Evolution

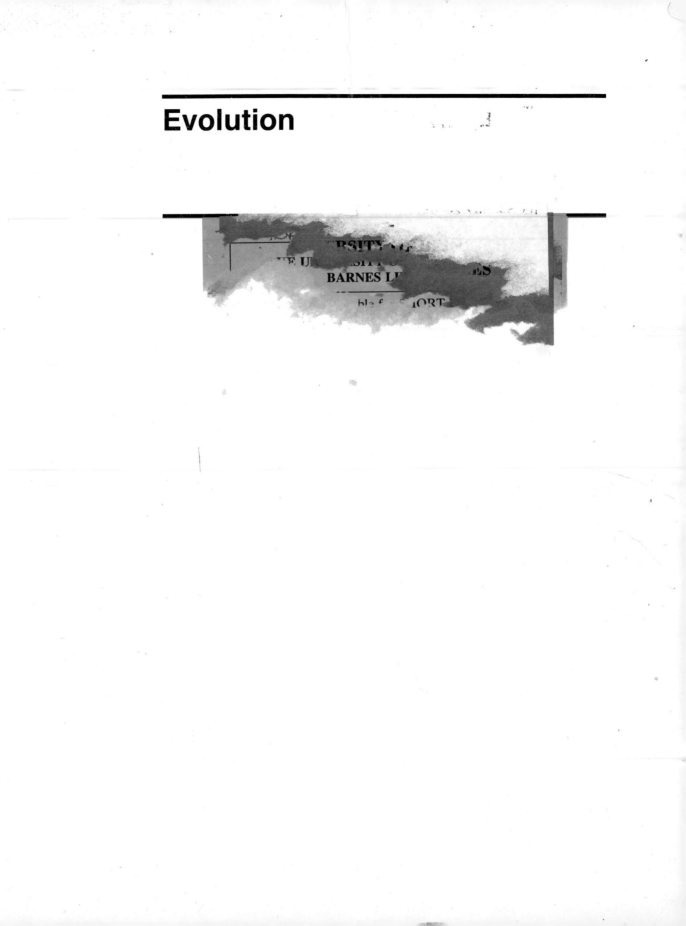

Evolution

MARK RIDLEY

Department of Biology
Emory University, Atlanta, Georgia

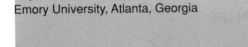

Boston
BLACKWELL SCIENTIFIC PUBLICATIONS
Oxford London Edinburgh
Melbourne Paris Berlin Vienna

©1993 by
Blackwell Scientific Publications, Inc.
Editorial offices:
238 Main Street, Cambridge,
  Massachusetts 02142, USA
Osney Mead, Oxford OX2 0EL,
  England
25 John Street, London, WC1N 2BL,
  England
23 Ainslie Place, Edinburgh EH3 6AJ,
  Scotland
54 University Street, Carlton, Victoria
  3053, Australia

Other Editorial Offices:
Librairie Arnette SA
1 rue de Lille
75007 Paris
France

Blackwell Wissenschafts-Verlag
Kurfürstendamm 57
10707 Berlin
Germany

Blackwell MZV
Feldgasse 13
A-1238 Wien
Austria

First published 1993

Set by Excel Typesetters Company,
Hong Kong

Printed and bound in the United States
of America by Hamilton, New York

94 95 96 97 5 4

Distributors:

USA
  Blackwell Scientific Publications, Inc.
  238 Main Street
  Cambridge, Massachusetts 02142
  (Orders: Tel: 800-215-1000
           617-876-7000)

Canada
  Oxford University Press
  70 Wynford Drive
  Don Mills
  Ontario M3C IJ9
  (Orders: Tel: 416 441-2941)

Australia
  Blackwell Scientific Publications Pty Ltd
  54 University Street
  Carlton, Victoria 3053
  (Orders: Tel: 03-347-5552)

Outside North America and Australia
  Marston Book Services, Ltd.
  P.O. Box 87
  Oxford OX2 0DT
  England
  (Orders: Tel: 0865 791155
           FAX: 0865 791927
           Telex: 837515)

Library of Congress
Cataloging-in-Publication Data
Ridley, Mark.
  Evolution/Mark Ridley.
    p.   cm.
  Includes bibliographical references and
  index.
  ISBN 0–86542–226–5
  ISBN 0–632–03481–5 (pbk)
  1. Evolution (Biology)   I. Title.
QH366.2.R524   1993
575—dc20

A catalogue record for this title is
available from the British Library

Contents

Preface

The theory of evolution is outstandingly the most important theory in biology, and it is always a pleasure to be a member, facing in either direction, of the class that is fortunate enough to be studying it. No other idea in biology is so powerful scientifically, or so intellectually stimulating. Evolution can add an extra dimension of interest to the most appealing sides of natural history—we shall see, for example, how modern evolutionary biologists tend to argue that the existence of sex is the most profound puzzle of all, and quite possibly a mistake that half the living creatures of this planet would be better off without. Evolution gives meaning to the dryer facts of life too, and it is one of the delights of the subject to see how there are ideas as well as facts within the disorienting technicalities of the genetics lab, and how deep theories about the history of life can hinge on measurements of the width of a region called "prodissoconch II" in the larval shell of a snail, or the number of ribs in a trilobite's tail. So great is the depth and range of evolutionary biology that in every other classroom on campus they must feel (as you can sort of tell) locked in with more superficial and ephemeral materials.

The theory of evolution, as I have arranged it here, has four main components. Population genetics provides the fundamental theory of the subject. If we know how any property of life is controlled genetically, population genetics can be applied to it directly. We have that knowledge particularly for molecules (together with some, mainly morphologic, properties of whole organisms), and molecular evolution and population genetics are therefore well integrated. Part 2 of the book considers them together. The second component is the theory of adaptation, and this is the subject of Part 3. Evolution is also the key to understanding the diversity of life, and in Part 4 we consider such topics as what a species is, how new species originate, and how to classify and reconstruct the history of life. Finally, Part 5 is about evolution on the grand scale; the fossil record is the main testing ground for these large evolutionary events, which occur over the geologic time scale of tens or hundreds of millions of years.

Controversy is always a tricky business to deal with in an introductory text, and evolutionary biology has more than its fair share of it. When I have come to a controversial topic, my first aim has been to explain the competing ideas in such a way that they can be understood on their own terms. In some cases, such as cladistic classification, I have then ventured to take sides, and in others, such as the relative empirical importance of gradual and punctuated change in fossils, I have not. I am well aware that not everyone will agree with the positions I have taken, or indeed with my decisions in some cases not to take a position; but the book's success mainly depends on how well it enables a reader who has not studied the subject much before to understand the various ideas and come to a sensible viewpoint about them.

The book is intended as an introductory text, and I have subordinated all other aims to that end. I have aimed to explain concepts, wherever possible by example, and with a minimum of professional clutter. The principal interest, I believe, of the theory of evolution is as a set of ideas to think about, and I have therefore tried in every case to move on to the ideas as soon as possible. The book is not a factual encyclopedia, nor (primarily) a reference work for research biologists. I have therefore sacrificed the usual convention, by which references are made through dates inserted in the text. Some references, it is true, can easily be made in this way; but others, particularly the general ones, cannot and these can easily end up spoiling the most powerful textual positions, such as the end of a paragraph or a section. The way I have things, those textual positions can be occupied by summary sentences and other more useful matter than directions to further reading. However, anyone who wishes to track down my sources will not have much difficulty. All the authors mentioned in the text are listed in the references at the end, and in the handful of ambiguous cases I have given the date of the reference in the text. The "further reading" section at the end of each chapter is the main vehicle for references. It should nearly always be possible to go directly to the appropriate source from there. I have referred to recent reviews when they exist, and the historical bibliography of each topic can be traced through them.

I started writing the book while I was a Research Fellow at St Catharine's College, Cambridge and completed it after moving to Emory University; I am most grateful to both institutions for the opportunity they provided. I am also grateful to Nick Barton, Paul Brakefield, Michael Ghiselin, Alan Grafen, Eddie Holmes, Bruce MacFadden, Chris Sanford, Randy Skelton, Michael Turelli, John Turner, and Elizabeth Vrba for their error-reducing comments on various parts, large and small, of the book. My thanks, too, to the tolerant and eagle-eyed members of ANT 385, BIO 462, BIO 470, and ANT 503 at Emory, who have cheerfully endured my chaotic manuscripts instead of a proper educational publication, and whose explicit (and implicit) experiences with the text have, more than anything else, inspired its final form.

Mark Ridley
Atlanta, 1992

Part 1
Introduction

Introduction

When Darwin put forward his theory of evolution by natural selection, he lacked a satisfactory theory of inheritance, and the importance of natural selection was widely doubted until it was shown in the 1920s and 1930s how natural selection could operate with Mendelian inheritance. The two key events in the history of evolutionary thought are therefore Darwin's discovery of evolution by natural selection and the synthesis of Darwin's and Mendel's theories—a synthesis variously called the modern synthesis, the synthetic theory of evolution, and neo-Darwinism. Chapter 1 discusses the rise of evolutionary theory historically, and introduces some of its main figures. During the twentieth century, the sciences of evolutionary biology and genetics have developed together and some knowledge of genetics is essential for understanding the modern theory of evolution. Chapter 2 provides an elementary review of the main genetic mechanisms. In chapter 3, we move on to consider the evidence for evolution—evidence that species have evolved from other, ancestral species rather than having separate origins and remaining forever fixed in form. The classic case for evolution was made in Darwin's *On the Origin of Species* and his general arguments still apply, but it is now possible to use more recent molecular and genetic evidence to illustrate them. Chapter 4 introduces the concept of natural selection. It considers the conditions necessary for natural selection to operate, and the main kinds of natural selection. One crucial condition is that the population should be variable—that individuals should differ from one another—and the chapter shows that variation is common in nature. New variants originate by mutation and the chapter reviews the main kinds of mutation, how mutation rates are measured, and discusses why mutation can be expected to be adaptively undirected.

The rise of evolutionary biology

1.1 Evolution means change in living things over long time periods

Evolutionary biology is a large science, and is growing larger. A list of its various subject areas could sound rather daunting. Evolutionary biologists now carry out research in some sciences, like molecular genetics, that are young and move rapidly, and in others like morphology and embryology that have accumulated their discoveries at a more stately speed over a much longer period. Evolutionary biologists work with materials as diverse as naked chemicals in test tubes, animal behavior in the jungle, and fossils collected from barren and inhospitable rocks.

However, a beautifully simple and easily understood idea—evolution by natural selection—can be scientifically tested in all these fields. It is one of the most powerful ideas in all of science, and is the only theory that can seriously claim to unify biology. It can give meaning to facts from the invisible world in a drop of rain water, or from the many colored delights of a botanic garden, or thundering herds of big game. The theory is also used to understand such topics as the geochemistry of life's origins and the gaseous proportions of the modern atmosphere. As Theodosius Dobzhansky, one of the twentieth century's most eminent evolutionary biologists, remarked in an often quoted but scarcely exaggerated phrase, "nothing in biology makes sense except in the light of evolution."

Evolution means change, change in the form and behavior of organisms between generations. (We have to say "change between generations" to exclude from the definition developmental change within the life of an organism.) Darwin called it "descent with modification." The forms of organisms, at all levels from DNA sequences to macroscopic morphology and social behavior, have been modified from those of their ancestors. But evolutionary modification in living things is of a particular kind. It depends on external environmental change and on random genetic innovation. At any moment, therefore, the form of future change is unpredictable, except conditionally. Moreover, the evolution of life has proceeded in a branching, tree-like pattern. The modern variety of species has been generated by the repeated splitting of lineages since the single common ancestor of all life. The changes of politics, of economics, of human history, technology, and even scientific theories, are sometimes loosely described as "evolutionary"; but this means mainly that they have changed through time, and perhaps not in a pre-ordained direction. Human ideas and institutions have sometimes split during their history; but this is not so characteristic and clear-cut a feature as it is in the history of life. Change, and splitting, provide two of the main themes of evolutionary theory.

5

1.2 Living things show adaptations

Adaptation is another of evolutionary theory's crucial concepts. Indeed, one of the main aims of modern evolutionary biology is to explain the forms of adaptation that we find in the living world. *Adaptation* refers to "design" in life—to those properties of living things that enable them to survive and reproduce in nature. The concept is easiest to understand by example. Many of the attributes of a living organism could be used to illustrate the concept of adaptation, because many details of the structure, metabolism, and behavior of an organism are well designed for life. Darwin's favorite particular example was the woodpecker. The woodpecker's most obvious adaptation is its powerful characteristically shaped beak. It enables the woodpecker to excavate holes in trees. They can thus feed on the year-round food supply of insects that live under bark, insects that bore into the wood, and the sap of the tree itself. Tree-holes also make safe sites to build a nest. Woodpeckers have many other design features as well as their beaks. Within the beak is a long, probing tongue, which is well adapted to extract insects from inside a tree-hole. They have a stiff tail that is used as a brace, short legs, and their feet have long curved toes for gripping onto the bark; they even have a special type of molting in which the strong central pair of tail-feathers (which are crucial in bracing) are retained and molted last. The beak and body design of the woodpecker is adaptive. The woodpecker is more likely to survive, in its natural habitat, by possessing them.

Camouflage is another, particularly clear, example of adaptation. Camouflaged species have color patterns and details of shape and behavior that make them less visible in their natural environment. Camouflage assists the organism in surviving by making it less visible to its natural enemies. Camouflage is adaptive. Adaptation, however, is not an isolated concept referring to only a few special properties of living things: in human beings alone, it applies to almost any part of the body. Hands are adapted for grasping, eyes for seeing, the alimentary canal for digesting food, legs for moving: all these functions assist us in survival. Although most of the obvious things we notice are adaptive, not every detail of an organism's form and behavior may be adaptive (see chapter 13). Adaptations are, however, so common that they have to be explained. Darwin regarded adaptation as the key problem that any theory of evolution had to solve. In Darwin's theory—as in modern evolutionary biology—the problem is solved by natural selection.

Natural selection means that some individuals in a population contribute more offspring to the next generation than do others. Provided that the offspring resemble their parents, the characters of individuals that leave more offspring than average will increase in frequency over time; the make-up of the population will then change automatically. Such is the simple, but immensely powerful, idea whose ramifying consequence we shall be exploring in this book.

1.3 A short history of evolutionary biology

We shall begin with a brief sketch of the historical rise of evolutionary biology. It has five main stages.

1 Evolutionary, and non-evolutionary, ideas before Darwin.
2 Darwin's theory (1859).

3 The eclipse of Darwin (c. 1880–1920).
4 The modern synthesis (1920s–1950s).
5 Developments within the modern synthesis (1960s–present).

1.3.1 Evolution before Darwin

The history of evolutionary biology really begins in 1859, with the publication of Charles Darwin's *On the Origin of Species*; but many of Darwin's ideas have an older pedigree. The most immediately controversial claim in Darwin's theory is that species are not fixed, but evolve into other species. Human ancestry, for instance, passes through a continuous series of forms tracing back to a unicellular stage. Species fixity was the orthodox belief in Darwin's time, though that does not mean that no one then or before had questioned it. Naturalists and philosophers a century or two before Darwin had often speculated about the transformation of species. The French scientist Maupertuis discussed evolution, as did *encyclopédistes* such as Diderot; Charles Darwin's grandfather Erasmus Darwin is another example. However, none of these thinkers put forward anything we would now recognize as a satisfactory theory to explain why species should change. They were mainly interested in the factual possibility that one species might change into another.

The French naturalist Jean-Baptiste Lamarck (1744–1829) was more influential. His most important book was the *Philosophie Zoologique* (1809), in which he argued that species change over time into new species. The way in which he thought species changed was importantly different from Darwin's and our modern idea of evolution, and historians prefer the contemporary word "transformism" to describe Lamarck's idea. (The historical change in the meaning of the term "evolution" is a fascinating story in itself. Initially it meant something more like what we mean by development (as in maturation from an egg to an adult) than by evolution: it meant an unfolding

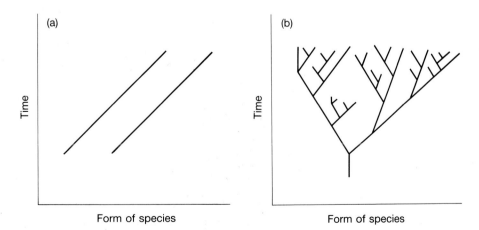

Figure 1.1 (a) Lamarckian "transformism," which differs in two crucial respects from evolution as Darwin imagined it. (b) Darwinian evolution is tree-like, as lineages split, and allows for extinction.

of predictable forms in a preprogrammed order. The course of evolution, in the modern sense, is not preprogrammed; it is unpredictable in much the same way that human history is unpredictable. The change of meaning occurred around the time of Darwin; he did not use the word in *On the Origin of Species* (1859), except in the form "evolved" which he used once as the last word in the book; but he did use it in *The Expression of the Emotions* (1872). It took a long time for the new meaning to become widespread.)

Figure 1.1 illustrates Lamarck's conception of evolution, and how it differed from Darwin's and the modern concept. Lamarck supposed that lineages of species persisted indefinitely, changing from one form into another; lineages in his system did not branch and did not become extinct. Lamarck had a two-part explanation of why species change. The principal mechanism was "internal forces": some sort of unknown mechanism within an organism causing it to produce offspring slightly different from itself, such that when the changes accumulated over many generations the line would be visibly transformed, perhaps enough to be a new species.

Lamarck's second (and possibly to him less important) mechanism is the one he is now remembered for: the inheritance of acquired characters. As an organism develops, it acquires many individual characters, due to its particular history of accidents, diseases, and muscular exercises. Lamarck suggested that a species could be transformed if these individually acquired modifications were inherited by the individual's offspring, and further modifications were added through time. In his famous discussion of the giraffe's neck, he argued that ancestral giraffes had stretched to reach leaves higher up trees; the exertion caused their necks to grow slightly longer; their longer necks were inherited by their offspring, who thus started life with a propensity to grow even longer necks than their parents. After many generations of neck stretching, the result was what we can now see. Lamarck

Figure 1.2 Charles Robert Darwin (1809–1882), painted in 1840.

described the process as being driven by the "striving" of the giraffe, and he often described animals as "wishing" or "willing" to change themselves. His theory has therefore sometimes been caricatured as suggesting that evolution happens by the will of the organism. However, the theory does not require any conscious striving on the part of the organism—only some flexibility in individual development and the inheritance of acquired characters.

Lamarck did not invent the idea of the inheritance of acquired characters. It is an ancient idea; Plato discussed it, for example. However, most modern thinking about the role of the process in evolution has been inspired by Lamarck, and the inheritance of acquired characters is now conventionally, if unhistorically, called "Lamarckian inheritance."

Lamarck, as a person, lacked the genius for making friends, and his main rival, the anatomist Georges Cuvier (1769–1832), knew how to conduct a controversy. Lamarck had broad interests, in chemistry and meteorology as well as biology, but his contributions did not always receive the attention he felt they deserved. By 1809, Lamarck had already persuaded himself that there was a conspiracy of silence against his ideas. The meteorologists ignored his scheme of weather forecasts,* the chemists ignored his chemical system, and when the *Philosophie Zoologique* was finally published, Cuvier saw to it that this, too, was greeted with silence. Actually, however, it was an influential book. It was at least partly in reaction to Lamarck that Cuvier and his school made a belief in the fixity of species a virtual orthodoxy among professional biologists. Cuvier's school studied the anatomy of animals to discover the various fundamental, unalterable plans according to which the different types of organism were designed, and Cuvier thus established that the animal kingdom had four main branches (*embranchements*): vertebrates, articulates, molluscs, and radiates. It was also Cuvier who established, contrary to Lamarck's belief, that species had become extinct (see chapter 22).

Lamarck's ideas mainly became known in Britain through a critical discussion in Charles Lyell's *Principles of Geology* (1830–1833). Cuvier's influence came more through Richard Owen (1804–1892), who had studied with Cuvier in Paris before returning to England. Owen became generally thought of as Britain's leading anatomist; and by the first half of the nineteenth century most biologists and geologists had come to accept Cuvier's view that each species had a separate origin, and then remained constant in form until it went extinct.

1.3.2 Charles Darwin

Meanwhile, Charles Darwin (Figure 1.2) was forming his own ideas. After graduating from Cambridge, he had traveled the world as a naturalist on

---

*Lamarck's weather forecasts were notorious among his Parisian contemporaries. Each year, starting in 1799, Lamarck published an *Annuaire Météorologique* which predicted (with scientific certainty) the weather for the next year. Unfortunately, some unforeseen accident always interfered, the weather turned out wrong, and it was necessary to modify the system in the next *Annuaire*. When Lamarck offered Napoleon a copy of the *Philosophie Zoologique*, the emperor turned it down, supposing it was Lamarck's latest volume of weather forecasts.

board the *Beagle* (1832–1837). He then lived briefly in London before settling permanently in the country. His father was a successful doctor, and his father-in-law controlled the Wedgwood china business; Charles Darwin was a gentleman of independent means. The crucial period of his life, for our purposes, was the year or so after the *Beagle* voyage (1837–1838). As he worked over his collection of birds from the Galápagos Islands, he realized that he should have recorded which island each specimen came from, because they varied from island to island. He had initially supposed that the Galápagos finches were all one species, but it now became clear that each island had its own distinct species. How easy to imagine that they had evolved from a common ancestral finch! He was similarly struck by the geographic variation of the ostrich-like birds called rheas in South America and it was probably these observations of geographic variation that first led Darwin to accept that species can change.

The next important step was to invent a theory to explain why species change. The notebooks Darwin kept at the time still survive. They reveal how he struggled with several ideas, including Lamarckism, but rejected them all because they failed to explain a crucial fact: adaptation. His theory would have to explain not only why species change, but also why they are well designed for life. In Darwin's own words (in his autobiography):

> It was equally evident that neither the action of the surrounding conditions, nor the will of the organisms [an allusion to Lamarck], could account for the innumerable cases in which organisms of every kind are beautifully adapted to their habits of life—for instance, a woodpecker or tree-frog to climb trees, or a seed for dispersal by hooks or plumes. I had always been much struck by such adaptations, and until these could be explained it seemed to me almost useless to endeavour to prove by indirect evidence that species have been modified.

Darwin came upon the explanation while reading Malthus's *Essay on Population*. He continued:

> In October 1838, that is fifteen months after I had begun my systematic enquiry, I happened to read for amusement 'Malthus on population', and being well prepared to appreciate the struggle for existence which everywhere goes on from long-continued observation of the habits of animals and plants, it at once struck me that under these circumstances favourable variations would tend to be preserved and unfavourable ones to be destroyed. The result of this would be the formation of a new species.

Because of the struggle for existence, forms that are better adapted for survival will leave more offspring and automatically increase in frequency in the population. As the environment changes through time, some new form will become best adapted to it and increase in frequency. And as the process continues, eventually (in Darwin's words) "the result of this would be the formation of a new species." This process provided Darwin with what he called "a theory by which to work." And he started to work. He was still at work, fitting facts into his theoretical scheme, 20 years later when he received a letter from another traveling British naturalist, Alfred Russel Wallace (Figure 1.3). Wallace had independently arrived at an idea very similar to

Figure 1.3 Alfred Russel Wallace (1823–1913), photographed in 1848.

Darwin's natural selection. Darwin's friends, Charles Lyell and Joseph Hooker (Figure 1.4), arranged for a simultaneous announcement of Darwin's and Wallace's idea at the Linnean Society in London in 1858; but by then Darwin was already writing an abstract of his full findings: that abstract is the scientific classic *On the Origin of Species*.

1.3.3 Darwin's reception

The reaction to Darwin's two connected theories—evolution and natural selection—was very mixed. Many biologists almost immediately came to accept that evolution was true; species were not separately created and not fixed in form. The new theory could in some cases make remarkably little

Figure 1.4 Joseph Dalton Hooker (1817–1911) on a botanical expedition in Sikkim in 1849. (From a sketch by William Tayler.)

difference to their day-to-day research. The kind of comparative anatomy practiced by the followers of Cuvier, including Owen, lent itself as well to the post-Darwinian search for pedigrees as to the pre-Darwinian search for "plans" of nature. The leading anatomists were by now mainly German; Carl Gegenbauer (1826–1903), one of the major figures, had soon reoriented his work to the tracing of evolutionary relationships between animal groups. The famous German biologist Ernst Haeckel (1834–1919) pursued the same line of research with spectacular vigor, as he applied his "biogenetic law"—the theory of recapitulation (which we shall meet in chapter 20)—to the unraveling of phylogenetic pedigrees.

Although some kind of evolution was thus widely accepted, it is doubtful whether many biologists shared Darwin's own idea. (In Darwin's theory, evolution is a branching tree, and it is largely a matter of accident what forms evolve at each stage.) There is nothing automatically progressive about Darwinian evolution: if later forms are "better" in some respect, that is simply the way things turned out; nothing guarantees the improvement. Most evolutionists of the late nineteenth and early twentieth centuries had a different conception of evolution from this. They imagined evolution instead as one-dimensional and progressive. They often concerned themselves with thinking up mechanisms to explain why evolution should have this unfolding, predictable, progressive pattern (Figure 1.5).

For while evolution—of a sort—was being accepted, natural selection was just as surely being rejected. People disliked the theory of natural selection for many reasons. It is not the purpose of this first chapter to explain the arguments in any depth, and what follows is only an introduction to the history of the ideas that subsequent chapters will explain.

One of the more sophisticated objections to Darwin's theory was that it

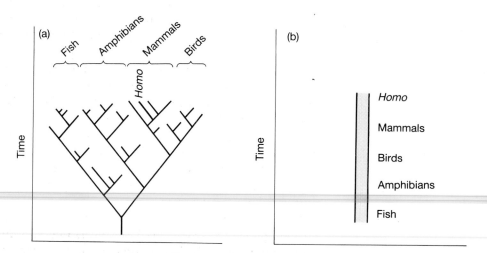

Figure 1.5 (a) Darwin's theory suggests that evolution has proceeded as a branching tree; note that it is arbitrary where *Homo* is positioned in the diagram—it does not have to be the right-hand extreme. The tree should be contrasted with the popular idea that evolution is a one-dimensional progressive ascent of life (b). In Stephen Jay Gould's words, Darwinian evolution is a bush, not a ladder (see Figure 1.1).

lacked a satisfactory theory of heredity. There were various theories of inheritance at that time, all of which are now known to be wrong. Darwin preferred a "blending" theory of inheritance, in which offspring blend their parental attributes; and one of the deepest hitting criticisms of his theory pointed out that natural selection could hardly operate at all if heredity blended (section 2.5, p. 32). At a more popular level, two of the main objections to natural selection then, as now, were that it explained evolution by chance and that there are gaps between forms in nature that could not be crossed if evolution was powered by natural selection alone. The anatomist St George Jackson Mivart (1827–1900), for instance, in his book *The Genesis of Species* (1871), listed a number of organs that would not (he thought) be advantageous in their initial stages. In Darwin's theory, organs evolve gradually, and each successive stage has to be advantageous in order for it to be favored by natural selection. Mivart retorted that although, for example, a fully formed wing is advantageous for a bird, the first evolutionary stage—a tiny proto-wing—might not be. Biologists who accepted the criticism sought to get round it by imagining processes other than selection that could work in the early stages of a new organ's evolution. Most of these processes belong to the class of theories of "directed mutation" or directed variation: they suggested that offspring, for some unspecified reason to do with the hereditary mechanism, consistently tend to differ from their parents in a certain direction. In the case of wings, the explanation using directed variation would say that the wingless ancestors of birds somehow tended to produce offspring with proto-wings, even though there was no advantage to it. (Chapter 13 deals with this general question; chapter 4 discusses variation.)

Lamarckian inheritance was the most popular theory of directed variation. Evolution is directed in this theory because the offspring tend to differ from their parents in the direction of characters acquired by their parents. If the parental giraffes all have short necks and acquire longer necks by stretching, their offspring have longer necks to begin with, before any elongation by stretching. (Darwin accepted that acquired characters can be inherited.) He even produced a theory of heredity ("my much abused hypothesis of pangenesis," as he called it) which incorporated the idea. In Darwin's time, the debate was about the relative importance of natural selection and the inheritance of acquired characteristics; but by the 1880s the debate had moved onto a new stage. The German biologist August Weismann (1833–1914) then produced strong evidence and theoretical arguments that acquired characteristics are not inherited, and the question became whether Lamarckian inheritance had any influence in evolution at all. Weismann initially suggested that practically all evolution was driven by natural selection, but he later retreated from this position.

Around the turn of the century, Weismann was a highly influential figure; but few biologists shared his belief in natural selection. Some, such as the British entomologist Edward Bagnall Poulton, were studying natural selection; but the majority view was that natural selection needed to be supplemented by other processes. An influential history of biology written by Erik Nordenskiöld in 1929 could even take it for granted that Darwin's theory was wrong. About natural selection, he concluded "that it does not operate in the

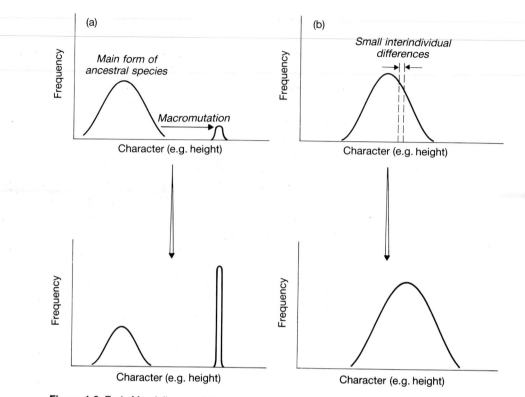

Figure 1.6 Early Mendelians and biometricians. (a) Early Mendelians studied large differences between organisms, and thought that evolution happened when a new species evolved from a "macromutation" in its ancestor. (b) Biometricians studied small interindividual differences, and explained evolutionary change by the transition of whole populations. Mendelians were less interested in the reasons for small interindividual variation. This figure is a simplification: no historical debate between two groups of scientists lasting for three decades can be fully represented in a single diagrammatic contrast.

form imagined by Darwin must certainly be taken as proved"; the only remaining question, for Nordenskiöld, was "does it exist at all?"

By this time, Mendel's theory of heredity had been "rediscovered." *Mendelism* (chapter 2) has been the generally accepted theory of heredity since the 1920s, and is the basis of all modern genetics. It was destined eventually to allow a revival of Darwin's theory; but its initial effect, in the first two decades of the century, was the exact opposite. The early Mendelians, such as De Vries and William Bateson, all opposed Darwin's theory of natural selection. They mainly did research on the inheritance of large differences between organisms, and generalized their findings to evolution as a whole. They suggested that evolution proceeds in big jumps, by macromutations. A *macromutation* is a large and genetically inherited change between parent and offspring (Figure 1.6a). (Chapters 7, 13, and 19 discuss various perspectives on the question of whether evolution proceeds in small or large steps.) Mendelism was not universally accepted in the period 1900–1920, however. Members of the other principal school, which rejected Mendelism, called themselves biometricians; Karl Pearson was one of the leading figures.

Biometricians studied small, rather than large, differences between individuals and developed statistical techniques for describing how character distributions passed from the parental to the offspring generation. They saw evolution more in terms of the steady shift of a whole population rather than the production of a new type from a macromutation (Figure 1.6b). Some biometricians were more sympathetic to Darwin's theory than were the Mendelians. Weldon, for instance, attempted to measure the amount of selection in crab populations on the seashore.

1.3.4 The modern synthesis

By the second decade of the century, research on Mendelian genetics had already become a major enterprise. It was concerned with many problems, most of which are more to do with genetics than evolutionary biology. But within the theory of evolution, the main problem was to reconcile the atomistic Mendelian theory of genetics with the biometrician's description of continuous variation in real populations. This reconciliation was achieved by several authors in many stages, but a 1918 paper by Ronald Aylmer Fisher is particularly important. Fisher demonstrated there that all the results known to the biometricians could be derived from Mendelian principles. The next step was to show that natural selection could operate with Mendelian genetics. The theoretical work was mainly done, independently, by R.A. Fisher, J.B.S. Haldane, and Sewall Wright (Figure 1.7). Their synthesis of Darwin's theory of natural selection with the Mendelian theory of heredity established what is known as *neo-Darwinism*, or the *synthetic theory of evolution*, or the *modern synthesis*, after the title of a book by Julian Huxley, *Evolution: the Modern Synthesis* (1942). The old dispute between Mendelians and Darwinians was brought to an end. Darwin's theory now possessed what it had lacked for half a century: a firm foundation in a well-tested theory of heredity.

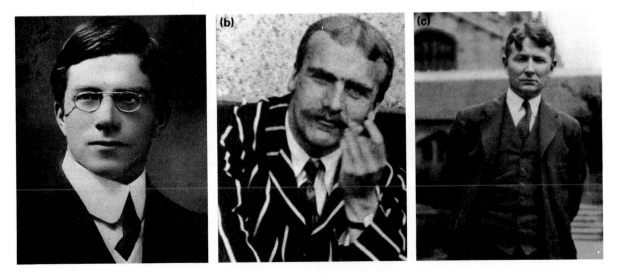

Figure 1.7 (a) Ronald Aylmer Fisher (1890–1962), seen in 1912, as a Steward at the First International Eugenics Conference. (b) J.B.S. Haldane (1892–1964), in Oxford in 1914. (c) Sewall Wright (1889–1988), photographed in 1928 at the University of Chicago.

The ideas of Fisher, Haldane, and Wright are known mainly from their great summary works, all written around 1930. Fisher published his book *The Genetical Theory of Natural Selection* in 1930. Haldane published a more popular book, *The Causes of Evolution* in 1932; it contained a long appendix under the title "A mathematical theory of artificial and natural selection," summarizing a series of papers published from 1918 onwards. Wright published a long paper on "Evolution in Mendelian populations" in 1931; unlike Fisher and Haldane, Wright lived to publish a four-volume treatise (1968, 1969, 1977, 1978) at the end of his career. These classic works of theoretical population genetics demonstrated that natural selection could work with the kinds of variation observable in natural populations and the laws of Mendelian inheritance. No other processes are needed. There is no need for the inheritance of acquired characters, or directed variation, or macromutations that produce extraordinary monsters. This insight has been incorporated into all later evolutionary thinking, and the work of Fisher, Haldane, and Wright is the basis for much of the material in chapters 5–9.

The reconciliation between Mendelism and Darwinism soon inspired new genetic research in the field and laboratory. Theodosius Dobzhansky (1900–1975), for example, began classic investigations of evolution in populations of fruitflies (*Drosophila*) after his move from Russia to the USA in 1927. Dobzhansky had been influenced by the leading Russian population geneticist Sergei Chetverikov (1880–1959), who had an important laboratory in Moscow until he was arrested in 1929. Dobzhansky, after he had emigrated, worked both on his own ideas and in collaboration with Sewall Wright. Dobzhansky's major book, *Genetics and the Origin of Species*, was first published in 1937 and its successive editions have been among the most influential works of the modern synthesis. We shall encounter several examples of Dobzhansky's work with fruitflies in later chapters.

Figure 1.8 Julian Huxley (1887–1975), photographed in 1918.

E.B. Ford (1901–1988) began in the 1920s a comparable, if more idiosyncratic, program of research in the UK. He studied selection in natural populations, mainly of moths, and called his subject "ecological genetics"; he published a summary of this work in a book called *Ecological Genetics* (1975). H.B.D. Kettlewell's (1901–1979) study of melanism in the peppered moth *Biston betularia* is the most famous piece of ecologic genetic research (section 5.6, p. 95). Ford collaborated closely with Fisher; their best known joint study was an attempt to show that the random processes emphasized by Wright could not account for observed evolutionary changes in the scarlet tiger moth *Panaxia dominula*. Julian Huxley (Figure 1.8) exerted his influence more through his skill in synthesizing work from many fields. His book *Evolution: the Modern Synthesis* (1942) introduced the theoretical concepts of Fisher, Haldane, and Wright to many biologists, by applying them to large evolutionary questions.

From population genetics, the modern synthesis spread into other areas of evolutionary biology. The question of how one species splits into two—the event is called speciation—was an early example. Before the modern synthesis had penetrated the subject, speciation had often been explained by macromutations or the inheritance of acquired characters. A major book, *The Variations of Animals in Nature* (1936), by two systematists, G.C. Robson and O.W. Richards, accepted neither Mendelism nor Darwinism. Robson and Richards suggested that the differences between species are non-adaptive and have nothing to do with natural selection. Richard Goldschmidt (1878–1958), most famously in his book on *The Material Basis of Evolution* (1940), argued that speciation is produced by macromutations, not the selection of small variants.

The question of how species originate is closely related to the questions of population genetics, and Fisher, Haldane, and Wright had all discussed it. Dobzhansky and Huxley emphasized the problem even more. They all

Figure 1.9 Ernst Mayr (1904–), on the right, pictured on an ornithologic expedition in New Guinea in 1928, with his Malay assistant.

reasoned that the kinds of changes studied by population geneticists, if they took place in geographically separated populations, could cause the populations to diverge and eventually evolve into distinct species (chapter 16). The classic work, however, was by Ernst Mayr: *Systematics and the Origin of Species* (1942). Like many classic books in science, it was written as a polemic against a particular viewpoint. It was precipitated by Goldschmidt's *Material Basis* but criticized Goldschmidt from the viewpoint of a complete and differing theory—the modern synthesis—rather than narrowly refuting him and it therefore has a much broader importance. Both Goldschmidt and Mayr (Figure 1.9) were born and educated in Germany and later emigrated to the USA; Mayr left in 1930 as a young man, but Goldschmidt was 58 years old and had built a distinguished career when he left Nazi Germany in 1936.

A related development is often called "the new systematics," after the title of a book edited by Julian Huxley (1940). It refers to the overthrow of what Mayr called the "typological" species concept and its replacement by a species concept better suited to modern population genetics (chapter 15). Species, on the typologic conception, had been defined as a set of more-or-less similar looking organisms, where similarity was measured relative to a standard (or "type") form for the species. The changes in gene frequencies analyzed by population geneticists take place within a "gene pool"—that is, a group of interbreeding organisms, who exchange genes when they reproduce. The crucial unit is the set of interbreeding forms, regardless of how similar looking they are to each other. The idea of a "type" for a species is meaningless in a gene pool containing many genotypes: no one genotype is more typical of the species than any other. Species thus came to be defined by interbreeding rather than by morphologic similarity. The modern synthesis had spread to systematics.

A similar treatment was given to paleontology by George Gaylord Simpson (Figure 1.10) in *Tempo and Mode in Evolution* (1944). Many paleontologists in the 1930s still persisted in explaining evolution in fossils by what are called "orthogenetic" processes; i.e. some inherent (and unexplained) tendency of a species to evolve in a certain direction. Orthogenesis is an idea related to the pre-Mendelian concept of directed mutation, and to the more mystical internal forces we saw in the work of Lamarck. Simpson argued that no observations in the fossil record required these processes. All the evidence was perfectly compatible with the population genetic mechanisms discussed by Fisher, Haldane, and Wright. He also showed how such topics as rates of evolution and the origin of major new groups could be analyzed by techniques derived from the assumptions of the modern synthesis (see chapters 19–22).

By the mid 1940s, therefore, the modern synthesis had penetrated all areas of biology. The 30 members of a "committee on common problems of genetics, systematics, and paleontology" who met (with some other experts) at Princeton in 1947 represented all areas of biology. But they shared a common viewpoint, the viewpoint of Mendelism and neo-Darwinism. A similar unanimity of 30 leading figures in genetics, morphology, systematics, and paleontology would have been difficult to achieve before that date. The Princeton symposium was published as *Genetics, Paleontology, and Evolution* (Jepsen *et al.*, 1949) and is now as good a symbol as any for the point at which

Figure 1.10 George Gaylord Simpson (1902–1984) shown with a baby guanaco in central Patagonia in 1930.

the synthesis had spread throughout biology. Of course, there remained controversy within the synthesis, and a counter-culture outside. Two eminent evolutionary biologists—the geneticist Muller, and the paleontologist Simpson—could still both celebrate the centenary of *On the Origin of Species* in 1959 with essays bearing (almost) the same memorable title: "One hundred years without Darwinism are enough."

1.3.5 Developments in the modern synthesis

The purpose of this book is to explain the main ideas of the modern synthesis, and how they are developing in recent research. Recent developments have been inspired by a number of sources, and molecular biology is one of the most important. Since the 1960s, molecular biology has been generating ever-increasing quantities of molecular evidence, particularly for protein sequences. In 1968, the Japanese population geneticist Motoo Kimura drew on the new molecular evidence to suggest his remarkably successful neutral theory of molecular evolution (chapter 7). He argued that most evolutionary change at the molecular level is not driven by natural selection, but consists of random change instead. Kimura's theory not only accounted well for the protein evidence that inspired it; but it turned out to account even better for the DNA sequences that started to accumulate through the 1980s, long after the neutral theory had been invented. The DNA sequences proved interesting in another respect too, as we shall see in chapter 10. The genome was found to have an architecture, with gene clusters and non-coding regions. These have their own peculiar modes of evolution, and the theory of population genetics has had to be extended to accommodate their non-Mendelian inheritance.

In the study of adaptation, a number of biologists in the 1960s realized that Darwin's theory of natural selection had been repeatedly misapplied (see chapter 12). Natural selection is often said to produce adaptations "for the good of the species" or "for the good of the group." In reality, natural selection probably mainly produces adaptations for the benefit of the individual organism, and this may have deleterious consequences at the level of the group or the species. When explanations in terms of species or group benefit were ruled out, a whole new set of problems, which had previously been thought to have been solved, were opened up: the function of sex is the outstanding example (chapter 11).

In comparative biology, the new systematics has been aggressively challenged, and a strong case can be made (chapter 14) that it only incompletely removed "typology" from classification. New ideas have multiplied in paleontology too. Paleontologists accept that the fossil record is (as Simpson showed) consistent with population genetics; but they increasingly doubt whether large-scale evolutionary events are simply magnified versions of the genetic processes that have been studied in the short term (see chapters 19–21). Large-scale evolutionary events may require explanations of their own, on their own level. There may, for example, be more to long-term trends, which persist over tens of millions of years, than natural selection working over a very long time. Likewise, very rare, but very catastrophic, geologic events (like mass extinctions, see chapter 22) may impose different forces from those in population genetics, and have a major influence on the course of evolution.

Evolutionary biologists differ in how they interpret these developments. Some see them as revolutionary ideas that are replacing the modern synthesis with a new and enriched theory of evolution. The neutral theory of molecular evolution, for example, was once proposed as a theory of "non-Darwinian evolution"; modern ideas on the fossil record have also been described as alternatives to Darwin's; and some members of the new schools of systematics have suggested that their ideas are revolutionary in another way—that they will lead classification back into a pre-Darwinian, pre-evolutionary phase after an anomalous century of unnecessary evolutionary entanglements. Other evolutionary biologists would describe these new ideas as interesting additions to the broad body of evolutionary thought. The difference in interpretation reflects a difference in how evolutionary theory is imagined as a whole. If the theory of evolution is a broad framework, new ideas will appear to readily fit into it; but if it is a relatively narrow set of ideas, a new hypothesis is more likely to appear to be a fundamental challenge.

1.4 Summary

1 Evolution means descent with modification, or the change in the form, physiology, and behavior of organisms over many generations of time. The evolutionary changes of living things occur in a diverging, tree-like pattern of lineages.

2 Living things possess adaptations; i.e. they are well designed in form, physiology, and behavior for life in their natural environment.

3 Many thinkers before Darwin had discussed the possibility that species change through time into other species. Lamarck is the best known. But in the mid nineteenth century most biologists believed that species are fixed in form.

4 Darwin's theory of evolution by natural selection explains evolutionary change and adaptation.

5 Darwin's contemporaries mainly accepted his idea of evolution, but not his explanation of it by natural selection.

6 Darwin lacked a theory of heredity. When Mendel's ideas were rediscovered at the turn of the century, they were initially thought to count against the theory of natural selection.

7 Fisher, Haldane, and Wright demonstrated that Mendelian heredity and natural selection are compatible; the synthesis of the two ideas is called neo-Darwinism or the synthetic theory of evolution.

8 During the 1930s and 1940s, neo-Darwinism gradually spread through all areas of biology and became widely accepted. It unified genetics, systematics, paleontology, and classical comparative morphology and embryology.

9 Since the 1960s, new ideas have challenged several strands within neo-Darwinism. These new ideas are seen by some as revolutionary, and by others as more-or-less enriching additions to the broad body of evolutionary thought.

1.5 Further reading A popular essay about the adaptations of woodpeckers is by Diamond (1990a). On the history: Bowler (1989) and Mayr (1982b) provide general histories of the idea of evolution. On Lamarck and his context, see Burkhardt (1977), Barthélemy-Madaule (1982), and Corsi (1988). On the nineteenth century history, see Ruse (1979). On Darwin's ideas, see Ghiselin (1969) and Mayr (1991). There are many biographies of Darwin; Desmond and Moore (1992) is a recent one. Darwin's autobiography is an interesting source. A pleasant (if more demanding) way to follow Darwin's life is through his correspondence. Much was published in *The Life and Letters* (Darwin, 1887) and *More Letters* (Darwin and Seward, 1903). A modern scholarly multivolume edition is underway (Burkhardt and Smith, 1985–). Bowler (1989) discusses and gives references about the reception and fate of Darwin's ideas. On the modern synthesis, see also Provine (1971) and Mayr and Provine (1980). There are biographies of many of the key figures: Box (1978) for Fisher, Clark (1969) for Haldane (see also Haldane, 1963 and Maynard Smith, 1987a), and Provine (1986a) for Wright. Huxley (1970, 1973) and Simpson (1978) wrote autobiographies. See Glass (1980) for Dobzhansky. The two Darwin centennial essays by Muller and by Simpson alluded to in the text are Muller (1959) and Simpson (1961a).

Gould's popular essays, which first appear in *Natural History* magazine and are then anthologized (Gould 1977b, 1980a, 1983a, 1985, 1991) introduce many aspects of evolutionary biology; Gould (1989) is about the popular misconception that evolution is a progressive ladder-climbing process.

Molecular and Mendelian genetics

2.1 Inheritance is caused by DNA molecules, which are passed from parent to offspring through the sperm and egg

In reproduction, each new individual develops from a single cell, which is produced by the fusion of sperm and egg from its parents. The cells of almost all multicellular organisms are *eukaryotic*; they have a complex internal structure, including internal organelles and a distinct region called the nucleus. Some unicellular organisms, such as bacteria, have *prokaryotic* cells, which are simpler and have no nucleus (Figure 2.1). In eukaryotes, the nucleus contains the hereditary material in the form of a molecule of DNA (deoxyribonucleic acid); in prokaryotes the DNA lies within the cell, but in no particular region. The DNA carries the information used to build a new body, and to differentiate its various body parts. Physically, the DNA is carried in structures called *chromosomes*, which can be seen through a light microscope. Each species characteristically has a certain number of chromosomes—each individual human has 46, for example. The finer structure of DNA is too small to be seen directly, but it can be inferred by X-ray diffraction. The molecular structure of DNA was elucidated by Watson and Crick in 1953.

The DNA molecule consists of a sequence of units; each unit, called a *nucleotide*, consists of a phosphate and a sugar group with a *base* attached. The alternating sugar and phosphate groups of successive nucleotides form the backbone of the DNA molecule. The full DNA molecule consists of two, paired complementary strands, each made up of sequences of nucleotides; the nucleotides of opposite strands are chemically bonded together. The two strands exist as a double helix (Figure 2.2).

2.2 DNA structurally encodes the information used to build the body's proteins

How does the DNA encode the information needed to build a body? The DNA in an individual human cell contains about 3×10^9 nucleotide units. This total length can be divided into regions called *genes* and spacer regions in between. Some genes are contiguous; others are separated by more-or-less lengthy spacer regions. Each gene encodes the information for a particular protein. To a crude but workable approximation, bodies are built from proteins; and different parts of the body have their distinct characteristics because of the kinds of proteins they are made of. Skin, for example, mainly comprises a protein called keratin; oxygen is carried in red blood cells by a protein called hemoglobin; eyes are sensitive to light because of pigment proteins such as rhodopsin (actually, rhodopsin is made up of a protein called opsin combined with a derivative of vitamin A); and metabolic processes are catalyzed by a whole battery of proteins called enzymes (cytochrome *c*, for instance, is a respiratory enzyme and alcohol dehydrogenase is a digestive

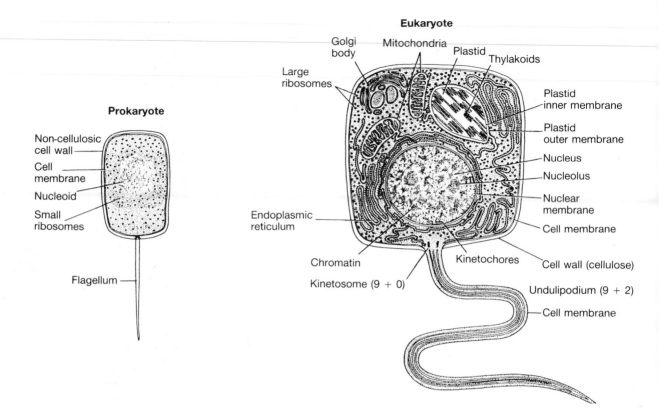

Figure 2.1 The cells of a body have a fine structure (or "ultrastructure") made up of a number of organelles. Not all the organelles illustrated here are found in all cells. Animal and fungal cells, for example, lack plastids; but all photosynthesizing organisms have them. Eukaryotes (i.e. all plants and animals) have complex cells with a separate nucleus. Within the nucleus the DNA is here illustrated in the diffuse form called chromatin; when the cell divides, the chromatin coalesces into structures called chromosomes. Prokaryotes, particularly bacteria, are simpler organisms and they lack a distinct nucleus; their DNA lies naked within the cell.

enzyme). It is by encoding proteins that DNA enables the egg cell to develop into an adult body.

Proteins are made up of particular sequences of amino acids. Twenty different amino acids are found in most living things; each amino acid behaves chemically in distinct ways, such that different sequences of amino acids result in proteins with very different properties. A protein's exact sequence of amino acids determines its nature. Hemoglobin, for example, has a sequence of 144 amino acids and insulin has a different sequence of 51 amino acids; the different behavior of hemoglobin and insulin are caused by the chemical properties of their different amino acids, arranged in their characteristic sequence. (As chapters 4 and 7 discuss in more detail, the particular sequence of any one protein can vary within and between species. Thus chicken hemoglobin differs from human hemoglobin, though it has the same general function in both groups. There are also different variants of hemoglobin within a species, a condition called *protein polymorphism*. However, the sequences of all variants of hemoglobin from the same or

(a) **Structure of single strand**

(b) **Structure of double strand**

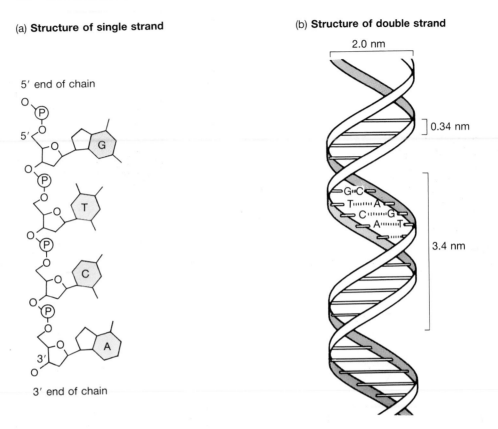

Figure 2.2 The structure of DNA. (a) Each strand of DNA is made up of a sequence of nucleotide units. Each nucleotide consists of a phosphate (P), a sugar, and a base (of which there are four types, here called G, guanine; C, cytosine; T, thymine; and A, adenine). (b) The full DNA molecule has two complementary strands, arranged in a double helix. nm = nanometer = 1×10^{-9} m.

different species are similar enough for them to be recognizably hemoglobin.)

The idea that one gene encodes for one protein is a simplification. The detail does not matter much here, but it is more exact to say that one cistron encodes for one polypeptide. (A cistron is defined as the length of DNA that codes for a polypeptide; a polypeptide is a chain of amino acids—the polypeptides in a body are of various lengths.) It is more exact because some proteins, such as hemoglobin, are assembled from more than one polypeptide, and the different polypeptides are encoded in separate (rather than contiguous) regions of the genome. Hemoglobin is made from four polypeptides, and the cistrons for two of them are a long way separated from the cistrons for the other two. (Thus, all proteins are polypeptides, but not all polypeptides are proteins. Each of the four polypeptides that make up hemoglobin are only parts of a protein.) There is no single "gene" encoding hemoglobin. However, for many purposes it is not grievously wrong to describe DNA as being made up of genes (and non-coding regions), and genes as coding for proteins.

How exactly do genes in the DNA encode for proteins? The answer is that the sequence of nucleotides in a gene specifies the sequence of amino acids in the protein. There are four types of nucleotide in the DNA. They differ only in the base part of the nucleotide unit; the sugar and phosphate group are the same in all four. The four are called adenine (A), cytosine (C), guanine (G), and thymine (T). Adenine and guanine belong to the chemical group called purines; cytosine and thymine are pyrimidines. In the double helix, an A nucleotide in one strand always pairs with a T nucleotide in the other; and a C always pairs with a G (as in Figure 2.2b). If the nucleotide sequence in one strand was . . . AGGCTCCTA . . . , then the complementary strand would be . . . TCCGAGGAT. . . . Because the sugar and phosphate are constant, it is often more convenient to imagine a DNA strand as a sequence of bases, like the . . . AGGCTCCTA . . . sequence above.

2.3 The information in DNA is decoded by transcription and translation

There are four types of nucleotide, but 20 different amino acids. A one-to-one code, in which one nucleotide coded for one amino acid, would therefore be impossible. In fact a triplet of bases encodes one amino acid; the nucleotide triplet for an amino acid is called a *codon*. The four nucleotides can be arranged in 64 (4 × 4 × 4) different triplets, and each one codes for a different amino acid. The relationship between triplet and amino acid has been deciphered and is called the *genetic code*.

The mechanism by which the amino acid sequence is read off from the nucleotide sequence of the DNA is understood in molecular detail. The full detail is unnecessary for our purposes, but we should distinguish two main stages. In the first, an intermediate molecule called messenger ribonucleic acid (mRNA) is transcribed from the DNA; the process is called *transcription*. mRNA has a similar structure to DNA, except that it is single-stranded and a

Table 2.1 The genetic code. The code is here expressed for mRNA. Each triplet encodes one amino acid (notice three triplets are "stop" codons, which signal the end of a gene)

First base in the codon	Second base in the codon				Third base in the codon
	U	C	A	G	
U	Phenylalanine	Serine	Tyrosine	Cysteine	U
	Phenylalanine	Serine	Tyrosine	Cysteine	C
	Leucine	Serine	Stop	Stop	A
	Leucine	Serine	Stop	Tryptophan	G
C	Leucine	Proline	Histidine	Arginine	U
	Leucine	Proline	Histidine	Arginine	C
	Leucine	Proline	Glutamine	Arginine	A
	Leucine	Proline	Glutamine	Arginine	G
A	Isoleucine	Threonine	Asparagine	Serine	U
	Isoleucine	Threonine	Asparagine	Serine	C
	Isoleucine	Threonine	Lysine	Arginine	A
	Methionine	Threonine	Lysine	Arginine	G
G	Valine	Alanine	Aspartic acid	Glycine	U
	Valine	Alanine	Aspartic acid	Glycine	C
	Valine	Alanine	Glutamic acid	Glycine	A
	Valine	Alanine	Glutamic acid	Glycine	G

base called uracil (U) is present instead of thymine (T). The DNA sequence AGGCTCCTA would therefore have an mRNA with the following sequence transcribed from it: UCCGAGGAU. The genetic code is usually expressed in terms of the codons in the mRNA (Table 2.1), though this is exactly equivalent to the complementary DNA sequence. The sequence UCCGAGGAU, for example, encodes for three amino acids: serine–glutamic acid–aspartic acid. The beginning and end of a gene are signaled by distinct base sequences, which provide a kind of punctuation in the DNA message.

Transcription takes place in the nucleus. After the mRNA molecule has been assembled on the gene, it then leaves the nucleus and travels to one of the structures in the cytoplasm called ribosomes (see Figure 2.1); ribosomes are made of another kind of RNA called ribosomal RNA (rRNA). The ribosome is the site of the second main stage in protein production: it is where the amino acid sequence is read off from the mRNA sequence and the protein is assembled. The process is called *translation*. The actual translation is achieved by yet another kind of RNA, called transfer RNA (tRNA). (That completes the three principal kinds of RNA: mRNA, rRNA, and tRNA.) There are 61 types of tRNA, one for each coding triplet. The tRNA has a recognition site, which binds to the complementary triplet in the mRNA, and has the appropriate amino acid attached at the other end. Thus protein assembly consists of tRNA molecules lining up on the mRNA at a ribosome. Other molecules are also needed to supply energy and attach the RNAs correctly. Figure 2.3 summarizes the transfer of information in the cell.

Different proteins have different numbers of amino acids, but a very approximate average figure is about 100–300 amino acids. Insulin (51 amino acids) is quite a short protein; long proteins may have well over 1000 amino acids. There may be approximately 100 000 genes in the human genome (and therefore a human body makes use of about 100 000 kinds of protein in its life). A 300 amino acid protein needs roughly 1000 nucleotides, which leads to the estimate that there are about 10^8 coding bases in the human genome. Because the total length of an individual's DNA is about 3×10^9 bases, about 90% of the DNA in the genome is of uncertain function (see chapter 10).

In addition to the DNA on the chromosomes in the nucleus, there are

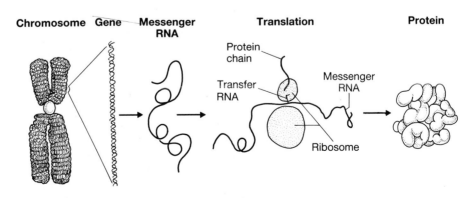

Figure 2.3 The transfer of information in a cell.

much smaller quantities of DNA in certain organelles in the cytoplasm (see Figure 2.1). Mitochondria—the organelles that control respiration—have some DNA, and in plants the organelles called chloroplasts which control photosynthesis also have their own DNA. Mitochondrial DNA is inherited maternally: mitochondria are passed on through eggs but not through sperm.

2.4 Organisms possess two sets of genes, which are inherited in characteristic Mendelian ratios

The DNA is physically carried on chromosomes. Humans, as noted above, have 46 chromosomes. However, the 46 consist of two sets of 23 distinct chromosomes. (To be exact, an individual has a pair of sex chromosomes—which are similar (XX) in females, but noticeably different (XY) in males—plus a double set of 22 non-sex chromosomes, called *autosomes*.) The condition of having two sets of chromosomes (and therefore two sets of the genes carried on them) is called *diploidy*. The figure given above of 3×10^9 nucleotides in the human genome is for only one of the sets of 23 chromosomes: the total DNA "library" of a human cell has about 6×10^9 nucleotides.

Diploidy is important in reproduction. An adult individual has two sets of chromosomes. Its gametes (eggs or ovules in the female, sperm or pollen in the male) have only one set: a human egg, for example, has only 23 chromosomes before it is fertilized. Gametes are said to be *haploid*. They are formed by a special kind of cell division, called *meiosis*; in meiosis, the double set of chromosomes divides to produce two gametes, each with one set. When male and female gametes fuse, at fertilization, the resulting *zygote* (the first cell of the new organism) has the double chromosome set restored, and it develops to produce a diploid adult: the cycle of genesis can then repeat itself.

Because each individual possesses a double set of chromosomes, it also possesses a double set of each gene. Any one gene is located at a particular place on a chromosome, called its *genetic locus*; an individual is therefore said to have two genes at each genetic locus in its DNA. One gene comes from its father and the other from its mother. The two genes at a locus are called a *genotype*. The two copies of a gene in an individual may be the same, or slightly different (i.e. the amino acid sequences of the proteins encoded by the two copies may be identical or differ to some extent). If they are the same, the genotype is *homozygote*; if they differ it is *heterozygote*. The different forms of the gene that can be present at a locus are called *alleles*. Genes and genotypes are usually symbolized by alphabetic letters. For instance, if there are two alleles at the genetic locus under consideration, we can call them *A* and *a*. An individual can then have one of three genotypes: it can be *AA*, or *Aa*, or *aa*.

Two more distinctions matter. The first is between genotype and *phenotype*. With two alleles at a locus, there are the three genotypes: *AA*, *Aa*, and *aa*. The genes will influence some property of the organism, and the property may or may not be easily visible. Suppose they influence color. The *A* gene might encode a black pigment and *AA* individuals would be black. *aa* individuals, lacking the pigment, would be (let us say) white. The coloration is then the phenotype controlled by the genotype at that locus: an individual's phenotype is its body and behavior as we observe them; this is distinct from its genotype, which is its set of genes. If we just consider *AA* and *aa* genotypes and phenotypes in this example, there is a one-to-one relationship between the two. Now we can consider two possibilities for the phenotype of the *Aa* genotype. One is that the color of *Aa* individuals is intermediate between the two homozygotes—they are gray; if this is the case, there are three phenotypes corresponding to the three genotypes, and there is still a one-to-one relationship between them. The second possibility is that the *Aa* heterozygotes resemble one of the homozygotes; they might be black, for instance. The *A* allele is then called *dominant* and the *a* allele *recessive*. (An allele is dominant if the phenotype of the heterozygote looks like the homozygote of that allele; the other allele in the heterozygote is called recessive.) If there is dominance, there will be only two phenotypes for the three genotypes and there is no longer a one-to-one relationship between them. (At different genetic loci, there can be any degree of dominance. Full dominance, in which the heterozygote resembles one of the homozygotes, and no dominance, in which it is intermediate between the two homozygotes, are extreme cases. The phenotype of the heterozygote could be anywhere between the two homozygotes. Instead of being either black or gray, it could have had any degree of grayness.)

The next important concept is the *Mendelian ratio*. Mendelian ratios express the proportions of different genotypes in the offspring of parents of particular combinations of genotypes. The easiest case is a cross between an *AA* male and an *AA* female (Figure 2.4a). After meiosis, all the male's gametes contain the *A* gene, as do all the female gametes. They combine to produce *AA* offspring. The Mendelian ratio is therefore 100% *AA* offspring. Now consider a mating between an *AA* homozygote and an *Aa* heterozygote (Figure 2.4b). Again, all the *AA* individual's gametes contain a single *A* gene. When a heterozygote reproduces, half its gametes contain an *A* gene, and half an *a*. The pair will produce either *AA* or *Aa* offspring in a 1:1 ratio. Finally, consider a cross between two heterozygotes (Figure 2.4c). Both male and

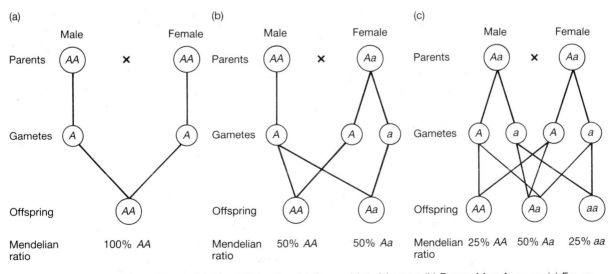

Figure 2.4 Mendelian ratios. (a) For an *AA* × *AA* cross. (b) For an *AA* × *Aa* cross. (c) For an *Aa* × *Aa* cross.

female produce half a gametes and half A gametes: if we consider the female gametes (eggs, or ovules), half of them are a, and half of these a eggs will be fertilized by a sperm, and half by A sperm; the other half of the eggs are A, and half of these A eggs will be fertilized by a sperm and half by A sperm. The resulting ratio of offspring is $\frac{1}{4} AA : \frac{1}{2} Aa : \frac{1}{4} aa$. The separation of an individual's two genes at a locus into its offspring is called *segregation*. The ratios of offspring types produced by different kinds of matings are examples of Mendelian ratios. They were discovered by Mendel in about 1856–1863. Mendel was a monk, later Abbé, in St Thomas's Augustinian monastery in what was then Brünn in Austria–Hungary and is now Brno in the Czech Republic.

Mendelian crosses can also concern more than one genetic locus. If the alleles at one locus are A and a, and at a second B and b, then an individual will have a double genotype, such as Ab/Ab (double homozygote) or Ab/ab (single heterozygote). It has a double set of genes at each locus, one set from each parent. The segregation ratios now depend on whether the genetic loci are on the same or different chromosomes. Recall that an individual human has a haploid number of 23 chromosomes and about 100 000 genes. That means there must be on average about 5000 genes per chromosome. Different genes on the same chromosome are described as being *linked*. Genes that are very close together are tightly linked, those further apart are loosely linked; those on separate chromosomes are unlinked.

The easy case is two unlinked loci. The genes at the two loci then segregate independently. Imagine a cross between an Ab/Ab male and an Ab/ab female. All the genes at the B locus are the same, while at the A locus the male is AA and the female is Aa: the ratio of offspring will be $\frac{1}{2} AAbb$ and $\frac{1}{2} Aabb$, which is a simple extension of the one locus case. If the B locus is heterozygous as well as the A locus, the ratios of B locus genotypes associated with each A locus genotype are exactly those you would predict from applying Mendel's principles independently to each locus. For instance, a cross between two Bb heterozygotes produces a ratio of offspring of $\frac{1}{4} BB : \frac{1}{2} Bb : \frac{1}{4} bb$. Then if the A locus and B locus are unlinked, the ratios of the three B genotypes with each of the A genotypes will be the same. In a cross between a male AB/Ab and a female AB/ab, there will be $\frac{1}{2} AA$ and $\frac{1}{2} Aa$ offspring: the ratios of B genotypes are the same with both: of the $\frac{1}{2}$ which are AA, $\frac{1}{4}$ are AB/AB, $\frac{1}{2}$ are AB/Ab, $\frac{1}{4}$ are Ab/Ab. Likewise for the $\frac{1}{2}$ Aa genotypes. The total offspring ratios are:

AB/AB	AB/Ab	Ab/Ab	AB/aB	AB/ab	Ab/ab
$\frac{1}{8}$	$\frac{1}{4}$	$\frac{1}{8}$	$\frac{1}{8}$	$\frac{1}{4}$	$\frac{1}{8}$

The segregation of unlinked genotypes is called *independent segregation*.

Figure 2.5 (*Opposite*) Recombination, seen at the level of (a) chromosomes; (b) genes; and (c) nucleotides. At recombination, the strands of a pair of chromosomes break at the same point and the two recombine. The post-recombinational sequence of genes, or nucleotides, combine one strand from one side of the break-point with the other strand from the other side. In (b) the gene sequence in chromosome 1 changes from ABC to ABc; in (c) the nucleotide sequence of the chromosome with bases A and T (stippled nucleotides) changes to A and A. (For the nucleotide sequence only one of the strands of the double helix is shown: each of the pair of chromosomes has a full double helix with complementary base pairs, as in Figure 2.2.)

(a) Recombination, at chromosome level

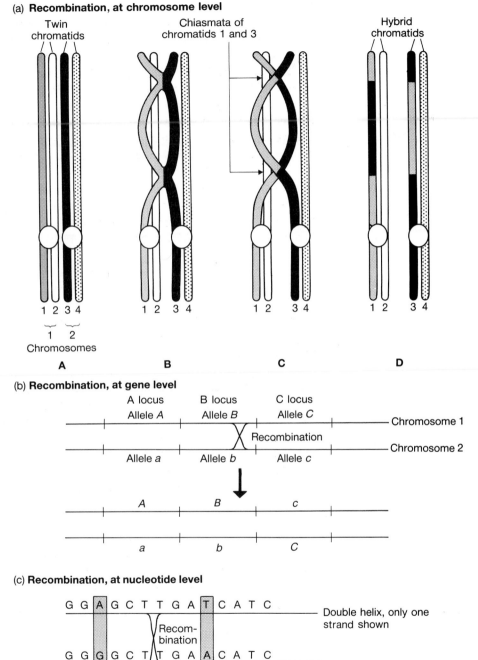

Twin chromatids

Chiasmata of chromatids 1 and 3

Hybrid chromatids

1 2 3 4   1 2 3 4   1 2 3 4   1 2 3 4

1   2

Chromosomes

A   **B**   **C**   **D**

(b) Recombination, at gene level

A locus   B locus   C locus
Allele *A*   Allele *B*   Allele *C*   — Chromosome 1
                                Recombination
                                                              — Chromosome 2
Allele *a*   Allele *b*   Allele *c*

A   *B*   *c*

a   *b*   *C*

(c) Recombination, at nucleotide level

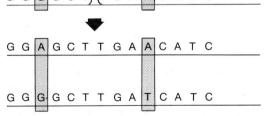

G G A G C T T G A T C A T C   Double helix, only one
                                     strand shown
                    Recom-
                    bination
G G G G C T T G A A C A T C

G G A G C T T G A A C A T C

G G G G C T T G A T C A T C

When the loci are linked on the same chromosome, they do not segregate independently. At meiosis, when haploid gametes are formed from a diploid adult, an additional process called *recombination* occurs. The pair of chromosomes physically line up and, at certain places, their strands join together and recombine (Figure 2.5). Recombination is a random process; it may or may not "hit" any point in the DNA. It occurs with a given probability, usually symbolized by r, between any two points on a chromosome. r can be defined between nucleotide sites or genes. We can then say that, if the A locus and the B locus are linked, the chance there will be recombination between them in an individual is r and the chance there is no recombination is $(1 - r)$.

The ratio of offspring genotypes for two linked gene loci depends on the rate of recombination between the two. If there is recombination (chance r), they segregate independently as if they were on separate chromosomes; if there is not (chance $(1 - r)$), they segregate as a unit. For a parent of genotype *Ab/aB*, the frequency of *AB* and *ab* gametes is r and of *Ab* and *aB* gametes is $(1 - r)$. These fractions can be used in the standard way; the principle is logically easy to understand, but the ratios can be laborious to work out in practice. The case of independent segregation corresponds to $r = \frac{1}{2}$. That is, when the A and B loci are on separate chromosomes, $r = (1 - r) = \frac{1}{2}$ and the *Ab/aB* parent produces *Ab*, *aB*, *AB*, and *ab* gametes in the ratio 1:1:1:1. One other detail should be noticed. For any two genes, recombination can "hit" more than once between them in an individual (this is called "multiple hits"). If there are two hits between a pair of loci, they cancel each other out and the chromosome has the same combination of genes as if there had been no recombination. It is more exact to say that the probability of recombination r above is equal to the probability there will be an odd number of hits, and the probability $(1 - r)$ is the chance of no hits plus the chance of an even number of hits.

The Mendelian ratios, in which paired diploid genes segregate into haploid gametes and the gametes of different individuals then combine at random, is the basis of all the theory of population genetics discussed in chapters 5–9.

When an organism reproduces, its DNA and genes are physically replicated. Normally an exact copy of the parental DNA is produced; but sometimes an accidental error occurs during replication. These errors are called *mutations*. The new sequence of DNA that results from the mutation may lead to the production of a different form of a protein, and the offspring will then differ from its parent because of the mutation. Various kinds of mutation can occur (section 4.6, p. 73).

2.5 Darwin's theory would probably not work if there was a non-Mendelian blending mechanism of heredity	As described in chapter 1, Mendel's theory of heredity plugged an important leak in Darwin's original theory, and the two theories together eventually came to form the synthetic theory of evolution, or neo-Darwinism. The problem was Darwin's lack of a sound theory of heredity; indeed it had even been shown in Darwin's time that natural selection would not work if heredity was controlled in the way that, before Mendel, most biologists thought it was. Before Mendel, most theories of heredity were "blending" theories. We can see the distinction in much the same terms as have just been

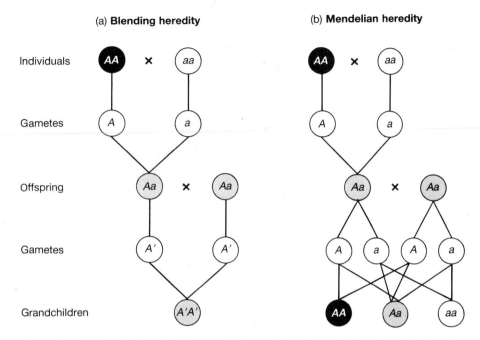

Figure 2.6 (a) Blending heredity. The parental genes for black (*A*) and white (*a*) color blend in their offspring, who produce a new type of gene (*A'*) coding for gray color. (b) Mendelian heredity. The parental genes are passed on unaltered by the offspring.

used for Mendelism (Figure 2.6). Suppose there is a gene *A* that causes its bearers to grow up black in color, and another gene *a* that causes its bearers to grow up white. We can imagine that, as in the real world of Mendelism, in our imaginary world of blending heredity, individuals are diploid and have two copies of each "gene." An individual could then either inherit an *AA* genotype from its parents and have a black phenotype, or inherit an *aa* genotype and have a white phenotype, or inherit an *Aa* genotype and have a gray phenotype. (Thus in the Mendelian version of the system, we should say there is no dominance between the *A* and *a* genes.)

The interesting individuals for this argument are the ones that have inherited an *Aa* genotype and have grown up to be gray in color. They could have been produced in a cross of a black and a white parent: then they will be gray whether inheritance is Mendelian (with no dominance) or blending. But now consider the next generation. Under Mendelian heredity, the gray *Aa* heterozygote passes on intact to its offspring the *A* and *a* genes it had inherited from its father and mother. Under blending heredity, the same is not true. An individual does not pass on the same genes as it inherited. If an individual inherited an *A* and an *a* gene, the two would physically blend in some way to form a new sort of gene (let us call it *A'*) causing gray coloration. Instead of producing $\frac{1}{2}$ *A* gametes and $\frac{1}{2}$ *a*, it would then produce all *A'* gametes. This makes a difference in the second generation. Whereas in Mendelian heredity, the black and white colors segregate out again in a cross between two heterozygotes, in blending heredity they do not—all the grandchildren are gray (see Figure 2.6).

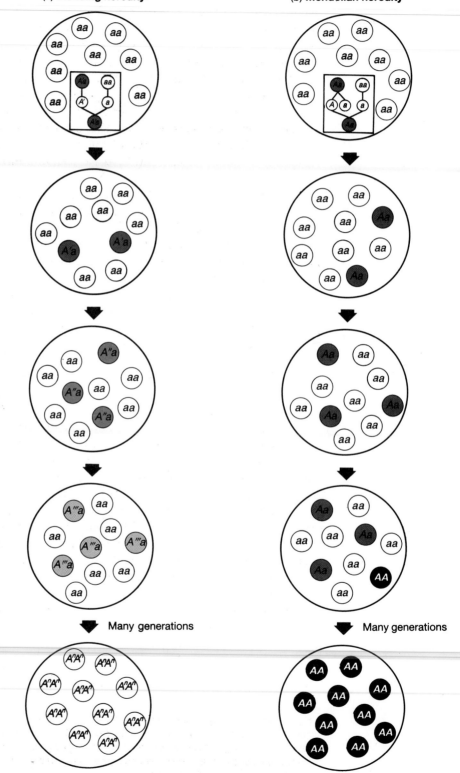

(a) Blending heredity

(b) Mendelian heredity

Many generations

Many generations

Mendelism is an "atomistic" theory of heredity. Not only are there discrete genes that encode discrete proteins, but also the genes are preserved during development and passed on unaltered to the next generation. In a blending mechanism, the "genes" are not preserved. The genes that an individual inherits from its parents are physically lost, as the two parental sets are blended together. In Mendelism, it is perfectly possible for the *phenotypes* of the parents to be blended in the offspring (as they are in the initial $AA \times aa$ cross in Figure 2.6), but the genes do not blend. Indeed, the phenotypes of real mothers and fathers often do blend in their offspring, and it was because they do that most students of heredity before Mendel thought that inheritance must be controlled by some blending mechanism. However, the case of heterozygotes that are intermediate between the two homozygotes (i.e. no dominance) shows that the blending of phenotypes need not mean blending of genotypes. In fact, the underlying genes are preserved. One way of expressing the importance of Mendelism for Darwin's theory is that it efficiently preserves genetic variation. In blending inheritance, variation is rapidly lost as extreme types mate together and their "genes" are blended out of existence into some general mean form; in Mendelian inheritance, variation is preserved because the extreme genetic types (even if disguised in heterozygotes) are passed down from generation to generation.

Why does this preservation of genes matter for Darwinism? Our full discussion of natural selection comes in later chapters, and some readers may prefer to return to this point after they have read about natural selection in more detail; but even after only the elementary account of natural selection in chapter 1, it is possible to understand why Darwin so to speak "needed" Mendel. Figure 2.7 illustrates the argument. Suppose that a population of individuals is white in color (and has the *aa* genotype) and heredity blends (in the manner of Figure 2.6). For some reason, it is advantageous for individuals in this population to be black in color: black individuals survive better and leave more offspring. Moreover, it is better to be a bit black (i.e. gray) than to be white. Suppose now that a single new gray individual somehow crops up by mutation, and it has an *Aa* genotype. This *Aa* individual will survive better than its *aa* fellow members of the population and produce more offspring. However, the advantageous gene cannot last long. In the first generation it produces A' gametes; these combine with *a* gametes (because everyone else in the population is white) and produce $A'a$ offspring. We can suppose these individuals are a bit less dark in color than the original *Aa* mutant; they have an advantage, but it is lower. The $A'a$ individual's genes in turn will blend and all its gametes will have an A'' gene. When that unites with an *a* gamete (because still almost everyone else is white) an $A''a$ offspring results, who are even less dark. It is only a matter of time until the original favorable mutation

Figure 2.7 (*Opposite*) Two populations with 10 individuals each (real populations would have many more members), one with blending heredity and the other with Mendelian heredity. (a) Under blending heredity a rare, new, advantageous gene is soon blended away. (b) Under Mendelian heredity a rare, new, favorable gene can increase in frequency and eventually become established in the population.

will be blended almost out of existence. The best result possible would be a population that was very slightly less white than it was to begin with. There is no chance that a population of black individuals could result from the mutation, because the original gene, which potentially could produce black individuals, will cease to exist after one generation. This objection was pointed out in Darwin's time by an engineer, Fleeming Jenkin. Darwin was very worried by Jenkin's argument and never did find a wholly satisfactory way round it.

Mendelism was what he needed. In the example just given, the original gray mutation will be in an Aa heterozygote, and fully half its offspring will be as gray as it—because they are also Aa heterozygotes. There is ample time for natural selection to increase the proportion of gray types, and eventually there would be enough of them for there to be a chance that two will mate together and produce some AA homozygotes among their offspring. It is now theoretically possible for a population of black AA individuals to result (Figure 2.6b). Thus natural selection is a powerful process with Mendelian heredity, because Mendelian genes are preserved over time; whereas it is at best a weak process with blending inheritance, because potentially favorable genes are diluted before they can be established.

2.6 Summary

1 Heredity is determined by a molecule called DNA. The structure and mechanisms of action of DNA are understood in detail.

2 The DNA molecule can be divided into regions called genes that encode for proteins. The code in the DNA is read off to produce a protein in two stages: transcription and translation. The genetic code has been deciphered.

3 DNA is physically carried on structures called chromosomes. Each individual has a double set of chromosomes (one inherited from its father, the other from its mother), and therefore two sets of all its genes. An individual's particular combination of genes is called its genotype.

4 When two individuals, of given genotypes, mate together the proportions of genotypes in their offspring appear in predictable Mendelian ratios. The exact ratios depend on the genotypes of the parents.

5 Different genes are preserved over the generations under Mendelian heredity, and this enables natural selection to operate. Before Darwin, it was generally (but wrongly) thought that genes blended rather than being preserved; if genes did blend, natural selection would be much less powerful than under Mendelian heredity.

2.7 Further reading

Any genetics text, such as Lewin (1990), Suzuki *et al.* (1989), Watson *et al.* (1988), and Weaver and Hedrick (1991), explains the subject in detail. Luria (1973) is a short introductory account. The classic statement of why Darwinism requires Mendelism, and does not work with blending heredity, is in the first chapter of Fisher (1930).

The evidence for evolution

3.1 A scientific argument can be given to demonstrate that evolution has happened

In this chapter, we shall be asking whether, according to scientific evidence, one species has evolved into another in the past, or whether each species had a separate origin and has remained fixed in form ever since that origin. For the purposes of argument, it is useful to have some articulate alternatives to debate between. We can discuss three theories (Figure 3.1): (a) evolution; (b) "transformism," in which species do change, but there have been as many origins of species as there have been species; and (c) separate creation, in which species originated separately and remain fixed. We shall thus be considering the evidence for two evolutionary claims: that species have changed, and that all species share a common ancestor. Separate creation and fixity of species are logically distinct, and it is worth considering how the evidence bears upon both questions.

Several lines of biological evidence suggest an evolutionary history for life. People differ in how persuasive they find the different lines of evidence, according to what they see as the main objections to the theory of evolution. Some, for instance, doubt whether species can change at all; others accept that evolution can happen on a small scale, within a species, but doubt whether a new species could be produced; yet others accept that the species barrier can be broken, but doubt whether evolution can produce major transitions, such as from reptiles into mammals.

Before we can discuss the evidence that evolution can produce new species, we should consider what a species is. The concept of a species is not straightforward (see chapter 15), but here it will suffice to define a species by two properties: reproductive isolation (only individuals, of opposite sex, belonging to the same species can interbreed together); and phenetic appearance (individuals belong to the same species if they look similar enough). The question of just how similar is similar "enough" can be answered formally, but here a less formal approach will suffice. Horses and zebra are (although they are closely related) different species, as are chimps and humans, and the difference between a zebra and a horse (or a chimp and a human) gives a rough idea of how different two groups of animals must look in order to count as different species. Any evidence that evolution can generate new species, therefore has to show that a population can be transformed enough to produce individuals that look different from their ancestors, and that a population can be produced that does not interbreed with forms like its ancestors.

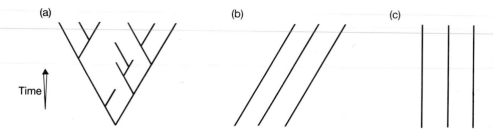

Figure 3.1 Three theories of the history of life. (a) Evolution; (b) transformism; and (c) separate creation. Evolution supposes that species have a common origin and do change, transformism that they have separate origins and do change, and separate creation that they have separate origins and do not change. Each line represents a species in time. If the line moves up vertically the species is constant, if it deviates left or right the species is changing in form.

3.2 On a small scale, evolution can be observed in nature

The time available for humans to observe evolution actually happening is only short, but a number of cases have been recorded in which a natural population of a species has changed in form over time. Most examples concern a phenomenon called melanism (i.e. black coloration) in moths. The most famous, and most thoroughly studied, example is the peppered moth *Biston betularia*; we shall examine it in detail in chapter 5. Here we can look at a different example, just to make the point that evolution can be see in action in nature.

The normal form of the arctiid moth *Panaxia dominula* has forewings that are mainly black, with white and yellow spots, and hindwings that are mainly red, with smaller black regions. A rare heterozygote, called *medionigra*, has more black on both fore- and hindwings. (The even rarer homozygous form of the gene producing the *medionigra* heterozygote is even blacker—in the homozygote, the hindwings are as much black as red and the forewings are almost completely black.) The frequency of the *medionigra* gene in a natural population of *Panaxia dominula* living near Oxford, in England, was measured

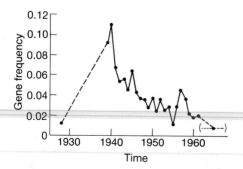

Figure 3.2 Changes in gene frequency in a natural population of the moth *Panaxia dominula* in England. The *medionigra* gene produces a darker colored moth than the normal form. The frequency before 1930 is uncertain. It was estimated from preserved collections, a method that almost certainly overestimates its real frequency: the increase from before 1930 to 1940 was probably much larger than the graph suggests. Data from Ford (1975).

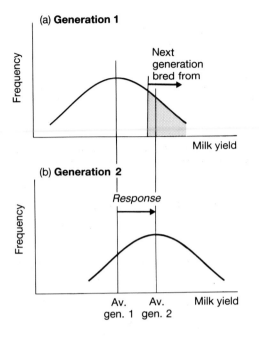

Figure 3.3 An artificial selection experiment. Generation 2 is formed by breeding from a selected minority (shaded area) of the members of generation 1. Here, for example, we imagine a population of cows and selectively breed for high milk yield. In nearly all cases, the average in the second generation changes from the first in the selected direction.

between 1930 and 1965; during this time, the gene frequency underwent a 10-fold increase followed by a similar decrease (Figure 3.2). The population, therefore, underwent evolutionary change. If the 10-fold increase between 1930 and 1940 had been continued for another century or so, the population would have evolved from a mainly black and red form, to a mainly melanic form. As it happened, the increase was not continued; it went into reverse. Evolutionary change in a natural population can therefore be seen even over as short a time period as a decade. Of the three alternatives in Figure 3.1, it supports both evolution and transformism (a,b), but not species fixity (c).

3.3 Evolution can also be produced experimentally

In a typical "artificial selection" experiment, a new generation is formed by allowing only a selected minority of the current generation to breed (Figure 3.3). The population in almost all cases will respond: the average in the next generation will have moved in the selected direction. The procedure is routinely used in agriculture: artificial selection has, for example, been used to alter the number of eggs laid by hens, the meat properties of bullocks, and the milk yield of cows. We shall meet several more examples of artificial experiments later (Figure 8.6, p. 201; also see particularly sections 9.7–9.8, pp. 226–233), but we can look at a curiosity here for purposes of illustration (Figure 3.4). In an experiment, rats were selected for increased or decreased susceptibility to dental caries on a controlled diet. As the graph shows, the rats could be successfully selected to grow better (resistant) or worse (susceptible) teeth. Evolutionary change can therefore be generated artificially.

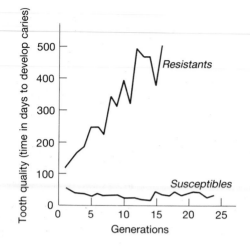

Figure 3.4 Selection for better and worse teeth in rats. Hunt *et al.* (1955) selectively bred from rats that developed caries later (resistants) or earlier (susceptibles) in life. The age (in days) at which the rats in the two lines developed caries was measured.

Artificial selection can produce dramatic change, if continued for long enough. A kind of artificial selection, for example, has generated almost all our agricultural crops and domestic pets. No doubt the artificial selection in these cases—begun thousands of years ago in some cases—employed less formal techniques than would a modern breeder. However, the longer time span has produced some telling results. The varieties of domestic dog are a clear example: within this single species, the difference between extreme forms, like a Pekingese and a St Bernard, is much greater than that between two real species in nature, such as a wolf (*Canis lupus*) and the silver-backed jackal (*Canis mesolemus*), or even two genera, such as either the wolf or jackal, and the African hunting dog (*Lycaeon pictus*). Darwin was impressed by the varieties of pigeon; and these illustrate the same point: on a small scale, species can be shown experimentally not to be fixed in form.

3.4 Natural species show extensive variation

At any one time and place, there do appear to be an array of distinct species in nature. This parochial perspective may be the reason why many people believe that species have separate origins and are fixed. In northern Europe, for example, two good species are the lesser black-backed gull (*Larus fuscus*) and the herring gull (*L. argentatus*). They do not interbreed, they choose different nest sites, and they look different. Any bird watcher can distinguish them. The lesser black-backed gull, as its name implies, has an almost black colored mantle, whereas the herring gull's is gray. They also differ in the color of their legs: the lesser black-backed gull has yellow legs, the herring gull has pink legs.

However, if we look further afield the impression of two distinct species becomes blurred. If we move to North America and look for the two gull species there, we find only one of them, the herring gull. (In fact there has been an increasing number of sitings of the lesser black-backed gull on the Atlantic coast in recent years, but we can ignore that here.) The North American herring gull looks much like the European; however, as we continue to follow it around the pole, from North America to Asia, it looks decreasingly like the European herring gull and increasingly like the

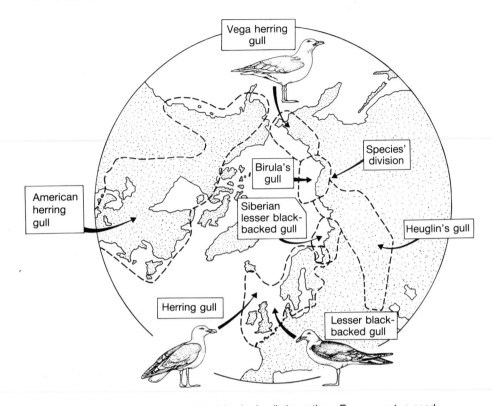

Figure 3.5 The herring and lesser black-backed gulls in northern Europe are two good species. Their geographic distribution is in the form of a ring around the North Pole, and if we trace either species around the ring (herring gull to the west, lesser black-backed gull to the east), we see a complete series of intermediates between it and the other species. Only at the endpoint of the ring, in Europe, are there two species.

European lesser black-backed gull. On the Siberian side of the Bering Strait, it is called the Siberian vega gull. The Siberian vega gull is classified in the same species as the herring gull, but it is noticeably different from the North American and European members of the species. It has almost yellow legs, and its mantle coloration is rather darker: in other words, it looks rather like a lesser black-backed gull. Across Siberia, the trend continues. At a location between central and north-east Siberia, the gulls' mantle coloration is dark enough for them to look more like European lesser black-backed gulls than like herring gulls; and the species is then classified as *Larus fuscus*. Hence, between Siberia and Europe, the coloration of the gull gradually changes until it merges into the lesser black-backed gull.

The herring gull and lesser black-backed gull are an example (not the only one) of a *ring species* (Figure 3.5). There is an almost continuous set of intermediates between the two species, and these intermediates happen to be arranged in a ring. At most points in the ring, there is only one species; but there are two where the endpoints meet. Ring species can provide important evidence for evolution. They show that intraspecific variation can be large enough to make two species; that the differences between species are the same in kind, though not in degree, as the differences between individuals, and populations, within a species.

Natural variation comes in all degrees. At the smallest level, there are slight differences between individuals. Populations of a species show rather larger differences, and species are more different still. Normally, although the geographic extremes of a species may look very different from each other, we do not know whether they are different enough to count as separate species in the reproductive sense, because we do not know whether they could interbreed. But in ring species the extremes meet, and we can see that they form two species. Those who deny that natural variation can be large enough to generate new species are put in some difficulty. They have to accept that at least some variation is natural: every individual is unique, and a creationist who says that all variation is due to separate creation would end up having to claim that every individual gull (and every individual human) is separately created, rather than naturally procreated: and we can *see* that this is false.

Once we accept that some variation is natural, we are at the top of a steep, slippery slope. To deny that natural variation can lead to new species requires an arbitrary decision about where evolution stops and separate creation starts. If evolution has produced the variation between gulls in Alaska and those in east Siberia, and between gulls in central Siberia and those in Europe, it is clearly absurd to suggest that the populations in east and central Siberia were separately created. The variation between any two points in the ring is of much the same kind. Ring species, therefore, show that there is a continuum from interindividual to interspecies variation. Natural variation is sufficient to break down the idea of a distinct species boundary.

The same argument, we shall see, can be applied to larger groups than species, and by extension to all life. The idea that nature comes in discrete groups, with no variation between, is a naïve perception. If the full range of natural forms, in time and space, is studied, all the apparent boundaries become fluid.

3.5 New species can be produced experimentally

The species barrier can be broken by experiment too. The varieties of artificially produced domestic animals and plants can differ in appearance at least as much as natural species; but they may be able to interbreed. Dog breeds that differ greatly in size probably in practice interbreed little, but it is still interesting to know whether we can make new species that unambiguously do not interbreed. It is possible to select directly for reduced interbreeding between two forms (section 16.4.2, p. 413). However, more extreme, and more abundant, examples of new, reproductively isolated species come from plants. The typical procedure is as follows. We begin with two distinct, but related species. The pollen of one is painted on the stigma of the other. If a hybrid offspring is generated, it is usually sterile: the two species are reproductively isolated. However, it may be possible to treat the hybrid in such a way as to make it fertile. The chemical colchicine can often restore hybrid fertility. It does so by causing the hybrid to double its number of chromosomes (a condition called polyploidy). Hybrids produced in this way can usually only interbreed with other hybrids, and not with the parental species. They are a new reproductive species.

The first artificially created hybrid polyploid species was a primrose *Primula kewensis*. It was formed by crossing *P. verticillata* and *P. floribunda*. *Primula kewensis* is a distinct species: a *P. kewensis* individual will breed with another *P. kewensis* individual, but not with members of *P. verticillata* or *P. floribunda*. *Primula verticillata* and *P. floribunda* have 18 chromosomes each, and simple hybrids between them also have 18 chromosomes. These hybrids are sterile. *Primula kewensis* has 36 chromosomes and is a fertile species. The procedure first successfully applied with *P. kewensis* is now a common method of producing new agricultural and horticultural varieties. Most garden varieties of irises, tulips, and dahlias, for example, are artificially created species; but their numbers are dwarfed by the huge numbers of artificial hybrid species of orchids, which it has been estimated are being formed at the rate of about 300 per month.

Polyploid hybrids are a major source of new natural, as well as artificial, plant species. It is possible to guess, from the number of chromosomes in groups of related species, which ones originated by hybridization. The hypothesis can be tested by artificially recreating the conjectural hybrid from its parents. This was first done for the mint *Galeopsis tetrahit*, which Müntzing in 1930 successfully created by hybridizing *G. pubescens* and *G. speciosa*. The artificially generated *G. tetrahit* can successfully interbreed with naturally occuring members of the species. It is quite common for a genus, or a family, of flowering plants to be made up of species with simple multiples of a basic number of chromosomes (N): the different species might have N, $2N$, $4N$, etc. chromosomes. The obvious interpretation is that many, or all, of the species with higher numbers of chromosomes have originated by polyploidization from species with fewer chromosomes. Chromosome counts have been made for many plant groups, and it has been estimated that as many as 70–80% angiosperm species are polyploids. Not all of these may have originated by the exact means described here, but they have probably all originated by some variant of it. In conclusion, it is possible to make new species, and by a method that has been highly important in the origin of new natural species.

3.6 Small-scale observations can be extrapolated over the long term

We have now seen that evolution can be observed directly on a small scale: the extreme forms within a species can be as different as two distinct species, and in nature and in experiments species will evolve into forms highly different from their starting point. It would be impossible, however, to observe in the same direct way the whole evolution of life from its common, single-celled ancestor a few billion years ago. Human experience is too brief. As we extend the argument from small-scale observations, like those described in moths, dogs, and gulls, to the history of all life we must shift from observation to inference. It is possible to imagine, by extrapolation, that if the small-scale processes we have seen were continued over a long enough period they could have produced the modern variety of life. The reasoning principle is called *uniformitarianism*. In a modest sense, uniformitarianism means merely that processes seen by humans to operate could also have operated when humans were not watching; but it also refers to the more controversial claim that processes operating in the present can account, by extrapolation over long periods, for the evolution of the earth and of life. This principle is not peculiar to evolution. It is used in all historical geology. When the persistent action of river erosion is used to explain the excavation of deep canyons, the reasoning principle again is uniformitarianism.

Differences, it may be argued, can be of kind as well as degree. For instance, many creationists believe that evolution can operate within a species, but cannot produce a new species. Their reason is a belief that species differences are not simply a magnified version of the differences we see between individuals. As a matter of fact, this particular argument is false. For the herring gull and lesser black-backed gull, we saw the smooth continuum of increasing difference, from the variation between individual gulls on any beach, to interregional variation, to speciation. Someone who permits uniformitarian extrapolation only up to a certain point in this continuum will inevitably be making an arbitrary decision. The differences immediately above and below the point will be just like the differences across it.

Analogous arguments to the one about species are sometimes made for higher taxonomic levels. It may be said, for example, that evolution is only possible within defined "types" (a type might be something like "dogs" or "cats," or even "birds" or "mammals"). But the evolutionist will advance the same counter-argument as for species. Nature only appears to be divided up into discrete "types" at any one time and place. Further study erodes away this impression. The fossil record contains a continuous set of intermediates between mammals and reptiles, and these fossils destroy the impression that "mammals" are a discrete type (section 20.1, p. 533); *Archaeopteryx* does the same for the bird type; and there are many further examples. In any case, if someone tries to argue that differences of kind arise at a certain level in the taxonomic hierarchy, they will be faced with these sorts of counter-example. If we draw on enough specimens from time and space, a strong argument can be made that organic variation is continuous, from the smallest difference between a pair of twins through to the whole history of life.

The argument for evolution does not have to rely only on small-scale observations and the principle of uniformitarianism. There are other kinds of evidence suggesting that all living things are descended from a common ancestor: the evidence comes from certain similarities between species, and from the fossil record.

3.7 There are homologous similarities between groups of living things

If we take any two living species, they will show some similarities in appearance. Here, we need to distinguish two sorts of similarity: *homologous* and *analogous* similarity. (In this chapter, these terms have a non-evolutionary meaning, which was common before Darwin. They should not be confused with the evolutionary meanings (see chapter 14). The non-evolutionary usage is needed here in order to avoid a circular argument: evolutionary concepts cannot be used as evidence for evolution.) An analogous similarity, in this non-evolutionary, pre-Darwinian sense, is one that can be explained by a shared way of life. Sharks, dolphins, and whales all have a hydrodynamic shape which can be explained by their habit of swimming through water. Their similar shape is analogous; it is a functional requirement. Likewise, the wings of insects, birds, and bats are all needed for flying: they too are analogous structures.

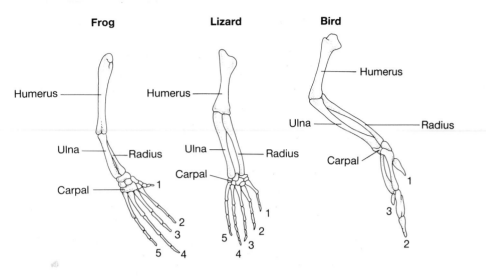

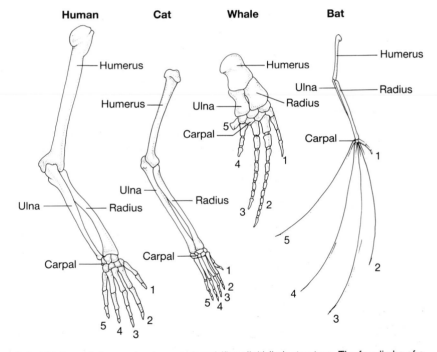

Figure 3.6 All tetrapods have a basic pentadactyl (five-digit) limb structure. The fore limbs of a bird, human, whale, and bat are all constructed from the same bones even though they perform different functions.

Other similarities between species are less easily explained by functional needs. The pentadactyl (five-digit) limb of tetrapods is a classic example (Figure 3.6). (Tetrapods are the group of vertebrates with four legs. Amphibians, reptiles, birds, and mammals are tetrapods; fish are not.) Tetrapods occupy a wide variety of environments, and use their limbs for many differing functions. There is no clear functional or environmental reason why all of them should need a five-digit, rather than a three- or seven- or 12-digit limb. And yet they all do. Bird wings are based on the five-digit limb, as are bat wings in a different way. Seal flippers are pentadactyl limbs, and the boneless hind fin of the whale conceals the vestiges of the characteristic five-digit pattern. Some mammals, such as horses, and some lizards have less than five digits in the adult stage; but even these develop from a five-digit precursor, losing a digit or two during development. In Darwin's words,

> What could be more curious than that the hand of man formed for grasping, that of a mole, for digging, the leg of a horse, the paddle of a porpoise and the wing of a bat, should all be constructed on the same pattern and should include similar bones and in the same relative positions?

The pentadactyl limb is a homology in the pre-Darwinian sense: it is a similarity between species that is not functionally necessary. Pre-Darwinian morphologists thought that homologies indicated a "plan of nature," in some more or less mystical sense; for evolutionary biologists, they are evidence of common ancestry. The evolutionary explanation of the pentadactyl limb is simply that all the tetrapods have descended from a common ancestor that had a pentadactyl limb and, during evolution, it has turned out to be easier to evolve variations on the five-digit theme, than to recompose the limb structure. If species have descended from common ancestors, homologies make sense; but if all species originated separately, it is difficult to understand why they should share homologous similarities. Without evolution, there is nothing forcing the tetrapods all to have pentadactyl limbs.

The pentadactyl limb is a morphologic homology: it has a wide distribution, being found in all tetrapods. At the molecular level there are homologies that have the widest distribution possible: they are found in all life. The genetic code is an example (see Table 2.1, p. 26). The translation between base triplets in the DNA and amino acids in proteins is universal to all life, as can be confirmed, for instance, by isolating the mRNA for hemoglobin from a rabbit and injecting it into the bacterium *Escherichia coli*. *Escherichia coli* does not normally make hemoglobin, but when injected with the mRNA will make rabbit hemoglobin. The machinery for decoding the message must therefore be common to rabbits and *E. coli*; and if it is common to them it is a reasonable inference that all living things have the same code. (Recombinant DNA technology is built on the assumption of a universal code.) Minor variants of the code, which have been found in mitochondria and in the nuclear DNA of a few species, do not affect the argument to be developed here.

Why should the code be universal? There are two possible explanations. One is that it is due to chemical constraint, the other is that it is a historical accident. In the chemical theory, each particular triplet would have some chemical affinity with its amino acid. The codon GGC, for example, would react with glycine in some way that matched the two together. Several lines of evidence suggest this is not so. One is that no such chemical relation has been found (and not for want of looking); and it is generally thought that one does not exist. Secondly, the triplet and amino acid do not physically interact

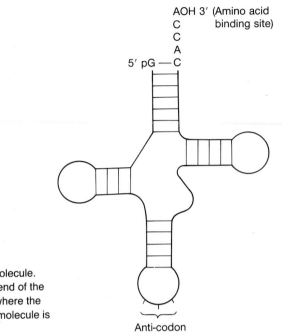

Figure 3.7 Transfer RNA (tRNA) molecule. The amino acid is held at the other end of the molecule from the anti-codon loop where the triplet code of the messenger RNA molecule is read.

in the translation of the code. They are both held on a transfer RNA (tRNA) molecule, but the amino acid is attached at one end of the molecule, while the site that recognizes the codon on the messenger RNA (mRNA) is at the other end (Figure 3.7). Finally, there are mutants that change the relationship between triplet code and amino acid (Figure 3.8). The mutants in question suppress the action of another class of mutants. Some of the triplets in the genetic code are "stop" codons: they act as a signal that the protein has come

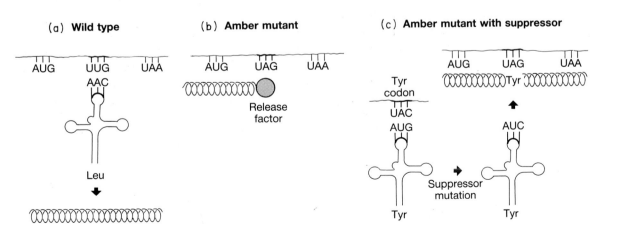

Figure 3.8 Mutations that suppress amber mutations suggest that the genetic code is chemically alterable. For example: (a) the normal codon is UUG and encodes leucine. (b) The UUG mutates to the stop codon UAG (this is called an amber mutation). (c) A tRNA for tyrosine mutates from AUG to AUC (which recognizes UAG) and suppresses the amber mutation by inserting a tyrosine.

to an end. If a triplet within a coding region mutates to a stop codon, the protein is not made. Examples of these mutations are well known in bacterial genetics—a mutation to the stop codon UAG, for example, is called an amber mutation. Now, once a bacterial culture with an amber mutation has been formed, it is sometimes possible to find other mutations that suppress the amber mutation: these mutants are normal, or near normal, bacteria. It turns out that the amber-suppressing mutants work by changing the coding triplet on a class of amino acid bearing tRNA to make it bind to UAG. The UAG codon then encodes an amino acid rather than causing transcription to stop. The fact that the relation between amino acid and codon can be changed in this way shows that the same genetic code has not been forced on all species by some unalterable chemical constraint.

If the genetic code is not chemically determined, why is it the same in all species? The most popular theory is as follows. The code is arbitrary, in the same sense that human language is arbitrary. In English the word for a horse is "horse," in Spanish it is "caballo," in French it is "cheval," in Ancient Rome it was "equus." There is no reason why one particular sequence of letters rather than another should signify that familiar perissodactylic mammal. Therefore, if we find more than one person using the same word, it implies they have both learned it from a common source. It implies common ancestry. When the *U.S.S. Enterprise* boldly descends on one of those extra-galactic planets where the aliens speak English, the correct inference is that the locals share a common ancestry with one of the English speaking peoples of the Earth. If they had evolved independently, they would not be using English.

All living species use a common, but equally arbitrary, language in the genetic code. The reason is thought to be that the code evolved early on in the history of life, and one early form turned out to be the common ancestor of all later species. (Notice that saying all life shares a common ancestor is not the same as saying life evolved only once.) The code is then what Crick called a "frozen accident." The original choice of a code was an accident; but once it had evolved, it would be strongly maintained. Any deviation from the code would be lethal. An individual that read GGC as phenylalanine instead of glycine, for example, would bungle all its proteins, and probably die at the egg stage.

The universality of the genetic code is important evidence that all life shares a single origin. In Darwin's time, morphologic homologies like the pentadactyl limb were known about; but these are shared between fairly limited groups of species (like all the tetrapods). Cuvier (section 1.3, p. 9) had arranged all animals into four large groups according to their morpho-logic homologies; and for this reason Darwin suggested there may have been a limited number, but perhaps more than one, common ancestor for modern living species. Universal molecular homologies, such as the genetic code, now provide the best evidence that all life has a single common ancestor.

Homologous similarities between species provide the most widespread class of evidence that living and fossil species have evolved from a common ancestor. The anatomy, biochemistry, and embryonic development of each species contains innumerable characters like the pentadactyl limb and the genetic code: characters that are similar between species, but would not be if the species had independent origins. The "argument from homology," however, is usually more persuasive for an educated biologist than for someone seeking immediately intelligible evidence for evolution. The most obvious evidence for evolution is that from direct observation of change; no one will have any difficulty in seeing how the examples of evolution in action, from moths and artificial selection, suggest that species are not fixed in form. The argument from homology is inferential, and more demanding. Some understanding of functional morphology or molecular biology is needed to appreciate that tetrapods would not share the pentadactyl limb, or all species the genetic code, if they originated independently.

But some homologies are immediately persuasive: the homologies, such as vestigial organs, in which the shared form appears to be positively inefficient. If we stay with the vertebrate limb, but move in from its extremities to the junction where it joins the spine, we find another set of bones—at the pectoral and pelvic articulations—that are recognizably homologous in all tetrapods. In most species, these bones are needed in order for the limb to be

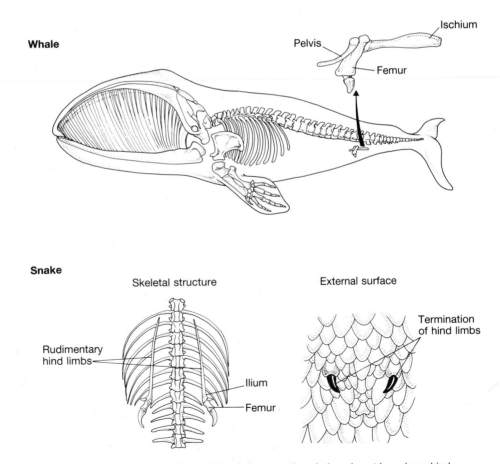

Figure 3.9 Whales have a vestigial pelvic girdle, even though they do not have bony hind limbs. The pelvic bones are homologous with those of other tetrapods. Snakes have vestigial hind limbs.

able to move; but in a few species the limbs have been lost. Modern whales, for instance, do not have hind limbs with bony supports. If we dissect a whale, we find at the appropriate place down the spine a set of bones that are clearly homologous with the pelvis of any other tetrapod (Figure 3.9). They are vestigial in the sense that they are no longer used to provide articulation for the hind limb; their retention suggests that whales evolved from tetrapods rather than being independently created. If an organ is called vestigial that does not mean it is necessarily functionless. Some of them may be truly functionless, but it is always difficult to confirm universal negative statements. Fossil whales called *Basilosaurus* living 40 million years ago, had functional pelvic bones and may have used them when copulating; and the vestigial pelvis of modern whales arguably is still needed to support the reproductive organs. However, that possibility does not count against the argument from homology: for why, if whales originated independently of other tetrapods, should whales use bones that are adapted for limb articulation in order to support their reproductive organs? If they were truly independent, some other support would be used.

In homologies like the pentadactyl limb and the genetic code the similarity between species is not actively disadvantageous. One form of genetic code would probably be as good as almost any other, and no species suffers for using the actual genetic code found in nature. However, there are some homologies that do look positively disadvantageous (sections 13.7–13.11, pp. 333–346). One of the cranial nerves goes from the brain to the larynx via a tube near the heart (see Figure 13.10, p. 344). In fish this is a direct route. But the same nerve in all species follows the same route, and in the giraffe it results in an absurd detour down and up the neck, so that the giraffe has to grow maybe 3–5 meters more nerve than it would with a direct connection. The "recurrent laryngeal nerve," as it is called, is surely inefficient. It is easy to explain such an efficiency if giraffes have evolved in small stages from a fish-like ancestor; but why giraffes should have such a nerve if they originated independently . . . well, we can leave that to others to try to explain.

3.8 Different homologies are correlated, and can be hierarchically classified

Different species share homologies, which suggests they are descended from a common ancestor; but the argument can be made both stronger and more revealing. Homologous similarities are the basis of biological classifications (see chapter 14): groups like "flowering plants" or "primates" or "cats" are formally defined by homologies. The reason homologies are used to define groups is that they fall into a nested, or hierarchical, pattern of groups within groups; and different homologies consistently fall into the same pattern.

A molecular study by Penny *et al.* illustrates the point, and shows how it argues for evolution. Different species can be more-or-less similar in the amino acid sequences of their protein, just as they can be more-or-less similar in their morphology. The pre-Darwinian distinction between analogy and homology is more difficult to apply to proteins. Our functional understanding of protein sequences is less well advanced than that of morphology, and it can be difficult to specify an amino acid's function in the way we can for the pentadactyl limb. Actually, the functions of many protein sequences are understood, but the chemistry takes a lot of explaining. For the argument

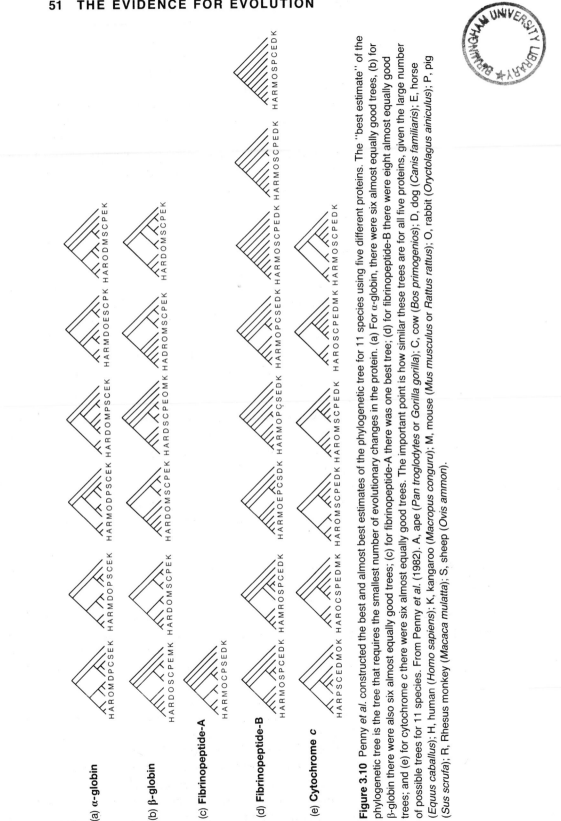

Figure 3.10 Penny *et al.* constructed the best and almost best estimates of the phylogenetic tree for 11 species using five different proteins. The "best estimate" of the phylogenetic tree is the tree that requires the smallest number of evolutionary changes in the protein. (a) For α-globin, there were six almost equally good trees, (b) for β-globin there were also six almost equally good trees; (c) for fibrinopeptide-A there was one best tree; (d) for fibrinopeptide-B there were eight almost equally good trees; and (e) for cytochrome *c* there were six almost equally good trees. The important point is how similar these trees are for all five proteins, given the large number of possible trees for 11 species. From Penny *et al.* (1982). A, ape (*Pan troglodytes* or *Gorilla gorilla*); C, cow (*Bos primogenios*); D, dog (*Canis familiaris*); E, horse (*Equus caballus*); H, human (*Homo sapiens*); K, kangaroo (*Macropus conguru*); M, mouse (*Mus musculus* or *Rattus rattus*); O, rabbit (*Oryctolagus ainiculus*); P, pig (*Sus scrufa*); R, Rhesus monkey (*Macaca mulatta*); S, sheep (*Ovis ammon*).

here, it only needs to be accepted that *some* of the amino acid similarities between species are not functionally necessary, in the same way that all tetrapods do not have to have five-digit limbs. There are a large number of amino acids in a protein, so this need not be controversial. If we accept that some amino acids are homologous in the pre-Darwinian sense, we can see how their distribution among species suggests evolution.

Penny *et al.* examined protein sequences in a group of 11 species. They used the pattern of amino acid similarities to work out the evolutionary "tree" for the species. Some species have more similar protein sequences than others, and the more similar species are grouped more closely in the tree (see chapter 17). The observation that suggests evolution is as follows. We start by working out the tree for one protein. We can then work it out for another protein, and compare the trees. Penny *et al.* worked out the 11 species tree for each of five proteins. The key observation was that the trees for all five proteins are very similar (Figure 3.10). For 11 species, there are 34 459 425 different possible trees, but the five proteins suggest trees that form a small subclass from this large number of possible trees.

The similarities and differences in the amino acid sequences of the five proteins are correlated: if two species have more amino acid homologies for one of the proteins, they are also likely to for the other proteins. That is why any two species are likely to be grouped together for any of the five proteins. If the 11 species had independent origins, there is no reason why their homologies should be correlated. In a group of 11 separately created species, some would no doubt show more similarities than others for any particular protein. But why should two species that are similar for, say, cytochrome c, also be similar for β-globin and fibrinopeptide-A? The problem is more difficult than that, because, as Figure 3.10 shows, all five proteins show a similar pattern of branching at all levels in the 11-species tree. It is easy to see how a set of independently created objects might show hierarchical patterns of similarity in any one respect. But these 11 species have been classified hierarchically for five different proteins: and the hierarchy in all five cases is similar.

If the species are descended from a common ancestor, the observed pattern is exactly what we should expect. All of the five proteins have been evolving in the same pattern of evolutionary branches, and we therefore expect them to show the same pattern of similarities. The hierarchic pattern of, and correlations among, homologies are evidence for evolution.

Consider an analogy. Consider a set of 11 buildings, each of which was independently designed and built. We could classify them into groups according to their similarities; some might be built of stone, others of brick, others of wood; some might have vaults, others ceilings; some arched windows, others rectangular windows; and so on. It would be easy to classify them hierarchically by one of these properties, such as building material. This classification would be analogous, in the study Penny *et al.*'s, to making the tree of the 11 species for one protein. The same buildings could then be classified by another property, such as window shape; this is analogous to classifying the species by a second protein. There would probably be some correlations between the two classifications of the buildings, because of functional factors. Maybe buildings with arched windows would be more likely to be built of brick or stone, than of wood. However, other similarities would just be non-functional, chance associations in the particular 11 buildings in the sample. Maybe, in this sample, the white-colored buildings also happened to have garages, whereas the red buildings tended not to. The argument for evolution concentrates on these "inessential," rather than functional, patterns of similarity.

The analogy of Penny *et al.*'s result in the case of the buildings would be as follows. We should classify 11 buildings by five independent sets of characters (like building material, window shape, type of garage, etc.). We should then look to see whether the five classifications all grouped the buildings in the same way. If the buildings were erected independently, there is no reason why they should show functionally unnecessary correlations. There would be no reason to expect that buildings that were similar for, say, window shape, would also be similar with respect to, say, number of chimney pots, or angle of roof, or the arrangement of chairs indoors.

Of course, some innocent explanation might be found for any such correlations. (Indeed if correlations *were* found in a real case, there would have to be some explanation.) Maybe they could all be explained by class of owner, or region, or common architects. But that is another matter; it is just to say that the buildings were not really independently created. If they were

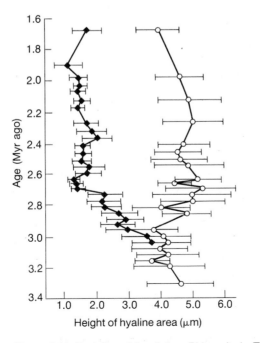

Figure 3.11 Evolution of the diatom *Rhizosolenia*. The form of the diatom is measured by the height of the hyaline (glass-like) area of the cell wall. ○—○ indicate forms classified as *R. praebergonii*, ◆—◆ indicate *R. bergonii*. Bars indicate the range of forms at each time. From Cronin and Schneider (1990).

independently created, it would be very puzzling if they showed systematic, hierarchical similarity in functionally unrelated characteristics.

In the case of biological species, we do find this sort of correlation between characters. Figure 3.10 shows how similar the branching patterns are for five proteins, and the same conclusion could be drawn from any well researched classification in biology. Biological classifications, therefore, provide an argument for evolution. If species had independent origins, we should not expect that, when several different (and functionally unrelated) characters were used to classify them, all the characters would produce strikingly similar classifications.

3.9 There is some fossil evidence for the transformation of species

Diatoms are single-celled, photosynthetic organisms that float in the sea and in fresh water. Many species grow beautiful glass-like cell walls, and these can be preserved as fossils. Figure 3.11 illustrates the fossil record for the diatom *Rhizosolenia* between 3.3 and 1.6 million years ago. About 3 million years ago, a single ancestral species split into two; and there is a comprehensive fossil record of the change at the time of the split. (We shall meet some similar examples in chapter 19.)

The diatoms in Figure 3.11 show that the fossil record can be complete enough to reveal the origin of a new species; but examples as good as this are rare. In other cases, the fossil record is less complete and there are large gaps between successive samples (see Appendix). There is then only less direct evidence of smooth transitions between species. The gaps are usually long, however (maybe 25 000 years in a good case, and millions of years in less complete records). There is enough time within one of the gaps for large evolutionary changes, and no one need be surprised that fossil samples from either side of a gap in the record show large changes.

In other respects, the fossil record provides important evidence for evolution. For someone who believes that the world has always been like it is now, the mere demonstration of bizarre extinct animals, like dinosaurs or the animals of the Burgess Shale, should make them think again. The mere existence of extinct species does not argue strongly for evolution, because an extinct species could just as well have been separately created as any modern species. Only a small change in the theory of separate creation would be needed (see Figure 3.1c). The faunas and floras of the past differ so much from those of today, both in what species were present and what were not, that all the species that have ever lived could not have been created at the same time: we

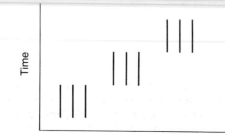

Figure 3.12 The theory of separate creation (see Figure 3.1c) can easily be modified to include extinct species and differences between present and fossil faunas. Each line represents a species in time.

should expect to find at least some modern species in the deep past if there had been a simultaneous mass creation. The creationist picture can be modified to accommodate this observation (Figure 3.12). Indeed, early paleontologists, who were well aware how different past faunas were from those of the present, suggested exactly this picture; they described the history of life as a succession of rounds of extinction followed by the creation of new species.

3.10 The order of the main groups in the fossil record suggests they have evolutionary relationships

The main subgroups of vertebrates on a conventional classification are: fish, amphibians, reptiles, birds, and mammals. It is possible to deduce that their order of evolution must have been fish then amphibians then reptiles then mammals; and not, for example, fish then mammals then reptiles then amphibians (Figure 3.13a). The deduction follows from the observation that an amphibian, such as a frog, or a reptile, such as an alligator, is intermediate in form between a fish and a mammal. Amphibians, for instance, have gills, as fish do; but have four legs, like reptiles and mammals, and not fins. If fish had evolved into mammals, and then mammals had evolved into amphibians,

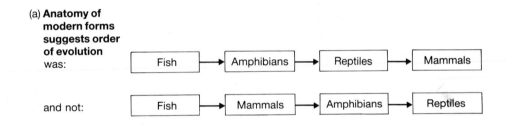

(a) **Anatomy of modern forms suggests order of evolution was:**

Fish → Amphibians → Reptiles → Mammals

and not:

Fish → Mammals → Amphibians → Reptiles

(b) **Order of main vertebrate groups in fossil records**

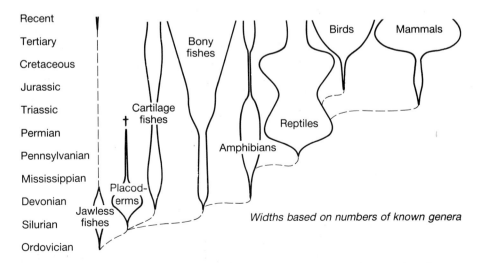

Figure 3.13 (a) Anatomic analysis of modern forms indicates that amphibians and reptiles are evolutionarily intermediate between fish and mammals. This order fits with (b) the geologic succession of the major vertebrate groups. The width of each group indicates the diversity of the group at that time. From Simpson (1949b).

the gills would have been lost in the evolution of mammals and then regained in the evolution of amphibia; which is much less likely than amphibia evolving from fish, retaining their gills, and the gills being lost in the origin of mammals. (Chapter 17 discusses these arguments more fully.) Gills and legs are just two examples: the full list of characters putting amphibians (and reptiles, by analogous arguments) between fish and mammals would be long indeed. The forms of modern vertebrates alone, therefore, enable us to deduce the order in which they evolved. (Strictly speaking, on the argument given here, it could also be that mammals came first and evolved into reptiles, the reptiles evolved into amphibians, and the amphibians into fish. However, we can extend the argument by including more groups of animals, back to a single-celled stage; the fish would then be revealed in turn as an intermediate stage between amphibians and simpler animals.)

The inference, from the modern forms, can be tested against the fossil record. The fossil record supports it: fish, amphibians, reptiles, and mammals, appear in the fossil record in the same order as they should have evolved (Figure 3.13b). The fit is good evidence for evolution, because if fish, amphibians, reptiles, and mammals had been separately created, we should not expect them to appear in the fossil record in the exact order of their apparent evolution. Fish, frogs, lizards, and rats would probably appear as fossils in some order, if they did not appear at the same time; but there is no reason to suppose they would appear in one order rather than another. It is therefore a revealing coincidence when they turn out in the evolutionary order. Similar analyses have been done with other large and well-fossilized groups of animals, such as the echinoderms, and have found the same result.

The argument can be stated another way. Haldane once said he would give up his belief in evolution if someone found a fossil rabbit in the Precambrian. The reason is that the rabbit, which is a fully formed mammal, must have evolved through reptilian, amphibian, and piscine stages and should not therefore appear in the fossil record 100 million years or so before its fossil ancestors. Creationists have appreciated the power of this argument. Various claims have been made for fossil human footprints contemporary with dinosaur tracks. Whenever one of these claims has been properly investigated, it has been exploded: some have turned out to have been carved fraudulently, others were carved as tourist exhibits, others are perfectly good dinosaur footprints. But the principle of the argument is valid. If evolution is correct, humans could not have existed before the main radiation of mammals and primates, and these took place after the dinosaurs had become extinct. The fact that no such human fossils have been found—that the order of appearance of the main fossil groups matches their evolutionary order—is the way in which the fossil record does provide good evidence for evolution.

3.11 Summary of the evidence for evolution

We have met three main classes of evidence for evolution: from direct observation, on the small scale; from homology; and from the order of the main groups in the fossil record. The small-scale observations work most powerfully against the idea of species fixity; by themselves, they are almost equally good evidence for evolution and for "transformism" (see Figure 3.1a,b). They show, by uniformitarian extrapolation, that evolution could have produced the whole history of life. Stronger arguments for large-scale evolution come from homologies and the fossil record. The geologic succession of the major groups and most classical morphologic homologies strongly suggest that these large groups have a common ancestor. The more recently discovered molecular homologies, such as the universal genetic code, extend the argument to the whole of life—and favor evolution (Figure 3.1a) over both transformism and separate creation (Figure 3.1b,c).

Such is the standard argument for evolution. Moreover, the theory of evolution can also be used to make sense of, and to analyze, a large array of additional facts. As we study the different areas of evolutionary biology, it is worth keeping the issue of this chapter in mind. How, for example, could we explain the molecular clock (section 7.11, p. 164) if species have independent origins? Or the difficulties of deciding whether closely related forms are different species (see chapter 15)? Or the relative changes in prezygotic and postzygotic reproductive isolation through time (section 16.8, p. 435)? Or the unique branching pattern of chromosomal inversions in the Hawaiian fruitflies (section 17.7, p. 456)?

3.12 Separate creation does not explain adaptation

There is another powerful reason why evolutionary biologists do not take the theory of separate creation seriously. Separate creation does not explain adaptation. Living things are well designed, in innumerable respects, for life in their natural environments; they have sensory systems to find their way around, feeding systems to catch and digest food, nervous systems to co-ordinate their actions. The theory of evolution has a mechanical, scientific theory for adaptation: natural selection.

Separate creation, by contrast, does not explain adaptation. When the species originated, they must have already been equipped with adaptations for life, because the theory holds that species are fixed in form after their origin. An unabashedly religious version of separate creation would attribute the adaptiveness of living things to the genius of God; but even this does not actually explain the origin of the adaptation, it just pushes the problem back one stage (section 13.1, p. 323). In the scientific version of the theory which we are concerned with here, supernatural events do not take place, and we are left with no theory of adaptation at all. Without a theory of adaptation, as Darwin realized (section 1.3.2, p. 10), any theory of the origin of living things is a non-starter.

3.13 Modern "scientific creationism" is scientifically untenable

That life has evolved is one of the great discoveries in all the history of science, and it is correspondingly interesting to know the arguments in favor of it. In modern evolutionary biology, the question of whether evolution happened is no longer a topic of research, because the question has been answered; but it is still controversial outside science. Christian fundamentalists—some of them politically influential—in the USA have supported various forms of the theory of separate creation and have been trying since the 1920s, sometimes successfully, sometimes unsuccessfully, to intrude them into school biology curricula.

What relevance do the arguments of this chapter have for these forms of the theory of separate creation? For a purely scientific form, the relevance is straightforward. Separate creation, as shown in Figure 3.1c, which simply suggests that species have had separate origins and been fixed since then, has been the subject of the whole chapter and we have seen that it is refuted by the evidence. The theory of separate creation in Figure 3.1c said nothing about the mechanism by which species originated and therefore need not assert that the species were created by God: a supporter of Figure 3.1c might merely say that species originated by some natural mechanism, the details of which are not yet understood. However, it is unlikely that anyone would now seriously support the theory of Figure 3.1c unless they also believed that the species originated supernaturally. Then we are not dealing with a scientific theory.

This chapter has confined itself to the scientific resources of logical argument and public observation. Scientific arguments only employ observations that anybody can make, as distinct from private revelations, and consider only natural, as distinct from supernatural, causes. Indeed, two good criteria to distinguish scientific from religious arguments are whether the theory invokes only natural causes, or needs supernatural causes too, and whether the evidence is publicly observable or requires some sort of "faith." Without these two conditions, there are no constraints on the argument. It is, in the end, impossible to show that species were not created by God and have remained fixed in form, because to God (as a supernatural agent) everything is permitted. It equally cannot be shown that the building (or garden) you are in, and the chair you are sitting on, were not created supernaturally by God 10 seconds ago from nothing: at the time, God would also have to have adjusted your memory and those of all other observers, but a supernatural agent can do that. This is why supernatural agents have no place in science.

Two final points are worth making. The first is that, although modern "scientific creationism" closely resembles the theory of separate creation in Figure 3.1c, it also possesses the added feature of specifying the time when all the species were created. Theologians working after the Reformation were able to deduce, from some plausible astronomical theory and rather less plausible Biblical scholarship, that the events described in Genesis chapter 1 happened about 6000 years ago; and fundamentalists in our own time have retained a belief in the recent origin of the world. A statement of creationism in the 1970s (and the one legally defended in court at Arkansas in 1981) included, as a creationist tenet, that there was "a relatively recent inception of the earth and living kinds." Scientists accept a great age for the earth because of radioactive dating and cosmological inferences from the background radiation. Cosmological and geologic time are important scientific discoveries, but we have ignored them in this chapter because our subject has been the scientific case for evolution: religious fundamentalism is another matter.

Finally, it is worth stressing that there need be no conflict between the theory of evolution and religious belief. This is not an "either/or" controversy, in which accepting evolution means rejecting religion. No important religious beliefs are contradicted by the theory of evolution, and religion and evolution should be able to coexist peacefully in anyone's set of beliefs about life.

3.14 Summary

1 A number of lines of evidence suggest that species have evolved from a common ancestor, rather than being fixed in form and created separately.

2 On a small scale, evolution can be seen happening in nature, such as in the color patterns of moths, and in artificial selection experiments, such as those used in breeding agricultural varieties.

3 Natural variation can cross the species border, for example in the ring species of gulls, and new species can be made artificially, as in the process of hybridization and polyploidy by which many agricultural and horticultural varieties have been created.

4 Observation of evolution on the small scale, combined with the extrapolative principle of uniformitarianism, suggest that all life could have evolved from a single common ancestor.

5 Homologous similarities between species (understood as similarities that do not have to exist for any pressing functional reason), suggest that the species descended from a common ancestor. Universal homologies—such as the genetic code—found in all living things suggest that all species are descended from a single common ancestor.

6 The fossil record provides some direct evidence of the origin of new species.

7 The order of succession of major groups in the fossil record is predicted by evolution, and contradicts the separate origin of the groups.

8 The separate creation of species does not explain adaptation; evolution, by the theory of natural selection, offers a valid explanation.

3.15 Further reading

Eldredge (1982), Ruse (1982), and Futuyma (1983) consider separate creation and the case for evolution at book length. Chapters 10–14 of *On the Origin of Species* (Darwin, 1859) are the classic account of the evidence for evolution. For examples of evolution in action, see Ford (1975), Endler (1986), and, for the tiger moth, Jones (1989). Gingerich *et al.* (1990) describe the Eocene whale with functional hind limbs. Gould (1989) describes the animals of the Burgess Shale. Wellnhofer (1990) describes *Archaeopteryx*. On adaptation, see Dawkins (1986). The broader issues are studied by Nelkin (1977, 1982) and in the anthology compiled by Zetterberg (1983). Larson (1989) describes the legal history of the evolution–creation issue.

Natural selection and variation

4.1 In nature, there is a struggle for existence

The Atlantic cod (*Gadus callarias*) is a large marine fish, and an important source of human food. It is also one of the most fecund of vertebrates. An average 10-year-old female cod lays about 2 million eggs in a breeding season, and large individuals may lay over 5 million (Figure 4.1a). Female cod ascend from deeper water to the surface to lay their eggs; but as soon as they are discharged, a slaughter begins. The plankton is a dangerous place for eggs. The billions of eggs released by cod alone are devoured by innumerable planktonic invertebrates, by other fish, and by fish larvae. About 99% of cod eggs die in their first month of life, and another 90% or so of the survivors die before reaching an age of 1 year (Figure 4.1b). A negligible proportion of the 5 million or so eggs laid by a female cod in her lifetime will survive and reproduce: an average female in the cod population will produce only two successful offspring.

This figure, that on average two eggs per female survive to reproduce successfully, is not the result of an observation: it comes from a logical calculation. Only two can survive, because any other number would be unsustainable over the long term. It takes a pair of individuals to reproduce. If an average pair in a population produce less than two offspring, the population will soon become extinct; if they produce more than two, the population will rapidly reach infinity—which is also unsustainable. Over a small number of generations, the average female in a population may produce more or less than two successful offspring, and the population will increase or decrease accordingly; but the long-term average must be two. Such is the reason for inferring that of the maybe 2 million eggs laid by a female cod in a season, 1 999 998 die before reproducing.

A life table can be used to describe the mortality of a population (Table 4.1). A life table begins at the egg stage and traces what proportion of the original 100% of eggs die off at the successive stages of life. In some species, mortality is concentrated early in life, in others mortality has a more constant rate throughout life. But in all species there is mortality, mortality which reduces the numbers of eggs down to a lower number of adults.

The condition of "excess" fecundity—females producing more offspring than survive—is universal in nature. In every species, more eggs are produced than can survive to the adult stage. The cod dramatizes the point in one way, because its fecundity, and mortality, are so high; but Darwin dramatized the same point by considering the opposite kind of species—one that has an extremely low reproductive rate. The fecundity of elephants is

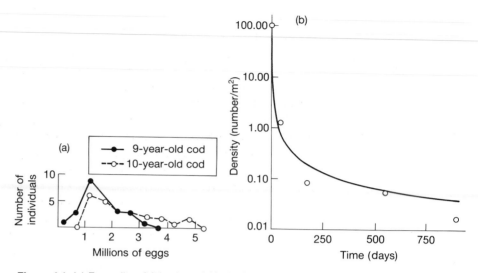

Figure 4.1 (a) Fecundity of Atlantic cod. Notice both the large numbers, and the variability between individuals. The more fecund cod lay perhaps five times as many eggs as the less fecund; much of the variation is associated with size, because larger individuals lay more eggs. From May (1967). (b) Mortality of cod in their first 3 years of life. From Cushing (1975).

low, but even they produce many more offspring than can survive. Here are Darwin's words:

> the elephant is reckoned the slowest breeder of all known animals, and I have taken some pains to estimate its probable minimum rate of natural increase; it will be safest to assume it begins breeding when thirty years old, and goes on breeding until ninety years old, bringing forth six

Table 4.1 A life table for the annual plant *Phlox drummondii* in Nixon, Texas. The life table gives the proportion of an original sample (cohort) that survive to various ages. A full life table may also give the fecundity of individuals at each age. From Leverich and Levin (1979)

Age interval (days)	Number surviving to day x	Proportion of original cohort surviving to day x	Proportion of original cohort dying during interval	Mortality rate per day
0–63	996	1.000	0.329	0.005
63–124	668	0.671	0.375	0.009
124–184	295	0.296	0.105	0.006
184–215	190	0.191	0.014	0.002
215–264	176	0.177	0.004	0.001
264–278	172	0.173	0.005	0.002
278–292	167	0.168	0.008	0.003
292–306	159	0.160	0.005	0.002
306–320	154	0.155	0.007	0.003
320–334	147	0.148	0.043	0.021
334–348	105	0.105	0.083	0.057
348–362	22	0.022	0.022	1.000
362–	0	0	—	

young in the interval, and surviving till one hundred years old; if this be so, after a period of 740 to 750 years there would be nearly nineteen million elephants alive, descended from the first pair.

In elephants then, as in cod, many individuals die between egg and adult; they both have excess fecundity. This excess fecundity is the reason for the ecological "struggle for existence." There are inadequate resources to support all the young that are born. There are only limited amounts of food and space in the world; a population may expand to some extent, but logically there will come a point beyond which the food supply must limit its further expansion. As resources are used up, the death rate in the population increases, and when the death rate equals the birth rate the population will stop growing. In theory, it could be purely a matter of luck which individuals survive and which do not; and luck in fact is an important influence on survival. But organisms also compete—both directly, for example by defending territories, and indirectly, for example by eating food that could otherwise be eaten by another individual. The actual competitive factors limiting the sizes of real populations make up a major area of ecological study, and in different species different factors have been shown to operate. What matters here, however, is the general point that the members of a population, and members of different species, compete in order to survive. The word "struggle" in the Darwinian phrase "struggle for existence" is metaphoric; it need not imply an actual physical fight over resources. However, organisms are built to do their utmost to survive and reproduce.

The struggle for existence takes place within a web of ecological relations. Above an organism in the ecological food chain there will be predators and parasites seeking to feed off it; and below it are the food resources it must in turn consume in order to stay alive. At the same level in the chain are competitors, which may be competing for the same limited resources of food or space. An organism competes most closely with other members of its own species, because they have the most similar ecological needs to its own; other species, in decreasing order of ecological similarity, also compete and exert a negative influence on the organism's chance of survival. In summary, organisms produce more offspring than (given the limited amounts of resources) can ever survive, and organisms therefore compete for survival. Only the successful competitors will reproduce.

4.2 Natural selection operates if certain conditions are met

The excess fecundity, and consequent competition to survive, in every species provides the precondition for the process Darwin called natural selection. Natural selection is easiest to understand, in the abstract, as a logical argument, leading from premises to conclusion. The argument, in its most general form, requires the four conditions listed below.

1 Reproduction. Entities must reproduce to form a new generation.

2 Heredity. The offspring must tend to resemble their parents: roughly speaking, "like must produce like."

3 Variation in characteristics of the members of the population.

4 Variation in the "fitness" of organisms associated with their various characteristics. In evolutionary theory, fitness is a technical term, meaning the probability that an individual will survive to reproduce. This condition therefore means that individuals in the population with some characteristics must be more likely to reproduce (i.e. have higher fitness) than others. (The evolutionary meaning of the term "fitness" differs from its meaning in athletics.)

If all these conditions are met for any property of a species, natural selection automatically results. If any are not, it does not. Thus entities, like planets, that do not reproduce cannot evolve by natural selection; entities that reproduce but without the characteristics of parents being inherited by their offspring also cannot evolve by natural selection. But when the four conditions apply, the organisms with the property conferring higher fitness will leave more offspring, and the frequency of that type of organism will increase in the population.

We can illustrate the process by a famous example. (The example will be discussed in detail in chapter 5; here it is only being used to show how the four conditions apply in a real case.) The usual form of the peppered moth *Biston betularia* in northern Europe has a "peppered" pattern of coloration (see Figure 5.4, p. 96). The moth rests on tree branches and its color pattern camouflages it against predatory attack. The camouflage only works against the right background. The light coloration of tree branches is mainly caused by lichens that grow there. Smoke pollution in the industrial revolution in the UK killed these lichens near to industrial areas, leaving tree branches black. At about this time, around 1830, a "melanic" form of the peppered moth became increasingly common in contemporary moth collections. The melanic form is camouflaged on dark tree branches. Through the nineteenth century, the melanic form increased in frequency until, near industrial regions, it was the most common type of the moth. The increase was almost certainly driven by natural selection. Observations made by Kettlewell reveal that birds are more likely to eat poorly camouflaged moths; that when both forms are released in industrial areas, the melanic moths are more likely to be recollected later; and that when both forms are released in non-industrial areas, the peppered forms are more likely to be recollected later.

The moths satisfy all four conditions for natural selection to operate. They reproduce; their color pattern is inherited (melanic moths are more likely to produce melanic offspring than are moths of the original peppered coloration); there is variation in their color patterns; and the different forms have different fitnesses (melanic forms are more likely to survive in industrial areas). In polluted areas, the melanic moths survived better, produced offspring like themselves, and increased in frequency: natural selection favored them.

4.3 Natural selection explains both evolution and adaptation

When the environment of the peppered moth changed, such that the background where the moths settled was dark rather than light colored, the population of moths changed in form over time. The moth population evolved. Natural selection produces evolution when the environment changes; it will also produce evolutionary change in a constant environment if a new form arises that survives better than the existing forms of the species. If the process that operated in the nineteenth century in a single species of moth were continued for the thousands of millions of years since life originated, much larger evolutionary changes could be accomplished. Indeed, one version of evolutionary theory maintains that the abstract process—natural selection—that caused the particular change in the peppered moth population is also responsible, in a general way and over a much grander time scale, for the whole diversification of life from a simple common ancestor about 3.5×10^9 years ago.

Natural selection not only produces evolution; it also produces adaptation.

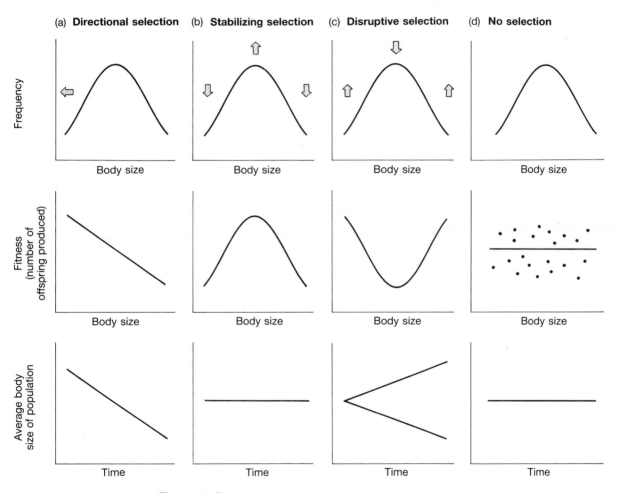

Figure 4.2 Three kinds of selection. The top line shows the frequency distribution of the character (body size). For many characters in nature, this distribution has a peak in the middle, near the average, and is lower at the extremes. (The normal distribution, or "bell curve," is a particular example of this kind of distribution.) The second line shows the relationship between body size and fitness within one generation, and the third shows the expected change in the average for the character over many generations (if body size is inherited). (a) Directional selection. Smaller individuals have higher fitness, and the species will decrease in average body size through time. Figure 4.3 is an example. (b) Stabilizing selection. Intermediate-sized individuals have higher fitness. Figure 4.4 is an example. (c) Disruptive selection. Both extremes are favored. If selection is strong enough, the population splits into two. Figure 4.5 is an example. (d) No selection. If there is no relation between the character and fitness, natural selection is not operating on it.

The camouflage of the peppered moth is a straightforward example of adaptation (section 1.2, p. 6): the color pattern of the moth, against the appropriate background, makes it less visible and less likely to be eaten by birds. The moth's survival is therefore increased. When the environment changed, the form of the adaptation needed in it changed too. In industrial regions, the peppered coloration was no longer adaptive. The action of natural selection to increase the frequency of melanic moths resulted in the moths becoming adapted to their environment. Over time, natural selection generates adaptation. The theory of natural selection therefore passes the key test set by Darwin (section 1.3.2, p. 10) for a satisfactory theory of evolution.

4.4 Natural selection can be directional, stabilizing, or disruptive

We can now distinguish three kinds of natural selection, according to their effect on a character—as an example we shall use body size. The first possibility is that smaller individuals have higher fitness (i.e. produce more offspring) than larger individuals. Natural selection is then *directional*: it favors smaller individuals and will, if the character is inherited, produce a decrease in average body size (Figure 4.2a). Directional selection could, of course, also produce an evolutionary increase in body size if larger individuals had higher fitness.

For example, pink salmon (*Onchorhynchus gorbuscha*) fished in the northwest Pacific have been decreasing in size in recent years (Figure 4.3). In 1954, fishermen started being paid by weight, rather than per individual, for the

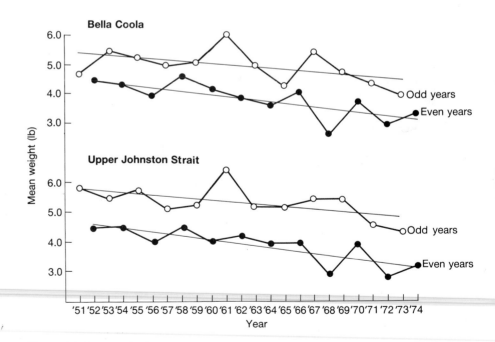

Figure 4.3 Directional selection by fishing on pink salmon, *Onchorhynchus gorbuscha*. The graph shows the decrease in size of pink salmon caught in two rivers in British Columbia since 1950, driven by selective fishing for the large individuals. Two lines are drawn for each river: one for the salmon caught in odd-numbered years, the other for even years. Salmon caught in odd years are consistently heavier, presumably because of the 2-year life cycle of the salmon. From Rìcker (1981).

salmon they caught and they increased the use of gill netting, which selectively takes larger fish. The selection for small size was intense, because fishing is thorough—about 75–80% of the adult salmon coming up the river were caught in these years. The selectivity of gill netting can be shown by comparing the average size of salmon taken by gill netting with those taken by an unselective fishing technique: the difference ranged from 0.3 to 0.48 lb (0.7 to 1.1 kg). Therefore, after gill netting was introduced, smaller salmon had a higher chance of survival. The average weight of salmon duly decreased, by about one-third, in the next 25 years.

A second (and in nature, more common) possibility is for natural selection to be *stabilizing* (Figure 4.2b). The average members of the population, with intermediate body sizes, have higher fitness than the extremes. Natural selection now acts against change in form, and keeps the population constant through time. Birth weight in humans is a good example of stabilizing selection; babies that are heavier or lighter than average do not survive as well as babies of average weight (Figure 4.4).

Thirdly, natural selection could favor both extremes over the intermediate types. This is called *disruptive* selection (Figure 4.2c). In nature, sexual dimorphism is probably a common example; but here we shall use an experiment by Thoday and Gibson on fruitflies as an example. Thoday and Gibson bred from fruitflies with high, or low, numbers of bristles on a certain region

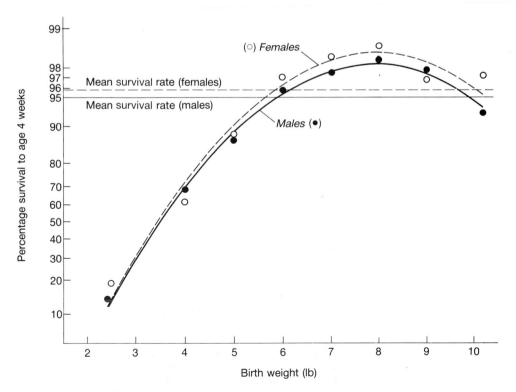

Figure 4.4 Stabilizing selection on human birth weight. Infants weighing 8 lb (18 kg) at birth have a higher survival rate than heavier or lighter infants. The graph is based on 13 700 infants born in a hospital in London, England, from 1935 to 1946. From Karn and Penrose (1951).

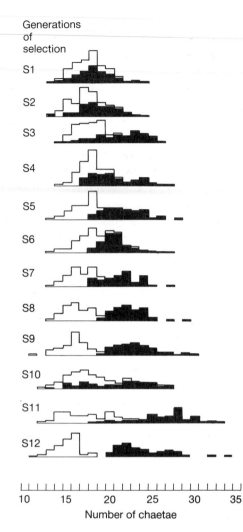

Generations
of
selection

S1

S2

S3

S4

S5

S6

S7

S8

S9

S10

S11

S12

10   15   20   25   30   35

Number of chaetae

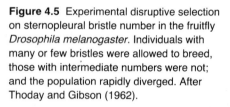

Figure 4.5 Experimental disruptive selection on sternopleural bristle number in the fruitfly *Drosophila melanogaster*. Individuals with many or few bristles were allowed to breed, those with intermediate numbers were not; and the population rapidly diverged. After Thoday and Gibson (1962).

of the body; individuals with intermediate numbers of bristles were pre-vented from breeding. After 12 generations of this disruptive selection, the population had noticeably diverged (Figure 4.5). Disruptive selection is of particular theoretical interest, both because it can increase the genetic diversity of a population (by frequency-dependent selection, section 5.12, p. 112) and promote speciation (see chapter 16).

A final theoretical possibility is for there to be no relation between fitness and the character in question; in which case natural selection is not acting on it (Figure 4.2d).

4.5 Variation in natural populations is widespread

Natural selection will operate whenever the four conditions (in section 4.2, p. 63), are satisfied. The first two conditions need little more to be said about them. It is well known that organisms reproduce themselves: this is often given as one of the defining properties of living things. It is also well known that organisms show inheritance. Inheritance is produced by the Mendelian process, which is understood down to a molecular level. Not all characteristics of organisms are inherited; natural selection will not adjust the frequencies of non-inherited characters. But many are inherited; on these natural selection can potentially work. The third and fourth conditions do need further comment.

How much, and with respect to what characters, do natural populations show variation and, in particular, variation in fitness? Let us consider biological variation through a series of levels of organization, beginning with the organism's morphology, and working down to more microscopic levels. The purpose of this section is to give examples of variation, and show how variation can be seen in almost all properties of living things.

At the morphologic level, the individuals of a natural population will be found to vary for almost any character we may measure. In some characters, like body size, every individual differs from every other individual; this is called continuous variation. Other morphologic characters show discrete variation: they fall into a limited number of categories. Gender is an obvious example; some individuals of a population are female, others male. This kind of categoric variation is found in other characters too. Within a population of the peppered moth, for example, there are two main wing color forms (the dark, melanic form and the light, peppered form). A population that contains more than one recognizable form is *polymorphic* (the condition is called polymorphism). There can be any number of forms in real cases, and they can have any set of relative frequencies. With gender, there are usually two forms; in the peppered moth, two main forms are often distinguished, though real populations may contain three or more. In the snail *Cepaea nemoralis*, which is polymorphic for its shell pattern (section 11.1, p. 267), there can be many more. As the number of forms in the population increases, the polymorphic, categoric kind of variation blurs into the continuous kind of variation.

Variation is not confined to morphologic characters. If we descend to a cellular character, such as the number and structure of chromosomes, we again find variation. In the fruitfly *Drosophila melanogaster*, the chromosomes exist in giant forms in the larval salivary glands and they can be studied with a light microscope. They turn out to have characteristic banding patterns, and

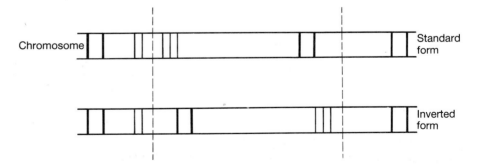

Figure 4.6 Chromosomes can exist in standard and inverted forms. It is arbitrary which is called "standard" and which "inverted." The inversion can be detected by comparing the fine structure of bands, as is diagrammatically illustrated here, or by the behavior of the chromosomes at meiosis.

chromosomes from different individuals in a population have subtly varying banding patterns. One type of variant is called an "inversion" (Figure 4.6), in which the banding pattern—and therefore the order of genes—of a region of the chromosome is inverted. A population of fruitflies may be polymorphic for a number of different inversions.

Chromosomal variation is less easy to study in species that lack giant chromosomal forms, but it is still known to exist. Populations of the Australian grasshopper *Keyacris scurra*, for example, may contain two (normal and inverted) forms for each of two chromosomes; that makes nine grasshopper forms in all because an individual may be homozygous or heterozygous for any of the four chromosomal types. As Figure 4.7 illustrates, the nine types of grasshopper differ in size and viability.

Chromosomes can vary in other respects besides inversion patterns. For example, individuals may vary in their number of chromosomes. In many species, some individuals have one or more extra chromosomes, in addition to the normal number for the species. These "supernumary" chromosomes, which are often called B chromosomes, have been particularly studied in maize and in grasshoppers. In the grasshopper *Atractomorpha australis*, normal individuals have 18 autosomes; but individuals have been found with from

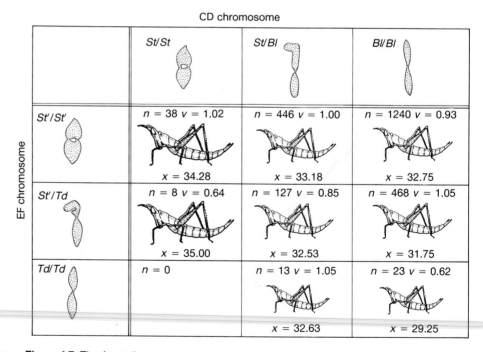

Figure 4.7 The Australian grasshopper *Keyacris scurra* is polymorphic for inversions in two chromosomes. The two chromosomes are called the CD and the EF chromosomes. The standard and inverted form of the CD chromosome are called *St* and *Bl*; the standard and inverted form of the EF chromosome are called *St'* and *Td*. *v* is the relative viability at a site at Wombat, New South Wales, expressed relative to the viability of the *St/Bl St'/St'* grasshoppers, which is arbitrarily set as 1. *x* is the mean live weight, and the picture illustrates relative sizes of the grasshoppers. From White (1973).

one to six supernumary chromosomes. The population is polymorphic with respect to chromosome number. Inversions and B chromosomes are just two kinds of chromosomal variation. There are other kinds too; but these are enough to make the point that individuals vary at the subcellular, as well as the morphologic level.

The story is the same at the biochemical level, such as for proteins. Proteins are molecules made up of sequences of amino acid units. A particular protein, like human hemoglobin, has a particular characteristic sequence, which in turn determines the molecule's shape and properties: but do all humans have exactly the same sequence for hemoglobin, or any other protein? In theory, we could find out by taking the protein from several individuals and then working out the sequence in each of them; but it would be excessively laborious to do so. There is a much faster method, called *gel electrophoresis*. Gel electrophoresis works because different amino acids carry different electric charges. Different proteins—and different variants of the same protein—have different net electric charges, because they have different amino acid compositions. If we place a sample of proteins (with the same molecular weight) in an electric field, those with the largest electric charges will move fastest. For the student of biological variation, the importance of the method is that it can reveal different variants of a particular type of protein. For a good example, we shall consider an enzyme called alcohol dehydrogenase, in the fruitfly.

Fruitflies, as their name suggests, lay their eggs in, and feed on, decaying fruit. They are attracted to rotting fruit because of the yeast it contains. Fruitflies can be collected almost anywhere in the world by leaving out rotting fruit as a lure; and drowned fruitflies are usually found in a glass of wine left out overnight after a garden party in the late summer. As fruit rots, it forms a number of chemicals, including alcohol, which is both a poison and a potential energy source. Fruitflies, and many other living species, deal with alcohol by means of a detoxifying enzyme called alcohol dehydrogenase. Gel electrophoresis reveals that in most populations of the fruitfly *Drosophila melanogaster*, alcohol dehydrogenase comes in two main forms. The two forms show up as different bands on the gel after the sample has been put on it, an electric current passed across it for a few hours, and the position of the enzyme has been exposed by a specific stain. The two variants are called slow (*Adh-s*) or fast (*Adh-f*) according to how far they have moved in the time. The multiple bands show that the protein is polymorphic; the enzyme called alcohol dehydrogenase is actually a class of two polypeptides with slightly different amino acid sequences. Gel electrophoresis has been applied to a large number of proteins in a large number of species; different proteins show different degrees of variability (see chapter 7). The point for now is that many of these proteins have been found to be variable: there is extensive variation in proteins in natural populations.

If there is variation in every organ, at every level, among the individuals of a population, it is almost inevitable that there will be variation at the DNA level too. The inversion polymorphisms of chromosomes that we met above are due to inversions of the DNA sequence. Variation of DNA has not been directly studied so extensively as morphologic and enzyme variation, because no equivalently rapid method of measuring DNA variation has been available. There are approximate methods, such as the use of restriction enzymes. Restriction enzymes break the DNA at the sites of characteristic sequences which are distributed through the DNA; any DNA molecule therefore is broken into a certain set of fragments. When the DNA of different individuals are treated with restriction enzymes, they are found to break into distinct sets of fragments. This must be because the individuals' DNA sequences differ, and the sequences recognized by the restriction enzymes are at different places along their DNA. The variation is called *restriction fragment length polymorphism*.

The most direct method of studying DNA variation is to sequence the DNA itself. Kreitman isolated the DNA encoding alcohol dehydrogenase from 11 independent lines of *D. melanogaster* and individually sequenced them all. The DNA was variable, as we should expect from the results with gel electrophoresis. The difference between *Adh-f* and *Adh-s* is always due to a single amino acid difference (threonine or lysine at codon 192); but that was not the only source of variation. The DNA is even more variable than the protein study suggests. For whereas at the protein level, only the two main variants would be found in the sample of 11 genes, at the DNA level, there were 11 different sequences with 43 different variable sites. At the DNA level, therefore, the greatest amount of variation is found. At the level of gross morphology, a *Drosophila* with two *Adh-f* genes is indistinguishable from one with two *Adh-s* genes; gel electrophoresis resolves two classes of fly; but at the DNA level, the two classes decompose into innumerable individual variants.

Therefore, when we move on to look at natural selection in more detail, we can assume that in natural populations the requirement of variation, as well as of reproduction and heredity, is met. However, if natural selection is to take place, there must not only be variation: the variation must be for fitness—in the degree to which individuals contribute offspring to the next generation. Fitness is more difficult to measure than a phenotypic character like body size, and there are far fewer observations of variation in fitness than variation in phenotype. However, there are still a good number of examples. We have met some already in section 4.4 (p. 66) and we shall meet more later in the book. These examples are enough to show that natural selection can operate, though they leave open the question of how often it does so. Is it the normal or the exceptional condition for individuals of a population to vary in their fitness? A plausible argument can be made that it is normal: we have to combine Darwin's argument (section 4.1, p. 63) that there is a struggle for existence, with the observation that phenotypic variation is common in nature. In almost every species, such a high proportion of individuals are doomed to die, that any attribute increasing their chance of survival, in a way that might appear trivial to us, is likely to result in a higher than average fitness; and any tendency of individuals to make mistakes, slightly increasing their risk of death, will result in lowered fitness. The struggle for existence, and phenotypic variation, are both universal conditions in nature; variation in fitness associated with some of those phenotypic characters is therefore also likely to be very common. The argument is one of plausibility, rather than certainty: it is not logically inevitable that a population showing variation in phenotype will also vary in fitness. If it does, however, natural selection will operate.

4.6 New variation is generated by mutation

The extent of variation in natural populations is so large that every individual must be genetically unique. Several processes can generate new variation in a population. Recombination between existing chromosomes produces new chromosomes with their own unique sequences, and many new genetic variants of a character like body size were probably generated by recombination. Migration is another important source of new genetic variation: when individuals arrive from distant parts they will often have different genotypes from the local population; they thus provide new genetic variation. Both recombination and migration work with existing allelic variation; they put existing variation into new genetic, or geographic, combinations. Important though this is, if there were no pre-existing allelic variation, recombination and migration would not generate new genetic variants. Recombination between identical chromosomes produces the same identical chromosomes over again. However, even in a population in which all copies of a chromosome were identical, new genetic variants would arise: they would be generated by *mutation*.

Mutations probably mainly arise as copy errors when DNA is replicated at

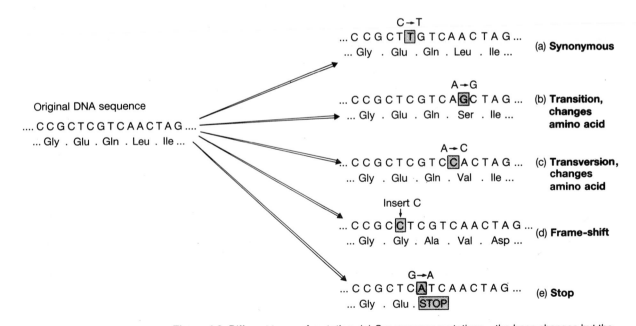

Figure 4.8 Different types of mutation. (a) Synonymous mutations—the base changes but the amino acid encoded does not. (b) Transition—change between purine types or between pyrimidine types. (c) Transversion—change from purine to pyrimidine or vice versa. (d) Frame-shift mutation—a base is inserted. (e) Stop mutation—an amino acid encoding triplet mutates to a stop codon. The terms transition and transversion can apply to synonymous or amino acid changing mutations, but it has only been illustrated here for mutations that alter amino acids. The base sequence here is for the DNA (with the 3′ end at left and 5′ end at right). The genetic code is conventionally written for the mRNA sequence; thus G has to be transcribed to C, etc. when comparing the figure with Table 2.1 (the genetic code, p. 26).

mitosis and meiosis. Copy errors can be of several kinds. One is *point mutation*, in which a base in the DNA sequence changes to another base. The 20 amino acids are encoded by 61 base triplets (Table 2.1, p. 26), and the effect of a mutation depends on what sort of base change it is (Figure 4.8a–c). Synonymous, or silent, mutations (Figure 4.8a) are mutations between two triplets that code for the same amino acid, and have no effect on the protein sequence; non-synonymous, or meaningful, point mutations do change the amino acid. Because of the structure of the genetic code, most synonymous mutations are in the third base position of the codon. About 70% of changes in the third position are synonymous; whereas all changes in the second, and most (96%) in the first position are meaningful. Another distinction for point mutations is between transitions and transversions. Transitions are base changes from one pyrimidine to the other, or from one purine to the other: between A and G, and between C and T. Transversions convert a purine base into a pyrimidine, or vice versa: from A or G to T or C (and from C or T to A or G). The distinction is interesting because transitional changes are much more common in evolution than transversions.

Successive amino acids are read from consecutive base triplets. If, therefore, a mutation *inserts* a base pair into the DNA, it can alter the meaning of every base "downstream" from the mutation (Figure 4.8d). These are called *frame-shift mutations*, and will usually produce a completely nonsensical, functionless protein. Another kind of "nonsense" mutation happens when a previously coding triplet mutates to a "stop" codon (Figure 4.8e); the resulting protein fragments will probably again be functionless.

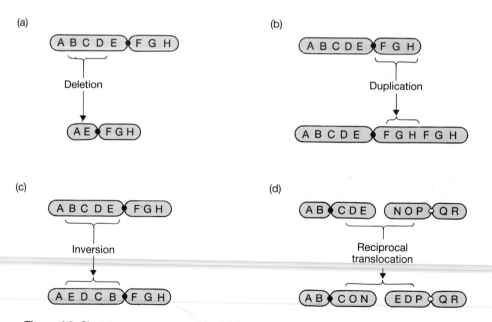

Figure 4.9 Chromosomes can mutate by (a) deletion; (b) duplication of a part; (c) inversion (see also Figure 4.7); or (d) translocation. Translocation may be either "reciprocal" (in which the two chromosomes exchange equal lengths of DNA) or "non-reciprocal" (in which one chromosome gains more than the other). In addition, whole chromosomes may fuse, and whole chromosomes (or the whole genome) may duplicate.

Mutations may involve whole chunks of chromosomes rather than single bases (Figure 4.9). A length of chromosome may be translocated to another chromosome, or to another place on the same chromosome, or be inverted. Whole chromosomes may fuse, as has happened in primate evolution; chimpanzees and gorillas have 24 pairs of chromosomes whereas humans have 23. Some or all of the chromosomes may be duplicated. The phenotypic effects of these chromosomal mutations are more difficult to generalize about. If the break-points of the mutation divide a protein, that protein will be lost in the mutant organism. But if the break is between two proteins, any effect will depend on whether the expression of a gene depends on its position in the genome. In theory, it might not matter whether a protein is transcribed from one chromosome or another; though in practice gene expression is probably at least partly regulated by relations between neighboring genes and a chromosomal mutation will then have phenotypic consequences.

4.6.1 The rates of mutation can be measured and estimated

What is the rate of mutation? A direct method of finding out is to observe the rate at which a visible mutant phenotype arises in a laboratory population of a species. The method has been used with many laboratory species, such as bacteria, yeast, *Drosophila*, and mice. The "multiple-locus technique" is a version of the method in mice. In one experiment, by Russell, a large number of females that were homozygous for seven recessive mutations were crossed with homozygous wild-type males. Almost all the offspring had the wild-type phenotype, but a few did not. Table 4.2 shows that observations of about 0.5 million offspring suggested a mutation rate of about 8×10^{-4}. It is not certain, however, that all the "mutant" offspring were produced by new mutation; some of the ostensibly homozygous male parents may have been heterozygotes. Table 4.3 summarizes some other measurements, by direct observation, of rates of visible mutation. Many of them are mutations from a rare, often dysfunctional, form back to the normal form of the gene. (Laboratory estimates of mutation rates may be biased, because the genetic changes that are most likely to have been noticed will be those with high mutation rates: changes with low mutation rates simply may not happen often enough to be seen.)

A memorable approximate figure from Table 4.3 is that mutation rates are about 10^{-6}; but the figures for different mutations show great variation. One reason is that the mutations are of different kinds. When the mutant needed

Table 4.2 Spontaneous mutation rates at seven gene loci in the mouse *Mus musculus*. From Wallace (1968)

Gender	Number of offspring	Number with mutations	Mutations per locus per gamete
Male	544 897	32	0.84×10^{-5}
Female	98 828	1	0.14×10^{-5}

Table 4.3 Mutation rates for various genetic changes

Organism	Character (gene → gene)	Rate
Bacteriophage (T$_2$)	Lysis inhibition, $rII \rightarrow rII^+$	1×10^{-3}–1×10^{-9} or less
	Host range, $h^+ \rightarrow h$	3×10^{-9}
Bacteria (*Escherichia coli*)	Lactose fermentation, $lac^- \rightarrow lac^+$	2×10^{-7}
	Phage T$_1$ sensitivity, T$_1$-$s \rightarrow$ T$_1$-r	2×10^{-8}
	Histidine requirement, $his^- \rightarrow his^+$	4×10^{-8}
	$his^+ \rightarrow his^-$	2×10^{-6}
	Streptomycin sensitivity, str-$s \rightarrow str$-d	1×10^{-9}
	str-$d \rightarrow str$-s	1×10^{-8}
Algae (*Chlamydomonas reinhardi*)	Streptomycin sensitivity, str-$s \rightarrow str$-r	1×10^{-6}
Fungi (*Neurospora crassa*)	Inositol requirement, $inos^- \rightarrow inos^+$	8×10^{-8}
	Adenine requirement, $ade^- \rightarrow ade^+$	4×10^{-8}
Corn (*Zea mays*)	Shrunken seeds, $Sh \rightarrow sh$	1×10^{-5}
	Purple, $P \rightarrow p$	1×10^{-6}
Fruitfly (*Drosophila melanogaster*)	Yellow body, $Y \rightarrow y$, in males	1×10^{-4}
	$Y \rightarrow y$, in females	1×10^{-5}
	White eye, $W \rightarrow w$	4×10^{-5}
	Brown eye, $Bw \rightarrow bw$	3×10^{-5}
Mouse (*Mus musculus*)	Piebald coat color, $S \rightarrow s$	3×10^{-5}
	Dilute coat color, $D \rightarrow d$	3×10^{-5}
Human (*Homo sapiens*)	Normal → hemophilic	3×10^{-5}
	Normal → albino	3×10^{-5}
Human bone marrow cells in tissue culture	Normal → 8-azoguanine resistant	7×10^{-4}
	Normal → 8-azoguanosine resistant	1×10^{-6}

to restore the normal gene is something as easy as a transitional point mutation, it has a high rate (10^{-3} or so); but when a particular frame-shift, or deletion, is needed, the rates are much lower (10^{-9} or less). There are other methods of estimating mutation rates too. Box 4.1 describes a laboratory method that can be used with bacteria, and chapter 5 (p. 109) and chapter 7 (p. 172) describe two other methods.

Mutation rates can be increased by exposing organisms to certain chemicals, called *mutagens*. Caffeine, as Figure B4.2 (in Box 4.1) shows, increases the rate of mutation in bacteria. Many other chemicals act as mutagens; and in some cases the reason is understood at the chemical level. Nitrous acid, for example, directly converts bases in the DNA; other mutagens are chemical analogs of bases and are incorporated into DNA. 5-Bromouracil, for example, is a thymine analog and increases the mutation rate as it is substituted for thymine bases in the DNA. X-rays also increase mutation rates, particularly the rates of breaks and rearrangements in chromosomes. These artificial

Box 4.1 Mutation rates in bacteria.

Here we consider a method of estimating mutation rates that can be used with the very large populations of bacteria that can be kept in the laboratory. Consider a genetic locus that can have two alleles, A

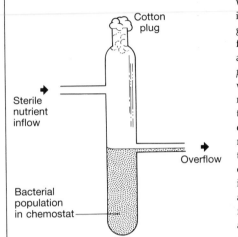

Cotton plug

Sterile nutrient inflow

Overflow

Bacterial population in chemostat

Figure B4.1 Diagram of a chemostat. The bacterial population grows until it reaches a stable number at which the rate of birth by cell division equals the rate of loss into the overflow.

or a; let the mutation rate from A to a be m and from a to A be n. Now suppose that nearly every individual in the population has allele A. If the frequency of allele A in the population in generation zero is written p_0, then $p_0 \simeq 1$. The frequency of a in generation zero, $q \simeq 0$. In the first generation, A will mutate to a with frequency $pm \simeq m$. The frequency of A after one generation (p_1) will therefore be $p_1 = p_0 - m$. Likewise the frequency of a will be $q_1 = q_0 + m$. We can ignore mutations from a to A because a is so rare that these mutations have negligible effect on gene frequencies compared with mutations from A to a. After t generations, the process will have occurred t times over, and the frequency of a will have increased, while q remains small, according to the relation $q_t = q_0 + tm$. The frequency of a should increase over time, and the rate of increase is a measure of the mutation rate m. The increase can be measured in a device called a *chemostat* (Figure B4.1). Figure B4.2 illustrates the result of one experiment, and the experiment suggests mutations rates of 0.24×10^{-8} and 2.2×10^{-8} before and after the addition of a mutagen.

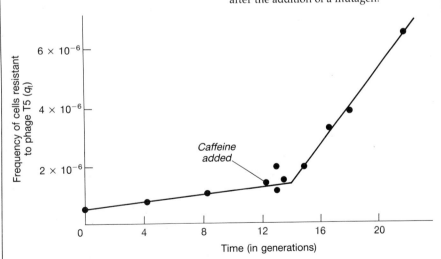

Caffeine added

Figure B4.2 The increase in frequency of gene for resistance to phage T5 in a bacterial chemostat. The phage was added at time 0. The spontaneous mutation rate equals the slope between 0 and 13 generations; the slope there is 1.3×10^{-8} hour^{-1}. Caffeine is a bacterial mutagen, and the mutation rate increased to 12×10^{-8} hour^{-1} after it was added. The generation time in the experiment was 5.5 hours. From Novick (1955).

influences on mutation rates are environmentally important, but may not have been of major importance in natural evolution.

4.6.2 Mutation is random with respect to the direction of adaptation

When the environment changes, as it did for British peppered moths in the nineteenth century, a new form of the species comes to be favored. According to Darwin's theory, natural selection will favor whatever form, among those represented in the species, fits the new environment best. The environmental change does not itself cause mutations of the right form to appear. New mutations of all sorts will be constantly arising, independently of what is required for adaptation to the current environment; selection works on this variation. Such is the Darwinian theory. The alternative would be some kind of *directed mutation*. If mutation is directed, then when the environment changed to favor melanic moths, melanic mutations would be generated by some mutational mechanism.

There are both factual and theoretical reasons to doubt that mutations are adaptively directed. The factual evidence comes from laboratory observations of spontaneous mutations. There is no evidence that these mutations take place toward the adaptive needs of the organism. Until recently, a classic experiment of Luria and Delbrück in 1943 was the "textbook" demonstration that mutation is undirected; but that particular experiment was effectively challenged by Cairns *et al.* in 1988 and is currently the subject of active research.

The strongest reason to doubt that mutations are adaptively directed is theoretical. In the case of the peppered moths, melanic moths probably already existed in the population at the time of the industrial revolution, and selection simply increased their frequency; but we can use the case as an example as if there were no melanic moths when the environment changed. The industrial revolution imposed on the moths an environment they had never encountered before. The environment (probably) was completely new. If there were no melanic moths in the population, a mutation, of melanic coloration, was then needed for the moths to survive. Could it arise by "directed" mutation? At the DNA level, the mutation would have consisted of a set of particular changes in the base sequence of a gene. No genetic mechanism has been discovered that could direct the right base changes to happen.

If we reflect on the kind of mechanism that would be needed, it becomes clear that an adaptively directed mutation would be practically impossible. The organism would have to recognize that the environment had changed, work out what change was needed to adapt to the new conditions, and then cause the correct base changes in the relevant parts of its DNA. It would have to do so for an environment the species had never previously experienced. It would be like humans describing subject matter they had never encountered before in a language they did not understand: like a seventeenth century American using Egyptian hieroglyphics to describe how to change a computer program. (Hieroglyphics were not deciphered until the discovery of the Rosetta Stone in 1799.) Even if it were just possible to imagine, as an extreme theoretical possibility, directed mutations in the case of moth coloration, the changes in the evolution of a more complex organ (like the brain, or circulatory system, or eye) would require a virtual miracle. It is for this reason that mutation is thought not to be directed toward adaptation.

Although mutation is random and undirected with respect to the direction of improved adaptation, that does not exclude the possibility that mutations are non-random at the molecular level. The higher frequency of transitions than transversions is a case in point. It is not true that changes from the base A, for example, are equally probable to G, T, or C; a change to T may be 10 times more probable. Other molecular biases are also known. One example is the tendency of the two base sequence CG to mutate, when it has been methylated, to TG. (The DNA in a cell is sometimes methylated, for reasons that do not matter here.) After replication, a complementary pair of CG on the one strand and GC on the other will then have produced TG and AC. Species with high amounts of DNA methylation have (probably for this reason) low amounts of CG in their DNA.

These molecular mutational biases are not the same as changes toward improved adaptation, however. You cannot change a light-colored moth into a dark one just by causing transitional, rather than transversional, mutations in its DNA, or by converting a proportion of its CG dinucleotides into TG. Some critics of Darwinism have read that the Darwinian theory describes mutation as "random," and have then trotted out these sorts of molecular mutational biases as if they contradicted Darwinian theory. But mutation can be non-random at the molecular level without contradicting the Darwinian theory. What Darwinism rules out is mutation directed toward new adaptation. Because of this confusion about the word "random," it is often better to describe mutation not as "random," but as "undirected" or "accidental" (which was the word Darwin used).

One last point should be made about assessing evidence for directed mutations in laboratory mutations. The mutations studied in the laboratory, as we saw, are likely to be a biased sample. They are likely to be the more frequent kinds of changes. Now, it is possible that some of these frequent changes are not spontaneous point (or chromosomal) mutations; some may be caused by the insertion of mobile genetic elements (section 10.6, p. 255). For some purposes, this does not matter. A spontaneous genetic insertion can be just as much a mutation as a change in a base or a chromosome. However, insertion elements may in some cases have an organized, probably adaptive, genetic structure; they may therefore be able to confer whole new adaptations on the organism into which they insert, like plasmids in bacteria. If the organism could in turn influence the chance that insertion elements will be mobilized inside itself, then a sort of directed mutation might appear to be operating. It is thus important, when assessing evidence for directionality in mutation, to know what the genetic basis of the alleged mutation is. The Darwinian claim that mutation is not adaptively directed applies only to spontaneous new mutations, not to the activation of previously adapted genetic sequences. This is a hypothetical possibility; but the claim that mutation is undirected is so fundamental to Darwinism that we must understand exactly what it means.

4.7 Summary

1 Organisms produce many more offspring than can survive, which results in a "struggle for existence," or competition to survive.

2 Natural selection will operate among any entities that reproduce, show inheritance of their characters from one generation to the next, and vary in "fitness" (i.e. their chance of surviving to reproduce) according to the characters they possess.

3 The increase in frequency of the melanic, relative to the light-colored, form of the peppered moth *Biston betularia* clearly illustrates how natural selection causes both evolutionary change and the evolution of adaptation.

4 Selection may be directional, stabilizing, or disruptive.

5 The members of natural populations vary with respect to characteristics at all levels. They differ in their morphology, their microscopic structure, their chromosomes, the amino acid sequences of their proteins, and in the DNA sequences.

6 New genetic variation originates by mutational changes in DNA.

7 Rates of mutation can be estimated by direct observation, or inference from gene frequencies in populations.

8 The direction of mutation must be random with respect to the direction of improved adaptation. No genetic mechanism could have the foresight to direct mutations towards novel adaptive requirements.

4.8 Further reading

Any ecology text, such as Begon *et al.* (1990), introduces life tables. For the theory of natural selection, see Darwin's original account (1859, chapters 3 and 4) and Endler (1986). Johnson (1976, 1979) introduces the kinds of selection. Law (1991) describes the selective effects of fishing. Travis (1989) reviews stabilizing selection. Variation and the genetics of mutation are described in all the larger population genetics texts, such as Hartl (1988) and Hartl and Clark (1989), Hedrick (1983), Spiess (1988), and Wallace (1981). Lewontin (1982) describes variation in humans; White (1973) and Dobzhansky (1970) describe chromosomal variation. Variation in proteins and DNA will be discussed further in chapter 7, which gives references.

For mutation, see genetics texts such as Lewin (1990) or Watson *et al.* (1988); Klekowski and Godfrey (1989) estimated the mutation rate of one set of genes in mangroves. On biases in the direction of mutation at the DNA level, see Golding (1987). References for the classic experiment on mutational directionality and the modern controversy are: Luria and Delbrück (1943), Cairns *et al.* (1988), Benson *et al.* (1988), and the references and discussion in Brenner (1992).

Part 2
Evolutionary Genetics

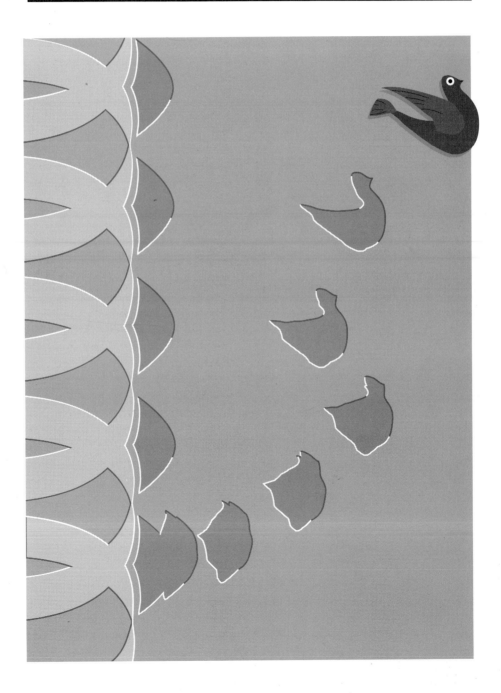

Evolutionary Genetics

The theory of population genetics is the most important, most fundamental body of theory in evolutionary biology. As Medawar once remarked, "it is only in genetics that there can be said to be a 'theoretical biology,' in the sense in which theoretical physics is so described." We shall start with the simplest cases and move on to the more complex. The simplest case is when natural selection is working on only one genetic locus, and one allele of higher fitness is being substituted for an inferior allele. We shall also look at how natural selection can maintain variation at a single locus, in three circumstances. Examples from ecological genetics will illustrate the ideas. Up to this point the theory will have been purely deterministic. Mendelian inheritance is not purely deterministic, however, as chapter 6 will discuss. The frequencies of genes can drift up and down between generations at random. A new set of possibilities is opened up, in which evolution is not controlled by natural selection; the "neutral theory" suggests that molecular evolution is mainly driven by random processes. We shall examine the evidence for and against the neutral theory and its selectionist alternative in chapter 7. We then move on to consider natural selection working simultaneously on more than one locus (chapter 8). Linkage between loci complicates the one-locus model. With more than one locus, there is the possibility of "coadaptation" between genes—this makes the existence of genetic recombination puzzling. It is a matter of controversy how important higher level interactions between gene loci are, and how far the one-locus model is an adequate description of the real world. As we move from two-locus evolution to multiple-locus evolution we abandon Mendelian exactitude and use a quite different method: quantitative genetics. In quantitative genetics (chapter 9), the relations between individuals and between successive generations are described approximately and abstractly. It is concerned with "continuous" characters, at the morphologic level. As we saw in chapter 4, morphologic characters show variation in natural populations, and we shall consider how to account for the level of variation that is observed. Finally, we consider in chapter 10 some of the special evolutionary properties of multigene families and of parasitic "selfish DNA."

The theory of natural selection

5.1 Population genetics is concerned with genotype and gene frequencies

There may be about 100 000 gene loci in the human genome. Let us focus on just one of them. Let us focus, in particular, on a locus at which there is more than one allele, because no evolutionary change can happen at a locus for which every individual in the population has two copies of the same gene. We shall be concerned in this chapter with models of evolution at a single genetic locus; these are the simplest models in population genetics. In many real cases, evolutionary change probably occurs at many loci simultaneously; but we can keep the problem at a manageable size by concentrating on only one of them.

The theory of population genetics at one locus is mainly concerned with understanding two closely connected variables: *gene frequency* and *genotype frequency*. They are easy to measure. The simplest case is one genetic locus with two alleles (*A* and *a*) and three genotypes (*AA*, *Aa*, and *aa*). Each individual has a genotype made up of two genes at the locus and a population can be symbolized thus:

Aa   AA   aa   aa   AA   Aa   AA   Aa

This is an imaginary population with only eight individuals. To find the genotype frequencies we simply count the number of individuals with each genotype. Thus:

Frequency of $AA = \frac{3}{8} = 0.375$

Frequency of $Aa = \frac{3}{8} = 0.375$

Frequency of $aa = \frac{2}{8} = 0.25$

In general we can symbolize genotype frequencies algebraically, as follows.

Genotype	*AA*	*Aa*	*aa*
Frequency	*P*	*Q*	*R*

P, *Q*, and *R* are expressed as percentages or proportions, so in our population, *P* = 0.375, *Q* = 0.375, *R* = 0.25 (they have to add up to one, or to 100%). They are measured simply by observing and counting the numbers of each type of organism in the population, and dividing by the total number of organisms in the population (the population size).

The gene frequency is likewise measured by counting the frequencies of each gene in the population. (Because each genotype contains two genes, there are a total of 16 genes per locus in a population of eight individuals.) In the population above,

Frequency of $A = \frac{9}{16} = 0.5625$

Frequency of $a = \frac{7}{16} = 0.4375$

Algebraically we can define p as the frequency of A, and q as the frequency of a. p and q are always called "gene" frequencies, but in a strict sense they are allele frequencies: they are the frequencies of the different alleles at one genetic locus. The gene frequencies can be calculated from the genotype frequencies:

$$p = P + \tfrac{1}{2}Q$$

$$q = R + \tfrac{1}{2}Q \qquad (5.1)$$

(and $p + q = 1$). The calculation of gene from genotype frequencies is highly important: we shall make recurrent use of these two simple equations in the chapter. Although the gene frequencies can be calculated from the genotype frequencies (P, Q, R), the opposite is not true: the genotype frequencies cannot be calculated from the gene frequencies (p, q).

Now that we have defined the key variables, we can see how population geneticists analyze changes in those variables through time.

5.2 An elementary population genetics model has four main steps

Population geneticists try to answer the following question: if we know the genotype (or gene) frequencies in one generation, what will they be in the next generation? It is worth looking at the general procedure before going into particular models. The procedure is to break down the time from one

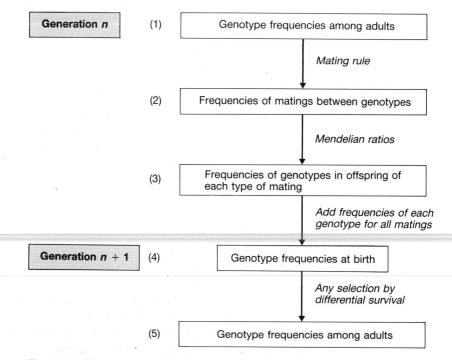

Figure 5.1 The general model of population genetics.

generation to the next into a series of stages, and work out how genotype frequencies are affected at each stage. We can begin at any arbitrarily chosen starting point in generation n and then follow the genotype frequencies through to the same point in generation $n + 1$. The general outline of a population genetics model is shown in Figure 5.1. The total genotype frequencies at each stage, and the total frequencies of matings, must add up to one. In words, the procedure is as follows. We start with the frequencies of genotypes among the adults in generation n. The first step is to specify how these genotypes combine to breed (called a mating rule); then the Mendelian ratios are applied (see chapter 2) for each type of mating; we then add the frequencies of each genotype generated from each type of mating to find the total frequency of the genotypes among the offspring at birth in the next generation. If the genotypes have different chances of survival from birth to adulthood, we multiply the frequency of each genotype at birth by its chance of survival to find the frequency among adults. When the calculation at each stage has been completed, the population geneticist's question has been answered. (Natural selection can operate in two ways, by differences in survival among genotypes or by differences in fertility. There are two theoretical extremes. Surviving individuals of all genotypes could produce the same number of offspring; natural selection would then operate only by differences in survival. Alternatively, individuals of all genotypes could have the same chance of survival, and natural selection could operate only by differences in the number of offspring they produce (i.e. their fertility). Many real cases presumably contain a mixture of selection by differences in survival and in fertility. The models we shall consider all express selection in terms of differences in chance of survival: this is not to suggest that selection always operates on survival; it is to keep the models simple and consistent.)

The model, in this general form, may look rather complicated. However, we can cut it down to size by making some simplifying assumptions. The first two simplifying assumptions to consider are random mating and no selection (i.e. no differential survival between stages 4 and 5).

5.3 Genotype frequencies in the absence of selection are given by the Hardy–Weinberg equilibrium

We can stay with the case of one genetic locus with two alleles (A and a). The frequencies of genotypes AA, Aa, and aa are still P, Q, and R. Our question is, if there is random mating and no selective difference among the genotypes, and we know the genotype frequencies in one generation, what will the genotype frequencies be in the next generation? The answer is called the *Hardy–Weinberg equilibrium*. Let us see what that means.

The calculation is shown in Table 5.1. Mating is random, and the mating frequencies are straightforward. To form a pair, we pick out at random two individuals from the population. What is the chance of an $AA \times AA$ pair? Well, to produce this pair, the first individual we pick has to be an AA and the second one also has to be an AA. The probability that the first is an AA is simply P, the genotype's frequency in the population. We are imagining an infinite population, thus the probability that the second one is AA is also P. The chance of drawing out two AA individuals in a row is therefore P^2. (The frequency of $Aa \times Aa$ and $aa \times aa$ matings are likewise Q^2 and R^2 respectively.) Similar reasoning applies for the frequencies of matings in

Table 5.1 Calculations needed to derive the Hardy–Weinberg ratio, for one locus and two alleles, A and a. (Frequency of AA = P, of Aa = Q, of aa = R.) The table shows the frequencies of different matings if the genotypes mate randomly, and the genotypic proportions among the progeny of the different matings

Mating type	Frequency of mating	Offspring genotypic proportions		
		AA	Aa	aa
AA × AA	P^2	1		
AA × Aa	PQ	$\frac{1}{2}$	$\frac{1}{2}$	
AA × aa	PR		1	
Aa × AA	QP	$\frac{1}{2}$	$\frac{1}{2}$	
Aa × Aa	Q^2	$\frac{1}{4}$	$\frac{1}{2}$	$\frac{1}{4}$
Aa × aa	QR		$\frac{1}{2}$	$\frac{1}{2}$
aa × AA	RP		1	
aa × Aa	RQ		$\frac{1}{2}$	$\frac{1}{2}$
aa × aa	R^2			1

which the two individuals have different genotypes. The chance of picking an AA and then an Aa (to produce an AA × Aa pair), for example, is PQ; the chance of picking an AA and then an aa is PR; and so on.

The genotypic proportions in the offspring of each type of mating are given by the set of Mendel's ratios for that cross. We can now work out the frequency of a genotype in the next generation by addition. We look at which matings generate the genotype, and add the frequencies generated by all the matings. Let us work it out for the genotype AA. AA individuals, Table 5.1 shows, come from AA × AA, AA × Aa (and Aa × AA), and Aa × Aa matings. We can ignore all the other types of mating. AA × AA matings have frequency P^2 and produce all AA offspring. AA × Aa and Aa × AA matings each have frequency PQ and produce $\frac{1}{2}AA$ offspring. Aa × Aa matings have frequency Q^2 and produce $\frac{1}{4}AA$ offspring. The frequency of AA in the next generation*, P', is then

$$P' = P^2 + \tfrac{1}{2}PQ + \tfrac{1}{2}PQ + \tfrac{1}{4}Q^2 \tag{5.2}$$

This can be rearranged to

$$P' = (P + \tfrac{1}{2}Q)(P + \tfrac{1}{2}Q) \tag{5.3}$$

$(P + \tfrac{1}{2}Q)$, we have seen, is simply the frequency of the gene A (p). Therefore

$$P' = p^2 \tag{5.4}$$

The frequency of genotype AA after one generation of random mating is equal to the square of the frequency of the A gene. Analogous arguments

* Population geneticists conventionally symbolize the frequency of variables one generation on by writing a prime. If P is the frequency of genotype AA in one generation, P' is its frequency in the next; if p is the frequency of an allele in one generation, p' is its frequency in the next generation. We shall follow this convention here.

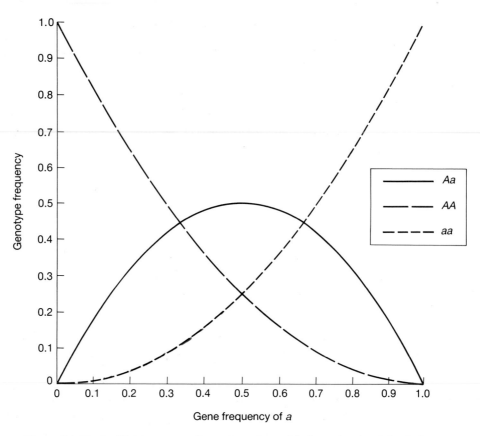

Figure 5.2 Hardy–Weinberg proportions of genotypes *AA*, *Aa*, and *aa* in relation to the frequency of the gene *a* (*q*)

show that the frequencies of *Aa* and *aa* are $2pq$ and q^2, respectively. The Hardy–Weinberg frequencies are then

$AA : Aa : aa$
$p^2 : 2pq : q^2$

They are reached, from any initial genotypic frequencies (the values of P, Q, and R therefore do not matter), after a single generation of random mating; they will then remain stable, provided that mating remains random and there is no selection. (The calculation also assumes that the population size is infinite. However, when evolutionary biologists talk about an infinite population they do not mean it is literally infinite. They mean it is large— large enough for the random effects discussed in chapter 6 to be unimportant.) Indeed, it is an unavoidable theoretical truth than the genotypes at a locus will have stable Hardy–Weinberg frequencies for as long as there is no selection, mating remains random, and the population is infinitely large. Figure 5.2 shows the Hardy–Weinberg proportions of the three different genotypes at different frequencies of the gene *a*; heterozygotes are most frequent when the gene frequency (*q*) is 0.5.

We saw in section 5.1 that it is not in general possible to calculate the

Box 5.1 The Hardy—Weinberg theorem for three alleles.

We can call the three alleles A_1, A_2, and A_3, and define their gene frequencies as p, q, and r, respectively. We form new zygotes by sampling two successive gametes from an infinite pool of gametes. So, if we first pick (with chance p) an A_1 allele from the gamete pool, the chance that the second allele is another A_1 allele is p, the chance that it is an A_2 allele is q, and the chance that it is a A_3 allele is r. So far, the frequencies of A_1A_1, A_1A_2, and A_1A_3 zygotes are p^2, pq, and pr; but there are other ways of forming A_1A_2 and A_1A_3 zygotes. The only way to form an A_1A_1 zygote is by picking two A_1 alleles, and the frequency of A_1A_1 genotype in the next generation will therefore be p^2.

Now suppose that the first allele we picked out had been an A_2 (which would happen with chance q). The chances that the second allele would be A_1, A_2, or A_3 would be p, q, and r, respectively; the only way to form an A_2A_2 zygote is by picking two A_2 alleles in a row, and the frequency of A_2A_2 in the next generation is therefore q^2. Finally, if we had picked (with chance r) an A_3 allele, there would be the same chances of combining it with an A_1, A_2 or A_3 allele. By addition, the total chance of forming an A_1A_3 zygote is $pr + rp = 2pr$; of forming an A_1A_2 zygote is $pq + qp = 2pq$; and of an A_2A_3 zygote is $2qr$. The complete Hardy—Weinberg proportions are:

$$A_1A_1 : A_1A_2 : A_1A_3 : A_2A_2 : A_2A_3 : A_3A_3$$
$$p^2 \quad 2pq \quad 2pr \quad q^2 \quad 2qr \quad r^2$$

$p + q + r = 1$, and the sum of the genotype frequencies is also one.

genotype frequencies in a generation if you only know the gene frequencies. We can now see that it is possible to calculate, from gene frequencies alone, what the genotype frequencies will be in the *next* generation, provided that mating is random, there is no selection, and the population in effectively infinite. If the gene frequencies in this generation are p and q, in the next generation the genotypes will have Hardy—Weinberg frequencies.

The proof of the Hardy—Weinberg theorem we have worked through was long-winded. We worked through it all in order to illustrate the general model of population genetics in its simplest case. However, a simpler proof can be given in terms of gametes. Diploid organisms produce haploid gametes. We could imagine that the haploid gametes are all released into the sea, where they combine at random to form the next generation. This is called "random union of gametes." In the "gamete pool" A gametes will have frequency p, and a gametes frequency q. Because they are combining at random, an a gamete will meet an A gamete with chance p and an a gamete with chance q. From the a gametes, Aa zygotes will therefore be produced with frequency pq, and aa gametes with frequency q^2. A similar argument applies for the A gametes (which have frequency p): they combine with a gametes with chance q, to produce Aa zygotes (frequency pq) and A gametes with chance p to form AA zygotes (frequency p^2). If we now add up the frequencies of the genotypes from the two types of gamete, the Hardy—Weinberg genotypic frequencies emerge. We have now derived the Hardy—Weinberg theorem for the case of two alleles; the same argument easily extends to three alleles (Box 5.1).

(Some people may be puzzled by the two in the frequency of the heterozygotes. It is a simple combinatorial probability. Imagine flipping two coins and asking what the chances are of flipping two heads, or two tails, or one head and one tail. The chance of two heads is $(\frac{1}{2})^2$ and of two tails $(\frac{1}{2})^2$; the chance of a head and a tail is $2 \times (\frac{1}{2})^2$. The "head" is analogous to genotype A, "tail" to a; two heads to producing an AA genotype, one head and one tail to a heterozygote Aa. The coin produces heads with probability $\frac{1}{2}$, and is analogous to a gene frequency of $p = \frac{1}{2}$. The frequency $2pq$ for heterozygotes is analogous to the chance of one head and one tail, $2 \times (\frac{1}{2})^2$.)

5.4 We can test, by simple observation, whether genotypes in a population are at Hardy–Weinberg equilibrium

The Hardy–Weinberg theorem depends on three main assumptions: no selection, random mating, and effectively infinite population size. In a natural population, any of these could be false; hence we cannot assume that natural populations will be at the Hardy–Weinberg equilibrium. In practice, we can find out whether a population has its genotypes at the Hardy–Weinberg frequencies simply by counting the genotypic frequencies. We first calculate the gene frequencies (p and q) from the genotype frequencies; then, if the observed homozygote frequencies equal the square of their gene frequencies (p^2 and q^2), the population is in Hardy–Weinberg equilibrium. If they do not, it is not.

The MN blood group system in humans is a good example, because the three genotypes are distinct and the genes have reasonably high frequencies in human populations. Three phenotypes, M, MN, and N are produced by three genotypes (MM, MN, NN) and two alleles (M and N) at one locus. The phenotypes of the MN group, like the better known ABO group, are recognized by injecting blood into a rabbit, to make an antiserum to the type of

Table 5.2 The frequencies of the *MM*, *MN*, and *NN* blood group genotypes in three American populations. (The figures for expected proportion and expected number have been rounded)

Population		*MM*	*MN*	*NN*	Total	*p*	*q*
African Americans	Observed no.	79	138	61	278		
	Expected proportion	0.283	0.499	0.219		0.532	0.468
	Expected no.	78.8	138.7	60.8			
European Americans	Observed no.	1787	3039	1303	6129		
	Expected proportion	0.292	0.497	0.211		0.54	0.46
	Expected no.	1787.2	3044.9	1296.9			
Native Americans	Observed no.	123	72	10	205		
	Expected proportion	0.602	0.348	0.05		0.776	0.224
	Expected no.	123.3	71.4	10.3			

Specimen calculation—for African Americans: Frequency of M allele $= 79 + (\frac{1}{2} \times 138) = 0.532 = p$

Frequency of N allele $= 61 + (\frac{1}{2} \times 138) = 0.468 = q$

Expected proportion of $MM = p^2 = (0.532)^2 = 0.283$
Expected proportion of $MN = 2pq = 2(0.532)(0.468) = 0.499$
Expected proportion of $NN = q^2 = (0.468)^2 = 0.219$

Expected nos = expected proportion $\times$ total no. (n)
Expected no. of $MM = p^2 n = 0.283 \times 278 = 78.8$
Expected no. of $MN = 2pqn = 0.499 \times 278 = 138.7$
Expected no. of $NN = q^2 n = 0.219 \times 278 = 60.8$

blood that was injected. If the rabbit has been injected with M-type human blood, it produces anti-M serum. Anti-M serum agglutinates blood from humans with one or two *M* alleles in their genotypes; likewise anti-N blood agglutinates the blood of humans with one or two *N* alleles. Therefore *MM* individuals' blood reacts with anti-M, *NN* individuals with anti-N, and *MN* individuals react with both.

Table 5.2 gives some measurements of the frequencies of the *MN* blood group genotypes for three human populations. Are they at Hardy–Weinberg equilibrium? In European Americans, the frequency of the *M* allele (calculated from the usual $p = P + \frac{1}{2}Q$ equation) is 0.54. If the population is at Hardy–Weinberg equilibrium, the frequency of *MM* homozygotes (p^2) will be $0.54^2 = 0.2916$ (1787 in a sample of 6129 individuals); the frequency of *MN* heterozygotes ($2pq$) will be $0.54 \times 0.46 = 0.497$ (3045 in a sample of 6129). As Table 5.2 shows, these expectations are very close to the observed frequencies. In fact all three populations are at Hardy–Weinberg equilibrium. As we shall see, when a population is not at Hardy–Weinberg equilibrium, the same calculations would not have correctly predicted the genotypic frequencies.

However, the importance of the Hardy–Weinberg theorem is not as an empirical prediction. We have no good reason to think that genotypes in natural populations will generally have Hardy–Weinberg frequencies, because it would require both no selection and random mating. Real populations may often experience selection and non-random mating. The interest of the theorem lies elsewhere. For empirical evolutionary biologists, it can act as a springboard, launching us toward the interesting problems. If we compare genotypic frequencies in a real population with the Hardy–Weinberg ratios and they deviate, it suggests something interesting (such as selection or non-random mating) may be going on, which would merit further research. For the theoretician, the theorem provides a useful short-cut. In the general model of population genetics (section 5.2 above) there were five stages, joined by four calculations. The Hardy–Weinberg theorem simplifies the

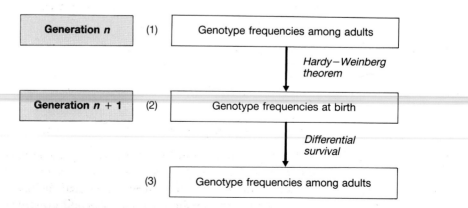

Figure 5.3 General model of population genetics simplified by the Hardy–Weinberg theorem.

model wonderfully. If we assume random mating, we can go directly from the adult frequencies in generation n to the genotype frequencies at birth of generation $n + 1$, collapsing three calculations into one (Figure 5.3). If we know the adult genotype frequencies in generation n (stage 1), we only need to calculate the gene frequencies: the genotype frequencies at birth in the next generation (stage 2) must then have Hardy–Weinberg frequencies, because the gene frequencies do not change between the adults of one generation and the newborn members of the next generation. A simple model of selection can concentrate on how the genotype frequencies are modified between birth and the adult reproductive stage (from stage 2 to stage 3 of Figure 5.3).

5.5 The simplest model of selection is for one favored allele at one locus

We shall start with the simplest case. It is the case of natural selection operating on only one genetic locus, at which there are two alleles, one dominant to the other. Suppose that individuals with the three genotypes have the following relative chances of survival from birth to the adult stage:

Genotype	Chance of survival
AA, Aa	1
aa	$1 - s$

s is a number between zero and one; it is called the *selection coefficient*. If s is 0.1 then *aa* individuals have a 90% chance of survival, relative to 100% for *AA* and *Aa* individuals. The chance of survival is the *fitness* of the genotype; we are assuming that all surviving individuals produce the same number of offspring. Obviously, selection will eventually eliminate the *a* allele and *fix* the *A* allele. (To "fix" a gene is genetic jargon for carrying its frequency up to one. When there is only one gene at a locus, it is said to be "fixed" or in a state of "fixation.")

How rapidly will the population change through time? To find out, we seek an expression for the gene frequency of A (p') in one generation in terms of its frequency in the previous generation (p). The difference between the two, $\Delta p = p' - p$, is the change in gene frequency between two successive generations. Now that selection is operating, the genotype frequencies will not have Hardy–Weinberg frequencies among the adults, because a disproportionate number of *aa* individuals will have died. We can, however, assume that mating is random; the three genotypes will have Hardy–Weinberg frequencies among newborn individuals. At birth:

Genotype	AA	Aa	aa
Frequency	p^2	$2pq$	q^2

The ratios among the adults, after selection, will be

Genotype	AA	Aa	aa
Relative frequency	p^2	$2pq$	$q^2(1 - s)$

Notice two things about the frequencies after selection. First, they do not add up to one. Because s is positive, the number of *aa* individuals has been reduced. The relative frequencies of *AA* and *Aa* have increased. To obtain an expression for the frequencies that does add up to one, we divide the relative frequencies by the total frequency of all the genotypes in the adult population. The total number is the sum of the three genotype frequencies, and is called the *mean fitness*. Here

$$\text{Mean fitness} = p^2 + 2pq + q^2(1 - s) = 1 - sq^2 \qquad (5.5)$$

We can then write down the exact frequencies among adults, after selection:

Genotype	*AA*	*Aa*	*aa*
Frequency	$\dfrac{p^2}{1 - sq^2}$	$\dfrac{2pq}{1 - sq^2}$	$\dfrac{q^2(1 - s)}{1 - sq^2}$

These frequencies must add up to one. The second point to notice is, as was asserted above, that the adult frequencies are not in Hardy–Weinberg ratio. If we tried to predict the proportion of *aa* from q^2, as in the MN blood group (section 5.4, p. 92), we should fail. The frequency of *aa* is $q^2(1 - s)/1 - sq^2$, not q^2.

What is the relationship between p' and p? Remember that the frequency of the gene *A* at any time is equal to the frequency of *AA* plus half the frequency of *Aa*. We have just listed those frequencies in the adults after selection:

$$p' = \frac{p^2 + pq}{1 - sq^2} \qquad (5.6)$$

($1 - sq^2$ is less than one, because s is positive, so p' is greater than p: selection is increasing the frequency of the *A* gene.) We can now derive a result for Δp, the change in gene frequency in one generation. The algebra looks like this:

$$\begin{aligned}
\Delta p = p' - p &= \frac{p^2 + pq}{1 - sq^2} - p \\
&= \frac{p^2 + pq - p + spq^2}{1 - sq^2} \\
&= \frac{p(p + q - 1 + sq^2)}{1 - sq^2} \\
&= \frac{spq^2}{1 - sq^2} \qquad (5.7)
\end{aligned}$$

For example, if $p = q = 0.5$ and *aa* individuals have fitness 0.9 compared with *AA* and *Aa* individuals ($s = 0.1$) then the change in gene frequency to the next generation will be $(0.1 \times 0.5 \times (0.5)^2)/(1 - 0.1 \times (0.5)^2) = 0.0128$; the frequency of *A* will therefore increase to 0.5128. We can use equation 5.7 to calculate the change in gene frequency between successive generations for any selection coefficient (s), and any gene frequency. The result in this simple case is that the *A* gene will increase in frequency until it is eventually fixed (i.e. has frequency of one). Table 5.3 illustrates how gene frequencies change for two selection coefficients. We can notice two points in the table. Firstly, the *A* gene increases in frequency more rapidly with a higher selection coefficient against the *aa* genotype. Secondly, the increase in the frequency of *A* slows down when it becomes common, and it would take a long time finally to eliminate the *a* gene. This is because the *a* gene is recessive. When *a* is rare it is almost always found in *Aa* individuals, which are selectively

Table 5.3 A simulation of changes in gene frequency for selection against the recessive gene *a*. The change between generation 0 and 100 is found by applying equation 5.7 100 times successively

Generation	Selection coefficient (s) = 0.05*		Selection coefficient (s) = 0.01*	
	p	*q*	*p*	*q*
0	0.01	0.99	0.01	0.99
100	0.44	0.56	0.026	0.974
200	0.81	0.19	0.067	0.933
300	0.89	0.11	0.15	0.85
400	0.93	0.07	0.28	0.72
500	0.95	0.05	0.43	0.57
600	0.96	0.04	0.55	0.45
700	0.96	0.04	0.65	0.35
800	0.97	0.03	0.72	0.28
900	0.97	0.03	0.77	0.23
1000	0.98	0.02	0.80	0.20

p, frequency of *A*; *q*, frequency of *a*.
* A selection coefficient of 0.05 (0.01) means *aa* individuals have a relative chance of survival of 95% (99%) against 100% for *AA* and *Aa*.

equivalent to *AA* individuals: selection can no longer "see" the *a* gene, and it becomes more and more difficult to eliminate them. Logically, selection cannot eliminate the one final *a* gene from the population, because if there is only one copy of the gene it *must* be in a heterozygote.

Haldane first produced this particular mathematic model of selection in 1924. It can be extended in various ways. However, for our purposes it is mainly important to see that an exact model of selection can be built and exact predictions made from it. The modifications for different degrees of dominance, and separate selection on heterozygotes and homozygotes, are easy, though they make the algebra more complex. The model is in many respects simplified, but it helps us to understand a number of real cases. We can use observations on genotype frequencies to estimate the genotype fitnesses, using the formula we derived in this section for gene frequency change as a function of fitness.

5.6 Industrial melanism in moths evolved by natural selection

The peppered moth *Biston betularia* provides perhaps the best known story in evolutionary biology. In collections made in Britain before the industrial revolution, the form of the moth was always a light, peppered color. The dark (melanic) form was first recorded in 1848 near Manchester; it then increased in frequency until it made up more than 90% of the populations in polluted areas in the mid twentieth century. In unpolluted areas, the light form remained common. Kettlewell explained the change as being due to visually hunting birds. The melanic form of moth was better camouflaged on the tree trunks in polluted areas, where soot killed the lichen; but the peppered form remained better camouflaged in unpolluted areas (Figure 5.4). Kettlewell finally convinced the skeptics that camouflage reduces predation

Figure 5.4 Peppered moths naturally settle on the undersides of twigs in higher branches of trees, not on tree trunks. Melanic forms are better camouflaged in polluted areas (compare (a) and (b)). Peppered forms are better camouflaged in unpolluted areas (d). Photographs from Brakefield (1987).

by birds when he put the two forms on trees in different areas, photographed birds in the act of taking the moths, and measured the rate at which birds took the two forms: light-colored moths were indeed taken more in polluted areas, and melanic moths in unpolluted areas. The evolution of melanic forms also took place in many other moth species through this period, but *Biston betularia* is the most thoroughly studied species.

What were the relative fitnesses of the genes controlling the melanic and light coloration during this evolutionary change? We first need to know the genetics of coloration in the moths, and then we need measurements of the frequencies of the different color forms for at least two instances; we can then estimate the gene frequencies from the genotype frequencies, and insert them into the formula for gene frequency change under selection to estimate the fitnesses of the light and melanic moths.

Genetic crosses initially suggested that the difference in color was controlled by one main locus; the original, peppered form (called *typica*) was one homozygote (*cc*) and the melanic form (called *carbonaria*) was another homozygote (*CC*). *C* is dominant to *c*. The first estimates of fitnesses were made by Haldane, using essentially the theory of the previous section. (The genetics is now known to be more complex than Haldane assumed: see below.)

So we shall assume that light-colored moths had genotype *cc* and melanic moths had genotypes *CC* or *Cc*. Now we need measurements of their frequencies at two known times. As we saw, the melanic form was first seen in 1848; but it was probably not a new mutation then. Although no earlier melanic specimens of the peppered moth are known, in other moth species there are melanic forms in pre-industrial forests; the melanic form of the peppered moth might therefore have been a rare form. Its frequency would probably be that of "mutation–selection balance." Mutation–selection balance means that the gene is disadvantageous and has a low frequency, determined by a balance between being formed by mutation and being lost by selection; a disadvantageous dominant gene in mutation–selection balance has a frequency of m/s, where m is its mutation rate and s its selective disadvantage (see equation 5.12, p. 108). The values of m and s are unknown for the gene in the eighteenth century, but Haldane guessed that $m = 10^{-6}$ and $s = 0.1$ for the melanic C gene. It would then have had a frequency of 10^{-5}. By 1898, the frequency of the light-colored (*typica*) genotype was 1–10% in polluted areas (it was not more than 5% near Manchester, for example, implying a gene frequency of about 0.2 for the *typica* gene and 0.8 for *carbonaria*). There would have been 50 generations between 1848 and 1898.

We now know all we need. If C was dominant, what selective coefficient would generate an increase in its frequency from 10^{-5} to 0.8 in 50 generations? The answer can be calculated from equation 5.6 (p. 94). This gives the gene frequency in one generation in terms of its frequency in the previous generation, but between 1848 and 1898 there would have been 50 generations. The formula therefore has to be applied 50 times over, which is most easily done by computer. A change from 10^{-5} to 0.8 in 50 generations, it turns out, requires $s = 0.33$: the peppered moths had two-thirds the survival rate of

Table 5.4 Theoretical changes in gene frequencies in the evolution of melanism in the peppered moth, starting with an initial frequency of *C* of 0.00001 (rounded to zero in the table). *C* is dominant, *c* is recessive: genotypes *CC* and *Cc* are melanic and *cc* is peppered in color

Generation	Year	Gene frequency	
		C	*c*
1848	0	0.00	1.00
1858	10	0.00	1.00
1868	20	0.03	0.97
1878	30	0.45	0.55
1888	40	0.76	0.24
1898	50	0.86	0.14
1908	60	0.90	0.10
1918	70	0.92	0.08
1928	80	0.94	0.06
1938	90	0.96	0.04
1948	100	0.96	0.04

melanic moths (Table 5.4). A frequency of 0.8 is reached in 40–50 generations. The calculations are crude; but they show how fitnesses can be inferred.

The estimate of fitness can be checked against other estimates. The gene frequency change is believed to be produced by differential survival in nature, and we can attempt to measure the rate of survival of the two types and see whether the peppered moths survive only two-thirds as well as melanic moths. Kettlewell measured survival rates by "mark–recapture" experiments in the field. He released melanic and light-colored peppered moths in known proportions in polluted and in unpolluted regions, and then later recaught some of the moths (which are attracted to mercury-vapor lamps). He then counted the proportions of melanic and peppered moths in the moths recaptured from the two areas. Table 5.5 gives some results for two sites, Birmingham (polluted) and Deanend Wood, an unpolluted forest in Dorset, England. The proportions in the recaptured moths are as we should expect: more light-colored *typica* in the Deanend Wood samples, more melanic *carbonaria* in the Birmingham samples. In Birmingham, melanic moths were recaptured at about twice the rate of light-colored ones, implying $s = 0.57$. This is a higher fitness difference than the $s = 0.33$ implied by the change in gene frequency. Even if we round $s = 0.57$ down to $s = 0.5$, Haldane calculated that the change in gene frequency from 10^{-5} to 0.8 would have been accomplished in 27 generations, rather than the 50 actually taken.

The discrepancy is unsurprising because both estimates are uncertain; it could have a number of causes. It may just be sampling error (the numbers in the mark–recapture experiment were small) and the frequency in 1848 may have been less than 10^{-5}. The genetics were probably also more complicated than we have assumed. The *carbonaria* gene that is now widespread is not the only melanic allele and the melanic allele that was spreading in the nineteenth century may not have been *carbonaria*. If the early melanic alleles had not been so well camouflaged as *carbonaria* they would have had a lower fitness

Table 5.5 Frequencies of three peppered moth forms in samples recaptured at two sites: Birmingham (polluted) and Deanend Wood, Dorset (unpolluted). *insularia* is an intermediate form between the normal light form (*typica*) and the melanic (*carbonaria*). The observed numbers are the actual numbers recaught; the expected numbers are the numbers that would have been recaught if all forms survived equally (i.e. proportions in released moths × no. moths recaptured). The recaptured moths at Birmingham were taken over a period of about 1 week, at Deanend Wood, over about 3 weeks. Data from Kettlewell (1973)

	Numbers recaptured		Relative survival rate	Relative fitness
	Observed	Expected		
Birmingham				
typica	18	35.97	0.5	0.43 (0.5/1.15)
insularia	8	8.57	0.93	0.81 (0.93/1.15)
carbonaria	140	121.46	1.15	1 (1.15/1.15)
Deanend Wood				
typica	67	54.3	1.23	1 (1.23/1.23)
insularia	5	2.3	—*	—*
carbonaria	32	47.4	0.68	0.55 (0.68/1.23)

*Data ignored as numbers are so small.

and not spread so fast; moreover, they might not have been fully dominant. If the heterozygotes were intermediate, the allele would not have spread so fast. Haldane also estimated how long it would take a melanic allele to increase from frequency 10^{-5} to 0.8 if it had the fitness advantage of *carbonaria* (Table 5.5) but produced intermediate heterozygotes. It would have taken 37 generations, which is more than the estimate of 27 given above. It is less than the observed 50 generations, but the discrepancy has been reduced.

There would also have been migration of light-colored moths from unpolluted areas into polluted areas, and the fitness difference in the polluted areas alone would therefore exaggerate the advantage of the dark form in the moth population as a whole. The fitness differences in the 1950s probably also differed from 1848 to 1898. After clean-air laws were passed in the twentieth century, industrial pollution decreased, and the frequency of the light form increased again in formerly more polluted areas; this factor, however, may aggravate the discrepancy. Whatever the cause, the two calculations do illustrate two important methods of estimating fitness. In summary, the early research on the peppered moth suggested that a rare gene (*carbonaria*) became advantageous in the nineteenth century. Moths possessing it were better camouflaged on trees and less frequently eaten by birds. It duly increased in frequency in polluted areas; but the original form remained common in unpolluted areas, where the light color was better camouflage. Finally, when smoke levels decreased in the twentieth century, the frequency of the light form increased again in some regions.

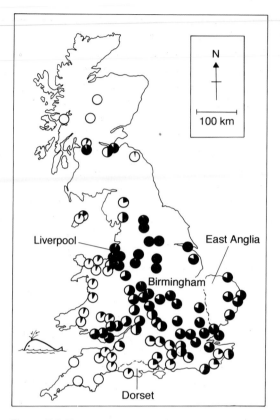

Figure 5.5 Frequency of melanic and peppered forms of the peppered moth in different parts of Britain. The filled-in part of each pie-diagram is the frequency of the melanic form in that area. Melanic moths are generally higher in industrial areas, such as central England; but note the high proportion in East Anglia, a non-industrial region. From Lees (1971).

What has recent research added to the story? One complication is that the genetic control is now known to involve at least three melanic alleles, not just one; but a more important discovery is that there are further selective forces at work. Melanic forms in other species—including beetles, pigeons, and cats—have increased in frequency in polluted regions even though they do not suffer significant predation by birds. In ladybird beetles (*Adalia bipunctata*), the melanic form absorbs solar radiation more efficiently in smoky places and has an advantage for that reason; it also has an advantage in breeding. But even if we concentrate on the peppered moth *Biston betularia*, the geographic distribution of the forms does not exactly fit the simple story. The melanic form, for example, has a frequency of up to 80% in East Anglia, where pollution is low (Figure 5.5). And in polluted areas, the dark form does not seem to have a high enough frequency. It never exceeded about 95% even though it was clearly better camouflaged and ought for that reason to have had a frequency of 100%.

Moreover, the decrease in frequency of melanic moths as pollution decreased seemed to move ahead of what would be predicted from their camouflage alone. Between 1960 and 1975, the frequency of melanic moths near Liverpool decreased from 95 to 82%, even though in 1975 they still appeared to be the better camouflaged form. There is also evidence that the forms differ in fitness independently of bird predation. In 1980, Creed *et al.* collected all the measurements that had been made on the moths' survival to adulthood in the laboratory. They analyzed the results of 83 broods, containing 12 569 offspring; the original measurements had been made by many different geneticists in the previous 115 years. The viability of light-colored homozygotes was found to be about 30% less on average than that of the melanic homozygote in the laboratory, where there is no bird predation—the reason for this is not known, but the fact alone implies there is some "inherent" advantage to the melanic genotype. The fitness advantage detected in the laboratory, although it is probably only approximately similar to that in nature, implies that melanic moths would replace light ones even without bird predation in polluted areas; and in unpolluted areas, light-colored moths remain only because birds eat more of the conspicuous melanic moths.

The final factor to be taken into account is migration. Male moths can fly long distances to find females, and male peppered moths may mate on average about 2.5 km away from where it is born. Migration may explain why melanic moths are found in some unpolluted areas like East Anglia and why light-colored moths persist in polluted areas where they are less well camouflaged. The three factors—bird predation, inherent advantage to melanic genotypes, and migration—were incorporated in a simulation by Mani in 1982 to try to explain the pattern of the gene in England. He could simulate the pattern most successfully with an "inherent" selective advantage (independent of bird predation) of 19% for the melanic homozygote, less than Creed *et al.*'s 30%, but the general success of Mani's simulation suggests that melanism in the peppered moth must be understood not only in terms of bird predation, but also in terms of migration and some other "inherent" selective advantage of the melanic moths.

In conclusion, although the industrial melanism of the peppered moth is a classic example of natural selection, it illustrates only imperfectly the simple one locus, two allele model of selection. Haldane made rough estimates of the difference in fitness between the two forms of moth using estimates of their frequencies at different times, and these estimates show in principle how the model can be used. But the selection is not as simple as once suspected. Bird predation is not the only factor, because the melanic form seems to have an inherent advantage, independent of location. It also seems that migration, as well as selection, is needed to explain the geographic pattern of gene frequencies.

5.7 Pesticide resistance in insects is an example of natural selection

Malaria is caused by a protozoan blood parasite (see section 5.11.1, p. 110), and humans are infected with it by mosquitoes (family Culicidae; genera include *Aedes*, *Anopheles*, and *Culex*). Transmission can therefore be prevented by killing the local mosquito population, and health workers have recurrently responded to malarial outbreaks by spraying insecticides such as DDT in affected areas. DDT, sprayed on a normal insect, is a lethal nerve poison.

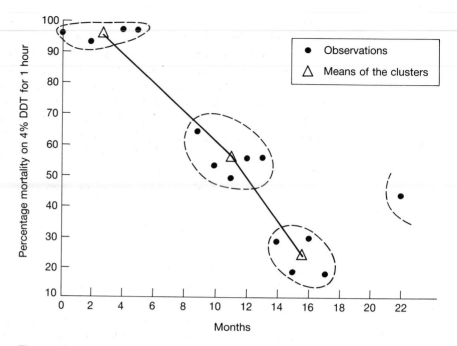

Figure 5.6 Increase in frequency of pesticide resistance in mosquitoes (*Anopheles culicifacies*) after spraying with DDT. A sample of mosquitoes was captured at each time indicated and the number that were killed by a standard dose of DDT (4% DDT for 1 hour) in the laboratory was measured. From Curtis *et al.* (1978).

When it is first sprayed on a local mosquito population, the population goes into abrupt decline. What happens then depends on whether DDT has been sprayed there before.

On its first use, DDT is effective for several years; in India, for example, it remained effective for 10–11 years after its first widespread use in the late 1940s. On a global scale, DDT was one reason why the number of cases of malaria reduced to 75 million or so per year by the early 1960s. But by then, DDT-resistant mosquitoes had already begun to appear. These were first detected in India in 1959, and they have increased so rapidly that when a local spray program is begun now, most mosquitoes become resistant in a matter of months rather than years (Figure 5.6). The malarial statistics reveal the consequence. The global number of cases had increased to over 200 million by 1972 and in parts of India the incidence of malaria increased 70-fold between the 1960s and 1970s. Pesticide resistance was not the only factor at work, but it was undoubtedly very important.

DDT becomes ineffective so quickly because, when it is sprayed, a strong force of selection in favor of DDT-resistant mosquitoes is immediately created. A graph such as Figure 5.6 allows a rough estimate of the strength of selection. As for the peppered moth, we need to understand the genetics of the character, and to measure the genotype frequencies at two or more times. We can then use the formula for gene frequency change to estimate the fitness. We have to make a number of assumptions. One is that resistance is controlled by a single allele (we shall return to that below). Another concerns the degree of dominance: the resistance allele might be dominant, recessive, or intermediate, relative to the natural "susceptibility" allele. The case of dominant resistance is easiest to understand. (If resistance is recessive we follow the same general method, but the exact result differs.) Let us call the resistance allele *R* and the susceptibility allele *r*. All the mosquitoes that die, in the mortality tests used in Figure 5.6, would then have been homozygous (*rr*) for susceptibility; assuming (for simplicity rather than accuracy) Hardy–Weinberg ratios, we can estimate the frequency of the susceptibility gene as the square root of the proportion of mosquitoes that die in the tests. The selection coefficients are defined as follows, where fitness is measured as the chance of survival in the presence of DDT:

Genotype	*RR*	*Rr*	*rr*
Fitness	1	1	$1 - s$

If *p* is the frequency of *R* and *q* is the frequency of *r*, the equation for gene frequency change is equation 5.7 on p. 94 (and also applied to the peppered moth): selection is working against a recessive gene. The graph (Figure 5.6) shows the decline in the initially common susceptible types. We need the expression for the change in *q* in one generation (Δq), rather than for Δp (as on p. 94). The decrease in *q* is the mirror-image of the increase in *p*, and we just need to put a minus sign in front of equation 5.7:

$$\Delta q = \frac{-spq^2}{1 - sq^2} \tag{5.8}$$

The generation time is about 1 month. (The generations of mosquitoes overlap, rather than being discrete as the model assumes; but the exact procedure is similar in either case, and we can ignore the detailed correction for overlapping generations.) Table 5.6 shows how the genotype frequencies were read off the graph (Figure 5.6) in two stages, giving two estimates of fitness. Again, the formula for one generation has to be applied recurrently, for 8.25 and 4.5 generations in this case, to give an average fitness for the genotypes through the period. It appears that in Figure 5.6 the resistant mosquitoes had about twice the fitness of the susceptible ones—this is very strong selection.

The genetics of resistance in this case are not known, and the one-locus, two-allele model is an assumption only; but they are understood in some

Table 5.6 Estimated selection coefficients against DDT-susceptible *Anopheles culicifacies*, from Figure 5.6. Selection coefficient *s* where relative fitness of susceptible type is $(1 - s)$. The estimate assumes the resistance allele is dominant. From Curtis *et al.* (1978)

Frequency of susceptible		Time	Selection
Before DDT	After DDT	(months)	coefficient (*s*)
0.96	0.56	8.25	0.4
0.56	0.24	4.5	0.55

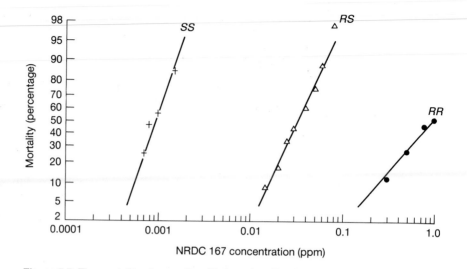

Figure 5.7 The mortality of mosquitos (*Culex quinquifasciatus*) of three genotypes at a locus when exposed to various concentrations of permethrin. The susceptible (*SS*) homozygote (+ —— +) dies at lower concentrations of the poison than the resistant homozygote (*RR*) (● —— ●). The heterozygote (*RS*) (△ —— △) has intermediate resistance. From Taylor (1986).

other cases. Resistance is often controlled by a single "resistance" allele. For example, Figure 5.7 shows that the resistance of the mosquito *Culex quinquifasciatus* to permethrin is due to a resistance (*R*) allele, which acts in a semi-dominant way, with heterozygotes intermediate between the two homozygotes. In houseflies, resistance to DDT is due to an allele called *kdr*; *kdr* flies are resistant because they have fewer binding sites for DDT on their nerve axons. In other cases, resistance may be due not to a new point mutation, but to gene amplification. *Culex pipiens*, for instance, in one experiment became resistant to an organophosphate insecticide called temephos because individuals arose with increased numbers of copies of a gene for an esterase enzyme that detoxified the poison. In the absence of temephos, the resistance disappeared, which suggests that the amplified genotype has to be maintained by selection. Gene amplification may be generally important in insecticide resistance; the semi-dominant behavior of resistance alleles in *C. quinquifasciatus* is, for example, compatible with this genetic mechanism. These few examples suggest that insects can evolve resistance by a large number of different mechanisms, and Table 5.7 summarizes the main means that have been identified so far.

When an insect pest has become resistant to one insecticide, the authorities often respond by spraying it with another. The evolutionary pattern we have seen here then usually repeats itself, and on a shorter time scale. On Long Island, New York, for example, the Colorado potato beetle (*Leptinotarsa septemlineata*) was first attacked with DDT. It evolved resistance to it in 7 years. The beetles were then sprayed with azinphosmethyl, to which it evolved resistance in 5 years; next came carbofuran (2 years), pyrethroids

Table 5.7 The main mechanisms of resistance to insecticides. From Taylor (1986)

Mechanism	Insecticides affected
Behavioral	
Increased sensitivity to insecticide	DDT
Avoid treated microhabitats	Many
Increased detoxification	
Dehydrochlorinase	DDT
Microsome oxidase	Carbamates Pyrethroids Phosphorothioates
Glutathione transferase	Organophosphates (O-dimethyl)
Hydrolases, esterases	Organophosphates
Decreased sensitivity of target site	
Acetylcholinesterase	Organophosphates Carbamates
Nerve sensitivity	DDT Pyrethroids
Cyclodiene resistance genes	Cyclodienes (organochlorines)
Decreased cuticular penetration	Most

(another 2 years), and finally pyrethroids with synergist (1 year). The decreasing time to evolve resistance is probably partly due to detoxification mechanisms that work against more than one pesticide. As more pests are sprayed with pesticides, more evolve resistance: according to a list in 1983, over 400 insect species were resistant to one or more pesticide. The list grows longer every year.

Insecticide resistance matters not only in the prevention of disease, but also in farming. Insect pests at present destroy about 20% of world crop production, and it has been estimated that in the absence of pesticides as much as 50% would be lost. Insect pests are a major economic and health problem. The evolution of resistance to pesticides causes misery to millions of people, whether through disease or reduced food supply. The fact that insects can rapidly evolve resistance is not the only problem with using pesticides against pests—the pesticides themselves (as is well known) can cause ecological side-effects that range from the irritating to the dangerous. However, the evolution of resistance to pesticides does provide a marvelously clear example of evolution by natural selection.

5.8 Fitness may be estimated by three main methods

The fitness of a genotype, in the theory and examples we have met, is its relative probability of survival from birth to adulthood. The fitness also determines the change in gene frequencies between generations. These two properties of fitness allow two methods of measuring it. One is to measure the relative survival of the genotypes within a generation. Kettlewell's mark–recapture experiment with the peppered moth is an example. If we assume that the relative rate of recapture of the genotypes is equal to their relative chance of survival from egg to adulthood, we have an estimate of fitness. The assumption may be invalid. The genotypes may, for instance, differ in their chances of survival at some stage of life other than the time of the mark–recapture experiment; if survival is measured in adult moths, any differences in survival at the egg or caterpillar stages will not be detected. Also, the genotypes may differ in fertility: fitnesses estimated by differences in survival can only be accurate if all genotypes have the same fertility. The second method is to measure changes in gene frequencies between generations. We then substitute the measurements into the formula that expresses gene frequency in one generation in terms of fitness and gene frequency in an earlier generation (equations 5.6 and 5.7). Both methods have been used in many cases; the main problems are the obvious difficulties of accurately measuring survival, and gene frequencies, respectively. Apart from them, in the examples we considered there were also difficulties in understanding the genetics of the characters: we need to know the genotypes underlying the observed phenotypes.

We shall meet a third method of estimating fitness below, in the case of sickle-cell anemia (Table 5.9, p. 112). It uses deviations of adult genotype frequencies from the Hardy–Weinberg ratios. It can only be used when the gene frequencies are constant between birth and adulthood, but genotypes have different survival. It therefore cannot be used in the examples of directional selection against a disadvantageous gene that we have been concerned with so far, because in them the gene frequency changes between birth and the adult stage.

5.9 Natural selection operating on a favored allele at a single locus is not meant to be a general model of evolution

Evolutionary change in which natural selection favors a rare mutation at a single locus, and carries it up to fixation, is one of the simplest forms of evolution. Sometimes evolution may happen that way. But the situation can be more complicated in nature. We have considered selection in terms of different chances of survival from birth to adulthood; but selection can also take place by "differential fertility," in which the genotypes—even after they have survived to adulthood—produce different numbers of offspring. Our model had random mating among the genotypes: but mating may be non-random. Moreover, the fitness of a genotype may vary in time and space, and depend on what genotypes are present at other loci (a subject we shall deal with in chapter 8). Much of evolutionary change probably consists of adjustments in the frequencies of alleles at polymorphic loci, as fitnesses fluctuate through time, rather than the fixation of new favorable mutations.

These complexities in the real world are important, but they do not invalidate—or trivialize—the one-locus model. For the model is intended as a model. It should be used as an aid to understanding, not as a general theory of nature. In science, it is a good strategy to build up an understanding of nature's complexities by considering simple cases first and then working toward the complex whole. Simple ideas rarely provide accurate, general theories; but they often provide powerful paradigms. The one-locus model is concrete and easy to understand. It is a good starting point for the science of population genetics. Indeed, population geneticists have constructed models of all the complications in the previous paragraphs, and those models are all developments within the general method we have been studying.

5.10 A recurrent disadvantageous mutation will evolve to a calculable equilibrial frequency

The model of selection at one locus revealed how a favorable mutation will spread through a population. But what about unfavorable mutations? Natural selection will act to eliminate any allele that decreases the fitness of its bearers, and the allele's frequency will decrease at a rate specified by equations; but what about a recurrent disadvantageous mutation that keeps arising at a certain rate? Selection can never finally eliminate the gene, because it will keep on reappearing by mutation. In this case, we can work out the equilibrial frequency of the mutation: the equilibrium is between the mutant gene's creation, by recurrent mutation, and its elimination by natural selection.

To be specific, we can consider a single locus, at which there is initially one allele, a. The gene has a tendency to mutate to a dominant allele, A. We must specify the mutation rate and the selection coefficient (fitness) of the genotypes. Define m as the mutation rate from a to A per generation; we ignore back mutation (though actually this assumption does not matter). The frequency of a is q, and of A is p. Finally, we define the fitnesses as follows:

Genotype	aa	Aa	AA
Fitness	1	$1 - s$	$1 - s$

How does evolution proceed? We shall try to find the equilibrial frequency of A (defined as p^*). To do so, we first find the relationship between the frequency of A in one generation (p) and in the next generation (p'): we can then reason that at equilibrium $p = p'$; the gene frequency is constant. We thus find p^* from the equation $p = p'$. The relation between p and p' is found in two stages; we first find the effect of selection on gene frequency (Table 5.8), much as in section 5.5; then we calculate the effect of mutation. Mutation increases the number of A genes by $m \times$ the number of a genes and decreases the number of a genes by the same amount; the total number of genes is unchanged. From Table 5.8, the frequency of A after selection (but before mutation) $= 2p(1 - s)/(2 - 2s(p + pq))$; and the frequency of a after selection $= 2q(1 - ps)/(2 - 2s(p + pq))$. Therefore the frequency of A after both selection and mutation, p', is:

$$p' = \frac{2p(1 - s) + 2mq(1 - ps)}{2 - 2s(p + pq)} \tag{5.9}$$

This simplifies to

$$p' = \frac{p - ps + mq - mpqs}{1 - sp(1 + q)}$$

At equilibrium, there will be only a few A genes: selection is working against A and mutation rates are low (10^{-6} or so). Therefore p^* will be low. Now, if p

Table 5.8 (a) Frequency of genotypes after selection, when A is dominant to a and AA and Aa have a selective disadvantage of $(1 - s)$ relative to aa. (b) Number of the genes A and a after selection. Note the number of genes is twice the number of genotypes

(a)

Genotype	Frequency after selection
AA	$p^2 (1 - s)$
Aa	$2pq (1 - s)$
aa	q^2

(b)

Gene	Number after selection
A	$2(p^2 + pq)(1 - s) = 2p(1 - s)$
a	$2pq(1 - s) + 2q^2 = 2q(1 - ps)$
Total	$2 - 2s(p^2 + 2pq) = 2 - 2s(p + pq)$

is low the denominator $(1 - sp(1 + q))$ will be approximately one, and mq in the numerator will be much larger than $mpqs$. The equation then simplifies to:

$$p' \approx p - ps + mq \tag{5.10}$$

at equilibrium,

$$p^* = p^* - p^*s + m(1 - p^*) \tag{5.11}$$

which rearranges to

$$p^* = \frac{m(1 - p^*)}{s} \approx \frac{m}{s} \tag{5.12}$$

The simple result is that the equilibrial gene frequency of the mutation is equal to the ratio of its mutation rate to its selective disadvantage. The result is intuitive: the equilibrium is the balance between the rates of creation and elimination of the gene.

The expression $p = m/s$ can allow a rough estimate of the mutation rate of a harmful mutation just from a measurement of the mutant gene's frequency. If the mutation is rare, it will be present mainly in heterozygotes, which at birth will have frequency $2pq$. If p is small, $q \simeq 1$ and $2pq \simeq 2p$. Define N as the frequency of mutant bearers, which equals the frequency of heterozygotes: i.e. $N = 2p$. As $p = m/s$, $m = sp$; if we substitute $p = N/2$, $m = sN/2$. If the mutation is highly deleterious, $s \simeq 1$ and $m \simeq N/2$. The mutation rate can be estimated as half the birth rate of the mutant type. The estimate is clearly approximate, because it relies on a number of assumptions. In addition to the assumptions of high s and low p, mating is supposed to be random. We usually have no means of checking whether this is the case.

Chondrodystrophic dwarfism is a dominant deleterious mutation in humans. In one study, 10 births out of 94 075 had the gene, a frequency of

10.6×10^{-4}. The estimate of the mutation rate by the above method is then $m = 5.3 \times 10^{-4}$. However, it is possible to estimate the selection coefficient, enabling a more accurate estimate of the mutation rate. In another study, 108 chondrodystrophic dwarves produced 27 children; their 457 normal siblings produced 582 children. The relative fitness of the dwarves was $(27 \times 457)/(108 \times 582) = 0.196$; the selection coefficient $s = 0.804$. Instead of assuming $s = 1$, we can use $s = 0.804$. Then the mutation rate $= sN/2 = 4.3 \times 10^{-4}$, a rather lower figure because with lower selection the same gene frequency can be maintained by a lower mutation rate.

For many genes, we do not know the dominance relationships of the alleles at the locus. A similar calculation can be done for a recessive gene, but the formula is different, and it differs again if the mutation has intermediate dominance. We can only estimate the mutation rate from $p = m/s$ if we know the mutation is dominant. The method is therefore unreliable unless its assumptions have been independently verified.

5.11 Selection can maintain a polymorphism when the heterozygote is fitter than either homozygote

We come now to an influential theory. We are going to consider the case in which the heterozygote is fitter than both homozygotes. The fitnesses can be written:

Genotype	AA	Aa	aa
Fitness	$1 - s$	1	$1 - t$

What happens here? Again, we look for the relationship between the gene frequency of A in one generation (p) and the next (p') and reason that at equilibrium $p = p'$. We find the expression for the next generation by the standard procedure, which we have used before: it is the frequency of AA individuals plus half the frequency of Aa individuals all divided by the mean fitness (which equals the total frequency of all the three genotypes after selection):

$$p' = \frac{p^2(1 - s) + pq}{p^2(1 - s) + 2pq + q^2(1 - t)} \tag{5.13}$$

which simplifies to

$$p' = \frac{p(1 - ps)}{1 - p^2 s - q^2 t} \tag{5.14}$$

At equilibrium $p' = p$, and

$$1 - ps = 1 - p^2 s - q^2 t \tag{5.15}$$

We then write $q = 1 - p$ and solve the equation for the equilibrial value of p. The result is

$$(p - 1)[(s + t)p - t] = 0 \tag{5.16}$$

The equation can be satisfied if $p = 0$, $p = 1$, or $p = s/(s + t)$. The former two equilibria are not particularly interesting; they merely say that the system is stable if either a or A is fixed, as it must be in the absence of mutation. The third, "internal" equilibrium is an equilibrium in which all three genotypes are present. The two homozygotes are at a disadvantage; they are selected against, but it is impossible to eliminate them because matings among heterozygotes generate homozygotes. The exact gene frequency at equilibrium depends on the relative selection against the two homozygotes. If, for instance, AA and aa have equal fitness, then $s = t$ and $p = \frac{1}{2}$ at equilibrium. If AA is relatively more unfit than aa then $s > t$ and $p < \frac{1}{2}$; there are fewer of the more strongly selected against genotypes.

When heterozygotes are fitter than the homozygotes, therefore, natural selection will maintain a polymorphism. The result was first proved by Fisher in 1922 and independently by Haldane. We shall come later to consider in more detail why genetic variability exists in natural populations, and *heterozygous advantage* will be one of several controversial explanations to be tested.

5.11.1 Sickle-cell anemia is a polymorphism with heterozygous advantage

Sickle-cell anemia is the classic example of a polymorphism maintained by heterozygous advantage. It is a lethal or near lethal condition in humans; it is responsible for about 100 000 deaths a year. It is caused by a genetic variant of α-globin. If we symbolize the normal hemoglobin allele by A and the sickle-cell hemoglobin by S, then people who suffer from sickle-cell anemia are SS genotype. Hemoglobin S causes the red blood cells to become curved and distorted ("sickle" shaped); they then block up the capillaries and are destroyed by phagocytosis, causing severe anemia. About 80% of SS individuals die before reproducing. With such apparently strong selection against hemoglobin S it was a puzzle why it persisted at quite high frequencies (10% or even more) in some human populations. The answer was first suggested by Haldane and confirmed by Allison.

Haldane actually discussed another anemic condition; but his argument works for sickle-cell anemia too. He compared a map of the incidence of malaria with a map of the hemoglobin S gene frequency (Figure 5.8): they are strikingly similar. It turned out that, although SS is almost always lethal, the heterozygote AS is more resistant to malaria than the normal homozygote (AA). AS red blood cells do not normally sickle, except if the oxygen concentration falls. When the malarial parasite *Plasmodium falciparum* enters a red blood cell it destroys (probably eats) the hemoglobin, which causes the oxygen concentration in the cell to go down and the cell sickles and is destroyed, along with the parasite. The human survives because most of the red blood cells are uninfected and carry oxygen normally. Therefore, where the malarial parasite is common, AS humans survive better than AA, which suffer from malaria. (Malarial resistance may not be the only advantage of the AS genotype.)

Once the heterozygote had been shown physiologically to be at an advantage, the adult genotypic frequencies can be used to estimate the relative fitnesses of the three genotypes. The fitnesses are:

Genotype	AA	AS	SS
Fitness	$1 - s$	1	$1 - t$

If the frequency of the gene A is p and of S is q, then the relative genotype frequencies among adults will be $p^2(1 - s):2pq:q^2(1 - t)$. If there were no

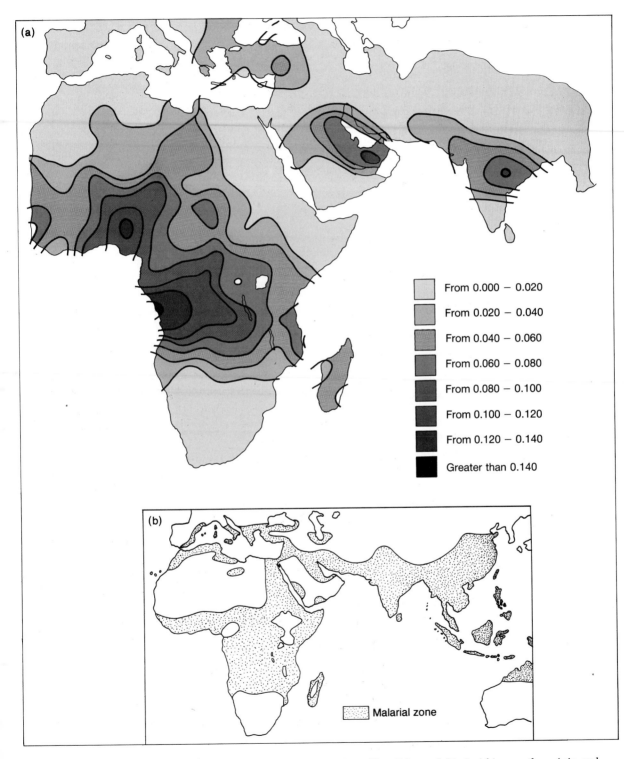

Figure 5.8 (a) Frequencies of gene for sickle-cell hemoglobin in Africa, southern Asia, and Europe. The frequencies match the incidence of malaria (b).

Table 5.9 Estimates of selection coefficients for sickle-cell anemia, using genotypic frequencies in adults. The sickle-cell hemoglobin allele is S, and the normal hemoglobin (which actually consists of more than one allele) is A. The genotypic frequencies are for the Yorubas of Ibadan, Nigeria. One small detail is not explained in the text. The observed expected ratio for the heterozygote may not be equal to one. Here it turned out to the 1.12. All the observed expected ratios are therefore divided by 1.12 to make them fit the standard fitness regime for heterozygote advantage. From Bodmer and Cavalli-Sforza (1976)

Genotype	Observed adult frequency (O)	Expected Hardy–Weinberg frequency* (E)	Ratio O/E	Fitness
SS	29	187.4	0.155	0.155/1.12 = 0.14 = 1 − t
SA	2993	2672.4	1.12	1.12/1.12 = 1.00
AA	9365	9527.2	0.983	0.983/1.12 = 0.88 = 1 − s
Total	12 387	12 387		

*Calculation of expected frequencies is done thus: gene frequency of S = frequency of SS + 1/2 (frequency of SA) = (29 + 2993/2)/12 387 = 0.123. Therefore frequency of A allele = 1 − 0.123 = 0.877. From the Hardy–Weinberg theorem, the expected genotype frequencies are $(0.123)^2 \times 12\,387$, $2(0.877)(0.123) \times 12\,387$, and $(0.877)^2 \times 12\,387$, for SS, SA, and AA, respectively.

selection ($s = t = 0$), the three genotypes would have Hardy–Weinberg frequencies of $p^2 : 2pq : q^2$.

Selection causes deviations from the Hardy–Weinberg frequencies. Take the genotype AA as an example. The ratio of the observed frequency in adults to that predicted from the Hardy–Weinberg ratio will be $(1 − s)/1$. The frequency expected from the Hardy–Weinberg principle is found by the usual method: the expected frequency is p^2, where p is the observed proportion of AA plus half the observed proportion of AS. Table 5.9 illustrates the method for a Nigerian population, where $s = 0.88$ and $t = 0.14$.

The method is only valid if the deviation from Hardy–Weinberg proportions is caused by heterozygous advantage. If heterozygotes are found to be in excess frequency in a natural population, it may indeed be because the heterozygote has a higher fitness, but there are other reasons. Disassortative mating, for instance, can produce the same result (in this case, disassortative mating would mean that aa individuals preferentially mate with AA individuals). But for sickle-cell anemia, the physiologic observations showed that the heterozygote is fitter and the procedure is well justified. Indeed, in this case, although it has not been checked whether mating is random, the near lethality of SS means that disassortative mating will be unimportant. It is also being assumed that surviving members of all the genotypes have the same fertility.

5.12 The fitness of a genotype may depend on its frequency

The next interesting complication is to consider selection when the fitness of a genotype depends on its frequency. In the models we have considered so far, the fitness of a genotype (1, $1 − s$, or whatever) was constant, regardless of whether the genotype was rare or common. It is also possible for the fitness

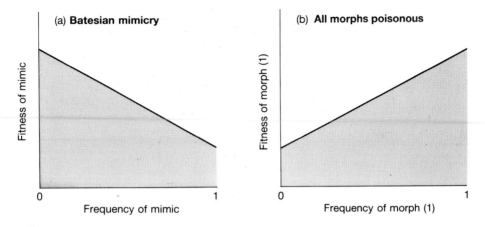

Figure 5.9 (a) With Batesian mimicry within a species fitnesses are negatively frequency dependent. (b) When all morphs of a species are poisonous, fitnesses become positively frequency dependent. In a graph of fitness against frequency, any line other than a horizontal one is a case of frequency-dependent fitness.

of a genotype to increase (positively frequency dependent) or decrease (negatively frequency dependent) as the genotypic frequency in the population increases.

Examples of both positive and negative frequency dependence can arise in systems of mimicry. Some poisonous species of animals, such as butterflies, have evolved recognizable color patterns, and predatory species, such as birds, learn to avoid feeding on these poisonous prey. Natural selection may then favor other non-poisonous butterflies that have a similar color pattern: they will be avoided by birds which have learned to avoid the poisonous types. This system is called Batesian mimicry. In many real cases, the mimic and model belong to separate species (and selection is strictly speaking number dependent rather than frequency dependent). But Batesian mimicry can also exist within a species, and the fitness of each type then depends on its frequency. Consider the non-poisonous mimics. When they are rare, birds will tend to avoid them, because they will have already encountered a poisonous butterfly of the same appearance. But when the non-poisonous type is common, the previous encounters of birds with butterflies of their appearance are more likely to have been rewarding; the birds will not avoid eating them, and their fitness will be lower. The fitness of the mimics is negatively frequency dependent (Figure 5.9a).

In other butterflies, such as in Central and South American *Heliconius*, there are several morphs within a species, each morph having a different color pattern. All the morphs are poisonous. In this case the fitness of each morph increases as its frequency increases. When a morph is common, it will be more likely that birds will have already learned to avoid them, whereas birds will not yet have learned to avoid a rare morph. An individual of a rare morph is therefore more likely to be the unlucky prey that educates the bird, and gets killed in the process. The fitness of each morph is positively frequency dependent (Figure 5.9b). With positively frequency-dependent

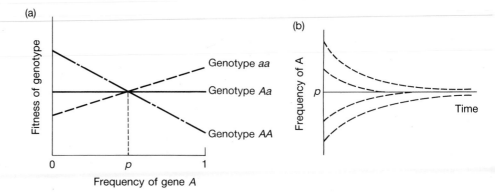

Figure 5.10 (a) *AA* and *aa* have negatively frequency-dependent fitnesses; that of *Aa* is frequency independent. To the left of the point (*p*) where the lines for *AA* and *aa* cross, *A* is favored and increases in frequency, vice versa to the right. (b) Natural selection will take the population from any initial gene frequency to the gene frequency at which the fitnesses of *AA* and *aa* are equal.

selection, a morph becomes more likely to be fixed as its frequency increases. We therefore do not expect to find polymorphisms. Indeed, in *Heliconius*, there is only a single morph in any one area; the different morphs are found in different places.

With negatively frequency-dependent fitnesses (as in Batesian mimicry), however, it is possible for natural selection to maintain a polymorphism. When a genotype is rare, it is relatively favored by selection and it will increase in frequency; as it becomes more common, its fitness decreases and there may come a point at which it is no longer favored (Figure 5.10). At that point, the fitnesses of the different genotypes are equal and natural selection will not alter their frequencies: they are at equilibrium. The sex ratio is another case in which selection is frequency dependent (see chapter 11).

It is in principle straightforward to modify one of the genetic models we have considered for the cases of frequency dependence. We just have to write the fitness of a genotype as a function of its frequency. One simple example is, for gene frequency of $A = p$, and of $a = q$:

Genotype	*AA*	*Aa*	*aa*
Fitness	$2(1 - p)$	*1*	$2(1 - p)$

Fitness is negatively frequency dependent here. The fitnesses of the genotypes are 2, 1, and 0 when *A* is rare ($p \approx 0$), and 0, 1, and 2 when *A* is common. However, in more general models, the algebra becomes more difficult than in the frequency-independent case. Frequency-dependent selection is most easily investigated by means of graphs like Figure 5.9.

In theory, the graphic relationship between fitness and frequency can have any shape, straight-lined or curved (Figure 5.11). The shape of the curve is theoretically important. It controls whether a polymorphism is possible, as well as the exact dynamics of gene frequency changes through time. The theoretical possibilities in models of this general kind are remarkably rich. They can, for instance, produce oscillations in gene frequency (section 11.3.3,

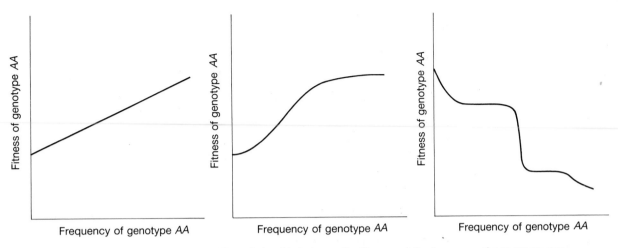

Figure 5.11 The relationship between the fitness and the frequency of a genotype can theoretically have any shape.

p. 271). The main point here, however, is the possibility that, with negatively frequency dependent fitnesses, natural selection can maintain a polymorphism.

5.13 Multiple niche polymorphism can evolve in a heterogeneous environment

Normal individual fruitflies (*Drosophila*) have orangey-red colored eyes. In some mutant forms, however, the red pigment is lacking and their eyes are white. These eye-color mutations affect the fruitfly's response to light. A normal *Drosophila*, with red eyes, is attracted to light, though repelled by very bright light; white-eyed mutants are more sensitive and (if given the choice) will settle in places of lower light intensity. Jones and Probert performed the following experiment with normal and white-eyed *Drosophila simulans*. They first set up cages, with a mixture of the two genotypes, in either white or red light; in both cases the white-eyed mutants were at a disadvantage and selection reduced their frequency (Figure 5.12a). They also placed the two types of fly together in a cage that was illuminated with red light in one half and white light in the other half. In this case the two genotypes were maintained (Figure 5.12a). As might be expected, the white-eyed flies concentrated in the red half of the cage, and the normal red-eyed flies in the half with white light; the flies show "habitat selection" (Figure 5.12b).

The polymorphism in the cage with both white and red light is a *multiple niche polymorphism*, a topic first discussed by Levene in 1953. We can call the parts of the cage with different degrees of illumination different "niches," analogous to distinct regions in the species' natural environment. When both red and white niches are present, each type of fly chooses to live in the type of niche where it has higher fitness. The result is for normal flies to go to the region with white light, leaving enough resources for the white-eyed mutants to survive where the light is red.

Up to two conditions are needed for a multiple niche polymorphism. In nature, the members of a species occupy a variety of niches, and these vary in

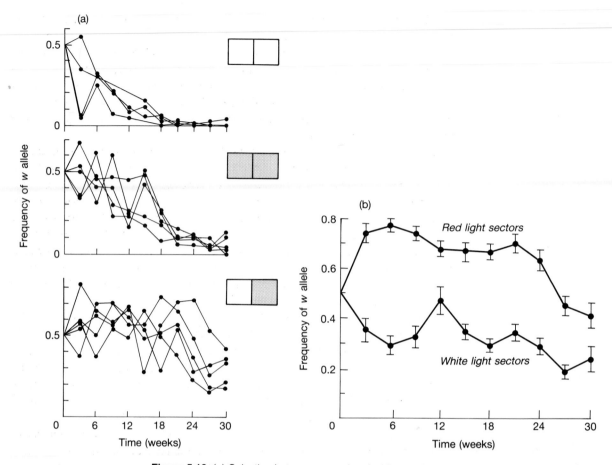

Figure 5.12 (a) Selection between normal and white-eyed genotypes of *Drosophila simulans* in three types of environment. In white light (top), the white-eyed genotype is removed by selection, as it is in red light (middle). But when the flies are placed in cages with red and white light, both genotypes persist (bottom). The experimental cages were divided in two by a barrier that the flies could walk through. In each case, the population was started with a mixture of the two genotypes, and the graphs plot the change in frequency of the white-eyed genotype. (b) In a cage with white light in one half and red light in the other (the bottom condition of Figure 5.12a), the normal red-eyed fruitflies concentrate in the white light and the white-eyed mutants concentrate in the red light. From Jones and Probert (1980).

their physical conditions (whether they are dark, light, dry, damp, etc.) and biologic conditions (what predators and parasites are locally active, what food is available). The first (and crucial) condition is that different genotypes should have different fitnesses in the various niches. In theory, this condition alone could produce a polymorphism (though no examples of this are known). The simplest case would have two niches, A and B, with genotypes *AA* and *Aa* better adapted to A and *aa* better adapted to B. The fitnesses are:

Genotypes	*AA*	*Aa*	*aa*
Fitness in niche A	1	1	$1 - s$
Fitness in niche B	$1 - t$	$1 - t$	1

It is in principle easy to write a model in which the frequency of niche A is N and of niche B is $(1 - N)$. If the genotypes experience niche types in these frequencies, the fitness of AA will be $[N + (1 - N)(1 - t)]$. Likewise for the other genotypes. The equations are fairly ugly and we need only notice the possibility that a polymorphism can evolve. The relative frequencies of the genotypes at equilibrium will depend on the proportions of the two niches: if A is more common gene A has a higher frequency and vice versa if niche B is more common.

A stable polymorphism can more easily evolve if there is *habitat selection* (as was the case in Jones and Probert's fruitflies): each genotype "chooses" the correct habitat to live in. Individuals with genotype AA or Aa would go and live in niche type A and those with aa pick niches of type B. A formal model would then have an even more complex expression for the fitness of a genotype, in terms not only of the frequency of each niche type but also the genotype's chance of living in it. However, the qualitative conclusion is again that multiple niche polymorphisms can arise, and they are more plausible if there is habitat selection as well as a relationship between a genotype's fitness and the niche it is living in.

Multiple niche polymorphism is a form of frequency dependent polymorphism. When the genotype AA is rare, it experiences relatively little competition in the niches for which it is well adapted, and increases in frequency. As AA becomes more common, it increasingly occupies all the niches of type A and has to make a living in B type niches too. Its advantage decreases as its frequency increases. The fitness of each genotype is negatively frequency dependent, in the manner of Figure 5.10a.

The purpose of sections 5.10–5.13 has been to illustrate the different mechanisms by which natural selection can maintain polymorphism. In chapter 6 we shall study another mechanism that can maintain polymorphism, a mechanism that does not rely on natural selection: genetic drift. Then, in chapter 7, we shall tackle the question of how important the mechanisms are in nature.

5.14 Subdivided populations require special population genetic principles

5.14.1 A subdivided set of populations have a higher proportion of homozygotes than an equivalent fused population: this is the Wahlund effect

So far we have considered population genetics within a single, uniform population. In practice, a species may consist of a number of separate populations, each more-or-less isolated from the others. The members of a species might, for example, inhabit a number of islands, with each island population being separated by the sea from the others; individuals might migrate between islands from time to time, but each island population would evolve to some extent independently. A species with a number of more-or-less independent subpopulations is said to have *population subdivision*.

Let us first see what effect population subdivision has on the Hardy–Weinberg principle. Consider a simple case in which there are two populations (we can call them population 1 and population 2), and we concentrate on one genetic locus with two alleles, A and a. Suppose allele A

Table 5.10 The frequency of genotypes *Aa* and *aa* assuming allele *A* has a frequency of 0.3 in population 1 and 0.7 in population 2. Calculated using the Hardy–Weinberg ratio

Genotype	Frequency		
	Population 1	Population 2	Average
AA	$(0.3)^2 = 0.09$	$(0.7)^2 = 0.49$	0.58/2 = 0.29
Aa	$2(0.3)(0.7) = 0.42$	$2(0.3)(0.7) = 0.42$	0.84/2 = 0.42
aa	$(0.7)^2 = 0.49$	$(0.3)^2 = 0.09$	0.58/2 = 0.29

has frequency 0.3 in population 1 and 0.7 in population 2. If the genotypes have Hardy–Weinberg ratios they will have frequencies, and average frequencies, in the two populations (we assume the two populations are of equal size) as shown in Table 5.10.

Now compare the average for the two with the Hardy–Weinberg genotype frequencies if the two populations were fused together. The gene frequencies of *A* and *a* in the combined population are (0.3 + 0.7)/2 = 0.5, and the Hardy–Weinberg genotype frequencies are:

Genotype	*AA*	*Aa*	*aa*
Frequency	0.25	0.5	0.25

Compare the frequency of homozygotes in the two cases. In the large, fused population there are fewer homozygotes than in the average for the set of subdivided populations. This is a general, and mathematically automatic, result. The excess of homozygotes in subdivided populations is called the *Wahlund effect*.

The Wahlund effect has a number of important consequences. One is that we have to know about the structure of a population when applying the Hardy–Weinberg principle to it. Suppose, for example, we had not known that populations 1 and 2 were independent. We might have sampled from both, pooled the samples indiscriminately, and then measured the genotype frequencies. We should find the frequency distribution for the average of the two populations (0.29:0.42:0.29); but the gene frequency would apparently be 0.5. There would seem to be more homozygotes than expected from the Hardy–Weinberg principle. We might suspect that selection, or some other factor, was favoring homozygotes. In fact both subpopulations are in perfectly good Hardy–Weinberg equilibrium and the deviation is due to the unwitting pooling of the separate populations. The moral is that we need to be on the lookout for population subdivision when interpreting deviations from Hardy–Weinberg ratios. A second consequence of the Wahlund effect is that when a number of previously subdivided populations merge together, the frequency of homozygotes will decrease. In humans, this can lead to a decrease in the incidence of rare recessive genetic diseases when a previously isolated population comes into contact with a larger population. The recessive disease is only expressed in the homozygous condition, and when the two populations start to interbreed, the frequency of those homozygotes goes down.

5.14.2 Migration acts to unify gene frequencies between populations

When an individual migrates from one population to another, it carries genes that are representative of its own ancestral population into the recipient population. If it successfully establishes itself and breeds it will transmit those genes between the populations. The transfer of genes is called *gene flow*. If the two populations originally had different gene frequencies and if selection is not operating, migration (or, to be exact, gene flow) alone will rapidly cause the gene frequencies of the different populations to converge. We can see how rapidly in a simple model.

Consider again the case of two populations and one locus with two alleles (A and a). Suppose this time that one of the populations is much larger than the other, say population 2 is much larger than population 1 (2 might be a continent and 1 a small island near it); then practically all the migration is from population 2 to population 1. The frequency of allele a in population 1 in generation t is written $q_{1(t)}$; we can suppose that the frequency of a in the large population 2 is not changing between generations and write it as q_m. (We are interested in the effect of migration on the gene frequency in population 1 and can ignore all other effects, such as selection.) Now, if we pick on any one allele in population 1 in generation $(t + 1)$, it will either be descended from a native of the population or from an immigrant. Define m as the chance that it is a migrant gene. (Earlier in the chapter, m was used for the mutation rate: now it is the *migration* rate.) If our gene is not a migrant (chance $(1 - m)$) it will be an a gene with chance $q_{1(t)}$, whereas if it is a migrant (chance m) it will be an a gene with chance q_m. The total frequency of a in population 1 in generation $(t + 1)$ is:

$$q_{1(t+1)} = (1 - m)q_{1(t)} + mq_m \qquad (5.17)$$

This can be rearranged to show the effect of t generations of migration on the gene frequency in population 1. If $q_{1(0)}$ is the frequency in the 0th generation, the frequency in generation t will be:

$$q_{1(t)} = q_m + (q_{1(0)} - q_m)(1 - m)^t \qquad (5.18)$$

(From $t = 1$ it is easy to confirm that this is indeed a rearrangement of the previous equation.) The equation says that the difference between the gene frequency in population 1 and population 2 decreases by a factor $(1 - m)$ per generation. At equilibrium, $q_1 = q_m$ and the small population will have the same gene frequency as the large population (Figure 5.13). In Figure 5.13, the gene frequencies converge in about 30 generations with a migration rate of 10%. Similar arguments apply if, instead of there being one source and one recipient population, the source is a set of many subpopulations, and q_m is their average gene frequency, or if there are two populations with each sending migrants to, and receiving them from, the other. Migration will generally unify gene frequencies among populations rapidly in evolutionary time. In the absence of selection, migration is a strong force for equalizing the gene frequencies of subpopulations in a species. Providing that the migration

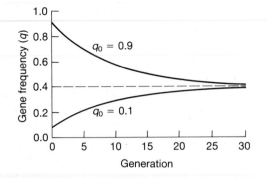

Figure 5.13 Migration causes the rapid convergence of gene frequencies in the populations exchanging migrants. Here a source population with gene frequency $q_m = 0.4$ sends migrants to two subpopulations, with initial gene frequencies of 0.9 and 0.1. They converge, with $m = 0.1$, onto the source population's gene frequency in about 30 generations.

rate is greater than zero, gene frequencies will eventually equalize. Even if there is only one successful migrant per generation, gene flow inevitably draws the population's gene frequency to the species' average. Gene flow thus acts to bind the species together.

5.14.3 The convergence of gene frequencies by gene flow is illustrated by the human population of the USA

The MN blood group system is controlled by one locus with two alleles (see p. 91). Frequencies of the M and N alleles have been measured, for example, in white and black Americans in Claxton, Georgia, and among West African blacks (which we can assume to be representative of the ancestral gene frequency of the black population of Claxton). The M allele frequency is 0.474 for the West Africans, 0.484 for the black Americans, and 0.507 for the white Americans. (The frequency of the N allele is $1 -$ frequency of the M allele.) The gene frequency among blacks in Claxton is thus intermediate between the frequencies for whites and for the West African sample. Individuals of mixed parentage were categorized as black and, if we ignore the possibility of selection favoring the M allele in the USA, we can treat the change in gene frequency in blacks as due to immigration of genes from the white population. The measurements can then be used to estimate the rate of gene migration. In equation 5.18, q_m is the gene frequency in the white population (the source of the "migrant" genes), q_0 is 0.474 (the original frequency in the black population), q_t is 0.484. As an approximate figure, we can suppose that the black population has been in the USA for 200–300 years, or about 10 generations. Then

$$0.484 = 0.507 + (0.474 - 0.507)(1 - m)^{10}$$

Which can be solved to find $m = 0.035$. That is, about 3.5% of the genes at the MN locus in the ancestry of the black population of Claxton have immigrated from the white population per generation. (Other estimates by the same method but using different gene loci suggest slightly different figures, more like 1%. The important point here is not the particular result; it is to illustrate how the population genetics of migration can be analyzed.) Notice again the rapid rate of genetic unification by migration: in only 10 generations, one-third of the gene frequency difference has been removed (after 10 generations the difference is $0.484 - 0.474$, against the original difference of $0.507 - 0.474$).

5.14.4 A balance of selection and migration can maintain genetic differences between subpopulations

If selection is working against an allele within one subpopulation, but the allele is continually being introduced by migration from other populations, it can be maintained by a balance of the two processes. The simplest case is again for one locus with two alleles. Imagine selection in one subpopulation is working against the a allele, and the fitnesses of the genotypes are

Genotype	AA	Aa	aa
Fitness	1	$1-s$	$1-2s$

The a allele has frequency q in the local population. Suppose that in other subpopulations, natural selection is more favorable to the gene a, and it has a higher frequency in them, q_m on average. q_m will then be the frequency of a among immigrants to our local population. The total change in gene frequency per generation is equal to the sum of changes due to selection and to migration, i.e.

$$\Delta q = \Delta q_s + \Delta q_m \tag{5.19}$$

In this case, the expression for the change in gene frequency due to migration is easily derived from equation 5.17. There an expression for q_{t+1} is given in terms of q_t. We write $\Delta q_m = q_{t+1} - q_t$. Then

$$\Delta q_m = -m(q_t - q_m). \tag{5.20}$$

(Notice that, in the case in which the allele is being maintained by migration, the gene frequency must be higher among the immigrants, and $q_t < q_m$. The expression in parentheses is then negative, and Δq_m is positive.) The change in gene frequency due to selection can be calculated by a simple extension of the methods used in sections 5.3 and 5.5. As usual, the frequency of a in the next generation is the frequency of aa plus half the frequency of Aa. For the effect of selection, we also calculate $\Delta q_s = q_{t+1} - q_t$.

In this case,

$$q_{t+1} = \frac{q_t(1 - q_t)(1 - s) + q_t^2(1 - 2s)}{\overline{w}} \tag{5.21}$$

($\overline{w}$ is the mean fitness of the genotypes $= p^2 + 2pq(1 - s) + q^2(1 - 2s)$, where $p = 1 - q$.) This expression does not simplify, but if we assume $\overline{w} \simeq 1$ (which means that s is small) then, approximately:

$$q_{t+1} = q_t - q_t s(1 - q_t)$$

and

$$\Delta q_s = -q_t s(1 - q_t) \tag{5.22}$$

We can now combine the expressions for selection and migration.

$$\Delta q = -q_t s(1 - q_t) - m(q_t - q_m) \tag{5.23}$$

As usual, we can find the equilibrium (q^*) by setting $\Delta q = 0$. The solution is a quadratic:

$$q^* = \frac{1}{2s}[(m + s) \pm ((m + s)^2 - 4msq_m)^{\frac{1}{2}}]$$

Biologically, it is interesting to distinguish three conditions. If migration is much more powerful than selection ($m \gg s$), the gene frequency in the subpopulation will equilibrate at approximately the frequency of the other populations sending migrants to it (q_m). If selection is much stronger than migration ($s \gg m$), the a allele is eliminated from the subpopulation. Migration is then not strong enough to unify the gene frequencies of the different subpopulations. There is regional variation, or "population differentiation": a is lost in the subpopulation in which it is selected against, but retained in the other population. The third condition is where $s \simeq m$. We can then substitute m for s (or vice versa) in the quadratic equation above and it simplifies to

$$q^* = 1 - \sqrt{(1 - q_m)} \tag{5.25}$$

For example, if $q_m = 0.4$, $q^* \doteq 0.225$. There will again be regional variation in gene frequency: the gene frequency is 0.4 in one population and 0.225 in the other. The difference in gene frequency between the two population is, not surprisingly, smaller than when there was less migration.

The main point of this section is to show how the relationship between migration and selection can be modeled. We have also seen how migration can be strong enough to unify gene frequencies between subpopulations, or if migration is weaker, the gene frequencies of different subpopulations can diverge under selection. This result will become important when we come to consider, in chapter 16, the relative importance of gene flow and selection in maintaining biological species.

5.15 Summary

1 In the absence of natural selection, and with random mating in an infinite population, the genotypic frequencies at a locus move in one generation to the Hardy–Weinberg ratio and then remain stable.

2 It is easy to observe whether the genotypes at a locus are in the Hardy–Weinberg ratio. In nature they will often not be, because the fitnesses of the genotypes are not equal, mating is non-random, and/or the population is small.

3 A theoretical equation for natural selection at a single locus can be written, by expressing the frequency of a gene in one generation as a function of its frequency in the previous generation. The relationship is determined by the fitnesses of the genotypes.

4 The fitnesses of the genotypes can be inferred from the rate of change of gene frequency in real cases of natural selection.

5 From the rate at which the melanic form of the peppered moth replaced the light-colored form, the melanic form must have had a selective advantage of about 50%.

6 The geographic pattern of melanic and light-colored forms of the peppered moth cannot be explained only by the selective advantage of the better camouflaged form. An inherent advantage to the melanic form, and migration, are also needed to explain the observations.

7 The evolution of resistance to pesticides in insects is in some cases due to rapid selection for a gene at a single locus. The fitness of the resistant types can be inferred, from the rate of evolution, to be as much as twice that of the non-resistant insects.

8 If a mutation is selected against but keeps on arising repeatedly, the mutation settles at a low frequency in the population. It is called selection–mutation balance.

9 Selection can maintain a polymorphism when the heterozygote is fitter than the homozygote, when fitnesses of genotypes are negatively frequency dependent, and when different genotypes are adapted to different niches (this third process is a special case of the second).

10 Sickle-cell anemia is an example of a polymorphism maintained by heterozygous advantage.

11 Subdivided populations have a higher proportion of homozygotes than an equivalent large, fused population.

12 Migration, in the absence of selection, rapidly unifies gene frequencies in different subpopulations; and it can maintain an allele that is selected against in a local subpopulation.

5.16 Further reading

There are a number of textbooks about population genetics. Bodmer and Cavalli-Sforza (1976), Crow (1986), Edwards (1977), Hartl (1988), and Maynard Smith (1968, 1989) are introductory. More comprehensive works include Cavalli-Sforza and Bodmer (1971), Crow and Kimura (1970), Hartl and Clark (1989), Hedrick (1983), Li (1976), Spiess (1989), and Wallace (1981). Dobzhansky (1970) is a classic study; Lewontin *et al.* (1981) contains Dobzhansky's most famous series of papers.

Ford (1975) and Sheppard (1975) include introductory accounts of the peppered moth; Kettlewell (1973) is the standard account by the main authority. Haldane (1924, 1958) are the classic papers on the selection coefficients. For more recent work, see Jones (1982), Brakefield (1987), and *Biological Journal of the Linnean Society*, vol. 39, pp. 301–371 (1990). On pests and pesticides, see Taylor (1986), Mallett (1989), Devonshire and Field (1991), and Raymond *et al.* (1991). See Endler (1986) on measuring fitness in general, and Primack and Kang (1989) for plants.

The various selective means of maintaining polymorphisms are explained in the general texts. In addition, on frequency-dependent selection, see Clarke (1979) and the special issue of *Philosophical Transactions of the Royal Society of London* B, vol. 319, pp. 457–640 (1988). On multiple niche polymorphism, the classic paper is by Levene (1953); Hedrick (1986), and Antonovics *et al.* (1988) are recent reviews. Models of migration are explained in the population genetics texts. For a probable example of balanced migration and selection, see Camin and Ehrlich's (1958) study of the water snakes of Lake Erie.

Random events in population genetics

6.1 Successive generations are a random sample from the parental gene pool

Imagine a population of 10 individuals, of which three have genotype *AA*, four are *Aa*, and three *aa*. There are 10 *A* genes in the population and 10 *a* genes; the gene frequencies of each gene are 0.5. We also imagine that natural selection is not operating: all genotypes have the same fitness. What will the gene frequencies be in the next generation? The most likely answer is 0.5 *A* and 0.5 *a*. However, this is only the most likely answer; it is not a certainty. The gene frequencies may by chance change a little from the previous generation. The change is possible because the genes that form a new generation are a *random sample* from the parental generation.

Random sampling starts at conception. In every species, each individual produces many more gametes than will ever fertilize, or be fertilized, to form new organisms. The successful gametes which do form offspring are a sample from the many gametes that the parents produce. If a parent is homozygous, the sampling makes no difference to what genes are passed on to the offspring; all of a homozygote's offspring inherit the same gene. However, sampling does matter if the parent is a heterozygote, such as *Aa*. It will then produce a large number of gametes, of which approximately half will be *A* and half *a*. (The proportions may not be exactly equal. Reproductive cells may die at any stage leading to gamete formation, or after they have become gametes; also, in the female, a randomly picked three-quarters of the products of meiosis are lost as polar bodies.) If that parent produces 10 offspring, it is most likely that five will inherit an *A* gene and five *a*. But because the gametes that formed the offspring were sampled from a much larger pool of gametes, it is possible that the proportions would be something else. Perhaps six inherited *A* and only four *a*, or three *A* and seven *a*.

In what sense is the sampling of gametes "random"? We can see the exact meaning if we consider the first two offspring produced by an *Aa* parent. When it produces its first offspring, one gamete is sampled from its total gamete supply, and there is a 50% chance it will be an *A* and 50% that it will be an *a*. Suppose it happens to be an *A*. The sense in which sampling is random is that it is no more likely that the next gamete to be sampled will be an *a* gene just because the last one sampled was an *A*: the chance that the next successful gamete will be an *a* is still approximately 50%. Coin flipping is random in the same way: if you first flip a head, the chance that the next flip will be a head is still 50%. The alternative would be some kind of "balancing" system in which, after an *A* gamete had produced an offspring, the next successful gamete would be an *a*. If reproduction was like that, the gene frequency contributed by a heterozygote to its offspring would always be

125

exactly 1:1, A:a. Random drift would then be less important in evolution. In fact reproduction is not like that. The successful gametes are a random sample from the gamete pool.

The sampling of gametes is not the only stage at which random sampling occurs. It can occur at any stage as the adult population of a new generation matures. Suppose, for example, that only 100 of 1000 juveniles in a species survive to become adults. If natural selection determined which 100 lived, and which 900 died, sampling would not be random. But if the average fitnesses of all genotypes are equal, the 100 would be a sample, independently of their genotypes, and there would then have been random sampling between juvenile and adult stages in the life cycle. In different species, random sampling may be important at different stages; but in principle it can operate whenever a smaller number of successful individuals (or gametes) are sampled from a larger pool of potential survivors. Now that we have established the meaning of random sampling, we can turn to its evolutionary effects.

6.2 The frequency of alleles with the same fitness will change at random through time in a process called genetic drift

If you are flipping a coin, it is possible, by chance, to flip 10 heads (or 10 tails) in a row. If there are two alleles at a locus, and they have the same fitness, random sampling can likewise cause their relative frequencies in a population to change. A population having two alleles with frequencies of 0.5 is most likely to have the same frequencies in the next generation, but they may by chance be higher or lower than 0.5. There is an equal chance that random sampling will shift the gene frequency up or down. The condition that genotypes AA, Aa, and aa leave on average the same number of offspring (they have identical fitness) is called selective *neutrality*. The random movement in gene frequencies between generations is called *genetic drift*, *neutral drift*, *random drift*, or (simply) *drift*. Genetic drift has important consequences for the random substitution of genes and the Hardy–Weinberg equilibrium.

The rate of change of gene frequency by random drift depends on the size of the population. Random sampling effects are more important in smaller populations. For example (Figure 6.1), Dobzhansky and Pavlovsky in 1957, working with the fruitfly *Drosophila pseudoobscura*, made 10 populations with 4000 initial members (large populations) and 10 with 20 initial members (small populations), and followed the change in frequency of two chromosomal variants for 18 months. The average effect was the same in small and large populations, but the variability was significantly greater among the small populations. An analogous result could be obtained by flipping 10 sets of 20, or 4000, coins. On average, there would be 50% heads in both cases, but the chance of flipping 12 heads and eight tails in the small population is higher than the chance of flipping 2400 heads and 1600 tails in the large. If a population is small, it is more likely that a sample will be biased away from the average by any given percentage amount; genetic drift is therefore greater in smaller populations. Indeed, because random effects only become strong in small populations, this chapter could as well have been called "population genetics in small populations" or "population genetics in finite populations" (when evolutionary biologists talk about finite populations they are referring

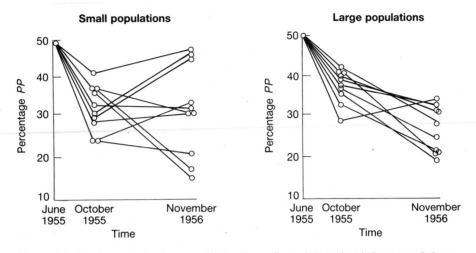

Figure 6.1 Random sampling is more effective in small populations than in large populations. Ten large (4000 founders) and 10 small (20 founders) populations of the fruitfly *Drosophila pseudoobscura* were created in June 1955 with the same frequencies (50% each) of two chromosomal inversions, *AP* and *PP*. After 18 months, the populations with small numbers of founders show a greater variety of genotype frequencies. From Dobzhansky (1970).

not to populations with any size less than infinity, but to populations that are small). The smaller the population, the more important are the effects of random sampling.

6.3 A small founder population may have a non-representative sample of the ancestral population's genes

A particular example of the influence of random sampling is given by what is called the *founder effect*. The founder effect was defined by Mayr (in 1963) as

the establishment of a new population by a few original founders (in an extreme case, by a single fertilized female) which carry only a small fraction of the total genetic variation of the parental population.

A population may be descended from a small number of ancestral individuals for either of two main reasons. A small number of individuals may colonize a place previously uninhabited by their species—the 250 or so individuals making up the modern human population on the island of Tristan da Cunha, for example, are all descended from about 20–25 immigrants in the early nineteenth century, and most are descended from the original settlers, one Scotsman and his family, who arrived in 1817. Alternatively, a population that is established in an area may fluctuate in size; the founder effect then occurs when the population passes through a "bottleneck," in which only a few individuals survive, and later expands again when more favorable times return.

If a small sample of individuals is taken from a larger population, there is a chance that an allele will be lost. In the special case of two alleles (*A* and *a* with proportions p and q), if one of them is not included in the founder population, the new population will be genetically monomorphic. The chance that an individual will be homozygous *AA* is simply p^2. The chance that two individuals drawn at random from the population will both be *AA* is $(p^2)^2$; in general, the chance of drawing N identical homozygotes is $(p^2)^N$. The founding population could be homozygous either because it is made up of N AA homozygotes or N aa homozygotes, and the total chance of homozygosity is therefore

$$\text{Chance of homozygosity} = [(p^2)^N + (q^2)^N] \qquad (6.1)$$

Figure 6.2 illustrates the relationship between the number of individuals in the founder population and the chance that the founder population is genetically uniform. The interesting result is that the founder effect is quite ineffective at producing a genetically monomorphic population. Even if the founder population is very small, with $N < 10$, it will usually possess both alleles. In general, the founder effect, whether by colonizations or population bottlenecks, is ineffective at reducing genetic variation. (An analogous calculation could be done for a population with three alleles, in which we calculated the chance that one of the three would be lost by the founder effect: the resulting population would not then be monomorphic, but have two instead of three alleles.)

However, the founder effect can have other interesting consequences. Although the sample of individuals forming a founder population are likely to have nearly all the ancestral population's genes, the gene frequencies may be peculiar. Isolated populations often have exceptionally high frequencies of otherwise rare alleles, and the most likely explanation is that the founding

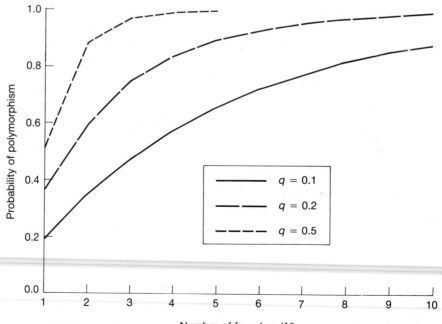

Figure 6.2 The chance that a founder population will be homozygous depends on the number of founders and the gene frequencies. If there is less variation and fewer founders, the chance of homozygosity is higher. Here the chance of homozygosity ($= 1 -$ probability of polymorphism) is shown for three different gene frequencies (q) at a two-allele locus.

population had a disproportionate number of those rare alleles. The clearest examples all come from humans. The Afrikaner population of South Africa is mainly descended from one shipload of immigrants who landed in 1652, though later arrivals have added to it. The population has increased dramatically since then to its modern level of 2500 000; the influence of the early colonists is shown by the fact that almost 1000 000 living Afrikaners have the names of 20 original settlers.

The early colonists included individuals with a number of rare genes. The ship of 1652 contained a Dutch male carrying the gene for Huntington's disease, a lethal autosomal dominant disease, and most cases of the disease in the modern Afrikaner population can be traced back to that individual. A similar story can be told for the dominant autosomal gene causing porphyria variegata. This condition is due to a defective form of the enzyme proto-porphyrinogen oxidase; carriers of the gene suffer a severe—even lethal—reaction to barbiturate anesthetics, and the gene was therefore not strongly disadvantageous before modern medicine. There are about 30 000 carriers of the gene in the modern Afrikaner population, a far higher frequency than in Holland, and all of them are descended from one couple, Gerrit Jansz and Ariaantje Jacobs, who emigrated from Holland in 1685 and 1688 respectively. Every human population has its own "private" polymorphisms, which were probably often caused by the genetic peculiarities of founder individuals.

Both of the examples we have just considered are for medical conditions. The individual carriers of the genes will have lower fitness than average, and selection will therefore act to reduce the frequency of the gene to zero. For much of the time, the porphyria variegata gene may have had a similar fitness to its allele; it may have been a neutral polymorphism until its "environment" came to contain (in selected cases) barbiturates. The gene for Huntington's disease will have been consistently selected against. Thus its present high frequency suggests that the founder population had an even higher frequency, because it will have probably been decreased by selection since then. Any particular founder sample would not be expected to have a higher than average frequency of the Huntington's disease gene, but if enough colonizing groups set out, some of them are bound to have peculiar, or even very peculiar, gene frequencies. In the case of Huntington's disease, the Afrikaner population is not the only one descended from founders with more copies of the gene than average. Some 432 carriers of Huntington's disease in Australia are descended from a Miss Cundick who left England with her 13 children; and a French nobleman's grandson, Pierre Dagnet d'Assigne de Bourbon, has bequeathed all the known cases of Huntington's disease on the island of Mauritius.

6.4 One gene can be substituted for another by random drift

The frequency of a gene is as likely to decrease as to increase by random drift; on average the frequencies of neutral alleles remain unchanged from one generation to the next. In practice, their frequencies drift up and down, and it is therefore possible for a gene to enjoy a run of luck and be carried up to a much higher frequency—in the extreme case, its frequency could after many generations be carried up to one by random drift (become fixed). Each generation, the frequency of a neutral allele has a chance of increasing, a chance of decreasing, and a chance of staying constant. If it increases in one generation, it again has the same chances of increasing, decreasing, or staying constant in the next generation; it thus has a small chance of increasing for two generations in a row (equal to the square of the chance of increasing in any one generation), a still smaller chance of increasing through three generations, and so on. For any one allele, fixation by random drift is very improbable. The probability is finite, however, and if enough neutral alleles, at enough loci, and over enough generations, are randomly drifting in frequency, one of them will eventually be fixed. The same process can occur whatever the initial frequency of the allele. It is less likely that a rare allele will be carried up to fixation by random drift than a common allele, because it would take a longer run of good luck; but it is still possible. Indeed, a unique neutral mutation has some chance of eventual fixation. Any one mutation is most likely to be lost; but if enough mutations arise, one will be bound to be fixed eventually.

So a gene can be substituted by random drift. What is the rate of this kind of neutral evolution? It can be shown, by a remarkably elegant argument, that the neutral evolution rate exactly equals the neutral mutation rate. The argument is as follows. In a population of size N at any one time there are a total of $2N$ genes at each locus. On average, each gene will contribute one copy of itself to the next generation; but because of random sampling, some genes will contribute more than one copy and others will contribute none. As we look two generations ahead, those genes that did not make it into the first generation cannot contribute copies to the second generation; and in the first generation some more genes will likewise "drop out," not be represented in the second generation, and be unable to contribute to the third. Each generation, some of the $2N$ original genes drop out in this way; and once a gene has failed to reproduce it is lost (Figure 6.3).

If we look far enough forwards there will eventually come a time when all the $2N$ genes are descended from just one of the $2N$ genes now. This is because in every generation, some genes will fail to reproduce; hence we must eventually come to a time when all but one of the original genes have dropped out. That one gene will have hit a long enough run of lucky increases and will have spread through the whole population. It will have been fixed by genetic drift. Because the process is pure luck, each of the $2N$ genes in the original population has an equal chance of being the lucky one. Any one gene, therefore, has a $1/(2N)$ chance of eventual fixation by random drift (and a $(2N - 1)/(2N)$ chance of being lost by it). Because the same argument applies to any gene in the population, it also applies to a new, unique, neutral mutation. When the mutation arises, it will be one gene in a population of $2N$ genes at its locus (i.e. its frequency will be $1/(2N)$), and it has the same $1/(2N)$ chance of eventual fixation as does every other gene in the population. It is therefore most likely that the mutation will be lost (probability of being lost $= (2N - 1)/(2N) \simeq 1$ if N is large); but it does have a small ($1/(2N)$) chance of success. That completes the first stage of the argument: the probability that a neutral mutation will be fixed is $1/(2N)$.

The rate of evolution equals the probability that a mutation is fixed multiplied by the rate at which mutations appear. We define the rate at which

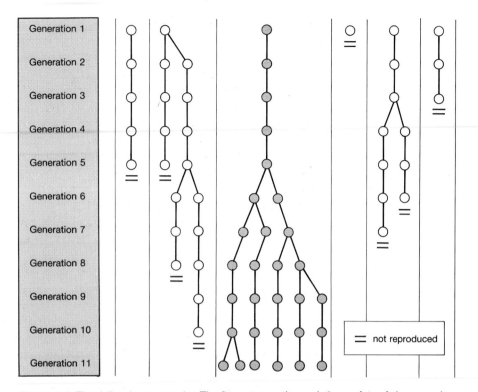

Figure 6.3 The drift to homozygosity. The figure traces the evolutionary fate of six genes; in a diploid species these would be combined each generation in three individuals. In every generation some genes may by chance fail to reproduce and others by chance may leave more than one copy. Because once a gene has failed to reproduce its line is lost forever, over time the population must drift to become made up of descendants of only one gene in an ancestral population a long time ago. In this example, the population after 11 generations is made up of descendants of gene number 3 (⬤) in generation 1.

neutral mutations arise as μ per gene per generation. (μ is the rate at which new selectively neutral mutations arise, not the total mutation rate. The total mutation rate includes selectively favorable and unfavorable mutations as well as neutral mutations. We are here considering only the effect of neutral drift.) At each locus, there are $2N$ genes in the population: the total number of neutral mutations arising in the population will be $2N\mu$ per generation. The rate of neutral evolution is then $1/(2N) \times 2N\mu = \mu$. The rate of neutral evolution is equal to the neutral mutation rate.

6.5 The Hardy–Weinberg "equilibrium" is not an equilibrium

Let us stay with the case of a single locus, with two selectively neutral alleles A and a. If genetic drift is not happening—if the population is infinite—the gene frequencies will stay constant from generation to generation and the genotype frequencies will also be constant, in Hardy–Weinberg proportions (section 5.3, p. 87). But in a real, finite population the gene frequencies can drift around. The average gene frequencies in one generation will be the same as in the previous generation, and it might be thought that the long-term average gene and genotype frequencies will simply be those of the Hardy–

Weinberg equilibrium, but with a bit of "noise" around them. That is not so, however. The long-term result of genetic drift is that one of the alleles will be fixed; the polymorphic Hardy–Weinberg "equilibrium" is unstable in a finite population.

Just to explain the process, suppose that a population is made up of five individuals, containing five A alleles and five a alleles (that is obviously a tiny population, but the same point would apply if there were 50 000 copies of each allele). The genes are randomly sampled to produce the next generation. Maybe six A alleles are sampled and four a alleles. This is now the starting point to produce the next generation; maybe six A and four a are drawn again. The fourth generation might be seven A and three a, the fifth six A and four a, the sixth seven A and three a, then seven A: three a, eight A: two a, eight A: two a, nine A: one a, and then $10\,A$. The same process could have gone off in the other direction, or started by favoring A and then reversed to fix a: random drift is directionless. However, when one of the genes is fixed, the population is homozygous and will stay homozygous (Figures 6.3 and 6.4).

The Hardy–Weinberg theorem is a good approximation for all but small populations, and retains its importance in evolutionary biology; but it is also true that, once we allow for random drift, the Hardy–Weinberg ratios are not an equilibrium. The Hardy–Weinberg ratios are for neutral alleles at a locus and the Hardy–Weinberg result suggested that the genotype (and gene) ratios are stable over time. However, in a small population, gene frequencies drift and change. Eventually one of the genes will be fixed, and only then will the system be stable. The true equilibrium of the Hardy–Weinberg system in a small population is, so to speak, at homozygosity. (This conclusion, some geneticists would say, is overdramatized. They would prefer to say that the Hardy–Weinberg equilibrium refers only to an infinite population, and events in finite populations have nothing to do with the Hardy–Weinberg equilibrium. The point is verbal merely: the important thing to remember is that the Hardy–Weinberg equilibrium does not apply in small populations.)

6.6 Neutral drift over time produces a march to homozygosity

Over the long term, pure random drift causes the population to "march to homozygosity" at a locus. The process by which this happens we have already considered (section 6.4) and illustrated (Figure 6.3). All loci at which there are several selectively neutral alleles will tend to become fixed for only one gene. It is not difficult to derive an expression for the rate at which the population becomes homozygous. First we define the degree of homozygosity. Individuals in the population are either homozygotes or heterozygotes. Let f = the proportion of homozygotes, and $H = 1 - f$ be the proportion of heterozygotes (f comes from fixation). Homozygotes here includes all types of homozygote at a locus: if, for example, there are three alleles A_1, A_2, and A_3, then f is the number of A_1A_1, A_2A_2, and A_3A_3 individuals divided by the population size; H likewise is the sum of all heterozygote types. N will again stand for population size.

How will f change over time? We shall derive the result in terms of a special case: a species of hermaphrodite in which an individual can fertilize itself. Individuals in the population discharge their gametes into the water and each gamete has a chance of combining with any other gamete. New individuals

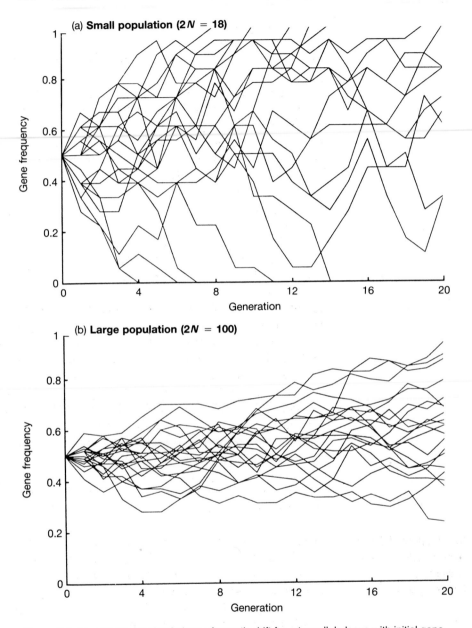

Figure 6.4 Twenty repeat simulations of genetic drift for a two-allele locus with initial gene frequency 0.5. (a) Small population ($2N = 18$); (b) larger population ($2N = 100$). Eventually one of the alleles drifts to a frequency of one. The other alleles are then lost. The drift to homozygosity is more rapid in a smaller population, but in any finite population without mutation homozygosity is the final result.

are formed by sampling two gametes from the gamete pool. The gamete pool contains $2N$ gamete types, where "gamete types" should be understood as follows. There are $2N$ genes in a population made up of N diploid individuals. A gamete type consists of all the gametes containing a copy of any one of these genes. Thus if an individual, with two genes, produces

200 000 gametes, there will be on average 100 000 copies of each gamete type in the gamete pool.

To calculate how f, the degree of homozygosity, changes through time, we derive an expression for the number of homozygotes in one generation in terms of the number of homozygotes in the generation before. We must first distinguish between a gene-bearing gametes in the gamete pool that are copies of the same parental a gene, and those that are derived from different parents. There are then two ways to produce a homozygote, when two a genes from the same gametic type meet or when two a genes from different gametic types meet (Figure 6.5); the frequency of homozygotes in the next generation will be the sum of these two.

The first way of making a homozygote is by "self-fertilization." There are $2N$ gamete types but, because each individual produces many more than two gametes, there is a $1/(2N)$ chance that a gamete will combine with another gamete of the same gamete type as itself: if it does, the offspring will be homozygous. (If, as above, each individual makes 200 000 gametes, there would be 200 000N gametes in the gamete pool. We first sample one gamete from it. Of the remaining gametes, practically 100 000 of them (99 999 in fact) are copies of the same gene. The proportion of gametes left in the pool that contain copies of the same gene as the gamete we sampled is 99 999/200 000N, or almost $1/(2N)$.)

The second way to produce a homozygote is by combining two identical genes that were not copied from the same gene in the parental generation. If the gamete does not combine with another copy of the same gamete type—

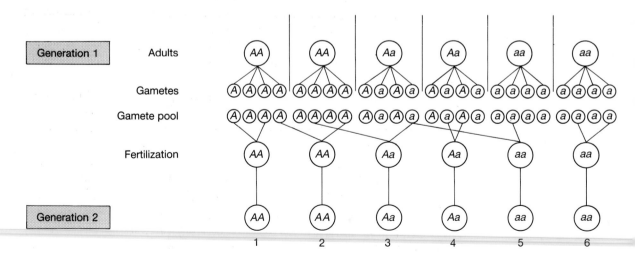

Figure 6.5 Inbreeding in a finite population produces homozygosity. A homozygote can be produced either by combining copies of the same gene from different individuals, or by combining two copies of the same physical gene. Here we imagine that the population contains six adults, which are potentially self-fertilizing hermaphrodites, and each produces four gametes. Homozygotes can then be produced by the kind of cross-mating assumed in the Hardy–Weinberg theorem (e.g. offspring number 2) or by self-fertilization (offspring number 1). Self-fertilization only necessarily produces a homozygote if its parent is homozygous (cf. offspring numbers 1 and 4).

chance $1 - (1/(2N))$— it will still form a homozygote if it combines with a copy made from the same gene but from another parent. For a gamete with an a gene, if the frequency of a in the population is p, the chance that two a genes meet is simply p^2. p^2 is the frequency of aa homozygotes in the previous generation. If there are two types of homozygote, AA and aa, the chance of forming a homozygote will be $p^2 + q^2 = f$. In general, the chance that two independent genes will combine to form a homozygote is equal to the frequency of homozygotes in the previous generation. The total chance of forming a homozygote by this second method is the chance that a gamete does not combine with another copy of the same parental gene, $1 - (1/(2N))$, multiplied by the chance that two independent genes combine to form a homozygote (f). That is, $1 - (1/(2N))f$.

Now we can write the frequency of homozygotes in the next generation in terms of the frequency of homozygotes in the parental generation. It is the sum of the two ways of forming a homozygote. Following the normal notation for f' and f (f' is the frequency of homozygotes one generation later),

$$f' = \frac{1}{2N} + \left(1 - \frac{1}{2N}\right)f \tag{6.2}$$

Since $H = 1 - f$, heterozygosity can be shown, by rearrangement of the equation, to decrease at the following rate:

$$H' = \left(1 - \frac{1}{2N}\right)H \tag{6.3}$$

That is, heterozygosity decreases at a rate of $1/(2N)$ per generation until it is zero. The population size N is again important in governing the influence of genetic drift. If N is small, the march to homozygosity is rapid. At the other extreme, we re-encounter the Hardy–Weinberg result. If N is infinitely large, the degree of heterozygosity is stable: there is then no "march to homozygosity."

Although it might seem that this derivation is for a particular, hermaphroditic breeding system, the result is in fact general (a small correction is needed for the case of two sexes). The march to homozygosity in a small population proceeds because two copies of the same gene may combine in a single individual. In the hermaphrodite, it happens obviously with self-fertilization. But if there are two sexes, a gene in the grandparental generation can appear as a homozygote, in two copies, in the grandchild generation. The process, by which a gene in a single copy in one individual combines in two copies in an offspring, is *inbreeding*. Inbreeding can happen in any breeding system with a small, finite population, and becomes more likely the smaller the population. With inbreeding, homozygosity is increased over the Hardy–Weinberg level generated by random mating, because it is impossible for two copies of the same gene to combine to form a heterozygote. The important point is that, in a finite population, homozygosity is increased relative to an infinite population: in an infinite population two copies of the same gene never fertilize each other; but in a finite population there is some chance they will.

6.7 The effect of new mutations

So far, it might appear that the theory of neutral drift predicts that populations should be completely homozygous. However, new variation will be contributed by mutation and the equilibrial level of heterozygosity will actually be a balance between the drift to homozygosity and the creation of heterozygosity by mutation. We can now work out what that equilibrium is. The *neutral* mutation rate is µ per gene per generation. (µ, as before, is the rate at which new selectively neutral mutations arise, not the total mutation rate.) To find out the equilibrial heterozygosity under drift and mutation, we have to modify equation 6.2 to account for mutation. If an individual was born a homozygote, and if neither gene has mutated, it stays a homozygote. (We ignore the possibility that mutation produces a homozygote, for example by a heterozygote *Aa* mutating to a homozygous *AA*; we are assuming that mutations produce new genes.) In order for a homozygote to produce all its gametes with the same gene, *neither* of its genes must have mutated. If either of them has mutated, the frequency of homozygotes is decreased. The chance that a gene has not mutated is $(1 - \mu)$ and the chance that neither of an individual's genes has mutated is $(1 - \mu)^2$.

Now we can simply modify the recurrence relationship derived above. The frequency of homozygotes will be as before, but decreased by the probability that they have mutated to heterozygotes:

$$f' = \left[\frac{1}{2N} + \left(1 - \frac{1}{2}N\right)f\right]\left(1 - \mu\right)^2 \tag{6.4}$$

Homozygosity (f) will now not increase to one. It will converge to an equilibrial value. The equilibrium is between the increase in homozygosity due to drift, and its decrease by mutation. We can find the equilibrial value of f from $f^* = f = f'$. f^* indicates a value of f that is stable in successive generations ($f' = f$). Substituting $f^* = f' = f$ in the equation gives (after a minor manipulation):

$$f^* = \frac{(1 - \mu)^2}{2N - (2N - 1)(1 - \mu)^2} \tag{6.5}$$

The equation simplifies if we ignore terms in μ^2, which will be relatively unimportant because µ is a low probability. Then

$$f^* = \frac{1}{4N\mu + 1} \tag{6.6}$$

The equilibrial heterozygosity ($H = 1 - f^*$) is:

$$H = \frac{4N\mu}{4N\mu + 1} \tag{6.7}$$

This is an important result. It gives the degree of heterozygosity that should exist for a balance between the drift to homozygosity and new neutral mutation. The expected heterozygosity depends on the neutral mutation rate and the population size (Figure 6.6). As the march to homozygosity is more rapid if the population size is smaller, it makes sense that the expected heterozygosity is lower if N is smaller. Heterozygosity is also lower if the

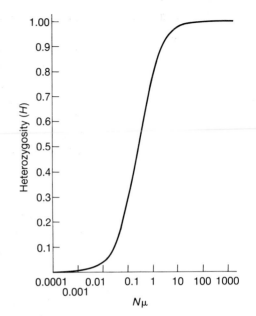

Figure 6.6 The theoretical relationship between the degree of heterozygosity and the parameter $N\mu$ (the product of the population size and neutral mutation rate).

mutation rate is lower, as we should expect. In sum, the population will be less genetically variable for neutral alleles when population sizes are smaller and the mutation rate lower.

6.8 Population size and effective population size

What is "population size"? We have seen that N determines the effect of genetic drift on gene frequencies. But what exactly is N? In an ecological sense, N can be measured as the number of adults in a locality. However, for the theory of population genetics with finite populations, that is only a crude approximation to the "population size," N, implied by the equations. What matters is the chance that two copies of a gene will be sampled as the next generation is produced, and this is affected by the breeding structure of the population. In a population of size N there will be $2N$ genes. The correct interpretation of N for the theoretical equations is that N has been correctly measured when the chance of drawing two copies of the same gene is $(1/(2N))^2$. If we draw two genes from a population at a locality, we may be more likely for various reasons to get two copies of the same gene than would be implied by the naïve ecological measure of population size. Population geneticists therefore often write N_e (for "effective population size") in the equations, rather than N. In practice, effective population sizes are generally lower than ecologically observed population sizes. The relation between N_e, the effective population size implied by the equations, and the observed population size N can be complex. A number of factors are known to influence effective population size.

1 *Sex ratio.* If one sex is rarer, the population size of the rarer sex will dominate the changes in gene frequencies. It is much more likely that identical genes will be drawn from the rarer sex, because fewer individuals are contributing the genes. Sewall Wright proved in 1932 that in this case:

$$N_e = \frac{4N_m \cdot N_f}{N_m + N_f} \qquad (6.8)$$

where N_m is the number of males, and N_f is the number of females in the population.

2 *Population fluctuations.* If population size fluctuates, homozygosity will increase more rapidly while the population goes through a "bottleneck" of small size. N_e is disproportionately influenced by N during the bottleneck, and a formula can be derived for N_e in terms of the harmonic mean of N.

3 *Small breeding groups.* If most breeding takes place within small groups, then the effective population size will be much smaller than if there is population-wide random mating (called "panmixis"). The degree of homozygosity will be nearer that predicted from the size of the breeding groups than from the total population size. Breeding groups can be small either because of geographic subdivision of the population or because of non-random mating. If certain types of individual consistently choose a particular kind of mate then the mating will be conducted through only a part of the whole population even without geographic subdivision. Either way, the chance that copies of the same gene will be "drawn" together is increased relative to the case with a large population. There are various ways of describing population subdivision, to derive expressions for N_e. Effective population size is mainly influenced by the degree of subdivision and the migration rate between the subgroups: the more migration, the more N_e tends to N.

4 *Variable fertility.* If the number of successful gametes varies between individuals (as it often does among males when sexual selection is operating, see chapter 11), the more fertile individuals will accelerate the march to homozygosity. Again, the chance that copies of the same gene will combine in the same individual in the production of the next generation is increased and the "effective" population size decreased relative to the total number of adults. Wright showed that if k is the average number of gametes produced by a member of the population and σ_k^2 is the variance of k, then

$$N_e = \frac{4N - 2}{\sigma_k^2 + 2} \qquad (6.9)$$

For $N_e < N$, the variance of k has to be greater than random. If k varies randomly, as a Poisson process, $\sigma_k^2 = k = 2$ and $N_e \simeq N$.

These are all quite technical points. The N_e in the equations for neutral evolution is an exactly defined quantity, but it is difficult to measure in practice. It is usually less than the observed number of adults, N. N_e is equal to N when the population mates randomly, is constant in size, has an equal sex ratio, and approximately Poisson variance in fertility. Natural deviations from these conditions produce $N_e < N$. How much smaller N_e is than N is difficult to measure, though it is possible to make estimates by the formulae we have seen. Other things being equal, species with more subdivided and inbred population structures have lower N_e than more panmictic species.

6.9 Summary

1 In a finite population, random sampling of gametes to produce the next generation can change the gene frequency. These random changes are called genetic drift.

2 Genetic drift has a larger effect on gene frequencies if the population size is small than if it is large.

3 If a small population colonizes a new area, it may have a reduced and non-representative sample of the ancestral population's genes.

4 One gene can be substituted for another by random drift. The rate of neutral substitution is equal to the rate at which neutral mutations arise.

5 In a finite population, in the absence of mutation, one allele will eventually be fixed at a locus. The population will eventually become homozygous. The Hardy–Weinberg equilibrium does not apply to finite populations. The effect of drift is to reduce the amount of variability in the population.

6 The amount of neutral genetic variability in a population will be a balance between its loss by drift and its creation by new mutation.

7 The "effective" size of a population, which is the population size assumed in the theory of population genetics for finite populations, should be distinguished from the size of a population that an ecologist might measure in nature. Effective population sizes are usually smaller than observed population sizes.

6.10 Further reading

Population genetics texts, such as those of Crow (1986), Crow and Kimura (1970), Ewens (1979), Gale (1990), Hartl and Clark (1989), Hedrick (1983), Spiess (1989), or Wallace (1981) explain the theory of population genetics for finite populations. Kimura (1983), Lewontin (1974), and Nei (1987) also explain much of the material. Wright (1969) is more advanced. Li (1977) is an anthology of many of the classic papers of stochastic population genetics. Kimura (1983) also contains a clear account of the parts of the theory most relevant to his neutral theory and discusses the meaning of effective population size. For the medical examples of the founder effect in humans, see Dean (1972) and Hayden (1981).

Molecular evolution and the neutral theory

7.1 Four main observations originally suggested, to some biologists, that most molecular evolution has been by neutral drift

In 1968 and 1969, Kimura, and King and Jukes, suggested that most evolutionary change at the molecular level is driven by random drift rather than natural selection. The theory of neutral genetic drift (see chapter 6) then took on a new importance in evolutionary biology. One gene can undoubtedly be substituted for another by genetic drift, but the process had previously usually been thought to explain only a minority of evolutionary change. Now Kimura, and King and Jukes, were suggesting that random drift causes more evolutionary change, as measured by numbers of changes in the DNA, than any other evolutionary process. The result was a controversy between two theories: according to the *neutral theory*, or *neutralism*, most evolutionary change in molecules is caused by random drift; and according to the "selectionist" alternative, most evolution is caused by natural selection. The neutral theory does not suggest that random drift explains all evolutionary change. Natural selection is still needed to explain adaptation. It is, however, possible that the adaptations we observe in organisms required only a minority of all the evolutionary changes that have taken place in the DNA. The neutral theory states that evolution at the level of the DNA and proteins, but not of adaptation, is dominated by random processes; most evolution at the molecular level would then be non-adaptive.

The controversy about the relative importance of natural selection and neutral drift has passed through several stages. In the 1960s and 1970s, for example, the controversy was mainly concerned, at the level of evidence, with proteins; but it has increasingly shifted to DNA sequences as these have become available in the 1980s and 1990s. There has also been movement in the concepts and theories; in many cases, the theoretical issues are exactly the same for DNA and for proteins, though the evidence for DNA has allowed tests that were impossible with protein sequences alone; but the important theoretical positions have been refined. To understand the subject now, we need to know about some of its earlier stages; and we shall therefore consider it, to some extent, in historical order.

We shall discuss the four main observations about molecular evolution that have been used to support the neutral theory. Three concern evolutionary changes between species. When, in the 1960s, the sequences of particular proteins became known for several species, biologists could work out the number of amino acid changes that occurred during the evolutionary divergence of a pair of species. For any two species, the approximate date of their common ancestor can be estimated from the fossil record. The rate of protein evolution can then be calculated as the number of amino acid differences

Protein sequence

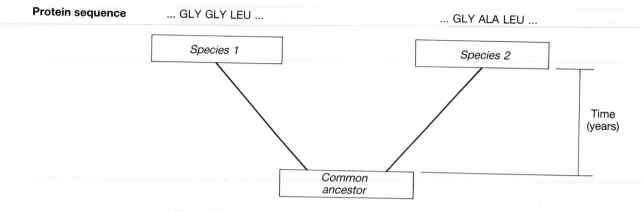

Figure 7.1 Imagine that some region of a protein has the illustrated sequences in two species. The evolutionary change has happened somewhere within the lineage connecting the two species via their common ancestor. An alanine has been substituted for a glycine in the lineage leading to species 2, or a glycine for an alanine in the lineage to species 1. Either way, the amount of evolution is one change, and it has taken place in twice the time (t) from the modern species back to their common ancestor; or, one change in $2t$ years.

between the protein of the two species divided by twice the time to their common ancestor (Figure 7.1). Table 7.1 lists the rate of evolution for a number of proteins. As can be seen, different proteins evolve at different rates: histones evolve slowly, fibrinopeptides rapidly; but an approximate memorable figure is that amino acids are substituted at a rate of around 1.5×10^{-9} per amino acid site per year. Kimura found this number remarkably high, if proteins evolved by natural selection. We shall come to the argument in a moment; but for now we need only note the observation: the rate of protein evolution has been high.

The rate calculation can be repeated for several pairs of species. The species pairs will have common ancestors of varying ages, and a graph like Figure 7.2 can be built up to show the rate of a protein's evolution in various lineages. The striking property of the graph is that it is almost a straight line: each protein appears to have evolved at a constant rate. The constant rate of

Table 7.1 Rates of evolution in eight proteins. The rates are expressed as average numbers of changes per amino acid site in 10^9 years. From Kimura (1983)

Protein	Rate (number of changes in 10^9 years)
Fibrinopeptides	8.3
Pancreatic ribonuclease	2.1
Lysozyme	2.0
α-Globin	1.2
Myoglobin	0.89
Insulin	0.44
Cytochrome c	0.3
Histone H4	0.01

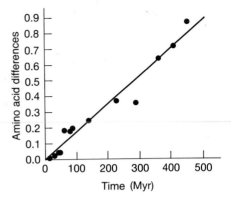

Figure 7.2 The rate of evolution of α-globin. Each point on the graph is for a pair of species, or groups of species, and the value for that pair was obtained by the method of Figure 7.1. From Kimura (1983).

protein evolution can also be demonstrated without the use of (probably inaccurate) dating from the fossil record. Graphs like Figure 7.2, with absolute times, do need fossil evidence; but there is another argument which does not (Box 7.1). Kimura thought it unlikely (for reasons we shall come to) that natural selection would keep evolutionary rates as constant as is shown in Figure 7.2: this was the second observation.

The third observation concerned the rate of evolution of different parts of a protein. Some parts of a protein appear functionally more important than others; in an enzyme, for instance, the active site is functionally more important than the outlying regions. It was observed that the functionally more important parts evolve at a slower rate. If natural selection works more on the functionally important parts, they should evolve faster, not slower; but neutral drift would be more powerful in the less important parts of the protein because there is a higher chance that a mutation there will be neutral.

The fourth observation is about variation within a species. In 1966, the method of gel electrophoresis (section 4.5, p. 71) had first been used to estimate the average amount of genetic variation, in humans by Harris and in the fruitfly *Drosophila* by Hubby and Lewontin. The amount of variation can be described by two main indexes. One is the proportion of heterozygotes, or heterozygosity (*H*, section 6.6, p. 135); the other is the percentage of polymorphic loci. If, say, 20 loci are studied by gel electrophoresis, and 16 show no variation and four have more than one band on the gel, then the percentage polymorphism would be 4/20 × 100 = 20%. In real populations, percentage polymorphisms are about 10–20% (Table 7.2). These levels of variation are quite high: too high, Kimura thought, to be accounted for by natural selection.

7.2 Neutral drift and natural selection can both hypothetically explain molecular evolution

The neutral theory proposes that the changes in protein sequences that have occurred during evolution were selectively neutral; the evolutionary substitution of one form of the protein for another was by genetic drift. Changes in the sequence of a protein are most likely to be selectively neutral if the two forms are functionally identical. However, the neutral theory does not draw on biochemical evidence about the functional properties of different protein forms. Instead it simply assumes that, for any one protein, there are many

Box 7.1 The relative rate test.

The relative rate test is a method of testing whether a molecule (or, in principle, any other character) evolves at a constant rate in two independent lineages. It was first used by Sarich and Wilson in 1973. Suppose we know the sequence of a protein in three species, a, b, and c, and we also know the order of phylogenetic branching of the three species (Figure B7.1). We can now infer the amount of change in the two lines from the common ancestor of a and b to the modern species (x and y in Figure B7.1). If the protein evolved at the same rate in the two lineages, the number of amino acid changes between the common ancestor and a (x) should equal the number of changes between the common ancestor and b (y); that is, x = y. x and y can be inferred by simple simultaneous equations. We know the differences between the protein sequences in a and b (k), b and c (l), and a and c (m). Thus:

$$k = x + y$$
$$l = y + z$$
$$m = x + z$$

We have three equations with three unknowns and can solve for x, y, and z. We then test whether the rates were the same by seeing whether by x = y. Notice that we do not need to know the absolute date (or the identity) of the common ancestors.

The relative rate test can only show that a molecule evolved at the same rate in the two lineages connecting the two modern species with their common ancestor. This does not prove that the molecule always has a constant rate; it does not, in other words, confirm the molecular clock. If the identity of relative rate is shown for many pairs of species, with common ancestors of very different antiquities, that is suggestive of (and consistent with) a molecular clock, but it is not conclusive evidence. We can see why in a counter-example (Figure B7.2). Suppose that a

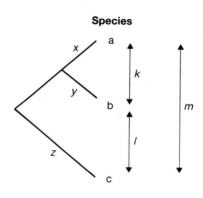

Figure B7.1 Phylogeny of three species: a, b, and c. k, l, and m are the observed number of amino acid differences between the three species. The amounts of evolution (x, y, z) in the three parts of the tree can be simply inferred, as the text explains.

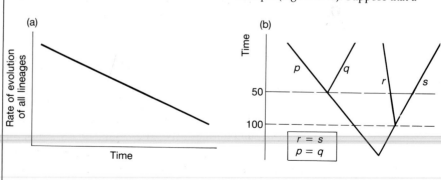

Figure B7.2 (a) The rate of evolution of a molecule has slowed down gradually through time; but the rate of evolution is always the same in all lineages at any one time. The molecule does not evolve like a molecular clock (which would show up as a flat graph of rate against time). Then, (b) for any pair of species, with common ancestors at any time, the amount of change will be the same in both lineages.

continued on page 145

molecule evolves at the same rate in all lineages at any one time, but that it has been gradually slowing down through evolutionary history. A pair of species with a common ancestor 100 million years ago will then show rate constancy according to the relative rate test (because the molecule evolves at the same rate in all lineages at any one time); and any other species pair, for instance with a common ancestor 50 million years ago, will also show relative rate constancy. However, there is no molecular clock because the rate slows down through time. The relative rate test will not detect that the more recent species pair have a smaller absolute number of changes: absolute dates would be needed for that. The same point would apply if there were any trend in evolutionary rate with time; it does not have to be directional. The molecule could speed up and slow down many times in evolution; but so long as the speeding up and slowing down apply to all lineages, the relative rate test will show equal rates of evolution in the two lineages. The relative rate test, therefore, cannot conclusively test the molecular clock hypothesis.

Further reading. Fitch (1976).

Box 7.1 (*continued*) The relative rate test.

Table 7.2 Amount of variation in natural populations. Variation can be measured as percentage of polymorphic loci (*P*) and average percentage heterozygosity per individual (*H*). Also given is the number of gene loci used to estimate *P* and *H*. From Nevo (1988)

Species	Number of loci	P (%)	H (%)
Phlox cuspidata	16	11	1.2
Liatris cylindracea	27	56	5.7
Limulus polyphemus	25	25	5.7
Balanus eburneus	14	67	6.7
Homarus americanus	28–42	18	3.8
Gryllus bimaculatus	25	58	6.3
Drosophila robusta	40	39	11
Bombus americanorum	12	0	0
Salmo gairdneri	23	15	3.7
Bufo americanus	14	26	11.6
Passer domesticus	15	33	9.8
Homo sapiens	71	28	6.7

equivalent forms. A mutation that changes a protein from one of these forms to another will not affect the organism's fitness. The mutation is neutral. If the neutral theory is right, almost all protein evolution has consisted of shuffling around between these functionally equivalent forms. Notice that this does not mean that neutralism rules out natural selection, nor that all mutations are neutral. Many (even most) mutations could be selectively deleterious, and simply be eliminated by selection soon after they arose. These deleterious mutations never produce an evolutionary change, and do not contribute to the molecular differences observed between species. The neutral theory applies to the changes that did actually happen: it suggests that most evolutionary changes were produced by drift between selectively equivalent forms.

The main alternative to the neutral theory is that molecular evolution is driven by natural selection. In this case, when a protein changes in evolution, it is because the later form was an improvement over the earlier, in the local environmental conditions. Mutations are fixed not by random drift but because they are favored by natural selection. If molecular evolution has been by natural selection, there will not be many equivalent forms of a protein: when a change happens it will be for the better or worse; it will not be neutral.

The essential difference between the selective and neutral theories concerns the frequency distribution of mutations of various selection coefficients (s). Selectionism states that neutral mutations are relatively rare and there are enough selectively advantageous mutations to account for protein evolution (Figure 7.3a). The neutral theory states the opposite (Figure 7.3b): there are many neutral, but few selectively advantageous, mutations. On the neutral theory, mutations are nearly always either disadvantageous ($s < 0$) or neutral ($s = 0$); the former are rapidly eliminated by selection and have no evolutionary effects; observable evolution is largely a matter of drifting between the neutral variants.

The important regions of Figure 7.3 are the frequencies of mutations with $s = 0$ and with $s > 0$. However, to the left of the neutral point in the graphs there is a region for selectively disadvantageous mutations. It is important to realize that the two theories do not disagree about this region. Both accept that the majority of mutations are deleterious and selected against: the mutants will be removed soon after they arise. Neutralists, in the sense intended in this book, do not deny natural selection; but they have a different use for it from selectionists. Selectionists use natural selection to explain both why mutations are lost (when they are deleterious), and are fixed (when they are advantageous); but neutralists use selection only to explain why deleterious mutations are lost: they use drift to explain why new mutations are fixed.

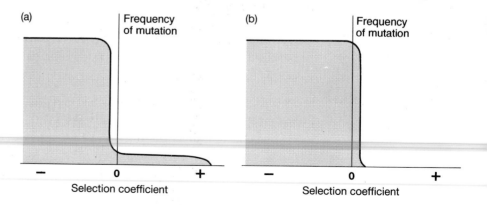

Figure 7.3 The neutral and selectionist theories postulate different frequency distributions for the rates of mutation with various selection coefficients. (a) According to the selectionists, exactly neutral mutations are rare and there are enough favorable mutations to account for all molecular evolution; whereas (b) neutralists believe there are many more neutral, and hardly any selectively favored, mutations.

This sort of neutralism (which we might call "true" neutralism) should not be confused with a misunderstanding of neutralism, according to which almost *all* mutations are neutral (Figure 7.4). Li has called this *pan-neutralism*. To an excellent approximation, no one believes pan-neutralism. There are good empirical reasons to reject it. It cannot explain why different genes evolve at different rates, or (see section 7.8.6) why different species have different biases in codon usage. Nor is it theoretically plausible. It is absurd to suggest that hardly any mutations are disadvantageous. Organisms, including their molecules, are adapted to their environments; we only need reflect on the efficiency of digestive enzymes—or any other biological molecule—in supporting life to realize that. If molecules are adaptations, many (or most) changes in them will be for the worse. This objection does not apply to the neutral theory, because the neutral theory recognizes that most mutations are deleterious and that molecules are adaptations: it merely suggests that there are many equally well adapted forms of a molecule, and evolution consists of neutral drift among them.

So the crucial difference between the selective and neutral theories is about the relative frequencies of neutral and selectively advantageous mutations. The most direct test between them would simply be to measure the frequencies of mutations of various fitnesses. Unfortunately that is impossible. Testing between the theories has been attempted by less direct means. One possibility has been to argue about which of the two frequency distributions (Figure 7.3a,b) is more realistic; but this has been inconclusive. Some people imagine that a protein could have many selectively equivalent forms; others that any change, no matter how small, will make some difference under some conditions. The main controversy has been concerned with other kinds of argument. The case for the neutral theory mainly consists of reasoning that the conditions needed for natural selection to produce the observed patterns of molecular evolution are false, or contradictory, whereas neutral drift runs into no such difficulties; and the case for the selective interpretation mainly consists of showing that the difficulties are illusory and that the neutral theory does not fit the facts.

We shall now consider the four observations of section 7.1 in more detail. We shall take them in three stages. We start with the rate of molecular

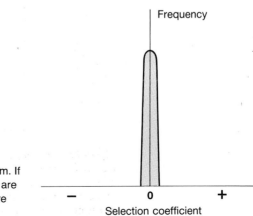

Figure 7.4 The theory of pan-neutralism. If pan-neutralism is correct, all mutations are selectively neutral (cf. neutralism, Figure 7.3b).

evolution and the level of polymorphism. Kimura suggested they raise a common difficulty for the theory of natural selection: a difficulty that can be expressed in terms of "genetic load." Genetic load is a general concept in evolutionary biology and we can take the opportunity to introduce it as a whole as well as in its neutralist application. We shall see how there are effective replies to Kimura's argument, and neither observation convincingly rules out natural selection. We then move on to the constancy of the rate, and see that the rate of molecular evolution is indeed strikingly more constant than that of morphologic evolution; but the form of the constancy, in protein evolution, is not exactly what the neutral theory predicts. After that we turn to the relationship between functional importance and evolutionary rate. For both the constancy of molecular evolutionary rates and their relationship with functional importance, evidence from DNA has added striking new tests, and suggests that the neutral theory may better apply to the evolution of some DNA sequences than to the proteins that first inspired it.

7.3 A population in which not all individuals have the optimal genotype is said to have a genetic load

At a particular time, there may be a single optimal genotype that an individual in a particular population could possess. We can avoid philosophic worries about imaginary omnipotent organisms (section 13.6, p. 332) by confining attention to the genotypes that actually exist in the population. (We also have to ignore frequency-dependent selection.) Suppose there are a variety of genotypes segregating out in the population, each with its characteristic fitness; one genotype has a higher fitness than the rest (at this particular time and place) and we call its fitness w_{opt}. We can also measure the average fitness of the whole population; it is just the fitness of each genotype multiplied by its frequency: it is called mean fitness and written $\bar{w}$. The general formula for genetic load (L) is as follows:

$$L = \frac{w_{opt} - \bar{w}}{w_{opt}} \qquad (7.1)$$

If all the individuals in the population have the optimal genotype, $\bar{w} = w_{opt}$ and the load is zero; if all but one have a genotype of zero fitness, $\bar{w} = 0$ and $L = 1$. Genetic load is a number between zero and one and it measures the extent to which an average individual in a population deviates from the best it could be. To be exact, the genetic load equals the chance that an average individual will die before reproducing because it possesses deleterious genes that exist in the population.

Genetic load can exist for several reasons. The balance between selection against a deleterious gene and its production by mutation is a simple example, called *mutational load*. At equilibrium, the deleterious gene has frequency m/s, where m is the mutation rate and s is the selective disadvantage of the mutation (section 5.10, see p. 108). What is the mutational load in this case? Suppose, for simplicity, there are two genotypes in the population, the majority type *aa*, which has relative fitness 1, and a rare dominant mutant type *Aa*, which has fitness $(1 - s)$; we can ignore *AA* because it will not be found if the mutant is rare. The frequency of *Aa* will then be m/s, and of *aa* $(1 - (m/s))$. $w_{opt} = 1$, because *aa* is the best genotype. $\bar{w}$ is the average fitness, i.e. the frequency of each genotype multiplied by its fitness:

$$\bar{w} = 1\left(1 - \frac{m}{s}\right) + (1 - s)\frac{m}{s} = 1 - m \tag{7.2}$$

We can substitute this result into the general formula for genetic load (equation 7.1), and

$$L = \frac{1 - (1 - m)}{1}$$

$$L = m \tag{7.3}$$

In this case, the mutational load equals the mutation rate. The load expresses the fact that individuals are dying because of the deleterious mutations that arise. The population carries a "load" of deleterious mutations, which reduces the average fitness of its members.

Genetic load has a real meaning and can be measured; but its main value is as a theoretical concept. We can express the consequences of many theoretical processes as a genetic load, and the concept can then be used as a heuristic reasoning device. Because a population contains only a finite number of individuals, there is a limit to the genetic load it can endure, and this fact can be used to estimate the upper limit on the processes that cause genetic loads. These kinds of argument make various assumptions, however, and we shall discuss them in due course; the point of this section is only to explain what a genetic load is.

7.4 Haldane described a "cost" of natural selection

When a gene is substituted by natural selection, the bearers of the inferior gene have to die at a higher rate, or have lower fertility, than bearers of the superior gene. The higher the intensity of natural selection, the higher the amount of selective death or infertility there must be in every generation. (It does not matter whether the selection is by differential mortality or fertility, or some combination of both; we shall discuss selection here in terms of mortality, but only as a shorthand—"mortality" stands for both real mortality and differential fertility.) Now, a population cannot tolerate an indefinitely large amount of selective death. If selection is too strong it will drive the population extinct. (Imagine a mutation that arose and caused all the rest of the population to have zero fitness. The population would become extinct: trivially because the single mutant would not find a mate; and more interestingly because the statistical probability that a population will become extinct increases as the population size decreases. Moreover, if two such mutants arose at different loci the population would necessarily become extinct.) Haldane's point was not that selection could never be very strong. It was that if it ever has been, the species would have become extinct. Species that have survived, and that we study now, cannot have suffered too great a selective cost during their evolutionary histories. Haldane referred to the selective death that must occur in order for a gene to be substituted as *the cost of selection* and in a 1957 paper he investigated what upper limit it would place on the rate of evolution.

The principle can be discussed most easily in a haploid case. Suppose there are two genes in a population, A and A', with fitnesses 1 and $(1 - s)$ and frequencies p and q $(= 1 - p)$, respectively. In any generation, of the q

A'-bearers, s will die without reproducing and $(1 - s)$ will survive like A-bearers. A proportion sq of the population die without reproducing because of selection at this locus. We can now define a ratio of the proportion of individuals in the population that survive to the proportion that die. sq die, and $p + q(1-s) = 1 - sq$ survive. The ratio in one generation is $sq/(1 - sq)$. The ratio is the same every generation until A' is eliminated. As selection operates in each generation, more and more selective death accumulates. We can now define the total cost of natural selection (C) as:

$$C = \sum \frac{sq}{1 - sq} \qquad (7.4)$$

where the summation is over all the generations it takes to fix the A gene. Table 7.3 gives some illustrative numbers. The most interesting property of Haldane's cost is actually not very relevant to the neutralist–selectionist controversy, but we can notice it here. It is that the total cost of selection is almost independent of the selection coefficient s when s is small (Table 7.3 illustrates the fact). The reason is as follows. If selection is stronger (s higher), more individuals die within each generation, but the gene itself is substituted more rapidly and the cost therefore has to be added over fewer generations. The two factors, cost within a generation and the number of generations taken to substitute the gene, approximately cancel; the total amount of selective death needed to substitute a gene is therefore more or less independent of the gene's selective advantage.

If a population is to maintain itself, the individuals that survive have to produce sufficient extra offspring to make up for those that die before reproduction. If, for example, the ratio $sq/(1 - sq)$ was 0.1/0.9, then each survivor would have to produce $1\frac{1}{9}$ offspring on average, which should be manageable. But there will be an upper limit to the cost of natural selection that can be born. If the ratio was 0.999/0.001, each survivor would have to leave 1000 *surviving* offspring, which would be much more difficult. Haldane's approximate estimate for the upper limit of cost that could be born was derived for a diploid rather than a haploid model. The algebra for the diploid case is in principle the same, but more complex. We shall not go into the details here. The upper limit suggested by Haldane was one gene substitution per 300 generations. He arrived at the figure by guessing the cost of a typical gene substitution was about 30 (cf. Table 7.3) and that a species

Table 7.3 The cost of natural selection (C), for various selection coefficients (s) including the limit for arbitrarily small s. The initial frequency of the favored allele is assumed to be 0.01

s	C
1.00	99
0.99	52
0.50	6.2
0.10	4.8
0.01	4.63
Limit	4.61

had about 10% excess reproductive capacity available for selective processes. Each gene substitution will then take 300 generations. These figures are very crude, because we have no evidence that a typical cost of a gene substitution is 20 or 40 rather than 30, but they illustrate Haldane's abstract argument.

The cost of selection can be expressed as a genetic load. While selection is fixing the favored allele, there will be some inferior, and some superior, alleles in the population. We can therefore define a *substitutional load* with the standard formula (equation 7.1). The mean fitness ($\overline{w}$) varies with gene frequency; it moves closer to w_{opt} as the favored allele moves to fixation. But whether the concept is formulated as a load or a cost, the key idea is that the substitution of one gene for another by selection requires death (or non-reproduction); without that death, gene frequencies cannot change.

7.5 The cost of selection was argued to imply that the rates of molecular evolution are too fast to be explained by selection

When the earliest measurements of the rate of molecular evolution became available, Kimura noticed that they implied much higher rates of change than Haldane's estimate of the maximum rate of evolution by natural selection, i.e. one substitution per 300 generations. The figures for rates of evolution in Table 7.1 are in units of changes per amino acid site per year. Kimura multiplied these numbers up to find the total number of changes in the genome per year. In a mammal there are perhaps 3×10^9 nucleotides in the genome. Kimura used an estimate (cf. Table 7.1) of one change in 28×10^8 years per amino acid site. Multiplying that rate by the number of amino acid triplets in the genome (10^9) gives an estimate of 0.357 changes per year, or about one change in the genome every 2–3 years. That is more than 100 times as fast as Haldane's upper limit. If Haldane's argument is correct, natural selection could not drive evolution as fast as it has been taking place in proteins. Although Haldane's estimate was very crude, the real rate is so much higher than the theoretical limit that the exact figures hardly matter. Kimura duly concluded that natural selection could not account for the majority of molecular evolutionary change.

With neutral drift, no difficulty arises. There is no cost of substituting one allele for another by random drift. Evolution by random drift is as rapid as the rate at which neutral mutations arise (see section 6.4, p. 130). That rate is unknown. And because it is unknown, the neutral theory can hypothetically explain almost any observed rates of evolution by postulating appropriate values for the neutral mutation rate μ. All we can conclude is that the rate of molecular evolution is not inconsistent with neutralism. The only upper limit on μ is the total mutation rate, because the neutral mutation rate must be less (probably much less) than the total mutation rate (see Figure 7.3). This upper limit, however, poses no problem, because the observed rates of molecular evolution are all much less than total mutation rates. Take, for instance, the figure for the rate of molecular evolution of one change in 28×10^8 years per amino acid site. A typical total mutation rate (section 4.6.1, p. 75) for a whole gene might be about 10^{-6} per generation, or 10^{-8} per amino acid site for a gene of 100 amino acids. This mutation rate would permit one change in 10^8 years, which is nearly 30 times as fast as the observed figure of one change in 28×10^8 years. The neutral theory is not contradicted by what we know about mutation rates. The neutralist argument, in summary, is that the observed rate of evolution is too high for evolution by natural selection, but not for evolution by neutral drift.

7.6 The concept of segregational load was used to argue that the degree of variation is too great to be maintained by selection

Gel electrophoresis revealed a large amount of protein variation in natural populations, even though the method does not detect all protein variation. For the discussion to follow, it is useful to have a rough figure for the amount of variation in a natural population. In a fruitfly (*Drosophila*), an approximate figure for the proportion of loci that are polymorphic may be around 30% and the average heterozygosity is about 10% (see Table 7.2); this is probably an underestimate because gel electrophoresis does not detect all protein variability. A low estimate of the number of loci in a fruitfly is 10 000; a population of fruitflies will therefore have about 3000 polymorphic loci. Their average heterozygosity will be 0.1/0.3 = 0.33 (0.1 is the average heterozygosity of the whole genome, but the heterozygosity is concentrated in 0.3 of the loci; the actually polymorphic loci will therefore have a heterozygosity of about 0.33). The first question is whether the selective and neutralist theories can account for this degree of variation.

At the time the controversy began, heterozygous advantage was thought to be the only general explanation of polymorphism. Dobzhansky had previously (e.g. in 1951) suggested that there are a large number of selectively maintained polymorphisms in natural populations. When large amounts of variation were indeed found, heterozygous advantage was the first explanation to be considered. For one or two loci, the idea is plausible. However, for large numbers of loci it soon runs into the problem of *segregational load*. Kimura and Crow first put forward this argument, in 1964. We start with the standard selection scheme for heterozygous advantage:

Genotype	*AA*	*Aa*	*aa*
Fitness	$1 - s$	1	$1 - t$

We have seen how to calculate the equilibrial gene frequency (section 5.11, p. 109) and it is then easy to calculate the proportions of the three genotypes at equilibrium. Thus:

The equilibrial gene frequencies at the locus are (*p* for gene *A*, *q* for *a*):

$$p = \frac{t}{s + t} \quad q = \frac{s}{s + t} \tag{7.5}$$

The proportion of heterozygotes at the locus is:

$$H = \frac{2st}{(s + t)^2} \tag{7.6}$$

The mean fitness of the population (for this one locus) is:

$$\bar{w} = 1 - \frac{st}{(s + t)} \tag{7.7}$$

(Mean fitness is found, as usual, by multiplying the frequency of each genotype by its fitness, i.e. $\bar{w} = p^2(1 - s) + 2pq + q^2(1 - t)$. To reduce that to the formula above, first multiply it out, then note that $p^2 + 2pq + q^2 = 1$, and substitute for p and q from the formulae in terms of s and t (equation 7.5). Note also that $p^2s + q^2t = ps = qt$. The algebraic simplification is then easy, as readers can confirm.) The formula for mean fitness can be further simplified if we substitute the formula for heterozygosity into the one for mean fitness. Then,

$$\bar{w} = 1 - H\left(\frac{s + t}{2}\right)$$

$$\bar{w} = 1 - H\bar{s} \tag{7.8}$$

where $\bar{s} = (s + t)/2$. If $\bar{s} = 0.1$, and $H = 0.33$, the average member of the population has a fitness of 0.967 as compared with one for a heterozygote. Mean fitness is less than one, because s and t are both positive. With heterozygous advantage, the population mean fitness is lower than it would be if all individuals were heterozygotes. $\bar{w}$ is lower if H is higher because when there is a higher proportion of heterozygotes, selection against the homozygotes must be stronger (higher $\bar{s}$). More homozygotes must die each generation and the population "suffers" more reduction in mean fitness than when the homozygotes and heterozygotes have similar fitnesses.

The load exists because with Mendelian heredity it is impossible to create a population of pure-breeding heterozygotes. Even in the extreme cases of $s = t = 1$, when all homozygotes die before reproduction, still the matings between heterozygotes will generate the inferior homozygous forms; in practice, s and t are less than one and the homozygote crosses contribute too. The population inevitably produces homozygotes every generation, and the average fitness of its members will therefore be below that of a heterozygote.

So far we have considered only one locus. But in a fruitfly population, there are about 3000 polymorphic loci, with heterozygosity averaging 0.33 each. If $\bar{s} = 0.1$ for each of them, then at every locus the fitness of an average individual is about 0.967 compared with one for a heterozygote. If the average fitness for all loci is found by multiplying up the average fitness of each of the 3000 polymorphic loci

$$\bar{w} = (1 - 0.33 \times 0.1)^{3000}$$

Natural selection is here assumed to operate independently on each locus; homozygotes at each of the 3000 loci are contributing independently to the loss of average fitness in the population. The resulting fitness of an average individual is very low. Indeed, with $\bar{s} = 0.1$ and $H = 0.33$ as in the formula, $\bar{w} = 10^{-43}$ (compared with $w = 1$ for an imaginary individual that was heterozygous at all 3000 loci). This is an absurdly low number. It means that if we compare the fitness of a heterozygote at all 3000 loci with the fitness of a random member of the population, the full heterozygote would be 10^{43} times fitter than the randomly picked individual, which is an obviously impossible requirement. As Lewontin put it, "One can hardly imagine a *Drosophila* female, no matter how many loci she was heterozygous for, laying 10^{43} eggs."

Something has gone wrong. Kimura's conclusion was that the error lies in the idea that polymorphism is maintained by natural selection. He deduced that natural selection cannot account for the observed levels of natural variation in proteins. As in the case of evolutionary rates, no such difficulty arises if the variability is due to neutral drift. As we have seen (section 6.7, p. 136), neutral drift will preserve a degree of heterozygosity. The heterozygotes are "transient." New alleles appear first as unique mutations, they then exist as transient polymorphisms for a while, until drift either fixes or eliminates the allele. The heterozygosity maintained is given by

$$H = \frac{4N\mu}{4N\mu + 1} \qquad (7.9)$$

As we saw when explaining the rate of evolution by neutral drift, almost any result could be explained if we put the right numbers into the formula. The same is true here. For example, $H = 9\%$ if $N = 2.5 \times 10^5$ and $\mu = 10^{-7}$. This μ may be approximately right, though the N is quite low. A population size of 2.5×10^5, however, is within the bounds of possibility, and a single observation of $H = 9\%$ certainly does not contradict the neutral theory.

Kimura's original argument, in summary, stated that an impossible segregational load would be run up if selection was maintaining the observed levels of heterozygosity. The observations, however, can be explained by neutral drift, with plausible values for the population size and neutral mutation rate. His argument, therefore, has much the same structure both for the rate of evolution and for the amount of heterozygosity.

Before we move on, let us notice another observation. The neutral theory would be more comfortable with rather higher heterozygosities than are found in nature. A figure of 9% heterozygosity, we have seen, is predicted by the neutral theory if the product of population size and mutation rate, $N\mu = 2.5 \times 10^{-2}$. Several critics have argued that natural values for $N\mu$ will be higher. The neutral theory should therefore predict higher heterozygosities than are observed. The population size of *Drosophila*, for example, may be 100 times larger than that required by the neutral theory ($H \simeq 10\%$ in *Drosophila*). Moreover, as heterozygosity has been measured in further species, it has become clear that they are consistently too low for the neutral theory (Figure 7.5). Because N and μ are both unknown numbers, this is not strong evidence against the theory; but it is a problem to worry about (Box 7.2 discusses one modification to the neutral theory to account for the "low" levels of heterozygosity, and one or two other awkward facts).

7.7 Natural selection can operate without running up an impossible genetic load if it is "soft" rather than "hard"

In this section, and the next two, we shall see three reasons to doubt whether the rates of evolution and levels of polymorphism pose any theoretical difficulty for natural selection. The first concerns a distinction between "hard" and "soft" selection. Haldane's "cost" of natural selection assumes that there are two kinds of death (or, in more abstract terms, population control processes). On average, each pair of individuals must produce two offspring, because natural populations cannot expand or contract indefinitely. However, females in all species produce more—usually many more—than two offspring. The excess fecundity, of course, is the reason for Darwin's "struggle for existence." It means that, whether or not natural selection is

Box 7.2 Slightly deleterious mutations in the neutral theory.

A neutral mutation has the same fitness as the existing allele, or alleles, at its locus. Its evolution is controlled by neutral drift. However, random drift not only influences the evolution of neutral genes. It influences all genes to a greater or lesser extent. If one allele has a very different fitness from another, their fate will mainly be governed by natural selection. (Even here, the fate of a new mutation is strongly influenced by chance; the chance that a mutation with selective advantage s will be lost by random drift is $1 - 2s$. A mutation with as high a selective advantage as 5% still has an approximately 90% chance of random loss.) As the selective difference between the two alleles decreases, neutral drift becomes more important. Besides the difference in fitness, the other factor determining the relative importance of selection and drift is population size; random drift is more important in smaller populations. The neutral theory therefore is not only relevant for neutral genes. To a rough approximation, genes with a selective advantage (or disadvantage) of $1/(2N)$ or less behave as effectively neutral genes: their evolutionary fate is influenced more by chance than by selection.

Ohta realized that a number of difficulties in the neutral theory could be overcome if many mutations are slightly deleterious rather than exactly neutral. Selection does work, feebly, against slightly deleterious mutations (SDM), but their dynamics are dominated by drift. Ohta tackled three difficulties in the neutral theory: the level of heterozygosity, which is often too low for the neutral theory; the absence of a generation time effect in the molecular clock; and a third which we have not discussed because the evidence is too controversial (it is the possibility that the molecular clock is too variable for one driven by neutral drift). Neutralism, as we saw, was criticized for predicting too high levels of heterozygosity (section 7.6). However, the equilibrial heterozygosity is lower if most mutations are slightly deleterious, rather than exactly neutral. The facts can then be accounted for.

The argument for evolutionary rates is more complex. Under pure neutrality, the rate of evolution is equal to the mutation rate. It is independent of population size. But for slightly deleterious mutations, the evolutionary rate is higher in larger populations. The exact formula for the relationship between slightly deleterious mutation rate, population size, and rate of evolution is complex, but it is not difficult to see how the dependence on population size arises. The number of mutations of different selection coefficients will be related to population size (N). As N increases, the number of mutations of each value of s increases, and there will be more slightly deleterious mutations in a larger population. Ohta made use of this result to explain why the molecular clock, at least for proteins, shows no generation time effect. Species with larger population sizes probably tend to have shorter generation times. This relationship, she simply says, "we know as an empirical fact," though actually it is questionable. But if it is accepted, species with shorter generation times will evolve more rapidly through the intermediate variable of higher population size. This will tend to make up for their lower (in absolute time) mutation rates.

Thus certain difficulties in the neutral theory can be removed, or at least reduced, by postulating a large class of slightly deleterious mutations. The theory can still reasonably be called the neutral theory, because chance still determines the fate of genes in evolution; but it is a modified version of the original theory (Figure B7.3). And it, in turn, can be criticized. Gillespie in particular looked in detail at the population sizes required for Ohta's theory to explain the observations. Ohta qualitatively observed that the dependence of evolutionary rates on population size will shift the predictions closer to the observations. But it is possible to calculate exactly what values of N are needed. Gillespie's original papers should be consulted for the arguments, but his conclusion doubted whether neutral theory and genetic fact could be reconciled by realistic population size fluctuations. Even if Ohta's theory does fit the facts, that would hardly confirm it, because it postulates a large class of slightly

continued on page 156

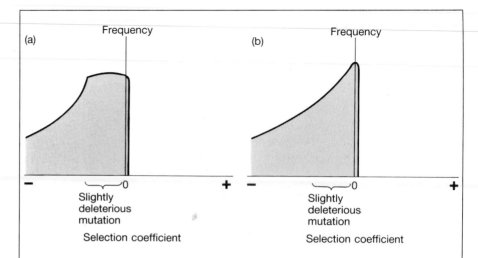

Figure B7.3 (a) According to the theory of slightly deleterious mutations, there is a large class of mutations with selection coefficients that are negative, but so slightly negative that they behave as if they were neutral. (b) The theory is a small modification of the original neutral theory, which postulated a large class of exactly neutral mutations (cf. Figure 7.1). The exact shape of the graphs to the left of the slightly deleterious region is unimportant; both theories suggest there are very few advantageous mutations, many neutral ones, and many deleterious ones; they differ in the frequencies of slightly deleterious mutations.

deleterious mutations for which there is no independent evidence. We therefore do not know how important slightly deleterious mutations are in evolution. In summary, Ohta suggested the neutral theory could explain the facts of protein polymorphism and the molecular clock better if there are a large number of slightly deleterious mutations. The idea is theoretically important, but how well it explains the facts is controversial.

Further reading. Ohta (1974, 1977) and Gillespie (1992).

Box 7.2 (*continued*) Slightly deleterious mutations in the neutral theory.

altering gene frequencies, many individuals must die every generation. In Haldane's formulation, when natural selection is actually substituting one gene for another, it does so by means of extra mortality, on top of the background mortality. This is called *hard selection*. If selection does indeed require extra mortality, it will tend to depress the population size. The rate of evolution then, has an upper limit: if selection is too strong it will depress the population size to zero.

However, the limit is relaxed under *soft selection* (Figure 7.6). Under soft selection, the selective deaths are substituted for non-selective background mortality. The possible rate of evolution is then much higher. If all the mortality suffered by a population was selective, evolution could proceed at an immense speed. Consider a species like the cod, in which a female can produce over 5000 000 eggs. In theory it is possible for a single mutation in a population of 2500 000 cod to increase from a frequency of 1/2500 000 to one

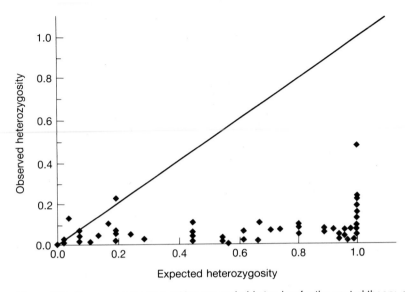

Figure 7.5 Observed heterozygosities are probably too low for the neutral theory, at least in its simplest form. Each point gives the observed heterozygosity (y-axis) for a species (total 77 species), plotted against the "expected" heterozygosity from estimates of the population size and generation length of the species and assuming a neutral mutation rate of 10^{-7} per generation. From Gillespie (1992).

(i.e. 2500 000/2500 000) in a single generation and for the population size to stay constant. No doubt this would not really happen: the intensity of selection is unrealistic and much of the mortality suffered by a cod population will inevitably be non-selective. But the point is theoretically interesting. It shows that rapid evolution can take place under natural selection provided that selection is soft: the selective mortality is being substituted for the natural mortality of the population. The population size then need not be decreased by the action of natural selection. Hard and soft selection are extremes of a continuum. In practice, natural selection in different cases could be any mixture of hard and soft selection. (The cod is a dramatic case, because its fecundity is so high; but in all species there are many more eggs laid than there are reproducing adults.)

The general concept of a cost of selection need not presuppose either hard or soft selection. There has to be selective death (or non-reproduction) for evolution by natural selection. But Haldane's (and Kimura's) use of the concept, to calculate an upper limit on the rate of evolution, does assume that selection is quite hard. They assumed only 10% selective non-reproduction could be endured. In the cod example, that would mean that, of the 4999 998 of one female's 5000 000 eggs that do not survive to reproduce, about 4949 999 were hard deaths that would happen in any case and only 49 999 offspring failed to reproduce because natural selection worked against them. There is considerable upside in the rate of evolution if selection really is softer than Haldane assumed.

In nature, when will selection be hard and when soft? The answer depends on the circumstances. A case of pure soft selection might arise as follows

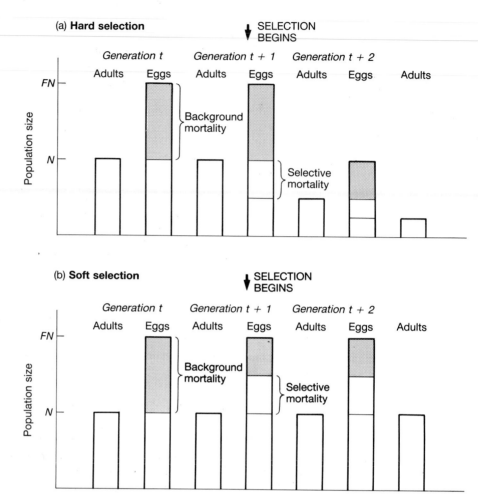

Figure 7.6 A population with N adults, of average fecundity $2F$ per female, produces FN eggs. By the ordinary ecologic processes of mortality, if the population size is approximately constant, the FN eggs must be reduced to N adults. $FN - N$ of the eggs die before reproducing. (a) In the extreme of hard selection, selective mortality is added to the background ecologic mortality. The population size is reduced below N adults while selection operates. Selection begins operating between generation t and generation $t + 1$. (b) In the extreme of soft selection, all selective mortality replaces some of the background ecologic mortality and the population size is not altered by the operation of selection.

(Figure 7.7). Imagine a population in which nearly all mortality is due to competition for food. Selection is taking place at a gene locus that influences feeding efficiency; a new mutation might be superior in territorial defence, for instance. The individuals that fail to defend territories, let us suppose, will starve. When the mutation arises the total space for territories does not change. All that will happen is that the bearers of the mutation will be underrepresented among the starving individuals. No extra mortality is incurred; the population size need not change as the gene was fixed. Selection has been purely soft.

(a)

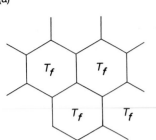

(b)

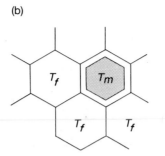

(c)

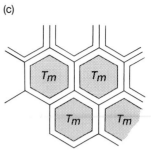

Figure 7.7 Theoretical examples of pure soft selection. The species is territorial. (a) The individuals (genotype T_f) divide up the available space and only those with territories breed; every one else dies without reproducing. (b) A mutation (T_m) arises, whose bearers are superior at holding territories; the double border symbolizes this superior ability. (c) The mutation spreads, but as it does so the population size remains constant; those with the genotype (T_f) being selected against are now concentrated in the non-territorial pool of non-breeders.

At the other, "hard" extreme, we could imagine that most of a species' mortality happens during the juvenile phase. Perhaps 99% of those born die before reproducing. In the absence of selection, the remaining 1% all breed and manage to maintain the population. Now suppose an advantageous mutation arises and influences the relative chance of adult survival. The mutation has the effect of reducing the survival of the rest of the adult population (for instance because the mutant individual takes more food). Extra mortality is now being added to the juvenile mortality; and the population will go into decline. Selection would have been soft if the selective mortality could have been shifted into the juvenile phase; but in this case the mutation influences only the adult, and the juvenile mortality happens for some inevitable ecological reason. Selection is hard.

The important point is that Haldane's argument assumed that selection is hard. In the case of hard selection, the limit on the amount of selective mortality that can be tolerated will be lower than under soft selection. In some natural cases, selection may be hard; but it does not have to be always. To the extent that selection is soft, the neutralist argument is weakened.

7.8 Natural selection can act jointly on many genetic loci

In the neutralist's calculations of segregational and substitutional loads, the high figures were obtained by multiplying figures for one gene by the number of genes, to estimate the load for the genome as a whole. The calculation assumes that selection operates independently at each genetic locus. In the case of the segregational load in *Drosophila*, the load was estimated by raising $(1 - Hs)$ to the power 3000, for the 3000 loci that may be polymorphic in a population of fruitflies. But does selection at each locus independently influence the individual's mortality?

Imagine a situation like Figure 7.8. A substrate can be successively metabolized by two enzymes in a metabolic sequence, provided that the individual has the enzymes. If the individual lacks enzyme 1, it cannot form the first stage product, and whether or not it has enzyme 2 will make no

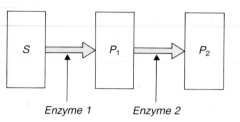

Figure 7.8 Theoretical example of a metabolic pathway in which a substrate S is successively degraded into product P_1, by enzyme 1, and then P_2, by enzyme 2. Suppose the enzymes are encoded by heterozygotes and it is advantageous for the organism to degrade S. Possessing the genotype for enzyme 2 is useless in an organism that lacks enzyme 1; the fitnesses of the enzymes encoding the two loci interact.

difference. Suppose that both enzymes are encoded by heterozygotes, and that it is advantageous to the organism to be able to metabolize the substrate S into the product P_2. There is now heterozygous advantage at each locus, but they do not contribute independently to the organism's fitness. The fitness of an organism lacking both heterozygotes is the same as that of an organism lacking either one of them.

In practice, some of an organism's loci will interact in the manner of Figure 7.8, and they will not independently influence fitness; other loci will be independent, and the fitness of multiple homozygotes will then be less than that of single homozygotes. It depends on the effect of the particular locus, and its relationship with other loci. Reasoning generally, it seems unlikely that a fruitfly is so atomistically organized that all 3000 polymorphic loci influence fitness independently. Its different metabolic systems interact. Kimura's calculation was therefore exaggerated. Heterozygous advantage could maintain many polymorphisms if each does not have an independent effect on the organism's fitness. This point was made by a number of population geneticists, including Sved *et al.* King, Milkman, and Maynard Smith, in 1967 and 1968.

However, the other extreme position is surely also false: a fruitfly is not so "holistically" organized that none of the 3000 loci have independent effects on fitness. The truth is somewhere between these two positions. The argument can be made formal in a "threshold" or "truncation" model of selection. The models propose that an organism's fitness is related to the number of loci at which it is heterozygous in the manner of Figure 7.9. The graph has a simplified, idealized form, and not much significance need be attached to the exact shape of the threshold function (an exact shape has to be specified for modeling). For the general point, the function could have any decreasing shape. Selection will have this threshold shape if, after an organism has more than a threshold number of heterozygous loci, the different loci stop having independent effects on fitness. The point is slightly different from the particular example in Figure 7.8, but the abstract idea is the same and it arguably may be quite realistic.

A relationship like Figure 7.9 is plausible if an organism has a number of partly independent metabolic subsystems, and the loci controlling enzymes within each subsystem interact in a "decreasing returns" manner. To have

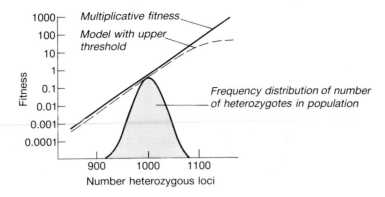

Figure 7.9 Fitness of an organism in relation to the number of loci at which it is heterozygous. The straight line (—) for fitness corresponds to the multiplicative assumption of Kimura's argument. There could instead be a non-linear, or "threshold," relationship (– – –). The *y*-axis is logarithmic.

one or two heterozygous loci within a subsystem is highly beneficial; but increasing numbers provide ever smaller advantages. Now imagine (as a thought experiment) taking a homozygous fruitfly and adding a few heterozygotes at random. (We are at the bottom left of the graph in Figure 7.9 and we are moving up the graph to the right.) These first few heterozygotes will probably mainly affect independent systems and have independent effects on fitness. Now we can make some more of the loci heterozygous. As the number of loci at which the fly is heterozygous increases, extra heterozygotes will increasingly be within metabolic systems that already have a number of heterozygotes, and the heterozygous loci will no longer all have independent effects. Clearly, in the case of the threshold model, it is wrong to estimate the total segregational load by raising $(1 - Hs)$ to the power of the number of loci at which there is heterozygous advantage; in the particular case of Figure 7.9, a step function is needed. Without entering into algebraic detail, the general point should be clear. With threshold selection, heterozygous advantage can maintain many polymorphisms without generating a paradoxical segregational load.

A similar argument can be made for the rate of evolution. When Kimura argued that evolution had proceeded too fast to have been driven by natural selection (reasoning from Haldane's cost of selection), he was assuming that selection ran up an independent cost at each of the loci it was operating on. Kimura calculated that one nucleotide is substituted every two or three generations, in which case changes must be taking place simultaneously at very large numbers of loci. If all the changes are by hard selection, and each one imposes an independent "cost" on the population, the paradox arises. However, each locus may not be contributing an independent increment to the total cost of selection. Selection need not be working independently at each locus. Again, there could be a threshold effect, such that if an organism has more than a threshold number of the favored genes, its fitness is no longer a power multiple of the number of favored genes (Figure 7.10). Many loci could then "pay" the cost of selection together in a block, and Kimura's

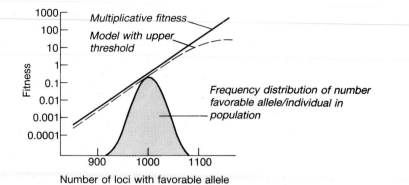

Figure 7.10 Assume that directional natural selection is operating at each of a number of loci. At each locus, some allele is being favored over others. Kimura's calculation of the cost of selection assumed that the cost could be multiplied up for the number of loci at which selection was operating. But it could be that the loci interact and there is a threshold relation. The x-axis differs from Figure 7.9, but the point is the same.

calculation would be invalid. If selection operates jointly on many loci, the loads implied by the observed rates of molecular evolution and protein polymorphism are no longer absurd.

7.9 Frequency-dependent polymorphism can exist without any genetic load

Although, in the mid 1960s, heterozygous advantage was the best known mechanism by which selection could maintain a polymorphism, it is not the only one. Frequency-dependent selection is an important alternative (section 5.12, p. 114). Does it generate a genetic load at equilibrium? With heterozygous advantage there is segregational load because inferior homozygotes are continually formed. Whether or not frequency-dependent selection results in a genetic load depends on whether all the genotypes at equilibrium have equal fitnesses. In a simple case with one locus, two alleles, and three genotypes, at equilibrium all the genotypes have equal fitness. They are equally the "best" genotype and the fitness of the best genotype is equal to the mean fitness of the population. Load, as defined by the standard formula (equation 7.1) is zero. Inferior genotypes are not generated each generation, and there is no need for selective death to maintain the polymorphism. There can be other frequency-dependent equilibria in which not all genotypes have equal fitness: then there is a genetic load. However, frequency-dependent selection can maintain an equilibrium with no genetic load. In theory it could maintain an indefinite amount of polymorphism without running into any difficulty over genetic load. (By the way, notice that just because, at equilibrium, genetic load is zero, that does not mean fitness $\bar{w}$ is maximized (i.e. $\bar{w} = w_{opt}$). With frequency dependence, the fitness of a genotype is at a maximum when it is rare; but it is logically impossible to have an equilibrium at which all the genotypes are rare. In general, under frequency-dependent selection the gene frequency that maximizes mean fitness is not the same as the equilibrial gene frequency.)

So far we have been concerned with theoretical questions. If we allowed (for sake of argument) that most polymorphisms were maintained by selection, the question would remain of the relative influence of heterozygous advantage, frequency dependence, and other forms of selection such as spatial and temporal environmental change. Heterozygous advantage is not now widely believed to be the main cause of polymorphism, because examples of it have accumulated too slowly. The main example, sickle-cell anemia, was one of the first polymorphisms to be understood in selective terms and it was natural to assume (tentatively) that the process might be general. Forty years later, it remains the clearest case. Other examples do exist; but Endler found only six in a review in 1986. Other, more abstract arguments have also been put forward to suggest that heterozygous advantage is likely to be rare; but they are not finally persuasive. Although the question has not finally been settled, few would now argue that heterozygous advantage is the main cause of protein polymorphism.

With frequency-dependent selection, the situation is even less clear. Some population geneticists, such as Clarke, have argued that it is a general cause of polymorphism. The possibility has not been widely accepted, but it cannot be ruled out. There are no general arguments against it, except perhaps a vague intuition that not enough biological interactions have the right characteristics to be negatively frequency dependent and to maintain 3000 polymorphisms in a fruitfly population. Some widespread biological processes such as the interaction of predators and prey and of parasites and their hosts (section 11.3.5, p. 275) produce frequency-dependent relationships, although not necessarily stable polymorphisms. The importance of frequency-dependent selection in nature remains an open question, but the point of this section is theoretical: with frequency dependence, polymorphism can be maintained without any selective load of deaths.

7.10 The first two observations are indecisive

Kimura's argument was that neutral drift easily explains the high rate of molecular evolution and high levels of polymorphism, whereas natural selection does not. The case against natural selection made unnecessary, and probably spurious, assumptions about genetic loads, and is unconvincing. Moreover, the explanation for the facts offered by the neutral theory itself is not powerful: it explains the observed evolutionary rates and heterozygosities by plugging conjectural values for N and μ into the equations for genetic drift. However, the actual values of N and μ are unknown, and a neutralist could "explain" almost any evolutionary rate or heterozygosity by positing suitable values for N and μ.

When natural selection is considered positively, as a possible explanation for molecular evolution, it turns out to be in similar position to the neutral theory. Selectionists too do not generally know the value of the variables that control the rate of evolution. The probability of fixation of a selectively advantageous mutation with selective advantage s is approximately $2s$, which makes the rate of evolution $2sm$ where m is the rate of selectively advantageous mutations. m is just as much of an unknown as μ and not much can be said about s apart from that it is probably low (1% or less). Like the neutralist, the selectionist can "explain" almost any observation by inventing suitable values for the unknown numbers, s and m. In short, the observations about rates of molecular evolution and levels of polymorphism are inconclusive.

7.11 The rates of molecular evolution were argued to be too constant for a process controlled by natural selection

The rate of molecular evolution, we have seen (Figure 7.2), is approximately constant. Kimura reasoned that constant rates are more easily explained by neutral drift than selection. Neutral drift has the property of a random process and its rate will show the variability characteristic of a random process. Neutral mutations crop up at random intervals, but if they are observed over a sufficiently long period the rate of change will appear to be approximately constant. Neutral drift will drive evolution at a fairly constant rate. Natural selection, Kimura argued, does not produce such constant change. Under selection, the rate of evolution is influenced by environmental change as well as the mutation rate; and it would require a surprisingly steady rate of environmental change, over hundreds of millions of years, in organisms as different as snails and mice and sharks and trees to produce the constant rate of change seen in Figure 7.2.

Moreover, if we look at characteristics, such as any adaptive morphologic characters, that have undoubtedly evolved by natural selection, they do not seem to evolve at constant rates. Kimura's discussion of the evolution of the vertebrate wing illustrates the argument. There was first a long period before the wing evolved, in which the vertebrate limb remained relatively constant (in the form of the tetrapod limb of amphibians and reptiles); then came a shorter period when the wing originated and evolved; finally, there was a long period of "fine-tuning" of a more-or-less finished wing form. The wing undoubtedly evolved under the influence of natural selection; and the rate of change in this case probably underwent large fluctuations between fast and slow evolution. The argument both suggests that the rate of molecular evolution is strikingly constant, and is also Kimura's reason for confining the neutral theory to molecules, and not applying it to the gross phenotypes of organisms. Molecular evolution appears to have a fairly constant rate, as would be expected for a random process; but the rate of morphologic evolution has a different pattern, and was probably produced by the non-random process of selection.

Molecular evolution may be constant enough to provide a *molecular clock* of evolution. Evolutionary change at the molecular level ticks over at a roughly constant rate, and the amount of molecular change between two species therefore measures how long ago they shared a common ancestor. (Molecular differences between species are therefore useful in phylogenetic inference—see chapter 17.) It has been questioned whether molecular evolution is as constant in rate as the neutralists claim; but the statistics are involved and inconclusive and we shall not enter into them here. It can reasonably be concluded at present that the rate of molecular evolution is strikingly constant, as compared with that of morphology.

Molecular evolution in "living fossils" provides an example both of the constant rate of molecular evolution and of the independence of molecular and morphologic evolution. The Port Jackson shark *Heterodontus portusjacksoni* is a "living fossil"—a species that closely resembles its fossil ancestors (there are fossils over 300 million years old that are very similar to the Port Jackson

Table 7.4 Amino acid differences between the α- and β-globins, for three species pairs. From Kimura (1983)

Species pair	Number of amino acid differences
Human α vs human β	147
Carp α vs human β	149
Shark α vs shark β	150

shark). Its molecules have been evolving very differently from its morphology. Hemoglobin duplicated into α and β forms before the ancestor of mammals and sharks, at the beginning of the chordate radiation. We can count the amino acid differences between α- and β-globin as a measure of the rate of molecular evolution in the lineages leading to the modern species. Table 7.4 reveals that changes have accumulated in the Port Jackson shark lineage at the same rate as the human lineage. The rate of molecular evolution in the two lineages is roughly equal. This is to be compared with the large difference in the rate of morphologic evolution in the two lineages: the Port Jackson shark lineage has hardly changed at all, but humans have evolved from fish-like ancestors, and passed through amphibian, reptilian, and several mammalian stages. Moreover, as Table 7.4 shows, human β-globin is as different from human α-globin as it is from carp α-globin. This is despite the fact that human α- and β-globin will have shared much more similar external selective pressures, as they have been locked in the same kind of organisms throughout evolution, than have human β-globin and carp α-globin. The result suggests that the α- and β-globin molecules have been accumulating changes independently, at roughly constant rates, regardless of the external selective cirumstances of the molecule. This in turn suggests that most of the evolutionary changes in the hemoglobin molecule have been neutral shifts among equivalent forms, of equal adaptive utility. While the rate of morphologic change varies greatly among the various evolutionary lineages of vertebrates, the rate of molecular evolution all seems to have been much more similar.

7.12 The protein molecular clock runs relative to absolute time, not generation time

It was first well established by Wilson and his colleagues, in 1977, for 12 proteins and 25 species-pair comparisons, that the rate of protein evolution is constant relative to absolute time, not the number of generations (Figure 7.11). For each protein, they selected pairs of species, one a mammal with a short generation time, the other a mammal with long generation time. They then estimated the evolutionary rates between the common ancestor of the species pair and each of the two modern species. If the proteins evolve at a rate proportional to absolute time, the rates should be equal for the two species; if the rate is relative to number of generations, the proteins should have changed more in the line to the modern species with a short generation time. The evidence clearly shows that it is absolute, not generation, time that matters. This result, for proteins, is not in doubt; evolution in the DNA tells a different story, as we shall see.

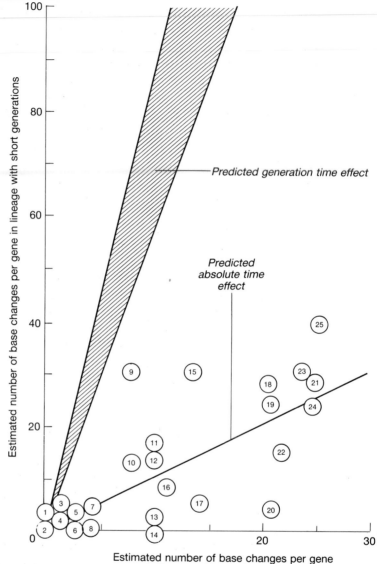

Figure 7.11 The molecular clock does not show a generation time effect for proteins. The graph shows the rate of evolution for 25 species pairs. Each pair was made up of one species (such as elephant or gray whale) with a long generation time and one (such as rabbit or mouse) with a short generation time. If there is no generation time effect, the rates of evolution in the two lineages of a pair should be the same, and the points will cluster around the 1:1 line. This is indeed what happens. If there were a generation time effect, evolution per year would be slower in the lineage with a longer generation time. For the particular species used in the test, the points would then have fallen in the region indicated on the graph. From Wilson *et al.* (1977).

According to the neutral theory, the rate of evolution is equal to the neutral mutation rate. The evolutionary rate will therefore be proportional to the time scale of the process generating mutations. This is normally thought to imply that evolutionary rates will be influenced by generation time. Most mutations probably arise as copy errors during mitotic or meiotic replication. The mutation rate should therefore be proportional to the number of DNA replications in the germ line. There are fewer such replications per unit of absolute time in species with longer generations, because the number of mitotic divisions in the germ line is much less than proportional to longevity. For visible mutation, mutation rates per generation are as a matter of fact not noticeably longer in species with longer generation times (Table 4.3, p. 76). If (as seems reasonable) neutral mutations are like other mutations, neutral evolution should be slower in species with longer generation times. Selectionists can therefore retort that neutralism, far from predicting the molecular clock, actually does not predict it: the neutralist's clock would be a quite different kind of clock from the one that has been found. It would tick in proportion to numbers of generations, not of years. (Box 7.2, however, suggests how the neutral theory can be modified to fit the facts.)

Selection probably provides a better explanation than neutral drift for why the clock depends on absolute time. Under neutral evolution, the rate of evolution is exactly the mutation rate; neutralism therefore strongly predicts a relation between number of generations and evolutionary changes. But under selection, the rate of evolution can be controlled by the pace of environmental change, which could depend on absolute time. Suppose, for instance, that two kinds of mammal, such as a mouse (generation time approximately 0.33 years) and an elephant (generation time approximately 33 years) were both separately evolving in relation to changes in parasites, such as bacteria, with short generation times. After 0.33 years, the bacterial population will have accumulated 0.33 years worth of change, which acts as an agent of selection on the mouse population; but after 33 years, the bacteria that parasitize elephants will have accumulated 100 times as much change, which will select 100 times as strongly on the elephants. The longer elephant generation time will to some extent be compensated by the stronger selection it will experience. The resulting rates of evolution in mice and elephants might be quite similar. The argument is simplified, but it suggests that natural selection can explain why the molecular clock keeps absolute, not generational, time.

7.13 DNA sequences can now be used to test for a generation time effect in the molecular clock

As an increasing number of DNA sequences have been elucidated, it has become possible to test for the constancy of molecular evolution in the DNA sequences. The genetic code contains 64 codons, of which 61 code for 20 amino acids (see chapter 2); the three-fold redundancy in the code means that not all base changes in the DNA cause amino acid changes in the protein. It is an almost inevitable theoretical consequence that there cannot be an evolutionary molecular clock both for base changes and for amino acid changes. We have seen that the clock for proteins keeps reasonably good time, though it ticks over according to absolute, not generational, time. What happens in the DNA?

The evidence suggests there is a generation time effect. We can distinguish

Table 7.5 Rates of evolution in silent base sites are faster in groups with shorter generation times. There are estimates for various pairs of species, and each estimate is an average for a number of proteins; the number of sites is the total number of base sites (for all proteins) that have been used to estimate the rate. The divergence times, which are in millions of years, are uncertain; a range of estimates (in parentheses) has been made. From Li *et al.* (1987)

Comparison	Number of proteins	Number of sites	Divergence time (Myr)	Rate ($\times 10^{-9}$ year)	Generation time
Primates					
Human vs chimpanzee	7	921	7 (5–10)	1.3 (0.9–1.9)	
Human vs orangutan	4	616	12 (10–16)	2.0 (1.5–2.4)	} Long
Human vs Old World monkey	8	998	25 (20–30)	2.2 (1.8–2.8)	
Artiodactyls					
Cow vs goat	3	297	17 (12–25)	4.2 (2.9–6)	} Medium
Cow/sheep vs goat	3	1027	55 (45–65)	3.5 (3–4.3)	
Rodents					
Mouse vs rat	24	3886	15 (10–30)	7.9 (3.9–11.8)	Short

the results for "silent" and "replacement" base substitutions. Silent base changes do not result in an amino acid change in the protein; the protein is the same after a silent mutation as before. Silent mutations, therefore, seem particularly likely to be neutral. Rodents, such as mice and rats, have shorter generation times than primates and artiodactyls (such as cows). For both silent substitutions and replacement substitutions, evolution is faster in rodents than in artiodactyls, and faster in artiodactyls than in primates; and the effect is particularly marked in silent sites (Table 7.5). Silent substitutions occur faster in species with shorter generation times, as the neutral theory predicted. Maybe, therefore, the silent parts of the DNA do evolve neutrally. (We shall have more to say about the rate of evolution in DNA sequences below; here we are concerned only with whether they show a generation time effect.)

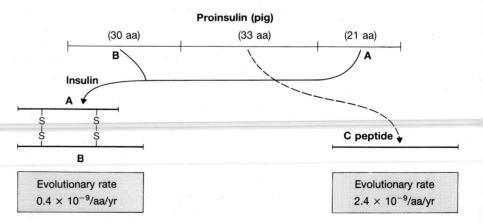

Figure 7.12 The insulin molecule is made by snipping the center out of a larger proinsulin molecule. The rate of evolution in the central part, which is discarded, is higher than that of the functional extremities. From Kimura (1983).

Table 7.6 Rates of evolution in the surface and heme pocket parts of the hemoglobin molecules. Rates are expressed as number of amino acid changes per 10^9 years. From Kimura (1983)

Region	α-Globin	β-Globin
Surface	1.35	2.73
Heme pocket	0.165	0.236

In conclusion, the absence of a generation time effect in the molecular clock for proteins suggests that neutral drift is not so important in their evolution as Kimura originally thought. For silent base changes in DNA, however, the clock depends on generation time and the neutral theory is much more plausible.

7.14 The more functionally constrained parts of proteins evolve at slower rates

A protein contains functionally more important regions (such as the active site of an enzyme) and less important regions. It has consistently been found that the rate of evolution in the functionally more important parts or proteins is slower. Insulin, for example, is formed from a proinsulin molecule by excising a central region (Figure 7.12); the central region is then discarded, and its sequence is probably therefore less crucial than that of the outlying parts which form the final insulin protein. The central part evolves six times more rapidly than the outlying parts. The same result has been found by comparing evolutionary rates in the active sites and in other regions of enzymes; the surface of a hemoglobin, for example, may be functionally less important than the heme pocket, which contains the active site. The evolutionary rate is about 10 times faster in the surface region (Table 7.6).

7.15 Both natural selection and neutral drift can explain the trend

The neutral explanation for the relationship between evolutionary rate and functional constraint is as follows. In the active site of an enzyme, an amino acid change will probably change the enzyme's activity. Because the enzyme is relatively well adapted, the change is likely to be for the worse. It may well spoil the enzyme's function. In the other parts of the molecule it may matter less what amino acid occupies a site, and a change is more likely to be neutral. The proportion of mutations that are neutral will be lower for the functionally constrained regions; therefore, if the total mutation rate is similar throughout the enzyme, the number of neutral mutations will be lower in the active site. The evolutionary rate will then be lower there too. The fit between prediction and fact is good and has been confirmed repeatedly. We shall see further examples, from DNA, later.

And what is the selective explanation? It has been argued that if all evolution is by natural selection there should either be no relationship between functional constraint and evolutionary rate, or the opposite relationship from that which is found. Natural selection would arguably be more important in the functionally more important parts of proteins, and evolutionary rates should therefore be higher there. The facts show the opposite. However, as Clarke pointed out, natural selection can explain the facts. The argument is often put in terms of a model that Fisher presented for adaptive evolution; the model illustrates why natural selection favors small changes more often than large changes (Box 7.3).

Mutations in a protein's active site will tend to have large effects; mutations in the outlying regions will have smaller effects. A change in amino acid in the active site is a virtual macromutation, which will almost always make things worse; natural selection will only rarely favor amino acid changes. But a similar change in the less functionally constrained parts may have more chance of being a small "fine-tuning" improvement which natural selection would favor. Selection will then more often favor changes in the less constrained regions of molecules, because there is more scope for fine-tuning in those parts. We could make an analogy with a radio, in which the more and less constrained regions of a protein are like two tuning knobs, one for fine adjustments and the other for large movements. Once a radio station has been located, we make more use of the fine adjustment knob, to keep track with the signal. Likewise, changes outside the active site are more frequent in evolution. Such is the selective explanation of the slower evolutionary rate in functionally constrained regions of molecules.

Kimura rejects the argument as "completely false." It ignores the probability that a mutation, even if it is advantageous, is eventually fixed. The chance of fixation of a favorable mutation is, as we have seen, about $2s$ and is therefore lower for a mutation of smaller s. The fine-tuning mutants will,

Box 7.3 Fisher's model of adaptive evolution.

Fisher's model is of general importance, but it is convenient to introduce it here. For almost any character, there will be a relationship between the form of the character and its fitness, as in Figure B7.4.

For the point that follows it does not matter whether the graph has multiple peaks, or what the detailed shape of the slope is: all that matters is that there is at least one fitness peak, with some "hills"

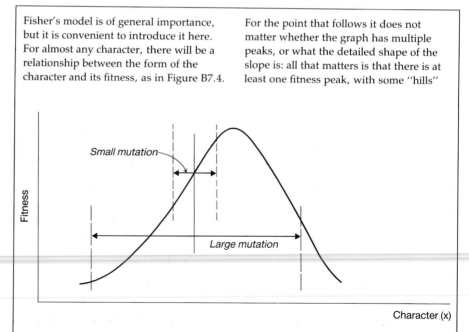

Figure B7.4 A general model of adaptation. For some trait (x), the fitness of an individual has an optimum at a certain value of x, and declines away from that point. There is then a "hill" of fitness values. A mutation which changes the value of x also changes its bearer's fitness.

continued on page 171

leading up to it from all sides. At the peak, the organism is better adapted than any slightly different form. The abstract idea can apply to any kind of character: a morphologic character like bone diameter, or a molecular one, such as a chemical property of an enzyme. We can assume that, in nature, the members of a species will be somewhere near the peak, because organisms, if not perfectly adapted, are at least fairly well adapted.

Mutations will arise at random with respect to the direction of improved adaptation. A mutation is equally likely in any direction from where the species is; one is equally likely "uphill" as "downhill." The important relationship is between the size of a mutation and the chance that it is an improvement. Consider first a large mutation. If it is directed downhill, it will make its bearer less adapted. More interestingly, if it is in the uphill direction, if it is a large enough mutation it will overshoot the peak and make the organism less adapted by taking it further down the hill on the other side. This is a well known argument, showing that adaptations will not usually evolve by macromutations: if you make a large enough random change in a well adjusted machine, the change will be for the worse. Now consider a small mutation. If it is downhill it will again make its bearer less adapted; but if it is uphill it will probably be an improvement. The probability depends on how near the peak the species already is. If it is far off, quite a large mutation uphill can be favored; but if it is close to the peak the mutation will have to be very small. During evolution, a species will wander up and down the hills around a peak because the peak will move as the environment changes. Fisher concluded

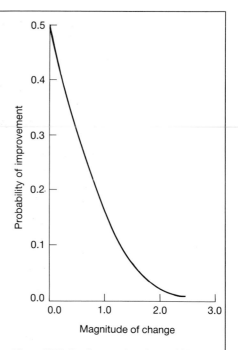

Figure B7.5 Smaller mutations have a higher chance of being selectively advantageous. Macromutations are almost never advantageous. Most evolutionary change, therefore, is achieved by substituting mutations of small effect.

that, in the limit of mutations of indefinitely small size, there is a 50% chance that they will be favorable; and as the magnitude of the mutation increases the chance it is favorable decreases to zero (Figure B7.5). If a character is related to its bearer's fitness as in Figure B7.4, smaller mutations are more likely to be advantageous.

Further reading. Fisher (1930) and Kimura (1983).

Box 7.3 (*continued*) Fisher's model of adaptive evolution.

Kimura reasons, have lower selective advantage, and the larger proportion of selectively advantageous small mutations will be canceled by their higher chance of random loss. However, things are not as clear as Kimura suggests. His argument works for a unique mutation. But if a favorable mutation occurs recurrently, it will eventually be fixed even if its chance of survival is low each time. Then the recurrent, advantageous mutations in unconstrained regions will eventually be fixed, and the recurrent disruptive changes in active sites will always be rejected. The rate of evolution will then be higher in the functionally less constrained parts of molecules, as is observed. The selective explanation is quite plausible. Both theories can explain the well-documented relationship between functional constraint and evolutionary rate.

7.16 Silent sites in the DNA evolve more rapidly than replacement sites

The same relationships between functional constraint and evolutionary rate have been found, and the same explanations defended, for DNA as for proteins. Two properties of DNA sequences are particularly interesting: the relationship between "silent" (or "synonymous") and replacement changes in the third codon position, and the evolutionary rate of *pseudogenes*. A pseudogene is a region of a DNA molecule that clearly resembles the sequence of a known gene, but differs from it in some crucial respect and probably has no function (Figure 7.13). Some pseudogenes, for example, cannot be transcribed, because they lack promotors and introns. (Promotors and introns are sequences that are needed for transcription, but are removed from the mRNA before it is translated into a protein. The pseudogene therefore may have originated in a mistaken reverse transcription of processed mRNA into the DNA.) Pseudogenes, once formed, are probably under little or no constraint and mutations will accumulate by neutral drift at the rate at which they arise. They will show pure neutral evolution in the pan-neutralist sense that all mutations are neutral. The neutral theory predicts that pseudogenes should evolve rapidly. And they do (Table 7.7). It is tempting to use rates of pseudogene evolution as direct estimates of the mutation rate, though there are grounds for caution, as we shall see.

A selectionist could, I suppose, argue that the fast evolution of pseudo-

	Met	Ala	Thr	Lys	Ala	Val	Cys	Val	Leu	Lys	Gly	Asp	Gly	Pro	Val
SOD-1	ATG	GCG	ACG	AAG	GCC	GTG	TGC	GTG	CTG	AAG	GGC	GAC	GGC	CCA	GTG
ψ69.1	ATA	ATG	ATG	AAG	GTC	ATG	TAC	ATG	TTG	AAG	GGC	CAG	AGC	CCG	GTG
	Ile	Met	Met	Lys	Val	Met	Tyr	Met	Leu	Lys	Gly	Gln	Ser	Pro	Val

	Gln	Gly	Ile	Ile	Asn	Phe	Glu	Gln	Lys			Glu	Ser	Asn	Gly
SOD-1	CAG	GGC	ATC	ATC	AAT	TTC	GAC	CAG	AAG	G	intron	AA	AGT	AAT	GGA
ψ69.1	CAG	GCG	A C	ATC	CAT	TT	GAG	CAG	AAG	G		AA		AAT	GAA
	Gln	Val	Thr	Ser	Ile										

	Pro	Val	Lys	Val	Trp	Gly	Ser	Ile	Lys	Gly	Leu	Thr	Glu	Gly	Leu
SOD-1	CCA	GTG	AAG	GTG	TGG	GGA	A GC	ATT	AAA	GGA	CTG	ACT	GAA	GGC	CTG
ψ69.1	CCA	TTT	ATG	GTG	T C	AGA	ATGC	ATT	ACA	GGA	TTG	ACT	GAA	CGC	CAG

	His	Gly	Phe	His	Val	His	Glu	Phe	Gly	Asp	Asn	Thr	Ala		
SOD-1	CAT	GGA	TTC	CAT	GTT	CAT	GAG	TTT	GGA	GAT	AAT	ACA	GCA	intron	
ψ69.1	CAC	AGA	TTC	CAT	GTT	CAT	CAG	TTT	GGA	G T	A T	AAC	ACA		

Figure 7.13 Sequences of part of a real human gene, superoxide dismutase (*SOD*-1) and the same part of a pseudogene (ψ69.1). The introns in *SOD*-1 have been excised, and various base changes have taken place, in the pseudogene. From Danciger *et al.* (1986).

Table 7.7 Pseudogenes evolve at about the same rate as silent base changes. Rates are expressed in numbers of base changes per 10^9 years. The comparisons are for various genes and pseudogenes in the globin gene family. From Li *et al.* (1987)

Species pair	Divergence time (Myr)	Evolutionary rate	
		Pseudogenes	Silent sites
Human vs chimpanzee	7	1.2	1.3
Human vs orangutan	15	1.0	2
Human vs rhesus monkey	25	1.5	2.2
Human vs owl monkey	35	1.6	—
Rhesus monkey vs owl monkey	35	1.9	—
Cow vs goat	17	2.7	4.2

genes is due to particularly frequent fine-tuning in these molecules, but I do not know of anyone who has said so. In the present state of knowledge, it would be a forced argument. Selectionists will be more tempted to concede pseudogenes as an exception, but an exception that tells us nothing about the reason for faster evolution in other cases of reduced functional constraint. They might add that no one has confirmed that pseudogenes are function-less—that claim is based only on negative evidence and intuition. However that may be, the neutralist explanation for rapid evolution in pseudogenes looks reasonable at present.

As we saw earlier (section 7.13), some base changes in the third position of a codon do not cause amino acid replacements, whereas others do. Silent sites should be less constrained than replacement sites, and Kimura predicted, before DNA sequences were available, that silent changes would evolve more rapidly. It has now been well confirmed that they do; evolution in silent sites runs on average at about five times the rate in replacement sites (Table 7.8). A few exceptions are now known; one is in the antigen recognition site of the

Table 7.8 Rates of evolution for meaningful (i.e. amino acid changing) and silent base changes in various genes. Rates are expressed as inferred number of base changes per 10^9 years. From Li *et al.* (1985)

Gene	Meaningful rate	Silent rate
β_2-Microglobulin	1.21	11.77
Albumin	0.92	6.72
Histone H4	0.027	6.13
Immunoglobin V_H	1.07	5.67
α-Globin	0.56	3.94
β-Globin	0.87	2.96
Parathyroid hormone	0.44	1.73
Average (38 proteins)	0.88	4.65

Note. The absolute numbers are imperfect because of difficulties in the fossil dating used in the rate calculations; but comparisons between genes are meaningful.

HLA genes (Figure 7.14). Whether the usual pattern of rapid change in silent sites is due to neutral drift or frequent selective fine-tuning has not been conclusively settled, but most people accept that neutral drift is the main force, if only because it is difficult to imagine why silent changes should be selected for so often.

If silent codon positions were selectively unconstrained, they should show pan-neutral evolution. They should then evolve in a different manner from proteins. Different proteins, as we have seen, evolve at different rates, which neutralists explain by different degrees of constraint (and selectionists by different intensities of selection). If the silent positions in the DNA encoding those same proteins are unconstrained, they should not show different evolutionary rates. They should evolve at the rate of a universal molecular clock, i.e. at their total mutation rate. The idea can be tested by comparing the rate of change in the silent sites of different genes, to see whether the evolutionary rate is constant in silent sites. Initially, Miyata and others suggested that the silent rate is indeed constant, but more recent evidence has not supported them (see right-hand column of Table 7.8, where the rates vary almost 10-fold between different genes); the evolutionary rate is less variable in silent than in replacement sites, suggesting a lower degree of constraint, but it still does vary. Bulmer *et al.* have also found evidence that evolutionary rates for a single gene can vary between different lineages, again suggesting they do not show pan-neutral evolution.

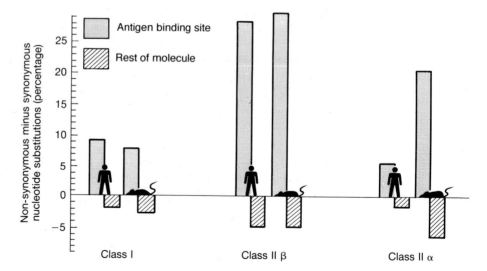

Figure 7.14 Evolutionary rates of synonymous (i.e. silent) and non-synonymous (amino acid changing) base changes in the human (HLA) and mouse (H-2) major histocompatibility loci. The *y*-axis is the non-synonymous minus the synonymous rate: the rate of synonymous changes is higher for most genes, which would give a negative value on the *y*-axis. This graph for the major histocompatibility loci breaks the regions of the molecules down into two parts: the antigen binding site (stipple shading) and the rest of the molecule (hatched shading). The regions outside the antigen binding site show the normal pattern of evolutionary rates. But in the antigen binding site, the rate of amino acid changing (non-synonymous) base substitutions is higher than the rate of synonymous changes. From Potts and Wakeland (1990); data from Hughes and Nei (1988, 1989).

There is other, more direct, evidence bearing on selective constraints in silent base sites. The evidence is of two kinds. The first concerns the relative rates of evolution in pseudogenes and silent positions. It was initially thought that pseudogenes evolve about twice as fast as silent codon positions, but the more recent evidence in Table 7.7 suggests that evolution is just as fast in silent codon positions as it is in pseudogenes. We can provisionally conclude that pseudogenes and silent sites evolve at about the same rate, though further evidence may suggest otherwise. The question is interesting, because when it was thought that pseudogenes evolve faster than any other DNA, it was tempting to argue that their evolution was completely unconstrained; their rate of evolution would then be a direct estimate of the total mutation rate. But now that we think pseudogenes evolve at the same rate as silent base changes, it is argued that evolution is probably to some extent constrained even in pseudogenes. They then should not be used to estimate the total mutation rate. The argument works because we have independent evidence that evolution is constrained in silent sites: this evidence is the topic of the next section.

7.17 Codon usages are biased

Biases in codon usage provide the second kind of evidence for constraints on silent sites. If all the silent alternative codons were functionally equivalent, we should expect only random variation in the frequency of those codons in a species. In fact, there are consistent biases. Table 7.9 gives evidence for the six arginine codons in *Homo sapiens*, *Drosophila*, and *Escherichia coli*, from the work of Grantham. AGG is the most common in humans, but the rarest in *Drosophila* and *E. coli*; CGC is most common in *Drosophila*, and CGU in *E. coli*. The relative rarity of, for example, AGA compared with CGU in *E. coli* suggests that mutations from CGU to AGA tend to be selected against. The changes are not all neutral. Pan-neutralism is not valid in this case. (Notice again that we are only concerned with the possibility that some codons, out of a synonymous set, are deleterious, and selected against; the non-random

Table 7.9 Frequencies of six arginine codons in the DNA of three species. The table gives the percentages of arginine amino acids that are encoded by each of the six codons in various numbers of genes in the species. From Grantham *et al.* (1986)

Codon	Frequency (%)		
	Human	*Drosophila*	*E. coli*
AGA	22	10	1
AGG	23	6	1
CGA	10	8	4
CGC	22	49	39
CGG	14	9	4
CGU	9	18	49
Total number of arginine codons	2403	506	149
Total number of genes	195	46	149

frequency distributions could then be explained. It is not an argument against Kimura's neutral theory. Logically, it could be that the non-random frequency distributions arise because of positive selection for certain codons: the neutral theory would then be wrong, but that is another matter.)

How could selection discriminate between silent codons? It is not really known, but there are two suggestions. One is that the nucleotide sequence controls the secondary structure of the DNA molecule; changes in nucleotides might then influence the molecular shape, which could make a difference to the organism's fitness. Silent substitutions would be as likely to influence structure as replacement ones, and selection would work on both for the same reason. The second factor is transfer RNA (tRNA). The different silent codons use, to some extent, different tRNA molecules. It has been observed, in yeast and in *E. coli*, that the frequency of use of the codons in a set of synonyms is correlated with the tRNA abundances in the cell (Figure 7.15). We should expect this relationship if the frequencies of condons were set by some other factor: the tRNA abundances would then adjust (by natural selection) to the quantity needed. But the relationship could also arise if the tRNA abundances were fixed by some other constraining factor; then the mutations among the silent codons would not be neutral, as mutations to codons whose tRNA was in short supply would be selected against, relative to those with abundant tRNA. These possibilities are theoretical only. It remains a puzzle how natural selection operates on silent codon changes.

Directional mutation pressure is another possible explanation for biases in codon usage. A tendency for A and T bases to mutate more often to G or C than do G to C to A or T would result in a build up of G and C bases. Codon biases would result automatically. It has indeed been suggested that GC is more stable than AT, and favored by mutation pressure; the task is then to

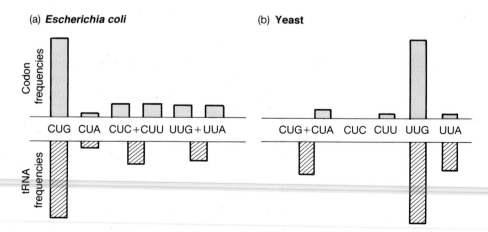

Figure 7.15 Relative frequencies of codons match tRNA abundance. (a) The stippled columns (above) are the relative frequencies of six leucine codons in *E. coli*; the hatched columns (below) are the relative frequencies of the corresponding tRNA molecules in the cell. The two sets of codons joined by a + sign are recognized by a single tRNA molecule. (b) Same relationship, but in yeast. Notice the different bias in codon usage in the two species, which reillustrates the point of Table 7.9. From Kimura (1983).

apply these chemical mechanisms to explain the differences between species in their codon biases (Table 7.9). In summary, the functionally more important parts of proteins (such as the active sites of enzymes) and DNA (such as replacement rather than silent sites) evolve more slowly than the less important parts. There are two explanations for the trend. It may arise because a higher proportion of mutations are neutral in the less important parts; this is the generally accepted explanation for the rapid rate of evolution for silent base changes. However, even these sites are not completely unconstrained; different taxonomic groups show different biases in codon usage and have different rates of evolution in silent sites. The second explanation for the trend is that natural selection works more readily on small changes than large, and a change by one amino acid is more likely to be a small change in a less important part of a molecule. It is not generally agreed whether selection or drift is the reason why the functionally important parts of proteins evolve slowly.

7.18 DNA sequences provide strong evidence for natural selection on protein structure

When, in chapter 4, we considered the evidence for biological variation, we noticed that many DNA sequence variants can be uncovered if a single gel electrophoretic class of a protein is sequenced at the DNA level (section 4.5, p. 72). This observation has important implications for molecular evolution. At the alcohol dehydrogenase (Adh) locus in the fruitfly, there are two electrophoretic classes (fast and slow). When Kreitman sequenced proteins from the two classes, he found that the amino acid sequence of each was uniform, but there were many variants in silent third codon positions. The combination of a fixed amino acid sequence and variable silent sites provides, as Lewontin has particularly emphasized, evidence that natural selection has been operating to maintain the enzyme structure.

There are two possible reasons why the enzyme sequence, at the amino acid level, should be fixed within each gel electrophoretic class. One is "identity by descent": all the copies of the Adh-f allele, for example, may be descended from an ancestral mutation, which had that sequence and has been passively passed from generation to generation. Eventually another mutation may arise and the Adh-f allele will then become two alleles; the constant sequence within a population merely indicates that not enough time has passed for such a mutation to occur. Alternatively, the Adh-f copies may all have the same sequence because the sequence is maintained by natural selection; when a mutation arises, selection removes it. The observed variability distinguishes between these two hypotheses. The variability in the silent base sites means that there has been time for mutations to arise in the molecule. If mutations have arisen in silent sites, they will surely have arisen in replacement sites too. Therefore, we can reason that the identity in amino acid sequence is unlikely to be identity by descent. Mutations in replacement sites have presumably not been retained because natural selection eliminated them.

If it had turned out that the Adh-f allele was fixed for one DNA sequence at all sites, replacement and silent, we could not know whether the uniformity was due to selection or identity by descent. We should be in the same position as we were when we only had the gel electrophoretic evidence.

Maybe the fixity would mean only that no mutations had occurred. The DNA sequences thus provide evidence for selection that could not have been obtained with amino acid sequences alone.

The absence of amino acid sequence variation within the Adh-f (and Adh-s) allelic class is particularly striking because 30% of the enzyme is made up of isoleucine and valine, which are biochemically very similar (and electrophoretically indistinguishable); a neutralist might have predicted that some of the valines could be changed to isoleucines, or vice versa. The only amino acid sequence variant is the one that causes the Adh-f/Adh-s polymorphism; and because there is abundant evidence that natural selection operates on that change, Lewontin concludes that there are actually no neutral amino acid sites in the 255-amino acid alcohol dehydrogenase enzyme of the fruitfly. Interestingly, that means that we could almost construct Figure 7.3 for alcohol dehydrogenase at the amino acid level: and the graph would be like Figure 7.3a. Natural selection is powerfully maintaining the amino acid sequence, while the silent base changes are probably neutral.

7.19 The evolutionary rate of a molecule is correlated with its heterozygosity

In the neutral theory, the explanation of both evolutionary change and of variability within a population is the same. All polymorphism is "transient" polymorphism; it exists because alleles are drifting up towards fixation or down to be lost. Selectionists, by contrast, usually offer different explanations for polymorphism and for evolutionary change. Polymorphism is explained by a balance of selective forces, such that the polymorphism is actively maintained; it is stable. Evolutionary change takes place when the selective forces on the population alter; the newly favored alleles then rapidly increase in frequency. Changes in selective forces, according to a common version of the selective position, are so rare that only a minority of polymorphisms will be transient. (Other selectionists could argue that selection in fact produces large amounts of transient polymorphism. The direction of selection may change, continually and rapidly, as happens in the parasitic theory of sex (see chapter 11).)

The unified explanation of both polymorphism and evolution in the neutral theory makes possible another test. The neutral theory explains the different evolutionary rate of different genes by their different neutral mutation rate: some genes are more selectively constrained than others; these more constrained genes have a lower neutral mutation rate, and evolve more slowly. Now, the neutral mutation rate not only controls the rate of evolution; it also controls heterozygosity (section 6.7, p. 136): genes with a higher neutral mutation rate have higher heterozygosity. If the neutral theory is correct, therefore, the evolutionary rate and the heterozygosity of genes should be correlated. It is less clear what the selectionist would predict. A correlation between heterozygosity and evolutionary rate for different molecules is certainly not a strong selective prediction. If, for example, polymorphism and evolution are produced by different processes, they would be correlated only by coincidence. Some other version of selectionism, perhaps, could be constructed to explain a correlation between heterozygosity and rate of evolution, but the correlation is not so automatically predicted as it is by the neutral theory.

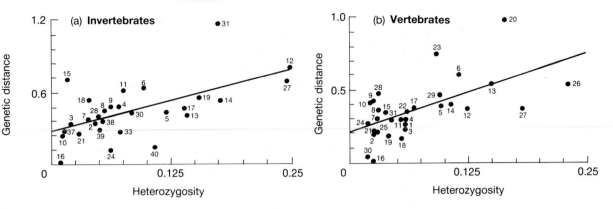

Figure 7.16 Proteins that have higher heterozygosities evolve faster. Each graph shows genetic distance on the *y*-axis and heterozygosity on the *x*-axis; each point is for a pair of species. Genetic distance measures the amount of genetic difference between the species. For formal definition, see Box 16.3, p. 436. (a) Invertebrate interspecies comparisons; (b) vertebrate interspecies comparisons. From Ward and Skibinski (1985).

Ward and Skibinski examined the relation for proteins in 1985. They broke down the published data into five classes: *Drosophila*, invertebrates (interspecific or intraspecific comparisons), and vertebrates (interspecific or intraspecific comparisons). In all five there was a positive correlation (Figure 7.16 shows the invertebrate interspecies and vertebrate interspecies comparisons). Some neutral models predict the exact relationships better than others, but the general result clearly fits the neutralist prediction.

More recently, in 1987, Hudson *et al.* carried out a smaller scale but better controlled version of the same test. They looked at the variability within a *Drosophila melanogaster* population, and the difference between two *Drosophila* species, for two regions of the DNA: the alcohol dehydrogenase gene and the 5′ flanking region next to the gene. In this case, the results do not fit the neutral prediction (Table 7.10). The two regions show similar levels of variability within a population, which the neutralist would explain by similar

Table 7.10 Variation and rate of evolution for two regions of the *Drosophila* genome. Hudson *et al.* (1987) measured the polymorphism within a *D. melanogaster* population by sequencing the DNA of 81 females; they measured the rate of evolution by comparing a single genome of *D. melanogaster* with a single genome of the closely related *D. sechellia*. Length is the total length of the region; number of sites compared is the number of DNA sites actually sequenced

	Length	Number of sites compared	Number of sites variable
5′ Region			
Within species	4000	414	9
Between species	4052	4052	210
Adh			
Within species	900	79	8
Between species	900	324	18

degrees of selective constraint. This, however, is contradicted by the relative rate of evolution in the two regions: the 5' flanking region has evolved four times as fast as the *Adh* gene itself. Thus, in alcohol dehydrogenase, variability and evolutionary rate are not correlated.

The neutral theory predicts that genes that are more variable in a population will evolve more rapidly; natural selection predicts no such general relationship. In Ward and Skibinski's comparison of many proteins, evolutionary rate and heterozygosity were correlated; but there was considerable scatter in the results. Hudson *et al.* carried out a more controlled comparison for the alcohol dehydrogenase gene in fruitflies and found similar levels of variation but different evolutionary rates in different regions of the gene. The neutral prediction was supported by one study, but not the other.

7.20 The analysis of DNA sequences is only the most recent stage in a long controversy

The controversy between the selective and the neutral theories has moved through several stages. The first stage was concerned with the amino acid sequences of proteins. Kimura suggested that the high rate, and constancy, of protein evolution, and high levels of heterozygosity, could not be explained by natural selection. His argument proved unconvincing and most biologists accept that natural selection could explain these observations. Moreover, further evidence suggests that the levels of heterozygosity are too low for the neutral theory and the molecular clock for proteins ticks according to generational, rather than absolute, time.

More recently, the availability of base sequences for DNA has added another dimension to all these tests. DNA contains both silent and replacement base positions. It seemed likely, or at least possible, that silent base changes—which do not alter the amino acid—would be neutral. Silent changes should have a higher rate of evolution than changes at replacement sites, and it has been well confirmed that they do (with a few interesting exceptions, such as the HLA genes). Moreover, the evolutionary rate for silent sites should be proportional to generation time; and the evidence suggests it is. Silent sites are not pan-neutral in their evolution, however; there is evidence for constraints from the different patterns they show in different taxonomic groups. The DNA sequence evidence has also provided strong evidence that selection maintains the amino acid sequence of enzymes.

The neutralist–selectionist controversy is thus not settled. However, it is unlikely that neutral drift is as important, and selection as unimportant, in protein evolution as Kimura originally suggested. His case against selection has been rebutted and the facts do not comfortably fit the neutral predictions. For silent sites in the DNA, however, the neutral theory is widely accepted: the theory made clear predictions, many years before the facts were discovered, and the predictions have turned out to be correct. The neutral theory therefore probably applies better to some types, or parts, of molecules than to others. But there remains ample room for disagreement about which molecular evolutionary changes were driven by selection, and which by neutral drift.

7.21 Summary

1 The neutral theory of molecular evolution suggests that molecular evolution is mainly due to neutral drift. The mutations that have been

substituted in evolution were selectively neutral with respect to the genes they replaced. The alternative is that molecular evolution is mainly driven by natural selection.

2 Four main observations were originally interpreted in favor of the neutral theory: molecular evolution has a rapid rate, its rate has a clock-like constancy, it is more rapid in functionally less constrained parts of molecules, and natural populations are highly polymorphic.

3 Kimura argued that the high rate of evolution, and the high degree of variability of proteins, would, if caused by natural selection, impose a high genetic load.

4 Neutral drift can drive high rates of evolution, and maintain high levels of variability, without imposing a genetic load.

5 Selection can operate without producing impossible genetic loads, and Kimura's original case for the neutral theory is no longer convincing.

6 The constant rate of molecular evolution gives rise to a "molecular clock".

7 Neutral drift should drive evolution at a stochastically constant rate; Kimura pointed to the contrast between the uneven rate of morphologic evolution and the constant rate of molecular evolution and argued that natural selection would not drive molecular evolution at a constant rate.

8 The molecular clock for proteins ticks over according to absolute time rather than generational time. But for silent changes in DNA, lineages with shorter generation times probably evolve faster. Neutral drift should cause the molecular clock to run according to generational, not absolute, time.

9 The neutral theory explains the higher evolutionary rate of functionally less constrained regions of proteins by the greater chance that a mutation there will be neutral.

10 Selectionists explain the higher evolutionary rate of functionally less constrained regions of proteins by the greater chance that a mutation there will be a small, rather than a large, change.

11 The four properties observed in the evolution of proteins (rate of evolution, constancy of rate of evolution, level of variability, relationship between functional constraint and evolutionary rate) have also been seen in DNA.

12 Pseudogenes and silent changes in third codon positions may be relatively functionally unconstrained. These parts of the DNA evolve faster than do the first two positions in codons, and meaningful third base changes. Neutralists attribute this high rate of evolution to enhanced neutral drift.

13 For amino acids encoded by more than one codon, there are consistent biases in the frequencies of the codons. Changes between the silent codons are therefore not completely unconstrained.

14 The neutral theory predicts a positive relationship between the degree of variability of a molecule and its rate of evolution. Genes with higher amounts of variation should evolve at a higher rate. Comparisons between proteins broadly support the prediction. A finer comparison for the DNA within the *Adh* gene and its surrounding regions contradicts it.

7.22 Further reading   Kimura (1983) is the principal reference on the neutral theory. He has written shorter introduction too (e.g. 1979, 1991). Gillespie (1992) is a modern review; Lewontin (1974) is less modern, though still an admirable introduction to the general issue. Li and Grauer (1991, chapter 4) introduce some of the material. The volumes edited by Selander *et al.* (1991), Clegg and O'Brien (1990), Nei and Koehn (1983), and the issue (entitled "The evolution of DNA sequences") of the *Philosophical Transactions of the Royal Society of London B*, vol. 312, pp. 189–354 (1986), all contain relevant papers. See also Nei (1987).

More specific references are as follows. For the general idea of genetic loads, and the cost of selection, a text book of population genetics, such as Crow and Kimura (1970) or Wallace (1981), should be consulted. Kimura (1983) and Lewontin (1974) deal with these topics in relation to neutralism. Haldane (1957) is the classic paper on the cost of selection. For threshold models, see Lewontin (1974) and Wills (1981).

The variability of rates provides a potentially important test, but all earlier attempts were unsettled by Bulmer (1988); see also Bulmer *et al.* (1991). Scherer (1990) reviews the molecular clock; and Czelusniak *et al.* (1982) and Wilson *et al.* (1977) contain other material discussed in the text.

Nevo (1988) reviews the natural levels of polymorphism detected by gel electrophoresis. On the controversy about the cause of polymorphism, see Ford (1975), Wills (1981), and Wright (1978). On biases in codon usage, see Grantham *et al.* (1986) and Bulmer (1987), as well as the general references given above. See Clark and Kao (1991) for another example like Figure 7.14. On the inference about alcohol dehydrogenase in fruitflies, see Lewontin (1985a,b, 1986), Kreitman (1983), and McDonald and Kreitman (1991). On the relation between polymorphism and evolutionary rate, see Kreitman (1987).

Two-locus and multi-locus population genetics

8.1 Mimicry in *Papilio* is controlled by more than one genetic locus

The swallowtails are a group of butterflies with a global distribution, and *Papilio* is the largest of the genera; their most striking character is a "tail" on the hindwing. They come in many colors—gorgeous greens, subtle shades of reds and orange, and marbled patterns in white and gray—but the most common type has stripes of black and yellow; the North American tiger swallowtail *Papilio glaucus* is easy to recognize by its tiger stripes, as it flutters through woodland lanes or humid valleys. Or rather, *most* tiger swallowtails are easy to recognize in this way. In part of the species' range (roughly, to the south-east of a line from Massachusetts to south Minnesota and from east Colorado to the Gulf Coast) the standard form of the tiger swallowtail lives alongside another form of the same species. This second form is black, with red spots on its hindwings, and is called *nigra*; it is only found in females. The *nigra* form is not poisonous, but mimics another species, the pipevine swallowtail *Battus philenor* which is poisonous. *Nigra*'s geographic distribution fits that of the pipevine swallowtail, and *nigra* is well protected there from predatory birds that have learned by stomach-churning experience not to eat butterflies looking like pipevine swallowtails. The tiger swallowtail, therefore, has a *mimetic polymorphism*. It has both the standard non-mimetic tiger morph of yellow and black stripes, and a black mimetic morph.

The tiger swallowtail *P. glaucus* comes to look almost simple when compared with the amazing array of female forms in the species *Papilio memnon*. *Papilio memnon* lives in the Malay archipelago and Indonesia; its male is again non-mimetic, though its color is deep blue rather than yellow and black stripes. However, instead of one mimetic female form, *P. memnon* females come in almost numberless variety. Their forewings show different geometric patterns of black and white; their hindwings, as well as varying in shape, can be colored in yellow, orange, or blood-red, and may or may not have a bright white "sunspot"; some have tails, others do not; the abdomen varies in color; and a spot at the butterfly's "shoulder" (i.e. at the base of the forewing near the head) called the epaulette, may be present in various shades of red. Clarke and Sheppard, who have studied the species, suggest that each female form mimics a different model (Figure 8.1 shows two examples: notice that one has a tail and the other does not). Their evidence is not strong, as it comes only from the geographic ranges of mimic and model, and from superficial similarity of appearance (which is not exact in all cases). Good evidence for mimicry requires experimental demonstration that birds that have learned to avoid the model will also then avoid the mimic; this has been done for the *nigra* form of *P. glaucus*, but not for *P. memnon*. However,

Figure 8.1 (b), (d), and (e) are three forms of *Papilio memnon*. (a) and (c) are the two other species, which are the models for forms (b) and (d). (e) is the rare, probably recombinant, form of *P. memnon* called *anura*, from Java. It is like the normal mimetic form called *achates* (b), but it lacks *achates*' tail. It may be a recombinant between *achates* (b) and a tailless form such as (d). From Clarke *et al.* (1968) and Clarke and Sheppard (1969).

we can accept as a working hypothesis that the apparently mimetic morphs of *P. memnon* indeed are mimetic. (*Papilus memnon* has yet further, non-mimetic forms too, but they are not essential here.)

Clarke and Sheppard were interested in the genetic control of this complex mimetic polymorphism. Crosses between the various morphs initially suggested that there is a single genetic locus with many alleles. When two forms are crossed, the offspring usually either all resemble one of the parents, or contain a mixture of the two parental types, as will happen with one locus and simple dominance relations among alleles. For instance, if one morph has genotype A_1A_1, another A_1A_2, and A_2 is dominant to A_1 then an $A_1A_1 \times A_1A_2$ cross produces the same two classes of offspring (A_1A_1 and A_1A_2) as were present in the parents.

However, the genetic story soon grew more complicated. In addition to the mimetic and non-mimetic morphs of *P. memnon*, all of which exist in reasonable frequencies in nature, some much rarer morphs have been found. An example, in Java, is the rare morph called *anura* (Figure 8.1e). A specimen found in Borneo was sent to Clarke and Sheppard in Liverpool. When it was crossed with a known *P. memnon* morph, it behaved like another allelic form of the mimicry "locus"; but a closer look at *anura* suggests a different interpretation. *Anura*'s morphology mixes patterns from two of the common morphs: it has the wing color pattern of the morph *achates* (Figure 8.1a,b), but it lacks *achates*' tail.

Clarke and Sheppard's interpretation is that *anura* is not actually an allelic variant, but a recombinant; and the mimetic patterns of *P. memnon* are not controlled by one locus but by a whole set of loci. If *anura* is a recombinant, then there must be at least one locus (call it T) controlling the presence (allele T_+) or absence (T_-) of a tail and at least one other locus (C) controlling the color patterns (C_1 for *achates*, and C_2, C_3, etc. alleles for other color morphs). *Achates* would have a genotype made up of one or two sets (depending on whether the alleles are dominant) of the two locus genotype T_+C_1, and *anura* would have T_-C_1, after recombination between a tailless morph and *achates*. The loci in question are so tightly linked that these recombinants practically never arise in the laboratory—which is why the different multi-locus genotypes appear, when crossed, to segregate like single locus genotypes. We can predict that if more than one locus really is involved, a sufficiently large number of crosses should be able to break one of the "alleles" (such as the *anura* "allele") into several real combinations of alleles at several loci. Darlington and Mather defined a set of genes that have so tightly linked loci that they behave like a single locus in a breeding experiment—as a *supergene*.

From *anura* alone, it seemed that there must be at least two loci controlling the mimetic polymorphism of *P. memnon*; but other rare types have also been found. Some, for example, combine the forewing color of one morph and the hindwing pattern of another, suggesting that there are separate loci controlling the color of fore- and hindwings. When all the inferred recombinants are considered together, there appear to be at least five loci in the mimicry supergene: T, W, F, E, and B. They control presence or absence of tail, hindwing pattern, forewing pattern, epaulette color, and body color, respectively. *Anura* is a recombinant between the T locus and the other four. The common morphs, which mimic natural models, should each consist of a particular set of alleles at the five loci. The morph mimicking model species number 1, for example, might have genotype $T_+W_1F_1E_1B_1/T_+W_1F_1E_1B_1$, and another morph (mimicking a second model) might have $T_-W_2F_2E_2B_2/T_-W_2F_2E_2B_2$ or $T_-W_2F_2E_2B_2/T_+W_1F_1E_1B_1$. The recombinant genotypes, such as $T_+W_1F_1E_2B_2$, do not exist naturally, except as very rare forms like *anura*.

The point to remember is that each of the morphs of *Papilio memnon* is thought to be controlled by a multi-locus genotype. It is not like the camouflage polymorphism in the peppered moth (section 5.6, p. 95), in which the different morphs are controlled by genotypes at one locus. A whole set of one-locus genotypes is needed to produce each of the swallow-tail butterfly morphs. The genetics are not definitively confirmed, but we can use the idea for purposes of discussion.

8.2 The genotypes at different loci in *Papilio memnon* are coadapted

How will natural selection act on a rare recombinant morph of *Papilio memnon*, such as *anura* in Java? Successful mimicry requires as complete a resemblance as possible between a mimic and its model. A potential mimic that mixes the patterns needed to mimic two species will mimic neither as successfully as a form that resembles one model in all respects. It will probably be selected against. *Anura* has the color pattern of *achates*, but will not mimic the model species of *achates* (Figure 8.1a) because it lacks a "tail" on its hindwings. The models of tailless morphs, in turn, have different color patterns, and *anura* will not mimic them either.

In general, natural selection will act against any recombinants between the mimetic five-locus genotypes. A five-locus genotype that mimics one model species in all five respects will be favored; but a swallowtail collage, mimicking one model in three aspects and another in two different aspects, will look like neither and be selected against. The genes at the five loci in this situation are said to be *coadapted*, or to show *coadaptation*, or to be part of a *coadapted gene complex*. Coadaptation means that a gene (or genotype), like T_+ (or T_+/T_+) is favored by selection if it is in the same individual as a particular gene (or genotype), such as W_1 (or W_1/W_1) at another locus but selected against when combined with other genes (or genotypes), such as W_2 (or W_2/W_2) at that locus: selection favors $T_+W_1F_1E_1B_1/T_+W_1F_1E_1B_1$ and $T_-W_2F_2E_2B_2/T_-W_2F_2E_2B_2$ individuals, but (if the alleles with the 2-subscript are dominant) works against $T_+W_2F_2E_2B_2/T_+W_1F_1E_1B_1$ individuals. It has not been empirically confirmed that selection works against the recombinant forms of *P. memnon*, but the argument is quite convincing. (The terms "supergene" and "coadapted gene complex" are related. Coadaptation refers to the fitness interactions among genes, and supergene to their linkage relationships. Extreme coadaptation is likely to lead, over evolutionary time, to the evolution of a supergene; but physical linkage and fitness relationships are conceptually distinct.)

8.3 Mimicry in *Heliconius* is controlled by more than one gene but not by a supergene

The butterflies of the genus *Heliconius* make an interesting comparison with *Papilio memnon*. In Latin America there are two species of *Heliconius*, *H. melpomene* and *H. erato*, both of which have multiple mimetic forms. The color patterns are again controlled by many loci: 15 in *H. erato* and 12 in *H. melpomene*. However, in both species the loci are scattered at random among the chromosomes rather than being tightly linked in a supergene. When two morphs of a *Heliconius* species are crossed, the offspring contain a kaleidoscopic variety of non-mimetic recombinant forms that resemble neither parent nor any known morph of the species.

Why do the genetics differ in *Heliconius* and *Papilio memnon*? The reason is probably geographic. At any one site in the range of *P. memnon*, there are often several morphs living side by side. Crosses between them will happen with high frequency naturally. But in *Heliconius* there is usually only one morph at any one place; the different morphs are geographically separated and will not interbreed in nature. The non-mimetic recombinant *Heliconius* are only generated when morphs from different places are put together in the laboratory. In *Heliconius* it does not matter if the mimicry genes are scattered around the chromosomes, because the non-mimetic progeny are not usually produced; but in *Papilio memnon* it does matter—if the mimicry genes were not linked in a supergene the recombinants would be produced, and be killed by predators.

8.4 Two-locus genetics is concerned with haplotype frequencies

The theory of population genetics for a single locus is concerned with gene frequencies; the analogous variable in two locus population genetics is *haplotype* frequency. (The term haplotype has two meanings. Here it refers to a combination of alleles at more than one locus. It is also used, in DNA sequencing, to refer to the base sequence of one of an individuals' two sets of DNA.) For two loci with two alleles each (A_1 and A_2, B_1 and B_2) there are four haplotypes, A_1B_1, A_1B_2, A_2B_1, and A_2B_2. A diploid individuals genotype will be something like A_1B_1/A_1B_2*; it has two haplotypes, one inherited from each parent, just as it is in a one-locus model it receives two genes per locus. If the A and B loci are on the same chromosome, each haplotype is a combination of genes on a chromosome; but haplotypes can also be specified for loci on different chromosomes. The frequency of a haplotype in a population can be counted as the number of gametes bearing a particular combination of genes. A haplotype can be specified for any number of loci. We shall mainly discuss two locus haplotypes, but the haplotypes in the *P. memnon* example had five loci, and the mimetic patterns of *Heliconius* are controlled by 12 or 15 gene loci. As the chapter will show, to understand the evolution of haplotype frequencies, we need some concepts that do not exist for gene frequencies. Two-locus population genetics is therefore not simply a doubled-up version of single-locus population genetics.

8.5 The frequencies of haplotypes may or may not be in linkage equilibrium

We can begin by asking a question like the one that led to the Hardy–Weinberg theorem for one locus. In the absence of selection, and in an infinite population with random mating, what will be the equilibrial frequencies of haplotypes? The question for multiple loci will lead us to another important concept, called *linkage equilibrium*.

The simplest case is for two loci with two alleles each. The crucial trick is to write the haplotype frequencies in terms of the gene frequencies at each locus, plus or minus a correction factor, called D. Let the gene frequency of A_1 be p_1, of A_2 be p_2, of B_1 be q_1, and of B_2 be q_2 in the population. Then:

---

*In this chapter, oblique strokes indicate diploid genotypes. Thus A_1/A_1 is a diploid genotype at one locus. The convention is to prevent confusion with haplotypes, which are written here without an oblique stroke, e.g. the A_1B_1 haplotype. A haplotype refers to the alleles at two (or more) loci that an individual received from one of its parents. A diploid individual has two haplotypes. Haplotypes have two different letters (for two loci), one-locus genotypes have only one letter. Diploid two-locus genotypes are also here written with an oblique stroke, e.g. A_1B_1/A_2B_2.

Haplotype	Frequency in population		
A_1B_1	a	$=$	$p_1q_1 + D$
A_1B_2	b	$=$	$p_1q_2 - D$
A_2B_1	c	$=$	$p_2q_1 - D$
A_2B_2	d	$=$	$p_2q_2 + D$

The total frequencies add up to one: $p_1q_1 + p_1q_2 + p_2q_1 + p_2q_2 = 1$, and the sum of the two $+D$ and two $-D$ factors is zero. The important term to understand is D; it is a measure of "linkage disequilibrium." Linkage equilibrium means that the alleles at the two loci are combined independently. The two B alleles would then be found with any one A allele (such as A_1) in the same frequencies as they are found in the whole population: if we take all the A_1 genes, q_1 of them are with B_1 genes and q_2 with B_2 genes; likewise, q_1 of the A_1 genes are with B_1 genes and q_2 with B_2. At linkage equilibrium, the frequency of the A_1B_1 haplotype is p_1q_1. D measures the deviation from linkage equilibrium. If $D > 0$ there are an excess of A_1B_1 (and A_2B_2) haplotypes: A_1 is more often found with B_1 (and less often with B_2) than would be expected if they combined at random. (D is conventionally defined by adding it to the frequencies of A_1B_1 and A_2B_2 and subtracting it from those of A_1B_2 and A_2B_1; it could equally well be defined the other way round. There are other possible measures of the non-random combination of genes, but all the points of principle can be made with D.)

Papilio memnon is an example of high linkage disequilibrium. If Clarke and Sheppard are correct, the allele T_- is almost always combined with the other alleles W_2, F_2, E_2, and B_2 rather than with W_1, or W_3, or W_4 (and equivalent alleles at the other loci). There is a large excess of the haplotypes $T_+W_1F_1E_1B_1$, $T_-W_2F_2E_2B_2$, $T_-W_3F_3E_3B_3$, etc., while haplotypes such as $T_+W_2F_2E_2B_2$, $T_+W_1F_2E_2B_2$, or $T_+W_1F_1E_2B_2$ are almost absent. The linkage disequilibrium in *P. memnon*, as we have seen, is caused by selection. In this section, however, we are asking how a set of haplotype frequencies should change through time in the absence of selection.

Let us return to the haplotype A_1B_1. It has frequency defined as a in one generation. What will its frequency be in the next generation? (We can use again the notation a' as the frequency of A_1B_1 one generation on.) In the absence of selection, the frequencies of each gene will be constant, but the frequencies of the haplotypes can be altered by recombination. The frequency of A_1B_1 cannot be altered by recombination in double or single homozygotes: the number of A_1B_1 haplotypes coming out of an A_1B_1/A_1B_1 individual, or of an A_1B_1/A_1B_2 individual, is the same as the number going in, whether or not there is recombination. The frequency can only be altered by recombination in the double heterozygotes A_1B_1/A_2B_2 and A_1B_2/A_2B_1. When recombination takes place in an A_1B_1/A_2B_2 individual, the number of A_1B_1 haplotypes is decreased; when it takes place in an A_1B_2/A_2B_1 individual, the number of A_1B_1 is increased.

The frequency of A_1B_1/A_2B_2 heterozygotes in the population is ad and of A_1B_2/A_2B_1 is bc. The frequency of recombination between the two loci is defined as r. (r can theoretically have any value up to a maximum of 0.5, if the loci are on different chromosomes; it is between 0 and 0.5 for loci on the

same chromosome, depending how tightly linked they are. See chapter 2.)
So:

$$a' = a - r(ad - bc)$$

Now, the expression $(ad - bc)$ is simply equal to the linkage disequilibrium D. (This is easy to confirm by multiplying out $ad - bc$ from the definitions above of a, b, c, and d.) If $D = 0$, i.e. if the genes are randomly associated, the haplotype frequencies are constant: $a' = a$. But if there is an excess of A_1B_1 haplotypes, the excess decreases by an amount rD per generation. The same relationship holds true for any successive pair of generations. We can see what is happening graphically if we substitute for a in the equation:

$$a' = p_1q_1 + D - rD$$
$$a' - p_1q_1 = (1 - r)D$$

The difference between a and p_1q_1 is the amount of excess of the A_1B_1 haplotype (i.e. the amount by which the frequency exceeds the random frequency). It is also equal to the linkage disequilibrium $(D = a - p_1q_1)$. Therefore:

$$D' = (1 - r)D$$

In the absence of selection and in an infinite, randomly mating population, the amount of linkage disequilibrium undergoes exponential decay at a rate equal to the recombination rate between the two loci (Figure 8.2). In other words, the difference between the actual frequency of a haplotype such as A_1B_1 (a) and the random proportion (p_1q_1) decreases each generation by a factor equal to the recombination rate between the loci.

Over time, any non-random gene associations will disappear; recombination will destroy the association. The higher the rate of recombination, the more rapid the destruction. The highest possible value of r is $\frac{1}{2}$, which is true when the two loci are on different chromosomes. Gene associations persist longer for tightly linked loci on the same chromosome, as we should intuitively expect.

The equilibrial haplotype proportions have $D = 0$. At equilibrium:

Haplotype	Equilibrial frequency
A_1B_1	$a = p_1q_1$
A_1B_2	$b = p_1q_2$
A_2B_1	$c = p_2q_1$
A_2B_2	$d = p_2q_2$

These are the haplotype frequencies we have met before and called linkage equilibrium. We can now see why it is called an equilibrium. In the absence of selection, the action of recombination will drive the haplotypes to these frequencies and then keep them there.

Recombination randomizes gene associations over time. If an excess of one haplotype, such as A_1B_1 exists, recombination will tend to break it down, and A_1 will end up with B_1 and B_2 in their population proportions (q_1 and q_2) and B_1 with A_1 and A_2 in their population proportions (p_1 and p_2). At linkage equilibrium, each of the two alleles at the A locus, A_1 and A_2, are then associated with B_1 in the same proportion.

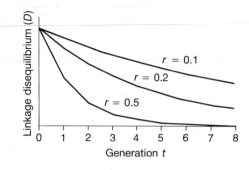

Figure 8.2 Non-random associations between alleles at different loci are measured by the degree of linkage disequilibrium (D). Recombination between the loci breaks down linkage disequilibrium, which decays at an exponential rate equal to the recombination rate between the loci.

Linkage equilibrium is, in a way, the analogy for a two-locus system of the Hardy–Weinberg equilibrium for the one-locus system. It describes the equilibrium that is reached in the absence of selection, and in an infinite, randomly mating population. Linkage equilibrium, however, is a property of haplotypes, not genotypes. A diploid individual has two haplotypes, and at the equilibrium the genotypes at each locus will be in Hardy–Weinberg proportions while the haplotypes are at linkage equilibrium. Notice also that whereas the Hardy–Weinberg equilibrium for one locus is reached instantly in one generation (section 5.3, p. 87), it takes several generations for linkage equilibrium to be reached. (The terms "linkage equilibrium" and "linkage disequilibrium" are not very satisfactory. They were first used by Lewontin and Kojima in 1960. "Linkage disequilibrium" can exist without linkage— among genes on different chromosomes—and it can also exist at equilibrium, as we shall see. It is, however, like the Hardy–Weinberg equilibrium, an equilibrium under certain specifiable conditions. The word linkage is avoided in certain other terms, such as "gametic phase equilibrium," which are also in use; but linkage disequilibrium is the most common term.)

Linkage equilibrium can also tell us whether the more complex two-locus theorem for one locus was the simplest model in single-locus population genetics; it illustrated how to construct a model with recurrence relationships for gene frequencies. The model of linkage equilibrium is, likewise, the simplest model for two loci and shows us how to construct a recurrence relationship for haplotype frequencies. Its second interest, also like the Hardy–Weinberg theorem, is that it provides a theoretical baseline telling us whether anything interesting is going on in a population. Deviations from Hardy–Weinberg proportions in a natural population suggest that selection, or non-random mating, or sampling effects may be operating. Likewise, if two loci are in linkage disequilibrium, we can also suspect that one or more of these variables are at work. If the first thing we had discovered about *P. memnon* had been its high linkage disequilibrium, we should have been led on to study how selection was operating on the loci, and perhaps ended up discovering the mimetic polymorphism. In fact, the direction of research in *P. memnon* was the other way round; but the general point, that linkage disequilibrium indicates something interesting, holds true.

Linkage equilibrium can also tell us whether the more complex two-locus theory is needed in a real case. To a rough approximation, the theory of population genetics for a single locus is satisfactory for populations in linkage equilibrium. It is when genes become non-randomly associated that a two-locus model is needed. The case we have been discussing can show why. At linkage equilibrium, A_1 and A_2 are equally associated with B_1. To understand evolution at the A locus we can then ignore the relative fitnesses of B_1 and B_2, because if B_1 is fitter than B_2, the association with B_1 will benefit A_1 and A_2 equally (and B_2 equally detract from them). But if A_1, for instance, is more associated with B_1 than is A_2 (i.e. there is linkage disequilibrium), then any advantage of B_1 over B_2 will passively give rise to an advantage to A_1. To understand frequency changes of A_1 we then need to know the relative fitnesses of B_1 and B_2, and the degree of asssociation A_1 has with them: we need a two-locus model.

8.6 The human HLA genes are a multi-locus gene system

The HLA system in humans is a set of linked genes on human chromosome 6; they control "histocompatibility" reactions. When an organ is transplanted from one individual to another, it is immunologically rejected by the recipient in a matter of days—skin grafts last about 2–15 days, for instance. The rejection implies that the immune system can distinguish between "self" and "foreign" cells, and the distinction is generally believed to be achieved by the products of the HLA genes. The best evidence for their role comes from the time course of kidney transplant rejection among siblings that either have or have not been matched for their HLA genes: for kidney transplants between HLA-matched siblings, over 90% of transplants still survive after 48 months; but among HLA-unmatched siblings, 90% survive for 4 months and only about 40% by 48 months.

The HLA system contains a number of genes (Figure 8.3); two loci, called HLA-A and HLA-B, are the main ones. Each HLA locus, in a human population, is highly polymorphic: at the B locus alone there will be perhaps 16 alleles with frequencies of 1–10% and many more rare alleles; for example, a sample of 874 individual Caucasians in France contained 31 different alleles at the B locus and another 17 alleles at the A locus. These are exceptionally

Figure 8.3 Genetic map of the human HLA loci on chromosome 6.

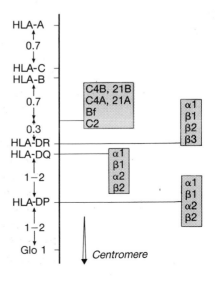

high degrees of variability. More typical loci (outside the HLA) might have one to five alleles, many less than the number found in the HLA. The reason for the high variability is uncertain; but it would allow the HLA genotype of an individual even in a large population to be unique, which is presumably important in the distinction of self from foreign cell types.

Particular HLA alleles are associated with particular diseases. The strongest association found so far is between ankylosing spondylitis and the allele B27; 90% of people with the disease have the B27 allele, against an allele frequency of only 7% in the population at large. The diversity of HLA types may reflect a history of coevolution between humans and disease agents. Disease agents may have tried to "fool" the immune system into treating the agent as part of the body, and the human population would then respond over evolutionary time by evolving new HLA alleles as new, reliable indicators of self. This would provide a further advantage to variability in the HLA loci. A heterozygous individual with two HLA proteins can compare itself with a possible invader in two ways: the invader has to match a homozygote only in one respect, but a heterozygote has to be matched in two independent respects. The HLA loci therefore probably show heterozygous advantage (section 5.11, p. 109), and the same process may have caused the exceptional pattern of evolution in silent and amino acid changing bases within codon triplets (Figure 7.14, p. 174).

The HLA system also provides examples of linkage disequilibrium. Particular combinations of genes are found in greater than random proportions. In north European populations, there is characteristically an excess of the *A1B8* haplotype. Figure 8.4 is a more general picture. It shows the linkage disequilibrium values for all combinations of *B* alleles and the allele *A1*. There could be an analogous graph for each *A* allele. In Figure 8.4, $D = 0.07$ for *A1B8*. If *A1* and *B8* combined in their population proportions, *A1B8* would have a frequency of about 0.023 (2.3%); but in fact it is found in about 9.3% of individuals ($0.093 - 0.023 = D = 0.07$). In all, the HLA system has about six clear cases of linkage disequilibrium; *A1B8* and *A3B7* are the most striking. The reason why these haplotypes are found in greater than random proportions is unknown, though it is generally believed to be due to past selection in favor of the gene combinations. But selection is not the only possible reason for linkage disequilibrium, as the next section will reveal.

8.7 Linkage disequilibrium can exist for several reasons

Recombination breaks down non-random gene associations, and yet in some cases like *P. memnon* and the HLA genes, non-random associations exist. What is causing the linkage disequilibrium? In these two cases, it is probably due to selection. If selection favors individuals with particular combinations of alleles, then linkage disequilibrium will evolve in the population. But selection is not the only process that can generate linkage disequilibrium, and a full study of a real case must examine all the possibilities.

There are three main factors other than selection. The first is that the equilibrium has not yet been reached. It takes a number of generations for recombination to do its randomizing work and, particularly for tightly linked genes (the *Papilio* supergene and the HLA loci are both tightly linked), linkage disequilibrium can persist for some time.

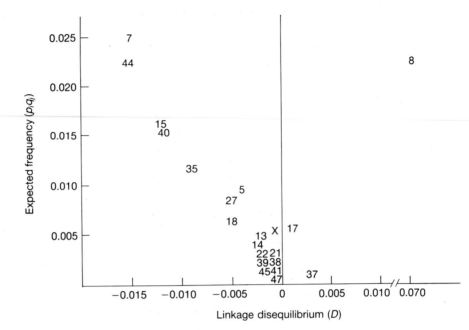

Figure 8.4 Degrees of linkage disequilibrium (D) of HLA B alleles with the $A1$ allele; an analogous graph can be drawn for every A allele. The $A1B8$ haplotype occurs at a much higher than random frequency. The y-axis is the expected frequency of the haplotype if the alleles were associated at random. Thus the observed frequency of a haplotype is the y-axis value plus (or minus) its x-axis value. From Hedrick *et al.* (1991a).

A second process is random drift in a finite population. Random processes have the interesting property that they cause persistent, not just transitory, linkage disequilibrium. If random sampling produces by chance an excess of a haplotype in a generation, linkage disequilibrium will have arisen. This is true for all four haplotypes: random sampling that produces an excess of any of them will disturb the state of linkage equilibrium. Any haplotype could be "favored" by chance, so the disequilibrium is equally likely to have $D > 0$ or $D < 0$. As a population approaches linkage equilibrium, all random fluctuations in haplotype frequencies will tend to be away from the linkage equilibrium values; if a population is well away from the point of linkage equilibrium, random sampling is equally likely to move it toward, as away from, the equilibrium. Most natural populations are probably near linkage equilibrium (see below, Figure 8.5, p. 197), and then the balance between the random creation of linkage disequilibrium and its destruction by recombination, in small enough populations, is such that linkage disequilibrium will persist.

The third possibility is non-random mating. If individuals with gene A_1 tend to mate with B_1 types rather than B_2 types, A_1B_1 haplotypes will have excess frequency over that for random mating. (The exact effect depends on whether it is homozygous A_1/A_1 individuals that mate non-randomly, or the homozygotes and the A_1/A_2 heterozygotes, and on whether they mate preferentially only with B_1/B_1 homozygotes, or with B_1/B_2 heterozygotes too. But the general effect of non-random mating on linkage disequilibrium is not complicated.)

The three processes other than selection probably account for some cases of linkage disequilibrium in nature. The process that has most interested evolutionary biologists, however, is natural selection. Let us now consider how we can model the effect of selection on haplotype frequencies.

8.8 Two-locus models of natural selection can be built

The effect of natural selection on haplotype frequencies in two-locus models, like its effect on gene frequencies in single-locus models, depends on the fitnesses of the genotypes. We have to write down the fitness of each genotype, and there are many possible ways in which it can be done. In one of the simplest two-locus models, the fitness of a two-locus genotype is the product of the fitnesses of its two single-locus genotypes; the model is realistic if the fitness effect of one locus is independent of the genotype at the other. Suppose, for example, that the A locus influences survival from 1 to 6 months, such that

Genotype	A_1/A_1	A_1/A_2	A_2/A_2
Chance of survival (to age 6 months)	w_{11}	w_{12}	w_{22}

and the B locus influences survival from ages 6 to 12 months

Genotype	B_1/B_1	B_1/B_2	B_2/B_2
Chance of survival (from 6–12 months)	x_{11}	x_{12}	x_{22}

The total chance of surviving from 1 to 12 months of age would then be the product of the two genotypes that an individual possessed because selection at 1–6 months of age is independent of selection at age 6–12 months:

	A_1/A_1	A_1/A_2	A_2/A_2
B_1/B_1	$w_{11}x_{11}$	$w_{12}x_{11}$	$w_{22}x_{11}$
B_1/B_2	$w_{11}x_{12}$	$w_{12}x_{12}$	$w_{22}x_{12}$
B_2/B_2	$w_{11}x_{22}$	$w_{12}x_{22}$	$w_{22}x_{22}$

These fitnesses are called *multiplicative*. An individual's fitness for its two genotypes is found by multiplying the fitnesses of each of its one-locus genotypes. The genotypes are independent, in the sense that the effect of one genotype on survival is independent of the other locus. An individual with the genotype A_1/A_1 has a chance of surviving from 1 to 6 months of w_{11} whether its genotype at the other locus is B_1/B_1, B_1/B_2, or B_2/B_2.

Additive fitness interactions are another possibility. Imagine now that the two loci encode two digestive enzymes. The enzyme produced by the A locus digests one kind of food, and that by the B locus another kind of food. The genotypes might enable a certain number of joules of energy to be obtained by feeding in a particular environment:

Genotype	A_1/A_1	A_1/A_2	A_2/A_2	B_1/B_1	B_1/B_2	B_2/B_2
Energy obtained	w_{11}	w_{12}	w_{22}	x_{11}	x_{12}	x_{22}

The fitness of an individual might be proportional to the total energy it obtains by feeding; its fitness for the two loci would then be found by adding those of its two genotypes together.

	A_1/A_1	A_1/A_2	A_2/A_2
B_1/B_1	$w_{11} + x_{11}$	$w_{12} + x_{11}$	$w_{22} + x_{11}$
B_1/B_2	$w_{11} + x_{12}$	$w_{12} + x_{12}$	$w_{22} + x_{12}$
B_2/B_2	$w_{11} + x_{22}$	$w_{12} + x_{22}$	$w_{22} + x_{22}$

Again, there is no interaction between the two loci. The fitness effect of a genotype at one locus does not depend on the other locus.

The next step is to derive a recurrence relationship between the frequency of a haplotype in one generation and in the next. However, the algebra is tiresome and we do not need to work through it here. In outline the procedure is the same as for the single-locus case, with the additional factor of recombination. The recurrence relationship for haplotype frequency takes account of the frequency and fitness of all the genotypes that a haplotype is found in. It also has to add and subtract the number of copies gained and lost by recombination: we multiply by $(1 - r)$ the frequency of the double heterozygotes containing the haplotype and by r that of the double heterozygote that can generate it if recombination occurs. The Mendelian rules are then applied, and the frequency of the haplotype in the next generation results.

Which selection schemes cause linkage disequilibrium? The question is important because, as we have seen, two-locus models are particularly needed when linkage disequilibrium exists. The two types of fitness (multiplicative and additive) that we have just considered both result in equilibrial haplotype frequencies with zero linkage disequilibrium. (The possibility only arises if the equilibrium has both loci polymorphic. If one gene is fixed at either locus, $D = 0$ trivially. The fitnesses, w_{11}, etc., as written above were frequency independent; a doubly heterozygous equilibrium then requires heterozygous advantage at both loci: $w_{11} < w_{12} > w_{22}$, $x_{11} < x_{12} > x_{22}$, see section 5.11, p. 109.) In conclusion, if ever linkage disequilibrium exists when fitnesses are independent at the two loci, it will decay to zero as the generations pass.

The more interesting case is when the fitnesses of the two loci interact *epistatically*. The selection in the mimetic polymorphism of *P. memnon* is epistatic. Epistatic interaction means that the fitness effects of a genotype depend on what genotype it is associated with at the other locus. We can reduce the situation in *P. memnon* to the most elementary theoretical form if we imagine there is one locus controlling whether the butterfly has a tail on its hindwing and a second controlling coloration. Let T_+ (presence of tail) be dominant to T_- (absence). At the other locus, C_1 is dominant, and C_1/C_1 and C_1/C_2 individuals have a color pattern that mimics a model species with a tail whereas C_2/C_2 individuals are colored like a model species that has no tail. The relative fitness of each genotype depends on what the genotype at the other locus is. For example, a T_+/T_+ genotype in the same butterfly as a C_2/C_2 will be less fit than a T_+/T_+ with C_1/C_1. The fitnesses can be written as follows (the simplification relative to the earlier fitness matrices arises because of dominance, and because there is one term for the fitness of both loci together rather than one term for each locus):

	T_+/T_+	T_+/T_-	T_-/T_-
C_1/C_1	w_{11}	w_{11}	w_{21}
C_1/C_2	w_{11}	w_{11}	w_{21}
C_2/C_2	w_{12}	w_{12}	w_{22}

In the case we discussed, w_{12} is the fitness of a butterfly with a tail and the color pattern of a tailless model. Therefore, $w_{12} < w_{11}$ and w_{22}. w_{21} is the fitness of a butterfly without a tail, but the color pattern of a tailed model. Therefore, $w_{21} < w_{11}$ and w_{22}. Selection now favors the T_+/T_a genotypes when they are with C_1/C_a, but not when with C_2/C_2, and T_-/T_- when it is with C_2/C_2 but not when with C_1/C_a (the a (any) implies it does not matter which gene is present, because of dominance). The fitness relationships are epistatic. There can now be a doubly polymorphic equilibrium. All four alleles will be present, and the haplotypes T_+C_1 and T_-C_2 will have "excess" frequencies. T_+C_2 and T_-C_1 haplotypes are selected against, because they often find themselves in poorly mimetic butterflies. Linkage disequilibrium ($D > 0$ in this case) exists at the equilibrium. And in general, selection can only produce linkage disequilibrium at equilibrium when the fitnesses of the genotypes at different loci interact epistatically. Not all epistatic fitness interactions generate doubly polymorphic equilibria with linkage disequilibrium; but all such equilibria do have epistatic fitnesses.

We have been discussing the different sorts of fitness interactions—multiplicative, epistatic, and so on—as properties of formal models. Real genes in real organisms will have fitness interactions too, and the more important question is what sort of interactions these are. There are cases like *Papilio* in which epistasis is present and powerful; but these may be isolated examples rather than representing a general condition. Evolutionary biologists are interested in whether fitness interactions between loci are generally epistatic, and generate strong linkage disequilibrium, or whether they are generally independent and generate linkage equilibrium. These two extremes roughly correspond to a more "holistic" and a more "atomistic" (or reductionist) school of thought, though that is not to say that they correspond to two clearly demarcated camps of biologists.

No general answer is yet available, but it is possible to make some observations. Different loci will tend to interact multiplicatively when they have independent effects on an individuals' survival and reproduction. This could be for loci that influence events at different times in an organism's life—though it is perfectly possible for such events to interact. Epistatic interactions are more likely for loci controlling closely interdependent parts of an organism. The extent to which we expect loci to interact epistatically or not will therefore loosely depend on how atomistic or holistic a view we have of the organism (see also section 7.8, p. 160).

Notice that epistatic fitness interaction is not the same as mere physiologic or embryological interaction. Fitness epistasis requires heterozygosity at two loci. Imagine a case in which the A locus controls, say, muscle strength and the B locus controls metabolic rate. Muscles and metabolism interact in a physiologic sense: when muscles are put to work, the metabolic rate goes up. However, if the population is fixed for homozygotes at both loci (all individuals are A_1B_1/A_1B_1) then there cannot be any fitness epistasis. Epistatic fitness requires heterozygosity at both loci, and the kind of fitness relations we saw in the *P. memnon* example. This is a special condition. Though epistasis is often called fitness "interaction," the term interaction is being used in a precise, not a colloquial, sense.

There is another, empirical method of answering the question of how common epistatic fitness interactions are in nature. Linkage disequilibrium is produced by epistatic selection, and the degree of linkage disequilibrium in a population can be measured. If it is high, then epistatic selection may be common. The argument works in one direction but not the other: because there are several possible causes of linkage disequilibrium (section 8.7), its existence does not demonstrate epistatic selection; however, if linkage disequilibrium is absent or low, we can infer that epistatic selection is unimportant in nature.

A few general surveys of the extent of linkage disequilibrium in natural populations have been made. They have used gel electrophoretic surveys of protein polymorphisms, to see directly whether genes at different loci are

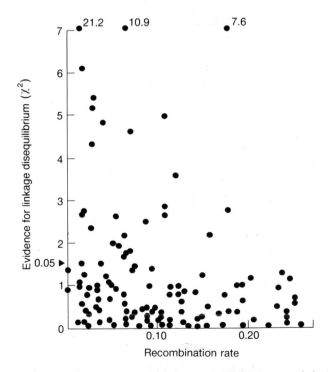

Figure 8.5 Linkage disequilibrium (on the *y*-axis) among gel electrophoretic samples of pairs of genes in *Drosophila*. It is plotted as a χ^2 value. The χ^2 value indicates how strong is the evidence for linkage disequilibrium: the higher χ^2, the more linkage disequilibrium between that pair of genes. (The χ^2 value corresponding to a statistical significance of 0.05 is indicated.) Most of the gene pairs have insignificant or low linkage disequilibrium (i.e. low χ^2). The *x*-axis is the rate of recombination between the pair of genes. From Langley (1977).

associated. Figure 8.5 illustrates some comprehensive results for the fruitfly *Drosophila*. Some evidence of linkage disequilibrium is found, but the results do not suggest high levels. One conclusion from this result would be that, although epistatic interactions are important in particular cases, like *Papilio*, they may not be of general importance in evolution.

Other biologists would disagree. They might be unconvinced by the evidence of Figure 8.5, perhaps calling it limited, or for a single species. The amount of interaction between loci that must go on during the development of a complex, organic body is so high that they would expect epistatic fitness interactions to be common. Such is the assumption of the school of thought that follows Wright, whose ideas we shall discuss at the end of the chapter.

8.9 Hitch-hiking occurs in two-locus selection models

When a gene is changing frequency at one locus over time, it can cause related changes at linked loci; events at linked loci can interfere with one another. Suppose, for instance, that directional selection is substituting one allele *A'* for another (*A*) at one locus, and there is a neutral polymorphism (*B,B'*) at a linked locus. Then whichever of *B* and *B'* happened to be linked with *A'* when it arose as a mutant will have its frequency increased. If the new mutant *A'* happened to arise on a *B*-bearing chromosome, then *B* will eventually be fixed together with the selected allele *A'* unless recombination splits them before *A* has been eliminated. The increase in the *B* allele frequency is due to *hitch-hiking*.

Another possibility is for the polymorphism at the B locus to be a selectively balanced polymorphism, due to heterozygous advantage. Suppose again that a selectively favored mutation *A'* arises at a linked locus, and that it happens to arise on the same chromosome as a *B* allele. Now the polymorphism at the B locus will interfere with the progress of *A'*. As *A'* increases in frequency by directional selection it will increase the frequency of *B* with it. Because it is linked to *B*, it will be more likely to find itself in *B/B* homozygotes than will the allele *A*, and less likely to be in *B/B'* heterozygotes. *B/B'* has higher fitness than *B/B* and the selection against *B/B* individuals will also work against the *A'* gene. Depending on the selection coefficients at the two loci, and the rate of recombination between them, the heterozygous advantage at the B locus can slow the rate at which *A'* is fixed. The *A'* gene will then have to wait for a recombinational hit between the two loci before it can progress to fixation.

Hitch-hiking has a further interest in relation to the neutral theory of molecular evolution. It alters the neutral theory's predictions for the level of heterozygosity. We saw in the last chapter (section 7.6, p. 154) that observed heterozygosities are lower than those the neutral theory predicts for plausible estimates of population size and mutation rate. Heterozygosities are not only too low, but also too uniform, given the variation in natural population sizes (see Figure 6.6, p. 137). Maynard Smith and Haigh pointed out that hitch-hiking could bring the neutral theory more closely into line with the observations. As a favored allele is fixed, the heterozygosity is reduced at any linked locus that has a neutral polymorphism; the more closely linked the polymorphic locus, the greater the reduction in its heterozygosity. We do not know to what extent hitch-hiking is the reason why heterozygosities deviate below the neutral theory's prediction; however, the hitch-hiking effect of a favorable gene must undoubtedly act to reduce heterozygosities at linked neutral loci.

8.10 Linkage disequilibrium can be advantageous, neutral, or disadvantageous

The linkage disequilibrium in *P. memnon*'s mimetic polymorphism is advantageous. Natural selection favors individuals with the gene associations like $T_-W_2F_2E_2B_2$, whereas it works against recombinants like $T_+W_2F_2E_2B_2$. An individual benefits from having the haplotypes that are in excess frequency in the population. Whole populations of *P. memnon* survive better than they would if the five loci were in linkage equilibrium.

In other cases, the opposite is true. We met an example in the previous section. As the favored allele A' increased in frequency, so the frequency of an allele, B, at a linked polymorphic locus was also increased by hitch-hiking. The polymorphism at the B locus was caused by heterozygous advantage, and the linkage disequilibrium built up by selection on the A locus (creating an excess of the $A'B$ haplotype) was therefore disadvantageous. The individuals on average have lower fitness than if there were linkage equilibrium between the A and B loci, because the increase in the $A'B$ haplotype reduces the proportion of B/B' heterozygotes. Natural selection will favor recombinant individuals that do not have the $A'B$ haplotype.

A third possibility is for linkage disequilibrium to be selectively neutral. An example of this was provided by the hitch-hiking of an allele at a neutral polymorphic locus with a selectively advantageous mutant at a linked locus. While the mutant is being fixed, linkage disequilibrium temporarily builds up between it and the alleles it happens to be linked to at nearby loci. It disappears when the mutant reaches a frequency of one.

The distinction between advantageous and disadvantageous linkage disequilibrium is crucial to understanding one of the major problems of evolutionary biology: why recombination exists.

8.11 Why does the genome not congeal?*

Genetic recombination undoubtedly exists, and yet it is not obvious why. The problem can be seen by considering two kinds of population near equilibrium, one at linkage equilibrium and the other with linkage disequilibrium due to epistatic selection (or coadaptation between genetic loci, like in *P. memnon*). The assumption that the populations are near equilibrium amounts to assuming that the environment has been constant for some time. In practice, environments may not be that constant. However, let us consider the consequences for recombination rates of assuming that they are. We shall then be able to work out more exactly what kind of environmental changes are required for recombination to be selected for. We shall also assume that mating is random and the population is infinite. What effect does recombination have in the first case, when there is linkage equilibrium? None whatever: the frequencies of A_1B_1, A_1B_2, A_2B_1, and A_2B_2 are repeated from generation to generation, whether or not there is recombination. If there is no recombination, the haplotype frequencies are necessarily constant. If there is

---

*In 1967 J.R.G. Turner published a strikingly titled paper, "Why does the genotype not congeal?" The paper is often mis-cited in the form of this section heading.

recombination, it creates and destroys each haplotype at the same rate: that is what is meant by linkage equilibrium. In the absence of linkage disequilibrium, therefore, recombination has no effect.

Suppose now that there is linkage disequilibrium, and it is due to coadaptation between loci. Let us suppose that A_1B_1 and A_2B_2 are coadapted pairs of alleles, like the tail genes and color genes in *P. memnon*, and are present in greater than random frequencies ($D > 0$). The actual amount of linkage disequilibrium will be a balance between the increase due to selection and the decrease due to recombination. In this case, recombination is disadvantageous. The lower the recombination rate, the stronger the association between the coadapted alleles that can be built up, and the fitter on average the individuals will be. There is selection to reduce the recombination rate between the loci.

The decrease in fitness due to recombination constitutes a genetic load on the population, called the *recombinational load*. It can be expressed in the standard formula (see equation 7.1, p. 148). The best genotypes here are A_1B_1 and A_2B_2. Let us suppose they both have the same fitness, w_{opt}. $\bar{w}$ is then less than w_{opt} by an amount proportional to the recombination rate, because it is recombination that is creating A_1B_2 and A_2B_1 haplotypes. (Sex does not help either, by producing A_1B_1/A_2B_2 heterozygotes, though this would not matter to the individual if there were dominance. In this idealized example, there is selection for speciation—the species should split into two forms, A_1B_1 and A_2B_2, that mate only among themselves. But that is another matter. The point here is that recombination, in so far as it exists, itself creates a genetic load.) A formal calculation of $\bar{w}$ and the exact genetic load is possible; here we only need to note that the disadvantage of recombination can be expressed as a recombinational load.

It might seem, then, that recombination either makes no difference or is disadvantageous. If there is linkage equilibrium, the genotype frequencies are unaltered by recombination; if beneficial associations between genes have built up, recombination tends to destroy them. So why does recombination exist? Why are recombination rates not reduced to zero? The question has not been answered with universal agreement. The vague answer is probably that recombination is favored because of environmental change; but it is still a

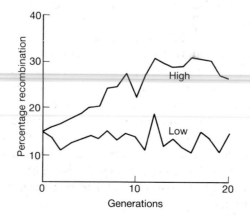

Figure 8.6 Artificial selection can increase (high) or decrease (low) the rate of recombination. These results are for the rate of recombination between two loci (called *Gl* and *Sb*) in laboratory stocks of the fruitfly *Drosophila melanogaster*. From Kidwell (1972).

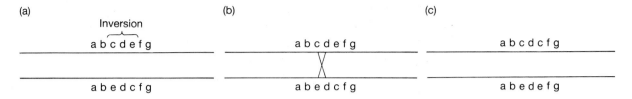

Figure 8.7 (a) The letters (a, b, c, etc.) are different genes along a chromosome. The genes c−d−e are an inversion polymorphism. (b) If recombination takes place within the inversion, the resulting chromosomes (c), which will become gametes, contain double copies of some genes but lack copies of other genes. Recombination within an inversion is therefore usually lethal. The only offspring of such an individual will be those in which recombination did not hit the inverted region.

problem to specify what kind of environmental change is needed for recombination to be advantageous, and to show that natural environments do change in the necessary way. But before we come on to environmental change, we should rule out one other possible explanation first.

Natural selection cannot work on something if it shows no genetic variation. If there were no genetic variation in recombination rates, it could not be reduced to zero. Is recombination rate genetically variable? The most convincing test is to carry out an artificial selection experiment. If a trait varies genetically, an attempt to change it by artificial selection should be successful; and if it does not, the attempt should fail. Several experiments have selected successfully for both higher and lower recombination rates in fruitflies (Figure 8.6). Absence of genetic variation is not the reason why recombination exists.

Moreover, in the case of linked groups of coadapted genes, recombination can be effectively eliminated by a chromosomal inversion. We do not know whether the *P. memnon* mimicry genes are contained within an inversion; but let us consider, for sake of argument, what would happen if they were. If a single recombination event occurs within an inverted chromosomal region, the resulting chromosomes lack certain genes and are therefore likely to be lethal (Figure 8.7); only when recombination (or, to be more exact, an odd number of recombination events) has not occurred in the inversion will the offspring be viable. The set of genes within the inversion are thus practically protected from recombination.

Chromosomal inversions are well known to exist. They are particularly easy to study in fruitflies (section 4.5, p. 69; section 17.7, p. 456) and Dobzhansky did much research on them. [He emphasized the idea that chromosomal inversions protect sets of epistatically interacting genes, and this is indeed an important principle of chromosomal evolution.] Notice, by the way, that the genes in an inversion will tend to behave as a "supergene," a set of linked genes that segregate as if they were a single Mendelian locus: the recombinants will often die and therefore not be seen.

Recombination can be reduced either by straightforward reduction of the rate of recombination, or by such means as the establishment of a chromosomal inversion. It is therefore unlikely that recombination persists in natural populations only because natural selection cannot eliminate it for want of genetic variation.

Recombination, we saw, is only disadvantageous when it disrupts coadapted sets of genes in linkage disequilibrium. Its disadvantage will depend on how much of this sort of linkage disequilibrium exists in nature; if the amount is low, the disadvantage will be small. As we have seen (Figure 8.5), such data as we have suggest quite low amounts of linkage disequilibrium. Coadapted gene complexes, in the strong sense of haplotypes in greater than random frequencies, may be exceptional in nature. To this extent, the problem of recombination is reduced.

But what positive advantage might recombination provide? The general condition for recombination to be favored is that disadvantageous gene associations should exist in the population. Recombination is selected against when it breaks up favorable gene combinations: it is selected for when it breaks up unfavorable gene combinations. Hitch-hiking is a case in point. When a selected polymorphic locus interferes with selection at a linked locus, recombination between the loci is advantageous. Strobeck *et al.* have demonstrated the advantage formally. The argument is fine in itself, but it may be a special circumstance. The process would be more common in a rapidly changing environment, because the fitnesses of different alleles would then be changing and selection would be attempting to adjust the frequencies of genes at many, potentially interfering, loci.

In a population changing under selection, recombination can be advantageous even without the hitch-hiking effect. Imagine that selection is operating because of environmental change. What kind of changes will favor recombination? Recombination, we have seen, is advantageous when it destroys disadvantageous linkage disequilibrium: the environmental changes must therefore be of a kind that produce disadvantageous linkage disequilibrium. Here is an example, invented by Maynard Smith, of how it could happen. Imagine several loci (let us say three), controlling adaptation to three different environmental variables, such as temperature, humidity, and feeding. Allele A_1 might be adapted to higher temperature, A_2 to low; B_1 to humid conditions, B_2 to dry; C_1 to feeding in one sort of environment, C_2 to another. Suppose that for several generations the conditions have favored the haplotype $A_1B_1C_1$, but that all six haplotypes have persisted at some frequency. The environment now changes, to favor $A_2B_2C_2$. Is recombination useful? No it is not: indeed, it is actively disadvantageous. It slows down the response to the environmental change. It is therefore an error to say that recombination is advantageous simply because of environmental change. Environmental change may often select against recombination.

However, recombination can be favored if the environmental change is of the right kind. Suppose instead that the environment had changed to favor $A_1B_2C_1$. If the existing haplotypes in the population are mainly $A_1B_1C_1$ and $A_2B_2C_2$, recombination is positively advantageous because it will split up the established haplotypes and increase the frequency of the newly favored one. The abstract condition is as follows. In order for recombination to be useful, the signs of the associations between environmental components must change between the generations. If the environment favors a positive correlation between the temperature and feeding strategy adaptations in one generation, it must be likely to change to favor a negative correlation in the next: then recombination will be maintained. It is not enough for the environment to change, it must change in such a way that new combinations of genes become favored.

Maynard Smith concluded that this condition makes recombination all the more puzzling. Different components of the environment will often change in a correlated way; if it is hot and dry one year, then if it is cooler next year it may also be wetter. Recombination between loci controlling adaptation to temperature and loci controlling adaptation to humidity is then disadvantageous. However, if we turn from physical to biological components of the environment, the right kind of change may be found. Hamilton and others have argued that the coevolutionary competition between parasites and hosts have the necessary property. Parasite and host exert contrary selection on each other. If one gene combination in the hosts is advantageous now, there will be strong selection on the parasite to overcome it, which will then lead to selection against the gene combination (section 11.3.5, p. 275).

Recombination, therefore, poses a problem. Its solution will probably be found in the types of environmental change going on in nature. For recombination to be advantageous, the environment has to change in such a way that ever new combinations of genes are favored: the problem is therefore to show that environments do change in this way. Currently, the most likely idea is that biological coevolution, such as between parasites and hosts, is the environmental component that changes in the necessary manner. But this is at present a suggestion, not an established theory; the question is still open. Whatever the answer turns out to be, it will probably contradict the initial assumption that populations are close to equilibrium. Persistent change, with a steady supply of transitorily favorable new alleles and gene combinations, may be a more accurate description of populations in nature.

8.12 Wright invented the influential concept of an adaptive topography

Wright's idea of an *adaptive topography* is particularly useful for thinking about complex genetic systems; but it is easier to begin with the simplest case. This is for a single genetic locus. The topography is a graph of mean population fitness ($\bar{w}$) against gene frequency (Figure 8.8). (Adaptive topographies can also be drawn for fitness in relation to genotype frequencies. They can even be drawn with phenotypic variables on the x-axis; see, for example, Raup's analysis of shell shape, discussed in section 13.6, p. 339.) We have repeatedly met the concept of mean fitness; it is equal to the sum of the fitnesses of each genotype in the population, each multiplied by its proportion in the population. In a case in which the genotypes containing one of the alleles have higher fitness than the others, the mean fitness of the population simply increases as the frequency of the superior allele increases and reaches a maximum when the gene is fixed (Figure 8.8a). That is fairly trivial. When there is heterozygous advantage, mean fitness is highest at the equilibrial gene frequency given by the standard equation (equation 5.16, p. 109); it declines on either side, where more of the unfavorable homozygotes will be dying each generation than at the equilibrium (Figure 8.8b). The graph is also called a *fitness surface*.

In these two cases, natural selection carries the population to the gene frequency where mean fitness is at a maximum. With one favorable allele, the

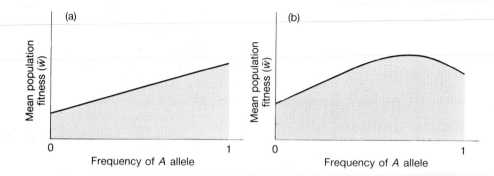

Figure 8.8 A fitness surface, or adaptive topography, is a graph of the mean fitness of the population as a function of gene, or in some cases genotype, frequency. (a) If genotype *AA* and *Aa* have higher fitness than *aa*, mean population fitness simply increases as the frequency of *A* increases. (b) With heterozygous advantage (fitnesses of genotypes *AA*:*Aa*:*aa* are $1 - s$:1:$1 - t$) mean population fitness increases to a peak at the intermediate frequency of *A* at which the proportion of heterozygotes is the maximum possible.

maximum mean fitness is where the allele is fixed—and natural selection will act to fix the allele. With heterozygous advantage, the maximum mean fitness is where the smallest number of homozygotes are dying each generation— and natural selection drives the population to an equilibrium where the amount of homozygote death is minimized. A question of interest in theoretical population genetics is whether natural selection always drives the population to the state at which the mean fitness is the maximum possible. There are cases in which it does not. Frequency dependence is an example we have already met (section 7.9, p. 162). If natural selection does not always maximize mean fitness, there is the further—and still unanswered— theoretical question of whether there is some other function that natural selection does always maximize; we shall not pursue the question here. And whatever the answer, there are many cases in which natural selection maximizes simple mean fitness, and for many purposes this is a useful assumption. We can then think of natural selection as a hill-climbing process, by analogy with the hills in the adaptive topography (Figure 8.8).

Now consider a second locus. Selection can be going on here too, and the fitness surface for the two loci might look like Figure 8.9;(a) shows a simple case in which one locus has heterozygous advantage and the other has a single favored allele. The idea of an adaptive topography can be extended to as many loci as interact to determine an organism's fitness, but the further loci have to be imagined, rather than drawn on two-dimensional paper.

Wright believed that, because the genes at different loci interact, a real multi-dimensional fitness surface would often have multiple peaks, with valleys between them (Figure 8.9b). The kind of reasoning involved is abstract rather than concrete. We imagine a large number of loci, many with more than one allele, and the alleles at the different loci interacting epistatically in their effects on fitness. Epistatic interactions, we now imagine, are common because organisms are highly integrated entities as compared with the atomistic chromosomal row of Mendelian genes from which organisms grow

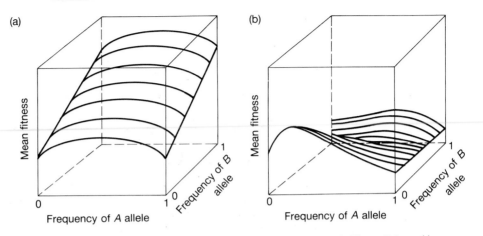

Figure 8.9 Fitness surface for two loci. (a) Combination of patterns in Figure 8.8a and b: heterozygous advantage at locus A and one allele has higher fitness than the other at locus B. (b) Two-locus fitness surface with two peaks.

up: the genes will have to interact to produce an organism. As we saw above, developmental interactions among genes do not automatically generate epistatic fitness interactions among loci. The extent to which undoubted developmental interaction will produce a multiply peaked fitness surface is therefore open to question; but the possibility is plausible. (Wright called the genes that interact favorably to produce an adaptive peak an "interaction system.")

In the coadapted supergenes of *P. memnon*, the mimetic genotypes occupy fitness peaks and the recombinants occupy various fitness valleys. The actual shape of the adaptive topography in nature is, however, a more advanced question than can be tackled here. The point of this section is to define what an adaptive topography is, and to point out that its visual simplicity can be useful in thinking about evolution when many gene loci are interacting.

8.13 The shifting balance theory of evolution

Wright used his idea of adaptive topographies in his general evolutionary theory. He imagined that real topographies would have multiple peaks, separated by valleys; some peaks would be higher than others. When the environment changed, and competing species evolved new forms, the shape of the adaptive topography for a population would change too. The surface would also change shape when a new mutation arose. A new allele at a locus may interact with genes at other loci differently from the existing alleles, and the fitnesses of the genes at the other loci will then be altered; genetic changes will take place at other loci to adjust to the new mutant. All the time, natural selection will be a hill-climbing process, directing the population up toward the current nearest peak. When the surface changes, the direction to the nearest peak may change, and selection will then send the population off in the new upward direction.

Natural selection, even in so far as it is a hill-climbing (i.e. mean fitness maximizing) process, is only a *locally* hill-climbing process. In theory, the local fitness peak could be in the opposite direction from a higher, or the global,

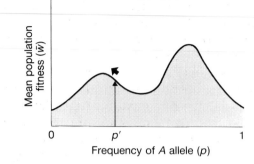

Figure 8.10 A two-peaked fitness surface with local and global maxima. Natural selection will take a population with gene frequency p' towards the local peak, away from the peak with highest average fitness.

peak (Figure 8.10). Natural selection, however, will direct the population to the local peak. Now suppose that the mean fitness of a population varies with the quality of its adaptations, such that a population with a higher mean fitness has better adaptations than a population with a lower mean fitness. Because natural selection seeks out only local peaks, it may not always allow a population to evolve the best possible adaptations. A population could be stuck on a merely locally adaptive peak. However, adaptation and mean fitness cannot always be equated in this way. In the simple case in which one allele is superior to another (Figure 8.8a) the organisms with the better genotype will also be better adapted. But when there is frequency dependence of fitness or when there is conflict between group and individual adaptations (see chapter 12) it is no longer true that the maximum of mean population fitness automatically corresponds to the best adaptation.

Wright was interested in how evolution could overcome the tendency of natural selection to become stuck at local fitness peaks. When fitness peaks correspond to optimal adaptations, the question is relevant to the evolution of adaptation; but when they do not, the question still has a technical interest in population genetics. Wright suggested that random drift could play a creative role. Drift will tend to make the population gene frequencies "explore" around their present position; the population could, by drift, move from a local peak to explore the valleys of the fitness surface. Once it had explored to the foot of another hill, natural selection could start it climbing uphill on the other side. If this process of drift and selection were repeated over and over again with different valleys and hills on the adaptive topography, a population would be more likely to reach the global peak than if it was under the exclusive control of the locally maximizing process of natural selection.

Wright included not only drift and selection within a population in his general theory. He also supposed that populations would be subdivided into many small local populations. Drift and selection would go on in each and the large number of subpopulations would multiply the chance that one of them within the whole population would find the global peak. If the members of a subpopulation at the highest peak were better adapted, they could produce more offspring; they would also send more emigrants to the other subpopulations. Thus the whole species would evolve to the higher peak. Wright's theory is thus an attempt at a comprehensive, realistic model of evolution. Everything is included: multiple loci, fitness interactions, selection within and between populations, drift, and migration. (The theory of adaptive peaks is also relevant to speciation: see section 16.4.4, p. 419).

(As always in science, complex realistic models are useful for some purposes and simple, idealized ones for others. Realistic models may describe nature more accurately, but they are less illuminating when it comes to explaining principles or studying the consequences of particular factors. For this reason, when in this book we are considering particular factors in isolation, we shall stay with simple and idealized models that concentrate on the factor of interest. The encumbrance of all the other factors that may influence gene frequencies in nature is needed when we are trying to describe the real world, but not when we are trying to understand arguments.)

Fisher generally disagreed with Wright over the importance of random drift in evolution. Fisher maintained that natural populations are generally too large for drift to be important; Wright retorted that Fisher had misunderstood the role of drift in the shifting balance theory, which was to assist, rather than counteract natural selection: much of their argument seems to have been at cross-purposes. Many of the points are interesting, however. Fisher, for instance, doubted whether natural selection actually would confine populations to local peaks. Fisher was pre-eminently a geometric thinker and he pointed out that, as the number of dimensions in an adaptive topography

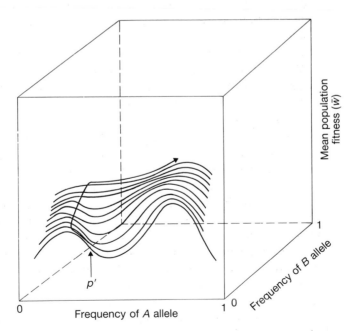

Figure 8.11 As extra gene loci are considered (extra axes in the adaptive topography) it becomes increasingly likely that what appeared to be a local maximum in fewer dimensions will turn out to be a hill-side or saddle-point in more dimensions. In this case, the fitness surface for the A locus at a gene frequency of zero for the B allele is the same as in Figure 8.10; but at other B gene frequencies, the local peak in the A locus fitness surface disappears. If the population started in the valley at a B gene frequency of zero and A gene frequency of p', natural selection would initially move gene frequencies to the local peak, but it would eventually reach the global peak by continuous hill-climbing.

increases, local peaks in one dimension tend to become points on hills in other dimensions, or saddle-points (Figure 8.11). In the extreme case, when there are an infinity of dimensions, it is certain that natural selection will be able to hill-climb all the way to the global peak without any need for drift. Each one- (Figure 8.8) or two-dimensional peak (Figure 8.9) will be crossed at the peak by an infinity of other dimensions, and it is highly implausible that the fitness surface will turn downhill in all of them at that point. This is a highly interesting argument, though it is of course purely theoretical; it refutes Wright's *theoretical* claim that natural selection will get stuck at local peaks. But it leaves open the empirical question of how important selection and drift have been in exploring the fitness surfaces of nature.

8.14 Summary

1 Population genetics for two or more loci is concerned with changes in the frequencies of haplotypes, which are the multi-locus equivalent of alleles.

2 Recombination tends, in the absence of other factors, to make the alleles of different loci appear in random (or independent) proportions in haplotypes. An allele A_1 at one locus will then be found with alleles B_1 and B_2 at another locus in the same proportions as B_1 and B_2 are found in the population as a whole. This condition is called linkage equilibrium.

3 A deviation from the random combinatorial proportions of haplotypes is called linkage disequilibrium.

4 The theory of population genetics for a single locus works well for populations in linkage equilibrium.

5 Linkage disequilibrium can arise because of non-random mating, random sampling, and natural selection.

6 For selection to generate linkage disequilibrium, the fitness interactions must be epistatic: the effect on fitness of a genotype (such as A_1/A_2) must vary according to the genotype it is associated with at other loci.

7 Pairs of alleles at different loci that cooperate in their effects on fitness are said to be coadapted. Selection acts to reduce the amount of recombination between coadapted genotypes.

8 When selection works on one locus, it will influence gene frequencies at linked loci. The effect is called hitch-hiking.

9 Recombination is selectively disadvantageous in so far as it breaks down favorable gene combinations. The existence of recombination suggests that natural populations must possess many unfavorable gene combinations, which it is advantageous to break up.

10 The mean fitness of a population can be drawn graphically for two loci; the graph is called a fitness surface or an adaptive topography.

11 Wright suggested that real adaptive topographies will have many separate "hills," with "valleys" between them.

12 Natural selection enables populations to climb the hills in the adaptive topography; but not to cross valleys. A population could become trapped at a local optimum.

13 Random drift could supplement natural selection by enabling populations to explore the valley bottoms of adaptive topographies.

14 It is questionable whether real adaptive topographies do have multiple peaks and valleys. They might have a single peak, with a continuous hill leading to it. Natural selection could then take the population to the optimum without any random drift.

·**15** Two-locus population genetics uses a number of concepts not found in single-locus genetics. The most important are: haplotype frequency, recombination, linkage disequilibrium, epistatic fitness interaction, hitch-hiking, and multiple-peak fitness surfaces.

8.15 Further reading

Ford (1975) and Sheppard (1975) both introduce mimicry in swallowtail butterflies. The multi-locus genetics of mimicry in *P. memnon* and in *Heliconius* is explained by Turner (1977, 1984); see also Turner (1985).

Population genetics for two loci, as for one locus, is introduced in such standard textbooks as Crow and Kimura (1970), Hartl and Clark (1989), and Hedrick (1983). Hedrick *et al.* (1978) is about multi-locus population genetics in particular. The HLA loci are introduced by Bodmer and Cavalli-Sforza (1976); Potts and Wakeland (1990) and Hedrick *et al.* (1991a) are more recent evolutionary reviews; for the extent of variation see Hedrick *et al.* (1991b).

Dobzhansky (1970) and Wright (1977) discuss chromosomal inversions and coadaptation. On hitch-hiking: Maynard Smith and Haigh (1974), Thomson (1977), and Kaplan *et al.* (1989). On recombination: Maynard Smith (1978a), Felsenstein (1985), several authors in Michod and Levin (1988), and Haig and Grafen (1991).

On Wright's "shifting balance" theory, see Wallace (1968), Wright (1970), and Wright's four volume treatise, especially vols 3–4 (1977–1978), Lewontin (1974, final chapter), Provine (1985, 1986a, 1986b), and Turner (1987) for the question of whether fitness surfaces will be multiply peaked.

Quantitative genetics

9.1 Climatic changes have driven the evolution of beak size in one of Darwin's finches

Fourteen species of Darwin's finches live in the Galápagos archipelago, and many of them differ most obviously in the size and shape of their beaks. A finch's beak shape, in turn, influences how efficiently it can feed on different types of food. Grant has been studying these finches since 1973, and his best evidence to demonstrate that beak size influences feeding efficiency comes from a comparison of two species, the large-beaked *Geospiza magnirostris* and the smaller *G. fortis*, feeding on the same kind of hard fruit.

The large-beaked *G. magnirostris* can crack the fruit (called the mericarp) of caltrop (*Tribulus cistoides*) transversely, taking on average only 2 seconds and exerting an average force of 26 kgf; it can then easily, in about 7 seconds, eat all the four to six seeds of the smashed fruit. The smaller *Geospiza fortis* are not strong enough to crack *Tribulus* mericarps and instead twist open the lower surface, applying a force of only 6 kgf and taking 7 seconds on average to reach the seeds inside; but only one or two of the seeds can be obtained in this way and it takes an average of 15 seconds to extract them. *Geospiza magnirostris* usually has an advantage with these large, hard types of food.

Smaller finches are probably more efficient with smaller types of food, but this is more difficult to show. Both large and small finches on the Galápagos do in fact eat small seeds, though there is an indirect reason (as we shall see) to believe that smaller finches do so more efficiently. From the evidence we have met so far, we should predict that natural selection would favor larger finches when large fruits and seeds are abundant. The prediction should apply both within and between species. A *G. magnirostris* looks like an enlarged *G. fortis*, and a larger individual *G. fortis* can probably deal with a large food item more efficiently than can a smaller conspecific, much as an average specimen of *G. magnirostris* is more efficient than an average *G. fortis*. When large seeds are common, we might expect the average size of a population of *G. fortis* to increase between generations, and vice versa when large seeds are rare—if beak size is inherited.

If beak size is inherited . . . but is it? For beak size to be inherited means that parental finches with larger than average beaks produce offspring with larger than average beaks. Grant measured the sizes of parental and offspring finches in several families and plotted the latter against the former (Figure 9.1). Large-beaked parental finches do indeed produce large-beaked offspring: beak size is inherited. It therefore makes sense to test the prediction that changes in beak size should follow changes in the size distribution of food items. The test was carried out by Grant on the species *G. fortis*, on one

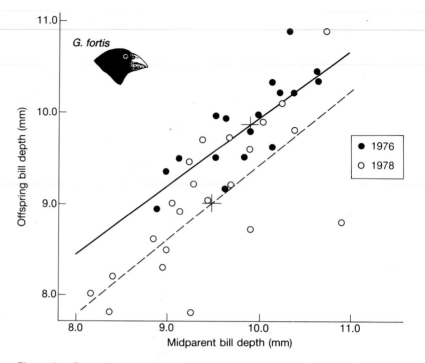

Figure 9.1 Parents with larger than average beaks produce offspring with larger than average beaks in *Geospiza fortis* on Daphne Major: beak size is inherited. From Grant (1986).

of the Galápagos islands, Daphne Major. Since the study began, this species has undergone two major, but contrasting, evolutionary events.

In the Galápagos, the normal pattern of seasons is for a hot, wet season from about January to May to be followed by a cooler, drier season through the rest of the year. But in early 1977, for some reason, the rain did not fall. Instead of the normal progression of wet season then dry, then wet then dry, the dry season that began in mid 1976 continued until early 1978: one whole wet season did not happen. The finch population of Daphne Major collapsed from about 1200 to about 180 individuals (Figure 9.2a), with females being particularly hard hit; the sex ratio at the end of 1977 was about five males per female. As the sex difference shows, not all finches suffered equally. Smaller birds died at a higher rate. The reason, again, lies in the food supply (Figure 9.2b). At the beginning of the drought, the various types of seeds were present in their normal proportions. *Geospiza fortis* of all sizes take small seeds, and as the drought persisted, these smaller seeds were relatively reduced in numbers; average available seed size became larger with time (Figure 9.2c). Now the larger finches were favored, because they eat the larger, harder seeds more efficiently; and the average finch size increased as the smaller birds died off. (Females died at a higher rate than males because females are on average smaller.) Size, as we have seen, is inherited. The differential mortality in the drought therefore caused an increase in the average size of finches born in the next generation. *Geospiza fortis* born in

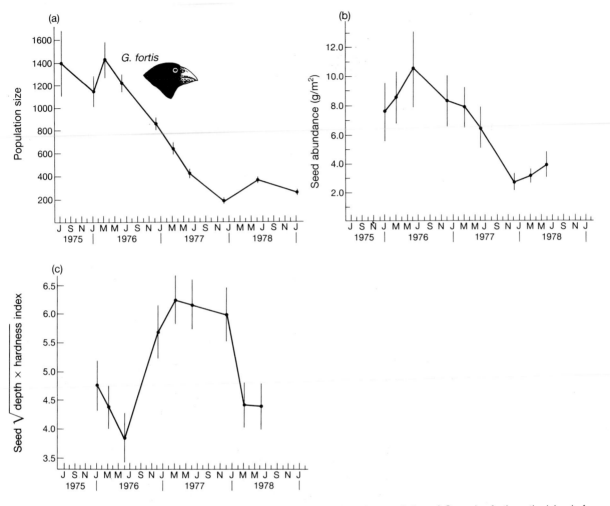

Figure 9.2 During a drought in 1976–1977 the population of *Geospiza fortis* on the island of Daphne Major in the Galápagos archipelago decreased (a), due to the decline of the food supply (b). (c) The average size of the seeds available as food increased during the drought. From Grant (1986).

1978 were about 4% larger on average than those born before the drought.

Four years later, in November 1982, the weather reversed. The rainfall of 1983 was exceptionally heavy and the dry volcanic landscape was covered with green in the periodic disturbance called El Niño. Seed production was enormous. The theory developed for 1976–1978 could now be tested. The conditions had reversed: the direction of evolution should go into reverse too. In the year after the 1983 El Niño event, there was an excess of small seeds: if the smaller finches can in fact exploit them more efficiently, the smaller finches should have survived relatively better. Grant again measured the sizes of *G. fortis* on Daphne Major in 1984–1985 and found that the smaller birds were indeed favored; finches born in 1985 had beaks about 2.5% smaller than those born before the El Niño downpours. His theory, that seed size controls beak size in these finches, was confirmed. The fluctuations in the direction of selection on beak shape—with smaller beaks favored in some years and larger ones in others—probably results in a kind of "stabilizing" selection over a long period of time, such that the average size of beak in the population is the size favored by long-term average weather. (Later in the chapter, we shall see how the degree of selection can be expressed more exactly; Figure 9.10 will show the results for 1976–1977 and 1984–1985.)

9.2 Quantitative genetics is concerned with characters controlled by large numbers of genes

The beak size of Galápagos finches is an example of a large class of characters that interest evolutionary biologists. It shows *continuous variation*. Simple Mendelian characters, like blood groups or the mimetic variation of *Papilio*, often have discrete variation; but many of the characters of species are like beak size in these finches: they vary continuously, and every individual in the population differs slightly from every other individual. There are no discrete categories of beak size within a species.

The other important point about beak size is that we do not know the exact genotype that produces any given beak size. We can, however, say something about the general sort of genetic control it may have. Characters like beak size, which has an approximately normal frequency distribution (i.e. a "bell curve"), are probably controlled by a large number of genes, each of small effect. The reason is as follows (Figure 9.3). Imagine first that beak size is controlled by a single pair of Mendelian alleles at one locus, with one dominant to the other, *AA* and *Aa* long and *aa* short: in this case, the population would contain two categories of individuals (Figure 9.3a). Imagine now that it was controlled by two loci with two alleles each. Beak size might

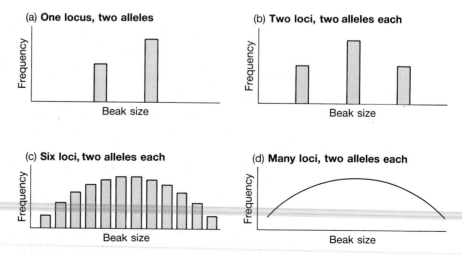

Figure 9.3 (a) The phenotypic character, beak size for example, is controlled by one locus with two alleles (*A* and *a*), where *A* is dominant to *a*. There are two discrete phenotypes in the population. (b) The character is controlled by two loci with two alleles each (*A* and *a*, *B* and *b*); *A* and *B* are dominant to *a* and *b*. There are three discrete phenotypes. (c) Control by six loci, with two alleles each. (d) Control by many loci with two alleles each. As the number of loci increase, the phenotypic frequency distribution becomes increasingly continuous.

now have a background value (say, 1 cm) plus the contribution of the two loci; an *A* adds 0.1 cm and a *B* subtracts 0.1 cm. *A* and *B* are dominant to *a* and *b*. Then *aaBb* and *aaBB* individuals have 0.9 cm beaks, *aabb*, *AaBB*, *AaBb*, *AABB*, and *AABb* have 1 cm beaks, and *AAbb* and *Aabb* have 1.1 cm beaks. Figure 9.3b is the frequency distribution if all alleles have frequency 0.5 and the two loci are in linkage equilibrium. It is starting to look like a normal distribution when the character is influenced by six loci (Figure 9.3c) and with enough loci the distribution will be normal (Figure 9.3d). Thus, when a large enough number of genes influence a character, it will have a continuous, normal frequency distribution.

The fact that multifactorial inheritance (i.e. the character is influenced by many genes) can generate a continuous frequency distribution had been anticipated by Mendel in his original paper in 1865; but it was not well confirmed until later work, particularly by East, Nilsson-Ehle, and others, in about 1910. Quantitative genetics is concerned with characters influenced by many genes, called *polygenic characters.* For a quantitative geneticist, five to 20 genes is a small number of genes to be influencing a character; whereas many quantitative characters may be influenced by more than a hundred, or even several hundred, genes. For characters influenced by a large number of loci, it ceases to be useful to follow the transmission of individual genes or haplotypes (even if they have been identified) from one generation to the next. The pattern of inheritance, at the genetic level, is too complex.

There is an additional complication. So far we have only considered the effect of genes. The value of a character, like beak size, will usually also be influenced by the environment in which the individual grows up. Beak size is probably related to general body size and all characters to do with bodily stature will be influenced by the amount of food an organism happens to find during its life. If we take a set of organisms with identical genotypes and allow some to grow up with abundant food and others with limited food, the former will end up larger on average. In nature, each character will be influenced by many environmental variables, some tending to increase it, others to decrease it. Thus if we take a class of genotypes with the same value of a character before the influence of the environment and add the effect of the environment, some of the individuals of each genotype will be made larger and others smaller in various degrees; and this will produce a further "spreading out" of the frequency distribution. Any pattern of discrete variation in the genotype frequency distribution is likely to be obscured by environmental effects and the discrete categories converted into a smooth curve (Figure 9.4).

The small effects of many genes and environmental variables are two separate influences that tend to convert the discrete phenotypic distribution of characters controlled by single genes into continuous distributions. If a character shows a continuous distribution, it in principle could be because of either process; quantitative genetics is mainly concerned with characters influenced by both. Quantitative genetics employs higher-level genetic concepts that are genetically less exact than those of one- or two-locus population genetics, but which are more useful for understanding evolution in polygenic characters. Instead of following changes in the frequency of genes

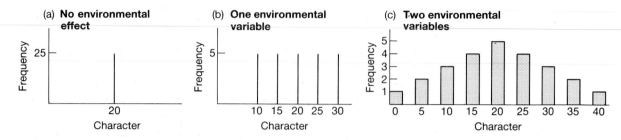

Figure 9.4 Environmental effects can produce continuous variation. (a) Twenty-five individuals in the absence of environmental variation all have the same phenotype, with a value for a character of 20. (b) Influence of one environmental variable. The variable has five states, and according to which state an organism grows up in its character becomes larger or smaller or is not changed. The five states change the character by $+10$, $+5$, 0, -5, and -10, and an organism has equal chance of experiencing any one of them. (c) Influence of a second environmental variable. This variable also has five equiprobable states, and they change the character by $+10$, $+5$, 0, -5, and -10. Of the five individuals in (b) with character value 10, one will get another -10, giving a value of 0, a second will get -5, giving 5, etc. After the influence of both variables, the frequency distribution ranges from 0 to 40 and is beginning to look bell-curved. With many environmental influences, each of small effect, a normal distribution would result.

or haplotypes, we now follow changes in the frequency distribution of a phenotypic character. Quantitative genetics is important because so many characters have continuous variation and multi-locus control.

9.3 Variation is first divided into genetic and environmental effects

Quantitative genetics contains an unavoidable minimum of formal concepts that we need to understand before we can put it to use: those formalities are the topic of this section and the next. To understand how a quantitative character like beak size will evolve, we have to "dissect" its variation. We tease apart the different factors that cause some birds to have larger beaks than others. Suppose, for example, that all the variation in beak size is caused by environmental factors—that is, all birds have the same genotype and they differ in their beak sizes only because of the different environments in which they grew up. Beak size could not then change during evolution (except for non-genetic evolution due to environmental change). For the character to evolve, it has to be at least partly genetically controlled. We therefore need to know how much beak size varies for environmental, and how much for genetic, reasons. However, even if different finches vary in their beak size for genetic reasons, that does not necessarily mean it can evolve by natural selection. As we shall see, we have to divide genetic influence into components that allow evolutionary change and those that do not.

In quantitative genetics, the value of a character in an individual is always expressed as a deviation from the population mean. Beak size will have a certain mean value in a population, and we talk about environmental and genetic influences on an individual as deviations from that mean. The procedure is easy to understand if we think of the population mean as a "background" value, and then the influences leading to a particular individual phenotype are expressed as increases or decreases from that value. Let us see how it is done. Suppose there is one locus with two alleles influencing beak depth. *AA* and *Aa* individuals' beaks are 1 cm from top to bottom, and *aa* individuals' are 0.5 cm; the environment has no effect. If the population average was 0.875 cm (as it would be for gene frequency of $a = 0.5$), then we should write the beak phenotype of *AA* and *Aa* individuals as +0.125 cm and that of *aa* as −0.375 cm. In general, we symbolize the phenotype by *P*: in this case, $P = +0.125$ for *AA* and *Aa* individuals and $P = −0.375$ for *aa*.

Clearly, the value of *P* for an individual with a certain genotype depends on the gene frequencies. When the frequency of *A* is 0.5, the genotypic effects are those just given. But when the frequency of *A* is 0.25, the population mean would be 0.8125 cm. For *aa*, *P* now is −0.3125 and for *AA* and *Aa*, $P = +0.1875$. In this example the phenotype is controlled only by genotype. We can symbolize the effect of the genotype by *G*. *G*, like *P*, is expressed as a deviation from the population mean. In this case, for an individual with a particular genotype (because the environment has no effect)

Mean of population + *P* = mean of population + *G*

The background population mean cancels from the equation, and can be ignored. We are then left (in this case in which the environment has no effect) with *P* equal to *G*.

The value of a real character will usually be influenced by the individual's environment as well as its genotype. The effect of the environment is expressed in the same way as for the genotype, as a deviation from the population mean. If an individual grew up in an environment causing it to grow a bigger beak than average, its environmental effect will be positive; and vice versa if it grew up in an environment giving it a smaller than average beak. The phenotype can then be expressed as the sum of environmental (*E*) and genotypic influences:

$P = G + E$

This, simple as it is, is the fundamental model of quantitative genetics. For any phenotypic character, the individual's value for that character (expressed, remember, as a deviation from the population mean) is due to the effect of its genes and environment.

We must look further into the genotypic effect. We need to consider both how to subdivide the genotypic effect, and why the subdivision is necessary. The main point can be seen in the one-locus example we have already used. The *A* gene is dominant, and both *AA* and *Aa* birds have 1 cm beaks ($P = +0.125$). (Because we are investigating the genotypic effect, it is simplest to ignore environmental effects, and then $P = G$.) Suppose we take an *AA* individual and mate it with another bird drawn at random from the population. The gene frequency is 0.5 and the random bird is *AA* with chance 0.25, *Aa* with chance 0.5, and *aa* with chance 0.25; but whatever the mate's genotype all the offspring will have beak phenotype $P = +0.125$ because *A* is dominant. Now suppose we take an *Aa* individual and mate it to a random member of the population. The average beak of their offspring can be calculated as $(0.25 \times +0.125) + (0.5 \times 0) + (0.25 \times −0.125) = 0$: the average offspring beak size is the same as the population average. Thus, for two genotypes with the same beak size ($P = +0.125$ cm), one produces offspring with beaks like their parent, the other produces offspring with beaks like the rest of the population.

So some genotypic effects are inherited by the offspring and some are not. We therefore divide the genotypic effect into a component that is passed on and a component that is not. The component that is passed on is called the *additive effect* (A) and the component that is not is called (in this case) the *dominance effect* (D). The full genotypic effect in an individual is the sum of the two:

$$G = A + D$$

The additive effect is the important one. The parent deviates from the population mean by a certain amount; its additive genotypic effect is the part of that deviation that can be passed on. However, when an individual reproduces, only half its genes are inherited by its offspring. The offspring inherit only half the additive effect of each parent. Thus the additive effect A for an individual is equal to twice the amount by which its offspring deviate from the population mean, if mating is random. For the AA parent, therefore, the additive effect is $= +0.25$. (The full quantitative genetics of the AA individuals is: $G = +0.125$, $A = +0.25$, and $D = -0.125$.) The offspring of Aa birds deviate by zero: their additive effect is twice zero—which is zero. (For Aa individuals, $G = +0.125$, $A = 0$, $D = +0.125$.) The exact values all depend on the gene frequency. The amount by which Aa heterozygotes deviate from the population mean is entirely due to dominance, and is not inherited by their offspring. The division of the genotypic effect into additive and dominance components tells us what proportion of the parent's deviation from the mean is inherited, and reveals how the non-inheritance of the Aa individuals' genotypic effect is due to dominance. (The negative value of the dominance effect of the AA homozygotes arises because these individuals do not deviate from the mean as much as they would in the absence of dominance—because the population mean has been "inflated" by the large beaks of Aa heterozygotes. Without dominance ($D = 0$), the population mean would have been 0.75, $G = A = +0.25$ for AA individuals, and $G = A = 0$ for heterozygotes. A non-zero dominance effect for any class of individuals indicates that dominance is influencing the character.)

In practice, quantitative geneticists do not know the genotypes underlying the characters they study; they only know the phenotype. They might, for instance, focus on the class of birds with 1 cm beaks ($P = +0.125$). We can work back from the genetics to find out what the additive effect is for this phenotypic class of birds. It contains individuals with both AA and Aa genotypes. Suppose we pick some birds with $P = +0.125$ and breed from them as before, by mating one bird with $P = +0.125$ to another bird drawn at random from the population. If the $P = +0.125$ bird has genotype AA, all its offspring will have $P = +0.125$; but if it has Aa then it can produce AA, Aa, or aa offspring. For a given gene frequency, we can calculate the average offspring phenotype. If the frequency of A is again 0.5, the class of birds with $P = +0.125$ contains one-third AA individuals and two-thirds Aa. The offspring of a $P = 0.125$ bird therefore have a one-third chance of having $P =$

+0.125 and a two-thirds chance of having $P = 0$. The average phenotype of the offspring of a $P = 0.125$ bird is $(\frac{1}{3} \times +0.125) + (\frac{2}{3} \times 0) = 0.04167$. And what is the additive effect? Again, only half the parent's additive effect is inherited by the offspring; for the birds with beaks $P = +0.125$, the additive effect is $2 \times 0.04167 = 0.08333$. Using $G = A + D$, $G = +0.125$, $A = +0.0833$, $D = +0.04167$ (Figure 9.5).

We have now seen what the additive effect of a phenotype is: but why is it so important? The answer is that once the additive effect for a character has been estimated, that estimate has much the same role in quantitative genetics as the exact knowledge of Mendelian genetics in a one- or two-locus case (see chapters 5 and 8). It is what we use to predict the frequency distribution of a character in the offspring, given a knowledge of the parents. In a one-locus genetic model, we know the genotypes corresponding to each phenotype, and can predict the phenotypes of offspring from the genotypes of their parents. In the case of selection, the gene frequency in the next generation is easy to predict if we know selection allows only *AA* individuals to breed. In

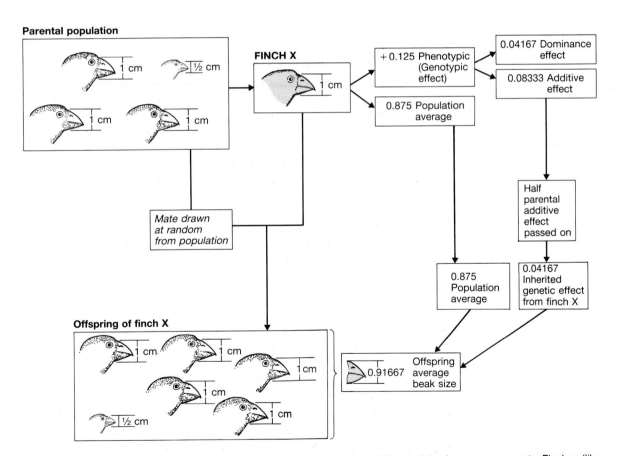

Figure 9.5 Analysis of genotypic effect into additive and dominance components. Finches (like 'finch X') with 1 cm beaks ($P = +0.125$) can have either of two genotypes, *AA* and *Aa*. Only part of the parental deviation from the mean appears in the offspring: that part is half the parental additive effect. See text for exact calculations.

two-locus genetics, the procedure is the same. If the next generation is formed from a certain mixture of Ab/AB and AB/AB individuals, we can calculate its haplotype frequencies if we know the exact mixture of parental genotypes.

In quantitative genetics, we do not know the genotypes. All we have is measurements of phenotypes, like beak size. But if we can estimate the additive genetic component of the phenotype, then we can predict the offspring in a manner analogous to the procedure when the real genetics is known. When we know the genetics, Mendel's laws of inheritance tell us how the parental genes are passed on to the offspring; when we do not know the genetics, the additive effect tells us what component of the parental phenotype is passed on. Estimating the additive effect is thus the key to understanding the evolution of quantitative characters. (In the example, the genetics was known: but that was in order to explain the concepts. In practice quantitative geneticists estimate the additive effects by breeding experiments. In the case of finches with 1 cm beaks in a population of average beak size 0.875 cm, the additive effect is estimated by mating 1 cm-beaked finches to random members of the population. The additive effect is then 2 times the offspring's deviation from the population mean.)

The genetic partitioning that we have made so far is incomplete. It applies reasonably well for one locus: in that case, dominance is the main reason why the genotypic effect of a parent is not exactly inherited in its offspring. When many loci influence a character, there can be effects due to *epistatic interactions* between alleles at different loci (see section 8.8, p. 196). Epistatic interactions are, like dominance effects, not passed on to offspring; they depend on particular combinations of genes and when the combinations are broken up (by genetic recombination) the effect disappears. They are not additive. An example of an epistatic interaction would be for individuals with the haplotype A_1B_2 to show a higher deviation from the population mean than the combined average deviations of A_1 and B_2; the extra deviation is epistatic. If the A and B loci are not linked, the extra epistatic effect will not be passed on, and the effect of the parental genotype in the offspring will be that of A_1 and B_2 separately, as each is inherited by different offspring. (If the two loci are linked the epistatic effect will be passed on with probability equal to 1 − recombination rate between the loci.) Other non-additive effects can arise because of gene–environment interaction (when the same gene produces a different phenotype in different environments) and gene–environment correlation (when particular genes are found more often than random in particular environments). A full analysis can take all these effects into account, but we do not need to enter into the details. In a full analysis, as in the simple one here, the aim is to isolate the additive effect of a phenotype. The additive effect is the part of the parental phenotype that is inherited by its offspring.

9.4 The variance of a character is divided into genetic and environmental effects

We return now to the frequency distribution of a character. We continue, as usual, to express effects as deviations from the population mean. If we consider an individual some distance from the mean, some of its deviation will be environmental, some genetic. Of the genetic component, some will be additive, some dominance, some epistatic. These terms have been defined so

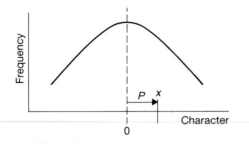

Figure 9.6 The x-axis of the continuous distribution for a character can be scaled to have a mean of 0. Consider the individuals (called x) with phenotype +P. Their phenotypic value are the sum of their individual combinations of E + A + D; the individuals with character x can have any combination of E, A, and D such that E + A + D = x. Any individual's deviation from the mean is due to its individual combination of environmental, additive, and dominance (and other, such as epistatic) effects.

that they add up to give the exact deviation of the individual from the mean. Any individual has its particular phenotypic value (P) because of its particular combination of environmental experiences and the dominance, additive, and interaction effects in its genotype (Figure 9.6). The different combinations of E, D, and A in different individuals are the reason why the character shows a continuous frequency distribution in the population.

The variation seen for the character in any one population could exist because of variation in any one of, or any combination of, these effects. Thus the individual differences could all be due to different environmental effects, with every individual having the same value for G. Or it could be 25% due to the environment, 20% to additive effects, 30% to dominance effects, and 25% to epistatic interaction effects. The proportion of variation due to the different effects matters when we wish to understand how a population will respond to selection. If all the variation exists because different individuals have different values of E, there will be no response to selection; but if the variation is mainly additive genetic variation, the response will be large. The proportion of the variation that is due to different values of A in different individuals tells us whether the population can respond to selection.

The variability in the population due to any particular factor, such as the environment, is measured by the statistic called the *variance*. (Box 9.1 explains some statistical terms used in quantitative genetics.) Variance is the sum of squared deviations from the mean divided by the sample size. So for all the values x of a character like beak size in a population, the variance of x is (see Box 9.1 for the notation):

$$V_x = \frac{1}{n} \sum (x_i - \bar{x})^2$$

We have seen how the total phenotype, genetic effect, environmental effect, and so on, can be measured for an individual; the measurements (P for phenotypic effect, etc.) are expressed as deviations from the mean. We can therefore easily calculate, for a population, the variances for all these factors:

BOX 9.1 Some statistical terms used in quantitative genetics.

The text mentions three main statistical terms. This box explains variance, but serves mainly as a reminder of the exact definition of covariance and regression; reference should be made to a statistical text for fuller explanation.

Variance. The variability of a set of numbers can be expressed as a variance. Take a set of numbers, such as 4, 3, 7, 2, 9. Here is how to calculate their variance.

1 Calculate the mean:

$$\text{Mean} = \frac{4 + 3 + 7 + 2 + 9}{5} = 5$$

2 For each number, calculate the square of its deviation from the mean. For the first number, 4, it is $(5 - 4)^2 = 1$. We do likewise for all five numbers.

3 Add up the sum of the squared deviations from the mean; for the five numbers, it is $1 + 1 + 4 + 9 + 16 = 31$.

4 Divide the sum by the sample size, n ($n - 1$ is sometimes used instead of n); $n = 5$ in this case.

$$\text{Variance} = \frac{31}{5} = 6.2$$

The general formula for the variance of a character X is

$$V_X = \frac{1}{n} \sum (x_i - \bar{x})^2$$

$\bar{x}$ is the mean. x_i is a standard notation for a set of numbers. Here we have five numbers. In terms of the notation, that means that i can have any value from one to five and is the value of the character for each i. Thus $x_1 = 4$, $x_2 = 3$, $x_5 = 9$. The summation in the general formula is for all values of i: here it is for all five numbers. If there had been 50 numbers, i would have varied from one to 50 and we should proceed as in the example for all 50 numbers. The variance describes how variable the set of numbers is: the higher the variance, the greater the differences among the numbers. If all the numbers were the same (all $x_i = \bar{x}$) then their variance is zero.

Covariance. Now imagine the individuals have been measured for two characters, X and Y. The covariance between the two is defined as

$$\text{cov}_{XY} = \frac{1}{n} \sum (x_i - \bar{x})(y_i - \bar{y})$$

Covariance measures whether, if an individual has a large value for X it also has a large value for Y. If the x_i and y_i of an individual are both large, the product $x_i y_i$ will also be large; if y_i is not large when x_i is, the product will be smaller. Generally, if X and Y covary, the product (and so the covariance) is large; and if they do not, the sum of the products will come to zero.

Regression. The regression, symbolized by b_{XY}, between characters X and Y is their covariance divided by the variance of X.

$$b_{XY} = \frac{\text{cov}_{XY}}{V_X}$$

Regressions are used to describe the slopes of graphs if X and Y are independent of each other, $b_{XY} = 0$; if there is a relationship between them, it can be positive or negative (Figure B9.1).

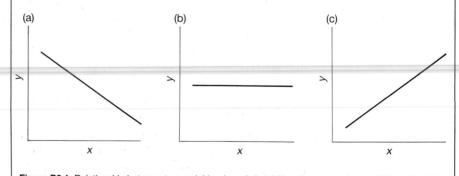

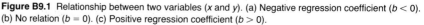

Figure B9.1 Relationship between two variables (x and y). (a) Negative regression coefficient ($b < 0$). (b) No relation ($b = 0$). (c) Positive regression coefficient ($b > 0$).

$$\text{Phenotypic variance} \quad = V_P = \frac{1}{n}\sum P^2$$

$$\text{Environmental variance} = V_E = \frac{1}{n}\sum E^2$$

$$\text{Genetic variance} \quad = V_G = \frac{1}{n}\sum G^2$$

$$\text{Dominance variance} \quad = V_D = \frac{1}{n}\sum D^2$$

$$\text{Additive variance} \quad = V_A = \frac{1}{n}\sum A^2$$

The important property of variances—and the reason why statisticians use them so much —is that they add up. Thus

$$P = G + E \quad and \quad V_P = V_G + V_E$$
$$G = D + A \quad and \quad V_G = V_D + V_A$$

For example, if all the variation in a population is due to different values of E in different individuals, $V_P = V_E$; or if $V_E = 14$, $V_D = 28$, and $V_A = 21$, then $V_P = 63$ and 33% of the variation is additive. In real cases, the variance terms can have any values (subject to the constraint that the equations hold true).

9.5 Relatives have similar genotypes, producing the correlation between them

In Figure 9.1, Grant measured the depth of the beak in many parent–offspring pairs in *Geospiza fortis*, and plotted the results on a graph. Each point is for one offspring and the average of its parents. Beak depth in *G. fortis*, like many characters in most species, shows a correlation between parent and offspring. This is an example of the correlation between relatives, for just as parents and offspring are similar to each other, so are siblings, and so are more distant relatives to some extent. Similarity among relatives may be either environmental or genetic, and the two effects may have different relative importance in different kinds of species. In humans, in which the young grow up together and both parents and offspring live in social groups, much of the similarity will be due to the common environment shared by relatives (i.e. there is gene–environment correlation). At the other extreme, in a species like a bivalve mollusc in which eggs are released into the sea at an early stage, relatives will not necessarily grow up in similar environments, and shared environments will not be such a strong influence on similarity among relatives. Darwin's finches are probably somewhere between these two extremes. The similarity among relatives that is caused by similar environments can readily be explained: in so far as relatives grow up in correlated environments, and there is environmental variation in a character, relatives will be more similar than non-relatives.

Of more interest to the evolutionary biologist is any similarity between relatives due to their shared genes. It is possible to deduce the correlation due to shared genes among any two classes of relatives from the variance terms we have already defined. We shall consider only one case, the correlation between parents and offspring, to see how it is done. We shall ignore epistatic interaction, to keep things simple. We can also assume that the environments of parent and offspring are uncorrelated, and ignore the environmental effects (because any environmental effect in the parent will not show up in the offspring: if the parent is larger than average because it chanced on a good food supply, that does not mean its offspring will too). Any correlation between parent and offspring will then be due to their genetic effects.

The genetic value of the character in the parent is, we have seen, made up of several components of which only the additive component is inherited by the offspring. When mating is random, half that additive component of the individual parent is diluted. At a locus, a parent has an additive deviation from the population mean in both its genes. When an offspring is formed, one of the parental genes goes into the offspring together with another gene drawn at random from the population (because we are assuming random mating). The average value of the character in the offspring is, as we saw above, half the additive value of the parent ($\frac{1}{2}A$); the average genetic value in the parent is $A + D$. The correlation between parent and offspring is the covariance between the two (Box 9.1):

$$\text{cov}_{OP} = \Sigma \tfrac{1}{2}A\,(A + D)$$

where the sum is over all offspring–parent pairs. The covariance can be re-expressed as:

$$\text{cov}_{OP} = \tfrac{1}{2}\Sigma A^2 + \tfrac{1}{2}\Sigma AD$$

$\frac{1}{2}\Sigma AD = 0$ because A and D have been so defined as to be uncorrelated: if an individual has a big value of A, we know nothing about whether its value of D will be big or small. That leaves

$$\text{cov}_{OP} = \tfrac{1}{2}\Sigma A^2 = \tfrac{1}{2}V_A$$

In words, the covariance of an offspring and one of its parents is equal to half the additive genetic variance of the character in the population.

The expression for the covariance between one parent and its offspring is true for each parent. It is a small step (though we shall not go into it here) to show that the same expression also gives the covariance between offspring and the midparental value: it is also $\frac{1}{2}V_A$. Other expressions can be deduced, by similar arguments, for the covariance between other classes of relatives (Table 9.1). The formulae are useful for estimating the additive variances of real characters. However, the estimates become most interesting, for the

Table 9.1 The covariance between several different classes of relatives

Relatives	Covariance
One parent and offspring	$\frac{1}{2}V_A$
Midparent and offspring	$\frac{1}{2}V_A$
Half siblings	$\frac{1}{2}V_A$
Full siblings	$\frac{1}{2}V_A + \frac{1}{4}V_D$

evolutionary biologist, when expressed in terms of the statistic called heritability.

9.6 Heritability is the proportion of phenotypic variance that is additive

The similarity between relatives in general, and between parents and offspring in particular, is governed by the additive genetic variance of the character. If a character has no additive genetic variance in a population, it will not be inherited from parent to offspring. For instance, many of the properties of an individual phenotype are accidentally acquired characters, such as cuts, scrapes, and wounds; if we measure these in parent and offspring they will show no correlation—V_A is zero. Moreover, some characters, such as the number of legs per individual in a natural population of, say, zebra, show practically no variation of any sort and for them V_A trivially is zero. Additive variance is therefore often discussed as a fraction of total phenotypic variance: and it is this fraction that is called the *heritability* (h^2) of a character

$$h^2 = \frac{V_A}{V_P}$$

Heritability is a number between zero and one. If heritability is one, all the variance of the character is genetic and additive. Given that $V_P = V_E + V_A + V_D$, all the terms on the right other than V_A must then be zero. In so far as the factors other than additive variance account for the variance of a character, heritability is less than one.

Heritability has an easy intuitive meaning. Consider a parent that differs from the population by a certain amount. If its offspring also deviate by the same amount, heritability is one; if the offspring have the same mean as the population, heritability is zero; if the offspring deviate from the mean in the same direction as their parents but to a lesser extent, heritability is between zero and one. Heritability, therefore, is the quantitative extent to which offspring resemble their parents, relative to the population mean.

How can we estimate the heritability of a real character? One method is to cross two pure lines. This is mainly of interest in applied genetics, where the problem might be to breed a new variety of crops; it has little interest in evolutionary biology. The two other main methods are to measure the correlation between relatives and the response to artificial selection. Figure 9.1 is an example which uses the correlation between relatives. The slope of the graph, which shows the beak size in offspring finches in relation to the average beak size of the two parents, is equal to the heritability of beak size in that population. The reason is as follows. The slope of the line is the regression of offspring beak size on midparental beak size. The regression of any variable y on another variable $x = \text{cov}_{xy}/\text{var}_x$ (Box 9.1). The covariance of offspring and midparental value $= \frac{1}{2}V_A$ (Table 9.1) and the variance of the midparental beak size is equal to $\frac{1}{2}V_P$. (It is half the total population variance because *two* parents have been drawn from the population and their values averaged: if Figure 9.1 had the value for one parent on the x-axis, its variance would be V_P.) The regression slope simply equals V_A/V_P, which is the character's heritability. For beak depth in *Geospiza fortis* on Daphne Major, the regression and therefore heritability is 0.79.

9.7 A character's heritability determines its response to artificial selection

How can quantitative genetics be applied to understand evolution? There are many ways, and we shall consider two of them here: directional selection and stabilizing selection (section 4.4, p. 66, explained the terms). This section will be concerned with *directional selection*, which has particularly been studied through artificial selection experiments. Artificial selection is important in applied genetics, as it provides the means of improving agricultural stock and crops.

If we wish to increase the value of a character by artificial selection, we can use any of a variety of selection regimes. One simple form is truncation selection: the selector picks out all individuals whose value of the character under selection is greater than a threshold value, and uses them to breed the next generation (Figure 9.7). What will be the value of the character in the offspring generation? First, we can define S as the mean deviation of the selected parents from the mean for the parental population; S is also called the "selection differential." The response to selection (R) is the difference between the offspring population mean and the parental population mean. In this case, calculating the response to selection is simply a matter of regressing (or correlating) the offspring value on the parental value, where the parents are the individuals that were selected to breed. As we saw in the previous section, the regression of offspring on midparent (b_{OP}) values for any character equals the heritability. Therefore:

$$R = b_{OP}S$$

or

$$R = h^2 S$$

This is an important result. The response to selection is equal to the amount by which the parents of the offspring generation deviate from the mean for their population multiplied by the character's heritability. (The response to selection or the parent–offspring regression can be used to estimate the heritability of a character; for a selected population, they are two ways of looking at the same set of measurements.)

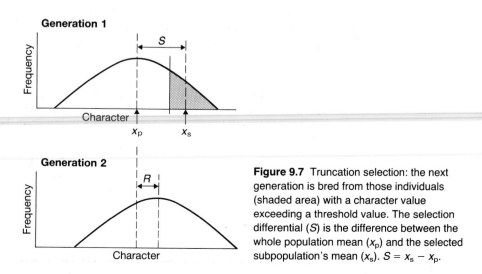

Figure 9.7 Truncation selection: the next generation is bred from those individuals (shaded area) with a character value exceeding a threshold value. The selection differential (S) is the difference between the whole population mean (x_p) and the selected subpopulation's mean (x_s). $S = x_s - x_p$.

A real example of directional selection may not have the form of truncation selection. In truncation selection, all individuals above a certain value for the character breed and all individuals below do not breed. All the selected individuals contribute equally to the next generation. It could be instead that there is no sharp cut-off, but that individuals with higher values of the character contribute increasing numbers of offspring to the next generation. However, the same formula for evolutionary response works for all forms of directional selection. The difference between the mean character value in the whole population and in those individuals that actually contribute to the next generation (if necessary, weighted by the number of offspring they contribute) is the selection differential and can be plugged into the formula to find the expected value of the character in the next generation.

A population can only respond to artificial selection for as long as the genetic variation lasts. Consider, for example, the longest running controlled artificial selection experiment. Since 1896, corn has been selected, at the State Agricultural Laboratory in Illinois, for (among other things) either high or low oil content. As Figure 9.8 shows, even after 76 generations the response to selection for high oil content was not exhausted. However, it will come to a stop eventually. As the corn is selected for increased oil content, the genotypes encoding for high oil content will increase in frequency and be substituted for genotypes for lower oil content. This process can only proceed so far. Eventually all the individuals in the population will come to have the same genotype for oil content. At the loci controlling oil content, no additive genetic variance will then be left; heritability will have been reduced to zero and the response to artificial selection will come to a stop. In the Illinois corn experiment, the process has not yet run its full course. The heritability of oil content in both the high and low selected lines has decreased through the experiment (Table 9.2), but it is still above zero—which is why the population is still responding to selection.

In other artificial selection experiments, the full process has been recorded. Figure 9.9 shows the response of a population of fruitflies to consistent directional selection for increased numbers of scutellar chaetae (i.e. bristles on a dorsal region of the thorax). Initially the population responded; then, as the additive genetic variation was used up (as its heritability declined), the rate of change slowed down to a stop in generations 4–14. It also appeared that, if

Table 9.2 The heritability of oil content in corn populations after different numbers of generations of artificial selection for high or low oil content, in the Illinois corn experiment (see Figure 9.8). The heritability declines as selection is applied. From Dudley (1977)

Generation	Heritability of oil content	
	High line	Low line
1–9	0.32	0.5
10–25	0.34	0.23
26–52	0.11	0.1
53–76	0.12	0.15

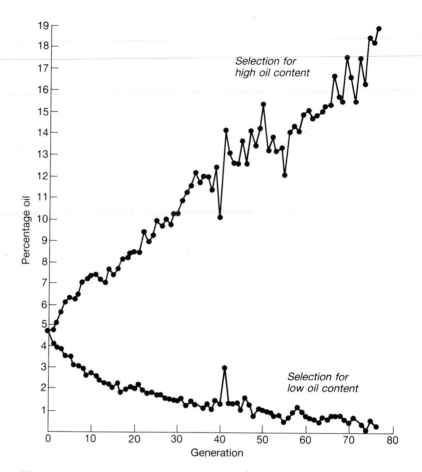

Figure 9.8 Response of corn (*Zea mays*) artificially selected for high or low oil content. The experiment began in 1896 when, from a population of 163 corn ears, the high line was formed from the 24 ears highest in oil content and the low line from the 12 ears with lowest oil content. From Dudley (1977).

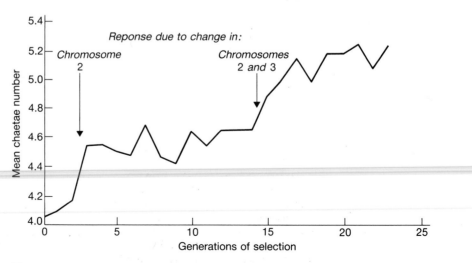

Figure 9.9 Artificial selection for increased numbers of scutellar chaetae in *Drosophila melanogaster*. The response took place in two rapid steps; the steps coincided with observable changes in the form of chromosomes 2, and 2 and 3, respectively; the changes are thought to have been recombinational events. From Mather (1943).

selection was still continued after the response stopped, the population suddenly started to respond again after an interval (in generations 14–17). The renewed bout of change is attributed to a rare recombinant or mutation that reinjected new genetic variation into the population.

The relationship between response to selection (R), heritability, and selection differential (S) enables us to calculate any one of the three variables if the other two have been measured. For example, we saw in chapter 4 that fishing has selected for small size in salmon, because larger fish are selectively taken in the nets. The selection differential S can be estimated from three measurements: the average size of salmon caught in the nets, the average size of salmon in the population at the mouth of the river (before it is fished), and the proportion of the population that is taken by fishing. All three have been measured and lead to the estimate that the salmon who survive to spawn are about 0.4 lb (0.9 kg) smaller than the population average. Figure 4.3 (p. 66) shows the response (R): the average size of the salmon decreased by about 0.1 lb between each 2 year generation. We can therefore estimate the heritability, $h^2 = 0.1/0.4 = 0.25$.

In the finches, Grant measured the response to selection and heritability for several characters related to body size, and used these to estimate selection differentials. We saw that in *Geospiza fortis* heritability of beak size is about 80%; and after the bout of selection for large size in 1976–1977, the finches were about 4% larger. Four per cent corresponds to an increase of about 0.47 standard deviations. (One standard deviation equals the square root of the variance for a population. See Box 9.1 for definition of variance.) We can then estimate the selection differential as $S \simeq 0.47/0.79 \simeq 0.6$ standard deviations. The results for several characters in three periods is shown in Figure 9.10. As the direction of selection reversed from favoring larger beaks between 1976 and 1978 and smaller beaks between 1983 and 1985, the selection differentials reversed from about +0.5 in 1976–1977 to about −0.2 in 1984–1985.

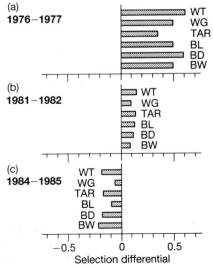

Figure 9.10 (a) In the drought of 1976–1977, *Geospiza fortis* individuals with larger beaks survived better and the average size of the finch population increased. (b) In the normal years of 1981–1982 there was a slight advantage to having a larger beak, but much smaller than during the drought. (c) After the 1983 El Niño, in 1984–1985, finches with smaller beaks survived better and the average size of the finch population decreased. WT, weight; WG, wing length; TAR, tarsus size; BL, bill length; BD, bill depth; BW, bill width. From Gibbs and Grant (1987).

(a)

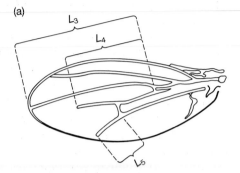

(b)

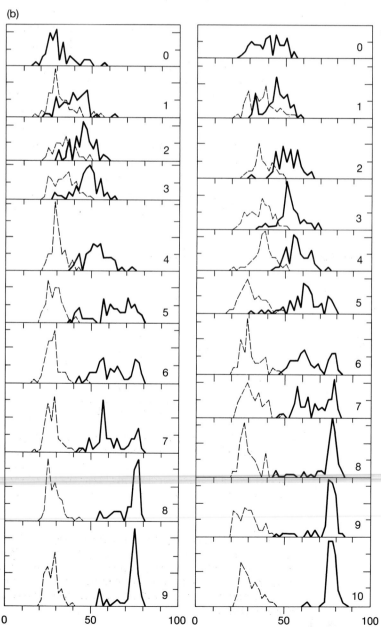

In summary, the response to directional selection is controlled by the heritability of a character. The relationship between the selection differential, response, and heritability can be used to estimate one of the variables if the other two are known. Sustained selection will use up the heritable variation, and the response will then come to a stop.

9.8 The relationship between genotype and phenotype may be non-linear, producing remarkable responses to selection

The remarkable bimodal response to artificial selection on wing vein length in the fruitfly shown in Figure 9.11 reveals that things can be more complicated than suggested by the results of the previous section. The key to understanding this shape of response is the relationship between genotype and phenotype. The typical response, such as that for oil content in Figure 9.8 or bristle number in Figure 9.9, results when there is an approximately linear relation between genotype and phenotype (Figure 9.12a). Genotype here is expressed as a metric variable. The easiest way to think of this is to imagine that the character is controlled by many loci; at each, some alleles (+) cause the phenotypic character to increase, and others (−) to decrease. The more + alleles an individual has, the higher its genotypic value (Figure 9.12a). An equivalent formulation is for the character to be controlled by a smaller number of loci, each with many alleles of varying power to increase the phenotypic character; an individual's genotypic value can be thought of as the sum of all its genotypic values at all relevant loci.

The approximately linear form of Figure 9.12a is not the only possible relationship between genotype and phenotype (cf. Figures 9.12b,c). The bimodal response in Figure 9.11 is thought to result from a threshold

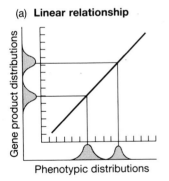
(a) **Linear relationship**

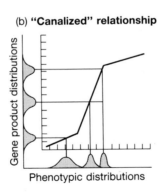
(b) **"Canalized" relationship**

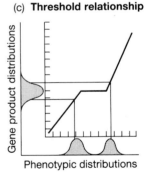
(c) **Threshold relationship**

Figure 9.12 Relationship between genotype and phenotype. The genotypic value is for a polygenic character, in which higher values might be produced by more + alleles. The stippled frequency distribution for the phenotype at the bottom is what would be observed, given the shape of the graph, if the genotypic frequency distribution were as illustrated. (a) Linear relationship. (b) "Canalized" relationship. (c) Threshold relationship.

Figure 9.11 (*Opposite*) (a) The main veins in the wing of *Drosophila*. Relative length of the fourth vein was measured by the ratio $L_4 : L_3$. (b) Artificial selection for relative length of fourth vein: — selection for long vein; --- selection for short vein. The series on the left is for females; males on the right. Each graph is a frequency distribution for the selected population. In both males and females, a bimodal distribution appears at vein lengths of about 60–80. From Scharloo (1987).

relationship between genotype and phenotype (Figures 9.12c and 9.13). In Figure 9.13, the graph has been rotated through 90° relative to the form in Figure 9.12; the *x*-axis (genotype) is drawn down the page on the left. The genotype is thought to control the amount of some vein-inducing substance. Vein length is shown across the top of the graph. The relationship between substance and vein length is hypothesized to contain a jump at vein length 60–80 (where the artificial selection response becomes bimodal).

Imagine the course of selection for longer wing veins with this threshold relationship between genotype and phenotype. The population starts at the top left of the graph with relatively short veins. Initially, there is some variation in the population for genotype, and an associated normal dis-

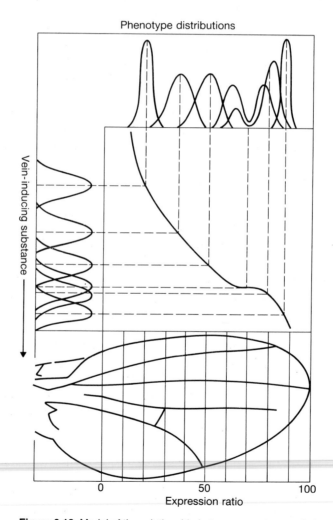

Figure 9.13 Model of the relationship between genotype and phenotype for *Drosophila* fourth wing vein. The genotype produces an amount of vein-inducing substance. The phenotype (vein length) is drawn along the top. The graph shows the relation between genotype and phenotype. The phenotype in the 60–80 range matches a threshold jump in the genotype–phenotype relationship; a bimodal phenotypic frequency distribution arises there. From Scharloo (1987).

tribution of vein lengths. Artificial selection produces flies with a higher concentration of the vein-inducing substance and with correspondingly longer wing veins. This process can continue until the population reaches a vein length of about 60–80. At that point, the normal (unimodal) variation for vein-inducing substance concentration generates a bimodal distribution of vein lengths. The bimodal distribution will later disappear, as the population passes beyond the jump in the genotype–phenotype mapping function. Hence the observed response to selection. The relationship of genotype and phenotype for vein length in Figure 9.13 is a hypothesis only; but it does show how in theory a bimodal response to selection could arise.

The main points are that when the genotype–phenotype relationship has the linear form of Figure 9.12a there is a simple response to artificial selection. The population changes until the genetic variation is used up. However, we have no reason to think that this is the typical genotype–phenotype relationship. When the relationship is more complex, the response to artificial selection can be interestingly different, as the bimodal response to selection on wing vein length in fruitflies illustrates.

9.9 How much genetic variability will polygenic characters have?

Natural populations have been undergoing the natural analogy of artificial selection for many generations. They can be thought of as being at the end of, rather than the beginning or anywhere else in, the selection experiment. For characters that directional selection has been working on, the heritability will have been reduced to a low level. We should expect that characters closely related to the reproductive fitness of individuals will show low heritability. (Incidentally, Figure 9.14, which is discussed more later, illustrates this point, as the heritabilities for life history characters are lower than those for morphologic characters.) High heritability can exist for characters that have little influence on the number of offspring its bearers produce. For characters that do influence fitness, higher heritabilities can be maintained if selection is balancing rather than directional; and if there is heterozygous advantage (section 5.11, p. 109) at a number of loci, significant heritability will be maintained.

For many characters, an intermediate rather than an extreme form is probably optimal: this is called *stabilizing selection*. (See section 4.4, p. 67, where Figure 4.4 illustrates how birth weight in humans is an example of stabilizing selection.) It is important to distinguish between stabilizing selection and heterozygous advantage. If, at one locus, *Aa* is intermediate in phenotype between *AA* and *aa*, then if there is stabilizing selection there is also heterozygous advantage. But if the heterozygote is not intermediate the identity breaks down. If *AA* is intermediate in phenotype between *Aa* and *aa*, then if there were stabilizing selection it would favor *AA*, not the heterozygote.

The difference between heterozygous advantage and stabilizing selection is more important for polygenic characters. Suppose the value of a character in an organism is influenced by a large number of loci. At each locus there are two possible alleles; one increases (+) the value of the character, the other decreases (−) it. An individual's haplotypes will then each be a series of alleles, and might for example be symbolized by −++−+−−++ (for 10 loci). An intermediate phenotype could be brought about by any genotype

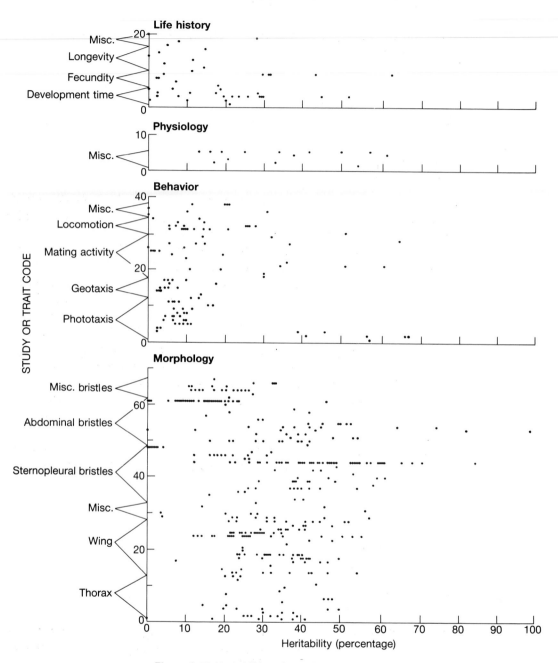

Figure 9.14 Heritabilities of quantitative characters in *Drosophila*. Roff and Mousseau (1987) compiled heritability estimates from 130 studies and divided them into morphologic, behavioral, and life history characters. The 130 studies are ordered arbitrarily up the *y*-axis: morphologic character 43, for instance, is "sternopleural bristle two sides" and morphologic character 61 is "abdominal bristles on one sternite." Each line consists of all the heritability estimates in that study. Some characters are the subject of more than one study and will have more than one line in the figure. Morphologic characters have heritabilities in the range 0.1–0.5; behavioral and life history characters have rather lower heritabilities, of about 0.1. From Roff and Mousseau (1987).

made up of half + genes and half − genes, such as any of the following (among others):

1 +++++++++
  − − − − − − − − −

2 ++++−−−−−
  − − − − −+++++

3 ++++−−−−−
  ++++−−−−−

Cases 1 and 2 would produce heterozygous advantage. But selection could also favor the intermediate genotype 3, and in a population containing only this haplotype there is no heterozygous advantage.

How much genetic variability do we expect for a character subject to stabilizing selection? In a population containing the four haplotypes, the three genotypes 1, 2, and 3 will all have the same fitness, and we might expect a population to retain considerable genetic variation as these genotypes interbreed and produce a variety of offspring types. However, over a long period of time, there is an advantage to genotypes like 3 that breed true, because all genotype 3's offspring have the optimal phenotype, whereas some of the offspring of genotypes 1 and 2 do not. In a population made up of these three genotypes, selection slightly favors genotype 3. If the environment were constant for a long time, always favoring the same phenotype, selection should eventually produce a uniform population with a genotype like 3.

We can take the argument a stage further. Genotype 3 is not the only true-breeding homozygote that can produce the intermediate optimal form. All the following do too:

4 +−+−+−+−+−
  +−+−+−+−+−

5 ++−−++−−+−
  ++−−++−−+−

6 +++−−−++−−
  +++−−−++−−

Suppose there was a population made up of genotypes 3–6, and selection still favors the intermediate phenotype. What will happen now? Evolution will again tend toward a population with only one genotype, and that genotype should be a multiple homozygote. The reason is that any one of the homozygotes that happens to have a slightly higher frequency than the others has an advantage. Suppose, for example, that genotype 4 had a higher frequency than 3, 5, and 6. All the genotypes will now be most likely to mate with a genotype 4. When genotype 4 mates with genotype 4, all their offspring have the favored phenotype, identical to their parents. But when genotype 3, 5, or 6 mate with genotype 4, the offspring contain potentially disadvantageous genotypes. The offspring of a mating between a genotype 5 and a genotype 4 will be +−+−+−+−+−/++−−++−−+− and has the favored intermediate phenotype. However, *its* offspring will contain disadvantageous recombinants. The end result is for selection to produce a uniform population, in which minority genotypes are selected against because they do not fit in with the majority form. Selection eventually reduces the genetic variability to zero, even with stabilizing selection.

The argument leads to the following conclusion. If we start with a set of genotypes which influence a character that is subject to stabilizing selection, and if the environment is constant for long enough, then selection eventually ought to produce a population containing only one, multiply homozygous, genotype. Alas, the conclusion is contradicted by observable facts. Heritabilities can be measured for real characters, and many show significant genetic variation. Figure 9.14 summarizes some measurements for *Drosophila*. It suggests that a typical figure for heritability is in the range 0.1–0.5. These estimates are for laboratory populations, and most reliable heritability estimates are unfortunately artificial; but there are measurements from nature too, such as Grant's for the Galápagos finch, and these natural observations fit in with the conclusion from Figure 9.14: real characters often have heritabilities higher than zero.

If stabilizing selection tends to eliminate heritable variation, why does it exist? The question is currently a matter of theoretical controversy. The traditional answer was that additional processes must be at work. Heterozygous advantage is an example. In the argument above we assumed that an intermediate phenotype could be equally well produced by a $+++---/$ $+++---$ homozygote as by a $+++++/-----$ heterozygote; then, as we saw, selection will fix the homozygotes. But if, for some reason, the heterozygotes were fitter, genetic variation would persist because of the segregation of recombinant genotypes from the heterozygotes. Just why heterozygotes should be fitter is another question; but if they are, it would explain the maintenance of genetic variation. A second possibility is that selective equilibrium is not reached because the phenotype favored by selection changes rapidly. Darwin's finches may be an example; the best beak size may change every few years, and genetic variation for beak size will persist because selection never favors one genotype for long enough to fix it in the population. Either of these arguments may be correct in particular cases; however, we should also consider a third possibility. Lande has argued that a balance of stabilizing selection and mutation alone could maintain the observed levels of genetic variation.

Lande's argument is plausible because in real populations, stabilizing selection does not work on a fixed initial set of genotypes. Mutation continually introduces new genetic variability. Although mutation rates for individual genes are low and the amount of genetic variability maintained at a single locus by mutation–selection balance is low (section 5.10, p. 107), for a polygenic character, mutation rates should be multiplied by the number of loci influencing the character. A character controlled by 500 loci will have 500 times the mutation rate of a one-locus character. The amount of variability that can be maintained is proportionally increased.

The exact arguments concerning whether mutation can, or cannot, account for the observed heritabilities are rather involved. Box 9.2 gives an outline of

Box 9.2 Can a balance of mutation and stabilizing selection explain genetic variation in quantitative characters?

Lande tackled the question by developing formulae derived by Kimura and by Latter. We need a formula for the rate at which variation is removed by stabilizing selection and another for the rate it is added by mutation; the balance of the two then gives the amount of genetic variation we should see if only these two processes are operating. The full argument is too long to present here, but we can follow the gist of it. The approach here is simplified from the work of Turelli. The strength of stabilizing selection can be expressed by the ratio V_S/V_E. V_E is as defined before; it is the environmental variance of a character. V_S is the phenotypic variance of the part of the population that has been selected to produce the next generation

(Figure B9.2). V_E is an upper limit on the strength of selection. The strongest possible selection happens when all the selected individuals that will produce the next generation have the same genotype; they would still show phenotypic variability equal to the environmental variance, V_E. If selection is less extreme, V_S will exceed V_E, because the selected individuals will contain more than one genotype. Turelli collected evidence to suggest that, in real cases of stabilizing selection, the ratio V_S/V_E ranged from 5 to 100; he used 20 as an approximate number.

Mutation will be generating variability in the population. Clayton and Robertson defined V_M as the variance due to

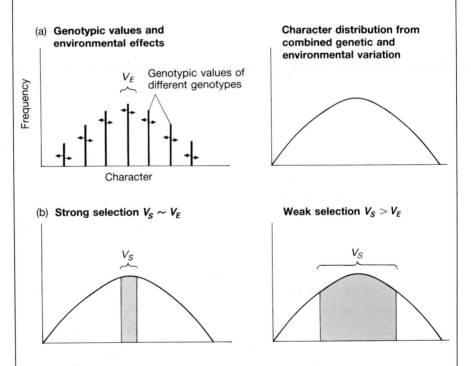

(a) **Genotypic values and environmental effects**

Character distribution from combined genetic and environmental variation

(b) **Strong selection $V_S \sim V_E$**

Weak selection $V_S > V_E$

Figure B9.2 The degree of stabilizing selection can be described by the ratio V_S/V_E. (a) The frequency distribution of a character exists because there are several (here seven) different genotypes, with different genotypic values, and the environment causes individuals with the same genotype to grow up with some deviation from the genotypic value. The genotypic and environmental effects combine to produce the normal character distribution. (b) If there is extremely strong selection, such that all the selected individuals have the same genotype, they will still have variance of V_E. A V_S/V_E ratio of one is therefore the strongest possible selection. Under weaker selection, the selected individuals will have some genotypic as well as the environmental variation: $V_S/V_E > 1$.

continued on page 238

mutation. It is made up of mutations of various effects on the character. If there are n loci, numbered from $1 \ldots n$, then mutation rate can be defined as μ_i and the average effect on the phenotype of a mutation as m_i at the ith locus (the effect, as usual, is measured as a deviation from the mean). The total variance in the character introduced by mutation is then, for a population of diploid individuals:

$$V_M = 2 \Sigma \mu_i m_i^2$$

the summation being over the n loci. (Therefore, if we think of μ and m as averages for all n loci, we can write the formula as $V_M = 2n\mu m^2$.) Lande collected several estimates of V_M; they suggested that

$$V_M \simeq 10^{-3} V_E$$

If the influence of mutation and stabilizing selection are put together, the amount of genetic variability is given by a formula that Kimura first derived. We shall not work through the derivation here, but the formula is

$$V_G \simeq \sqrt{2nV_M V_S}$$

The reader can confirm that the observed values of $V_M \simeq 10^{-3} V_E$ and $V_S = 20$ do indeed fit in with heritabilities of about 0.1–0.5. (Heritability = $V_G/(V_G + V_E)$.) We have now reached the conclusion of Lande's argument: realistic values of mutation and stabilizing selection in polygenic systems can alone account for the observed levels of heritability.

Turelli, however, noticed that Kimura's formula assumes mutations of very small effect. More exactly, it requires $m_i^2 \ll V_{G(i)}$: the effect of mutation at the ith locus is much smaller than the genetic variance at the locus. The condition can be tested crudely using Lande's own estimate of V_M

$\simeq 10^{-3} V_E$. Even a mutation rate as high as 10^{-4} implies $m_i^2 \simeq 10 V_{G(i)}$, and the mutational effects will be much larger with more plausible mutation rates of about 10^{-6}. The assumption is then wrong.

It is possible to re-derive Kimura's formula for $m_i^2 > V_{G(i)}$. Turelli has done so, and concludes that Lande's argument cannot be saved in this way. To explain the observed heritability values, very high mutation rates and numbers of loci are needed, such as 1000 loci with mutation rates of 10^{-5} each. More realistic mutation rates (about 10^{-6}) and numbers of loci (<100) imply much lower levels of genetic variability than are observed. Turelli therefore disagreed with Lande's conclusion.

Lande's general idea was to use estimates for mutation rates, the degree of stabilizing selection, and genetic variability, to see whether a balance between stabilizing selection and mutation could explain the amounts of genetic variability. It has been a fruitful line of research. The estimates used in the formulae are quite crude—they were made for artificial populations—and the conclusions may have to be modified. At present, it appears that the mutation–selection balance alone will not explain the high observed levels of heritability. Some further factor, presumably selection, is probably needed to account for the facts, and further theoretical work has been done on the nature of the selection that may be at work.

Further reading. For summaries see Barton and Turelli (1989), Bulmer (1989), and Turelli (1986, 1988). The original papers were Lande (1976) and Turelli (1984, 1985). See also Kondrashov and Turelli (1992).

Box 9.2 (*continued*) Can a balance of mutation and stabilizing selection explain genetic variation in quantitative characters?

them. The main conclusion from the box is tentative, but suggests that mutation alone is not enough. Some further process, such as heterozygous advantage, must be at work in natural populations to explain the facts shown in Figure 9.14.

In summary, with stabilizing selection, organisms containing a large number of alleles that increase (+) or decrease (−) the value of the character under selection are at a disadvantage. Selection favors genotypes with an intermediate number of + and − alleles. With this sort of genetic control, many kinds of genotype can produce the same phenotype. But over time, stabilizing selection will reduce the number of genotypes. Eventually it should reduce the number to one: all the individuals in the population would have the same homozygous multiple-locus genotype controlling the character. In fact characters in laboratory and natural populations (many of which are probably subject to stabilizing selection) have high amounts of genetic variability. Two reasons have been suggested. One (argued for by Lande) is that a balance between stabilizing selection and mutation alone can explain the observations. The other (argued for by Turelli and others) is that mutation alone is not enough, and further kinds of selection, such as heterozygous advantage or temporal changes in selective pressure, are needed to explain the maintenance of genetic variability in natural populations.

9.10 Conclusion

One- and two-locus population genetics is used for characters controlled by one or two loci and whose genetics is known. Quantitative genetics provides the techniques to understand evolution in characters that are influenced by a large number of genes, and for which the exact genotype (or genotypes) producing any given phenotype are unknown. It is possible that the majority of characters have this kind of genetics, in which case quantitative genetics would be appropriate for understanding the majority of evolution; at any rate, it is a highly important set of techniques. We have seen in this chapter how quantitative genetics divides up the variation in a character to recognize the component that controls how offspring resemble their parents; the component is called the additive genetic effect. The additive genetic effect plays the same role in quantitative genetics as a knowledge of Mendelian genetics in one- and two-locus population genetics. The response to selection can be analyzed by means of the heritability of a character, which is the fraction of its variation due to additive genetic effects. However, even with simple directional selection, the exact response depends on the underlying genetic control. For example, the possible threshold relationship between the genotype and phenotype for the wing veins of the fruitfly generates an interesting bimodal response to selection. Here, the heritability of the character would show strange changes as the character evolved. Directional selection unambiguously should continue to alter a character until its heritability is reduced to zero. With stabilizing selection, it might be thought that many genotypes could be maintained if they all produce the same intermediate phenotype. However, even here it can be argued that all but one of the genotypes should eventually be eliminated by selection. The argument appears to be contradicted by the facts, and it remains unsettled how mutation and the exact form of stabilizing selection stop populations from becoming genetically uniform.

9.11 Summary

1 Quantitative genetics, which is concerned with characters controlled by many genes, considers the changes in phenotypic and genotypic frequency distributions between generations, rather than following the fate of individual genes.

2 The phenotypic variance of a character in a population can be divided into components due to genetic, and to environmental, differences between individuals.

3 Some of the genetic effects on an individual's phenotype are inherited by its offspring; others are not. The former are called additive genetic effects; the latter are due to such factors as dominance and epistatic interaction between genes.

4 The heritability of a character is the proportion of its total phenotypic variance that is due to additive genetic effects.

5 The heritability of a character determines its evolutionary response to selection.

6 The additive genetic variance can be measured by the correlation between relatives, or by artificial selection experiments.

7 The response of a population to artificial selection depends on the amount of additive genetic variability and on the relationship between genotype and phenotype. If the relationship is non-linear, strange bimodal responses can arise.

8 Stabilizing selection acts to reduce the amount of genetic variability in a population. It has been controversially argued that the observed genetic variability of populations is a balance between mutation and stabilizing selection in polygenic characters; but some other factors may be needed to explain the observed amounts of genetic variability.

9.12 Further reading

Moving from the introductory to the advanced, the following are accounts of quantitative genetics: Falconer (1981), Lewontin's chapter in Suzuki *et al.* (1989), Bulmer (1980), and Wright (1969). Charlesworth (1980) shows how quantitative genetics can be applied to age-structured populations—as is necessary for such problems as life histories and senescence. Hill (1984) is an anthology of classic papers on quantitative genetics. The volumes edited by Pollak *et al.* (1977), Thompson and Thoday (1979), and Weir *et al.* (1988) contain many papers. On the finches, see Grant (1986, 1991). Mitchell-Olds and Rutledge (1986) review the application to natural plant populations.

On selection, see the references in chapter 4 for introductions. On the limits to the response, see Lee and Parsons (1968). Waddington (1957) and Rendel (1967) are classic discussions of the relationships between genotype and phenotype for polygenic characters. Scharloo (1987, 1991) reviews non-linear selection responses and canalization. See Box 9.2 for references on variation and stabilizing selection, together with Loeschcke (1987).

Genome evolution

10.1 Non-Mendelian processes must be added to classical population genetics to explain the evolution of the whole genome

The theory of population genetics, as we have discussed it in chapters 5–9, has been concerned with discrete genetic loci that are inherited according to Mendel's laws. The theory shows how mutation, selection, drift, migration, and linkage determine changes in gene frequencies. That theory was first developed in the 1920s and 1930s, even before the dawn of modern molecular genetics. Since the 1960s, however, it has become increasingly clear that only a part, probably a small part, of the DNA in an organism is made up of discrete genes, each of which codes for a protein; there are large amounts of non-genic DNA, and it is often arranged in the form of repeated unit sequences. The genes themselves, moreover, can have distinctive arrangements on the chromosomes; related genes, for instance, may be distributed in clusters in the DNA. These discoveries do not contradict the theory of population genetics; indeed, the theory of population genetics is being used to make sense of them. However, some extra concepts are needed. In this chapter we shall concentrate on two such concepts: concerted evolution and selfish DNA. They are related in that both depend on non-Mendelian heredity.

Any theory of how a genetic (or any other) phenomenon evolves needs two components. It must specify how the phenomenon originated, and then how it spread through the population. In the standard theory of evolution, phenomena originate as mutations at a locus and then spread by either natural selection or random drift. The new ideas in this chapter are mechanisms for the origin of variants, not for their spread through a population: they do not replace the theories of selection and drift.

10.2 Genes are arranged in gene clusters

How are the many genes of an organism arranged in its DNA? The theory of population genetics would all make good sense if single genes were scattered around arbitrarily (though we have seen in chapter 8 how selection can bring coadapted genes together into a supergene). Some genes may indeed exist by themselves as single copies, but many—perhaps the majority—of genes are arranged in groups of related genes. The groups are called *gene clusters*. Related genes may be arranged in more than one physical cluster and the whole family of related genes is called a *gene family*. Gene clusters have turned out to have peculiar evolutionary properties, which have been most studied in two examples: the ribosomal RNA genes and the globin gene family.

Eukaryotic ribosomes are made up of three classes of ribosomal RNA: 5S, 18S, and 28S. The 18S and 28S parts are formed from a large 45S precursor

(a) Eukaryotic ribosomal RNA genes: tandem repeats

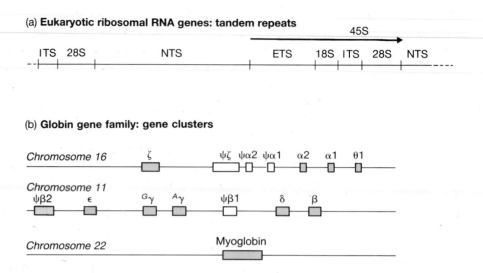

(b) Globin gene family: gene clusters

Figure 10.1 (a) Eukaryotic ribosomal RNA genes consist of a long row of tandem repeats of a unit genic organization, illustrated here. The transcribed part consists of the 18S and 28S genes together with an internal (ITS) and external transcribed spacer (ETS) region. The NTS part is a non-transcribed spacer. (b) The globin gene family consists of a number of clusters of related genes. In humans, illustrated here, there are three clusters. In other species, the globin gene in each cluster and the distribution of the clusters through the chromosomes differs from humans.

molecule which is transcribed from the DNA as a whole. Along the DNA, the 45S ribosomal genes are arranged in rows of multiple copies of a recognizable unit (Figure 10.1a); hundreds of these units are strung out in a row of *tandem repeats*. In the African clawed toad *Xenopus laevis* the 400–600 tandem repeats are all on one chromosome; in humans, approximately 300 repeats are scattered through five chromosomes. The organism presumably needs multiple copies of the gene in order to synthesize large quantities of the gene product. Other functional genes also exist in tandem repeats. The five histone genes, at least in *Drosophila* and sea urchins, have this structure; the histone genes of vertebrates are also repetitive, but dispersed.

The globin gene family has a slightly different arrangement. The hemoglobin molecule that transports oxygen in the blood is made up of a number of components. Human hemoglobin is a tetramer: in adults it is made up of two α-globins and two β-globins; in the fetus, of two α- and two γ-globins; and in the embryo, two ε- and two ζ-globins. The sequences of these five globin types are similar and the genes that code for them occur in clusters on the chromosomes. In humans, there are two main clusters of globin genes: the α-globin cluster on chromosome 16 and the β-globin cluster on chromosome 11; the myoglobin gene on chromosome 22 is also part of the same family (Figure 10.1b). Unlike the ribosomal RNA genes, the globins are not simple tandem repeats; the different globin genes are distinct genes with distinct sequences. But the ribosomal RNA genes and the globin genes share the common property of being linked in clusters.

10.3 Gene clusters probably originated by gene duplication

How could a cluster of genes, like the globin gene family, originate? For related genes strung out along a chromosome, duplication is the most likely mechanism. Gene clusters are so common that duplication must have been a highly important mechanism during evolutionary history. Duplications probably arise by "mistakes" in recombination; the process is called *unequal crossing over*. A misalignment happens, such that if there were initially two copies of a gene on each strand, then after crossing over there are three on one strand and one on the other (Figure 10.2a). It is particularly easy to envisage a misalignment if there are already a number of copies of a gene in a row. The sequences of the two strands are matched up when they align for recombination and a gene could easily align with a non-complementary copy in the series of repeats (Figure 10.2b).

In the human globin genes, duplication may have been facilitated by a non-coding repeat sequence between the globin genes. A repeat sequence has been found on either side of both γ-globins; if this existed before the duplication, it could have been the target for the mismatching that produced the duplication (Figure 10.2c). Certain variants of the human α-globin genes provide more evidence for the importance of unequal crossing over among

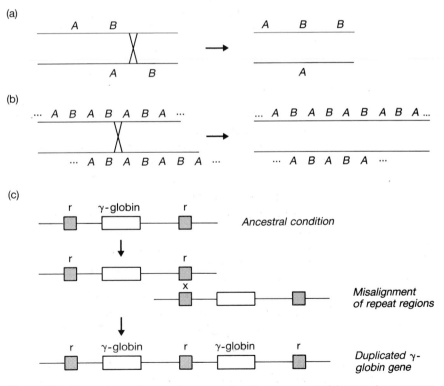

Figure 10.2 Unequal crossing over happens when the sequences of the two chromosomes are misaligned at recombination. (a) In a simple case, one chromosome with three copies of a gene and one with one copy could be generated from two chromosomes with two genes each. (b) In practice, misalignment is more likely if there is a long series of copies of similar sequences. (c) In the human γ-globin genes there is a repeat sequence (stippled box) on either side of the gene itself (clear box); this repeat could have been mismatched, causing duplication.

the globins. A normal human has two α-globin genes; but individuals can have three, or a single, α-globin genes. (Homozygotes for these chromosomes are one genetic cause of thalassemia.) The obvious explanation is that they were generated by unequal crossing over.

Once a duplication has originated in a single copy, it could spread through the population by drift or selection. Once the duplication had been fixed, the two genes could retain the same sequence (like in ribosomal RNA genes) if selection favors multiple copies of the gene. Or they could diverge to form a gene family (like the globins). Alternatively, if one of the copies of the gene became useless, selection would favor any mutation that switched it off.

Heterozygous advantage (section 5.11, p. 109) at a locus may set up selection for gene duplication. Suppose that selection at a locus favors heterozygotes because an individual is better off with two versions of a protein—perhaps because each is adapted to slightly different conditions. While these two versions are supplied by a single locus, disadvantageous homozygotes are generated when two heterozygotes breed together. If the gene were to duplicate, the two versions could be supplied by homozygotes for each at the two loci, and the disadvantageous segregational load (section 7.6, p. 152) disappears.

The human globin gene family, we saw, is arranged in two (or three, including myoglobin) clusters on two (or three) chromosomes. The kind of duplication illustrated in Figure 10.2 may well explain the origin of many genes within each cluster, but is less likely to explain how there is more than one such cluster. They are more likely to have originated by translocation, by the duplication of part or all of a chromosome, or by polyploidy (i.e. the doubling, or multiplying, up of the whole genome). For instance, suppose that one pair of chromosomes did not segregate equally at meiosis. One of the daughter cells would acquire a double set of the chromosome and the other would acquire none; the latter would probably die, because it is missing all the genes of the chromosome: the result is a mutation to a double set of the chromosome. As with gene duplication, after a chromosome has duplicated, its duplicated set of genes could be maintained, or diverge, or be suppressed, as drift or selective circumstances favored.

The globin gene family in two species of *Xenopus* support this interpretation. In *X. tropicalis*, unlike mammals, the α- and β-globin clusters are genetically linked; but *X. laevis* (a tetraploid species) has two sets of the linked α and β cluster. If the α set were suppressed on one chromosome and the β set on the other, the mammalian arrangement would have evolved.

The sequences of the human globin genes are known, and they can be arranged in a phylogeny (Figure 10.3) (chapter 18 discusses phylogenetic inference). The phylogeny suggests there have been at least seven duplications in the history of the molecule. Once we have the tree, it is easy to use the molecular clock (section 7.1, p. 143) to calibrate it and estimate the times of the duplications. The results are given in Figure 10.3.

These inferred dates pose an interesting paradox. If we compare the sequences of the human α_1- and α_2-globins, the split is apparently very recent; it may post-date the split between humans and the great apes and it certainly post-dates the split between the great apes and the rest of the

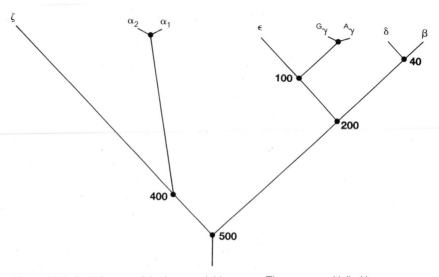

Figure 10.3 A phylogeny of the human globin genes. The genes multiplied by gene duplications, and the dates on the figure are the times of the duplications as inferred from the molecular clock. In fact, as the text explains, the inferred dates are untrustworthy. From Jeffreys *et al.* (1983).

primates. This being so, the primates outside the great apes should have only one α-globin. The prediction, however, is false: all the primates have two α-globin molecules. The full taxonomic distribution of the α-globins suggests that the gene had duplicated at least before the origin of the mammals (about 85 million years ago), and probably before the split between mammals and birds about 300 million years ago. And yet the similarity of the α_1- and α_2-globins in humans suggests the genes duplicated about 1 million years ago. Which of the figures, if either, is correct? The paradox is a common characteristic of evolution in gene families. It is called *concerted evolution*.

10.4 The genes in a gene family often evolve in concert

If each tandem repeat of the ribosomal RNA gene evolved independently by mutation, drift, and selection, the different genes might gradually diverge from each other over time (Figure 10.4). Evolutionary changes would accumulate in the different genes independently in different species, and we might therefore expect that the similarity between the genes in a gene family within a species would be about the same as the similarity between copies of the same gene in different species. This is only a rough prediction. The different genes in a gene family within a species have been separate since the duplication, whereas copies of the same gene in different species have been separate since the species split apart. If the duplication is in both species, then the speciation event will be more recent than the duplication; other things being equal, the copies of the same gene in different species should then be more similar than the duplicates within a species. On the other hand, there may be correlated selection pressures within a species, tending to make the duplicated copies within a species more similar.

Species A   *Species B*

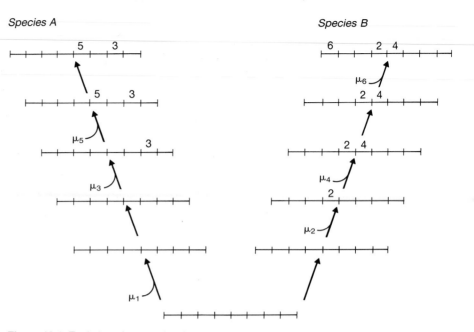

Figure 10.4 Evolution of a gene family in two evolutionary lineages (in separate species). If mutations arose independently at each locus, then to evolve by selection or drift, the different genes within a species should diverge at the same rate as the same gene in different species. From Arnheim (1983).

And what about the facts? The first study, by Brown *et al.* in 1972, was made for the ribosomal RNA genes of *Xenopus*. The 18S and 28S regions (Figure 10.1a) fitted the expected pattern. The sequences were almost identical in all the tandem repeats of both *X. borealis* and *X. laevis*, suggesting strong stabilizing selection. But for the non-transcribed spacer region (NTS) the pattern was different. Again, within a species the sequence was similar between tandem repeats; but this sequence differed in the two species. The NTS region of *X. borealis* and of *X. laevis* showed no more similarity than would two quite different genes—whereas within a species the NTS regions all had the same sequence. The NTS region showed concerted evolution (Figure 10.5).

The α-globin genes of primates described above illustrate the same principle. All primates, we saw, have two α-globins; we can therefore assume that the common ancestor of primates had two α-globin genes. The sequence of each α-globin gene differs between primate species; in the great apes, for example, any two species differ by about 2.5 amino acid substitutions in each gene. However, within a species the α_1- and α_2-globins differ by only about one-tenth of that amount (this has been shown not only in humans (Figure 10.3), but also in chimpanzees, gorillas, and rhesus monkeys). If one gene accumulates about 2.5 amino acid changes in the time between two species, then two different genes (α_1 and α_2) which have been separated for maybe 300 million years should have accumulated many more, maybe 100 times more, changes—if they have been evolving independently. The conclusion is that they have not evolved independently; they have evolved in concert.

(a) **Independent mutation at each locus**

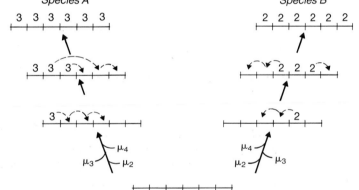

(b) **Lateral spread of mutations**

Figure 10.5 Concerted evolution. The different genes of a gene family tend to have similar sequences within a species, but are fixed for different sequences in different species. This could occur if there were many independent mutations at each locus (a), and selection favored the same sequence variant at every locus; or (b) mutations might be able to spread horizontally.

Why do gene families show concerted evolution? Neutral drift, acting independently at different loci, can be ruled out. If the α_1 and α_2 genes drifted at random independently, they should be much more different than they are within each species. In principle, selection can explain concerted evolution (Figure 10.5a). The same mutations would have to arise independently in both α_1- and α_2-globins in each species and then be fixed by selection, perhaps because in the conditions of humans one variant of the two genes was favored whereas in the conditions of chimpanzees another variant of both was favored. For the two globins in a number of primate species, this is a possibility. However, as the gene family showing concerted evolution grows larger, the selectionist explanation increasingly stretches our credulity. When we reach the ribosomal RNA genes, with their innumerable tandem repeats, it is beyond belief that the same mutation could have occurred independently at each locus and been fixed by selection. In this case, it is generally accepted, some genetic mechanism must act to homogenize the different members of the gene family. One sequence somehow is copied horizontally from one gene to the others. Mutations are not occurring independently at the different loci.

Two main mechanisms have been suggested: unequal crossing over and gene conversion. At the molecular level, the two are closely related; but at the chromosomal level they have different effects. We met unequal crossing over when we discussed gene duplication (Figure 10.2). The same process generates chromosomes with different numbers and combinations of genes; these can provide the mutational raw material for selection or drift to homogenize the gene family. Selection could directly favor more homogeneous gene clusters; or it could simply eliminate clusters with too many or too few copies of a type of gene (Figure 10.6).

In *gene conversion*, one of the alleles at a locus is converted into the other. Thus, when the heterozygote f_1f_2 segregates, instead of the Mendelian proportions of one f_1 for one f_2, two f_1 or two f_2 (and none of the other) would emerge. The same process could take place horizontally between the members of a gene family, with the gene variant at one locus being copied into the other locus. The mutational raw material for concerted evolution would again be produced. Gene conversion could be either biased or unbiased. Biased gene conversion means that it favors one variant of the gene rather than another; two f_1 genes might be produced more often than two f_2. Biased gene conversion is a kind of directed mutation, and it increases the chance that a gene cluster will be homogenized. Unbiased gene conversion means that the production of two f_1 or of two f_2 genes is equally likely.

Unbiased gene conversion and unequal crossing over do not by themselves produce concerted evolution, any more than undirected mutation alone produces other types of evolution. They account only for the origin of variants; they cause some individuals of a population to have more homogeneous gene families than other individuals. But we still need a process to explain how one of the gene family variants becomes fixed in the population. The process could be natural selection. Selection might favor similar genes in a gene cluster under some conditions. Alternatively, neutral drift could produce a "march to homozygosity" laterally between loci much as it can produce homozygosity at a single locus (section 6.6, p. 132). Either way, unequal crossing over and gene conversion act as lateral mutation processes, and make concerted evolution possible. The homogenization of gene families, producing concerted evolution, is a process Dover has called molecular drive.

So far we have considered concerted evolution in a single gene cluster. The process is not confined to gene clusters on one chromosome. The ribosomal RNA genes in primates (as we saw earlier) are in clusters on five chromosomes; but the genes on different chromosomes show concerted evolution like genes on the same chromosome. This is probably made possible when the regions of the chromosomes encoding the ribosomal RNA genes physically come into association. Unequal crossing over, or gene conversion, could then take place, just like that hypothesized to allow concerted evolution among genes on the same chromosome.

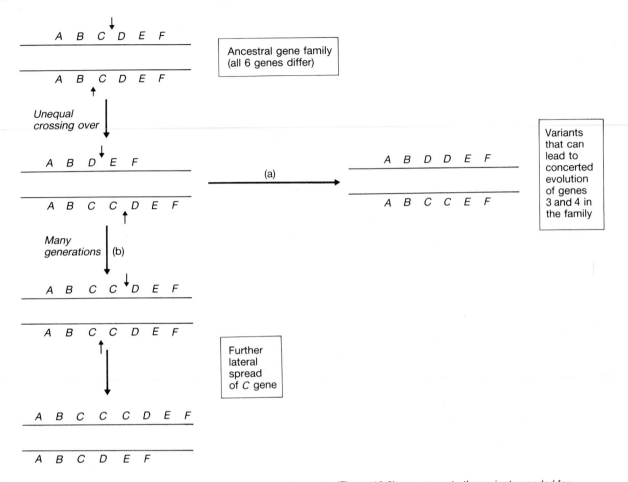

Figure 10.6 Unequal crossing over (Figure 10.2) can generate the variants needed for concerted evolution. The shuffling of genes among chromosomes can generate variants with more copies of one variant of a gene. Letters indicate genes, in a gene family, on a chromosome; arrows indicate the site of recombinational breaks. (a) Two successive unequal exchanges result in two chromosomes in which different variants have spread through two of the genes. (b) After one unequal cross over, the chromosome with two *C* genes increases in frequency until several generations later an individual with two copies of the chromosome is formed. Another unequal cross over could then form an offspring with three *C* genes; (a) and (b) are just two courses among many possibilities.

Concerted evolution is usually confined to only some of the genes in a gene family. For example, neither the α- and β-, nor the β- and γ-globins evolve in concert, even though they belong to the same gene family. The reason is that gene conversion only takes place between genes with some degree of sequence similarity. The chance of gene conversion is probably roughly proportional to the sequence similarity between two genes. A newly duplicated gene, therefore, might undergo one of two evolutionary courses. If gene conversion happens before the two genes have diverged much, they may evolve in concert. If, however, they diverge too far for gene conversion to be possible they will "escape" from each other; their future evolution will be independent. At a further stage, when they have diverged far enough, we should cease to classify them as members of the same gene family. Which of these two fates a pair of genes follows will be determined by the relative rates of gene conversion and of the mutations that break down sequence similarity. If the former is high relative to the latter, concerted evolution is likely; if the opposite is true, the genes will escape and diverge from each other. One additional complication is worth noticing. The chance of gene conversion is probably not simply proportional to sequence similarity. The insertion of a mobile genetic element into a gene can prevent future conversion of that gene, or of part of it to one side of the insertion. Thus Schimenti and Duncan have suggested that the insertion of an *Alu* element (see below) has prevented gene conversion, and concerted evolution, between certain genes in the β-globin families of cows and goats. The chance of concerted evolution is therefore controlled by both the sequence similarity of two genes and by whether or not a mobile genetic element inserts in them.

In summary, some of the genes in gene clusters evolve in parallel. This concerted evolution is probably not caused by selection and drift operating on independent mutations at each locus; it requires some mechanism for the same mutation to arise at the different loci. Gene conversion or unequal crossing over allow mutations to spread between loci, and selection or drift could then fix the mutations together. In the case of biased gene conversion, the mutational mechanism itself will help to fix the favored variant. The result would be the observed pattern of concerted evolution.

10.5 Not all the DNA codes for genes

So far we have been concerned with genes that code for proteins, or at least for RNA. The starting point for the second process is to observe that much of the DNA, in the organisms of many species, probably does not code for anything. Take humans as an example. The human genome is about 3×10^9 nucleotide pairs long, and geneticists disagree about what proportion of it is *informational* (i.e. is transcribed to produce proteins or regulate the production of proteins). Estimates are fallible, but the figure may be about 10–25%; the highest estimates go up to 50%. Even if 50% of the genome is informational, a large proportion of the genome is still doing something other than coding for, or controlling the production of, proteins. DNA that does not code for genes is called *non-informational* or *non-coding* DNA.

DNA reassociation experiments allow an estimate of the proportion of non-coding DNA. In these experiments, strands of DNA are allowed to join together (reassociate), and the rate at which they do so is controlled by their sequence similarity. The earliest work identified three classes of DNA. A part of the DNA reassociated quickly; this is mainly highly repetitive DNA, made up of large numbers of repeats of simple sequences. At the other extreme, "single copy" DNA joined together relatively slowly; this is presumably that part of the DNA which encodes for most of the genes. In between, there is a class of middle-repetitive DNA, which is repeated but not as much as the highly repetitive DNA. On the assumption that the single copy DNA is the coding part of the genome, we can use the proportion of single copy DNA (i.e. slowly reassociating DNA) as a first estimate of the informational DNA in the genome.

Table 10.1 lists the genome sizes, and proportion of single copy DNA, of a number of species. These numbers suggest, in two ways, that there is a large quantity of non-informational DNA. First, the figures for single copy DNA are often much less than 100%: only one-fifth of the DNA of the toad *Bufo bufo* is single copy DNA, for example. If only the single copy DNA is the coding part, then four-fifths of *B. bufo*'s DNA is non-informational. Second, there are large differences between species, which can hardly all be due to differences in the numbers of informational genes. It may need 15 times as many genes to build a human as a protozoan; but, in the words of Orgel and Crick, "it seems implausible that the number of radically different genes needed in a salamander is 20 times that in a man." The obvious deduction is that the differences are mainly in non-informational DNA. There is more to a genome than genes.

The experiments provide a second clue about the nature of the non-informational DNA: namely, that it is mainly repetitive DNA. Geneticists distinguish a number of kinds of repetitive DNA, according to the length of the repeated unit (whether it is long or short), the number of repeats, and

Table 10.1 Amount of DNA and percentage of single copy DNA (sc DNA) in various animal species. Amount of DNA is expressed as haploid DNA content, 1C, in picograms (10^{-12} g). For conversion: 1 picogram (pg) = 0.98×10^9 base pairs (bp); 1 bp = 1.02×10^{-9} pg. From John and Miklos (1988)

Invertebrates			Vertebrates		
Species	1C (pg)	% sc DNA	Species	1C (pg)	% sc DNA
Protozoa			Protochordata		
Tetrahymena pyriformis	0.2	90	*Ciona intestinalis*	0.2	70
Coelenterata			Pisces		
Aurelia aurita	0.7	70	*Scyliorhinus stellatus*	6.1	39
			Leuascus cephalus	5.5	44
Nemertini (Rhyncocoela)			*Raja montagui*	3.4	47
Cerebratulus	1.4	60	*Rutilus rutilus*	4.8	54
Mollusca			Amphibia		
Aplysia californica	1.8	55	*Necturus masculosus*	83.0	12
Crassostrea virginica	0.7	60	*Bufo bufo*	7.0	20
Spisula solidissima	1.2	75	*Triturus cristatus*	21.0	47
Loligo loligo	2.8	75	*Xenopus laevis*	3.1	75
Arthropoda			Reptilia		
Prosimulium multidentatum	0.18	56	*Natrix natrix*	2.5	47
Drosophila melanogaster	0.18	60	*Terrapene carolina*	4.1	54
Limulus polyphemus	2.8	70	*Caiman crocodylus*	2.6	66
Musca domestica	0.9	90	*Python reticulatus*	1.7	71
Chironomus tentans	0.21	95	Aves		
Echinodermata			*Gallus domesticus*	1.2	80
Strongylocentrotus purpuratus	0.9	75	Mammalia		
			Homo sapiens	3.5	64
			Mus musculus	3.5	70

whether the repeats are side by side (called *tandem repeats*) or scattered through the genome:

1 *Highly repetitive DNA* consists of long rows of repeated sequences, arranged side by side on the chromosome; they can be present in more than one block on the same or different chromosomes. Many of the repeated sequences are short: the simplest repeated unit is the dinucleotide AT; it is found, for example, in long stretches of poly (A–T) in the DNA of some crabs. In other cases, the repeated sequence may be longer—up to a thousand or so nucleotides.

2 *Dispersed middle-repetitive DNA* consists of longer sequences—a few hundred or so nucleotides—and is distributed throughout the genome, usually in single copies bound by other sequences rather than in tandem repeats. Three examples from humans are the sequences called *Alu* (which we shall discuss further below), *Kpn*, and poly (C–A); the three together make up about 20% of the human genome.

3 *Clustered, scrambled middle-repetitive DNA* consists of sequences of a few hundred or so nucleotides, found in clusters several thousand nucleotides long. Within a cluster there may be several unit sequences, which may be repeats or inverted repeats of one another. There may be several clusters in a genome, and they can share some of their sequences. The grasses (Gramineae) contain large amounts of these sequences; up to 80% of their genome may be made up of various families of middle-repetitive DNA, interspersed with one another and with single copy DNA.

4 *Tandem repeats of simple sequences.* Within introns, and elsewhere in non-coding DNA there can be short stretches of DNA made up of a few repeats of simple sequences, of about two to 10 nucleotides.

There are two main evolutionary hypotheses to explain all this repetitive, non-coding DNA. One is that it is functional, even though it does not actually encode genes. It may be needed for some regulatory or structural reason, for example; maybe it is needed to keep the genes apart or correctly configured in the DNA molecule's three-dimensional shape. Alternatively, the repetitive DNA may be selfish DNA—parasitic, or "junk," DNA. We shall concentrate on this second possibility here, not because it is better confirmed (which it certainly is not) but because it introduces a new evolutionary concept. If the repetitive DNA is functional, it will evolve just like genic DNA and requires no special discussion.

10.6 Repetitive DNA other than in gene clusters may be selfish DNA

In 1980, Doolittle and Sapienza, and Orgel and Crick, made more explicit an idea that many other geneticists had also thought about: the idea that some, or all, repetitive DNA may have no use for the organism. It may instead be *selfish DNA*. Selfish DNA is non-transcribed, non-informational, and contributes nothing to the well being of the organism; it is selectively neutral except for the energetic burden of replicating it. If it was excised, the organism would suffer no disadvantage. After a stretch of selfish DNA has originated, it is simply passively replicated and passed on from parent to offspring. Changes in its frequency in the population would be by drift. If a non-informational sequence was not neutral, perhaps because it interfered with the construction or metabolism of the organism, it would be selected against; likewise, if it accumulated to such an extent that the cell cycle was slowed down by the need to replicate it all, selection would probably act to reduce it. But provided the quantity of a particular sequence is not excessive, it is not transcribed, and it accumulates in parts of the genome where is does not interfere with gene regulation and transcription, there is no reason why selfish DNA should not evolve.

The sequences of selfish DNA could be of two kinds. The sequence itself might not influence the chance that it spreads in the DNA and is retained; it is then a passive kind of selfish DNA. It could accumulate as "junk" DNA in the genome. Alternatively, a particular sequence might be more likely to spread through the DNA; these sequences would be a more active, parasitic kind of selfish DNA, and would proliferate until checked because they interfered with vital DNA functions or became so abundant that their replication imposed a significant cost on the organism. For the possible examples of selfish DNA we shall discuss here, it is not known to which class they belong. When reading about selfish DNA, however, we should keep in mind the question not only of whether it is selfish DNA at all, but also of whether it is active or passive selfish DNA.

How might selfish DNA originate? In theory any kind of mutation could give rise to selfish DNA. A point mutation could inactivate a gene, converting it into a pseudogene (section 7.16, p. 172); if the inactivated gene was redundant and the new pseudogene was not in a position to interfere with anything important, the mutation could be fixed by drift and the pseudogene retained as selfish DNA. For repetitive DNA, other kinds of mutations would be needed. Unequal crossing over is probably important. Unequal crossing over produces a strand with an extra copy of a DNA sequence, and successive rounds of the process could build up long stretches of repeats. For the shorter repeated units, the related process of *slippage* may be more important. In slippage, a length of the DNA is copied twice over, perhaps because the first copy forms a loop and the replication machinery slips back and copies it all over again; likewise, a copy may be missed out (Figure 10.7). The number of repeats may thus increase or decrease. The key difference

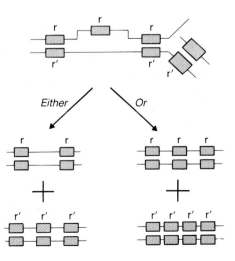

Figure 10.7 Slippage happens when a sequence of DNA is copied twice, generating a new strand with an extra copy of that sequence, or missed out, generating a new strand with one copy less. The row of repeats may therefore shrink or grow when the r and complementary r′ sequences are misaligned. The molecular details are unknown, but may involve the formation of a loop. Short repeats may be particularly vulnerable to slippage.

from unequal crossing over is that gain and loss are not matched. With unequal crossing over, if we start with two strands with two copies of a sequence unit each, we might end up with three repeats on one strand and one on the other; with slippage, the result might be three on one strand and two on the other.

Under both slippage and unequal crossing over, a sequence becomes more likely to be copied again by mutation as it becomes more numerous in the genome. Repeats of a sequence make sequence misalignment more probable, as (for instance) copy two on one strand aligns with copy three on the other; or copy eight on one strand with copy six on the other. The more copies of a sequence there are, the more likely this misalignment becomes. If, therefore, a genome contained two sorts of neutral, functionless sequence, one in single copy and the other in multiple repeats, then (other things being equal) the repeated sequence would be more likely to change its number of copies in the genome during evolution. If the highly repetitive rows of tandem repeats are indeed selfish DNA, then unequal crossing over, slippage, and drift will constantly be causing them to shrink and grow. At any one time, there will be some very long regions of repeats, and these are the ones we recognize as highly repetitive DNA. It could be that after a few million years, a currently long repeat will have shrunk and some other, previously shorter series will have lengthened; but the frequency distribution of lengths may remain probabilistically constant.

For scattered repetitive DNA, another process may be responsible. It is called *transposition*. Certain sequences of DNA can, under appropriate circumstances, copy themselves elsewhere in the genome. These mobile genetic sequences are called transposable elements, or (more informally) jumping genes. The molecular biology is complex, but we can distinguish two types of transposition (Figure 10.8): *conservative transposition*, in which the number of copies of the transposed sequence is the same before and after transposition; and *replicative transposition*, in which the number of copies increases. Replicative transposition can be achieved by the reverse transcription of an RNA intermediate (Figure 10.8b).

When replicative transposition takes place in an individual, it will clearly produce a scattered repeat sequence. It could then increase in frequency (by selection or drift) to result in the kind of pattern of scattered repeats we now observe. The best evidence that a scattered repeat sequence had originated by transposition would be to observe the sequence transposing. A more general method is to study the characteristics of the DNA sequences that are known to transpose, and those of the scattered repeats, to see whether the latter resemble the former. Geneticists currently distinguish three kinds of transposable DNA. The simplest are *insertion sequences*, which are short (700–2500 base pairs) and only contain the genetic instructions for transposition. *Transposons* are longer (2500–7000 base pairs) and are distinguished from insertion sequences by containing some other gene besides the instructions for transposition. More or less any gene may be incorporated into a transposon; the best known examples are genes for antibiotic resistance in bacteria. The gene in a transposon may be advantageous, disadvantageous, or neutral for the organism. Thirdly, there are various kinds of *retroelements*,

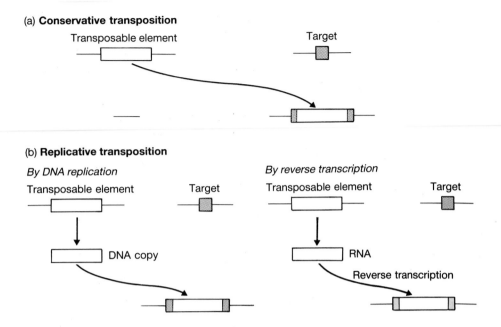

Figure 10.8 Transposition can be either (a) conservative, in which the original copy of the transposable element is lost as it jumps to a new site, or (b) replicative, in which extra copies of the transposable element are made. In some cases, replicative transposition is achieved by reverse transcription, in others it occurs by copying of the DNA.

which are distinguished by their containing the gene for reverse transcription in addition to the instructions for transposition. They are copied around the DNA by means of reverse transcription of an RNA intermediate. (Retroelements either may be able to live independently in the form of viruses, in which case they are "retroviruses," or they may only exist in the cell and have no independent life. The acquired immune deficiency syndrome (AIDS) virus is a retrovirus.)

Which sequences in our DNA may belong to these categories? It is likely that all (or most) organisms possess all three types. Transposition has been most studied in bacteria; the examples from multicellular organisms are less certain. A sequence called the *Alu* sequence is one likely candidate in mammals. *Alu* is the name for a characteristic sequence recognized by the restriction endonuclease *Alu 1*. The sequence is about 300 bases long and is found throughout the human genome; each of us contains about half a million copies. About 5% of our DNA consists of the *Alu* sequence. Few, if any, of the sequences are transcribed, and it may therefore have no function. Where does it all come from? One favored hypothesis is that it ultimately originates in the gene for a functional ribosomal RNA molecule called 7SL RNA. The gene for the 7SL RNA could have first been picked up by a retroelement and copied back into the DNA by reverse transcription; the many *Alu* sequences are believed to have been formed as second order derivatives of the 7SL RNA gene, by recurrent rounds of retroelemental reverse transcription from a number of reverse transcribed copies of the gene itself. The *Alu* sequence itself is not strictly a retroelement because it lacks the gene for reverse transcription; but it looks like a sequence that is derived from reverse transcription. Maybe it is picked up by a retroelement, and the gene for reverse transcription is lost when it is copied back into the DNA. It is not known to be functionless, but there is a strong suspicion that it is, and that it is an example of selfish DNA. The large numbers of copies of *Alu* suggests there is something about its sequence that makes it particularly likely to be copied; it would then be an example of active selfish DNA.

There are other examples of transposable sequences in eukaryotes. In the fruitfly *Drosophila*, there is a sequence called the *copia* element; it is almost certainly a retroelement, because it contains a gene similar to the gene for reverse transcription found in retroviruses. It may be that some transposable sequences have functions and are influenced by natural selection on the organism, whereas others do not have any function and form selfish DNA.

Whether or not transposable elements are functionally important, they can have interesting evolutionary consequences. The insertion of an *Alu* sequence, as we saw in section 10.4, can prevent gene conversion between a pair of genes. Concerted evolution between the two genes then becomes impossible: the consequence would be that duplicated genes in which an *Alu* sequence is inserted are more likely to show independent evolution than are genes without such an insertion. A second suggested consequence is in chromosomal evolution. Chromosomes may be particularly likely to rearrange around *Alu* sequences, and the numbers and positions of the sequence may therefore influence the rate and form of chromosomal evolution. Finally, at the level of the DNA sequence, when a transposon inserts into a new gene, it causes a mutation: the recipient gene has a new sequence after receiving the transposon. This may (or may not) cause a mutation at the phenotypic level; McClintock first showed that mutations can be caused by transposition, in maize. Transposition may differ from other kinds of mutation in that it may occur more at some sites in the genome than at others. Mutations caused by nucleotide changes are probably more randomly distributed.

10.7 Minisatellites are sequences of short repeats, found scattered through the genome

While studying the human myoglobin gene in the early 1980s, Jeffreys discovered a short sequence of repeated DNA within an intron. He used the short sequence as a "probe" to see whether the same sequence was present anywhere else in the genome. The probe hybridized at several regions. He then extracted the DNA from all these regions, and found that each consisted of a unit sequence in a row of tandem repeats. The number of repeats at each site is highly variable between individuals; at one site, for example, there might be 1250 repeats in one individual, five in another individual, and 470 in another (Figure 10.9). The unit sequence itself is variable, but a common core sequence has been characterized (Figure 10.10), and is 16 (or perhaps eight) bases long. The particular sequence in Figure 10.10 is probably not the only sequence that can form this sort of scattered short set of repeats, but it is the most studied. Jeffreys calls these sequences *minisatellites*; the name derives from the way they can be detected when DNA is centrifuged. (They are also called VNTR loci, for "variable number of tandem repeats.")

Why do minisatellites vary so much in the number of repeats? The

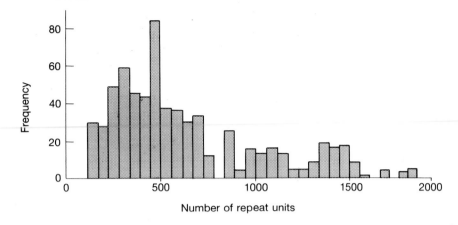

Figure 10.9 Frequency distribution of numbers of repeats at one minisatellite locus in a total of 344 individuals (688 gametes). From Jeffreys *et al.* (1988).

variation presumably originates by slippage and unequal crossing over, because these processes operate easily among tandem repeats (Figures 10.2 and 10.7). By following the pattern of minisatellites in a human pedigree, it is possible to estimate the rate at which new variants arise (the variants differ in the number of repeats of the unit sequence). The rate is high, up to 5% per gamete, and it varies among sites in a predictable way: more variable sites produce new variants at higher rates (Figure 10.11).

The high mutation rates of minisatellites are the reason why they can be used in *genetic fingerprinting*. New variants arise at a high enough rate for every individual (except monozygous twins) to have a unique frequency distribution, or "profile," of minisatellites. Genetic fingerprints are more forensically useful than real fingerprints: not only is an individual's genetic fingerprint unique, and identifiable from small quantities of body substances,

Clone	Repeat length	Number of alleles	Number of repeats	DNA sequence	
				G G A G G T G G G C A G G A A G	Myoglobin probe
1	62	6	40–20	A A G G G T G G G C A G G A A C	
2	32	1	6	G G A G G T G G G C A G G A A X	
3	64	5	18–10	T G G G G A G G G C A G A A A G	
4	17	1	14	G G A G G Y G G G C A G G A G G	Isolated clones
5	37	8	25–12	G G A G G A G G G C T G G A G G	
6	41	1	5	G G A – G T G G G C A G G C A G	
7	33	1	3	G G T G G T G G G C A G G A A G	
8	16	2	41, 29	A G A G G T G G G C A G G T G G	
				G G A G G T G G G C A G G A X G	Common core
				G C T G G T G G	*E. coli* χ

Figure 10.10 The unit sequence of the human minisatellite. Different minisatellites do not all have identical copies of the sequence, but there is a consensus sequence of 16 nucleotides. From Jeffreys *et al.* (1985).

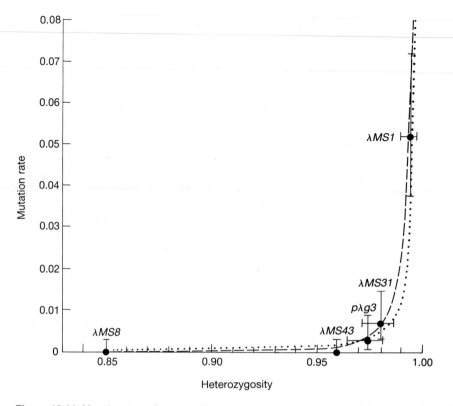

Figure 10.11 Mutation rates (i.e. rate of production of new length variants) for five minisatellite loci in humans. The more variable loci (those with higher heterozygosity) have higher mutation rates. From Jeffreys *et al.* (1988).

but it is also heritable. The profiles of a father and his children, for instance, are more similar than two random members of the same population. Genetic fingerprinting can therefore be used in paternity testing as well as in straightforward identification. Minisatellites similar to those in humans have been found in many other species, and the genetic fingerprinting probe has become an important method of tracing paternity in behavioral ecology.

Unequal crossing over, slippage, and drift will cause the minisatellites to expand and contract during evolution, but probably without affecting the fitness of the bearer. Unlike other kinds of repetitive DNA, the sequence of the minisatellite itself influences the rate of crossing over. In humans, it can increase the local recombination rate up to 700-fold. The reason may be as follows. The minisatellite unit sequence resembles an important sequence in the bacterium *Escherichia coli*, called the χ-sequence. In *E. coli*, the χ-sequence is thought to be recognized by one of the enzymes that control recombination. If the χ-sequence is a recombination site in *E. coli*, a similar sequence may have a similar function in humans. As the recombination rate would be increased in minisatellites, so would be their rates of slippage and unequal crossing over. If the number of repeats in a minisatellite was correlated with recombination rate, selection might eventually work against too lengthy regions of repeats: too high a recombination rate can be disadvantageous (section 8.11, p. 199). These ideas are uncertain, but they do illustrate how the sequence of a piece of DNA might cause it to behave as selfish DNA.

10.8 Selfish DNA may explain the C-factor paradox

Let us return finally to the observation (section 10.5) that there is more DNA in the cells of some species than is needed to code for their genes. The degree of excess is controversial, but no one doubts that there is at least some excess DNA in at least some species. (Likewise, no one doubts that at least some of the species differences in DNA content (Table 10.1), such as between humans and bacteria, are due to differences in phenotypic complexity.) The apparent excess of DNA is sometimes called the C-factor paradox. What is it all there for? Selfish DNA is an attractive, if hypothetical, solution: the excess DNA would then have no use or function; it would be selectively neutral, and passively replicated from generation to generation. The assumptions of the hypothesis are plausible: unequal crossing over, gene conversion, transposition, and neutral drift all do occur. Moreover, if the hypothesis is right, it would explain why this apparently functionless DNA actually is functionless. It could also explain why the degree of excess DNA differs so much among species, because the growth and shrinkage of repetitive DNA by mutation and drift, and its elimination by selection, will produce constant fluctuations in the amount of selfish DNA in a species; different species would be the still frames in a moving picture. It should be stressed that this is all hypothesis, even if theoretically attractive hypothesis.

The quantity of DNA in an organism's cells, however, is not a completely neutral character. Correlations at least (if not confirmed causal influences) have been found between genome size and the phenotype. Genome size is correlated with cell size, for example, in amphibians (Table 10.2) and other taxa; and in artificial polyploids, the cell size is proportional to the degree of ploidy. A more easily explained relationship is between genome size and the duration of the cell cycle (Table 10.3). Larger genomes may take longer to replicate and it therefore makes sense that in wheat taxa, mitosis lasts longer

Table 10.2 Genome size is related to cell size. Genome size and cell volumes are given for nine amphibian species, as measured by two sets of authors: Horner and Macgregor (H + M) and Olmo and Morescalchi (O + M). From John and Miklos (1988)

Species	Genome size (pg)		Nuclear volume (μm^3)		Cell volume (μm^3)	
	H + M	O + M	H + M	O + M	H + M	O + M
Xenopus laevis	3	3.1	10	29	250	295
Bombina variegata	6.9	10.5	60	171	650	1102
Plethodon cinereus	22.5	23.0	144	182	1476	1079
Triturus cristatus carnifex	24.5	21.8	210	91	1500	731
Triturus vulgaris	25.0	24.0	176	77	1663	449
Taricha granulosa	35.0	29.5	272	490	2090	1795
Notophthalmus viridescens	38.0	34.8	168	198	2204	1255
Taricha torosa	38.6	28.0	240	241	2232	1518
Taricha rivularis	38.9	30.0	232	119	2520	1828

Table 10.3 Mitosis takes longer in species with higher ploidy. Shown here are cell cycle times and ploidy for species in two genera of wheat (*Aegilops* and *Triticium*). 2C is the weight of the diploid quantity of DNA. G_1, S, G_2, and mitosis are the four stages of the cell cycle. From John and Miklos (1988)

Ploidy	Species	Duration (hours) 2C (pg)	G_1	S	G_2	Mitosis	Total cycle time (hours)	Duration of meiosis (hours)
$2x = 14$	A. squarrosa	11	1.7	6.2	3.0	0.54	11.4	
	T. monococcum	15	3.1	6.0	2.5	0.59	12.5	42
$4x = 28$	T. timopheevi	25	1.3	7.5	5.4	0.83	15.0	
	T. dicoccoides	25	2.7	7.6	4.4	0.76	15.5	
	T. dicoccum							30
$6x = 42$	T. spelta	38	6.3	9.3	2.6	0.93	19.7	
	T. aestivum	38	6.2	8.6	3.8	1.12	19.7	24

in species with larger genomes: though meiosis, for some unexplained reason, is shorter! Relationships have also been looked for with development time, gestation time, and whether (in plants) a species is an annual or perennial; these too show hints of trends, but the results are ambiguous. It seems likely that genome size has phenotypic consequences, but the relationships have not yet been well sorted out.

The causal significance of these correlations of genome size—if they exist—is another question. Cavalier-Smith has suggested that, when selection favors a longer cell cycle time, for example, it could be brought about by increases in the amount of repetitive junk DNA. This seems possible, though a longer cell cycle could presumably also be caused by other, more direct mechanisms. These are questions for the future. Meanwhile, the idea that excess DNA is due to passive accumulations of selfish DNA remains an attractive one.

10.9 Conclusion

Molecular biology is discovering many previously unsuspected genetic curiosities within the genome. The DNA of an organism is definitely not simply a chain of structural cistrons and regulatory genes. Evolutionary biologists want both to explain why the other kinds of genetic element exist, and why the genomes have the architecture that they do. They aim also to explore the evolutionary consequences of such non-Mendelian processes as transposition and unequal crossing over. The phenomena are sufficiently new and imperfectly described, and the kinds of theory they require are sufficiently novel, that favored hypotheses and the fit of fact and theory are changing almost by the day. It is an exciting area of the science.

10.10 Summary

1 Genes are usually arranged in clusters of related genes on the chromosome. The gene cluster may consist of tandem repeats, like the ribosomal RNA genes, or a linked group of related genes, like the globin genes.

2 Much of the non-informational DNA consists of repeated sequences. Several different kinds of repetitive DNA have been recognized.

3 Gene families originate by gene duplication, which in turn takes place by unequal crossing over or polyploidy.

4 The different genes in a gene family often show concerted evolution: the

genes at separate loci within a species are much more similar than the homologous copies of a gene in different species.

5 Concerted evolution among large numbers of genes is practically impossible to explain if mutations arise independently at each locus; it requires some mechanism for concerted mutations at all the loci.

6 Concerted mutation can occur by unequal crossing over or gene conversion. Concerted evolution happens when the more homogeneous variants produced by these processes are fixed by selection or drift. Gene conversion may be biased in favor of some sequences rather than others; the favored sequences would then proliferate by a form of lateral mutation pressure.

7 Gene conversion can occur between genes of similar sequence; when two genes have diverged more than a certain amount, concerted evolution will become unlikely.

8 Much of the genome does not code for information genes.

9 The large quantities of repetitive, non-informational DNA may be selfish DNA: it may be non-transcribed, have no function for the organism, and be replicated from generation to generation like a passive parasite.

10 The mechanism by which selfish DNA originates probably differs for different kinds of sequence. Tandem repeats may originate by unequal crossing over or slippage (specially for short repeats), scattered repeats by transposition.

11 Minisatellites (DNA consisting of a variable number of repeats of a characteristic short sequence) may be selfish DNA; they increase the rate of recombination, causing (by slippage and unequal crossing over) a high mutation rate for the number of repeats.

12 Much of the DNA in the genome in excess of that needed to code for genes may be selfish DNA.

10.11 Further reading

Maynard Smith (1989, chapter 11) and Li and Grauer (1991, chapters 6–8) introduce the subject. The volumes edited by Nei and Koehn (1983), Bendall (1983), and Selander et al. (1991), and the special issue of the *Philosophical Transactions of the Royal Society B*, vol. 312, pp. 189–354 (1986) contain many relevant papers. For background genetics, see any molecular genetics text such as Alberts et al. (1989), Watson et al. (1988), or Lewin (1990), or John and Miklos' (1988) more specialized book.

1 *Gene families*. Bodmer (1983) is a good place to begin. See Arnheim (1983) and Jeffreys et al. (1983) for concerted evolution in ribosomal RNA and globins. Hardison (1991) is a more advanced review for globins. Ohta (1988) reviews the theory. See also Dover (1986), Schimenti and Duncan (1984, 1985), and Walsh (1987).

2 *Selfish DNA and repetitive DNA*. Doolittle and Sapienza (1980) and Orgel and Crick (1980) are the classic references and good introductory reviews. On transposable elements in fruitflies, see Charlesworth and Langley (1991); on McClintock, see Keeler (1983). On minisatellites, see Jeffreys et al. (1985, 1988, 1991) and Wahls et al. (1990). On genetic fingerprinting, see Burke (1989) and Neufeld and Colman (1990); this fast-moving subject can be followed in the *Genetic Fingerprinting Newsletter*. Cavalier-Smith (1985) is about genome size.

Part 3
Adaptation and Natural Selection

Adaptation and Natural Selection

How can we find out what advantage (if any) an organism gains from possessing some characteristic—whether in its anatomy, physiology, or behavior? That is, how can we study adaptation? For many characters, such as muscles or digestive systems, it is well understood how they function as adaptations. For others, it is less well understood, or even not understood at all. Chapter 11 begins by posing this problem, and then discusses three main methods for dealing with it. The main part of the chapter is about three case studies which illustrate the methods; they are all drawn from the subject of sexual reproduction. The deepest problem—of why sex exists—is still mainly at the stage of developing a hypothesis. The theory of sexual selection is well developed, and the crucial empirical work is now beginning. In the theory of sex ratio, there is a good match of theoretical prediction and empirical tests. We then move on, in chapter 12, to ask in more theoretical terms what the entity is that adaptations evolve for the benefit of. Evolution by natural selection happens because adaptations benefit something, but what exactly—genes, whole genomes, individual organisms, groups of organisms, species, or what? This is the question of "What is the unit of selection?" Adaptations, the chapter suggests, usually benefit organisms, but there is a deeper criterion which can be used to understand the exceptions as well as the rule: more fundamentally, adaptations evolve for the benefit of genes. Only genes last long enough for natural selection to be able to adjust their frequencies over evolutionary time. Organismal adaptations usually result because gene reproduction is more closely tied to the reproduction of organisms than any other entity, and gene reproduction is maximized if adaptations are at the organism level. Thus far we have accepted that adaptations evolve by natural selection. Chapter 13 is about certain conceptual problems in the study of adaptation and begins by asking whether there are any explanations for adaptation other than natural selection. It argues there are not. However, some characteristics have probably evolved by processes other than natural selection, though they are not adaptations. Not all evolution proceeds by natural selection, but all adaptive evolution does. The chapter considers various kinds of constraint which can influence what characteristics evolve in a species, and how to test whether a constraint or natural selection is at work in a real case. The chapter also discusses how perfect the adaptations of living species are. Natural selection acts to improve adaptation, but there are various reasons why a state of perfection is not reached.

The analysis of adaptation

11.1 The way organisms are adapted may not be obvious

Cepaea nemoralis is a land snail, and the background color of its shell may be either yellow or some darker shade of brown or pink. On top of the background color are a number, usually between none and five, of dark bands. The snails are therefore highly variable in their external appearance; an individual snail may have any combination of background color and banding pattern. Why do *Cepaea* vary in this way? One possibility is that it does not matter, and Mayr remarked in 1942 that "there is no reason to believe that the presence or absence of a band on a snail shell would be a noticeable selective advantage or disadvantage."

Well, there is good reason now. So many selective factors have been shown to influence the pattern and coloration of *Cepaea* shells that the polymorphism has more recently been called "a problem with too many solutions." A series of ecological geneticists, particularly Cain and Sheppard, have studied the question over many years. The first suggestive observation was that the proportions of the different shell types vary from place to place. Initially, the geographic distribution of the shell types had seemed to be random, but it was soon found that the banded forms tend to be found in diversified habitats, such as mixed hedgerows, and the unbanded snails against more uniform backgrounds, such as those of dense woodlands. Perhaps the banding pattern is an adaptation for camouflage. Cain and Sheppard duly measured the rate at which the different snail types were eaten by birds. Birds, such as the thrush (*Turdus philomelos*), use stones as anvils to break open the snails, and it is possible to count the proportions of different snail types taken by the birds in the shell debris around an anvil; these proportions can then be compared with those in the local habitat. At one site near Oxford, England, for example, where the habitat is relatively uniform, Cain and Sheppard found that in the area as a whole 264 of a sample of 560 (47.1%) snails were banded; but 56.3% (486 out of 863) of the snails taken by thrushes were banded. The difference is statistically significant. In this habitat, the thrushes were taking the banded snails disproportionately. Similar results for other habitats also support the idea that differences in the color pattern and banding number serve as camouflage to protect the snails from bird predators.

The appearance of *Cepaea* is also influenced by the need for thermoregulation. How dark a shell is affects how quickly the snail warms up in sunshine, and it can be a matter of life and death for a snail if it has too many bands and lives on a sunny hill-slope. It can no longer be doubted that the color and banding patterns of *Cepaea* are adaptive, and adaptive in relation to more than one property of their local environment.

In 1942, as Mayr's remark shows, it was not obvious that shell pattern in *Cepaea* was an adaptation. It was certainly not obvious what it was an adaptation for. In this and the next two chapters, we shall be concerned with how we can find out why natural selection favors the particular characters that organisms possess. The first method, illustrated by *Cepaea*, is simply to look more closely at the character and see what consequences it has in the animal's life. There are other methods too, and we shall be discussing them in relation to three questions about sex. These questions are:

1 Why do organisms reproduce sexually?
2 Why do the sexes differ?
3 What sex ratio is favored by natural selection?

They are good examples because they are the subjects of active investigation, and reveal how evolutionary biologists actually carry out research into adaptation. Before we come to the three questions, we can look briefly at the abstract form of the methods they are going to illustrate.

11.2 Three main methods are used to study adaptation

The study of adaptation proceeds in three conceptual stages. The first is to identify, or postulate, what kinds of genetic variant the character can have. Sometimes, as in *Cepaea*, this is done empirically; other characters do not vary genetically and for them it is necessary to postulate appropriate theoretical mutant forms. For example, if we are studying why sex exists, we postulate a mutant form that reproduces asexually. The second stage is to develop a hypothesis, or a model, of the organ, or character's, function. Cain and Sheppard's first hypothesis about snail banding, for instance, was that it functioned as camouflage. Hypotheses are of varying quality, but the poorer sort can be improved on as work proceeds. A good hypothesis will predict the features of an organ exactly, and the predictions will be testable. In morphology, these predictions are often derived from an engineering model. For example, hydrodynamics is used to understand fish shape, while construction engineering is used for shell thickness in a mollusc: the costs of building a thicker shell have to be weighed against the benefits of reduced breakage, by wave action or predators. This sort of research carries on at all levels, from the simple and qualitative through to sophisticated algebraic modeling.

Stage three is to test the hypothesis's predictions. Three main methods are available. One is simply to see whether the actual form of an organ (or whatever character is under investigation) matches the hypothetical prediction; if it does not, the hypothesis is wrong somehow. A second method is to do experiments. It is only useful if the organ, or behavior pattern, can be altered experimentally. Almost any hypothesis about adaptation will predict that some specified form of an organ will enable its bearer to survive better than some other forms, but the alternatives are not always feasible: we cannot, for example, make an experimental pig with wings to see whether flight would be advantageous. When they are possible, experiments are a powerful means of testing ideas about adaptation. Animal coloration, for instance, has been studied in this way. Color patterns in some

Table 11.1 The wing stripe of some butterflies was painted over; controls were painted with transparent paint that did not affect their appearance. The numbers with intact wings at different times after the treatment was measured (table shows the last week in which wings were intact). The frequency distributions are not significantly different. From Silbergleid *et al.* (1980)

Age at capture (week)	Experimental		Control	
	N	%	*N*	%
0	81	83.5	88	90.7
1	14	14.4	6	6.2
2	2	2.1	2	2.1
3	0	0	1	1.0
4	0	0	0	0
5	0	0	0	0

butterfly species are believed to act as camouflage by "breaking up" the butterfly's outline. Silberglied *et al.*, in 1980 working at the Smithsonian Tropical Research Institution at Panama, experimentally painted out the wing stripes of the butterfly *Anartia fatima*. The butterflies with their wing stripes painted out showed similar levels of wing damage (which is produced by unsuccessful bird attacks) and survived equally well as control butterflies (Table 11.1); the wing-stripes therefore may not in fact be adaptations to increase survival. They may have some other signalling or reproductive function, though that would need to be tested by further experiments.

The *comparative method* is the third method of studying adaptation. It can be used if the hypothesis predicts that some kinds of species should have different forms of an adaptation from other kinds of species. Darwin's classic study of the relationship between sexual dimorphism and mating system is an example we shall discuss below. Some hypotheses predict that different kinds of species will have different adaptations, others do not. Darwin's theory of sexual dimorphism does; but, for example, an optical engineer's model of how the eye should be designed might specify just a single best design, with the implication that all animals with eyes should have that design. The comparative method would in that case be useless.

In summary, the three main methods of studying adaptation are to compare the predicted form of an organ with what is observed in nature, to alter the organ experimentally, and to compare the form of an organ in different kinds of species.

11.3 Example 1: the function of sex

11.3.1 Sexual and asexual reproduction should be distinguished

In sexual reproduction, a new organism is formed by the fusion of two gametes. It is to be contrasted with asexual reproduction, in which females produce offspring without any male contribution; the female's gametes develop directly into female offspring. The sex cells—eggs and sperm, pollen and ovules—are produced by a reduction division and the gametic fusion restores the original chromosome complement. In asexual reproduction, there is either no reduction division (apomixis) or the product cells of one individual's reduction division fuse together again (automixis); there are other, evolutionarily less interesting, methods of restoring diploidy, such as the doubling of the haploid chromosome set to produce a purely homozygous asexual offspring. The question to be discussed here is why sexual reproduction is so common: why has it not been replaced by asexual reproduction more often?

The simplest way in which sex could be lost would be for a mitotic division to be substituted for meiosis, and gametic fusion to be lost; the mitotically produced egg could then develop like a normal zygote. No complicated mutation would be needed. All the necessary mechanisms of mitosis already exist, and have much in common with meiosis; simply to substitute the one for the other is a less complex evolutionary step than must have taken place in the evolution of many other characters. The existence of a strange habit called pseudogamy partially compromises the argument (Box 11.1); but here we shall assume that mutations from sexual to asexual reproduction can easily occur. If the loss of sex would be relatively easy, why is sex maintained? How is sexual reproduction adaptive? (Genetic recombination is a related problem: see section 8.11, p. 199.)

11.3.2 Sex has a 50% cost

Although some species do reproduce asexually, the majority are sexual. So there is probably a selective advantage to sex in most species. The advantage, moreover, must be large, because it has to overcome an automatic two-fold *cost of sex*, relative to asexual reproduction.

What is this cost of sex? It is easy to understand by comparing the reproduction, over time, of a group of sexual females with a group of asexual females. We could start with a group of 100 asexual females and another group of 100 sexual females (if the sex ratio were 1:1, the sexual population would then have 200 members). There are 300 individuals in all, and one-third of them are asexual females, one-third sexual females, and one-third

Box 11.1 Pseudogamy.

If it is easy for a mutation to convert meiosis into mitosis, and thus a sexually reproducing into an asexually reproducing female, the form of reproduction called *pseudogamy* becomes puzzling. In some poecilid fish and a few odd insects, the sperm of a male is needed to stimulate the beginning of development, but no genetic material from the male enters the egg or contributes to the new individual. It is strange that the need for males in these species has not been lost. But it has not —and the need for their continued existence suggests that it is for some reason crucial to have development started off by sperm penetration. Triggering development in this way does not appear to be particularly fundamental, but this appearance cannot be correct, because pseudogamy does exist. The site of sperm penetration can provide an important spatial axis for development, and maybe it is for some reason difficult and complicated to substitute another mechanism of spatial control. Then the mutations needed to evolve from sexual to asexual reproduction would in reality be complex, and arise only rarely. The rarity of the necessary mutations could then be the reason why sex has not often been lost in evolution.

males. The members of the two groups, we suppose, are identical in all other respects: sexual and asexual individuals are equally good at finding food, avoiding enemies, and staying alive; they produce the same number of offspring, and those offspring have an equal chance of survival. We are considering only whether natural selection favors sexual or asexual reproduction.

Suppose, for simplicity, that each female produces two offspring. After one generation, the asexual group will have grown to 200 individuals, as each mother produces two daughters. The 100 sexual females will also produce 200 offspring, but only 100 of these will be daughters; the size of the sexual group will remain constant. Now, in this second generation, we have 400 individuals in all, and the proportion of asexual females has increased from one-third to one-half. After another generation, there will be 400 asexual females, 100 sexual females, and 100 males; the proportion of asexual females has grown to two-thirds. It will not be long before asexual reproduction has completely taken over. The clone of offspring from an asexual female multiplies at twice the rate of the progeny descended from a sexual female, and a sexual female has only 50% of the fitness of an asexual female. (We are assuming that the males make no energetic contribution to reproduction: there is no paternal care. In a species, like many bird species, in which the males defend and feed the young, the cost of sex will be less than 50%: the number of daughters will be more than half the number an asexual female could raise alone. However, in many species, the males do indeed contribute nothing more than sperm and sex has a full 50% cost.)

This 50% is a large cost. The problem of explaining sex is to find a compensating advantage of sexual reproduction that is large enough to make up for its cost. We are on the look-out for an extraordinarily large selective advantage: typical evolutionary events are thought to involve selective advantages of a few percent at most, and more often 1% or less. Consider this: a female who has survived to adulthood and is about to reproduce must be fairly well adapted to her environment. If she were to reproduce asexually, she would just make a copy of herself and produce a daughter as well adapted to the conditions of the next generation as she would be herself. If she reproduces sexually instead, she discards half her genes and produces an offspring by mixing the remaining half with other genes drawn from a stranger. If sex is to outweigh its two-fold cost, the sexual female must by this procedure expect to produce a daughter who will be twice as fit as a simple copy of herself. The problem, therefore, is not trivial. Indeed, G.C. Williams has described it as "the outstanding puzzle in evolutionary biology." Let us look at some of the possible solutions.

11.3.3 Sex can accelerate the rate of evolution

A population of sexually reproducing organisms can, under some conditions, evolve faster than a similar number of asexual organisms. Sexual reproduction can greatly increase the rate at which beneficial mutations, at separate loci, can be combined in a single individual (Figure 11.1). Suppose, for example, that a sexual and an asexual population are both fixed for genes A and B at two loci. In the environment where the two populations are living, mutations A' and B' are advantageous. A' and B' mutations would be likely

(a) **Asexual: high rate of favorable mutation**   (b) **Sexual: high rate of favorable mutation**

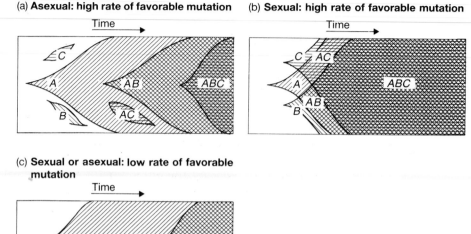

(c) **Sexual or asexual: low rate of favorable mutation**

Figure 11.1 Evolution in (a) asexual and (b) sexual population. The mutations *A*, *B*, and *C* are all advantageous. In the asexual population, an *AB* individual can only arise if the *B* mutation arises in an individual that already has an *A* mutation (or vice versa). In the sexual population, the *AB* individual can be formed by breeding of a *B* mutation-bearing individual with an *A* mutation-bearing individual; the second mutation of *B* is not needed. (c) If favorable mutations are rare, each will have been fixed before the next arises, and sexual populations do not evolve faster. The relative rates of evolution in asexual and sexual populations depends on the rate at which favorable mutations arise.

to arise initially in different individuals. The asexual population will then come to consist of *A′B* and *AB′* individuals, because the *A′* mutant cannot spread into the *AB′* clone or vice versa. *A′B′* individuals cannot appear until an *A* gene mutated to *A′* within the *AB′* clone (or *B* to *B′* in the *A′B* clone).

In the sexual population, evolution proceeds much faster. After *A′* and *B′* have arisen in different individuals, they can soon combine in a single individual by sex without waiting for the mutations to occur twice. Natural selection can therefore take the population from the state *AB* to *A′B′* faster than under asexual reproduction. This argument was first put forward by Fisher, who concluded that sexual populations have a more rapid rate of evolution than would an otherwise equivalent group of asexual organisms.

However, Fisher's conclusion depends on the rate of mutation. If favorable mutations are rare, each one will have been fixed in the population before the next one arises (Figure 11.1c). Sexual and asexual populations then evolve at the same rate. New favorable mutations will always arise in individuals that already carry the previous favorable mutation: they must do, because the previous favorable mutation is already in every member of the population. In terms of the example, the *B′* mutation will arise in an *A′B* individual in both sexual and asexual populations. However, if favorable mutations arise more frequently, then Fisher's argument works: the sexual population evolves faster. Each new favorable mutation will usually arise in an individual that does not already possess other favorable mutations, and the greater speed with which the different favorable mutations combine together causes the sexual population to evolve faster. The higher the rate at which favorable mutations are arising, the greater the evolutionary rate of a sexual relative to an asexual population.

11.3.4 Is sex maintained by group selection?

Perhaps the most common answer to the question of why sex exists is that it speeds up the rate of evolution. This is the "group selection" theory of sex. It accepts that sex is disadvantageous for the individual, because of its 50% cost; but claims that the cost is more than made up for by the faster evolution of the sexual population, or group. The sexual population would accumulate superior adaptations more rapidly and be able to outcompete an asexual population.

Group selection is a controversial subject, which we shall discuss in chapter 12. Here it is enough to note that most evolutionary biologists distrust theories that rely on advantages to the whole population. When individual and group advantage conflict, individual selection is usually more powerful. We should therefore not expect to find adaptations that are disadvantageous for the individual even if they do benefit the group. Although sexual populations may evolve faster than asexual ones, it is still true that sexual *individuals* reproduce more slowly than asexual individuals; asexuality, once it has arisen, will tend to take over sexual groups. Asexuality will not exist within a group only if it has not appeared by immigration or mutation. The reason to be suspicious of group selection is that it requires the rate at which asexual females arise in sexual groups to be very low.

The argument against group selection can at best show that adaptations for the benefit of the group are theoretically unlikely, not that they are theoretically impossible (see chapter 12). We should therefore also look at the facts. What do they suggest? We have seen that the effect of sex on the rate of evolution theoretically depends on the mutation rate, but what about populations in nature? Do natural populations occupy the "rare mutation" extreme in which sexual and asexual populations evolve at the same rate, or are favorable mutations frequent enough for sexual populations to evolve faster? The question cannot be answered directly, because we do not know the rate of favorable mutation; but the taxonomic distribution of asexual species allows an indirect inference.

The taxonomic distribution of asexuality shows it to be confined to the small twigs of the phylogenetic tree (Figure 11.2). It is usually found in single species, isolated within a mainly sexually reproducing genus; sometimes it is found in an entire genus, but only very rarely is it found throughout a larger phylogenetic group. The only such exceptions are a suborder of rotifers (the Bdelloidea) and of gastrotrichs (Chaetonoidea); all the former, and most of the latter, are believed to be asexual. The spindly taxonomic distribution of asexual reproduction suggests that asexual lineages have a higher extinction rate than sexual lineages (section 21.5.3, p. 584); asexual lineages usually do not last long enough to diversify into a genus or higher taxonomic level. If (as

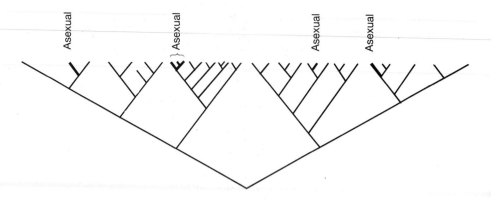

Figure 11.2 The taxonomic distribution of asexual reproduction is spindly; it is found in odd isolated taxa.

is likely), the difference in extinction rates is due to the slower evolution of asexual forms, then natural mutation rates must be high enough for Fisher's argument to apply.

To say that sexual populations have a lower extinction rate than asexual populations is to claim one thing: to say that sex exists because of its lower extinction rate is to claim something much stronger. It could be that sex exists in sexual species because sex is advantageous to the individuals of those species, and asexual reproduction exists in asexual species because it is advantageous to the individuals in those species. The different extinction rates would then be species-level consequences of different individual adaptations in the two types of species. By analogy, carnivores could have higher extinction rates than herbivores, but that would not mean that herbivory was disadvantageous to individual herbivores and maintained because of its advantage to the group. The taxonomic distribution of asexuality, therefore, although it is consistent with the group selectionist theory of sex, does not confirm it. The same pattern could have arisen if sex had an individual advantage.

Finally, Williams has put forward a general argument to doubt the importance of group selection in the maintenance of sex. It is his *balance argument*. Some species, such as many plants, aphids, sponges, rotifers, and water fleas (Cladocera), can reproduce either sexually or asexually according to the conditions. These species are called "heterogonic." Many heterogonic species time their sexual reproduction for periods of environmental uncertainty, and reproduce asexually when conditions are more stable; but that is not the important point here. What matters is that an individual can reproduce in either way. Therefore, when an aphid reproduces sexually, it must be advantageous to the individual, because if it was not the aphid could have reproduced asexually. Both sexual and asexual reproduction must have balanced advantages to maintain them in the species' life cycle: otherwise the inferior one would be lost.

The group selectionist, we recall, believes that sex is disadvantageous to the individual, and only advantageous to the group. But in aphids and other heterogonic species in which individuals have a choice, sex must have an individual advantage. The argument can be extended. If sex is advantageous in aphids, it is probably also advantageous to the individual in non-heterogonic species too. We have no good reason to think that sex is peculiar in aphids, or that special factors favor sex in heterogonic species. If we must find an individual advantage for sex in aphids, that same advantage will probably also exist in other species. If group selection can be ruled out for aphids, it can probably also be ruled out for other species.

Williams' argument is powerful, but not decisive. In most heterogonic species, the asexual and sexual propagules differ in other respects besides being asexual and sexual. For example, the Cladoceran sexual offspring form special winter eggs which are adapted for winter survival. Any Cladoceran that gave up sex would also lose its overwintering stage: in practice, the loss of sex while retaining the winter egg would need two mutations, one for the loss of sex and the other for transferring the winter egg phenotype to asexual eggs. So the balance argument is not perfectly clear-cut.

In summary, group selection will tend to favor sexual over asexual reproduction because sexual populations will have a lower rate of extinction. The taxonomic distribution of asexuality suggests that asexual populations tend to go extinct relatively quickly in evolution. However, biologists doubt whether group selection is the reason why sex exists, for two main reasons. One is a general disbelief in group selection; the other is Williams' balance argument. Neither of these objections is completely convincing, and group selection cannot finally be ruled out. However, the objections are strong enough to have inspired biologists to look for a short-term, individual advantage to sex. We shall consider two of the most influential modern ideas.

11.3.5 The coevolution of parasites and hosts may produce rapid environmental change

The group selectionist theory of sex required the environment to change, at least over the long term. In theory it only required new favorable mutations to arise at such a high rate that sexual groups evolve faster than asexual groups; but if we ask why new favorable mutations are arising, environmental change is much the most plausible reason. (If the environment was constant, most of the possible advantageous mutations would already have been fixed.) Now, some kinds of environmental change, like changes in the weather, are easy to observe and measure. But the environment has many less obvious features too, and we shall be concerned with some of these here. We have to think about them theoretically, because no one has yet discovered how to measure the crucial variables in nature.

The more rapidly the environment changes, the shorter is the time scale over which sex is advantageous. If the environment changes slowly, sex might be advantageous only on average every few hundred generations; and this is the kind of change supposed in the group selection theory of sex. If it changed rather faster, the advantage to sex might be felt perhaps every 50 generations. By extension, if the environment changes enough every generation, sex might be advantageous to the individual.

Sex undoubtedly could have an individual advantage, if environments change rapidly enough; the problem is to think up *how* environments could possibly be changing that rapidly. It is not difficult to believe that environments might change fast enough to make sex advantageous every few hundred years; but how could they be changing fast enough to make it advantageous every generation? Remember, the environment would have to be changing so rapidly that an average sexual female's daughters must be twice as fit as those of an average asexual female. We cannot take it for granted that ordinary environmental change will be enough: if we are to explain the existence of sex by environmental change, we have some work to do.

A recent, promising suggestion is that the coevolution between parasites and hosts may generate fast enough environmental change to make sex advantageous in the short term. The environment here, for the parasite, is the host's resistance mechanism and, for the host, the parasite's method of penetrating its defences. Several authors have suggested that *parasite–host coevolution* may be important in the maintenance of sex; Hamilton is the best known. The theory can be made more exact by a simple model. It is known that in some parasite–host relationships there are "gene-for-gene" matching systems such that one host genotype is adapted for resisting one parasite genotype, another host genotype for another parasite genotype, and so on. The best understood example is from wheat and parasitic rusts; similar selection may operate in the human HLA system (section 8.6, p. 191).

The simplest genetic model for parasite–host coevolution is haploid, with two alleles in each of the host and parasite species. One parasite allele is adapted to penetrate hosts with one of the host alleles, the other parasite allele to penetrate the other host allele (Table 11.2). Selection of this sort generates cyclic changes in gene frequency (Figure 11.3): as a genotype increases in frequency, its fitness (after a time lag) decreases. If parasite genotype P_1 is more common, host genotype H_1 will be favored and increase in frequency; the fitness of P_1 then goes down as more hosts are resistant to it. Then, as P_2 becomes more common, the fitness of H_1 decreases. When H_1 becomes rarer, the frequency of P_1 will in turn increase again. Cycles of gene frequency are driven by corresponding cycles of gene fitness.

Table 11.2 A simple model of gene-for-gene matching in a pair of host and parasite species. The numbers in the table are the fitnesses of the genotypes. (a) Fitness of parasite genotype in two types of host. (b) Fitness of host genotype against two types of parasite

(a)		Host genotype	
		H_1	H_2
Parasite	P_1	0.9	1
genotype	P_2	1	0.9

(b)		Parasite genotype	
		P_1	P_2
Host	H_1	1	0.9
genotype	H_2	0.9	1

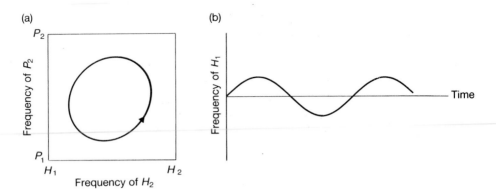

Figure 11.3 Frequency changes of host and parasite genotypes. (a) As H_2 becomes more common, there is selection to increase the frequency of P_2, which in turn selects against H_2, and H_1 increases in frequency. (b) Plotted against time, the frequency of each genotype oscillates cyclically.

We need a more complex, and more realistic, model to produce an advantage for sex. Imagine now that resistance and counter-resistance are controlled by two loci. Again, a haploid model is simplest. With two loci and two alleles at each, there are four haplotypes: AB, Ab, aB, and ab. There will be complementary sets in host and parasite; if $A_H B_H$, $A_H b_H$, $a_H B_H$, and $a_H b_H$ are the host genotypes, then we could write the parasite genotypes as $A_P B_P$, $A_P b_P$, $a_P B_P$, and $a_P b_P$. $A_H B_H$ and $A_P B_P$ are analogous to H_1 and P_1 in the previous model. As a concrete example, A_H and B_H might control two cell-surface molecules used by the parasite to penetrate the host. Hosts with allele A_H are efficiently penetrated by parasites with a_P, but not A_P; B_H hosts are penetrated by b_P parasites but not B_P. b_P parasites are therefore favored if the hosts are mainly B_H; likewise, b_H is a gene for resistance to b_P parasites. Haplotype frequencies at both loci will oscillate for the same reasons as did H_1 and P_1 in the simpler model. In the sexual parasites and in the sexual hosts, alleles at the two loci can recombine; in asexual parasites and hosts, they cannot. (This two-locus, haploid case is the simplest that can illustrate the important principle. More realistic models can have more than two loci, with more than two alleles each, and are diploid. Moreover, the host can be subject to attack by more than one species of parasite. These complications can make the conditions necessary for sex to be maintained over asexual reproduction easier.) A third locus determines whether reproduction is sexual or asexual. The locus controlling sex, we assume, has no effect on the organism's fitness: fitness is only influenced by the resistance genotypes. In practice, many other things besides parasite resistance will influence an organism's fitness; but we can reasonably ignore them while we are trying to understand the effect of parasite–host coevolution on the evolution of sex.

How can sex be advantageous in this model? The same argument can be given for either the host or the parasite: let us consider it from the viewpoint of the host. An asexual gene can be in a host with any of the four resistance genotypes. Over time, the fitness of that asexual gene will fluctuate along with that of the resistance genotype it is linked to (Figure 11.4). All four

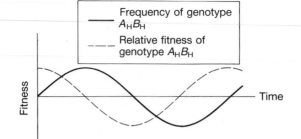

Figure 11.4 The frequency cycles of each genotype (such as the A_HB_H genotype in the host population) are driven by cyclical changes in the genotype's fitness.

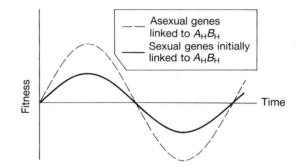

Figure 11.5 The fitness fluctuations of a gene for sex have lower amplitude than those for an equivalent gene for asexual reproduction. The fitness cycle of an asexual gene linked to only one of the resistance genotypes is shown: all four cycle, but not synchronously. The graph for the sexual gene shows the fitness cycles of the sex genes that were initially linked to the same resistance genotype as the asexual gene. Their fitness cycle depends on the rate at which they recombine with the other resistance genotypes, and the exact graph could have many different forms. It might even not cycle at all, and does not have to cycle synchronously with the initially equivalent set of asexual genes. However, the sex genes' fitness cycles will have lower amplitude than those of the genes for asexual reproduction.

resistance genotypes will undergo cycles, and a gene for asexuality experiences the ups and downs of the cycle of the resistance genotype it is bound up with. The asexual group as a whole contains four kinds of asexual gene, each undergoing a particular cycle of fitness.

The sexual gene has a different experience. Rather than being locked into one particular cycle, it is shuffled between the four resistance genotypes. The important consequence is that the sex gene undergoes fitness cycles of lower amplitude than the asexual gene (Figure 11.5). Imagine, for instance, that selection is favoring the A_HB_H genotype. In the asexual group, all the offspring of the favored A_HB_H individuals have the A_HB_H genotype; the frequency of the A_HB_H genotype in the next generation increases more than in the sexual group, because the offspring of a sexual A_HB_H individual do not all have the $A_H B_H$ genotype. (The exact genotypic frequencies in the sexual offspring depend on the frequencies in the population and on whether mating is random, but there will always be fewer of the parental genotypes in the offspring of a sexual female than in those of an asexual female.) The frequency of the $a_H b_H$ genotype therefore undergoes larger oscillations in the asexual than in the sexual group. Now, the genes both for asexual and for sexual reproduction are hitching an evolutionary ride with the resistance genotypes. The asexual gene hitching with $A_H B_H$, for instance, will increase in frequency when $A_H B_H$ is favored, but then decrease when $A_H B_H$ is selected against. The equivalent increases and decreases in frequency experienced by a gene for sex that starts out with $A_H B_H$ will have a lower amplitude. A sex gene in an $A_H B_H$ individual has a lower fitness than an equivalent asexual individual while selection favors $A_H B_H$, because the sexual $A_H B_H$ individual produces fewer $A_H B_H$ offspring. The fitness of the sexual gene is not so high during the up part of the cycle, and an individual with a favored genotype would do better to reproduce asexually. But as the cycle turns down, the opposite effect operates. Now the fitness of the sexual gene is less depressed than an asexual gene; it produces some favorable offspring, while the asexual female produces only disadvantageous $A_H B_H$ offspring. Through a full cycle, the fitness of the sex gene is damped down as it shuffles among the resistance genotypes at other loci.

Notice that the arithmetic average fitness of the sexual and asexual types in any one generation are the same. We have two equivalent "populations," each with four resistance genotypes experiencing the same selection. The average fitness of the population is just the fitness of the four genotypes multiplied by their frequencies; the arithmetic average is the same for the sexual and asexual populations. They differ in the amplitude of the fluctuations experienced by a sexual and an asexual gene linked initially to any one of the genotypes.

But this difference can be enough to make sex advantageous. The long-term change in a gene's frequency is determined by its geometric, not its arithmetic, mean. (The geometric mean of a set of n numbers is the nth root of their product. Thus the geometric mean of a, b, and c is $(a \times b \times c)^{\frac{1}{3}}$.) The propagation of a gene through a population is a geometric process, in which the number of copies of a gene after a number of generations is equal to the product (not the sum) of its reproduction in each generation. For two sets of numbers with the same arithmetic mean but different variabilities, the less variable set of numbers roughly speaking has a higher geometric mean. Suppose the sexual female always produced two offspring (this example is just to illustrate the numerical effect); after four generations the number of her descendants would be $2 \times 2 \times 2 \times 2 = 16$. Compare that with an asexual female with the same arithmetic average of two offspring per generation, but the number of whose offspring oscillates up and down in successive generations. After four generations, she might have produced (for instance) $2 \times 1.5 \times 2 \times 2.5 = 15$ descendants, lower than 16. The same numerical effect takes place for the more-or-less damped oscillations in Figure 11.5: the more variable fitnesses of the asexual gene will have a lower geometric mean. In biological terms, the sexual gene has an advantage by spreading its risks. It reduces the variability of its fitness changes between generations; it has a steadier fitness over the long term. The asexual gene has a high fitness while it is linked to the favored resistance gene, but its fitness will inevitably soon crash down again. Over the long term, it reproduces at a lower rate. The justification for risk spreading in human investment decisions is essentially the same.

The advantage to sex in the parasite–host model is quite subtle and an extreme case may clarify the principle. The frequencies of the haplotypes (A_HB_H, A_Hb_H, a_HB_H, and a_Hb_H) cycle, as we have seen; and the associated cycles in an asexual gene have greater amplitude than in a sexual gene. Imagine now that the cycles were so extreme that, from time to time, the fitness of a genotype became zero. In the asexual group, when the fitness of A_HB_H went to zero, all the asexual genes with that genotype would be eliminated. Some time later, the fitness of another genotype, such as A_Hb_H, would become zero, and all of them be eliminated. Before long all the genotypes would be eliminated: and the asexual gene would then be extinct. Once a genotype has been lost from an asexual group, it cannot be recreated by recombination; its loss is final. The same is not true with sex. When the fitness of A_HB_H becomes zero, all the sex genes with that genotype will be lost—but only temporarily, as A_HB_H will soon be reformed from the other genotypes. Even after the fitnesses of all four haplotypes has been zero, they will still all exist in the population. The sex gene will not be extinct.

In the extreme case where the fitness cycles go through zero, the advantage of sex is clear. By extension, if the cycles pass through a phase of very low fitness, there will be some chance that the asexual genotype will be lost, and as the cycles repeat each asexual genotype will inevitably be lost eventually. The loss of the sexual genotypes will again be only temporary. It is a small extension from this case to the geometric calculation above, which illustrates the same process in more general form.

For natural selection to favor sex, it is not enough that sexually reproducing females should have a higher fitness than asexually reproducing females. The fitness of sexual females must be quantitatively large enough to make up for the 50% cost of sex: sex must have twice the fitness of asexual reproduction. We have been concerned here with the biological process that produces an advantage to sex under parasite–host coevolution, but an exact model is needed to see whether the advantage is quantitatively large enough. Hamilton and others have simulated formal models of the type we have considered; and they have found that sex can be favored under some conditions, even including its 50% cost.

In summary, the advantage of sex in the parasite–host model arises from the following factors: (a) parasite–host coevolution with gene-for-gene relationships is cyclical; (b) a sexual gene linked to these cycles will undergo more damped oscillations of fitness than an asexual gene; (c) the evolutionary fate of a gene is determined by its geometric, not its arithmetic, mean; and (d) for two series of numbers with the same arithmetic mean, the series with the lower variability will have the higher geometric mean.

Whether the parasitic model is correct depends on how realistic its assumptions are. A few observations can be made. One is that parasite–host coevolution is almost universal in nature. Humans are parasitized by a long list of worms, flies, protozoa, bacteria, and viruses; all species, including bacteria, are parasitized by viruses. Whether the right kind of gene-for-gene matching systems, and cyclical coevolutionary dynamics, are also universal in nature is more difficult to say. However, although no positive evidence can be offered to suggest that they are, nor is there any negative evidence to show that they are not; the possibility remains open.

11.3.6 Sex may be favored by sibling competition

G.C. Williams, in his book *Sex and Evolution* (1975), suggested a series of models of sex, and Maynard Smith pointed out that the models share the essential feature of *sibling* (sib) *competition*. Sib competition means that a female's progeny compete among each other for survival.

Williams' "elm–oyster" model is an example. In the model, the environment is made up of patches, each big enough to support only one individual (Figure 11.6). At any one time, most patches are occupied; a patch only becomes empty when an adult elm tree dies. The seeds from many elm trees will then arrive at the patch, but only at most one can succeed. A seed's chance of success is assumed to depend on its genotype, because some genotypes will be better adapted to the patch's particular conditions than others. Different genotypes do better in different patches, but when a new patch opens up a parent cannot (so to speak) "predict" which genotype will do best in it. The model's final assumption is that each parent sends many seeds to the patch: this is what causes the sib competition. Many related seeds land in the patch, but only at most one will survive: they are competing among one another. The more seeds any one parent tree sends, the stronger the sib competition.

In these conditions, selection favors parents that produce a genetic variety of offspring. A sexual parent can do so, but an asexual one cannot. An asexual female, in Williams' analogy, is like someone who photocopies a lottery ticket: there is no advantage to entering multiple tickets in a lottery if they are all copies of the same original. Sending many seeds all with the same

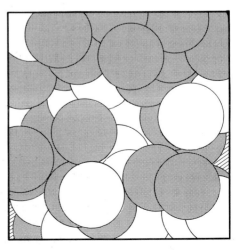

Figure 11.6 Sib competition model of sex. The environment is divided into patches, each of which can support a single individual. At any one time, some patches are occupied by sexual individuals (stippled circles), some by asexual individuals (unshaded circles), and a few are unoccupied (hatched). From Williams (1975).

genotype to the patch does not significantly increase the chance of success; sending seeds with different genotypes does. Sex can therefore be advantageous. Williams and Maynard Smith have both formally modeled the process and shown that, if sib competition is intense enough, the advantage of sex can outweigh its 50% cost.

Sib competition will undoubtedly operate in some cases. The main problem is whether it is a general, and powerful, enough process to explain the widespread existence of sex. Williams himself doubted this. Sib competition is most powerful in species with high fecundity, in which a parent sends many propagules to particular clearly defined places. Many species do not have that kind of life cycle. Some species, for example, discharge their eggs into the sea, where they become mixed up and sibs are no more likely to compete with sibs than with other, unrelated members of the species. Species with low fecundities pose another problem. If a parent produces one offspring at a time, and sends it alone to a new patch, there will be no sib competition at all. Other things being equal, sib competition will be reduced in species with lower fecundity. "Higher" vertebrates often have low fecundity, and Williams suggested that sex is now disadvantageous in them; sex persists, he thought, only because the mutations needed to lose it have not arisen.

But it might be that sib competition is very common. It requires only that more similar genotypes compete more strongly with each other than with other genotypes, and the competition can take place over a lifetime. Thus, in the sea, the apparently mixed up propagules may still show practical sib competition if different genotypes tend to exploit different resources or select their habitats differently; then more similar genotypes will end up competing more closely even if they have been mixed up in the plankton. And the fact that competition over a lifetime is what matters, rather than only that in the juvenile stages, could be important in higher vertebrates. Primates, including ourselves, usually produce one offspring at a time, but they mainly live in social groups. Related individuals can coexist in social groups for a long time. If more similar genotypes exploit more similar resources, or otherwise depress the reproduction of individuals with similar genotypes, sib competition could be operating in them too. Although biologists doubt that sib competition is a common enough process to explain the existence of sex, the process must operate to some extent; and there is still room for a variety of opinions about just how common it is in nature.

11.3.7 Conclusion: it is uncertain how sex is adaptive

There are other ideas besides the three we have discussed, and some of them are highly ingenious. However, the question of why sex exists remains an outstanding puzzle: evolutionary biologists are not confident the question has been satisfactorily answered. Maybe, as Maynard Smith once said, "some crucial aspect of the problem has been overlooked," and we need some radically new idea that has not yet been put forward or lies unappreciated. Alternatively, the gist of the answer may lie in the theories we have discussed and the problem is more one of showing how they apply in nature. When the question is answered, some other deep beliefs in evolutionary biology may have to be surrendered. If sex is an adaptation maintained by group selection, ideas on the importance of group selection (see chapter 12) will need to be modified. Continual, rapid, cyclical coevolution between parasites and hosts, or sib competition, may be more common than is customarily supposed; and both these theories suggest that the environments of offspring are so different from their parents that a genetic copy of the parent would have half the fitness of a set of offspring produced by mixing the mother's genes with those of another individual. That is dramatically rapid environmental change, much more rapid than anyone has evidence of; but it may yet turn out that it happens. The subject has moved fast in the past 20 years and the hope now is that this crucial question will be answered in the next decade or so of research. Important related questions wait upon the answer.

11.4 Example 2: sexual selection

11.4.1 Sexual characters are often apparently deleterious

For the most part, the characters of organisms are adaptive: they increase the organisms' chances of surviving to reproduce. However, there are some characters that do the opposite, and (as Darwin was well aware) natural selection does not explain why these characters exist. If a population contains some types with higher survival than other types, natural selection will fix the former and eliminate the latter.

Characters that reduce survival can be called "deleterious," or "costly." One large class of apparently costly characters are those found usually only in males and which Darwin called secondary sexual characters. The primary sexual characters are things like genitalia that are needed for breeding. The secondary sexual characters are not actually needed for breeding, but they function during reproduction. The peacock's "tail" (or, more exactly, train) is an example. In many other bird species too, the males have tails or other extravagantly developed and brightly colored structures. A peacock could inseminate a female just as well without his remarkable tail, and in that sense it is a secondary, not a primary, sexual organ. The peacock's tail almost certainly reduces the male's survival (though this has never actually been demonstrated): the tail reduces maneuverability, powers of flight, and makes the bird more conspicuous; its growth must also impose an energetic cost. Why are these costly characters not eliminated by selection?

11.4.2 Sexual selection acts by male competition and female choice

Darwin's solution was his theory of sexual selection. He defined the process by saying that it "depends on the advantage which certain individuals have over other individuals of the same sex and species, in exclusive relation to reproduction"; a structure produced by sexual selection in males exists not because of the struggle for existence, but because it gives the males that possess it an advantage over other males in the competition for mates. Darwin's idea is that the reduced survival of the peacocks with long, colorful tails is more than compensated by their increased "advantage in reproduction."

Darwin discussed two kinds of sexual selection. One is for males to compete among each other for access to females. *Male competition* can take the form of direct fighting, or it can be more subtle. Male insects, for instance, while copulating with a female have various means of removing any sperm she possesses from matings with previous males; one example is the damselfly *Calopteryx maculata* in which Waage showed that the male uses brushes and hooks on the penis to scrape out the sperm. There are many other delightful examples, but we shall not discuss them here because they do not pose deep theoretical questions. The situation is different for Darwin's other mechanism: *female choice*.

A structure like the peacock's tail cannot plausibly be explained by male competition. It would be no use in fighting—indeed it would reduce the male's fighting power—and no one has ever thought up a more subtle competitive function for the tail. Darwin instead suggested that the tail exists because females preferentially mate with males that have longer, brighter, or more beautiful tails. If they do, the mating advantage of males with longer tails will compensate a corresponding amount of reduced male survival.

Darwin's main argument for the importance of sexual selection was comparative. Sexual selection should operate more powerfully in polygamous than in monogamous species. In a polygynous species, in which several females mate with one male (and other males do not breed at all), a single male can potentially breed with more females than under monogamy; selection in favor of adaptations that enable males to gain access to females (whether by male competition or female choice) is proportionally stronger. Darwin therefore reasoned that secondary sexual characters would be more developed in polygynous, than monogamous, species. Polygynous species should have stronger *sexual dimorphism*.

Darwin's book *The Descent of Man, and Selection in Relation to Sex* (1871) contains a long review of sexual dimorphism in the animal kingdom; it is still the best (and classic) demonstration that sexual dimorphism is indeed mainly found in polygynous species. In polyandrous birds, such as phalaropes, sexual selection is "reversed": females compete for males, and it is the females that are the larger and brightly colored sex. There are exceptions, such as monogamous ducks that are also sexually dimorphic; Darwin had an additional theory for them. However, the main point is that Darwin's principal evidence for sexual selection came from a comparison of large numbers of species: the comparison showed that the species with brightly colored, large, or dangerously armed males are more often polygynous and the species in which males and females are more similar are more often monogamous.

11.4.3 Females may choose to pair with particular males

For Darwin, female choice of males was an assumption; he was mainly concerned to show that, if it exists, it can explain extraordinary phenomena like the peacock's tail. He did not have much to say about the prior question of why the female preference should ever evolve to begin with. Selection can work on a female preference just like on any other character. If females with one type of preference produce more offspring than females with another, selection will favor the more productive preference. The difficult case is in an extreme case like the peacock, in which the form of female choice appears to be disadvantageous to the female. Females are picking males that possess a costly character; that costly character will be passed on to their sons; the female preference therefore seems to be causing the females to produce inferior sons.

We can spell the problem out more fully, in terms of selection on a mutant non-choosy female. Suppose that peahens do prefer peacocks with dazzling tails, and a mutant female, who does not prefer these males, arises; she might mate at random, or prefer some other sort of male. What does selection do to this mutation? The mutant female will produce sons that do not possess the costly character, at least in so extreme a form; her sons will therefore survive better than average. So the mutant should be favored, the female preference should be lost, and the extreme male forms should disappear . . . or should it?

The mutant female will indeed produce sons that survive better than the population's average. But that, as Fisher first realized, is not enough to guarantee that it will spread. When the mutant female's sons grow up, with their inferior tails, they will be rejected as mates. The mutant female is a rare mutant, in a population where the majority of females prefer males with long tails, and this majority preference will work against the mutant's sons. Despite their superior survival, they will be condemned to celibacy. The randomly mating mutation, therefore, may not spread.

Fisher also discussed how the preference for a costly character could evolve to begin with. After the long male tail has evolved, it is costly; but at an earlier evolutionary stage, before the female preference arose, things might have been different. Male tails would have been shorter then. Suppose that, before some mutant female arose who picked longer-tailed males, most females picked their mates at random; suppose also that there was at that time a positive correlation between male tail length and survival (Figure 11.7a). Selection would then favor a mutant female with a preference for

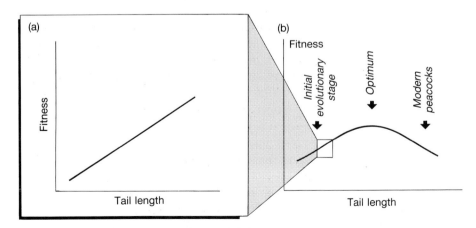

Figure 11.7 (a) An early stage in the evolution of a bizarre character such as the peacock's tail. Before females preferred to mate with long-tailed males, there might have been a positive correlation between tail length (then much shorter than in their descendants) and male fitness. (b) The full relationship between the degree of exaggeration of character (tail length) and survivorship. There is an intermediate optimum. Modern species like the peacock are toward the right of the graph.

males with longer tails; she would produce sons with longer than average tails and the associated higher survival. Then, as the mutation spreads, the males with longer tails would start to acquire a second advantage. There are increasing numbers of females in the population who prefer to mate with longer-tailed males, and the males so endowed will not only survive better but also enjoy an advantage in mating. The evolution of longer tails in males, and a mating preference for them in females thus come to reinforce each other, in what Fisher called a *runaway* process. Technically, they reinforce each other because the genes encoding them are in linkage disequilibrium (section 8.5, p. 187). The offspring of a female who mates preferentially with long-tailed males will possess both their mother's genes for choice and their father's genes for long tails. These two kinds of genes thus become non-randomly associated, and the genes for female choice increase in frequency by hitch-hiking with the advantageous genes for long tails in males.

If we consider a sufficiently wide range of tail lengths, the full relationship between tail length and male survival presumably shows some increase and decrease on either side of an optimum (Figure 11.7b). Eventually, powered by female choice, the average tail length in the population will reach the optimum; but evolution does not stop there. As the population evolves towards the optimal tail length, the longer-tailed males are still preferred. By now the female preference will have spread through the population and the majority of females will prefer long-tailed mates. Now the mating preference alone drives the evolution of longer tails. The preference may have become strong enough to compensate for lower male survival, and evolution will proceed into the interesting zone in which the male character, in a complete reversal of the original selective forces, evolves to become increasingly costly to its bearers. There are thus three stages in the evolution of the long tail: an initial stage in which long tails have only a survival advantage, and a second stage in which the survival advantage is supplemented by a mating advantage. As female choice grows more common and the tail length grows past the optimum for survival, and relative importance of the two advantages shifts over, until we reach a third stage at which further elongation is driven purely by female choice.

As the population evolves past the point of optimum tail length, the selective forces at work have become almost absurd. The males are evolving ever longer tails even though the original selective advantage has ceased, and then reversed. The female mating preference will also continue to strengthen, even though it now favors a character that decreases survival. The runaway process will only come to a stop when the death rate of males, due to their feathery excess, is so high that their success in mating no longer makes up for it. The tail length will then reach an equilibrium. That equilibrium, according to Fisher, is what we are now observing in birds like peacocks and birds of paradise.

The original problem was to explain the evolution of a set of apparently deleterious characters. Darwin's solution was that they could be maintained by female choice. He did not, however, explain why females should come to choose males with deleterious characters, nor why the choice would not be lost by natural selection. In Fisher's theory, when the choice first evolved, the male character was much smaller and choice then favored males with higher survival. Genes for choice could thus increase in frequency from being rare mutants to being the majority form in the population. Once nearly all the females in a population choose mates in a certain way, mutant females that pick some other sort of male are selected against (because of the effect on the kind of sons they produce). The cost of the male character at the final equilibrium serves no function for the female: it is maintained by female choice but is the useless end-product of an initially useful process.

11.4.4 Females may prefer to pair with handicapped males, because the male's survival indicates his high quality

We now turn to a second theory, in which the costliness of the male character is positively useful to the female. It is Zahavi's *handicap theory*. ("Handicap" is Zahavi's term for what we have been calling a costly or deleterious character—a handicap is a character that reduces survival.) The argument runs like this. Suppose that the males in the population vary in their quality. We shall concentrate on species, like peacocks, in which males contribute only sperm; in them, quality must mean genetic quality, because nothing else is transferred. We are thus assuming that some males have genes that confer higher fitness ("good genes") than do other males (who had "bad genes"); in practice there could be all degrees between good and bad, but the point can be explained in the simple dichotomous case.

If a female mates at random, her mates will have good and bad genes in the same proportions as the good and bad genes have in the whole population; if half the males in the population have good genes and half have bad, then 50% of her mates will have good genes and 50% bad. Now suppose that some of the males in the population possess a handicap—a character that reduces their survival. If only males with good genes can survive possessing a handicap, a female who mates preferentially with handicapped males will only mate with males with good genes (Table 11.3). The choice will be favored by selection if the advantage through the superior genes outweighs the cost of the handicap: then the net quality of the choosy female's offspring will be higher than those of the randomly mating female.

The handicap thus acts as an indicator of genetic quality: but why does the indicator have to be costly? The reason is that the cost guarantees that the

Table 11.3 The handicap principle. If only males with good genes can survive the possession of a handicap, females who mate with handicapped males will mate only with males who possess good genes. A female who mates with males lacking a handicap will mate with males possessing good and bad genes in their population proportions

	Males with bad genes	Males with good genes
Males without handicap	Alive	Alive
	In population proportions	
Males with handicap	Dead	Alive

indicator will be reliable. A male's genetic quality does not come written on him: it has to be inferred, and if females inferred it from an inexpensive signal, there would be selection on males to cheat. If females preferentially mated with males who merely said "I have good genes" (or rather, in a non-human species, something analogous to saying this) and rejected those that said "I have poor genes," mutant males who said the former independently of their true genetic quality would be favored. Words (and their analogs) are cheap. But if the criterion favored by females is costly, as growing a long and ostentatious tail is, then selection will less automatically favor cheats. In particular, if the cost of growing a handicap is less for a truly high-quality male than for a low-quality male, handicaps will be grown only by high-quality males and will be reliable signals for females to use. (This condition was met in the simple example in Table 11.3: the cost of the handicap for the males with bad genes was far higher than for males with good genes.)

So the reason for the costliness of the male character is completely different in Fisher's and in Zahavi's theories. In Fisher's theory, the cost arose as the end-product of a runaway process. To begin with, long tails were not costly, but as an open-ended female preference for males with longer tails was selected into the population, the tails evolved past their optimum and ended up reducing the survival of their bearers. In Zahavi's theory, the male character had to be costly from the start, and to remain costly as the female preference spreads. The function of the chosen male character is to indicate genetic quality at other loci, and it has to be costly in order to be reliable.

11.4.5 Female choice in Fisher's and Zahavi's models has to be open-ended, and this condition can be tested

There are two crucial means by which Fisher's and Zahavi's ideas can be tested. The first concerns the exact kind of female preference that they require. The preference has to be open-ended. We can distinguish between absolute preferences, which are of the form "mate preferentially with males whose tails are 30 cm long," and open-ended preferences, such as "mate preferentially with the male who has the longest tail you can find." In Fisher's theory, at the initial stage when the male character was positively correlated with survival, either an absolute or an open-ended preference could be favored. The average tail length might then have been 5 cm, and the longest tails in the population might have been 30 cm. If the mutant that was selected in happened to be one encoding an absolute preference, for males with 30 cm tails, then evolution would proceed until the average tail length was 30 cm and then come to a stop. Only if the preference is open-ended can there be an equilibrium with a costly male character. At the equilibrium, the lower survival of males with longer than average tails has to be compensated by a higher frequency of mating: it is not enough for the females to prefer average males, or mate at random: they must actively prefer males with longer than average tails.

Likewise, in the handicap theory, females must prefer males with the most costly handicaps. If the greater cost paid by the higher quality males is not compensated by higher mating success, a less costly handicap will evolve.

Therefore, in both theories, the female choice must be open-ended in a species with a costly male ornament. If we find evidence for such preferences, it suggests the male character indeed is maintained by female choice; but it does not tell us whether Fisher's runaway, or Zahavi's handicap, theory is at work.

The prediction has been tested in more than one species. One example to illustrate the procedure is Møller's study of barn swallows (*Hirundo rustica*). The two sexes are similar in the swallow, except for the outermost tail feather which is about 16% longer in the male than in the female. Møller tested whether females open-endedly prefer males with longer tails by experimentally shortening the tails of some males, by cutting them off with a pair of scissors, and elongating the tails of others, by sticking those severed tail feathers on to other intact males, with superglue that hardened in less than 1 second. He then measured how long it took the different males to find a mate: males with elongated tails mated faster (Figure 11.8a), resulting in higher reproductive success (Figure 11.8b).

Møller also confirmed that the male character is costly. Swallows molt in the fall and grow a new tail for the following breeding season. A male's new tail is on average about 5 mm longer than in the previous year, but the males whose tails were elongated grew a tail in the following year that was shorter than before the experimental treatment (Figure 11.8c). (Møller did not tamper with their tails in the year after the experiment.) Those males had enjoyed a good year during the experiment, but the extra effort of flying with an elongated tail exacted a physiologic cost. Next year the cost was paid: the males took longer to find a mate and their reproductive success decreased. In summary, Møller has shown that the sexually dimorphic tail feathers of swallows are maintained by female choice, that the choice is open-ended, and that the character chosen is costly.

11.4.6 Fisher's theory requires heritable variation in the male character, and Zahavi's theory requires heritable variation in fitness

In Fisher's runaway theory, the reason why females choose males with long tails at the final equilibrium point is that a mutant female who mated at random would have lower fitness because her sons would have shorter tails and be rejected as mates. This is only true if male tail length is heritable. If all the variation in tail length were environmental, and its heritability were zero (section 9.6, p. 225), the tail length of the mutant female's sons would be no shorter on average than those of the choosy females. If mate choice imposed any cost on a female at all, the randomly mating mutant would spread. Tail length, therefore, must be heritable or selection will favor the female who mates at random. This condition is testable, but has never been tested in a species with a costly and extravagant character like the peacock's tail.

In Zahavi's theory, the advantage of female choice does not depend on the inheritance of the male character, and choice could be maintained even if the heritability of tail length became zero. But the theory has an analogous condition. In species in which males transfer only sperm, female choice is for male genetic quality. There must be variation in male genetic quality: some males must have good genes, others bad. This is a condition called *heritability*

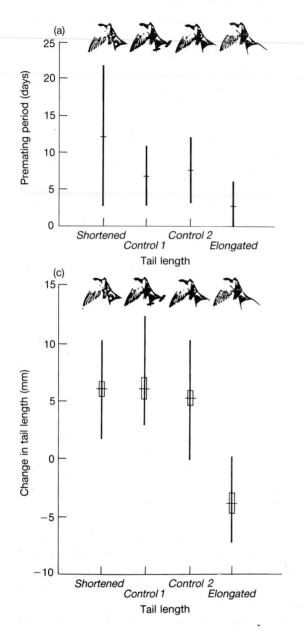

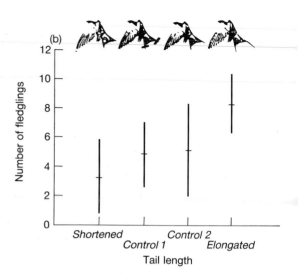

Figure 11.8 Barn swallows with longer tails are preferred by females, but the character is costly. Møller experimentally shortened some males' tails and elongated others; as one control he cut the males' tails off and then immediately stuck them back on again (control 1) and as another he left the males untreated (control 2). Males with elongated tails (a) obtain mates more quickly; (b) have higher reproductive success; but (c) next year grow a shorter tail while the other males grow a longer tail. (Møller also measured both the mating advantage and the cost of longer tails by other criteria too, and those results support the results illustrated here.) From Møller (1989).

of fitness; individuals of higher than average fitness (i.e. who produce more offspring than average) produce offspring who also have higher than average fitness. If the high quality males do not produce high quality offspring, there is no point in picking them as mates.

The conditions in the two theories face a common difficulty. While selection operates on any character, it reduces its heritability (e.g. the Illinois maize experiment, Table 9.2 and Figure 9.8, pp. 227–8). In a population in which some individuals possess good genes and others bad genes, selection acts to fix the good genes—and once it has done so there will be no variation in genetic quality left. Zahavi's theory then would not work. In fact it is not known how much variation in genetic quality exists in any species in nature, and no firm conclusion can be drawn. However, there are three arguments to be aware of. One is the possibility just noted, that Zahavi's theory may not work in species in which males transfer only sperm because there is not enough variation in genetic quality. Alternatively, there may be enough variation and the process may operate. Two reasons have been suggested. Mutation is one of them: although mutation is relatively ineffective at producing variation at a single locus (section 5.10, p. 107), it is much more important when we consider all the genetic loci in an organism (section 9.9, p. 223), and mutation alone might create new bad genes as fast as selection was eliminating them. A similar argument can be made in Fisher's theory if the male character, such as tail length, is influenced by a large number of genetic loci: again, mutation might create new genes for short tails as fast as selection was removing them. The second process creating new variation is environmental change. If the environment changes, what was a good gene a few generations ago may be a bad gene now, and selection may then never finally fix any one of the genes. As with the function of sex, one particularly important type of environmental change may be that due to the coevolution of parasites and their hosts.

11.4.7 Females may choose to pair with healthy, unparasitized males

We discussed above the conditions necessary for coevolution between parasites and hosts to cause unending cycles in the frequencies of genes for parasite resistance in the host population (Figures 11.3 and 11.4). At any one time, some individuals will be genetically resistant to the common parasites; they have good genes. Other individuals will be more easily parasitized, and have bad genes. Selection could then favor females who picked healthy, parasite-free males as mates, because they would pass on the males' genes for resistance to their offspring. Female choice of healthy males is a special case of female choice for good genes. This idea has been particularly supported by Hamilton.

If we allow that females might choose males that are resistant to parasites, what has that to do with brightly colored males? Hamilton and Zuk have suggested that these characters enable females to pick resistant mates. Bright plumage is a sign of health, because a parasitized male cannot produce as smart a plumage as can an unparasitized male. A similar argument can be made for other sexual displays and courtship behavior. As Hamilton and Zuk wrote,

> How could animals choose resistant mates? The methods used should have much in common with those of a physician checking eligibility for life insurance. Following this metaphor, the choosing animal should unclothe the subject, weigh, listen, observe vital capacity, and take blood, urine, and fecal samples. General good health and freedom from parasites are often strikingly indicated in plumage and fur, particularly when these are bright rather than dull or cryptic. The incidence of bare patches of skin, which may expose the color of blood in otherwise furry or feathered animals, and the number of courtship displays involving examination of male urine, are of interest in this regard. Vigor is often

conveyed by success in fights and by frequently exhausting athletic performances of many displaying animals.

The parasitic theory of sexual selection is a recent hypothesis and few tests of it have been performed so far; but it is proving to be a remarkably rich source of ideas. We can look at two tests, one comparative and the other experimental.

Hamilton and Zuk reasoned that, perhaps for external ecologic reasons, some bird species will suffer from parasites more than others. This parasitism will then set up selection on females to choose resistant mates, and the selection will be stronger in species suffering higher parasite loads than in species suffering lower loads. If male secondary sexual characters do enable females to choose resistant males, we should expect to find the relatively brightly colored males in the more strongly parasitized species. They collected evidence from the published literature of parasite loads and relative coloration of males and females in 7649 individuals from 109 bird species. They found a relation between the two (Figure 11.9). The result supports Hamilton's theory; bird species that suffer more from parasites do have brighter male coloration.

The experimental test is by Møller, again using the barn swallow. We saw above that female swallows prefer males with longer tails. In Hamilton's theory, tail length should then indicate health, and Møller measured the health of male swallows by counting the numbers of a blood-sucking mite (*Ornithonyssus bursa*) on the birds' heads. He found that males with longer tails indeed do carry fewer parasitic mites. But is the male's health passed on to his offspring? Møller placed 50 mites in the nests of swallows with all tail lengths, and counted the number of mites on the offspring in each nest at a

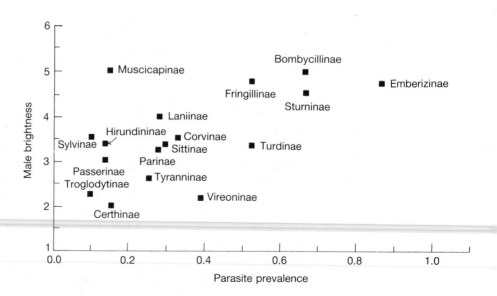

Figure 11.9 Bird species subject to higher degrees of parasitism tend to have more brightly colored males. The graph shows the relation for 17 subfamilies of North American passerines (based on data for 114 species). The scale of male brightness is a subjective ranking from 1–6. From Read (1988).

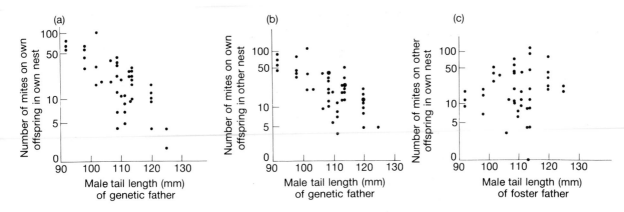

Figure 11.10 Male swallows with longer tails produce healthier offspring. Møller put 50 mites in many nests during the egg-laying period, and then counted the number of mites on the offspring at 7 days of age. He moved hatchling swallows between nests, to study the relationships between (a) the genetic father's tail length and offspring parasite load, when the offspring remain in his nest; (b) the genetic father's tail length and offspring parasite load, when the offspring are reared in another nest; and (c) the foster male's tail length and offspring parasite load in his nest, when the offspring were experimentally moved there. From Møller (1990).

later date. He found that offspring with longer-tailed fathers carried fewer mites (Figure 11.10a). He was also able to confirm that the effect was genetic, by two excellent controls.

A simple correlation between a father's tail length and offspring health could have a non-inherited cause: it could arise, for instance, if longer-tailed fathers were better at spotting parasites and removing them from their offspring or if they were better at foraging and better fed offspring are better at resisting parasites. Møller therefore cross-fostered baby swallows, by moving them between nests immediately after hatching (the adult swallows treat the foster offspring like their own). There was no relationship between the parasite load of young swallows moved to another nest and the tail length of the foster father at their nest (Figure 11.10c); health in young swallows, as measured in this experiment, is not caused by paternal skill. Moreover, the correlation between the genetic father's tail length and the offspring's health remained even when the offspring were reared in another nest (Figure 11.10b). Møller's results are strong evidence that parasite resistance in swallows is inherited. Altogether, Møller has shown that female swallows prefer males with longer tails, that tail length is costly for males, that natural tail length is an indicator of health, and that a male's health (or parasite resistance) is genetic and passed on to his offspring.

Møller's result (Figure 11.10) supports Zahavi's rather than Fisher's theory in the barn swallow. In Fisher's theory, the only advantage of female choice at equilibrium is the mating advantage of the sons. Sons with longer tails have lower survival, and there is no effect in daughters at all (because they lack the long tail). In Zahavi's theory, the good genes indicated by the male handicap will be passed on to both daughters and sons: and the fortunate young swallows descended from long-tailed males were in fact more resistant to parasites. If the Fisher process were all that were operating, we should not expect to see a correlation between the health of a female's offspring and the character of the male that she chooses.

11.4.8 Conclusion: the theory of sex differences is well worked out but incompletely tested

The theory of sexual selection is at a more advanced stage than the theory of why sex exists. The models, such as those of Fisher, Zahavi, and Hamilton, may be correct, and some work has been done to test them. The tests, however, are at an early stage. There are several pieces of evidence for open-ended female choice in species with extravagant, costly male characters. This suggests Darwin was right to explain those characters by female choice. But there has been less work on the other crucial theoretical variable: the inheritance of genetic quality. Møller's fine experiment suggests it can happen in one species, but skeptics will be waiting for repeats and further work.

If further work turns out to support Zahavi's and Hamilton's ideas, another whole set of questions will be opened up. For when we turn to points of detail, we only rarely know how sexual selection has operated. The innumerable more and less subtle details of sex differences in color, morphology, and behavior patterns are probably due in an abstract sense to sexual selection: but what exactly do the male characters we now see indicate, and why? For the swallow's tail, we now know it indicates (perhaps among other things) the number of ectoparasitic mites: but why has tail length in particular evolved to reveal this? And what about those more baffling male characters, such as the peacock's tail, or the ornaments of male birds of paradise? What do they reveal to the females of those species?

11.5 Example 3: the sex ratio

11.5.1 Natural selection usually favors a 50:50 sex ratio

The sex ratio is one of the most successfully understood adaptations. The main idea is again due to Fisher. In most species, the sex ratio at the zygote stage is about 50:50. Fisher explained the 50:50 sex ratio as an equilibrium point: if a population ever comes to deviate from it, natural selection will drive it back.

At first sight, the 50:50 sex ratio might seem inefficient. Most species do not have parental care and are not monogamous; a male can fertilize several females. It would be more efficient for the species to produce more females than males; the extra males are not needed to fertilize the females of the species and do not increase its reproductive rate. (This is another group selection argument: see section 12.2.5, p. 310.) However, imagine what would happen to a population with a persistently female biased sex ratio—one with four females for every one male, for instance. Each male in the population will fertilize on average four females. This condition could not be stable for long in evolution, because an average male is producing four times as many offspring as an average female. There is an advantage to being a male, and an advantage to a female who produces extra sons: sons have a higher reproductive success than daughters.

If a mutant female arose who produced only sons, the total reproductive success of her offspring would be 20/8 times that of an average female (the mutant produces five males, each with relative reproductive success four, for every one male and four females produced by the average female). The mutant would spread. As it did so, the population sex ratio would become less and less female biased. The same argument works in reverse for a population with a male biased sex ratio. The reproductive success of the average female is then higher than that of a male, and natural selection will favor mutant females that produce more daughters than sons. Only when the sex ratio is equal are the relative reproductive successes of the two sexes equal. At that point there is no advantage in producing more of one sex than the other. The 50:50 sex ratio is therefore the point that the population will, over evolutionary time, move to and then stay at. Any population that deviates from the 50:50 sex ratio will be shifted back to it by natural selection.

The fundamental reason why the 50:50 sex ratio is stable is that every organism has one father and one mother. All the females in the population together contribute the same number of genes to the next generation as all the males together; when members of one sex are in short supply, their average success must increase. A number of points should be made about Fisher's argument. One is that it usually applies to the sex ratio of zygotes. It is not affected by differential mortality. (It is not affected at all by differential mortality in species without parental care; but in species with parental care, any sex difference in mortality during the period of care will affect the ratio the parents should produce.) If males die at a higher rate, the adult populations will have more females than males, and will deviate from 50:50. Among the individuals that do survive to reproduce, the average male will have a higher reproductive success than the average female. But that does not mean there is selection to produce extra sons. As far as a mother is concerned, the extra reproductive success of her surviving sons exactly balances the zero reproductive success of those that die before reproducing. When she produces a son, she cannot know in advance whether he will be a survivor who will have higher than average success, or die and not reproduce at all. At birth, a male can only be expected to have the success of an average male, and the average male has the same reproductive success as the average female. The offspring of any one parent will suffer any sex differences in mortality in the same proportion as the population as a whole and there is nothing to be gained from producing more of the sex that will (in the adult stage) be in the minority. Any parent who did so would simply increase the average mortality rate among her progeny.

Such is Fisher's argument. It explains the normal case of the 50:50 sex ratio. Now let us look at a number of cases in which, suitably modified, it can explain deviations from that ratio.

11.5.2 Local mate competition causes deviation from a 50:50 sex ratio
Fisher's theory assumes that there is population-wide competition for mates. A typical male can expect to produce an average number of offspring equal to the total number of offspring produced by the population divided by the number of males in the population; the equivalent is also true for females.

This expectation is correct if an individual competes for mates with the whole of the rest of the population. But in some cases that condition does not apply.

Hamilton, in a paper in 1967, first pointed out that the sex ratio will become biased in species in which an individual competes against only a limited part of the rest of its population. The condition is called *local mate competition*. The pyemotid mites are an extreme case. In this group, brothers inseminate their sisters while they are still inside their mother. A mother therefore only needs to produce one or two males in a brood to inseminate all her daughters; the sex ratio is female biased. In *Pyemotes ventricosus*, for instance, the sex ratio of a brood contains on average four males and 86 females. The males are not competing with the whole population for mates; they compete only against their brothers, and a mother can reduce the fruitless competition by producing fewer sons.

Local mate competition is particularly common in parasitic Hymenoptera. Parasitic wasps insert their eggs in the early stages of other insects. In "gregarious" species, the wasp lays more than one egg in a host, and it is in gregarious species of parasitic wasp that female biased sex ratios are often found. Natural selection most strongly favors a female biased sex ratio when only one female lays her eggs in a host. If a second female also later lays her eggs in the same host, the competition among males becomes less local. (In the theoretical extreme case in which all the females in a wasp population lay all their eggs in the same host, Fisher's condition of population-wide competition for mates would be restored.)

Hamilton accordingly predicted that the second parasite of a host should produce more sons than the first parasite. The prediction has been tested

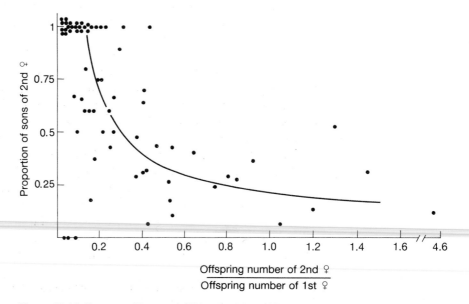

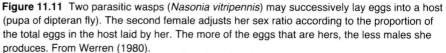

Figure 11.11 Two parasitic wasps (*Nasonia vitripennis*) may successively lay eggs into a host (pupa of dipteran fly). The second female adjusts her sex ratio according to the proportion of the total eggs in the host laid by her. The more of the eggs that are hers, the less males she produces. From Werren (1980).

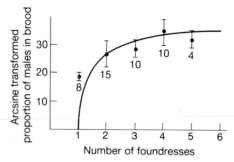

Figure 11.12 More than one fig wasp (*Blastophaga*) female may lay eggs into a fig. The later females to lay eggs in a fig produce increasing proportions of males. From Herre (1985).

(Figure 11.11). Indeed, it has been tested with quantitative precision. Parasitic wasps can tell (using chemical senses in the ovipositor) whether an individual host has already been parasitized. It is not difficult to calculate the exact optimal sex ratio for each female wasp; but we shall not work through the calculation. The main result is that the second female's best sex ratio depends on the relative number of eggs laid by herself and by the first female. If the second female contributes only a few eggs, most of them should be male; and as she lays further eggs, an increasing proportion should be female. Werren has tested the theory in *Nasonia vitripennis*, a parasite of maggots of common blow-flies. These wasps mate almost immediately after emerging from the maggot; the males wait on the outside for their sisters to emerge and then copulate with them. A female *N. vitripennis* lays about 30 eggs in a host, of which about two or three (8.7% to be exact) will be males if the host has not already been parasitized; later females lay more males. The observations cluster neatly around the theoretically predicted line. Werren's work is a good example of a quantitative experimental test of a theory about adaptation: the theory makes an exact prediction, and it has been tested experimentally.

The work of Herre, on fig wasps, provides a further test of Hamilton's theory. Fig wasps lay their eggs inside figs and, in some species, several females lay their eggs in a single fig. Herre's test relies on the same general principle as Werren's, namely that earlier females should produce fewer sons than later ones. In the three fig wasp species in Herre's study, the offspring all mate together inside the fig, before leaving it. Unlike *Nasonia* and its maggot, in Herre's species up to six "foundress" females may lay eggs in a fig. As each successive female lays her eggs, the degree of mate competition becomes less localized, and Hamilton's theory predicts that the successive females should lay increasing proportions of male eggs, though at a decreasing rate. Herre's observations support the prediction (Figure 11.12). Again, the test is quantitative. Local mate competition is proving a fertile testing ground for the theory of sex ratio.

11.5.3 Trivers and Willard identified another case in which sex ratio should deviate from 50:50

In polygynous species, a small number of males do a disproportionate amount of the breeding. Many males die before reaching the breeding stage and even then some males fail to breed. We saw above that, in Fisher's theory, a female cannot gain by producing more males, even if only a minority of them survive to breed. By producing more males, the average female would be as likely to produce more of the failures as of the successes.

However, there is a circumstance in which it can be advantageous to bias the sex ratio. It is when a female "knows" that she is not an average female, and that her offspring will not be average sons and daughters. Consider the case of the red deer *Cervus megalocerus*, which has been studied in detail by Clutton-Brock and his team on the island of Rhum, off Scotland. The breeding system is polygynous, as females (called hinds) live in groups, each of which is defended by a single male. Within a group of hinds, there is a dominance hierarchy. The more dominant hinds are better fed and stronger, and they can better provide for their offspring. Their offspring in turn grow up to be stronger than average. In these circumstances, as Trivers and Willard pointed out in 1973, there can be selection on higher ranking females to produce more sons, and on lower-ranking females to produce more daughters. A high-ranking female can expect to produce successful sons, and a low-ranking female to produce sons that fail to breed. Females can only make use of this fact if they in some sense "know" their social rank, but there is nothing particularly strange about assuming that they can. A hind can just as well "know" her social rank (i.e. adjust her behavior in relation to it) as she can "know" when she is hungry or when her blood has a high carbon dioxide concentration. Clutton-Brock tested Trivers and Willard's theory in red deer. The data are noisy, but higher-ranking hinds do produce a significantly higher proportion of sons than do low-ranking hinds (Figure 11.13). The scatter of the points suggests that other factors are influential, but the trend supports the theory. The test is not yet particularly quantitative. The theory does make an exact prediction. The graph for sex ratio should cross the 50% male threshold at that social rank at which a hind produces on average fitter sons than daughters. Clutton-Brock's data are consistent with the prediction, but the scatter is too great to say that the theory has been quantitatively confirmed.

The relationship between dominance and sex ratio, and local mate competition, are but two examples in which deviations from a 50:50 sex ratio have been successfully predicted. We shall not discuss any further examples here, for these two are enough to illustrate how the theory of adaptive sex ratio can be remarkably successful in explaining both the normal 50:50 sex ratio and deviations from it. The theory has suggested various different kinds of test using all three methods; it is quantitative in its predictions; and the key variable—sex ratio—is easy to measure. The sex ratio will therefore probably stay in the vanguard of evolutionary research for some time yet.

11.6 Different adaptations are understood in different levels of detail

We have looked at the function of sex, sexual selection, and sex ratio, as three related examples of research into adaptation. In each case, the research has advanced to a different stage. The problem of sex is still more-or-less unsolved and the main work is taking place at the imaginative level; the aim is to think up a reason why natural selection could favor sex and to build a model to show that the hypothetical advantage to sex is large enough to outweigh its two-fold cost. In the case of sexual selection, the main ideas—of

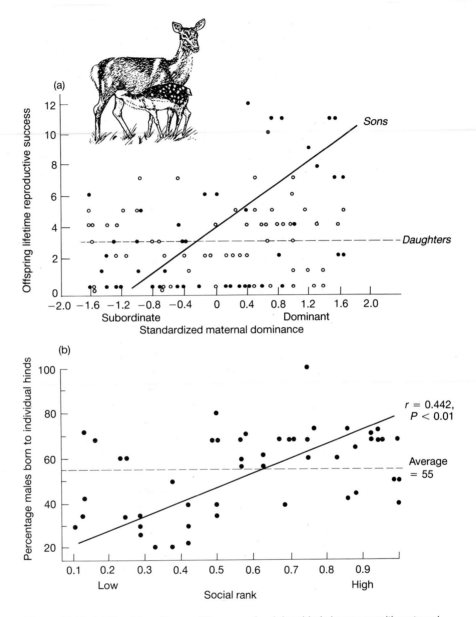

Figure 11.13 (a) The future fitness of the sons of red deer hinds increases with maternal dominance; whereas maternal dominance has no effect on the future fitness of daughters. (b) Dominant mothers bias the sex ratio of their progeny toward sons. From Clutton-Brock *et al.* (1984).

Fisher and Zahavi—provide a satisfactory abstract solution and are at least not known to be false; the full repertoire of techniques—model building, experiment, the comparative method—are being used. At a detailed natural history level, there are still many unanswered questions. We do not know whether the abstract ideas correctly explain the full natural variety of sexual behavior and dimorphism; and the ideas are not easy to test. The theory of sex ratio is still further advanced. The relationship between facts and theories is good. We have not only a general abstract theory as for sexual selection; the theory also makes quantitative predictions and suggests a number of types of test in special cases. Several of the tests have been followed up, and the fit of results to predictions suggests that the theory stands a good chance of being correct.

11.7 Summary

1 For many characters, it is not obvious how (or whether) they are adaptive.

2 Adaptation can be studied by comparing the observed form of an organ with a theoretical prediction, by experimentally altering the organ, and by comparing the form of the organ in many species.

3 Sex has a 50% fitness disadvantage relative to asexual reproduction.

4 Sexually reproducing populations will evolve faster than a set of asexual clones, provided that the rate of favorable mutation is high enough.

5 The taxonomic distribution of asexual reproduction suggests that asexual forms have a higher extinction rate than sexual forms. However, it is generally doubted that sex is maintained by group selection.

6 Two modern theories of why sex exists propose that it is favored by: (i) the coevolutionary arms race of parasites and hosts; and (ii) sib competition. The problem of why sex exists has not been finally solved.

7 Males in many species have bizarre and deleterious secondary sexual characters; the peacock's train is an example.

8 Darwin explained the evolution of strange secondary sex characters by sexual selection: the characters reduce their bearers' survival, but increase their success in reproduction; sexual selection in most species works by male competition and by female choice.

9 The greater sexual dimorphism of polygynous species than monogamous species suggests the importance of sexual selection.

10 The preference of females for males with deleterious characters is theoretically puzzling. It may be explained by Fisher's theory, in which deleterious characters were formerly advantageous and are maintained by majority preference, or by a Zahavi's handicap theory, in which the costly character indicates superior genetic quality.

11 In the barn swallow, females choose males with longer tail feathers, and tail length indicates health. Males with longer tails produce healthier, more parasite-resistant offspring.

12 The sex ratio is usually 50:50 because the reproductive success of all the males in a population must equal the reproductive success of all the females; if the population sex ratio deviates from 50:50, natural selection favors individuals that produce more offspring of the rarer sex.

13 The theory of sex ratio has correctly predicted when the ratio should differ from 50:50. It has been tested quantitatively and experimentally in the case of local mate competition.

14 The function of sex, sexual selection, and the sex ratio are three of the most important areas of research into adaptation. They have reached different stages of theoretical advance.

11.8 Further reading

Curio (1973), Harvey and Pagel (1991), Maynard Smith (1978b), Parker and Maynard Smith (1990), and Rudwick (1964), are reviews, general and particular, of the methods of studying adaptation. On *Cepaea*, see Ford (1975), Sheppard (1975), Jones *et al.* (1977).

On sex, see Williams (1975), Maynard Smith (1978a), Bell (1982), Hamilton *et al.* (1990), and Michod and Levin (1988). There are more theories of sex than there was space for in the text: see Kondrashov (1988) on deleterious mutations, Grafen (1988) on centrosomes, and Bernstein *et al.* (1988) on DNA repair. See also Kirkpatrick and Jenkins (1989) on evolutionary rates and Schmitt and Antonovics (1986) on sib competition.

The classic reference on sexual selection is Darwin (1871). The subject has been introduced more recently by, for example, Ghiselin (1974), Trivers (1985), Krebs and Davies (1987), and Dawkins (1976); Bradbury and Andersson (1987) is a conference. See Eberhard (1985) on genitalia, Zahavi (1975) and Grafen (1990a,b) on handicaps, and Kirkpatrick and Ryan (1991) and Maynard Smith (1991) for theoretical reviews. The influence of parasites on mate choice was originally discussed by Hamilton and Zuk (1982) (for further analysis, see Read and Harvey (1989)); it was also the topic of an introductory review by Read (1988) and a conference organized by G. Hausfater and R. Thornhill published in *American Zoologist*, vol. 30, pp. 225–352, (1990). See Hamilton (1990). Smith and Montgomerie (1991) confirm Møller's (1988, 1989, 1990) work with North American barn swallows. Baker and Parker (1979) question whether bright coloration in birds is costly.

On sex ratio, Fisher (1930) is the classic reference; but Dawkins (1976), Krebs and Davies (1993), and Trivers (1985) are more accessible introductions. Charnov (1982) is a fundamental modern study, which discusses the extension to sex change—on which see also Policansky (1982). For local mate competition, the key original references are Hamilton (1967), Werren (1980), and Herre (1985). On other deviations: Trivers and Willard (1973), Trivers and Hare (1976), Clutton-Brock *et al.* (1984), Gomendio *et al.* (1990); and Bull and Charnov (1988) for a review.

The units of selection

12.1 For the benefit of which level in the biological hierarchy of levels of organization does natural selection produce adaptations?

It is a familiar idea that life can be divided into a series of levels of organization, from nucleotide to gene, through cell, organ, and organism, to social group, species, and higher levels. Which, if any, of these levels does natural selection act on and produce adaptations for the benefit of? In a fairly superficial analysis, the answer does not matter. If an adaptation benefits an individual, it will often also benefit its species at a higher level and, at a lower level, all the parts that make up the individual. But there can be conflicts between these levels; in some cases, what benefits an organism may not also benefit its species, and in these cases the evolutionary biologist needs to know which level natural selection most directly benefits. The question therefore matters when we are studying particular adaptations. If we are seeking to understand why an adaptation evolved, we need to know what entities adaptations in general evolve for the benefit of. The question also has a more general, almost philosophic, interest: the theory of evolution should include a precise, and accurate, account of why adaptations evolve.

The issue can be made clearer in an example. Let us consider the adaptations that can be seen when lions go on a hunt. Lions often hunt alone, but they can improve their chance of success by hunting in a group. Here is part of a description of a group of hunting lions, by Bertram.

> When prey have been detected, a wildebeest herd perhaps, the lions start to stalk towards them. As they get close, they take different routes, some going on straight ahead, and some to the sides, so the prey herd is approached by lions stalking them from different directions.... Eventually one lion gets close enough to make a rush at a wildebeest, or else a lion is detected by the prey.

Then the trap is sprung. Panicked wildebeest run in all directions, some of them into the reach of other lions. The cooperative behavior of the hunting party is here an adaptation for catching food; but it is not the lions' only feeding adaptation. The lions' muscular jaws and limbs, their teeth and five senses, all contribute to the success of the hunt. Lions are well adapted for feeding: although some hunts are unsuccessful, and individual lions may starve to death, the lions in the Serengeti Plains of Kenya studied by Schaller and by Bertram spend about 20 hours a day in rest or sleep, and only 1 hour a day on average in hunting. Visitors tend to think lions are lazy.

When a lion hunt is successful, there are benefits for all but the highest biologic levels of organization. The individual lions obviously benefit, as does the pride. Each time a hunt is successful, there will be a small incremental increase in the species *Felis leo*'s chance of survival, or avoiding extinction.

The survival probability will also be increased, if by a smaller amount, for the genus *Felis* and the cat family Felidae. The hunt's effect at higher levels will depend on exactly what prey the lions caught. Almost all the Serengeti lions' food is made up of other mammals, so when we reach the class Mammalia, the effect of the hunt has probably become neutral: the lion's gain is the wildebeest's or the zebra's loss and the chance of survival of the class Mammalia is more or less unaffected. The beneficial effect of the hunt spreads downwards as well as up from the individual lion. As the survival of the lion is increased, so is the survival of its constituents. The lion's organs, cells, proteins, and genes, are all more likely to survive every time a lion hunt is successful. (Though if we trace the effect down through the nucleotides and their constituent atoms, it again disappears. A lion's atoms survive just as well whether it is alive and well fed or dead due to starvation.)

The levels of organization, from gene through individual lion to Felidae, are to a large extent bound together in their evolutionary fate, and what benefits one level will usually also benefit the others. However, this is not always so. Male lions can only join a pride by forcibly evicting the incumbent males. In the fight, lions may get killed or wounded, and in any case lions have a low rate of survival after they have been evicted from a pride. These fights have losers as well as winners: here the benefit of winning is confined to the individual (or male coalition) level and below; the lion species does not benefit. The survival of the species may be little affected by the death of male lions, because the mating system is polygynous and has plenty of males to spare; but the effect is clearly not positive.

Different adaptations, therefore, have different consequences for different units in nature. At one extreme, there may be adaptations—an improved DNA replication mechanism perhaps—that could benefit all life; but most adaptations will benefit only a smaller subsample of living things. Because the levels of living organization are bound together, if natural selection produces an adaptation to benefit one level, many other levels will benefit as a consequence. The question in this chapter is whether natural selection really acts to produce adaptations to benefit one level, with benefits at other levels being incidental consequences, or whether it acts to benefit all levels. And if it benefits one level, which is it? In evolutionary biology, this question is expressed as "What is *the unit of selection*?"

We shall seek to answer it by looking at a series of adaptations that appear to benefit different levels in the hierarchy of biological organization. Some adaptations seem to be in the interests of individual genes, at the organism's expense, others benefit organisms at the group's expense, others may benefit higher levels. When we have seen the examples, we can discuss generally which of the types of adaptation we should expect to see most often in nature.

12.2 Natural selection has produced adaptations that benefit various levels of organization

12.2.1 Segregation distortion benefits one gene at the expense of its allele

With normal Mendelian segregation at a genetic locus, on average half of an organism's offspring inherit one of the alleles and the other half the other allele. Mendelian segregation is, so to speak, "fair" in its treatment of genes: genes emerge from Mendelian segregation in the same proportions as they went in. There are, however, some curious cases in which Mendel's laws are broken: in which one of the alleles, instead of being inherited by half of a heterozygote's offspring, is consistently over-represented. The *segregation distorter* gene of *Drosophila melanogaster* is an example. (The phenomenon in general is also called *meiotic drive*.)

The segregation distorter gene was first found in *Drosophila* stocks from Wisconsin and Baja California. The gene is symbolized by *sd*, and we can call the other, more normal alleles at the locus "+"; a heterozygote for the segregation distorter is then *sd/+*. The majority (90% or more) of offspring from male heterozygotes have the *sd* gene. Female heterozygotes have normal Mendelian segregation. (The genetics are in reality more complex. Segregation distortion is controlled not by one locus, but by at least three tightly linked gene loci on the second chromosome. One locus, *sd*, interacts with another locus *Rsp* such that if the allele at *Rsp* is *Rsp*^{s} the sperm develop abnormally but if it is *Rsp*^{i} they develop normally (*Rsp*^{i} is therefore the wild-type condition). A third locus influences the strength of the effect. The consequence is that a disproportionate number of *Rsp*^{i} individuals are produced from matings of male heterozygotes. For simplicity, we shall treat the case as if it were controlled by one locus, and write the heterozygote for the segregation distorter gene as *sd/+*.)

A segregation distorter gene can have a great selective advantage. The allele that gets into more than half of a heterozygote's offspring will automatically increase in frequency and should spread through the population. Once the allele is fixed, the effect would disappear, because segregation is normal in homozygotes. In the case of the segregation distorter gene complex in *Drosophila*, however, other things are not equal. The abnormal sperm are infertile, and the total fertility of male heterozygotes is accordingly lowered. The fertility of an *sd/+* male is about half that of a normal male. The effect of the lowered fertility on selection at the *sd* locus is complex, and depends on whether the reduction in fertility is more or less than 50% and what effect the reduced fertility has on the number of offspring produced. There are, however, at least some segregation distorter alleles that produce enough copies of themselves in heterozygotes to have an automatic selective advantage. They produce more copies of themselves than would a normal Mendelian heterozygote; their increase in frequency up to fixation is then inevitable.

Segregation distortion sets up an interesting selection pressure in the rest of the genome. On average, all other genes at other loci suffer a disadvantage because of it. In the absence of the *sd* gene, a gene at any other locus has a 50% chance of getting into a particular gamete; this is because gametes are haploid whereas the individual is diploid. But segregation distortion produces a further reduction of about 50% in the chance that genes at other loci are passed on. In an *sd/+* heterozygote, if a gene at any other locus gets into a sperm, there is a 50% chance it is an *sd* sperm, which will be fertile, and a 50% chance it is a + sperm, which will be infertile (we are ignoring linkage between the *sd* locus and the other loci under consideration). Genes at other loci are all net losers: if they are in the same sperm as the favored *sd* allele, things are normal; if they are in the shrivelled disfavored sperm, they die. Selection acts at other loci to restore the status quo. Alexander and Borgia described genes like *sd* as *outlaw genes*, because at most other loci there is selection to suppress them.

We can imagine the following series of evolutionary events. At each locus, selection favors mutant alleles that can get into more than 50% of the offspring; whenever such an allele arises it has an automatic advantage. Once it has arisen, the next stage will depend on the fertility of its bearer. If it has no effect on, or increases, fertility, its frequency should increase to one. If it reduces fertility, alleles at other loci that suppress the effect will be favored; in this case, the outlaw gene will either be suppressed, or increase to a fixation, depending on how quickly a suppressor allele arises. In either case, segregation distortion will be evolutionarily short-lived. The allele responsible will either be fixed (and segregation distortion then disappears) or suppressed. At any one time, only a few cases should exist. This probably explains why only a few have been found. Meiosis generally does appear to be fair; alleles usually are equally represented in the offspring. It is important to remember, however, that this fairness exists despite selection at each locus for alleles that subvert the Mendelian ratios.

12.2.2 Selection may sometimes favor some cell lines at the expense of the rest of the body

In organisms like ourselves, a new individual develops from an initial single-cell stage; and that single cell derives from a special cell line, the germ line, in its parents. This kind of life cycle is called *Weismannist*, after the German biologist August Weismann, who first expounded the distinction between germ and somatic cell lines. In a Weismannist organism, most cell lines (the soma) inevitably die when the organism dies; reproduction is concentrated in a separate germ line of cells.

The separation of the germ line limits the possibilities for selection at the suborganismic level, between cell lines. One cell may mutate and become able to outreproduce other cell lines and (like a cancer) proliferate through the body. But this "adaptation" will not be passed on to the next generation unless it has arisen in the germ line. Any somatic cell line comes to an end with the organism's death. For this reason, cell selection is not important in species like ourselves.

However, Buss, in his book *The Evolution of Individuality* (1987), has pointed out that Weismannist development is relatively exceptional among multicellular organisms (Table 12.1). We tend to think of it as common because vertebrates, as well as the more familiar invertebrates like arthropods, develop in a Weismannist manner. However, more than half the taxa listed in Table 12.1 have somatic embryogenesis; a new generation may be formed from cells other than those in specialized reproductive organs. The most striking examples are from plants. Steward, for instance, in a famous experiment, grew new carrots from single cells taken from the root of an adult plant.

In a species in which new offspring can develop from more than one cell lineage, selection between cell lines becomes possible. When the organism is conceived it will be a single cell, and for the first few rounds of cell division the organism will probably remain genetically uniform. No selection can take place between cell lines if they are all genetically identical. Eventually a mutation may arise in one of the cells. If the mutation increases the cell's rate of reproduction, the cell line will cancerously proliferate at the expense of

Table 12.1 The modes of development in different groups of living things. In the cellular differentiation column, + means that it is present in all the species that have been studied in the group and +/− means present in some species and absent in others. In the developmental mode column, s means new organisms can develop from the "somatic cells" of their parent; e means epigenetic development; p means preformationistic development; u means unknown. From Buss (1987)

Taxon	Cellular differentiation	Developmental mode
Protoctista		
Phaeophyta	+/−	s
Rhodophyta	+/−	s
Chlorophyta	+/−	p
Ciliophora	+/−	s
Labyrinthulamycota	+/−	s
Acrasiomycota	+/−	s
Myxomycota	+/−	s
Oomycota	+	s
Fungi		
Zygomycota	+	s
Ascomycota	+	s
Basidiomycota	+	s
Deuteromycota	+	s
Plantae		
Bryophyta	+	s
Lycopodophyta	+	s
Sphenophyta	+	s
Pteridophyta	+	s
Cycadophyta	+	s
Coniferophyta	+	s
Angiospermophyta	+	s
Animalia		
Placozoa	+	s
Porifera	+	s
Cnidaria	+	s
Ctenophora	+	p
Mesozoa	+	p
Platyhelminthes	+	s, e, p
Nemertina	+	e
Gnathostomulida	+	u
Gastrotricha	+	p
Rotifera	+	p
Kinorhyncha	+	u
Acanthocephala	+	p
Entoprocta	+	s
Nematoda	+	p
Nematomorpha	+	u
Bryozoa	+	s
Phoronida	+	s
Brachiopoda	+	u
Mollusca	+	e, p
Priapulida	+	u
Sipuncula	+	u
Echiura	+	u

Table 12.1 (*continued*)

Taxon	Cellular differentiation	Developmental mode
Annelida	+	s, e, p
Tardigrada	+	p
Onychophora	+	p
Arthropoda	+	e, p
Pogonophora	+	u
Echinodermata	+	e
Chaetognatha	+	p
Hemichordata	+	s, e
Chordata	+	e, p

other cell lines in the organism. In a Weismannist species, that cell line will die when the organism dies; but if any cell line stands a chance of producing offspring, the mutant cell line would increase its chance of founding a new offspring. Explained in this way, the process would be detrimental to the organism; but it could be more useful. Whitham and Slobodchikoff have argued that in plants this kind of somatic selection enables the individual to adapt to local conditions more rapidly than would be possible with strictly Weismannist inheritance.

The process is at present more of a theoretical possibility than a confirmed empirical fact; but it may well be important in non-Weismannist species. It may also have been important in the non-Weismannist ancestors of such modern Weismannist forms as arthropods and humans. Buss has developed the idea that cell selection can explain certain features of embryology in Weismannist species (Box 12.1).

12.2.3 Natural selection has produced many adaptations to benefit organisms

We do not need to consider an example of organismal adaptation here: most of the adaptations described elsewhere in the book, from the beak of the woodpecker (section 1.2, p. 6) and the Galápagos finches (section 9.1, p. 211), to the color patterns of *Biston betularia* (section 5.6, p. 95), of the mimetic *Papilio* (section 8.1, p. 183), and of *Cepaea* (section 11.1, p. 267), are all adaptations that benefit the individual organism. It can hardly be doubted, therefore, that organismal adaptations exist, and natural selection can favor them.

12.2.4 Natural selection working on groups of close genetic relatives is called kin selection

In species in which individuals sometimes meet one another, as in social groups, individuals may be able to influence each other's reproduction. Biologists call a behavior pattern *altruistic* if it increases the number of offspring produced by the recipient and decreases that of the "altruist." (Notice that the term in biology, unlike in human action, implies nothing about the altruist's intentions: it is a motive-free account of reproductive

Box 12.1 The course of development.

The modern taxonomic distribution of non-Weismannist development (Table 12.1) suggests that such embryologically well-studied species as *Drosophila*, frogs, chicken, and humans, all of which have separate germ lines, evolved from ancestors in which new individuals could develop from any of a number of parental cell lines. Buss points out that many features of development in Weismannist species may be explicable by their ancestral history. If (as is true in a non-Weismannist species) any cell line can found a new organism, cell lines are selected to proliferate within the body and increase their chance of ultimate reproduction. Buss suggests that many features of developmental control, such as inhibition between cell lines, are what should be expected to evolve in a non-Weismannist state.

Buss's most striking interpretation is for gastrulation. During development, cells divide until they form a hollow sphere of cells. In the ancestral state, this sphere would probably have lived in the plankton and the cells would have had flagellae for swimming. Now, there is a well-confirmed generalization (discovered by Margulis) that cells cannot both have a flagellum (or cilium) and divide. If the cells were to proliferate they would have to be without flagellae. If they divided on the surface of the sphere, they would reduce its swimming efficiency. The obvious place to divide is inwards, into the center of the sphere; it may be for this reason that cells divide at the surface and invaginate in the process of gastrulation. The original reason for gastrulation no longer applies in modern forms, in which the embryo does not swim in the sea; but once a step has been acquired in development it may be difficult to lose, because small perturbations early in development can have magnified effects later in the adult. Evolution of the early stages of development is expected to be conservative, and gastrulation may be a vestige of ancestral selective conditions.

consequences.) Can natural selection ever favor altruistic actions, that decrease the reproduction of the actor? If we take a strictly organismic view of natural selection, it would seem to be impossible. And yet, in a growing list of natural observations, animals behave in an apparently altruistic manner. The altruism of the sterile "workers," in such insects as ants and bees, is one undoubted example; here the altruism is extreme, as in some species the workers do not reproduce at all. We shall consider a less extreme example.

The Florida scrub jay *Aphelocoma coerulescens* often breeds in groups containing more than just a male–female pair. The species has been studied by Woolfenden, and he has found that the breeding pair are assisted by a number of helpers. These helpers may defend the nest, or bring food for the young, but they usually do not incubate the eggs. Groups with helpers reproduce more successfully—by 33–80%—than those without (Table 12.2), and there is an additional advantage in adult survival, presumably because in the groups with helpers the parents do not have to work so hard to look after the young. But none of the extra offspring are those of the helpers themselves: the helpers do not breed. Helpers are the altruists, who pay costs, but to increase the reproduction of other birds. By following breeding groups through many reproductive cycles, Woolfenden found that the helpers are usually males, and the sons of the breeding pair. The helpers stay at their parental nest and help to rear later broods.

The breeding pair and the helpers are, therefore, genetically related. It can

Table 12.2 Numbers of offspring reared by scrub jay pairs with and without helpers. From Woolfenden and Fitzpatrick (1984)

	No helper	Helper present	Average number of helpers
Inexperienced pairs	1.24	2.2	1.7
Experienced pairs	1.8	2.38	1.9

be advantageous to the helpers to increase the reproduction of genetically related individuals. When an animal looks after its own young, in ordinary parental care, it is increasing the chance of survival of an individual (its offspring) to which it is 50% genetically related. As Hamilton first pointed out, an individual helping its parents to produce extra siblings has exactly the same genetic consequences. Its siblings, like its offspring, have a 50% chance of sharing its gene (or 25% for half-sibs). This type of selection, in which organisms are selected to produce more relatives other than offspring, is called *kin selection*. It takes place among groups of related individuals and results in the evolution of adaptations for the benefit of the kin group.

In the scrub jay, the selective situation is complex and kin selection is not the only factor. Breeding territories are in short supply, and the males may have no breeding opportunities elsewhere. Male helpers may not be sacrificing themselves in quite the way a naïve reading of Table 12.2 might suggest. The males may not have a straight choice between producing their own offspring (and obtaining the number of offspring for inexperienced pairs without helpers) or helping their parents to produce extra offspring. If a male has no opportunity to breed independently, the chance that altruistic helping will evolve is greatly increased. There are other factors too, but the main point here is that kin selection can result in the evolution of adaptations that are not in the individual's interest, but are in the interest of its group of kin as a whole.

12.2.5 The question of whether group selection ever produces adaptations for the benefit of groups has been controversial, though most biologists now think it is only a weak force in evolution

A group adaptation is a property of a group of organisms that benefits the survival and reproduction of the group as a whole. Adaptations produced by kin selection—such as helping in family groups of birds—will satisfy that definition, but we are concerned here with group adaptations that did not evolve by kin selection. If any exist, they will have come into existence by selection between groups: groups possessing the group adaptation would have become extinct at a lower rate, and sent out more emigrants, than groups lacking it. The group adaptation would have been favored by the differential reproduction of whole groups.

Many characteristics are beneficial at the group level, but also benefit all the individuals in the group; this would trivially be true, for example, of an improvement in the hunting skill of a lion: after the improvement has spread by individual selection, all the individuals in the group, and the group as a whole, will be better adapted. That is just to restate the earlier point that adaptations that have evolved for the benefit of one level of organization can incidentally benefit higher levels. The controversial group adaptations are those that benefit the group but not the individual. A hypothetical example is Wynne-Edwards' theory, put forward in *Animal Dispersion in Relation to Social Behaviour* (1962), that animals restrain their reproduction in order not to over-eat the local food supply. If all the individuals in a group reproduce at the maximum rate, their offspring might over-eat the food supply, and the group would then become extinct. They could avoid this fate by collectively restraining their reproduction, to maintain the balance of nature. Natural selection on individuals does not favor reproductive restraint. An individual that increases its reproduction will automatically be favored relative to individuals that produce fewer offspring. Within a group, if some individuals produce more offspring than others, the former will proliferate. But can individual selection within the group be overcome by selection between groups?

The question is highly important, both conceptually and historically. It is important historically because vague group selectionist thinking—particularly in the form of statements like "adaptation X exists for the good of the species"—was once common. It is now more usual (though by no means universal) for biologists to believe that group selection is a weak and unimportant process. There are both theoretical and empirical reasons for this. Empirically, there are no definite examples of adaptations that need to be interpreted in terms of group advantage. As Williams argued in 1966, the characteristics that Wynne-Edwards had suggested evolved to regulate population size can all be explained as adaptations that benefit individuals. Individuals generally reproduce at the maximum rate they can; the only obvious exceptions concern genetically related individuals, and can be explained by kin selection. Moreover, living things have characteristics that contradict the theory of group selection. The 50:50 sex ratio, which we discussed in section 11.5 (p. 294), is a case in point. In polygynous species, it is inefficient for the population to produce 50% males, most of whom are not needed. The widespread existence of the 50:50 sex ratio suggests that group selection has been ineffective on this trait.

Group selection is also implausible in theory. Consider a population containing two genotypes, one for an altruistic, group adaptive trait like reproductive restraint (call them A individuals) and another selfish (S) type that reproduces as fast as it can. The population is made up of a number of groups, which can contain any proportion of A and S individuals; groups with mainly altruistic members we call altruistic groups and those with mainly selfish members, selfish groups. The altruistic groups will become extinct at a lower rate; group selection favors altruism. Individual selection favors the selfish individuals within all groups; within each group S individuals increase in frequency. An altruistic group may temporarily contain no selfish members, but as soon as it is "infected" with one the selfish trait will proliferate and become fixed in the group. What result should we expect to find? It depends on the balance between the two processes. In theory, we can imagine a rate of group extinction so high that altruists will predominate: just imagine, for the sake of argument, what would happen if all groups with more than 10% selfish types instantly became extinct. All the groups we should see would clearly have at least 90% altruists. But that is only a thought experiment. The interesting question is what we should expect to happen naturally.

The reason most biologists suppose that group selection is a weak force in opposition to individual selection stems from the slow life cycles of groups as compared to individuals. Individuals die and reproduce at the rate of once per generation, and many individuals can move between groups within a generation; groups become extinct at a much slower rate. The amount of time it will take for selfish individuals to infect and proliferate in a group is a small part of the group's lifespan; at any one time, therefore, individual adaptations will predominate.

There are many models of group and individual selection; Maynard Smith has suggested they can be reduced to a common form (Figure 12.1). The groups are now supposed to occupy patches in nature. As before there are patches occupied by altruistic (A) and by selfish (S) groups; there are also empty patches (E). A selfish group in the model drives itself extinct by over-eating its patch's resources. The result of the model depends on whether a selfish group can "infect" an E or A patch before itself becoming extinct. Maynard Smith defines m as the number of successful migrants produced by one S group, on average, between its origin and extinction. (Successful means that the migrant establishes itself in another group and breeds.) If $m = 1$ the system will be stable; if $m < 1$ the selfish groups decrease in number and if $m > 1$ they increase. In other words, a selfish group only needs to produce more than one successful emigrant during its existence for the selfish trait to take over. This is a small number. So small that we can expect selfish individual adaptations to prevail in nature. Group selection, we conclude, is a weak force. It only works if migration rates are implausibly low and group extinction rates implausibly high. It is also not needed to explain the facts.

I have stated the case against group selection in stark terms, but only to make the arguments clear; the matter has not been settled finally, and group selection probably operates sometimes. The process is theoretically possible. Moreover, group selection can have evolutionary consequences even if it never over-rides individual selection. In chapter 21 we shall discuss, using sex as an example, how individual selection can establish different adaptations in different groups (or species, in the example of sex) and those different adaptations could result in different rates of group extinction or

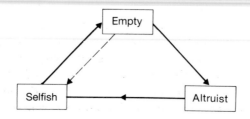

Figure 12.1 Maynard Smith's formulation of group selection models. A patch may change state in the direction of the arrows. From Maynard Smith (1976).

expansion. In these cases, there is no conflict between individual and group selection; individuals act in their own selfish interest in all the groups (or species). However, the kinds of group (or species) that become extinct less often will increase in frequency, and the increase is due to selection between groups. This idea is uncontroversial: the real "group selection controversy" in evolutionary biology was concerned with whether group selection could ever cause individuals to sacrifice their own reproductive interests for those of its group.

The conclusion of the controversy was that in nature group selection was rarely likely to over-ride individual selection, and establish individually disadvantageous behavior. In the laboratory, however, conditions can be made extreme enough for it to do so. Let us finish by looking at such a case: Wade's experiment on flour beetles, *Tribolium castaneum*. The life cycle of the flour beetle takes place, from egg to adult, in stored flour. They are pests, but they have also become a population biologist's standard experimental animal, especially in Chicago. Wade set up an experiment to illustrate Wynne-Edwards' hypothesis of reproductive restraint. The experiment had the following design (Figure 12.2a). In each of three experimental treatments, there were 48 different colonies of *Tribolium*. Each colony was allowed to breed for 37 days. Then 48 new colonies, of 16 beetles each, were set up from the progeny of the old ones. Wade artificially selected for groups that had showed a low (or high) fecundity (the third treatment was a control). He selected for low fecundity by forming a new generation of colonies from *Tribolium* colonies that had a low population density at the end of the 37 days; and for high fecundity by forming each new generation from colonies that had high population densities. He repeated a number of rounds of the process. Not surprisingly, the reproductive rate in the "low" lines decreased more than in the "high" lines (Figure 12.2b). The decrease in the low lines is due to group selection. Presumably, within the 37 days of any one cycle, the beetle types with high fecundity were increasing within each colony relative to the less fecund beetles. However, between cycles, Wade's group selection for low fecundity more than outweighed the individual selection and average fecundity declined. The group selection was strong enough to work.

In a way, the group selection structure of the experiment is superfluous. It would be perfectly possible simply to breed selectively from beetles with lower fecundity; the artificial selection would then reduce fecundity without the rigmarole of keeping them in groups for 37 days. But Wade's purpose was to illustrate group, not artificial, selection and his experiment does so. It has alternating rounds of individual and group selection and the experimental group selection is strong enough to produce the effect that Wynne-Edwards thought to be common in nature.

The fact that group selection can be carried out in an experiment does not bear upon how important it is in nature. Biologists doubt group selection for theoretical reasons, and because of the kinds of adaptation seen in nature; these reasons were not affected by Wade's experiment. The experiment is instructive, however, as it shows what group selection means.

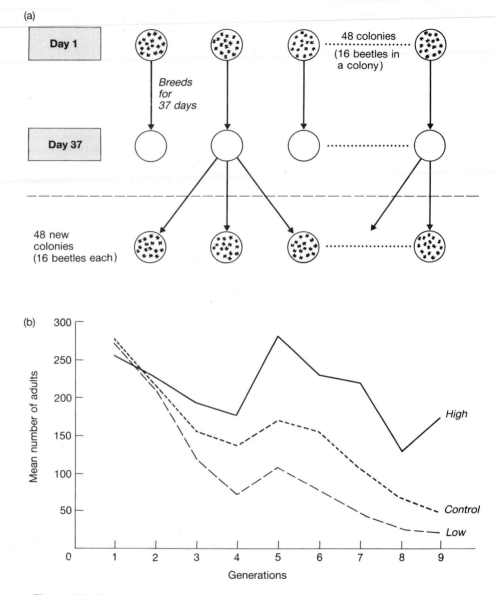

Figure 12.2 Wade's experiment with *Tribolium* beetles. (a) Experimental design. A set of 48 colonies were bred for 37 days, and then a new round of colonies was formed from the colonies that had grown to a low (or high) population density. (b) Results. Population densities in lines selected for high or low population densities and in unselected controls. Results are means for 48 colonies. From Wynne-Edwards (1986), using data of Wade.

12.2.6 Levels at which characters show "heritability" between generations are the levels at which adaptations evolve in the hierarchy of levels of organization

Adaptations can exist for the benefit of genes, cells, organisms, kin, or groups of unrelated individuals. Gene adaptations like segregation distortion are rare; cell line adaptations are very rare, at most, in Weismannist species, but may be found in non-Weismannist species such as plants; organismic adaptations are common; the number of examples of kin selected adaptations is increasing; and group adaptations are probably rare.

Why is it, then, that adaptations are mainly at the organismic level, with a few additional cases for groups of kin? In the five previous sections, we have already discussed the answer to this question, but a more general answer can be given. Maynard Smith has pointed out that the units in nature that show adaptations are the units that show heritability (section 9.6, p. 225). Mutations that influence the phenotype of a unit (whether a cell, organism, or group) must be passed on to the offspring of that unit in the next generation; if this happens natural selection can act to increase the mutation's frequency. Organisms show heritability in this sense. A finch with an improved beak shape, caused by a genetic change, will on average produce offspring with the improved beak shape. Natural selection can work on finches. But groups do not usually show this sort of heritability. A genetic variant that increases a group's chance of success tends not to be inherited by future groups. Immigration contaminates the group's genetic composition, such that heritability from generation to generation is low. Thus altruistic groups do not exclusively generate descendant altruistic groups, and selfish groups generate selfish groups. Migration from selfish groups causes altruistic groups to become selfish. Thus a group in one generation will only be genetically correlated with the group of its offspring in the next generation when there is practically no migration: then group selection works.

The same point can be made about kin selection and selection among cell lines. Kin selection operates because an offspring kin group genetically resembles the parental kin group. Cell selection, in Weismannist species, tends not to operate because somatic cells, although they are inherited down a cell line during one organism's brief life, are not passed on from an organism to its offspring; but when they are (in non-Weismannist species), cell selection and the evolution of cell line adaptations become theoretically more plausible. For gene adaptations, such as segregation distortion, the same basic argument applies.

In summary, we should expect to find adaptations existing for the benefit of those units in nature that show heritability. Adaptations will therefore usually benefit organisms. The cases of adaptations that benefit higher or lower levels of organization can be understood in the same general terms, because they only evolve in circumstances when groups, or parts, of organisms show heritability from one generation to the next. We can now give our first answer to the question of what is the unit of selection. The general answer is "that entity that shows heritability"; more specifically, it is usually the organism, with some interesting exceptions. This first answer specifies the units in nature that should possess adaptations.

12.3 Another sense of "unit of selection" is the entity whose frequency is adjusted directly by natural selection

Natural selection over the generations adjusts the frequencies of entities at all levels. We have implicitly seen this in the example of the lion hunt. If the lions of one pride become more efficient at hunting, perhaps because of some new behavioral trick, natural selection will favor them. If the trick is inherited, that type of lion will increase in frequency relative to other types of lion. All things associated with the trick will increase in frequency too. The type of lion, its type of neurons, of proteins, and their encoding genes would all increase in frequency relative to their alternatives. When the hunting success of lions as a whole increases, the frequency of lions in the ecosystem will increase too; and lions might, over geologic time, come to replace other competing predators on the plains. The question in this section is whether natural selection directly adjusts the frequency of any of these units—nucleotides, genes, neurons, individual lions, lion pride, the lion species?

The answer was most clearly given by Williams (1966), and Dawkins (1976). It is at least implicit in all theoretical population genetics and indeed in the previous section of this chapter. For natural selection to adjust the frequency of something over the generations, the entity must have a sufficient degree of permanence. You cannot adjust the frequency of an entity between times t_1 and t_2 if between the two times it has ceased to exist. A characteristic that is to increase in frequency under natural selection therefore has to be inherited.

We can work through the argument in terms of the example of an improvement in lion hunting skill. (We shall express it in terms of selection on a mutation: the same arguments apply when gene frequencies are being adjusted at a polymorphic locus.) When the improvement first appeared, it was a single genetic mutation. At a physiologic level, the mutation would produce its effect by making some minor change in the lion's developmental program. After the mutation has appeared, there is a "pool" of two types—the new mutation, and all the rest (i.e. all alleles of the mutation, and the behavior patterns they produce). There will, of course, be genetic variation at loci other than the one where the mutation arose, but that variation can be ignored because it will be randomly distributed among the mutant and non-mutant types. The lions with the mutation will survive better and produce more offspring. Natural selection is starting to work. Now we can ask what natural selection is adjusting the frequency of. Is it lions? Lion genomes? Or the mutation?

Williams' and Dawkins' answer is the gene: the particular mutation that produces improved hunting. Natural selection cannot work on whole lions because lions die: they are not permanent. Nor can it work on the genome. The mutant lion's offspring inherit only genetic fragments, not a copy of a whole genome, from their parents. Meiotic recombination breaks the genome. In Williams' expression, "meiosis and recombination destroy genotypes [i.e. genomes] as surely as death." What matters, in the process of natural selection, is that some of the lion's offspring inherit the mutation. These offspring in turn produce more offspring, and the gene increases in frequency. The gene can increase in frequency because it is not (like the genome) fragmented by meiosis, or (like the phenotype) returned to dust by death. The gene, in the form of copies of itself, is potentially immortal, and is at least permanent enough for it to be possible to alter its frequency in successive generations.

The objection may be made that recombination breaks genes as well as genomes. Recombination strikes at almost random intervals in the DNA and therefore could strike within the mutation we are concerned with. A little reflection, however, shows that is irrelevant. The information of the gene, not its physical continuity, is what matters. Consider the length of the chromosome

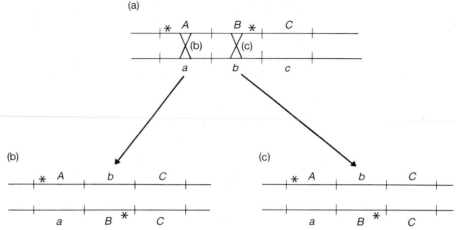

Figure 12.3 (a) Three genes along a chromosome. Loci A and B are heterozygous. *Indicates where the nucleotide differs from the other allele. Now consider the effect of recombination on the structure of the gene at the B locus. (b) Recombination in the neighboring gene does not affect the structure of the *A* genes; (c) nor does intragenic recombination. The same pair of *A* and *a* gene sequences come out of recombination as were present before.

containing the gene and its mutant form; there will usually be a number of polymorphic loci around the mutant locus (Figure 12.3a). Now consider what happens when recombination strikes either in a neighboring gene or in the gene itself. Nearby recombination breaks the information in the chromosome —which is just to repeat the point already made, that recombination destroys the genome (Figure 12.3b). Recombination within the gene does not usually alter the outcome (Figure 12.3c). If the locus was homozygous before the mutation, all the gene except for the mutant base pair will be identical in the original and mutant forms. Intragenic recombination therefore produces exactly the same result within the gene as no recombination; it only alters the combinations of genes.

Intragenic recombination can destroy the heritable information in a gene in one special circumstance. If the locus was heterozygous before the mutation and recombination occurs between the mutant site and the other site that differs between the two strands, the products of recombination differ from the initial strands (Figure 12.4). Clearly, this could happen. When it does, the length of DNA whose information is inherited is shorter than a gene. For this reason, if we take a long enough view, the only finally permanent units in the genome are nucleotide bases; recombination does not alter them. However, this long view is of little interest. We are concerned with the time scale of natural selection. It takes a few thousand generations for a mutation's frequency to be significantly altered (section 5.5, p. 93) and, over this time, genes, but not genomes or phenotypes, will be practically unaltered. Genes will then act as units of selection: they will be permanent enough to have their frequency altered by natural selection.

Williams defined the gene to make it almost true by definition that the gene is the unit of selection. He defined the gene as "that which segregates and

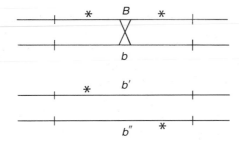

Figure 12.4 When intragenic recombination happens between two alleles with different heterozygous nucleotides, it does break the gene structure. *Indicates where the nucleotide sequence differs from the allele. The gene sequences coming out of recombination do differ from the initial sequences.

recombines with appreciable frequency.'' The gene on this definition need not be the same as a cistron (i.e. the length of DNA encoding one protein, or polypeptide). It is instead the length of chromosome that has sufficient permanence for natural selection to adjust its frequency: longer lengths are broken by recombination and shorter lengths have no more permanence than the gene (for the reason shown in Figure 12.3). The gene on Williams' definition is what Dawkins calls *the replicator*. In practice, the replicator (or Williams' gene) does not consistently correspond to any particular length of DNA. When selection is taking place at one locus, a cistron at a neighboring locus will to some extent (depending on the amount of recombination) have its frequency adjusted as a consequence. In a population genetic sense, this is hitch-hiking (section 8.9, p. 198) and builds up linkage disequilibrium between genes. The same will be true of loci further down the DNA from the selected locus: the hitch-hiking effect is gradually reduced by recombination with distance, but there is no clear cut-off. This poses no problem for Williams' definition of the gene. The neighboring allele that is hitch-hiking with the selected mutation is, in Williams' definition, part of the gene that is having its frequency altered. Williams' gene has a statistical reality, because shorter lengths of DNA are more permanent and longer lengths less so. The random hits of recombination will generate a frequency distribution of lengths of genome lasting for different periods of evolutionary time. The average length that survives long enough for natural selection to work on has been defined by Williams and Dawkins as the gene. Population geneticists have scolded them from time to time for assuming a one-locus zero linkage disequilibrium view of evolution; but the dispute is a matter of definition, not substance. The critics are identifying gene with cistron. It would be interesting to know whether the gene in Williams' sense is also a physical cistron; but it is a secondary question and has nothing to do with the fundamental logic of Williams' and Dawkins' argument.

We must discuss one further matter before considering the significance of the gene as a unit of selection. Critics, such as Gould, have objected that gene frequencies change between generations only in a passive, ''book-keeping'' sense; the frequency changes provide a record of evolution, but are not its fundamental cause. True natural selection, the critics would say, happens at the level of organismic survival and reproduction; the actual selection in the lion example happens when a lion catches, or fails to catch, its prey. The differential hunting success drives the gene frequency changes, and it is a mistake to identify the gene frequency changes as causal. Williams and

Dawkins, however, do not deny that whatever ecological processes are causing differential organismic survival also produce gene frequency changes within a generation. What they deny is that this ecological interaction of organisms means that natural selection directly adjusts the frequencies of organisms over the evolutionary time scale of many generations.

There is an easy philosophic method of deciding whether natural selection works on genes, or larger phenotypic units. We consider a phenotypic change such as a new hunting skill, and ask whether natural selection can work on it if it is produced genically and if it is produced non-genically. The case we discussed above was genic: the advantageous new hunting behavior was caused by a genetic mutation. Now suppose that the same advantageous phenotypic change was caused by a non-heritable phenotypic change instead, such as individual learning, or some developmental accident in the lion's nervous system. This thought experiment provides a test case between the organismic (phenotypic) and the genic accounts of evolution. In the genic case, we know, natural selection favors the improved hunting type and the gene for it increases in frequency. But what happens in the phenotypic case? The answer is too obvious to labor over. The individual lion with improved hunting ability will survive and produce more offspring than an average lion; but no evolution, or natural selection in any interesting sense, will occur. The trait will not be passed on to the next generation. Natural selection cannot directly work on organisms.

The change in gene frequency over time, therefore, is not just a passive book-keeping record of evolution. Genes are crucial if natural selection is to take place. The need for inheritance, and the fact that acquired characteristics are not inherited, gives the gene a priority over the organism as a unit of selection. Whenever a gene is being selected, it produces a phenotypic change and the frequency of different organismal types will change along with the gene frequency. But the change in organism frequency is a consequence of the change in gene frequency: it is the gene frequency that natural selection is actually working on and this is why Williams and Dawkins maintain that the gene is the unit of selection.

The argument is more a matter of logic than a testable claim that could be refuted by fact or experiment. By no means all biologists accept Williams' and Dawkins' argument. It has been criticized for the two reasons we have seen. I believe these criticisms have been misdirected. It is a confusion about the definition of "gene" to argue that hitch-hiking and linkage disequilibrium mean that selection adjusts the frequency of larger units than the gene; and the claim that natural selection really works on organisms rather than genes overlooks the importance of heredity.

Why does the argument matter? Its importance is to tell us what entities adaptations exist for the good of. Evolutionary biologists work on particular characters (like banding patterns in snails, and sex), trying to work out why the characters exist. The ultimate, abstract answer is that any adaptation exists because it increases the reproduction of the genes encoding it, relative to that of the alleles for alternative characters. The genes that exist in nature are the genes that in the past have out-reproduced alternative alleles. Natural selection will always favor a character that increases the replication of the genes encoding it.

It is important to know what the ultimate beneficiaries of adaptations are. When we are trying to explain the existence of particular characters, we need to know whether a proposed explanation is correct. The argument that genes are units of selection provides the fundamental logic that is used to find out. We imagine different genetic forms of the character, and the correct explanation must specify how the genes for the observed form of the character will out-reproduce other genetic types. In practice, there may be several possible hypotheses, and they can be tested between by the methods of chapter 11; but before those methods are applied, we have to ensure that the hypotheses make theoretical sense. We can rule out a hypothesis about adaptation before the practical testing stage if it contradicts the theory of gene selection.

12.4 The two senses of "unit of selection" are compatible; one specifies the entity that generally shows phenotypic adaptations, the other the entity whose frequency is generally adjusted by natural selection

We have now specified what the unit of selection is in two different senses. They have sometimes been confused, but many evolutionary biologists now appreciate the distinction. The two have been given names; Hull, for instance, distinguishes between the *interactor* and the *replicator* and Dawkins between *vehicles* and *replicators*. It is most important, however, to realize that there are two distinct issues and to understand the arguments used in the two cases.

Adaptations evolve because the genes encoding them out reproduce the alternative genes. In this sense, adaptations can only evolve if they benefit replicators. Genes do not, however, exist nakedly in the world; and the kinds of adaptations that evolutionary biologists seek to understand, such as social behavior, beak shape, or flower coloration, are not simple properties of genes. They are phenotypic properties of higher level entities (whole organisms, or societies). We therefore also have to ask which higher level entities should benefit from the natural selection of replicating genes. This leads to the discussion of section 12.2, and we have seen that adaptations should benefit entities that show heritability—which is usually organisms. This conclusion follows automatically from the fact that selection works on replicating genes, because the heritability in question is due to the propagation of genes: in other words, those entities will show adaptations that propagate genes efficiently.

12.5 Summary

1 Adaptations evolve by means of natural selection. When natural selection acts, it alters the frequencies of entities at many levels in the hierarchy of biological levels of organization. It also produces adaptations that benefit entities at many levels.

2 The discussion of units of selection aims to find out on which level natural selection directly acts, and which ones it affects only incidentally.

3 Evolutionary biologists are interested in what the unit of selection is both in order to understand why adaptations evolve and also in order that, when they study adaptations, they can concentrate on theoretically sensible hypotheses.

4 We can find out which level of organization shows adaptations by considering a series of adaptations at gene, cellular, organismic, and group levels and asking which evolves most often.

5 Segregation distortion is an adaptation of a gene against its allelic alternatives. Examples of this kind are rare.

6 In Weismannist organisms, with separate germ and somatic cell lines, selection between cell lines is a weak force. But many species do not have separate germ lines and in these we expect cell lines to evolve adaptations enabling them to proliferate at the expense of other cell lines. No clear examples are known, but Buss has suggested that the embryology of modern Weismannist species can be explained by a history of cell selection.

7 Adaptations are common at the level of organisms. When genetic relatives interact, adaptations may evolve for the benefit of kin groups (kin selection).

8 Group selection, in which selection produces adaptations for the benefit of groups of unrelated individuals, is thought to be a weak force.

9 Adaptations are possessed by the levels in the hierarchy of life that show heritability, in the sense that genetic changes are inherited by the progeny at that level. Group selection is weak because of the low genetic correlation (heritability) between succeeding generations of groups.

10 Natural selection only adjusts the frequencies of entities that are sufficiently permanent over evolutionary time. It therefore fundamentally adjusts the frequency of small genetic units. This small genetic unit is called the replicator. The gene can be so defined to be the unit of selection; but it is then not necessarily always a cistron in length.

11 Adaptations evolve because they increase the replication of genes. The replication of genes, in the real world, is enhanced by adaptations that benefit entities that show heritability.

12 The question of whether natural selection adjusts the frequencies of genes or of organisms is distinct from the question of the relative power of individual, kin, and group selection.

12.6 Further reading

The chapter follows the work of Dawkins (1976, 1982), Maynard Smith (1987b), and Williams (1966, 1985, 1992). Hull (1988), Sober (1984), and Brandon (1990) are more philosophic. Lewontin (1970), Alexander and Borgia (1978), and Wright (1980) are further general references. On segregation distortion, see Crow (1979) and a special issue of the *American Naturalist*, vol. 137, no. 3, pp. 281–456 (1991); Dawkins (1976, 1982) argues that the fairness of meiosis is necessary for the evolution of complex life forms, a theme not explored here. On cell selection, see Buss (1987) and Wolpert (1990), and Maynard Smith *et al.* (1985, pp. 281–282) for a possible example in the *bobbed* mutation of *Drosophila*; Cosmides and Tooby (1981), Eberhard (1980, 1990), and Nordström and Austin (1989) discuss various subcellular selective phenomena that would make a rung between this chapter's sections on gene and cellular selection. One example is the cytoplasmic factor that is currently spreading through *Drosophila simulans* in California (Turelli and Hoffmann, 1991). On kin selection, the fundamental works are by Hamilton (1964, 1972). For more introductory accounts see Krebs and Davies (1993), Grafen (1984), Trivers (1985), and Dawkins (1976); Grafen (1985) is also important. The literature on group selection is vast: Wynne-Edwards (1962, 1986) is the principal advocate; more recently, see Wade (1976, 1978) and Wilson (1980, 1983). Williams (1966) is a particularly important critic.

Maynard Smith (1976) and many of the references in Wilson (1983) discuss the models. Critics of gene, or replicator, selection are cited in Dawkins (1982) and Sober (1984); Gould (1980a, chapter 8, 1983b) is an example.

Adaptive explanation

13.1 Natural selection is the only known explanation for adaptation

The fact that living things are adapted for life on earth is sufficiently obvious that philosophers did not have to wait for Darwin to point it out. Adaptation was already a familiar fact, and much use was made of it by the school of thought called *natural theology*. Natural theologians explained the properties of nature theologically (i.e. by the direct action of God). They were highly influential from the eighteenth century until Darwin—John Ray and William Paley were two important thinkers of this type. Darwin himself was much influenced by the examples of adaptation, such as the vertebrate eye, discussed by Paley. Paley explained adaptation in nature by the creative action of God: when God miraculously created the world and its living creatures, he/she miraculously created their adaptations too. The key difference between natural theology and Darwinism is that the former explains adaptation by supernatural action, and the latter by natural selection.

Natural theology and natural selection are not the only two explanations that have ever been put forward for adaptation. The inheritance of acquired characters ("Lamarckism") suggests that the hereditary process produces adaptations automatically; and there are other theories which suggest that the hereditary mechanism itself produces designed, or directed, mutations and adaptation results as the consequence. These theories differ from Darwinism, in which variation is not directed toward improved adaptation: mutation is undirected and it is selection that provides the adaptive direction in evolution (section 4.6.2, p. 78).

It is one of the most fundamental claims in the Darwinian theory of evolution that natural selection is the only satisfactory explanation for adaptation. The Darwinian, therefore, has to show that the alternatives to natural selection either do not work or are scientifically unacceptable. Let us consider the natural theologians' supernatural explanation first. We can accept that an omnipotent, supernatural agent could create well-adapted living things: in that sense the explanation works. However, it has two defects. One is that supernatural explanations for natural phenomena are scientifically useless (section 3.13, p. 57). The second is that the supernatural Creator is not explanatory. The problem is to explain the existence of adaptation in the world; but the supernatural Creator already possesses this property. Omnipotent beings are themselves well-designed, adaptively complex, entities. The thing we want to explain has been built into the explanation. Positing a God merely invites the question of how such a highly adaptive and well-designed thing could in its turn have come into existence. Theological sophistry about the perfect simplicity of God and the

323

inexplicability of the First Cause can be ignored here: the problem is to *explain* adaptive complexity. The first alternative to natural selection, therefore, is a viciously circular argument, and unscientific.

The "Lamarckian" theory—the inheritance of acquired characters—is not unscientific. (I put "Lamarckian" in quotes because, as we saw in chapter 1, the inheritance of acquired characters was not specially important in Lamarck's own theory; nor did he invent the idea. However, the inheritance of acquired characters has generally come to be called Lamarckism; we can conveniently follow normal usage, outside purely historical discussion.) Since Weismann, in the late nineteenth century, it has generally been accepted that acquired characters are, as a matter of fact, not inherited. However, facts are always liable to revision and there may be some factual exceptions to Weismann's general doctrine. Even if there are no factual exceptions, it is still important to know whether the Lamarckian theory can account for adaptation in principle.

And can it? Consider the adaptations of zebra to escape from lions. Ancestral zebras would have run as fast as possible to escape from lions. In doing so, they would have exercised and strengthened the muscles used in running. Stronger legs are adaptive as well as being an individually acquired character: if the acquired character was inherited, the adaptation would be perpetuated. Superficially, this looks like an explanation, whose only defect is that as a matter of fact acquired characters happen not to be inherited.

Now let us imagine (for the sake of argument) that acquired characters are inherited, and look more closely at the explanation. The adaptation arises because zebra, within their lifetimes, become stronger runners. However, when muscles are exercised they do not become stronger by some automatic physical process; they might just as well become weaker, because they are used up. Muscle strengthening in an individual zebra requires explanation: it cannot be taken for granted. Muscles, when exercised, grow stronger because of a pre-existing mechanism which is adaptive for the organism. But where did that adaptive mechanism come from? The Lamarckian has no answer, and this is the crucial failing of the theory. For a complete explanation, it would have to fall back on another theory, such as God or natural selection. In the former case it would run into the difficulties we discussed above, and in the latter case it is natural selection, not Lamarckism, that is providing the fundamental explanation of adaptation. Lamarckism could work only as a subsidiary mechanism; it could only bring adaptations into existence in so far as natural selection had already programmed the organism with a set of adaptive responses. Pure Lamarckism does not by itself explain adaptation.

All theories of directed or designed mutation have the same problem. A theory of directed mutation, if it is to be a true alternative to natural selection, must offer a mechanism for adaptive change that does not fundamentally rely on natural selection to provide the adaptive information. Most alternatives to natural selection do not explain adaptation at all. For example, earlier this century some paleontologists, such as Osborn, were impressed by long-term evolutionary trends (see chapter 21). The titanotheres are a classic example. Titanotheres are an extinct group of Eocene and Oligocene perissodactyls (the mammalian order which includes horses); in a number of lineages, the earlier

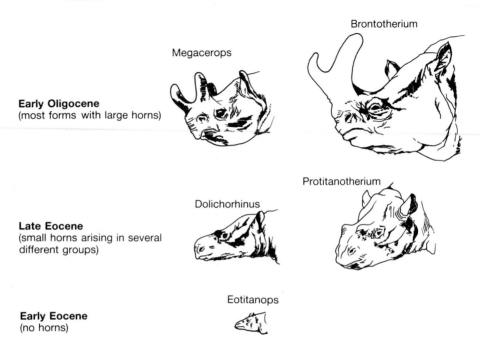

Brontotherium

Megacerops

Early Oligocene
(most forms with large horns)

Protitanotherium

Dolichorhinus

Late Eocene
(small horns arising in several
different groups)

Eotitanops

Early Eocene
(no horns)

Figure 13.1 Two lineages of titanotheres, showing parallel body size increase and the evolution of horns. Only two of many lineages are illustrated. From Simpson (1949).

forms lacked horns whereas later ones had evolved them (Figure 13.1). Osborn, and others, believed that the trend was *orthogenetic*: it arose not because of natural selection among random mutations but because titanotheres were mutating in the direction of the trend.

The idea could explain a simple, adaptively random trend. If it made no difference to a titanothere what size its horns were, then a trend toward larger horns might be generated by directed mutation. In fact the horns almost certainly were adaptive, and for trends toward undoubted adaptive complexity, such as the evolution of mammals from mammal-like reptiles (section 20.1, p. 533), we have to ask how "orthogenetic" mutations could keep on occurring in the direction of adaptive improvement. If it is replied that variation just happens to be that way, then adaptation is being explained by chance—and chance alone cannot explain adaptation, almost by definition. Theories of directed variation therefore generally boil down to supernatural intervention or natural selection. They are either unscientific or not genuine alternatives to Darwinism. Directed mutation and the inheritance of acquired characters not only do not operate in fact, but they are also unsatisfactory in theory. We can conclude that natural selection is the only available explanation for adaptation.

13.2 Pluralism is appropriate in the study of evolution, not of adaptation

So natural selection is our only explanation for adaptation. It should be emphasized that this statement applies only to *adaptation* and not to evolution as a whole. Biologists, such as Gould and Lewontin, have pointed out that Darwin himself did not rely exclusively on natural selection, but admitted other processes too; and they urge that we should accept a pluralism of evolutionary processes, rather than relying exclusively on natural selection. For evolution as a whole, this is a sensible idea. In chapter 7, for instance, we saw that many evolutionary changes in molecules may take place by neutral drift. The molecular sequences among which drift takes place are then not different adaptations; they are different variants of one adaptation, and natural selection has nothing to say about why one organism has one sequence variant, and another organism has another. We need drift as well as selection in a full theory of evolution.

The fact that processes beside natural selection can cause evolutionary change does not alter the argument of section 13.1: it just goes to show that not all evolution need be adaptive. This being so, we should be pluralists about evolution; but when we are studying adaptation, it is sensible to concentrate on natural selection. (Critics might agree, but retort that some evolutionary biologists exaggerate the amount of evolution that is adaptive rather than non-adaptive, and mistake cases of the latter for the former. That, however, is to change the subject. We shall have more to say about it later. The point being made here is that natural selection is the only known explanation of adaptation; how common adaptations are is another matter.)

13.3 Natural selection can, in principle, explain all known adaptations

The argument so far has been negative: we have ruled out the alternatives to natural selection, but we have not made the positive case for natural selection itself. We have seen before (see chapter 4) that natural selection can explain adaptation; but we can also ask a stronger question: can it explain *all* known adaptations?

The question is important historically, and it often still rises in popular discussions of evolution. The case against selection would run something like this. There is no doubt that natural selection explains some adaptations, such as camouflage. However, the adaptation in this case, as well as in other famous examples of natural selection, is simple. In the peppered moth it is just a matter of adjusting external color to the background. The problem arises in complex characters, which are adapted to the environment in many interdependent respects. Darwin's explanation for complex adaptations is that they evolved in many small steps, each analogous to the simple evolution in the peppered moth; that is what Darwin meant when he called evolution gradual. Evolution has to be gradual because it would take a miracle for a complex organ, requiring mutations in many parts, to evolve in one sudden step. If each mutation arose separately, in different organisms at different times, the whole process becomes more probable (Box 7.3, p. 170).

Darwin's gradualist requirement is a deep property of evolutionary theory. The Darwinian should be able to show for any organ that it could, at least in principle, have evolved in many small steps, with each step being advantageous. If there are exceptions, the theory is in trouble. In Darwin's words, "if it could be demonstrated that any complex organ existed which could not possibly have been formed by numerous successive slight modifications, my theory would absolutely break down."

Darwin argued that all known organs could have indeed evolved in small steps. He took examples of complex adaptations and showed how they could

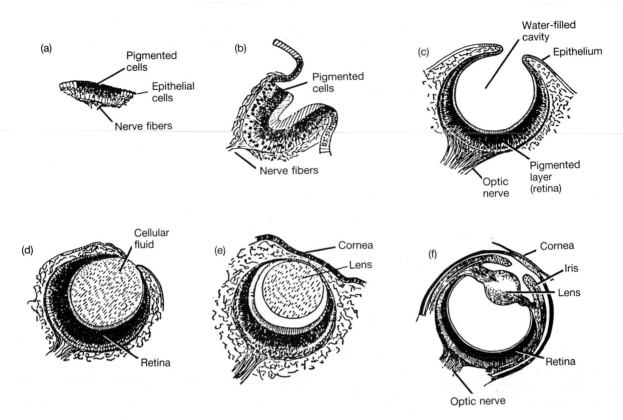

Figure 13.2 Stages in the evolution of the eye, illustrated by different modern molluscs. (a) A simple spot of pigmented cells. (b) Folded region of pigmented cells, which increases the number of sensitive cells per unit area. (c) Pin-hole camera eye, as is found in *Nautilus*. (d) Eye cavity filled with cellular fluid rather than water. (e) The eye is protected by adding a transparent cover of skin; and part of the cellular fluid has differentiated into a lens. (f) Full, complex eye, as found in octopus and squid.

have evolved through intermediate stages. In some cases, such as the eye (Figure 13.2), these intermediates can be illustrated by analogies with living species, in other cases they can only be imagined. Darwin only had to show that the intermediates could possibly have existed. His critics had the more difficult task: they must show that the intermediates could not have existed. It is very difficult to prove negative statements. Nevertheless, many critics suggested, for various adaptations, that natural selection cannot account for them. These types of adaptation can be considered under two headings.

13.3.1 Coadaptations

Coadaptation here refers to complex adaptations, the evolution of which would have required mutually adjusted changes in more than one of their parts. (Coadaptation is a popular word: it has already been used in a different sense in chapter 8, and will be used in a third sense in chapter 21!) In a historic dispute in the 1890s, Herbert Spencer and August Weismann discussed the giraffe's neck as an example. Spencer supposed that the nerves, veins, and bones and muscles in the neck were each under separate genetic control. Any change in neck length would then require independent, simultaneous changes of the correct magnitude in all the parts. As it were, a change in the length of the neckbones would malfunction without an equal change in vein length, and evolution by natural selection in one part at a time would be impossible. The example is unconvincing now because of the obvious retort that the lengths of all the parts could be under common genetic control.

The other standard example of a complex coadaptation was the eye. When one eye part, such as the distance from the retina to the cornea, changes during evolution, changes in other parts, such as lens shape, would be needed at the same time. Because of the improbability of simultaneous correct mutations in both parts at the same time (the argument goes), a complex, finely adjusted engineering device like the eye could not have evolved by natural selection. The Darwinian reply (illustrated in Figure 13.2) is that the different parts could evolve independently in small steps: it is not necessary for all the parts of an eye to change at the same time in evolution.

13.3.2 Functionless, or disadvantageous, rudimentary stages

An organ has to be advantageous to its bearer at all stages in its evolution if it is to be produced by natural selection. Some adaptations, it is said, although undoubtedly advantageous in their final form, could not have been when in a rudimentary form: "What is the use of half a wing?" is a familiar example. The anatomist St George Jackson Mivart particularly stressed this argument in his *The Genesis of Species* (1871). The Darwinian reply has been to suggest ways in which the character could have been advantageous in rudimentary form. In the case of the wing, partial wings might have broken the force of a fall from a tree, or proto-winged birds might have glided from cliff tops or between trees—as many animals, such as flying foxes, do now. These early stages would not have required all the muscular back-up of a full, final wing. The concept of preadaptation (see below) provides another solution to the problem.

These debates have disappeared from modern evolutionary biology, though their vestiges do give rise to misunderstanding. Evolutionary biologists are sometimes accused of making speculative, even fanciful, suggestions about the stages through which individual adaptations could have evolved. But the critics overlook the original point of the discussion. The speculations are not the prize specimens of evolutionary analysis. They were put forward solely to refute the suggestion that we cannot imagine how such-and-such a character could have evolved in small, advantageous steps. To refute that, it is only necessary to imagine a series of steps. It is not being claimed that the series is particularly profound or realistic, or even very probable. The long evolutionary history that precedes any complex modern adaptation will appear, with hindsight, to be an improbable series of accidents: the same point is as true for human history as evolution. Given the state of our knowledge at any one time, for some characters we can reconstruct their evolutionary stages with some rigor (see chapter 17), but for others we cannot —and for these it is only possible to make guesses to illustrate possibilities, not conduct a rigorous scientific investigation.

It is fair to conclude that there are no known adaptations that definitely could not have evolved by natural selection. Or (if the double negative is confusing!), we can conclude: all known adaptations are in principle explicable by natural selection.

13.4 Adaptation can be defined either historically or by current function

The Darwinian concept of adaptation, and the natural theologian's concept of design, differ not only in how they explain adaptation (or designfulness). Design, in natural theology, was a purely contemporary concept—it referred to the way organisms are now—whereas adaptation, in the theory of evolution, is at least partly a historical concept. Each adaptation will have passed from an initial stage, through intermediate stages, to the form we now see; adaptations were not suddenly created from nothing.

As we look at the historical stages in the evolution of an adaptation, it is possible that at every stage the organ served the same function (the eye is probably an example—all its stages probably had visual functions), or the earlier stages could have had different functions from the later ones. The classical Darwinian term for the second possibility is *preadaptation*. The earlier stage is described as a preadaptation for the later stage. A possible example is as follows. The most widely accepted theory of how vertebrates invaded the land suggests that lobe-finned fish originally walked on the bottom of lakes or streams with their fins, and by degrees came to walk on the muddy surrounds at the water edge, and finally on dry land as the necessary adaptations for air breathing and water retention evolved. Now, there are two main groups of bony fish (Figure 13.3): the ray-finned Actinopterygians (containing almost all the common fish that we think of as "fish") and the lobe-finned Sarcopterygians (a few oddities, including lungfish and the coelocanth). In a ray-finned fish, the skeleton of its fins are a ray of similar bones, and they are mainly moved by muscles with one end in the fin and the other end inside the body. In lobe-finned fish, there is a single main rod of skeletal support in the fin, and its movements are partly controlled within the fin. The difference is interesting because all tetrapods (amphibians, reptiles, birds, mammals) evolved from the lobe-finned fish, and they did so almost certainly because a lobe fin could evolve into a tetrapod limb, whereas a ray fin could not. The lobe-finned fish are preadapted for walking on land; ray-finned fish are not.

Preadaptation does not imply any futuristic or anticipatory faculty in evolution. Lobe-finned fish did not evolve their skeleton in order that they could later give rise to the tetrapods. All that is happening is that sometimes, by chance, an organ that works well in one function turns out to work well in another function after relatively little adjustment. Fins in both fish groups first evolved for swimming; then in some lobe-finned species, they probably came to be used for scuttling around near the seashore or on the bottom of rivers or lakes; it was then only a small change for them to walk on land. Whatever the details were, it is a reasonable inference that the lobe-finned skeleton was, unlike a ray fin, preadapted to evolve into a tetrapod limb. The term preadaptation is applied when a large change in function is accomplished with little change of structure.

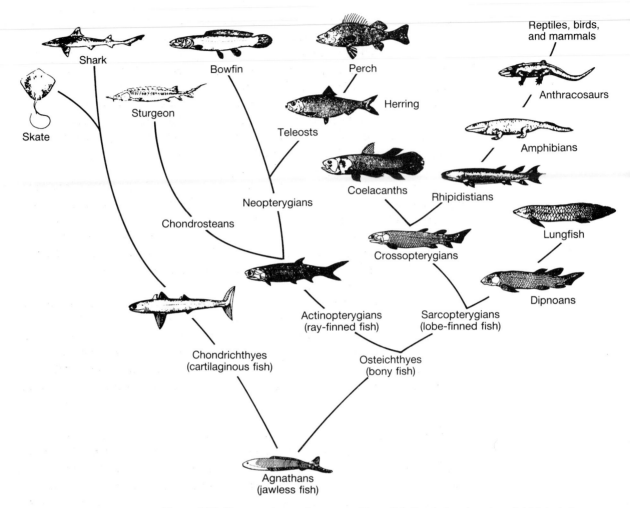

Figure 13.3 There are two main groups of bony fish: the Actinopterygians (which include nearly all modern fish) and the Sarcopterygians (from which we, and all other tetrapods, evolved). Cartilaginous fish are a related group.

More recently, Gould and Vrba suggested adding another term: *exaptation*. When an organ changes its function but retains a fairly constant structure, they would call the later stage an exaptation. The tetrapod limb is an exaptation for terrestrial locomotion. The organ was a preadaptation before the function change, and an exaptation afterwards. Actually, Gould and Vrba's concept was broader. They defined exaptations as any character that performs a function different from the reason why it originally evolved. It thus includes not only preadaptations, but also the theoretical possibility of an initially neutral character that later evolves a positive function.

There are two ways of defining adaptation. A broader use is closer to the pre-Darwinian concept of design. This broad meaning defines as an adaptation any character which helps its bearer to survive and reproduce; and the way in which a character is adapted is the way in which it now helps its bearer to survive and reproduce. By this definition, the tetrapod limb is an *adaptation* for walking: the definition includes characters both that serve the same function as they originally evolved to carry out and characters that have changed their function and are advantageous in some evolutionarily new respect. This definition is useful in the study of adaptation and is probably the more widely accepted among biologists who are actively doing research on adaptation. It is a practically useful criterion because it is possible to ask what difference it would make to the organism if some change were made to the character in question: for example (see chapter 11), what difference would it make if sex was changed to parthenogenesis, or the sex ratio from 50:50 to 40:60? In more formal, genetic terms, we can ask whether mutations that cause a change in the character would spread. This question can be asked independently of the functional history of the organ: we could ask it for a lobe fin in a lobe-finned fish as well as for a limb in a modern horse. If we are interested in how an organ is adapted for life in its environment now, there is no reason to distinguish characters that have changed function in the past from those that have not.

The second, narrower definition of adaptation confines it to organs that are still serving the function they originally evolved to do. By this definition, to say that a limb is an adaptation for terrestrial walking is not only to say that a limb is used for walking, but also that it originally evolved for terrestrial walking: so the tetrapod limb is then not an adaptation. Gould and Vrba favor the narrower definition. They would call adaptations those organs that are still serving their original function, and exaptations those organs that have changed function.

Nothing much depends on which definition is used, so long as neither of two crucial points is overlooked. One is that most study of adaptation is not concerned with the past history of functional changes, but with how natural selection maintains the adaptation in its modern form. It only confuses the issue if the problem of historical reconstruction is added to the problem of current function. They are separate problems. Biologists studying the current function of organs are not obliged to show that the organ has always served the same function. We therefore need a term for the broad sense. The second point is that organs undoubtedly can change their function during their history, and the reason why natural selection is maintaining an organ now may not be the reason that it evolved to begin with. These changes are worth knowing about. In summary, therefore, adaptation can be defined in either of two ways. It does not matter which definition is used, but the distinction between them is important.

13.5 The function of an organ should be distinguished from the effects it may have

A character of an organism can have beneficial effects that it is not useful to call adaptive. Some consequences follow from the laws of physics and chemistry without any need for shaping by natural selection. Here is a quotation from Williams (1966):

> Consider a flying fish that has just left the water to undertake an aerial flight. It is clear that there is a physiological necessity for it to return to the water very soon; it cannot long survive in the air. It is, moreover, a matter of common observation that an aerial glide normally terminates

with a return to the sea. Is this the result of a mechanism for getting the fish back into water? Certainly not; we need not invoke the principle of adaptation here. The purely physical principle of gravitation adequately explains why the fish, having gone up, eventually comes down.

The flying fish is not adapted to obey the law of gravity. When evolutionary biologists seek to understand how a character is adaptive, they consider the likely reproductive success of mutant, altered forms of the character. We can imagine many changes in the shape of the flying fish; but none of them will prevent it from returning to the sea. Even though returning to the sea is a biologic necessity, there has not been any selection in the past between some types of fish that did return to the sea and some types that did not, with the former surviving and reproducing better.

Two points can be made. One is that a thought experiment about alternative forms of a character is only sensible if the alternatives are plausible. Fish that disobey gravity are not. Imagining alternative forms of a character is not absurd, but it can be taken to absurd extremes. In real cases, the alternatives are usually plausible and may even be known to exist. There is nothing absurd in postulating a melanic form of the peppered moth, because it can be seen in nature; and it is reasonable to consider how natural selection would act on an asexual mutant of a sexual species, because asexual forms are known in many taxa. The second point is that not all the beneficial consequences of a character are correctly called adaptations. A character is an adaptation in so far as natural selection is maintaining its form in modern populations and (on the narrower definition) originally brought it into existence. Beneficial consequences that are independent of natural selection are not adaptations. The point is obvious in practice, but must be borne in mind in conceptual discussion.

13.6 Adaptations in nature are not perfect

Natural selection has brought into existence creatures that in many respects are marvellously well designed. The designs, however, are generally imperfect, and for a number of reasons. One reason we discussed in chapter 12. It may not be possible for an adaptation to be simultaneously perfect at all levels of organization: an adaptation like segregation distortion is good for the distorter gene but not for the organism; and birth control may be good for the population but not the individual. Most of the familiar examples of adaptation benefit the organism. They will therefore be (at best) imperfect at the genic, cellular, and group levels. However, we can still ask whether organismal adaptations are perfect even for the organism.

The quality of adaptation will progressively improve for as long as there is genetic variation to work on. If some genetic variants in the population produce a better adaptation than others, natural selection will increase their frequency. Although this process must always operate in the direction of improvement, it has never reached the final state of perfection. As Maynard Smith remarked, ''if there were no constraints on what is possible, the best phenotype would live for ever, would be impregnable to predators, would lay eggs at an infinite rate, and so on.'' What are the constraints that prevent this kind of perfection from evolving?

13.7 Adaptations may be imperfect because of time lags

Many flowering plants produce fruits, in order to induce animals to act as dispersal agents. The fruits of different species are adapted in various ways to the particular animals they make use of: fruits have to be attractive to the animal, but also protect the seed from destruction by the animal's digestive system; they also must remain in the animal's gut for about the right amount of time to be dispersed an appropriate distance from the parent and then be properly deposited, and this can be achieved by laxatives in the fruit. Many details are known about the ways in which individual fruits are adapted to the habits and physiology of the animal species that disperse them. Over evolutionary time, plants presumably have adapted the form of their fruits to whatever animals are around, and when the fauna changes, the plants will evolve (or rather "coevolve", section 21.3, p. 564), in time, to produce a new set of adapted fruits.

Natural selection, however, takes time, and there will be a period after a major change in the fauna during which the adaptations of fruits will be out of date, and adapted to an earlier form of dispersal agent. Janzen and Martin have argued that the fruits of many trees in the tropical forests of Central America are "neotropical anachronisms": they are anachronistically adapted to an extinct fauna of large herbivores (Figure 13.4). Until about 10 000 years ago North and Central America had a fauna of large herbivores comparable in scale to that of Africa in recent times. Just as Africa has elephants, giraffes, and hippopotamuses, in Central America there were giant ground sloths, an extinct giant bear, an extinct species of large horse, mammoths, and a group of large relatives of mastodons called gomphotheres. These mammals are all now gone, but the species of trees that they used to walk beneath still

Figure 13.4 The fruits of (a) *Crescentia alata* (Bignoniaceae) and (b) *Annona purpurea* (Annonaceae) are two examples of fruits that were probably eaten by large herbivores that recently became extinct. The larger fruits in (a) are about 20 cm long; the fruit in (b) is nearer 30 cm long. Both trees were photographed in Santa Rosa National Park, Costa Rica. (Photographs by Dan Janzen.)

remain. In the tropical forests of Costa Rica, some trees still drop large and hard fruits in great quantities. It accumulates, and much of it rots, at the base of the trees, and those that are moved by small mammals such as agoutis are not moved far. Here is how Janzen and Martin describe the fruiting of the large forest palm *Scheelea rostrata*: "in a month as many as 5000 fruits accumulate below each fruit-bearing *Scheelea*-palm. The first fruits to fall are picked up by agoutis, peccaries, and other animals that are soon satiated . . . The bulk of the seeds perish directly below the parent." The fruits seem overprotected, with their hard external coverings; they are produced in excessive quantities; and they are not adapted for dispersal by small animals like agoutis. It looks like a case of maladaptation: "a poor adjustment of seed crop size to dispersal guild." However, the fruits make sense if they are anachronistic adaptations to the large herbivores that have so recently become extinct. The fruits' large size would have been appropriate for a gomphothere, and the hard external cover would have protected the seeds from the gomphothere's powerful crushing teeth. Some 10 000 years has not been long enough for the trees to evolve fruits appropriate to the more modestly sized mammals that now dwell among them.

The principle illustrated by the fruits of these Central American plants is a general one. Adaptations will often be imperfect because evolution takes time. The environments of all species change more-or-less continually because of the evolutionary fortunes of the species they compete, and cooperate, with. Each species has to evolve to keep up with these events, but at any time they will lag some distance behind the optimal adaptation to their environment. (The degree of lag can be quantified as a "lag load," section 22.3, p. 596.) Adaptation will be imperfect because natural selection cannot operate as fast as the environment of a species changes.

13.8 Genetic constraints may cause imperfect adaptation

When the heterozygote at a locus has a higher fitness than either homozygote, the population evolves to an equilibrium at which all three genotypes are present (section 5.11, p. 109). A proportion of the individuals in the population must therefore have the deleterious homozygous genotypes. This is an example of a *genetic constraint*. It arises because the heterozygotes cannot, under Mendelian inheritance, produce purely heterozygous offspring: they cannot breed true. In so far as heterozygous advantage exists, some members of natural populations will be imperfectly adapted. The importance of heterozygous advantage is controversial; but there are undoubted examples such as sickle-cell hemoglobin—and sickle-cell anemia indeed is a practical manifestation of imperfect adaptation due to genetic constraint.

The balanced lethal system of the European crested newt *Triturus cristatus* is a more dramatic example. Members of the species have 12 pairs of chromosomes, numbered from one to 12, one being the longest and 12 the shortest. In 1980, Macgregor and Horner found that all individual crested newts of both sexes are "heteromorphic" for chromosome 1: an individual's two copies of chromosome 1 are visibly different; they look different under the microscope. They named the two types of chromosome 1, 1A and 1B (the same two types are found in every individual). Meiosis, they found, is normal: an individual produces equal numbers of gametes with 1A chromosomes as with 1B chromosomes. There is also little, if any, recombination between the two chromosomes.

The puzzle is why there are no chromosomally homomorphic newts, with either two 1A or two 1B chromosomes. Macgregor and Horner carried out breeding experiments, in which they crossed two normal individuals, and counted the proportion of eggs that survived. In every case, approximately half the offspring died during development. It is almost certainly the homomorphic individuals that die off, leaving only the heteromorphs. The reason why half the offspring die is as follows. The adult population has two types of chromosome, each with frequency 0.5. If we write the frequency of the 1A chromosome as p and of the 1B chromosome as q, $p = q = 0.5$. By normal Mendelian segregation, and the Hardy–Weinberg principle, the proportion of homozygotes (or homomorphs) is $p^2 + q^2 = 0.5$. Each generation, therefore, the heterozygous newts mate together and produce half homozygous offspring and half heterozygous offspring: and then all the homozygotes die. The system looks almost incredibly inefficient, because half the reproductive effort of the newts each generation is wasted; but the same sort of inefficiency exists, to some extent, at any genetic locus with heterozygous advantage.

If in humans a new hemoglobin arose that was resistant to malaria and viable in double dose, or in the crested newt a new chromosome 1 arose that was viable as a homomorph, it should spread through the population. Presumably the inefficiency remains only because no such mutations have arisen.

Could a system with heterozygous advantage easily evolve into a pure breeding genotype with the same phenotypic effect? Gene duplication (section 10.3, p. 243) suggests that it could. Imagine that the relevant hemoglobin gene duplicated in an Hb_+/Hb_s individual, to become Hb_+Hb_+/Hb_sHb_s. Genetic recombination could then produce an Hb_+Hb_s chromosome, and that chromosome should be able to achieve anything that an Hb_+/Hb_s heterozygote can. The chromosome would also breed true, once it had been fixed. We might expect therefore that the existing Hb_+/Hb_s system will evolve to a pure Hb_+Hb_s/Hb_+Hb_s system. Some "dosage compensation" might be needed after the gene had duplicated; but that should be no difficulty because regulatory devices are common in the genome. The apparent ease of this evolutionary escape from heterozygous advantage and segregational load is one possible explanation for the (apparent) rarity of heterozygous advantage. However that may be, the existence of some cases of heterozygous advantage suggests that natural populations can be imperfectly adapted because a superior mutation has not arisen.

13.9 Developmental constraints may cause imperfect adaptation

A nine-penned discussion (Maynard Smith *et al.*, 1985) of *developmental constraints* gave the following definition: "a developmental constraint is a bias on the production of variant phenotypes or a limitation on phenotypic variability caused by the structure, character, composition, or dynamics of the developmental system." The idea is that different groups of living things evolved distinct developmental mechanisms, and that the way an organism develops will influence the kinds of mutation it is likely to generate. A plant, for example, may be likely to mutate to a new form with more branches than would a vertebrate (in the vertebrate, new "branches" might be extra legs, or perhaps having two heads), because it is easier to produce that kind of change in the development of a plant. The rates of different kinds of mutation (or of "production of variant phenotypes" in the quoted definition) vary among the groups.

Developmental constraints can arise for a number of reasons. *Pleiotropy* is an example. A gene may influence the phenotype of more than one part of the body. A trivial instance would be that genes influencing the length of the left leg probably also influence the length of the right leg. The growth of legs probably takes place through a growth mechanism controlling both legs. This mechanism does not have to be inevitable for a constraint to exist. Perhaps some rare mutants do affect the length only of the right leg. A developmental constraint exists whenever there is a tendency for mutants (in this example) to affect both legs, and the tendency is due to the action of some developmental mechanism.

Pleiotropy exists because there is not a one-to-one relationship between the parts of an organism that a gene influences and the parts of an organism that we recognize as characters. The genes divide up the body in a different way from the human observer. Genes influence developmental processes, and a change in development will often change more than one part of the phenotype. Much the same reasoning lies behind a second sort of developmental constraint. New mutations often disrupt the development of the organism. A new mutant, with an advantageous effect on one character, may also disrupt other parts of the phenotype. The disruptions will probably be disadvantageous; but if the mutant has a net positive effect on fitness, natural selection will favor it. In some cases, the disruption can be measured by the degree of asymmetry in the form of the organism: in a species with bilateral symmetry, any deviation from bilateral symmetry in an organism is a measure of how well regulated its development was. Mutations can therefore cause *developmental asymmetry*.

The Australian sheep blowfly *Lucilia cuprina* provides an example. It is a pest, and farmers spray it with insecticides. The flies, as we should expect (section 5.7, p. 102), soon respond by evolving resistance. One such resistance mutation was studied by Clarke and McKenzie, and they found that, when it first appeared, it produced developmental asymmetry as a by-product. Presumably, the disruption of development is deleterious, though not so deleterious that the mutation is selected against: the advantage in insecticide resistance more than makes up for a little developmental disruption.

When the mutation appears, it will increase in frequency by selection. But selection at other loci will then start to act, to reduce the new mutation's deleterious side-effects while maintaining its advantageous main effect. That is, selection will make the new mutation fit in with the blowfly's pre-existing developmental mechanism. Genes at other loci that can restore the former symmetric development, while preserving the insecticide resistance, will be favored. These are called *modifier genes*, and the type of selection is called *canalizing selection*. Over time, in the sheep blowfly, the resistance mutation was modified such that it no longer disrupted development. Clarke and

McKenzie were able to show that the modification was caused by genes at loci other than the mutation-carrying locus. (This is important because, just as there is selection at other loci to reduce the deleterious side-effects of the mutation, so selection at that locus will favor other mutations that can produce insecticide resistance without harmful side-effects.) It is probably common, given the extent of genetic interaction in development, for new mutations to disrupt the existing developmental pattern. Canalizing selection, to restore developmental regulation with the new mutation, is therefore likely to be an important evolutionary process.

Another sort of developmental constraint can be seen in the "quantum" growth mechanism of arthropods. Arthropods grow by molting their exoskeleton and then growing a new, larger one. They do not grow while the exoskeleton is hard. The arthropod growth curve shows a series of jumps, often with a fairly constant size ratio of 1.2–1.3 (i.e. the animal is 1.2–1.3 times larger after the molt than it was before). There are various models of how body size can be adaptive; for example, it influences thermoregulation, competitive power, and what food can be taken. But none of these factors can plausibly explain the jumps in the arthropod growth curve. If, for example, the body size of an arthropod was adapted to the size of food items it fed on, it would hardly be likely that the distribution of sizes of food items in its environment set up a selection pressure for quantum growth. The explanation for the quantum jumps is a developmental constraint: growth, by molting, is dangerous and to grow with a smooth curve would require frequent risky molts. It is better to molt more rarely and grow in jumps.

Developmental constraints have been suggested as an alternative explanation to natural selection for two main natural phenomena. One is the persistence of fossil species for long periods of time without showing any change in form (section 19.6.3, p. 520). The other is the variety of forms to be found in the world. We can imagine plotting a *morphospace* for a particular set of phenotypes and then filling in the areas that are and are not represented in nature. Raup's analysis of shell shapes is an elegant example. Raup found that shell shapes could be described in terms of three main variables: translation rate, expansion rate, and distance of generating curve from coiling axis (Figure 13.5). Any shell can be represented as a point in a three-dimensional space, and Raup plotted the regions in this space that are occupied by living shells (Figure 13.6).

Large parts of the shell morphospace in Figure 13.6 are not occupied. There are two general hypotheses to explain why these forms do not exist: natural selection and constraint. If natural selection is responsible, the empty parts of the morphospace are regions of maladaptation. When these shell types arise as mutations, they are selected against and eliminated. Alternatively, the empty parts could be regions of constraint: the mutations to produce these shells have never occurred. If the constraint was developmental, it would mean that for some reason, it is developmentally impossible (or at least unlikely) for these kinds of shells to grow. The non-existent shells would be embryologic analogies for animals that disobey the law of gravity: they are shells that break the (unknown) laws of embryology. The absence of these shells would then no more be due to natural selection than is the absence of

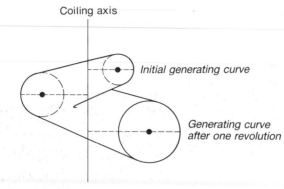

Coiling axis

Initial generating curve

Generating curve
after one revolution

Figure 13.5 The shape of a shell can be described by three numbers. The translations rate (T) describes the rate at which the coil moves down the coiling axis: T is zero for a flat planispiral shell, and has an increasingly positive number for increasingly elongated shells. The expansion rate (W) describes the rate at which the shell size increases; it can be measured by the ratio of the diameter of the shell at equivalent points in successive revolutions; $W = 2$ in the figure. The distance from the coiling axis (D) describes the tightness of the coil; it is the distance between the shell and the coiling axis, and in the figure it is half the diameter of the shell. See Figure 13.6 for many theoretically possible shell shapes with different values of T, W, and D. From Raup (1966).

animals that break the law of gravity. Just as natural selection and constraint are hypotheses to explain the absence of any form from nature, so they can both hypothetically explain the forms that are present. Faced with any form of organism, we can ask whether it exists because it is the only form that organism possibly could have (constraint), or whether selection has operated in the past among many genetic variants and the form we now observe is the one that was favored. If the form of an organism is the only one possible, an analysis that treated it as an adaptation would be misdirected. In some cases we can be more certain that variation is strongly constrained than in others. If the constraint is the law of gravity, adaptation is a fanciful hypothesis; but if the constraint is a conjectural piece of embryology, adaptation is much more worthy of investigation.

How can we test between selection and constraint? Maynard Smith and his eight co-authors listed four general possibilities. The first is the use of adaptive prediction. If a theory of shell adaptation predicted accurately and successfully the relationship between shell form and environment—which forms should be present, and which absent, in various conditions—then, in the absence of an equally exact embryologic theory, that would count in favor of adaptation and against developmental constraint. Vice versa, a successful, exact embryologic theory would be preferred to an empty adaptive theory. The second test is a direct measure of selection. In the case of the shell morphospace, this would mean somehow making the naturally non-existent shells experimentally, and testing how selection then worked on them (section 11.2, p. 268). We then find out by observation whether there is negative selection against these forms.

Thirdly, we can measure the character's heritability. If a constraint is preventing mutation in a character, it should not be genetically variable.

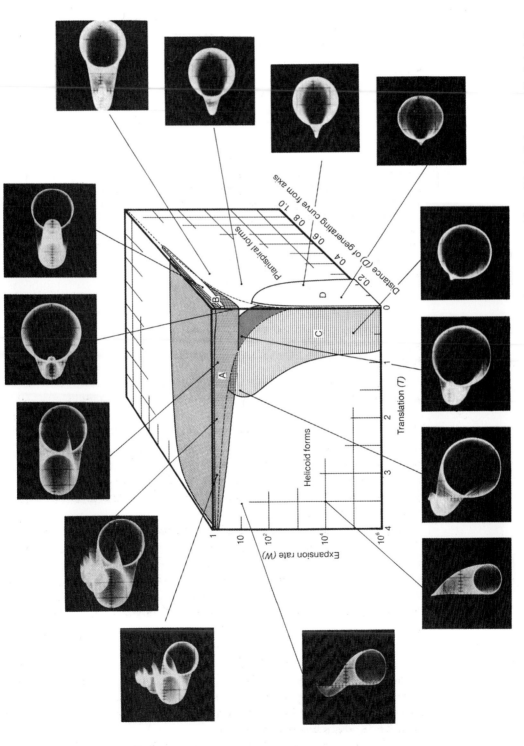

Figure 13.6 The three-dimensional cube describes a set of possible shell shapes. Around the outside of the figure, 14 possible shell shapes are illustrated; they were drawn by a computer. Only four regions in the cube are actually occupied by natural species: they are the regions marked A, B, C, and D. All other regions in the cube represent theoretically possible but naturally unrealized shell shapes. The space is called a morphospace. From Raup (1966).

Genetic variability can be measured, and the constraint hypothesis will be refuted for any character that shows significant heritability. As it happens, this kind of evidence suggests that the gaps in the shell morphospace are not caused by developmental constraint. The heritability of a number of shell properties has been measured, and significant genetic variation found. Shell shape, therefore, is at least to some extent unconstrained. (Depending on the exact constraint hypothesis, the measurement of heritability might be empirical overkill. If the hypothesized constraint purportedly makes any variation impossible, it could be refuted simply by demonstrating phenotypic variation: developmental constraints do after all apply to phenotype, not genotype. But in some cases, it might be necessary to test for genetic, as well as phenotypic, variation.)

Finally, comparative evidence from many species may be useful. It has particularly been used for pleiotropic developmental constraints. When more than one character is measured, and the values for the two characters in different organisms are plotted against each other, a relationship is nearly always found. This has been done most often for body size together with another character, and the relationships are then called *allometric* (Darwin referred to it as the "correlation of growth"). Graphs of tooth size, for example, or of testis size, against body size in different species of primates show positive correlations (Figure 13.7); these graphs are two-dimensional morphospaces, and are analogous to Raup's more sophisticated analysis for shells.

Again in the case of allometry, the observed distribution of points might be due either to adaptation or constraint. It might be adaptive for an animal with a large body to have large teeth, or large testes (or whatever is measured). Or it might make no difference what size an animal's teeth are, and changes in tooth size would simply be the correlated consequences of changes in body size (or vice versa): mutations altering one of the characters would in that case be constrained also to alter the other. Huxley was an influential early student of allometry, and he liked to explain allometric relationships by the hypothesis of constraint: "whenever we find [allometric relationships], we are justified in concluding that the *relative size* of the horn, mandible, or other organ is automatically determined as a secondary result of a single common growth-mechanism, and *therefore is not of adaptive significance*. This provides us with a large new list of non-adaptive specific and generic characters." (Huxley, 1932).

Correlated changes undoubtedly can happen for embryologic reasons, but that does not mean all allometric relations are non-adaptive. The idea can be tested by looking at the trends in more detail. The relationship between tooth and body size in primates is an example: it is allometric, but tooth size increases at a lower rate than body size; large primates therefore have proportionally smaller teeth than small primates. Lewontin for this reason once argued that "it would be erroneous to argue that for some special adaptive reason gorillas have been selected for relatively small teeth." However, when the interspecific differences are examined closely, relationships can be found between tooth size and other properties of the species' biology. Fruit-eating primate species have relatively smaller teeth than leaf eaters, and

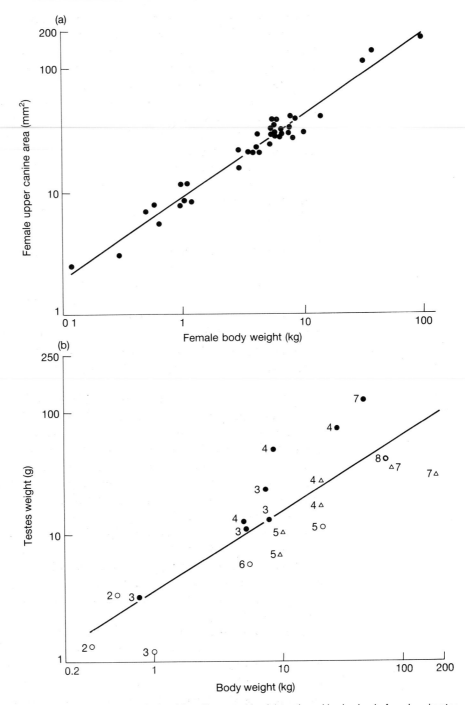

Figure 13.7 Allometric relationships, illustrated by (a) tooth and body size in female primates (each point is a different species), and (b) testis and body size in male primates. From Harvey *et al.* (1978) and Harcourt *et al.* (1981).

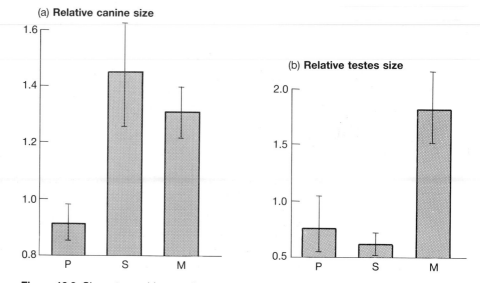

Figure 13.8 Characters subject to allometric relationships may also be influenced by other variables. (a) The relative tooth size of male primates (male tooth size divided by female tooth size for that body size) is larger in polygamous than in monogamous species; as is relative testis size (b). P, S, and M are three social systems: P are pair-living, or monogamous; S and M are polygamous—S species have a single male in the group, M species have more than one male. From Clutton-Brock and Harvey (1984).

monogamous species show less sexual dimorphism in canine size than do polygynous species; similar relationships are also found within the allometric relationship between testis size and body size (Figure 13.8). This suggests that allometric relationships can be broken, and they are influenced by natural selection, with different forms being favored under different conditions. This argument applies the first type of test—adaptive prediction—to the case of allometry.

In conclusion, developmental constraints have been actively discussed since the beginnings of the modern synthesis in evolutionary biology. Not much is known about how embryology constrains mutation, but the general idea is plausible. In particular cases, the alternatives of selection and constraint can be tested between. The way an organism develops will undoubtedly influence the kinds of mutations that can arise in it. The interesting problems begin when we try to move from this general claim to an exact demonstration in a real case; the attempts so far, as in the example of allometry, have not been finally convincing.

13.10 Historical constraints may cause imperfect adaptation

Evolution by natural selection proceeds in small, local steps; each change has to be advantageous in the short term. As Wright emphasized in his shifting balance model (section 8.13, p. 205), natural selection may climb to a local optimum, where the population may be trapped because no small change is advantageous, though a large change could be. As we saw, selection itself (when considered in a fully multi-dimensional context) or neutral drift, may lead the population away from local peaks; but it also may not. Some natural

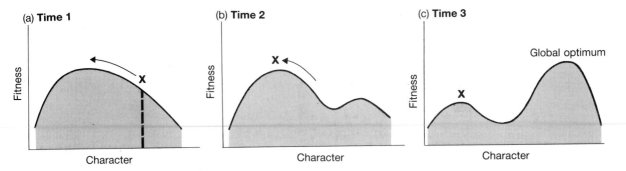

Figure 13.9 Historical change in adaptive topography leaves a species stranded on a local peak. (a) Initially, there is a single optimum state for a character, and the population (**x**) evolves to that peak. (b) As the environment changes through time, the adaptive topography changes. The species has now reached the optimum. (c) The topography has changed, and a new global peak has arisen. The species is stuck at the local peak, because evolution to the global peak would traverse a valley: natural selection does not favor evolution toward the global peak.

populations now may be imperfectly adapted because the accidents of history pointed their ancestors in what would later become the wrong direction (Figure 13.9).

The recurrent laryngeal nerve provides an amazing example. The laryngeal nerve is, anatomically, the fourth vagus nerve, one of the cranial nerves. These nerves first evolved in fish-like ancestors. As Figure 13.10a shows, successive branches of the vagus nerve pass, in fish, behind the successive arterial arches that run through the gills. Each nerve takes a direct route from the brain to the gills. During evolution, the gill arches have been transformed; the sixth gill arch has evolved in mammals into the ductus arteriosus, which is anatomically near to the heart. The recurrent laryngeal nerve still follows the route behind the (now highly modified) "gill arch": in a modern mammal, therefore, the nerve passes from the brain, down the neck, round the dorsal aorta, and back up to the larynx (Figure 13.10b). In humans, the detour looks absurd, but is only a distance of a foot or two. In modern camels, the nerve makes the same detour, but it passes all the way down and up the full length of the camel's neck. The detour is almost certainly unnecessary and probably imposes a cost on the camel (because it has to grow more nerve than necessary and signals sent down the nerve will take more time and energy). Ancestrally, the direct route for the nerve was to pass posterior to the aorta; but as the neck lengthened in the camel's evolutionary lineage, the nerve was led on a detour of increasing absurdity. If a mutant arose in which the nerve went directly from brain to larynx, it would probably be favored (though the mutation may be unlikely, if it would require a major embryologic reorganization); the imperfection persists because such a mutation has not arisen (or it arose and was lost by chance). The imperfection arose because natural selection operates in the short term, with each step taking place as a modification of what is already present. This process can easily lead to imperfections due to historical constraint—though most will not be as dramatic as the camel's recurrent laryngeal nerve.

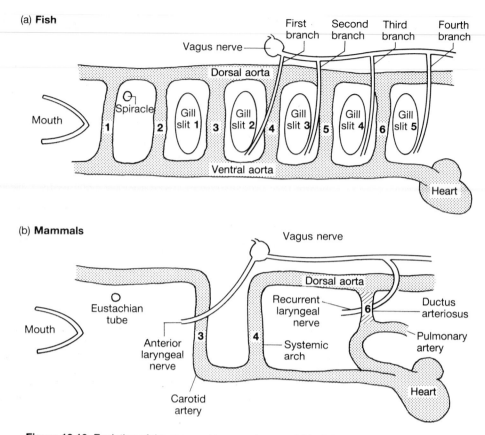

Figure 13.10 Evolution of the recurrent laryngeal nerve. (a) In fish, the vagus nerve sends direct branches between successive gill arches. (b) In mammals, the gill arches have evolved into a very different circulatory system. The descendant nerve of the fish's fourth vagus now passes from the brain, down to the heart (in the thorax) and back up to the larynx.

A similar historical contingency may produce not actual imperfection, but differences between populations or species that are not adaptively significant. In an adaptive topography with several adaptive peaks, there may be more than one of similar height. The camel's laryngeal nerve looks like a case in which a local peak is clearly lower than the global peak, and it is therefore recognizably an imperfect adaptation. If there were several peaks of similar height, one would not be recognizably inferior to the others. Imagine now that the ancestors of a number of different populations started out near different future peaks. If they then experienced the same external force of selection, each one would still evolve to its nearest peak. The different populations would then evolve different adaptations: but because of their different starting conditions, not due to adaptation to different environments (Figure 13.11).

Kangaroos and a placental herbivore, such as a gazelle, have different methods of locomotion, and Maynard Smith suggested they are a possible example. The two forms are ecologically analogous. Kangaroo hopping is no

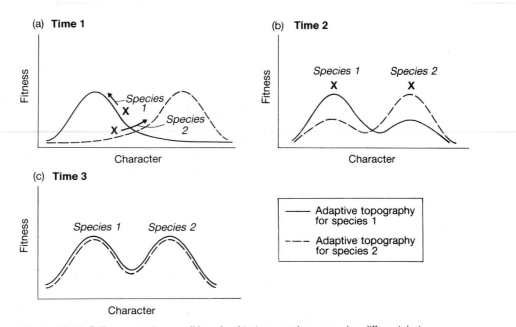

Figure 13.11 Different starting conditions lead to two species occupying different, but equivalent, adaptive peaks. (a) The adaptive topographies for two species differ, each evolves to its own peak. (b) The adaptive topographies now change, until (c) they become identical for the two species; but each species remains on its own peak. At stage (b) the species difference was adaptive: it was better for species 1 to be on its peak, and species 2 at its. By (c) the species difference is non-adaptive: either species would be equally well adapted on either peak.

better or worse a way of moving than running on four legs. The lineage leading to kangaroos improved one method of moving, while that leading to gazelles concentrated on the other. The difference is probably mainly a historical accident. If the argument is correct, the distant ancestors of kangaroos faced different selective conditions from those of gazelles; the adaptations fixed in those ancestors then influenced subsequent evolution such that now, even though kangaroos and gazelles occupy similar ecological niches, the mutations influencing locomotion that are favored in the two groups are completely different. The example illustrates a different idea from the camels' recurrent laryngeal nerve. Neither kangaroo nor gazelle is claimed to be imperfectly adapted; it is only the difference between the two that may be an historical accident. In the camel lineage, a similar kind of historical accident has generated actual imperfection in its laryngeal nerve. Whether historical accident leads to imperfection, or a neutral difference between lineages, depends on whether a global peak stays during evolution as a global peak or evolves into a local peak. In either case, past evolutionary events can lead to the establishment of forms that cannot be explained by a naïve application of the theory of natural selection. Adaptation has to be understood historically.

13.11 An organism's design may be a trade-off between different adaptive needs

Many organs are adapted to perform more than one function, and their adaptations for each are a compromise. If an organ is studied in isolation, as if it were an adaptation for only one of its functions, it may appear poorly designed. Consider how the mouth is used for feeding and breathing in different groups of tetrapods. In mammals, the nose and mouth are separated by a secondary palate, and the animal can chew and breathe at the same time. The earliest tetrapods, some modern reptiles, and all modern amphibians, lack a secondary palate and have only a limited ability to eat and breathe simultaneously. A boa constrictor, for example, has to stop breathing while it goes through the complex motions of swallowing its prey—a process that can take hours. The mouth of any such species that cannot breathe while it is feeding may, if it is judged only as an adaptation for feeding, appears inefficient compared with the mammalian system; the snake's mouth is a compromised adaptation for feeding. Of the reptilian groups, only crocodiles have a full secondary palate like mammals (it is presumably useful in crocodiles as it enables them to breathe air through the nose while the mouth is underwater), and reptilian feeding systems can be understood as compromised in varying degrees by the need to breathe.

Trade-offs do not only exist in organ systems. In behavior, an animal has to allocate its time between different activities, and the time allocated to foraging (for example) might be compromised by the need to spend time on other demands. Trade-offs exist over the whole lifetime too: an individual's life history of survival and reproduction from birth to death is a trade-off between reproduction early in life and reproduction later on. At any one time, an animal may appear to be producing less offspring than it could; but that does not mean it is poorly adapted, because it may be conserving its energies for extra reproduction later. In summary, the adaptations of organisms are a set of trade-offs between multiple functions, multiple activities, and the possibilities of the present and the future. If a character is viewed in isolation it will often seem poorly adapted; but the correct standard for assessing an adaptation is by its contribution to the organism's fitness in all the functions it is employed in, through the whole of the organism's life.

13.12 Conclusion: constraints on adaptation

In conclusion, we can draw two morals for the study of adaptation. The first is to emphasize the distinction between an adaptively insignificant difference between species and an imperfect adaptation within one species. The distinction corresponds to the two main ways of studying adaptation itself: evolutionary biologists are concerned with understanding both why different species have different adaptations and how adaptations function within each species. Let us see how the different kinds of imperfection could upset the methods of analyzing adaptation that we discussed in chapter 11.

The comparative method could be misled by cases of adaptively insignificant differences between species. If the different forms of the adaptation are selectively neutral, or are equivalent locally adaptive peaks that different species evolved by historical accident, then attempts to correlate the differences with ecologic circumstances should be unsuccessful. However, the fact that there can be several equivalently good forms of an adaptation does not, by itself, disturb the study of the character. The possibility of multiple adaptive forms should emerge from the analysis. If there is an optimal form of an enzyme, then it is no less an optimal form if 100 different amino acid sequences can realize it in practice. The problem for studies of particular adaptations within a species comes from the other source of imperfection. If the perfect form of the character has not arisen for reasons of history, embryology, or the genetic system, or because the environment has changed recently, then the character itself will be imperfectly adapted. If we try to predict the form of the character by an analysis of optimal adaptation, the prediction will be wrong.

What should the investigator do when a prediction turns out wrong? There is a danger that an analysis purely in terms of adaptation will produce spurious results. Any particular character could have evolved as an adaptation for any of a large number of reasons. Body size, for example, may be adaptive for thermoregulation, storing food, subduing prey, fighting other members of the same species, or other factors. If we assume that body size is an adaptation, we should then begin by picking on one factor, building a model, and seeing whether it predicts body size correctly. If it fails, we could move on to another factor: if, for instance, we started with a thermoregulatory model and it failed, we could move on to a diet model . . . and so on. This method, however, if carried far enough, will almost inevitably find a factor that ''predicts'' body size correctly. Eventually, by chance, a relationship will be found, if enough other factors are studied. One will be found even if body size is a neutral character.

The solution to the problem can be stated in a conceptually valid, but not always practically useful, form. The solution is the chapter's second moral. The methods of studying adaptation work well *if we are studying an adaptation*. If the character under study is an adaptation then it must exist because of natural selection and it is correct to persist in looking for the reason why it is favored. If body size is an adaptation, there will be an adaptive model for it that is correct. However, if the character (or different forms of it) is not favored by natural selection, the method breaks down. Methods of studying adaptation should therefore be confined to characters that are adaptive. Probably, in practice, they mainly are. Adaptation can be a self-evident property of nature, and it would be absurd to claim that no properties of living things are adaptive. While ''obvious'' adaptations are studied, the subject should be philosophically non-controversial.

However, there is still plenty of room for controversy. Biologists differ widely in how prevalent, and how perfect, they believe adaptation to be in nature. Some biologists believe that natural selection has fine-tuned the details, and established the main forms, of organic diversity. Others think that the main forms may be historical accidents and the fine details due to neutral drift. Not surprisingly, the evolutionary biologists who study adaptation tend to be among the former and those who criticize it among the latter. But this difference of opinion is not about the fundamental coherence of the methods; it is about the range of their application. History has—so far—been more on the side of the adaptationists. One character after another that had been written off as non-adaptive, has turned out on proper analysis to be controlled by natural selection. Allometry is one case in point; snail shell banding is another (section 11.1, p. 267). But the evidence of history, though it may encourage the study of adaptation, certainly does not prove that it is right to assume that all the characters of all organisms are adaptations.

13.13 How can we recognize adaptations?

The study of adaptation could be made foolproof if we had a criterion to distinguish in advance between adaptive and non-adaptive characters. Criteria for recognizing adaptations do exist. Adaptations can be recognized as characters that appear too well fitted to their environments for the fit to have arisen by chance; they are characters that help their bearers to survive and reproduce; they are purposive and often complex; they are the sorts of characters that before Darwin would have suggested the existence of God. No one will doubt that some adaptations, like the vertebrate eye, fit this definition. But there can be a problem with doubtful borderline cases. Small differences in a character in different species or individuals can be particularly difficult. There is little doubt that the vertebrate brain is an adaptation; but it may make no adaptive difference whether an individual has a brain of 250 or 300 cm^2. The criteria just listed will not tell us whether we should assume that such a difference is adaptive.

In theory, adaptation is a clear and objective concept. If a character is an adaptation, then natural selection will work against mutant alternatives. This criterion is theoretically objective because a mutant either will or will not spread; but its use is mainly theoretical. Only rarely can we study the fate of mutants. The criterion can sometimes be applied in practice by measuring the fitness of variants of a character. If variation in a character is significantly related to an organism's survival and well-being, natural selection must automatically be operating on it; the favored forms of the character are adaptations. When the method can be used, it is unambiguous; but it is not always possible to use it. It takes a lot of work to measure reproductive success in a character that does vary. Moreover, some characters do not vary in an easily measurable way. The vertebrate eye is undoubtedly an adaptation; but nobody has ever correlated variation in its optical properties with survival and reproductive success. A third problem is that a character could still be adaptive even if its relationship with reproductive success was statistically undetectable. Natural selection can theoretically work on a character over millions of years and produce major changes through selection coefficients of 0.001 or less; it would be practically impossible to detect this amount of selection in a modern population with the normal resources of an evolutionary biologist. Forces that are important in evolution can in some cases be impossible to study directly because they are so small. A direct measurement of reproductive success is most likely to demonstrate that a character is adaptive if the selection coefficient is large; but these will tend to be the obvious characters in any case. The method will be less useful for characters whose adaptive status is controversial.

If no detailed study of the relationship between a character and fitness has been performed, or if one is not possible, then we have to look at the character itself. The eye is a clearly adaptive: the criteria (beneficial, purposive, etc.) reveal this, without any need to measure fitness. But the criteria are not foolproof, because of the borderline cases noted above. The ambiguity in these borderline cases is not just a practical problem; it is inherent in the concept of adaptation itself, because natural selection can favor simple as well as complex characters—and simple characters can also arise by chance. We might make an analogy with the uncertainty in the definition of "design" in human fabrications. If we were to travel round the world and guess which objects were brought about by human design, we should see many obvious cases, such as architecture and engineered objects; and many non-obvious cases, such as heaps of earth. Earth could have been heaped up for a special purpose, such as for a burial mound, or it could have just accumulated there by natural accident. We should not expect the distinction between designed and non-designed entities to be clear in either the case of natural adaptation or of human fabrications. Heaps of earth can be brought about by natural landslides or by human agency. We cannot always tell which cause operated just by looking at the result. There is an objective distinction between the two causes, but it is historical: either the heaps of earth were constructed by human agency or they were not. But the history is unobservable, and when we have to make the distinction purely using modern observable evidence, there will be difficult borderline cases. Likewise, body coloration may be a simple adaptation, brought about by natural selection; or it may be non-adaptive and brought about by chance, as may be the case for the red color of the sediment-dwelling worm *Tubifex* (visual factors are not important in the sediment at the bottom of the water column). Again, either natural selection is favoring the body coloration or it is not; but if we try to decide whether it is just from looking at the character, the answer may not be clear. The clear theoretical meaning of adaptation implies that it cannot have a universal, foolproof, practical definition.

13.14 Summary

1 Three theories have been put forward to explain the existence of adaptation: supernatural creation, Lamarckism, and natural selection. Only natural selection works as a scientific theory.

2 Natural selection is not the only process that causes evolution, but is the only process causing adaptation.

3 Natural selection, at least in principle, can explain all known adaptations. Examples of coadaptation and useless incipient stages have been suggested but they can be reconciled with the theory of natural selection.

4 Adaptation can either be defined historically or by current function: we can say either that an adaptation is a character that evolved by natural selection to perform its modern function, or one that evolved by natural selection whether or not its modern function is the same as the one it first evolved to perform.

5 Not all the effects of an organ will have evolved as adaptations by natural selection. Some will be inevitable consequences of the laws of physics.

6 Adaptations cannot be simultaneously optimal for all the levels of organization in life. What is optimal for the organism may not be optimal for its population.

7 Adaptations may be imperfect because of time lags: a species may be adapted to its past environment, because it takes time for natural selection to operate.

8 Adaptations are imperfect because the mutations that would enable perfect adaptation have not arisen. The imperfections of living things are due to genetic, developmental, and historical constraints, and to trade-offs between competing demands.

9 For particular characters, adaptation and constraint can be alternative explanations. Likewise, differences in the form of a character between species may be due to adaptation to different conditions or to constraint. Forms that are not found in nature may be absent because they are selected against or because a constraint renders them impossible.

10 Adaptation and constraint can be tested between by several methods: by the use of predictions from a hypothesis of adaptation or constraint, by direct measurement of selection, by seeing whether the character is variable, and whether the variation is heritable, and by examining comparative trends.

11 The methods of analyzing adaptation are valid when applied to adaptive characters and interspecific trends; they might be misleading for non-adaptive characters and trends.

12 Biologists disagree about how exact, and how widespread, adaptation is in nature.

13 There are criteria to distinguish adaptive from non-adaptive characters. Measurement of selection provides an objective criterion, but is not always practical; other criteria, such as non-randomness and purposiveness are often useful, but may become subjective in borderline cases.

13.15 Further reading

Williams (1966) is a classic work on adaptation. Gould and Lewontin (1979) is an influential paper that criticizes the way adaptation has often been studied; Cain (1964) vigorously argues the opposite. Dawkins (1982, 1986) argues that only natural selection can explain adaptation. See Simpson (1944, 1947, 1953) on orthogenesis. Frazzetta (1975) and Wake and Roth (1989) discuss complex adaptations. See Ridley (1982) on the controversy about coadaptation. On preadaptation, see also the popular essay by Gould (1977b, chapter 12). Gould and Vrba (1982) is the reference for exaptation; Pinker and Bloom (1990) interestingly discuss whether human language is exaptive or adaptive.

Dawkins (1982, chapter 3), Oster and Wilson (1978, chapter 8), Krebs and Davies (1987, chapter 15) introduce constraints on perfection; see also Alexander (1985), Maynard Smith (1978b), Maynard Smith and Vida (1990) and Williams (1992).

On fruits, see Janzen and Martin (1982) in particular, and Janzen (1983) in general. Diamond (1990b) discusses some other ghostly adaptations. The remarkable genetic constraint in the crested newt is described in Macgregor and Horner (1980), Sims *et al.* (1984), and Horner and Macgregor (1985). On developmental constraint, Maynard Smith *et al.* (1985) is the major review. For the example of canalizing selection, see Clarke and McKenzie (1987) and Jones (1987); on developmental asymmetry see also Leary and Allendorf (1989) and Hoffmann and Parsons (1991). Chapter 9 has further references for canalizing selection, and chapter 21 for studies like Raup's on snails. For recent work on allometry, taking various theoretical positions, see Gould (1966), Lewontin (1980), Clutton-Brock and Harvey (1984), Harvey and Krebs (1990) and Harvey and Pagel (1991).

Part 4
Evolution and Diversity

Evolution and Diversity

Darwin closed *On The Origin of Species* with the following words.

> There is grandeur in this view of life, with its several powers, having been originally breathed into a few forms or into one; and that, whilst this planet has gone cycling on according to the fixed law of gravity, from so simple a beginning endless forms most beautiful and most wonderful have been, and are being, evolved.

Part 4 of this book is about how the theory of evolution can be used to understand the diversity of life or, in Darwin's words, the "endless forms most beautiful." We begin with the principles of classification. Classification might seem a dry subject, but the relationship between classification and evolution raises deep theoretical issues. Chapter 14 considers three main schools of classification and argues that classification is most objective when it represents only the branching relationships (or phylogeny) of species and ignores how similar the species look to one another. We then turn in chapter 15 to the question of what a species is. In evolutionary biology, species can be understood as gene pools—sets of interbreeding organisms—and these are important units because, in the theory of population genetics, natural selection adjusts the frequency of genes in gene pools. The millions of species now inhabiting this planet have, as Darwin said, a common ancestor, and the multiplication in the number of species has been generated as single species have split into two. Speciation (chapter 16) requires special circumstances, and we shall consider both the geographic and genetic conditions in which it takes place. Chapter 17 describes how the phylogenetic relationships of species, and higher taxa, can be reconstructed. The history of species cannot be simply observed, and phylogenetic relationships have to be reconstructed from clues in the molecules, chromosomes, and morphology of modern species (and in the morphology alone of fossils). Phylogenetic reconstruction is a crucial part of classification if it aims to represent phylogenetic relationships, and chapters 14 and 17 therefore depend on each other. Finally, the theory of speciation, as well as classification and phylogenetic reconstruction are all needed in evolutionary biogeography (chapter 18)—the use of evolutionary theory to understand the geographic distribution of species.

Classification and evolution

14.1 Biologists classify species into a hierarchy of groups

Biologists have so far described about one and a half million species of living plants and animals, and perhaps a further quarter of a million extinct fossil species. Describing a species is a formalized activity, in which the taxonomist has to compare specimens from the new species and other, similar species, and then explain how the new species can be distinguished; the description also has to be published. Describing species is the most important task of taxonomists, but it has no particular connexion with evolutionary biology. The evolutionary interest of classification begins at the next stage. Biologists do not think of their million and a half described species simply as a long list, beginning with the aardvark, working through buttercup, honeybee, and starfish, to end with zebra. Since Linnaeus, species have been arranged in a hierarchy (Figure 14.1). Species are grouped in genera; the gray wolf species *Canis lupus* and the golden jackal *Canis aureus*, for example, are grouped in the genus *Canis*: genera are grouped into families; the wolverine genus combines with several other genera, such as the fox genus *Vulpes*, to make up the family Canidae: several families combine to make up an order (Carnivora, in this example), orders to make a class (Mammalia), classes a phylum (Chordata), and phyla a kingdom (Animalia).

Each species, therefore, is a member of a genus, a family, an order, and so on. Now, if we take one and a half million species and seek to arrange them into a classification, there are a large number of ways in which the arrangement could be made. A classification does not even have to be hierarchic: chemists, for example, classify elements by the periodic table, which is not hierarchic. Why biological classification should be hierarchic is an interesting question in itself, and we shall consider it later in this chapter; but to begin with we shall accept that the classification is hierarchical and ask what the exact form of the hierarchy is.

14.2 There are phenetic and phylogenetic principles of classification

In biology, two main methods are used to classify species into groups: the *phenetic* and the *phylogenetic* methods. (Some would prefer to say phenotypic rather than phenetic throughout this chapter.) In the phenetic method, species are grouped according to their phenetic attributes. Species that are physically similar are put together in a group, species that are physically more different are put in different groups. Almost any physical attributes can be used. Fossil vertebrates can be classified phenetically by the shape of their bones; modern species of fruitfly by the pattern of their wing venation. Species can be grouped according to the number, shape, or banding pattern of chromosomes, by the immunologic similarity of their proteins, or by any

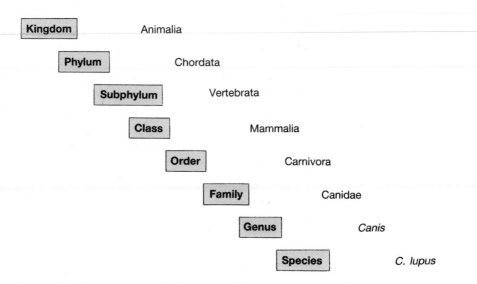

Figure 14.1 Each species in a biological classification is a member of a group at each of a succession of more inclusive hierarchic levels. Shown here is a fairly complete classification of the gray wolf *Canis lupus*. This way of classifying living things was invented by the eighteenth century Swedish biologist who wrote under the latinized name Carolus Linnaeus.

other measurable phenotypic property.

The characters used in classification are found by dividing the organism up into observable units, like bones, wing veins, and blood proteins. The division is arbitrary in the sense that if we look at an organism in different ways, or at different developmental stages, apparently distinct characters may merge or blur together. Pheneticists therefore aim to classify species according to a large number of characters; any accidents due to particular arbitrarily chosen characters should then be diluted out.

As an example, consider the phenetic relationships of a rhinoceros, a butterfly, and a beetle. The greater apparent similarity of a butterfly to a beetle than to a rhinoceros is due to many characters. A pheneticist might measure, for instance, body size, wing attributes, number of legs, physical composition of body parts, as well as such present/absent characters as whether or not the organism has wings; a full phenetic study might measure perhaps a hundred such characters in all. The hundred separate measures next have to be combined into an aggregate measure of similarity. There are, as we shall see, various ways in which this can be done; but Table 14.1 illustrates the principle by the measure called "mean character distance," in which the 100 measures are simply added up and averaged. (The aggregate measure of similarity is sometimes called *overall morphologic similarity*.) Phenetic similarity can be used to classify living things at all hierarchic stages. At the lowest taxonomic level, the more similar organisms are grouped together as conspecifics; the average similarity of the species is then used to group the species into genera, and then the genera into families, and so on. (An exact criterion is needed for the degree of similarity needed to define a species, a genus, and so on. One can easily be devised by working

Table 14.1 The phenetic similarity between two organisms, or between two groups of organisms, can be measured by averaging their similarities with respect to many characters. The table shows how aggregate (overall) morphologic similarity could be estimated for six properties of a beetle and a rhinoceros. This example uses the method of *mean character distance*, which is not the only method of measuring similarity; Figure 14.5 illustrates another method. It is conventional to use the values 0 and 1 for absence and presence of discrete characters. (Note the numbers are illustrate only—there are of course many different types of beetle)

Character	Rhinoceros	Beetle	Difference
Body length (cm)	400	2	398
Presence of wings	0	1	−1
Number of legs	4	6	−2
Chitin in surface (%)	0	90	−10
Presence of tusk	1	0	1
Annual fecundity	1	500	−499
Mean character distance			−18.833

from established classificatory groups. The criterion is then that a genus is a group of species showing the amount of overall morphologic similarity typical of some other, pre-established "reference" genera.)

Notice that nothing has been said about evolution. Phenetic classification is non-evolutionary. The same procedure could be applied to any set of objects, non-living or living: biological species, languages, furniture, clouds, and songs can all be classified according to their patterns of phenetic similarity.

The phylogenetic principle, however, is evolutionary. Only entities that have evolutionary relationships can be classified phylogenetically: the clouds in the sky, for instance, cannot be classified phylogenetically (unless any of them were formed by the division of ancestral clouds). The phylogenetic principle classifies species according to how recently they share a common ancestor. So, two species that share a recent common ancestor will be put in the same genus, and two species sharing a more distant common ancestor might be put in different genera, but in the same family. As the common ancestor of two species becomes more and more distant, they are grouped further and further apart in the classification: in the end, all species are contained in the largest phylogenetic category—the set of all living things—which contains all the descendants of the most distant common ancestor of life.

In most real cases in biology, the phylogenetic and phenetic principles give the same result. If we consider how to classify a butterfly, a beetle, and a rhinoceros, the butterfly and beetle are more closely related both phenetically and phylogenetically (Figure 14.2a). In other cases the principles can disagree. Adult barnacles superficially look rather like limpets and if we were to classify an adult barnacle, limpet, and lobster phenetically we might well put the barnacle and limpet together even though the lobster and barnacle share a more recent common ancestor and are grouped together phylo-genetically (Figure 14.2b). Groups like reptiles can also be given different phenetic and phylogenetic classifications. Some descendants (such as birds) of the common ancestor of the group have evolved rapidly and left a rump of

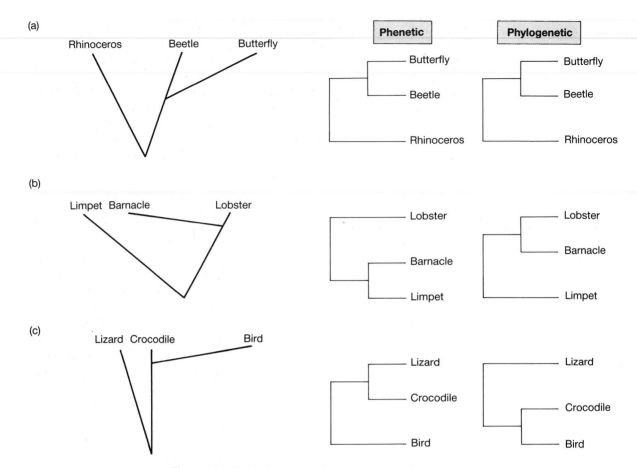

Figure 14.2 The phenetic and phylogenetic principles of classification may (a) agree, or (b and c) disagree.

quite distantly related groups that resemble one another phenetically (Figure 14.2c).

We shall be more concerned in this chapter with the question of how to group species into hierarchies than with the formalities of how to name the hierarchic levels. Figure 14.2b says nothing about whether the group of barnacle, lobster, and limpet should be a class or a phylum, and the barnacle and lobster an order or a class, or whatever. The naming of levels is largely a practical matter, which evolutionary theory has nothing to say about; but the prior question—of which classification (Figure 14.2a or 14.2b) is correct—is a matter of principle. Evolutionary theory does have something to say about that.

14.3 There are phenetic, cladistic, and evolutionary schools of classification

The phenetic and phylogenetic principles are the two fundamental types of biological classification, but there are more than two schools of thought about how classification should be carried out. We can discuss three main schools. The most influential modern school of phenetic classification is the *numerical taxonomy* that has been particularly defended by Sneath and Sokal; the terms "phenetics," "numerical phenetics," and "numerical taxonomy" are used almost interchangeably in modern biology. Phylogenetic classification has been defended by the German entomologist Hennig and his followers; Hennig called it *phylogenetic systematics*, but *cladism* (from the Greek κλαδoς for a branch) is now the more common term. The third school to be discussed uses a synthesis, or mixture, of phenetic and phylogenetic methods and is often called *evolutionary taxonomy*. In the reptilian example (Figure 14.2c) evolutionary taxonomy prefers the phenetic classification; in the barnacle example (Figure 14.2b), the phylogenetic. This school's best known advocates include Mayr, Simpson, and Dobzhansky.

14.4 A method is needed to judge the merit of a school of classification

How should we decide which school of classification, if any, is the best? To do so, we need a criterion to judge them against, and we shall use the *objectivity* criterion for this purpose. An objective classification is one that represents a real, unambiguous property of nature; it is to be contrasted with subjective classification, in which the classification represents some property arbitrarily chosen by the taxonomist. I might arbitrarily choose, for instance, to classify species into one group if I discovered them on a Monday or Tuesday and another group if I discovered them between Wednesday and Friday. The classification would then be subjective because I should have no method of justifying the choice—except my personal whim or convenience. If challenged about why I did not instead have one group for days beginning with the letter T and another group for all other days, I should have no principled argument to confound the critics. The underlying classificatory principle—time of discovery—is ambiguous because it can be applied in a number of conflicting ways; and it is unreal because it does not correspond to any natural relationship of the organisms being classified. The same problem, more interestingly, also arises if I chose to classify birds by the lengths of their beaks (or any other arbitrarily picked character). Again, it would be perfectly possible to do so; but without an external reason for the choice it is subjective. The objectivity test, therefore, is to ask whether a classificatory system has some compelling justification, external to the method it uses, for classifying in the way that it does.

A second distinction is between *natural* and *artificial classification*. To understand the difference, we must further distinguish between the characters of the organisms used to construct a classification and the characters that are not. Then, in abstract terms, a natural classification is one in which the members of a group resemble one another not only in the characters that define the group (as they must, by definition) but also for many other non-defining characters too. An artificial classification is one in which the members of a group only resemble each other in the defining characters; they show no similarities for non-defining characters (Figure 14.3 gives an imaginary example).

Objective, and natural, classifications have closely related advantages over subjective and artificial ones. The advantage of objective classification is that different, rational people, working independently, should be able to agree that it is the way to classify. The results should then be relatively stable and repeatable. If a classification is only valid for the few characters that define it

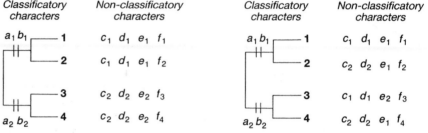

(a) **Natural classification**

Species   Characters and character states

Species	Characters and character states
1	a_1 b_1 c_1 d_1 e_1 f_1
2	a_1 b_1 c_1 d_1 e_1 f_1
3	a_2 b_2 c_2 d_2 e_2 f_2
4	a_2 b_2 c_2 d_2 e_2 f_2

(b) **Artificial classification**

Species	Characters and character states
1	a_1 b_1 c_1 d_1 e_1 f_1
2	a_1 b_1 c_2 d_2 e_1 f_2
3	a_2 b_2 c_1 d_1 e_2 f_3
4	a_2 b_2 c_2 d_2 e_1 f_4

Classificatory characters — *Non-classificatory characters*

$a_1 b_1$ ── 1 c_1 d_1 e_1 f_1
─── 2 c_1 d_1 e_1 f_2

─── 3 c_2 d_2 e_2 f_3
$a_2 b_2$ ── 4 c_2 d_2 e_2 f_4

Classificatory characters — *Non-classificatory characters*

$a_1 b_1$ ── 1 c_1 d_1 e_1 f_1
─── 2 c_2 d_2 e_1 f_2

─── 3 c_1 d_1 e_2 f_3
$a_2 b_2$ ── 4 c_2 d_2 e_1 f_4

Figure 14.3 Natural and artificial classifications differ according to whether the members of a group tend to share characters that have not actually been used to define the group. (a) Natural classification. Suppose a group of species 1 and 2 have been defined by characters a_1 and b_1, and a group of species 3 and 4 by characters a_2 and b_2. The members of the groups share similar states in the other characters. (b) Artificial classification. The groups made up of species 1 and 2 and of species 3 and 4 are again defined by characters a_1 and b_1 and a_2 and b_2 respectively; but in this case the members of a group do not share the same states for other characters.

(i.e. it is artificial), then when other taxonomists come along with other characters, they will want to alter it; if the classification is natural, it is less likely to need to be changed. Similarly, if several taxonomists working independently on the same group of species all strive for a natural classification they are more likely to come up with similar results than if they all are content with artificial classification.

Now let us see how well the three classificatory schools meet the criteria of objectivity and naturalness. To do so, we shall have to look in more detail at how the three schools actually operate.

14.5 Phenetic classification uses distance measures and cluster statistics

The modern forms of phenetic classification are numeric and multivariate, and they were developed in reaction to the uncertainties and imprecision of evolutionary classification. Evolutionary classification, whether the pure cladistic kind or the mixed evolutionary taxonomy of Mayr and others, requires a knowledge of phylogeny. We shall discuss in detail in chapter 17 how phylogenies can be inferred. Here, all we need to know is that, although the phylogenetic relationships between species can often be inferred, the inferences can sometimes be highly uncertain. For many groups of living things, hardly anything is known about phylogeny, and an evolutionary classification of such a group will inevitably be poorly supported by evidence. Numerical phenetics aimed to avoid all the evolutionary uncertainty by classifying only by phenetic relationships, and by using quantitative techniques to measure them. The classification would follow automatically, and therefore (it was thought) objectively, from the phenetic measurements. Let us consider the methods in some more detail, and see how well these aims can be achieved.

Numerical phenetic classifications are based not on one or two characters, but on whole suites of characters aggregated into a single measure. The reason is that different individual characters show different distributions among species and therefore tend to produce different classifications. Consider the birds, and some reptilian groups, such as crocodiles, lizards, and turtles (Figure 14.4). Crocodiles are more similar to lizards and turtles

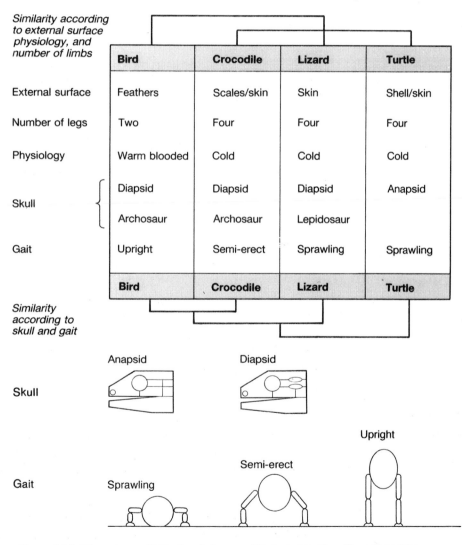

Similarity according to external surface physiology, and number of limbs

	Bird	**Crocodile**	**Lizard**	**Turtle**
External surface	Feathers	Scales/skin	Skin	Shell/skin
Number of legs	Two	Four	Four	Four
Physiology	Warm blooded	Cold	Cold	Cold
Skull	Diapsid	Diapsid	Diapsid	Anapsid
	Archosaur	Archosaur	Lepidosaur	
Gait	Upright	Semi-erect	Sprawling	Sprawling
	Bird	**Crocodile**	**Lizard**	**Turtle**

Similarity according to skull and gait

Skull

Anapsid   Diapsid

Gait

Sprawling   Semi-erect   Upright

Figure 14.4 Character conflict in the phylogeny of birds and reptiles. The gait and the anatomy of the skull link crocodiles and birds; leg number, physiology, and external surface link the reptilian groups. The anapsid skull has no openings, apart from the eye socket; the key feature of the diapsid skull is a single upper temporal opening, though most diapsids have an additional lower opening too. Archosaurs and lepidosaurs differ in their skulls (to be exact, lepidosaurs lack a lower temporal arch).

than to birds if we look at their external surfaces, number of legs, and cold-blooded physiology; but crocodiles and birds have anatomically more similar skulls than either have to lizards and turtles. The different types of character conflict. This is a common problem. If we classify using one sample of characters, we produce one classification; if we use another sample, we produce another. The pheneticists' solution is to aggregate all the characters into one grand measure of similarity; the idiosyncracies of particular samples should then disappear. The resulting classification groups the units according to their whole phenotype.

Statistically, it is easy to average over many characters. We saw above how it could be done for a beetle and a rhinoceros (Table 14.1). The same procedure can be expressed in a graph. We shall start with the simple case of two characters; the extension to further dimensions is easy. Suppose then that we wish to classify a group of fly species, and we have measured two characters, such as the length of a certain wing vein, and the length of the tibia of the hind leg. The average for each species can be represented as a point (Figure 14.5a). For any pair of species, the average difference in their

(a) **Phenetic measurements for five species**

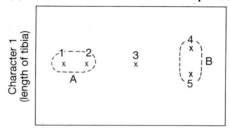

(b) **Nearest neighbor**   (c) **Average neighbor**

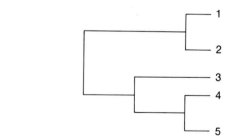

Figure 14.5 (a) The phenetic similarity between species can be expressed graphically. Suppose five species have been measured for two characters, length of a wing vein and length of tibia. The x-axis is the measurement of each species for length of a wing vein, the y-axis for length of tibia. The distance between two species on the graph is the phenetic distance between them. (Notice that the distance on the graph is different from the measure of mean character distance in Table 14.1.) (b) The phenetic classification by the single nearest neighbor technique puts species 3 with the group (cluster A) which has the nearest individual neighboring species (species 2). (c) But if it uses the nearest average neighbor technique it puts species 3 with the group (cluster B) that has the nearest average for all its species. Species 4 and 5 have a nearer average distance. (The averages are simply the averages of the distances of species 1 and 2, and of species 4 and 5, to species 3.)

wing vein lengths is the distance between their points on the x-axis, and the difference between the lengths of their tibiae is the distance on the y-axis: if we used either character by itself, the classifications would differ; species 1 and 3 for instance have identical tibial, but different wing vein, lengths. The aggregate difference for the two characters can be measured simply by the *distance* between the two species in the two-dimensional space. The species are then classified by putting each with the species, or group of species, that it is the shortest distance from.

If we measured a third character, such as pulse interval between the sounds in the courtship song, it could be drawn as a third dimension into the paper; now each species would be represented by a point in the three-dimensional space; the aggregate distance between the species could be measured as before by the distance between the species' points. We could likewise measure dozens of characters and measure the distance between the species by the appropriate line through hyperspace. Numerical taxonomists recommend measuring as many characters as possible—even hundreds—and classifying according to the aggregate similarity for all of them. The more characters that are measured, the more likely it is that peculiar individual characters will be averaged out, and the better founded and more natural the classification will be.

Are phenetic classifications artificial or natural? With numerical taxonomy, the school has pursued natural classification so far that it almost explodes the distinction. In the extreme case of a classification based on all possible characters, the classification would be both completely natural and completely artificial: the distinction referred to the difference between defining and non-defining characters, and as more and more characters are used to define the groups the number of non-defining characters decreases. However, for characters that were not used in the original classification, we have no reason to suppose they will fit in the same groups. A numerical phenetic classification is therefore natural in the sense that it is true of a very large number of characters; it must be, because all the characters it is true for were used to define it. But it may well be artificial for other, yet-to-be studied, characters.

Is a numerical phenetic classification objective or subjective? Objective classifications, remember, must represent some unambiguous property of nature. The phenetic classification itself represents the measure of aggregate morphologic similarity for large numbers of characters. The question, then, is whether there is some property of nature, some hierarchy of real phenetic similarity, that the measures of aggregate morphologic similarity may reasonably be said to be representing.

We can start by looking more closely at the statistical methods used in numerical taxonomy. In Figure 14.5 there were five species with two characters. To form Figure 14.5b, we group each species with its phenetically nearest neighbor. Two clusters—of species 1 and 2, and 4 and 5—immediately form. But to which of these clusters should we join species 3? The nearest species is 2. If (Figure 14.5b) we join species 3 to the cluster with the nearest single nearest neighbor we put it with cluster A (the nearest neighbor to species 3 in cluster A is species 2, whereas in cluster B it is species 4 and 5 equally). However, if we had calculated the average distance of each cluster as a whole, the answer is the opposite (Figure 14.5c); it follows from the geometry of Figure 14.5 that cluster B rather than cluster A has the nearer average neighbor to species 3.

The "nearest neighbor" and "average neighbor" are both examples of cluster statistics. They are not the only ones, but they are enough to make a point of principle. We have here, within the phenetic philosophy, managed to produce two different classifications. If the numerical phenetic claim to repeatability and stability is to be upheld, it must have some way of deciding which of the two is the correct phenetic classification. To do so, it would need some higher criterion to fall back on. The problem is it does not have one. The higher criterion would presumably be *the* hierarchy of aggregate morphologic similarity, but that hierarchy does not exist in nature independently of the statistics that measure it. And—as Figure 14.5 shows—different statistics produce different hierarchies.

So there is an essential degree of subjectivity in the phenetic philosophy. If its classifications are to be consistent, it must pick on one statistic, such as the average neighbor statistic, and stick to it. Classification would then be repeatable, but at a price. The consistency does not follow from the phenetic system itself; it is imposed by the taxonomist—subjectively. In practice, numerical taxonomists have never been able to agree on which statistic to use, and this is one reason why the school has lost much of its influence since its origin in the early 1960s.

Moreover, the choice of cluster statistic is not the pheneticists' only subjective decision. The measure of distance poses an analogous problem. We have seen two measures. The one used in Figure 14.5 is *Euclidean distance*: the straight line between two points; in two dimensions it is measured by Pythagoras' theorem. The other (used in Table 14.1) is *mean character distance* (MCD); this is the average distance between the groups for all characters measured. Thus, in two dimensions, if species 1 and 2 differ by x units in character A and y units in character B, then MCD = $(x + y)/2$ and Euclidean distance = $\sqrt{x^2 + y^2}$. Again, the two distance measures (and these are not the only two) can give different hierarchies and the pheneticist is faced with a subjective choice between them.

Phenetic classification, therefore, even in its modern numerical form, is not objective. It can produce classifications, but classifications that lack a deep philosophic justification. Let us see how the introduction of evolution into classification can help with the problem. We shall start with purely phylogenetic cladistic classification and then examine the synthetic evolutionary school.

14.6 Phylogenetic classification uses inferred phylogenetic relationships

14.6.1 Hennig's cladism classifies species by their phylogenetic branching relationships

Phylogenetic classifications group species solely according to recency of common ancestry. When a species splits during evolution it will usually form two descendant species, called *sister species*, and in a cladistic classification sister species are classified together. The branching hierarchy of ancestral relations is a unique hierarchy, extending back to the beginning, and

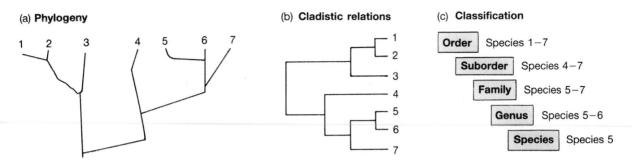

Figure 14.6 There is a simple relationship between the phylogenetic (cladistic) classification of a group of species, and their phylogenetic tree. (a) The phylogenetic tree of seven species. (b) Their cladistic classification. (c) The formal Linnaean classification, for species five as an example.

including all, of life. The phylogenetic hierarchy is easily converted into a classification (Figure 14.6).

The advantage of the phylogenetic system should be apparent. The phylogenetic hierarchy exists independently of the methods we use to discover it, and it is unique and unambiguous in form. When different techniques for inferring phylogenetic relationships disagree, there is always the external reference point to appeal to. When we cannot work out the phylogeny of some group or other, we do at least know that a solution exists to aim for. With the phenetic system there is no such external solution. There is no single natural phenetic hierarchy analogous to the phylogenetic hierarchy.

When a pair of species, like 5 and 6 in Figure 14.6b, are classified together cladistically, that means they share a more recent common ancestor than either does with any other species. Cladistic relationships are fundamentally ancestral relationships. In practice, the inference of ancestral relationships (i.e. the phylogeny in Figure 14.6a) can be difficult. Here we only need to know that the inferences can be made, and the inferred phylogeny can then be the starting point for phylogenetic classification. We must, however, refer forward to make one important point (chapter 17 discusses it in detail). The main evidence for phylogenetic relationships comes from a particular kind of character called shared derived characters. If we take all the characters shared between species, they can be divided into three types (Figure 14.7 and Table 14.2). A first division is into *analogies* and *homologies*: a homology is a character shared between species that was also present in their common ancestor; an analogy is a convergent character—one that is shared between species but that was not present in their common ancestor. (Note that this meaning differs from the pre-Darwinian meaning in chapter 3.) Homologies in turn are divided into shared derived homologies and shared ancestral homologies: a derived homology is one that is unique to a particular group of species (and their ancestor) and a shared ancestral homology is one that is found in the ancestor of a group of species and some, but not necessarily all, of its descendants. Now, of these three kinds of shared character, it is only the shared derived homologies that indicate phylogenetic groups: analogies

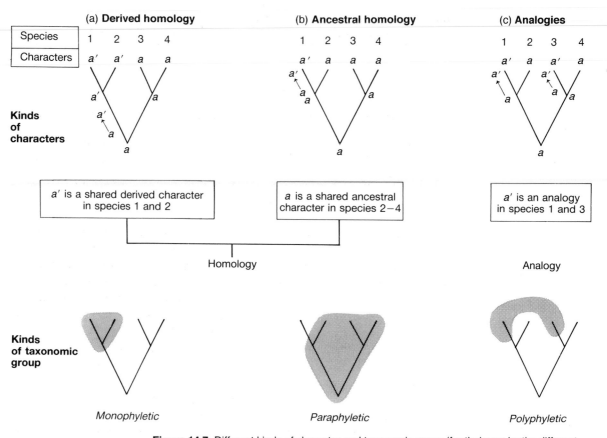

Figure 14.7 Different kinds of character and taxonomic group (for their use by the different schools of classification see Table 14.2). Homologies are characters shared between species that were present in the common ancestor. They can be derived or ancestral. (a) Shared derived homologies are found in all the descendants of the common ancestor, and are distributed in monophyletic groups. (b) Shared ancestral homologies are found in some but not all the descendants of the common ancestor, and are distributed in paraphyletic groups. (c) Analogies (convergent characters) are characters shared between species that were not present in the common ancestor; analogies fall into polyphyletic groups.

indicate convergent groups and homologies indicate not only phylogenetic groups but a special kind of non-phylogenetic group too. Here, the importance of this distinction between character types is to reveal the relationship between cladistic and phenetic classification. Numerical phenetic classification groups species using as many characters as possible, and averaging over them regardless of their evolutionary meaning. Cladists select, from all the characters shared between species, one special class of characters—the shared derived characters—and use only these to group the species; all other kinds of character are ignored in cladistic classification.

Are cladistic classifications natural? In a natural classification, the members of a group resemble one another for non-defining as well as defining characters. To answer the question, we should distinguish shared derived characters from the other kinds of character (i.e. shared ancestral and convergent characters). The cladistic classification should be natural for all

Table 14.2 Phenetic, evolutionary, and cladistic classification can be distinguished by the characters they use to define groups, and the kinds of groups they recognize

| | Groups recognized | | | Characters used | | |
| | | | | | Homologies | |
Classification	Monophyletic	Paraphyletic	Polyphyletic	Analogies	Ancestral	Derived
Phenetic	Yes	Yes	Yes	Yes	Yes	Yes
Phylogenetic	Yes	No	No	No	No	Yes
Evolutionary	Yes	Yes	No	No	Yes	Yes

shared derived characters: the shared derived characters not formally used to define the classification should also fall into the same groups, because all evolutionary changes occur in the same phylogenetic tree. If two sets of shared derived characters seem to fall into contradictory groups, then at least one of them must have been mistakenly analyzed: it is impossible for correctly analyzed shared derived characters to contradict each other. However, shared ancestral and convergent characters need not show the same pattern, and the cladistic classification will therefore not be natural with respect to them. Cladistic classifications are natural for cladistic characters, but may be artificial for non-cladistic characters.

The cladistic philosophy is beguilingly simple, and it may be worth pausing to consider what it implies. The complete shift from phenetic to phylogenetic classification produces results that, at first sight, can seem strange. Cladistic classifications are peculiar in two main ways: they exclude "paraphyletic" groups and they reclassify species even when there is scarcely any change of gene pool or phenetic appearance. To understand the first point we need to know some technical terms.

14.6.2 Cladists distinguish monophyletic, paraphyletic, and polyphyletic groups

The groups of cladistic classifications are *monophyletic* in the sense that they contain all the descendants of a common ancestor: the group has a common ancestor unique to itself (Figure 14.7a). Cladism rejects *paraphyletic* and *polyphyletic* groups. A paraphyletic group contains some, but not all, of the descendants from a common ancestor (Figure 14.7b). The members that are included are the forms that have changed little from the ancestral state; the excluded species are those that have changed more: a paraphyletic group therefore contains the rump of conservative descendants from an ancestral species. *Polyphyletic* groups are formed when two lineages convergently evolve similar character states (Figure 14.7c). The key difference between paraphyletic and polyphyletic groups is that paraphyletic groups contain their common ancestor, whereas polyphyletic groups do not.

As Figure 14.7 shows, the different kinds of group are defined by the different kinds of characters. Shared derived homologies fall into monophyletic groups; shared ancestral homologies fall into both monophyletic and paraphyletic groups; analogies fall into polyphyletic groups. Paraphyletic groups are defined when some descendants of a common ancestor retain their ancestral characters, while others evolve new derived character states. The difference between paraphyletic and polyphyletic groups therefore arises because ancestral characters would have been present in a group's common ancestor, but convergent characters would not.

Cladistic classifications only include monophyletic groups because only they have the unambiguous hierarchic arrangement of the phylogenetic tree. Only monophyletic groups are formed in the cladistic conversion of a phylogenetic tree into a classification (Figure 14.6). Monophyletic groups are defined unambiguously by their branching relationships: they contain all the branches below a given ancestor; nothing has to be said (or indeed be known) about the phenetic evolution of species within each branch. But in order to define paraphyletic and polyphyletic groups, we do need to know about phenetic similarity. We have to decide which species to include and which to exclude, and this is done by including in paraphyletic or polyphyletic groups only phenetically similar species. Because of the subjectivity of measures of phenetic similarity, the tree of life cannot be unambiguously divided into paraphyletic, or polyphyletic, groups.

Taxonomists had known about evolutionary convergence and polyphyletic groups for a long time before cladism, but paraphyletic groups proved a more insidious problem. Hennig was the first to recognize them clearly, and his work was not widely known about until it was translated into English in 1966. We can see how paraphyletic groups tend to crop up by means of an earlier example: reptiles (see Figure 14.2c). The phylogenetic relationships of the main vertebrate groups are probably those of Figure 14.8 (we return to this example in chapter 17). In conventional classifications, mammals, birds, and reptiles are given equal taxonomic rank as classes. What has happened is that two groups, mammals and birds, have independently undergone relatively rapid phenetic evolution and have come to look very different from reptiles. The different reptilian lineages have changed more slowly and been left looking more like each other than like birds or mammals. Crocodiles and lizards, for instance have cold blood, scales, four legs, walk with a reptilian gait; birds are hot blooded, have feathers, two legs and two wings, and they

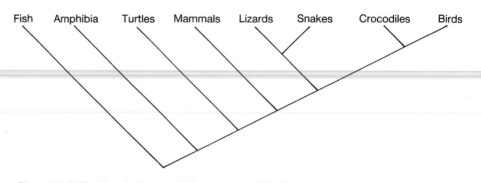

Figure 14.8 Phylogeny of main vertebrate groups. Reptiles are paraphyletic group, made up of turtles, lizards, snakes, and crocodiles in this picture.

fly. Yet crocodiles share a more recent common ancestor with birds than with lizards. The characters of crocodiles and lizards (scales, etc.) are ancestral for the group as a whole; and the paraphyletic group was formed because ancestral characters were used for definition. Paraphyletic groups become a danger whenever one or more subgroups have evolved relatively quickly and left their former relatives behind. If the cladistic philosophy is accepted, paraphyletic groups have to be ruled out. They are defined phenetically (by shared ancestral characters) and their recognition is inevitably subjective.

The cladistic classification of the tetrapods is quite unfamiliar, even counter-intuitive. The Reptilia were recognized in almost every formal classification before cladism; but cladism rules them out. There are many other examples of paraphyletic groups in non-cladistic classifications; fish are one of them. The tetrapods evolved from one particular group of fish, the lobe-finned fish (section 13.4, p. 329). If we consider the relationships between any tetrapod (such as a cow), any lobe-finned fish (such as a lungfish), and any ray-finned fish (such as salmon), the cow and the lungfish share a more recent common ancestor than do the lungfish and the salmon. The category "fish" (containing the lungfish and salmon, but excluding the cow) does not exist in cladism.

Some cladists have been rather fanatical in their insistence on ruling out Pisces and Reptilia. In a way, what we do in practice in these cases does not matter much. Their paraphyletic status is well known, and can do little damage; it is not worth getting worked up about. In other, less well known, cases it is more important to avoid paraphyletic groups. If a classification contains an unspecified mixture of monophyletic and paraphyletic groups, the evolutionary information in the classification becomes muddled. The beauty of a purely phylogenetic classification is that there can be no doubt what the branching relationships of its taxa represent (Figure 14.6). But if taxonomists define some relationships phenetically and others phylogenetically, it is no longer possible to say what any particular relationship means. The branching relationships are obscured and lost.

A second counter-intuitive feature of cladism is that a taxon can change names if it splits and one of the new lineages has hardly changed phenetically from its ancestor, or even not at all. The topic is discussed in the next chapter (section 15.3, p. 399).

14.6.3 A strictly cladistic classification could theoretically have an impractically large number of levels

If we could count all the branch points between the earliest common ancestor of all modern species and the present, the number would be very high. It would be far higher, by orders of magnitude, than could be fitted conveniently into the Linnaean system. The Linnaean hierarchy has perhaps seven main levels (kingdom, phylum, class, order, family, genus, species); these can be multiplied by adding in super-, sub-, and infralevels (e.g. superfamily, suborder, etc.—some are used conventionally, others are not) and extra levels, such as the tribe (between family and order). But this sort of system clearly could not accommodate the thousands of levels that the complete cladistic hierarchy would require.

Figure 17.8 (p. 459) shows the phylogeny of the Hawaiian picture-wing fruitflies. A cladistic classification of one part of it—such as the *Drosophila adiostola* species group—would need the invention of several (five in this case) new levels in the Linnaean hierarchy between genus and species (Figure 14.9). There are three things to say. One is that no interesting principles are at stake. The phylogenetic hierarchy can easily be divided up into the Linnaean levels with more than two members per group, as a matter of convenience. The division will be subjective, but that is unimportant because the division is only being done for convenience. It is the relationship of the cladistic hierarchy and the phylogenetic tree that is objective, and nothing fundamental is lost if the cladistic hierarchy is subjectively divided to produce a Linnaean naming scheme.

Secondly, the problem is not acute, in practice. We know very little about the phylogenetic relationships of living things. The situation in the Hawaiian fruitflies is exceptional; the crabs (Brachyura) are closer to the rule. Schram's *Crustacea* distinguishes about 50 families of crabs, but so little is known about their phylogeny that a classification can do little more than list the families alphabetically within about four tentative groupings. In the present state of

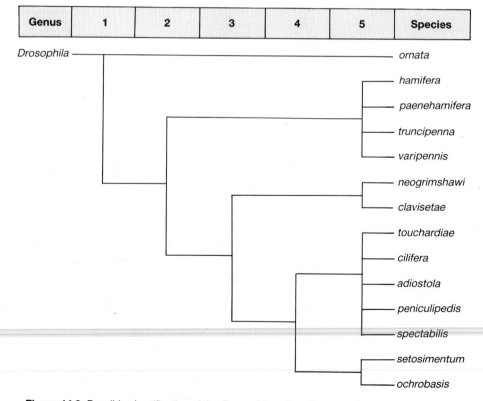

Figure 14.9 Possible classification of the *Drosophila adiostola* group of species of fruitflies. (Taken from the larger phylogeny of Hawaiian *Drosophila* in Figure 17.8; they are species 3–16 in that Figure.) These 14 species are only a small part of the Hawaiian *Drosophila*, which in turn are only a (large) part of the worldwide *Drosophila* fauna. And yet what we know about their phylogeny would require at least five new levels between the genus and species level.

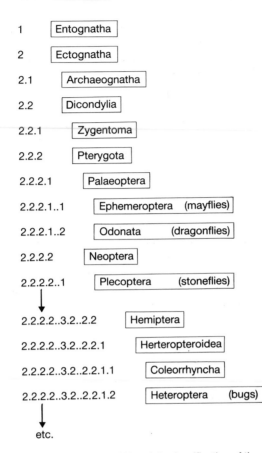

1	Entognatha
2	Ectognatha
2.1	Archaeognatha
2.2	Dicondylia
2.2.1	Zygentoma
2.2.2	Pterygota
2.2.2.1	Palaeoptera
2.2.2.1..1	Ephemeroptera (mayflies)
2.2.2.1..2	Odonata (dragonflies)
2.2.2.2	Neoptera
2.2.2.2..1	Plecoptera (stoneflies)
2.2.2.2..3.2..2.2	Hemiptera
2.2.2.2..3.2..2.2.1	Herteropteroidea
2.2.2.2..3.2..2.2.1.1	Coleorrhyncha
2.2.2.2..3.2..2.2.1.2	Heteroptera (bugs)

etc.

Figure 14.10 Part of Hennig's classification of the insects. The state of phylogenetic theory for the insects is too advanced for the limited number of Linnaean terms. Extra Linnaean terms could be invented, but Hennig instead classified each group by a series of numbers. The last number in the series identifies sister groups (e.g. 2.2.2.1..1 and 2.2.2.1..2, the mayflies and dragonflies, share a common ancestor). The order of splitting can then be traced back through the series of numbers. Only an illustrative part of Hennig's full scheme is shown here.

our ignorance, only four taxonomic levels are needed: infraorder (Brachyura), three sections, two subsections, and 50 families. Only when we find out more about the relationships of the families will extra levels be needed.

Thirdly, when we do know—or have some idea about—the fine branching of phylogeny, it is perfectly possible to devise a classificatory scheme to represent it. Hennig, for instance, in his final work on the phylogeny of the insects, produced at least hypotheses down to a fine branching level in several groups. He represented the phylogeny by a numerical scheme (Figure 14.10). The phylogenetic information in Figure 14.10 cannot all be represented in a simple Linnaean system. This is not in the least bit surprising. Linnaeus devised his system for approximately 10 000–20 000 named species. About a million species have been named now, 50–100 times the figure in the late eighteenth century. The expansion of our knowledge requires an expanded classificatory scheme to accommodate it.

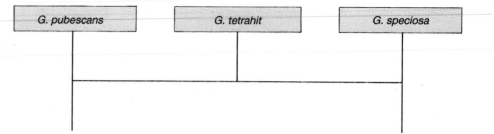

Figure 14.11 Origin of new species by hybridization. The mint species *Galeopsis tetrahit* originated by hybridization of two other species, *G. pubescens* and *G. speciosa* (see section 3.5, p. 43).

14.6.4 Hybridization can create difficulties in phylogenetic classification

The Linnaean system is well suited to represent cladistic relationships for groups of species that have evolved via a phylogenetic tree. There are two cases in which this condition may be violated. One is *hybridization* (Figure 14.11), which may be particularly common in flowering plants. The cladistic relationships of hybrid species are peculiar. If the new hybrid species and its parents belong to the same genus, as is the case in *Galeopsis* (Figure 14.11) and the other known botanical examples, no great naming difficulties arise. A deeper problem would arise if the parents of the hybrid were widely different forms. If (and this is a purely imaginary example) one parental species was from the Compositae, and the other from the Rosaceae, their hybrid would cladistically belong to two higher categories. Its phylogenetic relationships could not be represented in the Linnaean system.

The difficulty with hybrids can be thought of as a limit on cladism: the cladistic philosophy, it can be said, works well for normal phylogenies, but becomes inapplicable in cases like hybridization. The cladist could then reply that hybridization between different higher categories is unknown, and in any case introduces comparable problems in evolutionary (see below), if not phenetic, classification. However, there is a radical cladistic alternative— which is to apply the logic of cladism uncompromisingly. The fundamental demand of cladism is that the classification matches the phylogeny. If the phylogeny has overlapping, not hierarchic, relationships, then the classification should too. The hybrid ought to be classified, in a non-Linnaean manner, as a member of more than one higher category.

Although full hybridization between members of different higher groups is not known, gene transfer can take place. This is the second condition in which Hennig's conventional phylogenetic pattern may be violated. Gene transfer between species is specially common in bacteria, but there are a few odd cases (such as a hemoglobin-like protein in clover) suggestive of gene transfer between other higher taxa. If only a minority of a species' genes originated in this way, the phylogenetic tree would not be seriously spoiled; but as the process becomes more common, then it starts to alter in an important manner (Figure 14.12). The relationships of species are no longer simply hierarchic; part of a species ancestry is from one branch, part from another. In the extreme case where gene transfer is common, a large number

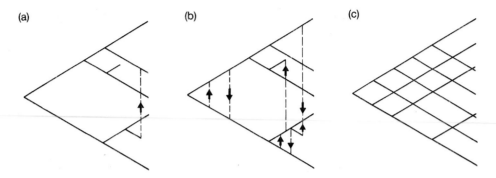

Figure 14.12 Gene transfer between species and partially anastomosing phylogenetic tree. (a) If an occasional gene (– – –) is transferred between distantly related species, the phylogenetic tree is not much upset. (b) As the process becomes more common, the phylogeny starts to break down. (c) The extreme case in which all the "species" behave as one large species.

of species—even all bacteria—could be behaving as one gigantic species. In the two extreme cases, of common and of rare gene transfer, classification is straightforward. The former makes the whole pool of forms into a single reproductive species (see sections 15.2.2–15.2.4, pp. 387–393); the latter implies a normal cladistic hierarchy. Intermediate cases could be more ambiguous. In practice, each bacterial species does seem to have its own characteristic gene pool. Bacterial gene transfer is almost entirely of plasmid genes: chromosomal genes can transfer, but the frequencies are very low indeed. It is therefore reasonable to treat separate bacteria species as species with cladistic relationships like other living things.

The closely related problems of hybridization and gene transfer are, for the moment at least, not serious. They can easily be accommodated in existing classificatory methods. However, as new facts are uncovered, difficult cases could arise, most probably in prokaryotes and plants. We shall then need the first principles of classification in order to decide how to classify them. Though neither hybridization nor gene transfer is practically insuperable, both issues illustrate the strong relationship, within cladism, between phylogeny and classification.

14.7 Evolutionary classification is a synthesis of the phenetic and phylogenetic principles

Evolutionary classifications incorporate both phenetic and phylogenetic elements. The school therefore describes itself as synthetic, drawing on the advantages, and avoiding the shortcomings, of the two purer schools. However, for the same reason it is liable to be criticized for doing the opposite— for retaining the philosophic shortcomings of pheneticism and adding to them the practical uncertainties of cladism. But first, what is evolutionary taxonomy? It is a school that recognizes both paraphyletic and monophyletic groups. In terms of the kinds of characters used to infer phylogeny (Figure 14.7; chapter 17), it forms groups by homologies rather than analogies; but does not distinguish ancestral from derived homologies. Referring back to Figure 14.2, evolutionary classification picks the phenetic classification in the reptilian case (Figure 14.2c) but the phylogenetic where there is convergence

(Figure 14.2b). (In Figure 14.2a it makes no difference, and it can be said to pick either one or the other, or both.)

Is a classification defined by homologies natural or artificial? Homologies can be either ancestral or derived, and as we saw, different shared derived homologies cannot (if correctly identified) contradict one another; however, different ancestral homologies can end up distributed in contradictory groupings. An evolutionary classification will therefore be natural in so far as shared derived homologies were used to construct it and to test its naturalness. If it contains shared ancestral homologies, or if defining shared derived homologies are compared with non-defining shared ancestral homologies, the classification will be less natural. There is no reason to suppose that non-defining analogies will fall into the same groups as an evolutionary classification. An evolutionary classification is natural in so far as it is cladistic.

How can the evolutionary taxonomist's synthesis of phenetic and phylogenetic principles be justified? The evolutionary school predates both modern phenetics and cladism and the main original discussions of the school therefore did not meet the pheneticists' and cladists' arguments head on; moreover, no complete modern evolutionary taxonomic defence against numerical and cladistic taxonomy exists. As such, the school differs from the other two, which were conceived partly in opposition to evolutionary taxonomy and made their objections to it clear. Nevertheless, a case can be made.

Evolutionary taxonomists disagree with pheneticists for much the same reason as we discussed above, though they expressed the argument differently. They criticized phenetic (or "morphologic") systems for being *idealistic*; that is, for supposing that a phenetic classification represents some ideal morphologic relationship between species. The ideal relationship would be some "idea" or "plan" in nature. An example is the pre-Darwinian theory that classifications represent the thoughts of God; the idea according to which the idealist tries to classify is then an idea existing in the mind of God but manifesting itself in (and inferrable from) God's creations. It is difficult for a modern scientist to make much sense of these old arguments, but "divine" taxonomy at least provides a concrete example of what idealism means. Modern numerical phenetics dropped the idealist philosophy of earlier phenetic classifications, but the same criticism—that there is no "ideal" phenetic plan of nature for the pheneticist to aim at—applies. The argument we considered in section 14.5 said that, in the absence of any natural hierarchy of phenetic similarity, phenetic classification lapses into subjectivity. The idealist error is the other side of the same coin. Idealists believe that a phenetic hierarchy exists "out there", but offer no good reason for their belief. When evolutionary taxonomists criticized phenetic classification for committing the error of idealism, they meant that no real phenetic hierarchy exists in nature, and a system which assumes such a hierarchy will be fundamentally subjective. Phenetic classifications try to group species according to a relationship—the ideal morphologic system—that evolution does not produce. Evolution does not produce one particular privileged phenetic hierarchy, which is more real than all other phenetic hierarchies.

Phenetic idealism can be avoided if taxonomists represent evolution, not phenetic similarity. But what does "evolution" mean? Evolutionary taxonomists exclude polyphyletic, but not paraphyletic, groups from classification. How do they interpret evolution to allow paraphyletic as well as monophyletic groups? To see why, we must look at what evolutionary taxonomists say about cladism.

Evolutionary taxonomists criticize cladism for its unnecessary puritanism. Cladism, as we have seen, leads to what at first sight can appear bizarre conclusions, such as the destruction of the Reptilia. It does so because of its distinction between paraphyletic and monophyletic groups. If both kinds of group are allowed in classification, the cladistic novelties do not arise. Evolutionary classification duly allows paraphyletic groups, and thus seeks to avoid the apparent perversity of cladism.

That is how evolutionary classification sees itself. But cladists, and pheneticists, see it rather differently. According to a cladist, the argument that evolutionary taxonomists accept against the pheneticist's polyphyletic groups works just as well against paraphyletic groups. What is sauce for the polyphyletic goose is sauce for the paraphyletic gander. If you accept paraphyletic groups, you must accept polyphyletic ones too, or be inconsistent. Paraphyletic groups are defined phenetically, just like polyphyletic groups. If some of the descendants of a common ancestor are to be excluded from a group, it has to be decided how many such descendants are to be left out, and the decision is phenetic and arbitrary (Figure 14.13); paraphyletic groups are not formed by phylogenetic relationships—as Figures 14.6 and 14.7 illustrated earlier. The standard phenetic problem then re-enters: according to some measures of phenetic similarity, one paraphyletic group will seem appropriate; according to another, another will. The choice between them is subjective—or idealist. When paraphyletic groups are admitted, the argument against pheneticism is lost. If phenetic criteria can be used in the case of paraphyletic groups, why not for polyphyletic ones? Paraphyletic groups presuppose idealism just as much as polyphyletic groups. (Box 14.1 discusses a method by which we might objectively define non-monophyletic groups.)

There is a further problem. Classifications that mix more than one type of information are less informative than classifications that represent only a single property. Consider an analogy with a library card index system. Suppose that a diligent librarian had read all the books in the library and devised a coded system for abstracting the facts he or she had found in them. Suppose also that this coding system happened to resemble closely the shelf mark codes. An evolutionary classification is like a card index that gives, for each book, a coded reference number, but does not tell you whether the numbers are the shelf mark or the abstracting codes. The card index reference code itself tells you neither where to find the book on the shelf, nor what the book's contents are. Given the reference code, you have to do further work to interpret it. Likewise, an evolutionary classification is constructed with both phenetic and phylogenetic information, and if you are only given the classification, you do not know which groups are phenetic and which phylogenetic. Classification by a pure system does not have this drawback.

Evolutionary taxonomy mixes phenetic and phylogenetic methods, but in a consistent and principled manner. It defines groups by homologies and

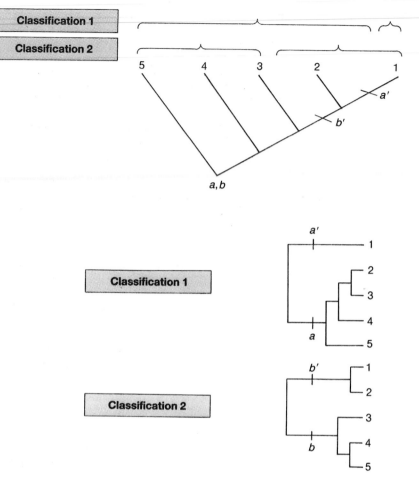

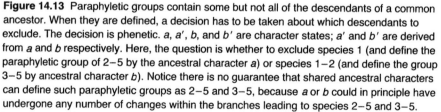

Figure 14.13 Paraphyletic groups contain some but not all of the descendants of a common ancestor. When they are defined, a decision has to be taken about which descendants to exclude. The decision is phenetic. *a*, *a'*, *b*, and *b'* are character states; *a'* and *b'* are derived from *a* and *b* respectively. Here, the question is whether to exclude species 1 (and define the paraphyletic group of 2–5 by the ancestral character *a*) or species 1–2 (and define the group 3–5 by ancestral character *b*). Notice there is no guarantee that shared ancestral characters can define such paraphyletic groups as 2–5 and 3–5, because *a* or *b* could in principle have undergone any number of changes within the branches leading to species 2–5 and 3–5.

excludes analogies; it therefore does not recognize polyphyletic groups. It allows phenetic groups such as reptiles and fish, but classifications in biology have a strong practical purpose and there is a case not to disband these long-established groups if a convincing reason can be found for keeping them. However, it is questionable whether the evolutionary taxonomist's reason is convincing enough.

Box 14.1 Adaptationist classification.

When evolutionary taxonomists define paraphyletic groups, the decision about which groups (like birds and mammals) to hive off from paraphyletic rumps (like the reptiles) is most easily understood as a phenetic decision. If it is, the general criticism of phenetic classification applies. However, if it is not a purely phenetic decision, evolutionary taxonomy may escape that criticism. One rather speculative criterion has been suggested; it comes from the idea of adaptation.

Birds do not differ from reptiles only phenetically; birds have evolved a set of adaptations for flying: they have achieved what is sometimes called an *adaptive breakthrough*, in the sense that birds have broken through during their evolution to a different way of life from reptiles. Adaptation is a more promising concept than phenetic difference for an objective system of classification. Adaptations fit ecological niches, and ecological niches exist—outside the classificatory system— in nature. If some method could be found of measuring adaptive differences, it might also be useful in classification. Groups could then be defined by shared adaptations, rather than shared ancestry or phenetic similarily. Fish might then be defined as "the group of vertebrates with adaptations for swimming." (Some other criterion would have to be added to exclude swimming reptiles and mammals; but if some definition along these lines is possible, it could be used in adaptationist

classification.) At present this is no more than an idea for the future. We should need for a start some method of measuring adaptive, as distinct from phenetic, similarity. No formal method has been proposed, and it might turn out to be impossible. We should note, however, that the abstract proposal exists, and that it is distinct from phenetic and phylogenetic classification. It can be called *adaptationist classification*.

The use of shared adaptations to define groups has been most often proposed by evolutionary taxonomists, such as Mayr. He suggested it could be used to define paraphyletic higher taxa, such as birds. "In the history of the vertebrates" Mayr has written "we know many such cases of the formulation of new grades, such as the sharks, the bony fishes, the amphibians, reptiles, birds, and mammals" (a *grade*, in Mayr's words, is a classificatory group "characterized (=defined) by a well integrated adaptive complex"). By itself, this is not an effective defence. It works just as well for the polyphyletic groups of phenetic classification as the paraphyletic groups of the evolutionary taxonomist. Convergence is usually caused by adaptation: sharks and dolphins have converged on the same hydrodynamic adaptations for the same environment. Classification by shared adaptations, if it can be made practical, is as likely to end up justifying a future version of phenetic, as of evolutionary, classification.

14.8 The principle of divergence explains why phylogeny is hierarchic

All three schools of classification—phenetic, cladistic, and evolutionary—aim at hierarchic classification. In the case of cladism, that is unsurprising. The phylogenetic tree is a hierarchy and phylogenetic classification will be hierarchic too. It is less obvious whether a phenetic classification has to be hierarchic. There are an infinity of phenetic patterns in nature; some indeed are nested hierarchies; but others are overlapping hierarchies, or non-hierarchic networks. If we aim at a phenetic classification, we have no strong reason to classify hierarchically. Biological classifications are hierarchic because evolution has produced a tree-like, diverging, hierarchic pattern of similarities among living things.

But why should this be? The question has an important place in the history of Darwin's thinking. He thought up natural selection in the late 1830s, as a natural explanation (or "secondary law") for adaptation and evolution.

As environments change, and competing species change, species will evolve new adaptations. By itself, this theory does not account for the tree-like, divergent course of evolution. He was well aware that evolution had steered such a course; the hierarchic structure, of groups within groups in classification, had been established as fact in the early nineteenth century—by (among others) Geoffroy St Hilaire's morphologic, and Milne-Edwards's embryologic, work. So striking a fact had to be fitted into the theory. Darwin recalled in his autobiography that in the early (1844) version of his theory:

> I overlooked one problem of great importance, and it is astonishing to me how I could have overlooked it and its solution. This problem is the tendency in organic beings descended from the same stock to diverge in character as they become modified. That they have diverged greatly is obvious from the manner in which species of all kinds can be classed under genera, genera under families, families under suborders and so forth . . . The solution occurred to me long after I had come to Down. The solution, I believe, is that the modified offspring of all dominant and increasing forms tend to become adapted to many and diversified places in the economy of nature.

This was Darwin's *principle of divergence* (Figure 14.14). He needed to explain

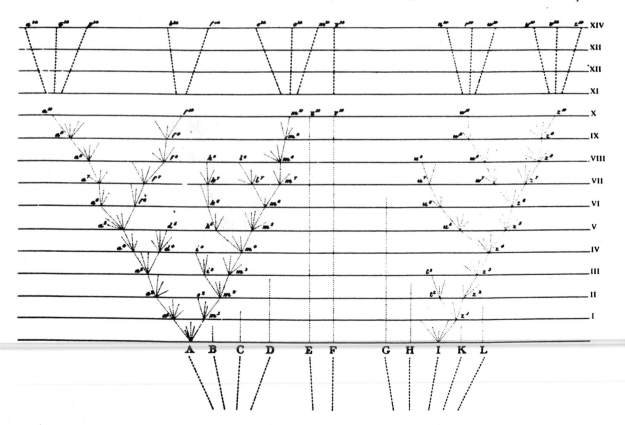

Figure 14.14 The divergent pattern of evolution. Each lineage tends to evolve away from related lineages, producing a tree-like pattern. Darwin suggested that the pattern evolves because more closely related forms compete more closely than distantly related lineages. From Darwin (1859).

why species apparently pushed one another apart in evolution. The mechanism he suggested was the relative strength of intraspecific competition. An individual of a species will compete against other members of its own species, and against members of other species in its own genus, and even perhaps against members of other families. The competition within a species will be strongest because the individual will encounter more members of its own species, and they will be more similar to it, exploiting more similar resources. One way to avoid competition is to become different from the competitors; hence there will be a force pushing similar competing types apart in evolution. Competition between similar individuals will make for the evolution of new adaptations in each that reduce the intensity of competition; divergence will result. The process is not inevitable. It depends on the contingencies of competition in particular cases; but provided that on average more similar individuals compete more closely, divergence will take place. Such was Darwin's theory.

A full explanation must tell us both why related species are pushed apart and also why highly unrelated species do not evolve to be very similar. Another argument, by Grafen, complements Darwin's and completes the explanation. Grafen asked why it is that more closely related species are more similar to each other than are more distantly related species. His answer was that speciation often takes place when a subpopulation from an existing species occupies a new, or unfilled, ecologic niche. This niche is most likely to be invaded by a species that already exploits a similar niche, because (in Grafen's words) "such a species can survive better initially in the vacant niche, and evolve sooner to exploit it fully, than a species in a more distant niche." As the population in the new niche speciates, the result will be two closely related, similar-looking species in two similar niches. This is probably the main reason why phylogenetically related species tend to resemble one another more closely than do distantly related species.

Grafen's argument employs a second Darwinian principle that is needed to explain divergence in nature: gradualism. Evolution proceeds in relatively small stages. If evolution often proceeded by large jumps, we might not see a smooth pattern of divergence. Species compete most against close relatives, and progressively less against more and more distant relatives, because the need to evolve in small stages forces them to. A species cannot rapidly escape the sphere of its present competitors' competition. A macromutational jump could lift a species out of the orbit of competition with close relatives, and it could then evolve independently of them; it might then be as likely to converge as to diverge (Figure 14.15).

The contingent, exploratory nature of evolution is also influential. Even if two species were not driving each other apart by competition, they could still be expected to diverge. They will encounter different environmental challenges, and generate different favorable variants (because favorable variants are rare it is not likely that the same one would coincidentally arise in both). As the different variants are fixed, the species will move apart.

Finally, the divergence of species can take place by a simple random walk of neutral evolution. Darwin may not have appreciated this, for he wrote in *On The Origin of Species* that:

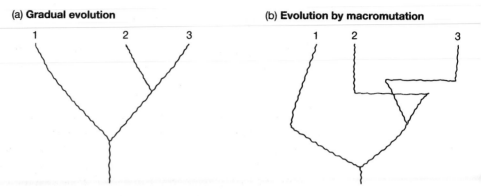

Figure 14.15 (a) Gradual evolution is more likely to be tree-like and divergent than is (b) evolution with macromutational jumps. The macromutations in species 1 and 3 take them out of competition with the other lineages, after which they are as likely to evolve toward as away from their former competitors, or not to change at all.

mere chance, as we may call it, might cause one variety to differ in some character from its parents, and the offspring of this variety again to differ from its parent in the very same character and in a greater degree; but this alone would never account for so habitual and large an amount of difference as that between varieties of the same species and species of the same genus.

However, two samples taken from a set of forms and allowed to evolve independently at random will soon diverge, in the way that evolution occurs in the neutral theory of molecular evolution (see also Figure 21.4, p. 562). Again, the reason is that the same mutations are not likely to occur in both lineages; random drift will tend to fix different amino acids in different lineages. Random divergence can have played no more than a part in evolutionary divergence, because it cannot account for adaptation and much evolution is clearly adaptive. The branching pattern Darwin was concerned with was at the level of gross morphology. At that level, the microscopic scale and rarity of evolutionary significant mutations, combined with the stronger competition among more similar forms, remains the most plausible principle of divergence.

14.9 Conclusion

If the arguments explained in this chapter are correct, cladism is theoretically the best justified system of classification. It has a deep philosophic justification which phenetic and partly phenetic systems lack. Cladism is objective, and objective classifications are preferable to subjective ones. Moreover, it is the only school that in theory will produce perfectly natural classifications: all sets of correctly identified shared derived characters must fall into the same phylogenetic hierarchy. Different sets of both shared ancestral homologies and of analogies can end up distributed in contradictory groups of species, and classifications that use these characters can therefore be artificial. But despite these theoretical advantages, cladism can run into practical problems. The uncertainties of phylogenetic inference make cladistic classifications liable to frequent revision. True cladists are not very worried about that, and remark that all healthy theories are modified as new facts come in.

Many biologists, however, would draw a different conclusion. They are biologists who are less anxious about the philosophic coherence of a classificatory school, and more concerned either that its practical procedures should match its theoretical ambitions (in which case they may prefer phenetic classification, with its clearly articulated techniques) or that not too much violence should be done to existing classification (in which case they will prefer evolutionary classification). At all events, the most important matter is to understand the arguments that have been used for and against each school; the exact conclusion we draw here is less important: there is no orthodoxy among evolutionary biologists about the best kind of classification.

14.10 Summary

1 There are two main principles, and three main schools, of biological classification: phenetic and phylogenetic principles, and phenetic, cladistic, and evolutionary schools. The schools differ in how (if at all) they represent evolution in classification.

2 Phenetic classification ignores evolutionary relationships and classifies species by their similarity in appearance; cladism ignores phenetic relationships and classifies species by their recency of common ancestry; evolutionary taxonomy includes both phenetic and phylogenetic relationships.

3 Phylogenetic inference is uncertain and phenetic classification has the advantage that it is not subject to revision when new phylogenetic ideas are put forward.

4 Phenetic classification is ambiguous because there is more than one way of measuring phenetic similarity and the different measures can disagree.

5 Cladism is unambiguous because there is only one phylogenetic tree of all living things.

6 Evolutionary taxonomy avoids some of the extraordinary properties of cladism. But it suffers from the ambiguity of phenetic taxonomy, and its argument for excluding one kind of phenetic relationship (convergence) works equally well against the kind of phenetic relationship (differential divergence) that it includes.

7 Living things show a diverging, tree-like pattern of relationships, probably because competition is stronger between more similar forms, evolution proceeds in small stages, and variation is undirected.

14.11 Further reading

I have previously discussed many of the points in this chapter, at an introductory level and at greater length, in Ridley (1986). Wiley (1981) is a textbook, written from a cladistic viewpoint; and Crowson (1970) is an accessible introduction, from a more-or-less cladistic viewpoint. Goto (1982) is also introductory, and more practical.

Sneath and Sokal (1973) is the standard work on numerical taxonomy. Sokal (1966) is a clear introduction. See Johnson (1970) for the main criticism. Hennig (1966) is the classic work on cladism—but he is not an easy read! Hennig (1981) offers further thoughts on the subject. Mayr (1969, 1976, 1981), Dobzhansky (1970), and Simpson (1961b) are the key works by the key evolutionary taxonomists. The controversy between the schools appears mainly in the pages of the journal *Systematic Zoology* in the 1970s and, more recently, increasingly in *Cladistics* and *Biology and Philosophy*. Hull (1988) gives an excellent history of the modern phenetic and cladistic movements, to illustrate his evolutionary philosophy of science.

Brooks and Wiley (1984, 1988; the former is more concise), Eldredge (1985a), Friday (1987), and Salthe (1985) discuss the principle of divergence. Browne (1983) and Ospovat (1981) are historically interesting. See also Grafen (1989).

The idea of a species

15.1 In practice species are recognized and defined by phenetic characters

Biologists almost universally agree that the species is a fundamental natural unit. When biologists report their research, they identify their subject matter at the species level and communicate it by a Linnaean binomial such as *Haliaeetus leucocephalus* (bald eagle) or *Drosophila melanogaster* (fruitfly). However, biologists have not been able to agree on exactly what a species is, or how species should abstractly be defined. The controversy is theoretical, not practical: there is no doubt about how particular species are defined in practice. Taxonomists practically define species in terms of easily identifiable morphologic characters. If one group of organisms consistently differs from other organisms, it will be defined as a separate species; the taxonomist, to give the species a formal definition, has to search for reliable characters— characters that are possessed by all members (or as many of them as possible) of the species to be defined, and not by others.

Almost any phenotypic character may end up being useful in the practical recognition of species. Let us consider just one example. Figure 15.1 shows the adult bald eagle (*Haliaeetus leucocephalus*) and golden eagle (*Aquila chrysaetos*), seen from below. A bird guide, such as Peterson's *Birds*, gives a number of characters by which the two species can be told apart. In the adult, the bald eagle can be recognized by its white head and tail, and its massive yellow bill; there is more of a problem telling apart the immature bald eagle and the adult golden eagle, but the immature bald eagle usually has enough white color on its wing linings for it to be recognizable. Therefore, we can define the bald eagle as a North American eagle with a white head and tail in the adult. (Strictly speaking, the characters used to recognize species are often diagnostic rather than defining characters. Box 15.1 explains the difference between the two.)

So species, in practice, are recognized and defined by phenotypic characters. Practical problems do arise. On the theory that species have evolved from a common ancestor, we should expect there to be some awkward intermediate cases in which it is unclear whether two populations have diverged enough to count as separate species. It is variation that poses most of the practical problems of species recognition. Phenotypic characters often vary between the individuals of a population at any one place, and also vary between populations living in different places. The taxonomist's aim, in looking for characters by which to recognize species, is to find a character that does not vary much between individuals and provides a reliable criterion. For most species it has been possible to find a character—or group of characters—that is reasonably reliable, though it may take an expert to be

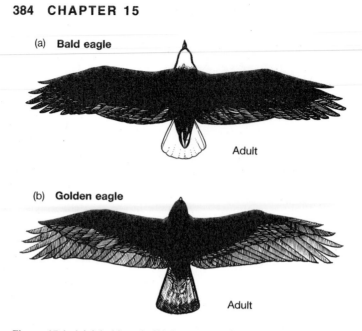

Figure 15.1 Adult bald eagle (*Haliaeetus leucocephalus*) and golden eagle (*Aquila chrysaetos*), seen from underneath. The species can be distinguished by their pattern of white coloration.

able to recognize them; but in extreme cases the problem can become almost vicious. Geographic variation causes the most difficulties. If a species varies geographically, a good character for species recognition in one place may become useless in another place. In ring species, in which the geographic extremes of a species overlap because the geographic distribution is ring shaped, there are two species where the ranges overlap, but they are connected by a continuously varying series of intermediates (see section 3.4, p. 40). It would be practically impossible to produce a non-arbitrary phenotypic character to divide the ring into two species, and there would not be any theoretical point in trying.

These practical difficulties have brought us to the theoretical question of the species concept. Species are in practice mainly recognized by phenotypic characters; but the interest of the subject for evolutionary biologists lies elsewhere. In this chapter we shall discuss whether there is some deeper theoretical concept beneath the individual characters used to recognize individual species. Is the bald eagle just the set of eagles that live in North America and have white heads and tails—or is there more to it? But first we should ask what it would mean for the species of bald eagles to have both a superficial, practical and a deeper, theoretical definition. A thought experiment should make the relationship clear. Imagine a bald eagle that lacked its diagnostic characters, but was otherwise unchanged: imagine that a bald eagle with a good white head and tail produced a nest of eagles of some different color pattern. Would it have given birth to a new species? If the color of the head and tail was all there was to being a member of *H. leucocephalus*, then the answer would clearly be yes. However, if there is a

Box 15.1 Description and diagnosis in formal taxonomy.

The point of the example of the two eagle species is merely the very simple one that species are defined in practice by particular characters. We should also notice a terminological formality, because when someone talks about the definition of a species they can mean one of two things. We have to distinguish a formal *description* of a species from a *diagnosis*. The formal description of a species has to satisfy certain rules, and the characters that in a formal sense "define" the species are those characters mentioned in the first formal description. In theory, the characters mentioned in the formal definition could be practically very difficult to use, but they still in a legalistic sense define the species.

Now, it could be that the characters in a species' formal definition are some arcane details of its genitalia, which can be observed through a microscope on a dead specimen in a museum. Taxonomists do not on purpose pick obscure characters to put in their definitions, but if the only distinct characters that the species' first taxonomist noticed were obscure ones then they will provide its formal definition. This is inconvenient, at the least, and subsequent taxonomists will then try to find other characters that can be more easily worked with. These useful characters, if they are not in the formal description, provide what is called a diagnosis. A diagnosis does not have the legalistic power of a description to determine which names are attached to which specimens; but it is more useful in the day to day practical taxonomic task of recognizing which species specimens belong to. As research progresses, better characters (i.e. more characteristic of the species, and more easily recognized) may be found than those in the first formal description. The formal definition then loses its practical interest, and the characters given in a work like Peterson's *Birds* are more likely to be diagnostic than formally defining.

When an evolutionary biologist discusses the definition of species, the formal distinction between description and diagnosis is beside the point. All that matters is that phenotypic characters are used to recognize species, as in the eagles. It is worth knowing about the distinction, however: both in order to avoid unnecessary muddles, and for other reasons—taxonomic formalities are important in the politics of conservation, for instance. Nevertheless, strictly speaking, we can ignore this distinction in the species controversy of evolutionary theory.

more fundamental definition of the species and the coloration was picked only as a practically useful marker, then the answer would be no. Indeed, the birth of the brood without the white coloration would mean that that character had become a less useful character, and taxonomists should start looking for a more reliable character to recognize the species.

Evolutionary biologists have discussed several possible theoretical definitions of the species. In this chapter, they are divided into two main classes. The first kind of definition aims to identify the members of a species at a particular time: there are millions of organisms in the world and we wish to know which ones belong to the same species. This kind of non-temporal (also called non-dimensional) definition is practically useful in the study of modern species. A field naturalist wishes to know which eagles are *H. leucocephalus* now, and is not much interested in eagles a million years in the past or future. The second kind of definition takes account of the temporal dimension of evolution; it aims to specify which individuals are members of a species not only at one time, but throughout all time.

15.2 Some species concepts define species at a point in time

Evolutionary biologists have suggested three main theoretical kinds of non-temporal species concept. One suggests that species exist because of interbreeding: we shall discuss two versions of it below—the biological and recognition species concepts (we can use the term "reproductive species concept" to refer to them jointly). A second suggests that species exist because selection maintains variation in the tight clusters we recognize as species: we shall discuss this as the ecological species concept. We shall do no more than mention the third possibility here. It suggests that species exist because of genetic constraints on variability (section 13.8, p. 334): in this theory, living things exist in discrete clusters because the mutations that would produce intermediates cannot arise. This would form a sort of structuralist species concept. No one has seriously developed it, and it is only mentioned here for completeness. In addition to these theoretical definitions, it has also been suggested that the species concept should not be tied to any particular theoretical idea: which brings us to the phenetic species concept.

15.2.1 The phenetic species concept

The *phenetic species concept* applies phenetic classification (section 14.5, p. 360) to the species category. Informally, the phenetic species concept defines a species as a set of organisms that look similar to each other and distinct from other sets. More formally, it would specify some exact degree of phenetic similarity, and similarity would be measured by a phenetic distance statistic. An exact definition would therefore sound more like "a species is a set of organisms not more than x phenetic distance units apart" or as "the set of organisms separated by a phenetic distance of at least y units from the nearest distinct set," and the value of x or y could be chosen to produce species much like those defined by earlier methods.

In practice, the pheneticist measures as many characters as possible in as many organisms as possible, and then recognizes phenetic clusters by multivariate statistics. The species are then the smaller clusters that approximate to the level of similarity typical of what would, before numerical techniques were available, have been called a species. There can be no doubt that species can be defined in this way; the techniques of phenetic measurement exist and are known to work.

The phenetic concept can be thought of as a simple extension of the way species are recognized in practice. Species are recognized by morphologic characters; the pheneticist then says that species *are* groups of individuals with certain morphologic characters. The phenetic species is properly defined by a large number of characters; the diagnostic characters like "white head and tail," which are used for rapid identification, are then proxies for the underlying multicharactered clusters. After the clusters have been found by multivariate numerical techniques, it will always be possible to find particular characters that discriminate conveniently between the clusters: they can then be used in practical species identification.

Another way of thinking about the phenetic concept, particularly emphasized by its critics, is as an updated, numerical form of the earlier *morphological* or *typological species concept*. Species had traditionally, particularly before the modern neo-Darwinian synthesis, been defined by reference to a morphologic type: a species would have been imagined (more-or-less articulately) to possess a particular "type" or "ideal" form, and the members of the species could then be defined as those individuals sufficiently similar to the type.

The phenetic species concept suffers from the general difficulty of all phenetic classification. We have worked through the argument before (chapter 14) and need only repeat it in outline here. The phenetic concept lacks a sound philosophic basis. It can necessitate subjective and arbitrary decisions. If we take a set of organisms, one phenetic measure may divide them up into one pattern of species, another into a different pattern. If the phenetic concept was sound, it would then have a criterion for deciding which pattern was correct; but it does not, and cannot, have one. Such a criterion would require there to be a real phenetic pattern of species in nature before the phenetic measures were made. If we pursue the argument, the real pattern will probably turn out to be some pattern of morphologic types. Neo-Darwinians have dismissed typological classification because there is no reason to suppose that any ideal pattern of morphologic types exists in nature (sections 14.5 and 14.7, pp. 360 and 373). Evolution does not make any variants in a population more typical or more real than others. The theory of evolution provides no deep support for the phenetic species concept. Species are identified phenetically, but that is only for practical convenience; the underlying theoretical concept of species is something other than a phenetic cluster.

If the phenetic species concept was valid, species definition would have little, or nothing, to do with evolution. Species can be phenetically defined quite regardless of how evolution happens—indeed it would be possible if evolution had not happened at all. This in itself would have been a reason for suspecting the concept, because (as we have seen) species are highly important units in the theory of evolution. It should be possible to define species in a theoretically powerful manner, in which the relationship of the species concept and the theory of evolution is clear. In science, definitions of terms are always most interesting in so far as they place the term within a scientific theory. Concepts outside theories are dull. The next two kinds of instantaneous species definition both draw on the theory of evolution.

15.2.2 The biological species concept

The *biological species concept* defines species in terms of interbreeding. Mayr (1969), for instance defined a species as follows: "species are groups of interbreeding natural populations that are reproductively isolated from other such groups." The concept actually predates Darwin—it was the concept used by John Ray in the seventeenth century, for instance—but it was strongly advocated by several influential founders of the modern synthesis, such as Dobzhansky, Mayr, and Huxley, and it is the most widely accepted species concept today, at least among zoologists. It is important because it places the taxonomy of natural species within the conceptual scheme of population genetics. A community of interbreeding organisms is, in population genetic terms, a gene pool. In theory, the gene pool is an abstract conception of a set of reproducing genetic units, within which gene frequencies can change. In the biological species concept, gene pools become more-or-less identifiable as species. The identity is imperfect, because species and populations are often subdivided; but that is a detail.

The biological species concept explains why the members of a species resemble one another, and differ from other species. When two organisms breed within a species, their genes pass into their combined offspring; as the same process is repeated every generation, the genes of different organisms are constantly shuffled around the species gene pool. The separate family lineages (of parent, offspring, grandchildren, etc.) soon become blurred by the transfer of genes between the lineages. The shared gene pool gives the species its identity. By contrast, genes are not (by definition) transferred to other species, and different species therefore take on a different appearance. The movement of genes through a species by migration and interbreeding is called *gene flow*.

And how, on the biological species concept, should the taxonomist's method of defining species be interpreted? Taxonomists actually identify species by morphology, not reproduction. Using the biological species concept, the taxonomist's aim should be, as far as possible, to define species reproductively; the justification for defining species morphologically is that the morphologic characters shared between individuals are indicators of interbreeding. When taxonomists can study interbreeding in nature they should do so and define the array of interbreeding forms as a species. With dead specimens in museums, morphologic criteria should be sought that mimic the interbreeding principle. Taxonomists should therefore seek morphologic criteria which define a species as a set of forms that appears to have the kind of variation that an interbreeding community would have. The morphologic characters of species are then indicators of interbreeding, as estimated by the taxonomist. Thus eagles with white heads and tails are one interbreeding unit: eagles with the color pattern of the golden eagle belong to another.

Taxonomists can estimate the degree of morphologic variation corresponding to a reproductive species. Interbreeding does give a species a degree of morphologic uniformity, and someone experienced with a group can estimate what the typical degree of uniformity is for a reproductive species in that group. But members of a species are by no means all uniform; biological species are *polytypic*—they have many (or no) morphologic types. Different biological species may have different degrees of morphologic variation, and the biological species concept is therefore practically as well as conceptually distinct from the phenetic species concept.

It is also possible for species to differ reproductively but not morphologically. They are then called *sibling species*. The classic example of a pair of sibling species is *Drosophila pseudoobscura* and *D. persimilis*. The two are phenetically almost indistinguishable. If a numerical phenetic cluster analysis was performed on them they would show up as a single phenetic species. They indeed *are* a single phenetic species. But reproductively they are separate. If flies from a *persimilis* line are put with flies from a *pseudoobscura* line they do not interbreed. Supporters of the biological species concept such as Dobzhansky used sibling species to criticize the principle of phenetic classification, but the criticism should not be well taken. Sibling species merely show that phenetic and biological species are not the same. To criticize the phenetic species concept for failing to distinguish *D. pseudoobscura* from *D. persimilis* assumes that the biological species concept has priority and the phenetic concept is to be judged by how well it recognizes biological species. But pheneticists will not see things that way. They might just as well turn the argument round and criticize the biological species concept for ruining a perfectly good phenetic species by splitting it in two. The main point is that phenetic and reproductive units can differ in nature. For the biological species concept, phenetic characters matter only in so far as they indicate inter-breeding. In fact species do often form phenetic clusters, and the biological species concept has, in gene flow, an explanation for this too.

15.2.3 Interbreeding between species is prevented by isolating mechanisms

Why is it that closely related species, living in the same area, do not breed together? The answer is, it is prevented by *isolating mechanisms*. An isolating mechanism is any property of the two species that stops them from interbreeding. Several main types of isolating mechanism are distinguished; Table 15.1, taken from Dobzhansky, gives one classification. Species may not interbreed because they live in different habitats, have a different courtship, their gametes do not fuse, or hybrid offspring fail to reproduce. Examples are known of all the mechanisms listed in Table 15.1, with the possible exception of mechanical isolation: no definite case is known of a pair of species that do not interbreed because copulation is mechanically impossible. (There are of course many cases in which copulation between species *is* impossible, but some prior factor always seems to isolate the species before any mechanical difficulties come into play.) Let us look at three examples.

Table 15.1 Dobzhansky's classification of reproductive isolation mechanisms. From Dobzhansky (1970)

---

1 *Premating* or *prezygotic* mechanisms prevent the formation of hybrid zygotes
 (a) *Ecological* or *habitat isolation.* The populations concerned occur in different habitats in the same general region
 (b) *Seasonal* or *temporal isolation.* Mating or flowering times occur at different seasons
 (c) *Sexual* or *ethological isolation.* Mutual attraction between the sexes of different species is weak or absent
 (d) *Mechanical isolation.* Physical non-correspondence of the genitalia or the flower parts prevents copulation or the transfer of pollen
 (e) *Isolation by different pollinators.* In flowering plants, related species may be specialized to attract different insects as pollinators
 (f) *Gametic isolation.* In organisms with external fertilization, female and male gametes may not be attracted to each other. In organisms with internal fertilization, the gametes or gametophytes of one species may be inviable in the sexual ducts or in the styles of other species

2 *Postmating* or *zygotic* isolating mechanisms reduce the viability or fertility of hybrid zygotes
 (g) *Hybrid inviability.* Hybrid zygotes have reduced viability or are inviable
 (h) *Hybrid sterility.* The F_1 hybrids of one sex or of both sexes fail to produce functional gametes
 (i) *Hybrid breakdown.* The F_2 or backcross hybrids have reduced viability or fertility

---

The first concerns isolation by time of breeding. Two closely related toads *Bufo fowleri* and *B. americanus* have overlapping ranges in central and eastern USA. *Bufo americanus* breeds earlier than *B. fowleri,* and little interbreeding takes place between them through most of the geographic range. However, in Michigan and Indiana they do interbreed and a third type, a hybrid, is found. The hybrid is fertile and as vigorous as either parent species; at another site, in New Jersey, only the hybrid is found. One interpretation, due to Blair and others, is that recent human induced changes have created new habitats that both species live in, whereas previously there was some separation by habitat. The toads have now been attracted to a common habitat, where they interbreed. The normal isolation would then have been partly temporal and partly by habitat.

In many species of frogs and toads, males attract females by croaking. It has been shown experimentally that females who have the croaks of males of their own species and croaks of other species played to them approach the loudspeaker playing their own species' croaks. The species are reproductively isolated by courtship. (The next section gives further details of this sort of experiment, but for orthopteran insects.)

Breeding time and courtship are examples of *prezygotic isolation mechanisms*: they prevent the two species from ever breeding together. However, if interbreeding takes place, the species may be isolated if the viability or fertility of the hybrid offspring is so low that the interbreeding is practically fruitless; the species are then maintained by a *postzygotic isolation mechanism*. One common, but incomplete means of postzygotic isolation is the phenomenon called *Haldane's rule*. Haldane observed that, in crosses in which only one sex of offspring is inviable or infertile while the other sex develops as normal, it is regularly the heterogametic sex that has reduced viability or fertility (the heterogametic sex is the one with two different sex chromosomes, i.e. in mammals, the males, whose genotype is XY, whereas the female genotype is XX). Since the tendency was noticed by Haldane, it has been supported by a large amount of evidence (Table 15.2); 20 mammalian crosses between closely related species result in offspring in which only one sex is infertile or inviable, and in 19 of them that sex is the male.

In particular cases, isolation is not likely to be completely due to only one

Table 15.2 Support for Haldane's rule. Asymmetry in the column "hybridizations with asymmetry" means that one sex is affected more than the other with respect to the trait, such as fertility. Many species of butterflies, moths, and mosquitoes are also known to follow the same rule. From Coyne and Orr (1989b)

Group	Trait	Hybridizations with asymmetry	Number obeying Haldane's rule
Mammals	Fertility	20	19
Birds	Fertility	43	40
	Viability	18	18
Drosophila	Fertility and viability	145	141

factor from Dobzhansky's list. It may be caused by a mix of several prezygotic and postzygotic factors. Moore estimated the relative importance of four different types of isolating mechanism among six congeneric North American frog (*Rana*) species. His results are given in Table 15.3. *Rana sylvatica*, for instance, is 100% isolated from *R. clamitans* by hybrid inviability and 30% by different ecology. The totals add up to more than 100% because species are isolated by more than one mechanism—there is some isolatory overkill. Notice the asymmetries between some pairs of species. The geographic range of one species may be contained within another, for example. The more local species will then be 0% isolated from the widely distributed species, which will, in turn, be isolated somewhere in the 0–100% range from the local species. From the table, it appears that 74% of the range of *R. palustris* is outside that of *R. pipiens* but none of *pipiens* is outside *palustris*. (Box 16.1, p. 420, describes how asymmetric isolation can be used to infer the direction of speciation.)

Table 15.3 Moore's estimates of the relative power of geographic isolation (G), ecological isolation (E), seasonal isolation (S), and hybrid inviability (D) among six species of North American frogs. From Moore (1949)

Females	Males					
	Rana sylvatica	*Rana pipiens*	*Rana palustris*	*Rana clamitans*	*Rana catesbeiana*	*Rana septentrionalis*
Rana sylvatica		G 29	G 61	G 59	G 68	G 80
		E 70	E 40	E 30	E 70	E 60
		S 60	S 100	S 100	S 100	S 100
		D 100	D 100	D 100	D 100	D ?
Rana pipiens	G 59		G 74	G 67	G 62	G 88
	E 70		E 70	E 70	E 85	E 80
	S 60		S 40	S 100	S 100	S 100
	D 100		D 0	D 100	D 100	D 100
Rana palustris	G 13	G 0		G 3	G 23	G 72
	E 40	E 70		E 60	E 70	E 50
	S 95	S 40		S 95	S 100	S 100
	D 100	D 0		D 100	D 100	D ?
Rana clamitans	G 28	G 0	G 24		G 18	G 22
	E 30	E 70	E 60		E 30	E 20
	S 100	S 100	S 95		S 50	S 0
	D 100	D 100	D 100		D 100	D 100
Rana catesbeiana	G 51	G 0	G 47	G 29		G 93
	E 70	E 85	E 70	E 30		E 30
	S 100	S 100	S 100	S 50		S 50
	D 100	D 100	D 100	D 100		D ?
Rana septentrionalis	G 0	G 0	G 37	G 36	G 79	
	E 60	E 85	E 50	E 20	E 30	
	S 100	S 100	S 100	S 0	S 50	
	D ?	D 100	D 100	D 100	D 95	

Complete isolation = 100; absence of isolation = 0.

15.2.4 The recognition species concept

Within a single habitat in the USA, as many as 30 or 40 different species of crickets may be breeding. The male crickets broadcast their songs and are approached by females; interbreeding is confined within a species because each species has its own distinctive song and females only approach males that are singing their species song. This can be confirmed using the simple experiment which was described above for toads. Female crickets approach loudspeakers playing cricket song, and it can be confirmed that the crickets are of the species whose song is being played. A still better experiment is to put female crickets on a kind of Y-shaped maze (Figure 15.2); the song of the female's own species can be played from one side and the song of another species from the other. The females almost invariably turn toward the song of their own species. (The direction from which the two species song are played can be reversed, to control for any tendency of crickets to turn to the left or right.)

The female cricket recognizes the song of males of her own species and will breed only with a male who sings that song. The song, and the female recognition of it, constitutes a mate recognition system. The species has a specific mate recognition system (SMRS). Paterson has particularly emphasized that species can be defined as a set of organisms with a common method of recognizing mates (a shared SMRS). This is the *recognition species concept*. In practice, the recognition concept should define very similar, or identical, species to the biological concept and it is possible to think of them as two versions of a general reproductive species concept. An isolation mechanism to keep species apart and a recognition mechanism to ensure breeding takes place within a species are, to a large extent, the two sides of the same coin.

However, Paterson has suggested that there are a number of advantages in defining species by a shared SMRS rather than interbreeding and reproductive isolation. One is practical: SMRSs can often be observed in a few, even dead, specimens whereas interbreeding is more difficult to observe and is usually

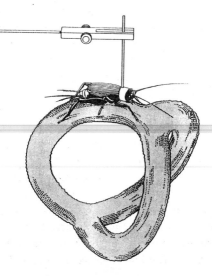

Figure 15.2 A Y-shaped maze for testing the acoustic preference of a female cricket. As the cricket walks, the Y-shaped maze moves beneath her feet. Loudspeakers can play the songs of different species from either side. From Bentley and Hoy (1974).

inferred indirectly. A second possible advantage is more theoretical. The recognition concept may represent more accurately what happens when a new species originates. The crucial event for the origin of a new species, according to Paterson, is the evolution of a new mate recognition system. He doubts whether the evolution of isolating mechanisms are so important. Dobzhansky used to argue that isolating mechanisms will evolve under natural selection to prevent interbreeding among incipient species: the isolating mechanisms are then a part of the causal process leading to new species. Paterson doubts whether isolating mechanisms are ever directly favored during speciation, and suspects they generally evolve as an incidental by-product of evolutionary changes taking place for other reasons during the divergence that leads to speciation, or later. Paterson's preference for his recognition concept therefore stems in part from his view of the relative importance of mate recognition systems and isolating mechanisms in speciation: the former he sees as causal and the latter more of a consequence. The recognition concept then better represents the causal process of speciation.

We shall return in chapter 16 to the question of whether isolating mechanisms evolve during speciation. For the purposes of the present chapter, isolating mechanisms and specific mate recognition systems can be considered as two ways of defining species as reproductive communities; the biological and recognition concepts are two parts of a more general reproductive species concept. It is worth noting that the biological species concept does not particularly have to be defined in terms of isolating mechanisms, although it was by Dobzhansky and by Mayr (in 1969, influenced by Dobzhansky). For the fundamental biological species concept, the crucial factor is interbreeding, and whether this is brought about by a shared mate recognition system or an isolating mechanism is a secondary matter.

15.2.5 The ecological species concept

The animals or plants that are members of one species are relatively similar to each other, compared with the members of other species. When the members of a species interbreed, they are breeding with other individuals similar to themselves. Why should they choose to do so? In a proximate sense, as the biological species concept identifies, it is because isolation and mate recognition mechanisms lead them to choose this kind of mate. But isolation and mate recognition mechanisms can evolve under selection just like any other character. If individuals confine themselves to mating with only a limited class of other individuals, there must be a selective advantage to the limitation. Dobzhansky's table of isolating mechanisms (Table 15.1) suggests one reason for the selective advantage: the lower half of the table lists various kinds of postzygotic isolation, in which the hybrid offspring are inviable or infertile. The mule is an example of hybrid infertility; it is a hybrid between a male ass (*Equus africanus*) and a female horse (*Equus caballus*) and is somatically vigorous, but sterile. There will be strong selection for horses and asses not to interbreed, because they produce sterile offspring. The sterility of the mule, however, looks rather like a contingent genetic accident: it could be that a small genetic change would allow a mule to breed. The mule may be a special case, which does not tell us the general reason why selection limits the degree of interbreeding. The *ecological species concept* points to a reason that may be more general.

It is that selection limits the degree of interbreeding in order to prevent the production of relatively maladaptive forms. The main reason why hybrids are inferior when two species are now crossed is that the genes of the two parental species are not coadapted (section 8.2, p. 186). Different genes have accumulated in the two species over time, and they do not interact well when they are mixed in a hybrid offspring. However, this does not tell us how selection would have operated on interbreeding in the early stages of speciation, before different genes had accumulated in the two separate species. At that time, it may have been that crosses between species produced forms that were not adapted to exploit natural resources efficiently, or failed to survive for some other ecological reason. In nature, only circumscribed niches are available for exploitation. The availability of resources, the physical conditions, the local distribution of parasites, predators, and competing species, all limit the ways of making a living. To exploit a niche, a species will need a particular set of adaptations, of behavior, morphology, and physiology. The ecological species concept supposes that ecological niches in nature occupy discrete zones, with gaps between. In other words, an organism whose attributes adapt it for life in one of the gaps between niches will be maladapted: it will be fit to exploit resources that do not exist, or to avoid non-existent parasites or predators. An ecological species can be defined as a set of organisms exploiting a single niche.

The idea that nature is divided into *adaptive zones* has a broader application. Simpson suggested that higher taxa, such as the Mammalia, occupy an adaptive zone; that is, there is an abstract space, set by the resources and competitors in nature, within which the mammalian sort of body plan is adaptive. The mammalian adaptive zone is more inclusive than a species niche, but we can imagine a hierarchy of adaptive zones corresponding to the successive taxonomic levels. Simpson developed the idea from Wright's more formally defined adaptive topography (section 8.12, p. 203, see also Raup's analysis of snail shells, Figure 13.6, p. 339); the adaptive topography is a surface of the average fitness of a population plotted on axes of gene frequency, whereas an adaptive zone has individual fitness plotted for the form of morphology; but the two are closely related. The common thread linking the ideas is that certain sets of adaptations work well in nature, but others do not.

The ecological species concept, therefore, can explain why the degree of interbreeding is limited in nature. If the concept is correct, crosses between species would tend to produce ecologically inferior forms; at least some of the offspring would be fitted to the gaps between species' adaptive zones—the valleys in the adaptive topography. The biological and ecological concepts are closely related; but there is room for considerable difference between them, in two areas. One is the question of whether one better explains why species form discrete phenetic clusters; the other, whether one provides a superior species concept. According to the biological species concept, as we saw, species form discrete units because of gene flow. The ecological species concept emphasizes selection instead: species form discrete units because selection favors certain forms and removes forms that are intermediate between species. In a particular real case, if enough work was done, the controversy could be settled empirically. For a set of populations, we could measure the degree of morphologic uniformity, and work out the extent to which migration (gene flow) and selection were maintaining it. In practice, extensive enough measurements have hardly ever been made, and then only for unnatural systems. As the next section describes, the controversy has had to be studied using less direct, and less conclusive, kinds of evidence.

15.2.6 Selection and gene flow can both explain the integrity of species

The reproductive and ecological aspects of species are probably correlated in nature: if the members of a species are interbreeding with too wide a range of mates, and producing maladaptive offspring, selection should adjust their mating preferences. Because of the correlation, it is difficult to test between the ideas, to find out whether gene flow or selection provides a better explanation of species integrity. Gene flow (migration) can rapidly unify the gene frequencies of separate populations if selection is weak (section 5.13, p. 122). But if selection is strong it can in theory keep populations distinct, and it is an empirical question how important the two processes are in nature. Here we shall look at a few test cases, in which the processes come into conflict. The reproductive and ecological concepts make different predictions when: (a) interbreeding is present, but selection favors divergence; and (b) when interbreeding is absent but selection favors uniformity. Ehrlich and Raven discussed examples of each in a classic paper in 1969, in which they strongly criticized the idea that species are held together by gene flow.

Selection can produce divergence despite gene flow

Bradshaw carried out a major ecological genetic study of the plants, particularly the grass *Agrostis tenuis*, on and around spoil-tips in the UK. Spoil-tips are deposited from metal mining and contain high concentrations of such poisonous heavy metals as copper, zinc, or lead. Few plants can grow on them. Some plants have colonized them, however, and the grass *A. tenuis* has been studied most closely. *Agrostis tenuis* has colonized the spoil-tips by means of genetic variants that are able to grow where the concentration of heavy metals is high; around a spoil-tip, therefore, there is one class of genotypes growing on the tip itself, and another class in the surrounding area. Natural selection works strongly against the seeds of the surrounding forms when they land on the spoil-tip: they are poisoned. There is also selection against the resistant forms off the spoil-tips; the reason is less clear, but there is probably some cost, or adaptive trade-off, in possessing a detoxification mechanism. Where the mechanism is not needed the grass is better off without it.

Populations of *A. tenuis* show divergence, in that there are markedly different frequencies of genes for metal resistance on and off the spoil-tips. The pattern is clearly favored by natural selection: but what about gene flow? The biological species concept predicts that gene flow will be reduced; otherwise the divergence could not have taken place. In fact, gene flow is large. Pollen blows in clouds over the edges of the spoil-tips and interbreeding between the genotypes is extensive. In this case, selection has been strong enough to overcome gene flow.

The conditions on a spoil-tip are exceptional and the example is more useful to illustrate a possibility than to determine the normal pattern in nature. Most other examples of populations that have diverged in the face of gene flow are also for exceptional conditions. This is what we should expect. If gene flow and selection come into conflict there is immediate selection to alter reproductive behavior. There are signs that this is already happening in *A. tenuis*. The flowering times of the resistant and normal types are already significantly different, which will reduce the amount of interbreeding. If the selection for and against heavy metal tolerance in close proximity continues, we should expect reproductive isolation to evolve. The grass will speciate. The conflict between gene flow and selection will be short lived, because either the gene flow pattern, or the selection regime, will alter.

Selection can produce uniformity in the absence of gene flow

Ehrlich and Raven also discussed a number of cases in which a species is subdivided into several geographically separate populations. The gene flow between the populations was arguably trivial or zero, yet the populations were remarkably uniform. Unfortunately, all the evidence was unquantitative. The recurring pattern in one example after another was suggestive at least—even persuasive for some—but each example could be criticized. The checkerspot butterfly *Euphydryas editha*, which Ehrlich has extensively studied himself, illustrates the point.

> Colonies of the butterfly occur scattered throughout California, many of them separated by distances of several kilometers and some by gaps of nearly 200 kilometers. It has been demonstrated that there is almost no gene flow in this species over as little as 100 meters. For this reason, there seems no possibility that gene flow "holds together" its widely scattered populations.

The argument may well be true. However, it should also be noted that a skeptic would want more exact evidence about morphologic similarity, gene flow, and selection in relation to morphology.

Another example comes from the work of Ochman *et al.* on the snail *Cepaea nemoralis* in the Pyrenees. The snail rarely lives above 1500 meters in the mountains, and never above 2000 meters, because of the cold. In the Pyrenees, it lives in neighboring river valleys separated by mountains: where those mountains are higher than 1500 meters, gene flow between valleys will be absent—and there is probably little gene flow even between the valleys with lower mountains. If gene flow is required to maintain the integrity of the species (i.e. the similarity of gene frequencies), populations in different valleys should have diverged. Ochman *et al.* observed the shell morphologies (which vary in banding pattern, see section 11.1, p. 267) and measured gene frequencies by gel electrophoresis in 197 populations. Shell morphology showed little geographic differentiation: the frequency distribution of color forms and banding patterns was much the same in all populations. For protein polymorphisms, the situation was more complex. The frequencies of different protein forms had diverged between different areas, but in a pattern that transcended the mountainous barriers to gene flow. Ochman *et al.* recognized three main areas, within which protein form frequencies were relatively constant; but within each of the areas there are several valleys separated by mountains. Gene flow cannot explain the uniformity within each of the three areas, and particularly not within the high central area (Figure 15.3). *Cepaea* in the Pyrenees, therefore, is another case in which selection maintains uniformity between populations when there is no gene flow.

Further evidence comes from asexual species. There is no reason why only sexual, and not asexual, forms should inhabit niches, and selection should therefore maintain asexual species in integrated clusters much like sexual species. But if gene flow is more important in holding a species together, asexual forms should have diverged more than related sexual species. Unfortunately, the evidence published so far is indecisive. Many authors—especially critics of the biological species concept, Simpson being an example—have asserted that asexual species form integrated phenetic clusters just like sexual species. Maynard Smith (1986), by contrast, has pointed to the example of hawkweed (*Hieracium*). It reproduces asexually and is highly variable, such that taxonomists have recognized many hundreds of "species" and no two taxonomists agree on how many forms there are. But then an analysis of rotifers, by Holman, seems to show the opposite. He compared bdelloid rotifers, which are asexual, with monogont rotifers, which can reproduce sexually, and argued that species of the asexual bdelloids are more consistently recognized by different taxonomists than are the species of the sexual monogonts. Asexual species, therefore, are a potentially interesting test case; but the evidence that has so far been assembled does not point to a definite conclusion.

Selection and gene flow are probably not usually opposed forces in nature. Gene flow, tending to break down differences between two forms, and natural selection, working on reproductive behavior, should normally unify the ecological and reproductive species concepts. The few examples we have looked at here show that natural selection can be powerful enough to overcome gene flow, and to produce integrity without the help of gene flow; but the examples only illustrate a possibility, not prove a generalization. By themselves, they do not demonstrate that the ecological is a more powerful species concept than the biological, and the controversy between these two concepts will continue to inspire research in evolutionary biology. (Similar issues have been raised in relation to sympatric and allopatric, and parapatric and allopatric, speciation, which we discuss in the next chapter.)

15.2.7 A pluralistic species concept
There seem to be two possible conclusions to the controversy between ecological and reproductive species concepts. One would be to accept that one or the other of the processes is fundamentally more important. If it were shown, for example, that species of a similar degree of integrity are formed in the absence of gene flow and in its presence, or (on the other hand) that populations diverge in the absence of gene flow even when selection favors

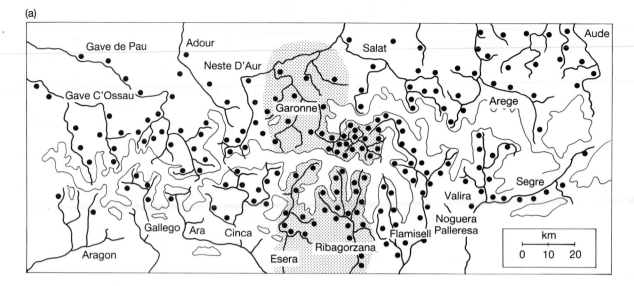

(a)

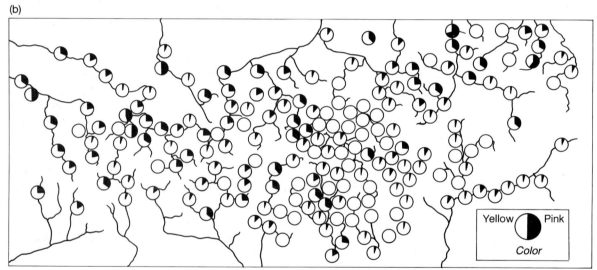

(b)

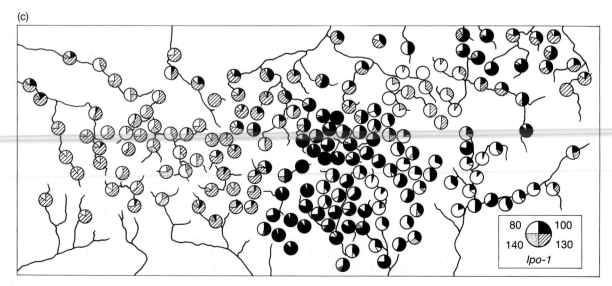

(c)

uniformity, then the ecological or reproductive concept, respectively, would be supported. Alternatively, we might be led to conclude that both factors work in different degrees in different taxa to produce the kind of phenetic clusters we call species. Some species may be more ecological, others more reproductive. We should then need more than one species concept. Cain, for example, in 1954, suggested that we need at least three. More recently, Mishler, Donoghue, and Brandon have favored a "pluralistic" species concept, which would explicitly recognize that no single concept accounts for all species. The evidence on gene flow and selection in different species hardly warrants any firm conclusion. Even if we cannot identify the relative importance of the different processes—or show that one process is overwhelmingly important—we probably have identified the main processes at work. Interbreeding and selection within adaptive zones are probably the reason why species form phenetic clusters in nature.

15.3 The cladistic species concept defines species throughout their evolutionary history

The species concepts we have discussed so far aim to define how, and explain why, individuals should be grouped into species at one time instant. Species last for longer than an instant. They last for many generations and another species concept may be needed to include all the members of a species throughout its existence. In this section, we shall see that all five instantaneous concepts—the phenetic, biological, recognition, ecological, and pluralistic species concepts—lead to undesirable conclusions if applied in evolutionary time; another concept is therefore needed. The problem can best be seen by looking again at the material we discussed in chapter 14, on phenetic and phylogenetic classification.

Consider the branching diagrams in Figure 15.4. The species concepts we have discussed so far have been "horizontal." At any time, they seek to define which individual organisms belong in which line—which organisms are in species 2 and which in 3. The question now is "vertical": what length of line should be included in one species? We start by asking whether the horizontal concepts can answer the question: we shall see that their answers are unsatisfactory and then move on to consider a different kind of answer. In the case of Figure 15.4, we can ask for each of the five concepts (phenetic, biological, recognition, ecological, pluralistic) how much of the line 1–2 will be included in a single species. The x-axis has been drawn as representing phenetic distance; but it will represent all five horizontal concepts if only certain sets of phenetic forms can interbreed and those sets fit neatly into the available ecological niches.

Figure 15.3 (*Opposite*) (a) Map of the Pyrenees, showing sites where the snail *Cepaea nemoralis* was sampled, and river valleys. Rivers are separated by high ground and mountains, and the shaded regions indicate heights exceeding 1500 meters. (b) Shell morphology (in this case, background color), shows little geographic variation; whereas (c) protein polymorphism falls into three main areas. The map is for the four alleles of one enzyme, indophenol oxidase (*Ipo-1*) The similarity within mountainous areas is unlikely to be maintained by migration (gene flow). From Ochman *et al*. (1983).

(a)   (b)

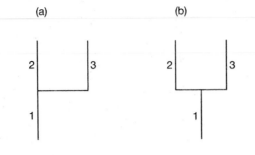

Figure 15.4 Two types of speciation. (a) The ancestral species does not change phenetically (or reproductively or ecologically) after a daughter species evolves. Species 1 and 2 are phenetically identical but cladistically different. (b) The ancestral species changes after the evolution of the new species; species 1 and 2 are both phenetically and cladistically different.

The horizontal concepts are then unambiguous in Figure 15.4. In Figure 15.4a, for all five concepts, the line 1–2 is only one species because the individuals throughout the lineage are a phenetic, ecological, and reproductive unit. (Let us ignore the pedantic point that individuals from different times cannot interbreed; we are only interested in whether they could interbreed if they were contemporary. Formally, this implies that the reproductive concept is defined in terms of isolation or mate recognition systems rather than interbreeding.) In Figure 15.4b, the lineage 1–2 would on all five concepts probably contain two species, because at the time of the split the population evolved to a different phenetic (and therefore reproductive and ecological) form. So for the horizontal concepts, the pattern in Figure 15.4a differs from that of 15.4b: in the former there are two species, in the latter there are three. The fact that a species has branched off between 1 and 2 in Figure 15.4a is irrelevant to the species definition on all the horizontal concepts.

Now let us turn to a gradually changing lineage, as in Figure 15.5. How do the horizontal concepts divide up this lineage? All five will divide the lineage into more than one species, if the lineage is long enough: the longer the lineage is, the more species they will recognize. However, difficulties arise when we ask exactly how the division is to be carried out. The phenetic species concept will run into the usual problem. Any division of the line is arbitrary, because the dividing point depends on the phenetic distance measure: different distance statistics will divide the lineage in different ways. And what about the ecological and reproductive species concepts? Do they

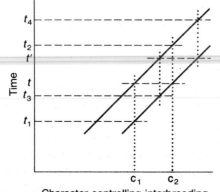

Figure 15.5 A species that changes gradually through time. Is the lineage one species, or more? And if it is more than one, how should it be divided? The thickness of the lineage corresponds to the array of forms that exist, and interbreed, at any one time. Thus at time t, the array of forms from c_1 to c_2 recognize each other as mates and can interbreed; forms outside that range do not interbreed. If we used the range c_1–c_2 to define a species, we should include the populations between t_1 and t_2 in a single species, and start a new species after t_2.

provide non-arbitrary criteria to divide the lineage? If we take a lineage at any one time, it will contain a range of forms that can interbreed. It might naïvely be thought that the length of the lineage to be included as a species can be objectively defined by looking forward and backward in the lineage and including in a species the length corresponding to the array of forms that could interbreed. But this definition is arbitrary, because the place we start determines what is included in the species (Figure 15.5). If we pick time t as the starting point, we define the length from t_1 to t_2 as a species. But this definition is arbitrary, because if we had started instead at time t', the biological species would have been from t_3 to t_4. The content of the species depends on the starting point, and because there is no objective way to pick that, the whole definition of the species becomes arbitrary. The same point will apply if the width of the lineage refers not to interbreeding but to an ecological niche or adaptive zone. Therefore, none of the horizontal species concepts provide an objective criterion for dividing a temporal lineage into species.

An objective solution can be found in Hennig's cladistic principle of classification (chapter 14). The lineages between branch points exist objectively; it is only when we try to divide up a single branch that we run into the problem of arbitrary division. We can define as a *cladistic species* that set of organisms in a lineage between two branch points. (To be more exact, we need to take account of extinction and the possibility that a species is still alive. A fuller definition of a cladistic species is that a species is the set of organisms between two branch points, or between one branch point and an extinction event or a modern population.) Therefore, in Figure 15.4 the two patterns are cladistically identical: in Figure 15.4a and b there are three species and species 1 gives birth to two new species at the branch point. Whereas on the horizontal concepts there are two species in Figure 15.4a and three in Figure 15.4b, on the cladistic concept there are three in both. This is because the cladistic concept recognizes species by branch points, independently of how little or how much change takes place between them. For the same reason, in Figure 15.5, there is only one cladistic species regardless of the extent of phenetic, reproductive, or ecological change.

There is no orthodoxy among evolutionary biologists about how to treat cases of branching without change (Figure 15.4a) and extensive change without branching (Figure 15.5). Our main purpose here is to use the extreme theoretical cases to illustrate the different concepts. I would point out, however, that the objections to the cladistic treatment of the two cases are more-or-less phenetically inspired, and lead to the general phenetic difficulties: arbitrary and subjective classification. In Figure 15.4, there could be any number of intermediate stages between the two extremes, and if someone wants to recognize two species at one extreme and three at the other, they will have to make a subjective choice about how much change is needed for a new species to be recognized. The cladistic treatment requires no subjectivity: when there is a branch point, the ancestral species goes extinct and two new species are born. It can seem strange at first, but it has the advantage over the alternatives of objectivity.

Moreover, the cladistic species concept and the non-temporal biological (or ecological) concepts do not have to be opposed: indeed, they can help each other. The cladistic concept is not a replacement for the instantaneous concepts. By itself, the cladistic concept offers no account of what a lineage or a branch point is, whereas the biological and ecological concepts both offer plausible accounts. By themselves, the biological and ecological concepts are inadequate when applied through evolutionary time; they can therefore benefit from the cladistic concept.

The cladistic species concept, as defined here, is not the only temporal species concept. One other has been influential. It is Simpson's *evolutionary species concept*; it defines a species as "a lineage (an ancestral-descendant sequence of populations) evolving separately from others and with its own unitary evolutionary role and tendencies." The definition is vague, but of all the concepts in this chapter it is probably closest to the ecological concept, with an added time dimension. (Box 15.2 describes an interesting philosophic consequence of the temporal species concepts for the nature of species in biology.)

In summary, we need both a definition of species at a time instant, and a definition for all evolutionary time. At any one time, species may be defined by interbreeding (biological and recognition concepts) or by a shared ecological niche. If these concepts are applied through time, they lead to arbitrary divisions of lineages; they are therefore best supplemented by a temporal concept such as the cladistic species concept for their most general, temporal, form.

15.4 Taxonomic concepts may be nominalist or realist

15.4.1 The species category

When we classify the natural world into units such as species, genera, and families, are we imposing categories of our own devising on a seamless natural continuum, or are the categories real divisions in nature? The problem is an old one. It applies to all taxonomic categories, but has particularly been discussed in the case of the species category. The idea that species are artificial divisions of a natural continuum is called *nominalism*; the alternative, that nature is itself divided into discrete species, is called *realism*.

The question is often posed in phenetic terms, but the answer depends on what species concept we are using. If we use a phenetic species concept, the correct question to ask is as follows. If we measure many individual organisms phenetically, do they fall into clusters or are they evenly scattered through the morphospace? Do the species exist in the frequency distribution for individual variation, or do they have to be imposed on a continuum? The answer is that clusters are indeed usually found, especially if we sample from only a limited geographic area. (Discrete clusters would not be found, for example, in a sample from the full geographic range of the ring species *Larus argentatus* and *L. fuscus*.)

The most striking evidence that species exist as phenetic clusters comes from folk taxonomy. People working independently of Western taxonomists usually have names for the species living in their area, and we can look whether they have hit on the same division of nature into species as have Western taxonomists working with the same raw material. Some tribes, it seems, do classify species much like Western taxonomists. The Kalám of New Guinea, for instance, recognize 174 vertebrate species, all but four of which

Box 15.2 Species in evolutionary theory are ''individuals'' rather than ''classes.''

The species concepts we have discussed turn out to have an interesting philosophic consequence. Species have often, particularly by philosophers, been treated as classes. A class is, in this sense, a set: if an object has the defining property of a set, then it is a member of the set. Gold, for example, is the set of atoms that have atomic number 79: any atom that has atomic number 79 belongs in the class of gold atoms. It is tempting to believe that species are classes in the same sense. The complete definition is rather a mouthful, but can be listed. For example (and it is not important to understand the following anatomic detail, which is illustrative merely!), the diagnostic characters in Scott's *Butterflies of North America* imply that the yellow and black colored swallowtail *Papilio machaon* would be the class of butterflies with six functional legs, one pair of tarsal claws, and a single anal vein in the hindwing (all of which defines the family Papilionidae), a bent-down, or ''fluted,'' shape in the hind margin of the male's hindwing (which defines the genus *Papilio*, in North America), and whose eyespot on the upper side of the hindwing touches the wing edge (the eyespot in other species has other positions).

On the phenetic species concept, species are classes. An organism is a member of a phenetically defined species if it has the phenetic characters needed to make it a member of the species. However, as we have seen, the phenetic species concept is not generally accepted. Species are instead thought of as ecological or reproductive units at one time and as cladistic lineages through time. On these concepts, species are not classes, but ''individuals.'' The argument is mainly due to Ghiselin and Hull.

The term ''individual'' does not here have its normal biological meaning. Biologists use individual as a synonym for organism, but in logic an individual is a particular, or unique entity. Particular organisms are good examples of individuals in the logician's sense. Charles Darwin was an individual—the person who was born in 1809, went to University at Edinburgh and Cambridge, sailed in the

Beagle, moved to Down House in Kent in 1839, wrote *On The Origin of Species*, and died in 1882. As an individual, he was a member of many logical classes, such as the class of Cambridge undergraduates and the class of residents in Kent. Organisms are not the only logical individuals, however. The book you are reading is an individual book—it had a physical beginning at the printing press and will physically come to an end sooner or later—and is a member of the class of all books. Mars is an individual, and a member of the class of planets, and so on. The difference between a class and an individual is that a particular individual is born and dies, whereas a class is spatiotemporally unrestricted. Whereas this copy of the book will come to an end (i.e. ''die''), the class of all books exists forever (indeed, the class exists whether or not any actual books exist).

Biological species, according to Ghiselin and Hull, have the property of being individuals, or particulars, in the logician's sense. A species, like species 1 in Figure 15.4, comes into existence at a speciation event, lasts for a while, and then becomes extinct at the next speciation event. The important point is that an organism is a member of species 1 by virtue of being in that lineage between two speciation events and not because of any phenetic defining attribute it may possess. It will often be true that some phenetic attributes will be shared by all the organisms in a lineage. These are the diagnostic characters used by taxonomists to recognize species. However, the possession of common characters is a contingent matter, and if there are no common characters, that does not affect in the least the fundamental species definition; it just makes the species more difficult to recognize. Likewise, even if there are phenetic characters held in common between all the members of a species at one time, those characters may be lost as the lineage evolves; as we have seen, this makes no difference to the species definition. With the cladistic species concept, a species lineage may undergo an indefinitely large amount of change without changing into a new

continued on page 404

species, provided only that it does not split.

Species are, in this respect, analogous to individual organisms. An individual organism, like Charles Darwin, undergoes extensive phenetic change between the single egg stage (when a mammal resembles a protozoan) and old age. The phenetic change is irrelevant to his definition. Charles Darwin is the individual between one particular event of birth and one particular event of death, regardless of his phenetic appearance between. An individual organism can undergo extreme phenetic change (from a caterpillar to a butterfly, for example) without changing into a new organism. So it is with species. They can undergo phenetic change without becoming new species.

It follows that the definition of the swallowtail butterfly *Papilio machaon* as a set, at the beginning of this section, must be wrong. The reason is that, although *Papilio machaon* may be definable that way now, if one of them were to mutate to a new wing color pattern (while still interbreeding with the others) it would be in the same lineage even though it violated the set or class definition of the species. The set definition is just a contingent set of criteria, which may happen to work now but do not truly define the species. Likewise, it might have been possible to define the individual Charles Darwin by certain phenotypic characters at a certain time (e.g. in 1840 he might have been defined as the set of all people who are male, live in Upper Gower Street, Bloomsbury, are members of the

Geological Society of London, are clean-shaven, and wear a blue waistcoat when sitting for their portrait); but that does not mean he could *really* have been defined by this, or any other, set of phenotypic characters. When he moved to Kent and grew a beard, he was still Charles Darwin, as he would remain whatever his phenotype.

Species, then, are particular individuals, not classes. The main point here is to understand the difference between an individual and a class, and why all the species concepts except the phenetic concept imply that species are individuals. We might also notice that the philosophic distinction is important. For example, scientific laws can be expressed for classes, but not for individuals. If species are individuals, laws cannot be written about them. The statement "all swans are white" may sound like a scientific law, but it cannot be one. In terms of the discussion above, the reason why it cannot be a law is as follows. If swans were defined phenetically, and their white color was part of the definition, the generalization would be tautologically valid. However, on the biological, ecological, and cladistic species concepts, swans could evolve (without a split in the lineage) to be non-white. The new non-white swans will still be part of the swan lineage. The generalization "all swans are white," therefore, can only be a contingent empirical truth, not the kind of necessary truth that qualifies as a scientific law.

Further reading. Ghiselin (1966, 1987) and Hull (1978, 1988).

Box 15.2 (*continued*) Species in evolutionary theory are "individuals" rather than "classes."

correspond to species recognized by Western taxonomists. Other tribes in other places do not agree quite so closely with museum classifications; but a reasonable conclusion is that folk taxonomists and professional Western taxonomists agree to a large extent.

However, the fact that independently observing humans see much the same species in nature does not show that species are real rather than nominal categories. The most it shows is that all human brains are wired up with a similar perceptual cluster statistic. That cluster statistic, however, has no objective priority over any other cluster statistic, and it would be easy (for the reasons discussed in chapter 14) to invent another statistic that classified the birds of New Guinea differently from the Kalám or the taxonomists of a natural history museum. In this sense, the question of whether phenetic species exist nominally or really is meaningless. For the same set of observations, one cluster statistic might sense a continuum while another cluster statistic sensed discrete clusters. Phenetic taxonomy certainly in practice does reveal clusters, and cross-cultural anthropology supports these clusters in some cases; but those clusters have no firm theoretical justification. They suffer from the general problem of phenetic classification.

The other species concepts have sounder theoretical foundations, and the question of nominal and real classification can be answered more conclusively for thém. For the reproductive species concepts (in both the biological and recognition versions), the question is whether species exist as reproductive communities. It could be that there is a gradual blurring off of interbreeding as we move to more and more distant forms; but in practice, though mating is often non-random within a species, interbreeding within a species is typically extensive and hybrids between species are rare. For example, in a sample of 725 copulating pairs of two similar cicada species (*Magicicada*) breeding in an area near Chicago, Dybas and Lloyd found only seven heterospecific pairs; and even those seven pairs probably did not produce offspring. The reproductive criterion is clear-cut; real species do exist in the reproductive sense. The same is probably true for the ecological species concept. In this case, however, we usually have no external measure of the adaptive zones in nature and we cannot say for sure whether they are discrete and separated by gaps, or blur into each other. Actually, the strongest reason to think that niches are discrete is that species form phenetic clusters—but this argument could only persuade someone who accepts the ecological species concept to begin with: others might interpret the evidence differently.

The controversy between realistic and nominalistic concepts applies more to instantaneous than temporal classification. But there is a strong argument that species really exist in the cladistic sense. The reality of ecological and reproductive lineages at any one time will guarantee that the phylogenetic hierarchy has real branches in it, and that there are cladistic species. Indeed, to deny that there are real cladistic species would be to deny the existence of a phylogenetic tree of life. We can therefore conclude that species, in the reproductive, (probably) the ecological, and the cladistic senses are real and not nominal units in evolutionary theory.

15.4.2 Categories above the species level

Those evolutionary biologists who support the biological species concept characteristically differ from those who support the ecological species concept in their attitude to the reality of taxa above the species level. The biological (and recognition) species concepts can apply to only one taxonomic level. If species are defined by interbreeding, then genera, families, and orders must exist for some other reason. Mayr has been a strong supporter of the biological species concept and (in 1942, for example) duly reasoned that species are real, but that higher levels are defined more phenetically and have less reality; higher levels are relatively nominalistic. Dobzhansky and Huxley held a similar position.

Simpson, however, favored a more ecological theory of species. The ecological concept can apply in much the same way at all taxonomic levels. If there is a narrow adaptive zone for the lion, there can be a broader zone for the genus *Felis*, and a still broader zone for the Mammalia. Adaptive zones could have a hierarchic pattern corresponding to (and causing) the taxonomic hierarchy. All taxonomic levels could then be real in the same way. The relative reality of the species, and of higher taxonomic levels, is therefore part of the larger controversy between the ecological and reproductive species concepts.

15.5 Conclusion

In evolutionary biology, the interesting questions about species are theoretical. The practical question of which actual individuals should be classified into which species can on occasion be awkward; but biologists do not tie themselves in knots about it. The majority—perhaps over 99.9%—of specimens can easily be fitted into conventionally recognized species and do not raise even practical problems. Other specimens can be identified after a bit of work, or even left on one side until more is learned about them.

The more interesting question is why variation in nature comes arranged in the clusters we recognize as species. There are several possible answers, as we have seen. Different horizontal (non-temporal) species concepts follow from different ideas about the importance of interbreeding (or gene flow) and natural selection. It is sometimes possible to test between them, but the results so far have not been sufficient to confirm any one concept (or any plurality of concepts) decisively. However, there is general agreement that phenetic distinction alone is not an adequate concept, and that the key theoretical processes are interbreeding and ecological distinction.

15.6 Summary

1 Species in practice are defined by easily recognizable phenotypic characters that reliably indicate what species an individual belongs to.

2 The phenetic species concept defines a species as a set of organisms that are sufficiently phenetically similar to one another.

3 The biological species concept defines a species as a set of interbreeding forms. Interbreeding between species is prevented by isolating mechanisms.

4 The recognition species concept defines a species as a set of organisms with a shared specific mate recognition system: different members of the species recognize one another (by an evolved method of communication) as potential mates.

5 The ecological species concept defines a species as a set of organisms adapted to a particular ecologic niche. Interbreeding between species is selected against because hybrids would not be adapted to a natural adaptive zone.

6 The biological species concept explains the integrity of species by gene flow, the ecological concept by selection. The two processes are usually correlated, but can be tested between in special cases. Selection can be strong enough to overcome gene flow, and selection can maintain a species' integrity in the absence of gene flow.

7 The cladistic species concept defines a species as the members of an evolutionary lineage between two branch points. If the biological or ecological

species concept is applied temporally, it can lead to arbitrary divisions of a lineage.

8 The question of whether species exist as real, or only nominal, phenetic clusters has no deep theoretical meaning. Species, however, do realistically exist in the sense of the biological species concept.

9 Using the biological species concept, species can be real, but not the higher taxonomic levels. With the ecological species concept, all taxonomic levels from species to kingdom can have a similar degree of realism.

15.7 Further reading

Mayr (1963) is the classic account of the species in evolutionary biology. Mayr (1969) is more introductory, and Mayr (1976, 1981, 1988) update his views on various details. Dobzhansky (1970), Huxley (1942), Cain (1954), and Simpson (1961b) also contain classic material. Wiley (1981) contains a more recent account. See Raven (1976) and Levin (1979) on species in plants. Hull (1965) and Ghiselin (1984) explain the distinction between the theoretical definition and practical recognition of species.

On the phenetic species concept, see Sneath and Sokal (1973) and many of the references in chapter 14. On the biological concept, see the Mayr references. For isolating mechanisms see the books by Mayr and Dobzhansky. On plants, see Levin (1978). See Paterson (1978, 1985) on the recognition concept; Coyne *et al.* (1989) and Mayr (1988) criticize it. On the ecological concept, see Van Valen (1976). The controversy between selection and gene flow was stimulated particularly by Ehrlich and Raven (1969); see also Jackson and Pounds (1979) and Slatkin (1985). On heavy metal tolerance in plants, see Antonovics *et al.* (1971), Bradshaw (1971), Ford (1975)—and Antonovics *et al.* (1988) for more recent related work. Mishler and Donoghue (1982) and Mishler and Brandon (1987) favor the pluralistic, and Ridley (1989) the cladistic, concepts. See Van Valen (1971) on adaptive zones, and Berlin (1973) and Gould (1980a) on folk taxonomy.

Speciation

16.1 How can one species split into two reproductively isolated groups of organisms?

The crucial event, for the origin of a new species, is reproductive isolation. The members of a species do usually differ genetically, ecologically, and in their behavior and morphology (i.e. phenetically) from other species, as well as in who they will interbreed with, and it may be that none of these criteria provide a universal species definition, applicable to all animals, plants, and microorganisms (chapter 14). However, many species, if not all, differ by being reproductively isolated. To understand in these cases how a new species comes into existence, we need to understand how a barrier to interbreeding can evolve between the new species and its ancestors.

Evolutionary biologists have studied the circumstances, and process, of speciation in a number of ways. They have considered whether new species evolve in geographically isolated populations; whether species become reproductively isolated while geographically separated, or if the process only starts there and is completed if the incipient species later re-encounters its ancestors. They have considered whether the populations that give rise to new species are peculiar in any way, for instance by being small. The genetics of speciation has been studied, to see what kinds of genetic changes give rise to reproductive isolation. And work has also been done on how long it takes for one species to split into two. All these topics are considered in this chapter. However, we shall not discuss again one particular kind of speciation: instantaneous speciation by polyploidy (see section 3.5, p. 42). The process is important in plants, but probably does not operate much in other taxa. (All species that have evolved by this mechanism are examples of sympatric speciation, a topic to be discussed later in this chapter.)

16.2 A newly evolving species could theoretically have an allopatric, parapatric, or sympatric geographic relationship with its ancestors

In the abstract, the following series of events must happen in the origin of a new species. We start with a single species, made up of a set of interbreeding organisms. A genetic variant (or several variants) must spread through part of the species, and the bearers of this genetic variant must mate only (or preferentially) with other bearers of the same genetic variant. If the reproductive isolation starts as a mating preference, it must eventually tighten up to exclusive breeding with its own type. Once it has done so, the species will have split into two: from one initial population, two separate interbreeding populations will have evolved. At some stage along the way, further phenetic, ecologic, and behavioral differences associated with the reproductive isolation may also evolve. The big question about speciation is why, and in what circumstances, the genetic variant causing reproductive isolation should evolve. The first matter to consider is whether the new

(a)

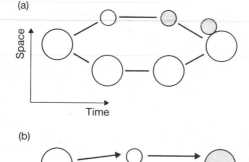

(b)

(c)

Figure 16.1 Three main theoretical types of speciation are distinguished, according to the geographic relationships of the ancestral species and the newly evolving species. (a) Allopatric speciation: the new species forms geographically apart from its ancestor. (b) Parapatric speciation: the new species forms in a contiguous population. (c) Sympatric speciation: a new species emerges from within the geographic range of its ancestor.

species evolves in geographic isolation from its ancestor, called *allopatric speciation*, or in a geographically contiguous population, called *parapatric speciation*, or whether a new species can evolve within the geographic range of its ancestor, called *sympatric speciation* (Figure 16.1).

16.3 Geographic variation is widespread and exists in all species

Johnston and Selander measured 16 morphologic variables in house sparrows (*Passer domesticus*) sampled from 33 sites in North America. The 16 characters could be reduced to a single abstract character of "body size" (to be statistically exact, this character was the first principal component). In Figure 16.2 the average body size of house sparrows in plotted on a map: two things should be noticed. The most important, for our purposes, is simply that the characters vary in space: house sparrows from one part of the continent differ from those in other parts. Almost every species that has been studied in different places has been found to vary in some respect. Not all characters vary, of course (humans have two eyes everywhere), but populations always differ in some characters. Moreover, if we look at enough species, and enough populations of them, we can usually find a case in which any particular character will vary geographically; this is particularly true if we concentrate on metric variation. Different populations differ in size, in color, in physiology, in their chromosomes, and in their behavior and courtship patterns; and geographic variation has been found at the genetic level, for gel electrophoretically assessed gene frequencies.

Mayr, most powerfully in his book *Animal Species and Evolution* (1963; chapter 11), has collected more evidence about geographic variation than anybody else and he concludes that "every population of a species differs from all others," and "the degree of difference between different populations of a species ranges from almost complete identity to distinctness almost of species level." Moreover, "all characters employed to distinguish species from each other are known also to be subject to geographic variation."

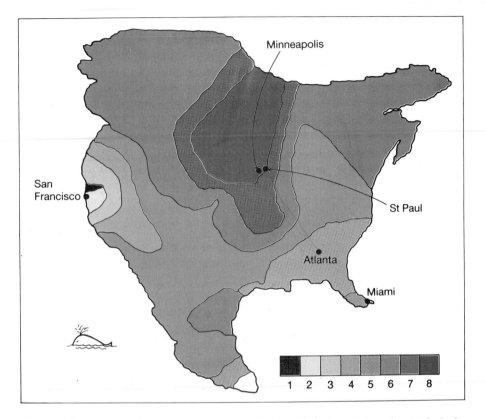

Figure 16.2 Size of male house sparrows in North America. Size is measured as a principal component score, derived from 16 skeletal measurements. The score of 8 is for the largest birds, the score of 1 is the smallest. From Gould and Johnston (1972), corrected from Johnston and Selander (1971).

That is, species differ by body size and proportions, and these also vary intraspecifically; species differ in their chromosomes, and these vary intraspecifically; species differ in their courtship, and genital morphology, and these also can vary within a species from place to place.

The second point to notice about Figure 16.2 is that the form of the geographic variation is explicable. House sparrows are generally larger in the north, in Canada, than in Central America. The generalization is imperfect (compare, for instance, the sparrows of San Francisco and Miami); but in so far as it applies, it illustrates *Bergman's rule*. Animals tend to be larger in colder regions, presumably for reasons of thermoregulation. Geographic variation in this species is therefore adaptive: the form of the sparrow differs between regions because natural selection favors slightly differing shapes in different regions.

If we drew a line on Figure 16.2 from Atlanta to the Twin Cities (Minneapolis and St Paul), or from the Twin Cities to San Francisco, and looked at the size of sparrows along it, we should have an example of a *cline*. A cline is a gradient of continuous variation, in a phenotypic or genetic character, within a species. Clines can arise for a number of reasons. In the house sparrows,

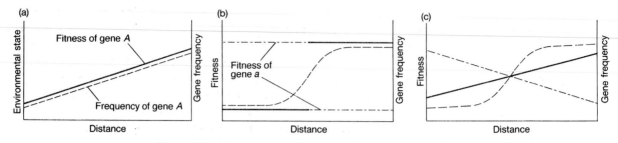

Figure 16.3 (a) A cline can arise in a continuous environmental gradient. The house sparrow example (Figure 16.2) probably has polygenic inheritance; the *y*-axis would more appropriately express the proportion of genes for a larger body size than the average for the USA. (b) A cline can also result when natural selection favors different genotypes in different discrete environments and there is gene flow (migration) between them; (c) is as (b) except that the environment changes gradually rather than suddenly. The figures are drawn for one locus with two alleles (*A* and *a*), but the concepts apply more generally.

the reason is probably that natural selection favors a slightly different body size along the gradient; sparrows are continuously adapted to their local environments (Figure 16.3a). A cline can also arise if two forms are adapted to two different environments separated in space, and migration (gene flow) takes place between them (Figure 16.3b,c); we shall return to the theory of clines later (section 16.5.1), but the important point at this stage in the argument is that geographic variation exists and often in the form of a continuous cline.

16.4 Allopatric speciation

16.4.1 Speciation may occur when a barrier is intruded within the continuous geographic variation

The extremes of geographic variation within a species can be almost as different from each other morphologically as are two closely related species, as was remarked in the passage quoted above from Mayr. Normally, however, we do not know whether the extremes can interbreed, because they never meet each other. An interesting test case is provided by ring species (section 3.4, p. 40). In ring species, the populations of a species are arranged geographically in a ring, and the endpoints do meet. When they meet, the two forms behave as perfectly good, reproductively isolated, biologic species. Two classic examples are the ring made by the herring and black-backed gulls around the North Pole and the ring of species of great tit (*Parus major major* to *P. major cinereus*) in Asia; they are not the only examples, however. Neither is thought to have evolved by one population gradually spreading out through space until it met itself again: the history of the subpopulations is difficult to reconstruct; but they do show how geographic variation within a species might lead to the origin of new species.

One species could split into two if a physical barrier divided its geographic range. The barrier could be something like a new mountain range or river cutting through the formerly continuous population. If the barrier is large enough, gene flow between the two groups would cease and each would evolve independently. Over time, different alleles would be fixed in each, whether because of the hazards of mutation and drift, or because selection favored different characters in the two. Whether two full species would evolve depends on whether the populations persist long enough for them to diverge to that degree. Suppose that the two populations have diverged to approximately the difference in genetics and morphology that we see between two species. The two might at this point meet again, either because the barrier between them disappeared or because of migration. If they had diverged enough, they might not recognize each other as members of the same species and would not interbreed: they would have speciated, and be isolated by a prezygotic isolation mechanism. There would be two species where there was formerly one. (The same process could occur with any kind of isolation mechanism (section 15.2.3, p. 389); the species might, for instance, mate but the offspring not develop. Isolation is then postzygotic.) Alternatively, the two populations might meet up again before they had diverged as far as separate species. They might recognize each other as conspecifics, interbreed, and their offspring develop normally; the two populations would then soon merge back into one. Or one of the two new species might be better adapted than the other, and drive it extinct.

16.4.2 When the diverged populations remeet, reproductive isolation may be reinforced by natural selection

When the two populations come back into contact, the reproductive isolation between them might be incomplete and they might exert a selection pressure on each other. The two forms might initially prefer to mate with their own type, but occasionally mate with the other type. If the hybrids between the two were at a disadvantage, selection would act to increase the reproductive isolation because each form would do better not to mate with the other and produce disadvantageous hybrids: speciation would be speeded up by selection in sympatry. The process is called *secondary reinforcement*. It is secondary if the reproductive isolation has partly evolved allopatrically, and is then reinforced on secondary contact. The process by which selection increases reproductive isolation, independently of the history of the populations, is simply called *reinforcement*. (It is also possible for selection to act to reduce the partial reproductive isolation, if the hybrids are as fit—or fitter—than the two original types.)

Reinforcement is a controversial theory. Many influential biologists, from Wallace to Fisher and Dobzhansky and their followers, have argued that it often operates during speciation. It is certainly possible in theory: if an individual has a choice between two sorts of potential mate, and would produce more offspring by mating with one sort than the other, selection undoubtedly will favor a mating preference. It will act to reinforce the prezygotic reproductive isolation.

Moreover, the process can be simulated by artificial selection. It is a simple matter of selecting for assortative mating. (A population in which each individual preferentially mates with another of its own genotype shows *assortative mating*; this is to be contrasted with random mating.) Figure 16.4 illustrates the results of an experiment by Kessler. *Drosophila pseudoobscura* and *D. persimilis* are sibling species (section 15.2.2, p. 388), and their degree of prezygotic ethologic isolation depends in the laboratory on temperature. At low temperature, females of each species will mate with males of the other

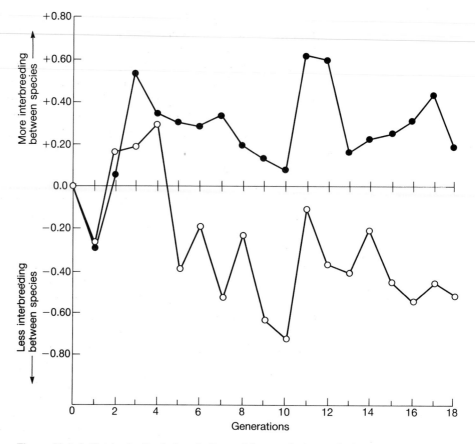

Figure 16.4 Artificial selection in female *Drosophila pseudoobscura* for increased (low isolation, solid circles) and decreased (high isolation, open circles) tendency to mate with male *D. persimilis*. The *y*-axis is an index of frequency of heterospecific mating. When the index is positive, females are more likely to mate with heterospecific males than are control females; when it is negative they are less likely to. From Kessler (1966).

(heterospecific) as well as their own (homospecific) species. Kessler watched the behavior of the females closely; he selected for increased ethologic isolation by breeding from females that he saw reject a heterospecific male and then mate with a homospecific male; he selected for decreased isolation by the reciprocal procedure. He obtained significant changes in prezygotic isolation in 18 generations. Selection can alter the mating preferences of females for different kinds of males.

However, the theoretical conditions for speciation to take place by reinforcement are difficult. That it can be imagined, and even acted out in an artificial selection experiment, does not mean it is generally important in nature. In real cases, there will be a time window after secondary contact during which reinforcement must happen, for the following reason. The simplest case in which selection favors reinforcement has two alleles A and a at a locus, and genotypes AA and aa have higher fitness, while Aa is selected against. Natural selection then favors AA types that mate preferentially with other *AA* types; and *aa* types that mate only with other *aa*. The result could be speciation, with one species having genotype *AA* and the other species *aa*. However, before speciation is complete, the selection against heterozygotes will also act to remove the less common of the two alleles (if *a*, for example, is rare it will tend to find itself more in *Aa* than in *aa* individuals and will accordingly be selected against). There will be an evolutionary race between speciation and the loss of the allele. The latter process will often be faster because it needs no new genetic variation, whereas speciation cannot happen without genetic variation for mating preferences. A more complex model might consider genetic differences like *AA* and *aa* at many loci, but here again selection will tend to remove the rarer genotypes until the population is uniform. Moreover, recombination between the genes controlling the mating preference and the genes under selection will frustrate the evolution of reinforcement.

In the case of two populations that have diverged partly in allopatry and then re-encounter each other, there is an additional factor: gene flow. Selection might reinforce any isolation between the populations. But until isolation is complete, gene flow will be acting to equalize their gene frequencies. Once the two populations become genetically the same, there can no longer be selection to decrease breeding between them.

Selection for reinforcement is therefore likely to be strongest immediately after the two populations meet. If the necessary genetic variation in mating preferences is present, and selection is strong enough, the two species may completely split; but, on theoretical grounds, a gradual slide back into a single species is also possible. We shall see below (sections 16.4 and 16.10) that in addition to the difficult theoretical conditions, the evidence for reinforcement is imperfect. A theory of speciation, therefore, can avoid a theoretical and empirical minefield if, while not excluding the possibility of reinforcement, it nevertheless does not depend on it. The allopatric theory has this virtue. In the allopatric theory, it could be that reproductive isolation only ever evolves in allopatry (and when it does not the two populations simply collapse back into one when they meet); alternatively, it could be that partial reproductive isolation sometimes evolves allopatrically and is then reinforced on secondary contact. Either possibility fits in with the general theory of allopatric speciation. The two alternatives—parapatric and sympatric speciation—depend strongly on the theory of reinforcement: indeed, they require it, and if reinforcement does not operate, neither do they.

16.4.3 Allopatric speciation may take place in peripherally isolated populations

The model of allopatric speciation discussed above proposed that some barrier physically divided the species' geographic range. Bush calls it speciation by subdivision, and Mayr the "dumb-bell" model; its characteristic is that the ancestral population is divided into two fairly equal halves (Figure 16.5a). It can be contrasted with the *peripheral isolate* model. Here the idea is that a small population, at the extreme edge of the species' range, is separated off (Figure 16.5b). The same sequence of divergence and, it may be,

(a)

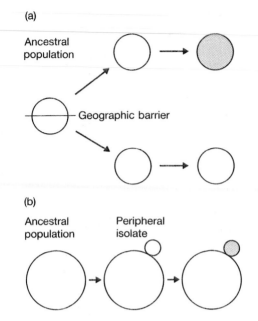

Ancestral population

Geographic barrier

(b)

Ancestral population

Peripheral isolate

Figure 16.5 Two models of allopatric speciation. Stippling indicates evolution of new species. (a) The dumb-bell model in which the ancestral species is divided into two roughly equal halves, each of which forms a new species. (b) The peripheral isolate model, in which the new species forms from a population isolated at the edge of the ancestral species range.

subsequent meeting of the two populations could then take place as in speciation by subdivision.

What are the relative frequencies of allopatric speciation via peripheral isolates and via subdivision? It has been argued, particularly by Mayr, that the process via peripheral isolates has been much more common in evolution. This is for two reasons. One is that it may be physically more probable that a small population would be isolated at the edge of a species range than that a barrier would divide the whole of a species range. The other comes from the pattern of geographic variation in modern species. It is thought to be particularly common for isolated populations at the edge of a species range to be distinct in form, whereas the individuals in the main part of the species range show less variation. Distinct peripheral isolates have often been observed on islands. Here is an example, in Mayr's (1954) own words.

> Let us look, for instance, at the range of the Papuan kingfishers of the *Tanysiptera hydrocharis-galatea* group [Figure 16.6]. It is typical of hundreds of similar cases. On the mainland of New Guinea three subspecies occur which are very similar to each other. But whenever we find a representative of this group on an island, it is so different that five of the six Papuan island forms were described as separate species and four are still so regarded.

The kingfishers on the peripheral islands have diverged more than would be predicted from the degree of variation on New Guinea. We should bear in mind that the phenomenon, of phenotypically distinct, peripherally isolated populations, has only ever been illustrated by examples, not supported by systematic study; but we can provisionally accept its reality. It has two possible explanations. One is local adaptation: the conditions on the island may be such that natural selection favors a distinct phenotype there. The second, favored by Mayr, is isolation: gene flow from the rest of the species

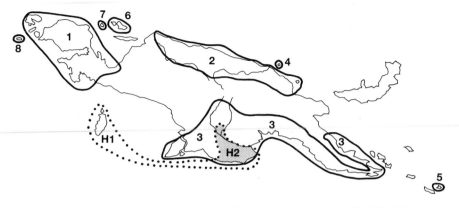

Figure 16.6 Kingfishers of the *Tanysiptera hydrocharis-galatea* group in the New Guinea region. The three forms on mainland New Guinea (1,2,3) are almost indistinguishable. The five island forms (4–8) are strikingly distinct and most were originally described as species. H1 and H2 make up the form *hydrocharis*. The Aru Islands (where H1 now lives) and southern New Guinea (H2) once formed a separate island and *hydrocharis* evolved there. When southern New Guinea joined the main island (dotted line), the south-east New Guinea subspecies (3) invaded the area, but did not interbreed with *hydrocharis*. *Tanysiptera hydrocharis* had therefore evolved into a full species while isolated. From Mayr (1942).

may be reduced on the island, allowing the population there to diverge. Whatever its explanation, if peripherally isolated populations are particularly likely to diverge from the form of their ancestors, they may indeed be a common stage on the way to speciation.

16.4.4 Much work has been done on the theory of allopatric speciation via peripheral isolates

Speciation by peripheral isolation may have a number of characteristics. The speciating population is likely to be small and at the edge of the species' range. It may live in relatively extreme physical conditions. Mayr, and his followers, have accordingly suggested that the speciating population may evolve rapidly (because the population is small) and by drift as well as selection. Because the peripheral isolate has been cut off accidentally, it may have a non-representative sample of the ancestral species' genes. In the *founder effect* (section 6.3, p. 127), the peripheral population is formed by a small number of founder individuals, who lack some of their ancestors' genes. This is a controversial idea, because it implies that speciation might be non-adaptive. It is an old criticism of natural selection that many species differences are non-adaptive, and if they indeed are, then a theory of non-adaptive speciation will be needed. Founder effect speciation, with random genetic loss and drift, would be one such theory. However, hypotheses of non-adaptation are fallible (section 11.1, p. 267), and there is no convincing evidence that any differences between species are non-adaptive. A final suggestion made by Mayr is that speciation might be accompanied by a *genetic revolution*: an extensive reorganization of the gene pool, which takes place in the extraordinary genetic and environmental conditions of the peripherally isolated population.

All these ideas have been influential (see, for example, the theory of punctuated equilibrium in chapter 19); but their importance in evolution is more difficult to assess. One hope might be that we could test them with facts. We might measure the amount of genetic variation, and the population sizes, of recently evolved species to see whether they are small. The snag is that both these variables can change rapidly, and we can only observe them in modern species: this is at best an unreliable guide to the population size when a species originated. However, one of the most important pieces of evidence against the founder effect is that recently evolved species, such as the fruitflies of Hawaii which have been suggested as examples of speciation by the founder effect, show extensive genetic variation. This does not decisively refute the founder effect, because the variation could have built up since the species originated. For the case of population size, there is a second possible source of evidence in the fossil record. One particularly detailed study (section 19.5.2, p. 514) of the Pleistocene snails of Lake Turkana by Williamson did not support the founder effect: in those snails, the speciating population contained "many millions of individuals." However, even that is only one result, and supporters of the founder effect will not be troubled by a few counter-examples. More systematic studies have been attempted, using the geographic ranges of recently evolved pairs of species, and they do not suggest that speciating populations tend to be particularly small. In summary, there is little evidence for the founder effect, but the evidence is not of decisive quality.

Likewise, we might try to test whether genetic revolutions occur at speciation by measuring the genetic differences between recently evolved species; but the test faces the same snag. The gel electrophoretic evidence does usually shows large differences between species, larger than the differences between populations within species (Table 16.1). This weakly supports the idea that extensive genetic change happens at speciation: if the

Table 16.1 Genetic identities of local populations and closely related species. A genetic identity can be between zero and one; when it is one, two samples are genetically identical (i.e. two samples from different populations are no more different than two samples from the same population). Genetic identity is the inverse of genetic distance. (Box 16.3, p. 436)

| Group | Identity among | |
	Local populations	Species
Drosophila (fruitfly)	0.97	0.35, 0.56*
Lepomis (fish)	0.97	0.54
Anolis bimini group (lizards)	0.97	0.21
Anolis rocquet group (lizards)	0.99	0.67
Thomomys (mammals)	0.93	0.85
Sigmodon (mammals)	0.98	0.76
Lupinus (lupin)	0.96	0.35
Stephanomeria (Compositae)	0.98	0.95

*Identities for subspecies and semispecies, respectively, within the *D. willistoni* species group.

differences between species had often been small, it would have counted against the idea. However, the test is weak because the large differences did not necessarily evolve at speciation. Indeed, proteins may evolve at a constant rate by the molecular clock (section 7.11, p. 164), and most of the changes would then probably have accumulated with each lineage after speciation. The idea of genetic revolutions, therefore, is also difficult to test, though in this case the weak evidence is more in its favor.

Meanwhile, population geneticists have done extensive theoretical work to investigate the logic of the idea. What effect, to begin with, should small population size have on genetic variation? The founder effect would be most likely to operate if the colonizing founder population by chance lacked one of the ancestral population's genes. If the founding population contains all the ancestral population's genes, and only differs in gene frequency, natural selection could adjust the frequencies back to the best adapted state. But if a gene is missing from the sample, natural selection cannot take the population back to the ancestral species' state, and a precondition may be established for the populations to diverge. However, the random loss of genes is theoretically unlikely except in very small samples and many theoreticians doubt the importance of the process (chapter 6). A further possibility is that sexual selection (section 11.4, p. 283) could operate differently in isolated populations, and a behavioral analogy of the genetic founder effect has been used to explain certain asymmetries in the interspecific mating preferences of Hawaiian fruitflies (Box 16.1).

Next, what relationship should we expect between population size and evolutionary rate? Mayr suggested that evolution is rapid in small populations. The rate of evolution within a population by random genetic drift is independent of population size (section 6.4, p. 129). For natural selection, there is no general result. If the rate of evolution is limited by the number of favorable mutations, evolution is faster in larger populations, which have a higher total mutation rate. If the rate of evolution is limited by selection pressure, there is no relationship between evolutionary rate and population size. (In the special case in which evolutionary change requires inbreeding, it is faster in smaller populations: see section 16.11 below.)

Population size also has a complicated relationship with evolutionary rate in the shifting balance theory of evolution (section 8.13, p. 205). An adaptive landscape can have multiple peaks. Wright argued that natural selection will drive evolution uphill to the nearest peak, but random drift is needed to explore the surface and drive evolution across the valleys. It has been suggested that speciation could take place when a population crosses a valley to a new peak in the landscape: this is called speciation by *peak shifts*. It has been suggested that a peak shift might be more likely in a peripherally isolated population than in the larger main population of the species.

What effect does population size have on the ability of a population to evolve from one adaptive peak to another? The answer depends on the shape of the landscape, as the theoretical work of Wright and Barton has revealed. Consider the adaptive landscape with twin peaks in Figure 16.7. It has one peak higher than the other; let us imagine that the population is initially at the lower peak. What chance is there of reaching the higher peak? In a large

Box 16.1 Asymmetric speciation.

The phylogeny of the Hawaiian drosophilid flies is exceptionally well understood (section 17.7, p. 457). If the cytogenetic evidence used to reconstruct their phylogeny is combined with geologic evidence about the ages of the islands, it is possible to make reasonably confident deductions, in some cases, about which species are ancestral to which. Kaneshiro carried out an interesting experiment on two ancestor–descendant species pairs: *Drosophila differens* and *D. planitibia*, which live on the islands of Molokai and Maui, and are close to the ancestors of *D. heteroneura* and *D. silvestris*, which live on Hawaii. (Figure 17.8, p. 459, gives a map and a phylogeny of these species: the four species are numbers 28, 29, 30, and 31 on the phylogeny.) For a pair of species, Kaneshiro placed a female of both species with a male of one of the species, and observed whether or not the female mated according to whether the male was the same species as her (homospecific) or the other species (heterospecific). The result was asymmetric (Table B16.1): females of the ancestral species in both cases mate preferentially with males of their own species; but descendant species females showed no preference. This result could only be found in an experiment; in nature the ancestral and descendant species are geographically isolated from each other.

Kaneshiro explained the result in terms of changes in courtship during the speciation event. The courtship of male fruitflies varies greatly among the hundreds of species, but is a complicated affair, lasting 15 minutes or more. It is made up of particular movement patterns (the courtship dance), sound production, stroking the female in particular ways, and even courtship feeding in some species. It thus contains many elements. Kaneshiro suggests that if the two species originated in founder events, by chance all the males of the new species might have lacked certain elements in the courtship. If the females of the founder population tolerated the new type of courtship, the species would survive (and if they did not, it would suffer extinction by sexual abstention).

Now imagine what will happen when (long after the speciation event) males and females of the ancestral and descendant types are combined in Kaneshiro's experiment. The ancestral females are given a choice between the males they normally meet, who perform the normal complete courtship and other males who perform an incomplete courtship display; the females will prefer the former. The same choice is also given to the descendant females. They have not been selected to discriminate against ancestral male courtship, which contains all the elements of the descendants' courtship, with a few extra bits. The descendant females should show no preference, or

Table B16.1 A male of one species is placed with two females, one of the same species and another of a different species. The number of homospecific and heterospecific matings are recorded. Females of the ancestral species prefer not to mate with males of a descendant species; but females of a descendant species mate with males of their own, or of an ancestral, species. *D, Drosophila differens; H, D. heteroneura; P, D. planitibia; S, D. silvestris*. From Kaneshiro (1989)

Male	Female	Number of matings	
		Homospecific	Heterospecific
D	D, H	24	25
H	D, H	33	0
P	P, S	20	12
S	P, S	32	5

continued on page 421

might still find the ancestral courtship more stimulating. Hence the results of Table B16.1.

Kaneshiro's model requires two main conditions. First, during speciation some elements of courtship should be lost by the males of the incipient descendant species. The founder effect is the most likely reason; selection could logically do the same, but it is difficult to imagine why it would. The second requirement is no gene flow between the two species after they split. The courtship change does not result in reproductive isolation, and if the ancestral and descendant populations mixed they would soon genetically merge, and the novel reduced form of courtship would be lost. But if they are isolated on separate islands (as is the case in the two *Drosophila* species pairs studied by Kaneshiro), then there is no reason why they should not behave as separate species.

If the theory is accepted, it can be applied backwards. It can be used to infer the direction of evolution when the patterns of reproductive isolation are known. Kaneshiro and Kurihara tried this for the populations of *D. silvestris* on the island of Hawaii. The species exists as several different populations in the fragmented rainforest of the island. It was not known whether the populations on the north and west of the island were relatively ancestral or descendant, and Kaneshiro and Kurihara crossed flies from these populations both among themselves and with foreign populations. They found asymmetric patterns which led to the inference "that the population of Hualalai is the oldest population, and from these two separate lineages gave rise to the remaining five populations" (Figure B16.1).

The model requires particular conditions, and these conditions do not apply to all speciation events. It is therefore not surprising that the kind of asymmetry of mating preferences that the model predicts is not found universally. Watanabe and Kawanishi, for instance, found that in the *D. melanogaster–D. simulans* species pair (in which *D. simulans* is thought to be the descendant), *D. simulans* males are more likely to mate

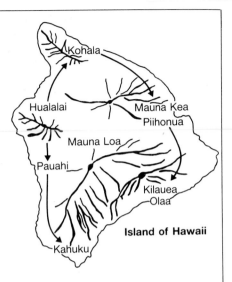

Figure B16.1 Evolution of *Drosophila silvestris* on Hawaii.

with *D. melanogaster* females than are *D. melanogaster* males with *D. simulans* females. This is the opposite of Kaneshiro's pattern. But there is no evidence that the requirements, of a founder event and subsequent absence of gene flow, are met in this case; Kaneshiro's model is therefore not expected to apply. The restricted applicability of the model means that care must be exercised when applying it in reverse; the species in question should be known with reasonable certainty to have evolved by the Kaneshiro mechanism. Moreover, the conditions of the model posit historical events that are always difficult (and usually impossible) to study; so it is not likely to become a general method of inferring evolutionary polarities. However, it is an interesting theory, and should inspire further work, both to establish the conditions, and the generality, of its operation and also to find out whether the process it postulates does actually occur. Is it true, for example, that the courtship of the descendant species like *D. heteroneura* is simpler than that of ancestors like *D. differens*?

Further reading. Giddings and Templeton (1983), *Evolutionary Biology* vol. 21 (1987), chapters 1–3, and Kaneshiro (1989).

Box 16.1 (*continued*) Asymmetric speciation.

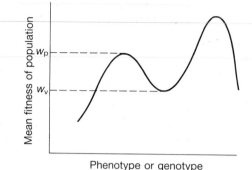

Figure 16.7 Adaptive landscape with two peaks, one higher than the other. If a population occupies the lower peak, the chance of a peak shift to the higher peak depends mainly on the depth of the valley in a large population, but on the distance between the peaks in a small population.

population, selection is relatively powerful compared with random drift and the chance of crossing the valley is dominated by the valley's depth (i.e. the ratio w_v/w_p): the lower the ratio, the less likely the population is to descend into the valley. In fact, Wright showed that the chance of being in the valley approximately equals $(w_v/w_p)^{2N}$. The chance that the population will enter the valley is the rate limiting step in the large population, and the time (T) taken to cross between the peaks is therefore the inverse of that key probability; it is given by the approximate formula $T \simeq (w_p/w_v)^{2N}$.

However, in a small population random drift becomes more powerful compared with selection. Now drift will more often take a population down the nearby valleys, and the controlling variable becomes the distance between the two peaks: the closer the neighboring peak is, the more likely it is to be reached. Therefore, in large populations the chance of a peak shift is controlled by the depth of the valley; and is more likely if the valley is shallow: in a smaller population, as in a founder event, the depth of the valley is less important and the chance of a peak shift is controlled by the distance between the peaks. The effect of population size depends on the shape of the adaptive landscape.

The general conclusion is that the effect of population size on evolutionary rate is a complex theoretical question: many different relationships could exist, depending on mutation rates, selection coefficients, and the shape of the adaptive landscape. Both the evidence and the theoretical work, therefore, suggest that it would be premature to draw any firm conclusion about the importance of Mayr's peripheral isolate and founder effect theory of speciation in nature. We do not know whether speciating populations are small, large, or of no particular size; nor do we know how often speciation proceeds non-adaptively by drift; nor do we know what proportion of genetic differences between species arise during speciation. The questions are important, however, and Mayr's writings have stimulated a lot of work on them.

16.4.5 Conclusion

Allopatric speciation is an important evolutionary process. Ring species provide clear evidence for it, but the evidence that the process is common comes from the widespread existence, and extent, of geographic variation in all species, and the theoretical possibility that reproductive isolation could evolve as part of the genetic divergence in an isolated allopatric population. Geographic variation also provides the best evidence for the peripheral isolate version of the theory of allopatric speciation; observations of isolated populations that differ from the rest of the species suggest that peripheral isolates may be a common route to speciation. The main controversies about the role of geography in speciation concern not whether allopatric speciation happens, but the details of how it operates. There is also the question of whether it is the exclusive mode of speciation, or whether speciation can also take place parapatrically, or even sympatrically.

16.5 Parapatric speciation

16.5.1 Speciation begins with the evolution of a hybrid zone

In parapatric speciation, the new species evolve from contiguous populations, rather than completely separate ones (as in allopatric speciation). The crucial evidence concerns a kind of geographic variation called a *hybrid zone*. A hybrid zone is an area of contact between two noticeably different forms of a species, at which hybridization takes place. The forms on either side of the zone may be different enough to have been classified as separate species, and the hooded crow (*Corvus corone*) and carrion crow (*Corvus cornix*) in Europe are a classic example (Figure 16.8). The hooded crow is distributed more to the west, the carrion crow to the east; the two species meet along a line in central Europe. At that line—the hybrid zone—they interbreed and produce hybrids. The hybrid zone for the crows was first recognized phenotypically, because the hooded crow is gray with a black head and tail, whereas the

Figure 16.8 Hybrid zone between the hooded crow (*Corvus corone*) and carrion crow (*C. cornix*) in Europe. From Mayr (1963).

carrion crow is black all over. The two species are now known to differ in many other respects too. The fact that the crows interbreed in the hybrid zone mean that speciation between them is incomplete. Figure 16.9 illustrates another example, from Harrison's work on the crickets (*Gryllus*) of Connecticut. In the crickets, the hybrid zone is not a simple line as it is in the crows; there is a mosaic of zones where hybrids meet. Hybrid zones have been described in all the main groups of living things that have been studied by population geneticists, and for gel electrophoretic, chromosomal, and phenotypic characters. The actual zones vary considerably in form in different cases: some, like the composite weed *Gaillardia pulchella* which forms hybrid zones in Texas, are narrow and the transition takes place in a few meters; others, like the hybrid zone between forms of the bishop pine *Pinus muricata* in California, are a few kilometers wide.

Hybrid zones are examples of *stepped clines*. We saw above that continuous

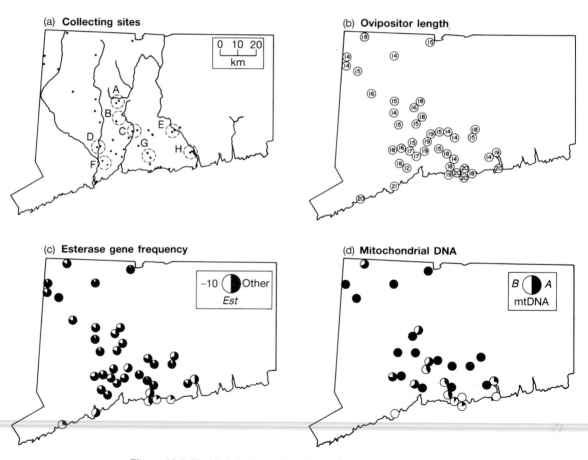

Figure 16.9 Mosaic hybrid zone in crickets *Gryllus firmus* and *G. pennsylvanicus* in Connecticut. (a) Collecting localities; (b) average ovipositor length; (c) frequencies of the Est^{-10} allele; (d) frequencies of two mitochondrial DNA genotypes called *A* and *B*. The samples are almost pure *G. firmus* (long ovipositor, high frequency of Est^{-10} allele, *B* mitochondrial DNA) at the coast and *G. pennsylvanicus* (short ovipositor, low frequency of Est^{-10} allele, *A* mitochondrial DNA) inland. In between there is a mosaic pattern, rather than a simple line, of hybrid zones. From Harrison and Rand (1989).

clines are a common kind of geographic variation, and that they will evolve when different genotypes are favored in different places and individuals migrate between the areas (Figure 16.3b,c). A simple model would be as follows. Suppose there are three genotypes at a locus, *AA*, *Aa*, and *aa*, and *AA* is increasingly favored in one direction on the geographic gradient and *aa* in the other (Figure 16.3c). The heterozygote is on average selected against. The cline will be steeper (or narrower, depending on how you look at it) if the selection against heterozygotes is stronger, and if there is less migration. To be exact, if h is the selection against *Aa* heterozygotes, and if s is the standard deviation of the distance to which genes disperse per generation (i.e. the average distance between a reproducing offspring and its parent) then the width of the cline $= s/\sqrt{h}$. (Width can be defined as 1/maximum slope of the graph of gene frequency against space: if the change in gene frequencies is sudden, the slope is steep, and the cline is narrow.) The formula can be tested against natural observations. It is almost always true (see below) that hybrid heterozygotes are selected against in hybrid zones, though the exact disadvantage is not known in most cases. We can test for the predicted positive relationship between degree of dispersal and width of cline, on the assumption that different intensities of selection will only contribute noise rather than spoiling the trend. Figure 16.10 illustrates the

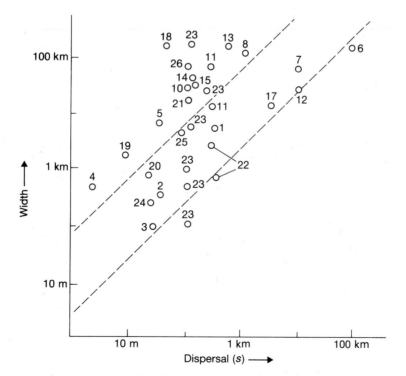

Figure 16.10 Species with higher rates of dispersal form wider hybrid zones. The graph shows the relationship between dispersal (s) and width (w) for hybrid zones in 26 genera. Different points with the same number mean more than one species has been studied in a genus. For the exact meaning of s and w, and the theoretically predicted relationship, see section 16.3. The two dashed lines are for selection against the heterozygote of $h = 10^{-5}$ and $h = 10^{-1}$. From Barton and Hewitt (1985).

relationship for 26 hybrid zones and shows that species with higher dispersal rates do indeed form wider hybrid zones. The kind of cline that will evolve depends on the rate of dispersal and the pattern of the environment.

We can now see how these ideas relate to the theory of speciation. Suppose that a population initially existed in an area to which it was well adapted, and that it then started to expand into a contiguous area in which the environment favored a different form. If the transition between the two environments was sudden, a stepped cline will evolve at the border. As selection worked on the population in the new area, different genes would accumulate in it and the two populations would diverge to become adapted to their respective environments. If they diverged almost to be different species, the border would be recognized as a hybrid zone. The two populations would have separated while they were geographically contiguous, along an environmental gradient. This is the first stage of parapatric speciation.

The two populations on either side of the hybrid zone, on this interpretation, have diverged without any period of geographic separation. The contact at the zone is called primary. The allopatric theory of speciation would have a different interpretation of the observed hybrid zones. It suggests the two forms first diverged allopatrically; their ranges then changed, and they spread up to each other's borders. Prezygotic reproductive isolation, however, had not evolved, and they could interbreed and form hybrids. Hybrid zones are then cases of secondary contact: they exist because populations have diverged allopatrically and subsequently come back into contact.

Neither mechanism is theoretically controversial: a stepped cline can evolve along an environmental gradient and populations can diverge by allopatry and then change their ranges. However, there can be empirical controversy about any particular hybrid zone, and it has turned out to be difficult, or impossible, to tell whether the forms of most real hybrid zones diverged allopatrically or within a continuous population. One piece of evidence that has been argued to favor a secondary explanation of hybrid zones is the coincidence of clines in many characters and genes at the same place. At a typical hybrid zone, the frequencies of many allozymes, morphologic markers, and mitochondrial DNA genotypes all change through the same region (e.g. Figure 16.9). If selection were acting on each independently (the argument runs), it would be unlikely that it would favor a change in all of them at exactly the same site; but if the populations diverged in all the characters allopatrically, they would all change in space at the same hybrid zone. However, the argument is unconvincing, because if the environment changes abruptly at the location of a hybrid zone, it will place selection on many features of the organisms at the same point in space. However, although the evidence is indecisive, many hybrid zones probably are secondary. The ice ages may explain the crows, and other examples: species ranges generally changed at that time (section 18.4, p. 483); and it is possible that populations would have been isolated in refuges, and then expanded later when the climate became warmer.

16.5.2 Hybrid zones may evolve into species barriers by reinforcement

Will reinforcement operate in a hybrid zone? If the hybrids are not selected against, it will not, and the hybrid zone could be stable indefinitely.

However, if the hybrids are disadvantageous, natural selection will act to reinforce the reproductive isolation between the two forms. If reinforcement is effective, two full species can evolve. The full process of parapatric speciation, therefore, is for a cline first to evolve, without any allopatric phase, as far as a hybrid zone, in which the hybrids are selected against, and then the two different forms are selected to mate only with individuals of their own type.

The difficulty with reinforcement, we saw above, is that it has to occur in a limited time. If there are two populations, containing genotypes that when crossed produce inferior hybrids, reproductive isolation is only favored while the populations remain distinct; but if they are not yet isolated, either natural selection will eliminate one of the genes, or gene flow will merge the two populations. In the theory of parapatric speciation, this problem is solved by the stable cline. A cline like Figure 16.3b will be maintained indefinitely by natural selection, and reinforcement has a much longer time to operate. The stable cline is an essential feature of parapatric speciation; without it, there would probably not be enough time for reinforcement.

So the theoretical possibility of reinforcement exists: but what evidence is there for it in modern hybrid zones? We first have to establish that the hybrids in hybrid zones are actually disadvantageous, and the evidence suggests that they usually are—though there are some exceptions. Barton and Hewitt had in 1989 reviewed 170 reported hybrid zones and concluded that most of them were *tension zones*. A tension zone is a hybrid zone in which the hybrids are selected against, and in which the two forms on either side of it are adapted to different environments. The hybrid zone exists because of dispersal of individuals out of the area to which they are best adapted.

This being so, natural selection should have favored mating preferences. Within the zone, the mating preference should be stronger (because of reinforcement) than outside it where only one form is found. Curiously, however, there is rather little evidence of reinforcement. A general mating preference exists in most cases; in the crickets studied by Harrison, for example, each species prefers a mate of its own species. What is in short supply is evidence that the mating preferences are any stronger in the hybrid zone. In a 1985 review of 32 hybrid zones in which there was known to be partial isolation, Barton and Hewitt found evidence of reinforcement in only three of them. It is not known why reinforcement of prezygotic isolation is not more powerful. Maybe hybrid zones are not tension zones at all, and the apparent evidence that hybrids are selected against is misleading in some way. In some cases, as Moore (1977) has argued, hybrids may be favored. However that may be, the fact that interbreeding persists is probably the explanation for another property of hybrid zones: their stability over time. Three chromosomal forms of the grasshopper *Warramaba* meet in hybrid zones on Kangaroo Island, South Australia, and the hybrid zones are thought to have been stable for 8000–10 000 years.

In summary, the theoretical argument for the two stages of parapatric speciation—divergence along an environmental gradient, and reinforcement—is persuasive; but its application to natural evidence is less complete. It can usually be argued that real hybrid zones are secondary, not primary, and further work is needed to sort out how reinforcement is (or is not) operating within them.

The same evidence of geographic variation as was used to support the theory of allopatric speciation can be used for parapatric speciation too. Both processes require geographic variation; in both it is the source material that is converted into speciation. In the case of particular modern species, however, it is difficult to decide whether any given pair of species evolved by parapatric or allopatric speciation. The key difference between the two processes is in the geographic distribution during a short phase in the past; and that evidence has now been lost down the stream of history. Modern geographic distributions can be suggestive, but not decisively so: the parapatric theory predicts that recently formed pairs of species will have contiguous geographic distribution, whereas the allopatric theory does not particularly predict this (though it does not rule it out). The relative importance of the two processes is unlikely to be settled by biogeographic evidence. A solution is more likely to come from theoretical work, and further research on hybrid zones.

16.6 Sympatric speciation is theoretically possible

In sympatric speciation, a species splits into two without any separation of the ancestral species' geographic range. It has been a source of persistent controversy in evolutionary biology whether sympatric speciation ever happens. Mayr has particularly cast doubt on it, and in doing so has stimulated others to work out conditions under which it may be possible. There are many models and here we shall consider one; a general model, by Seger, which clarifies why some models apparently permit sympatric speciation whereas others do not.

The main problem is familiar. Speciation is most likely to be favored when there are two genotypes in the population, and cross-mating between the genotypes produce deleterious hybrids; selection will then favor individuals that mate only with other individuals of the same genotype (i.e. mate assortatively). However, natural selection will act to eliminate all but one of the genotypes, and gene flow to unify a population genetically and to break down any incipient difference between subpopulations. The conditions for speciation become easier if selection is forcibly maintaining the genetic diversity in the population (as it does in the stable stepped cline of a hybrid zone): then natural selection and gene flow will not pre-empt the speciation process. If the mutations needed to cause reproductive isolation can arise, speciation will be possible. In short, we need a stable polymorphism, and a stable polymorphism in which assortative mating would be favored by selection.

Seger suggests we imagine a bird species in which (like the Galápagos finches, section 9.1, p. 211) beak size determines the type of food the bird can eat. If the food is seeds, then there will be some distribution of seed sizes in the environment. Suppose also that there are several genotypes influencing beak size in the bird population. The fitnesses of the genotypes will be negatively frequency dependent (section 5.12, p. 112), because as there are more birds with a certain size of beak they will compete with each other for food and lower one another's fitness. There will be a stable polymorphism of beak genotypes. The question is how the seed size distribution will influence

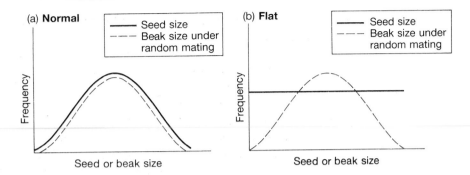

Figure 16.11 Two theoretical resource distributions. (a) Seed sizes have a normal distribution: random mating among bird genotypes will match the distribution. (b) Seed sizes have a flat distribution: random mating among bird genotypes will produce too few birds with extreme beak sizes, and those birds will have a higher fitness because of reduced competition for food.

the evolution of the bird population.

The simplest case is for the seed sizes to have a normal distribution (Figure 16.11a). The bird population can then match the resource distribution by random mating among the genotypes controlling beak size. The population will evolve to a state at which all the beak sizes have equal fitnesses; then there is no advantage to any beak size over any other, and the population will be stable.

The interesting cases are where the resource distributions are not normal. Seger first considers the case of a flat distribution, in which there are equal quantities of all seed sizes (Figure 16.11b). With random mating of birds, the beak size distribution will be normal as before; but now the different genotypes will not have equal fitness. Birds with extreme beak sizes will experience less competition for food, and have higher fitness. In this circumstance, selection will favor assortative mating for beak size. Birds at the edge of the distribution by mating assortatively would produce more offspring like themselves, and therefore with higher fitness, than if they mated at random. As selection fixes assortative mating, the population will split into two new species: one with long beaks, the other with short. Box 16.2 explains Seger's model.

The general point of the model is as follows. When random mating can produce as good, or a better, fit of beak sizes to the seed resource distribution than can assortative mating, then selection will not produce speciation. Sympatric speciation is therefore particularly unlikely when the resources to which the genotypes are adapted are normally distributed. But if assortative mating produces a better fit, a condition for sympatric speciation exists. Selection now favors assortative mating and can be strong enough (depending, as Box 16.2 shows, on the form of choice) to produce full speciation. We have been discussing Seger's model in terms of seeds and beaks; but it could apply to any resource and to any genotypically controlled character related to the resource. When the phenotypic distribution can be brought more into line with the resource distribution by assortative mating, sympatric speciation is a possibility.

Box 16.2 Seger's model of sympatric speciation.

Suppose that, in our model bird population, beak size is controlled by two independently segregating gene loci, each with two alleles. Let us confine ourselves to the haploid case for simplicity. (The general principle is identical in the more realistic diploid case.) Of the two alleles, A_1 at locus A and B_1 at locus B each contribute a phenotypic value of 1; and A_2 and B_2 each contribute 2. (These could be beak lengths in centimeters, and an A_1B_2 haploid individual's beak would be 3 cm long.) There are thus three genotypic values (Table B16.2). However, the phenotype is influenced by the environment as well as the genotype, and Seger supposed that the environment can make the phenotype deviate by +1 or −1 from its genotype, with probability 0.25, for sake of example. Table B16.2 then gives the distribution of beak sizes (phenotypic values) for each genotype.

The population with these beak sizes is exploiting a certain frequency distribution of resources. We can define seed size such that it, like beak size, varies from 1 to 5 units; a bird with beak size 1 then exploits seeds of size 1, beak size 2 exploits seeds size 2, and so on. Seger considered four resource distributions: binomial, uniform, overdispersed, and underdispersed (Table B16.3). A uniform distribution is just an extreme form of an overdispersed distribution. Now we are ready to ask the interesting question. Suppose there is another locus, controlling mate preference. Birds with genotype C_1 (choosy) mate assortatively with respect to phenotypic beak size. Assortative mate choice could happen in a number of ways. Under what Seger called genotypic choice, a C_1 female would mate assortatively only with a male with the same genotype as herself; another possibility is phenotypic choice, in which a female picks a male with the same phenotype as herself. An $A_1B_1C_1$ female, for instance, would choose a male with beak size 1 with chance 0.25, with beak size 2 with chance 0.5, and with beak size 3 with chance 0.25. Genotype C_2 mates randomly. The condition for sympatric speciation is that the population should be split into two, with no interbreeding between them.

With genotypic choice, Seger found (as expected) that with a binomial resource distribution, choice is neutral; random mating would remain. But with both under- and overdispersed resource distributions, C_1 is fixed and speciation happens. These are the conditions of mismatch between the seed distribution and the beak distribution, where assortative mating is favored.

The difference between genotypic and phenotypic choice is instructive. With an underdispersed resource distribution, choice is favored under Seger's genotypic choice model but eliminated in his phenotypic choice model. The reason lies in the kind of offspring produced in the two cases. If the seed size distribution is underdispersed, there is an excess of medium-sized seeds. With genotypic

Table B16.2 Beak lengths, expressed as genotypic and phenotypic values (see chapter 8), in Seger's model. For example, birds of genotype A_1B_1 have a genotypic value of 2, and the influence of the environment results in those birds growing up with a phenotype (beak length) of 1 with probability 0.25, beak length 2 with probability 0.5 and beak length 3 with probability 0.25

Genotype	Genotypic value	Phenotypic values				
		1	2	3	4	5
A_1B_1	2	0.25	0.5	0.25		
A_1B_2	3		0.25	0.5	0.25	
A_2B_1	3		0.25	0.5	0.25	
A_2B_2	4			0.25	0.5	0.25

continued on page 431

Table B16.3 Four types of seed size distribution. In the uniform distribution, for example, seeds of all sizes 1–5 have relative frequency 0.2

Distribution	Class				
	1	2	3	4	5
Binomial	0.0625	0.25	0.375	0.25	0.0625
Uniform	0.2	0.2	0.2	0.2	0.2
Overdispersed	0.075	0.25	0.35	0.25	0.075
Underdispersed	0.05	0.25	0.4	0.25	0.05

choice, the genotypes that are adapted to exploit these seeds, A_1B_2 and A_2B_1, can mate assortatively and enjoy the abundant resources. The beak size distribution is brought more into line with the seed distribution by the assortative genotypic mating. But with phenotypic choice, the beak distribution, instead of becoming more peaked is actually flattened. A_1B_2 genotypes, for instance, mate with all the other genotypes and produce all the possible genotypes among their offspring. The beak and seed size distributions are therefore not aligned, and selection does not favor assortative mating. With a relatively flat (overdispersed) resource distribution, assortative mating is selectively favored with both genotypic and phenotypic choice. Full speciation, however, only results if choice is genotypic. Again, under phenotypic choice, matings between all genotypes continue to take place even at equilibrium; speciation is impossible. We can conclude that it is possible to build a model in which sympatric speciation will occur, and whether or not it does occur in the model depends on the exact assumptions.

Further reading. Seger (1985).

Box 16.2 (*continued*) Seger's model of sympatric speciation.

The conditions producing a multiple niche polymorphism (section 5.13, p. 115) and the conditions of Seger's model are closely related. A multiple niche polymorphism will be stable within a species provided that either there are no genes (like gene C_1 in Box 16.2) enabling assortative mating, or random mating generates genotypes in the same frequencies as their niches exist in the environment. If the niches are in frequencies that cannot be matched by random mating, then selection will tend to produce sympatric speciation. The general conditions for Seger's model to apply are that the genotypes are adapted to different resources and the limited resources generate frequency-dependent selection. Then if the resources are not in the frequencies of the genotypes generated by random mating, sympatric speciation becomes a possibility. How common these conditions are in nature is not the issue here (though they arguably could be common): the point of this section is to clarify what conditions are needed for sympatric speciation to take place.

16.6.2 Two species of green lacewings may have split sympatrically

A study by Tauber and Tauber provides a possible example of sympatric speciation, in which the genetics are also understood. They worked on the green lacewings *Chrysoperla carnea* and *C. downesi*, which live in North America, and are similar in appearance, but have subtle differences in their coloration. *Chrysoperla carnea* is light green in spring and early summer and changes to brown in the fall; *C. downesi* is a darker green all year. Their colors are adapted, as camouflage, to their habitats. *Chrysoperla carnea* lives in fields and meadows in the summer and moves to deciduous trees in the fall, whereas *C. downesi* mainly lives on conifers. The geographic ranges of the two species are sympatric: the range of *C. downesi* is completely contained within that of *C. carnea*. The Taubers know of no evidence that the two species' ranges were ever separated in the past.

The species are separated by breeding season as well as habitat. *Chrysoperla carnea* breeds in winter and again in the summer; *C. downesi* breeds only in the spring. The breeding time is controlled by a photoperiodic response, and by bringing the lacewings into the laboratory and keeping them on artificial light–dark cycles, the Taubers were able to make the two species breed at the same time. The species could hybridize successfully. From genetic experiments, they found that the species differed in three main genetic loci. One locus controlled color: the *C. carnea* color phenotype was produced by one homozygote (G_1G_1) and the *C. downesi* phenotype by another homozygote (G_2G_2); heterozygotes (G_1G_2) were intermediate in color. The fact that no lacewings with the hybrid coloration were found in nature suggests that the two are "good" species that do not interbreed naturally. The other two loci controlled the photoperiodic response and there was dominance at both. A lacewing with the two recessive homozygotes breeds (like *C. downesi*) in the spring; and one with a dominant allele on at least one of the loci breeds in the winter and summer, like *C. carnea*.

The Taubers suggest that speciation in these lacewings took place as follows. The first stage was the establishment of a polymorphism within a species. One morph was adapted to live on conifers and the other in the deciduous habitat: there was a multiple niche polymorphism. The stable polymorphism is necessary, or selection and gene flow will reduce the species back to a single genotype. Once the polymorphism existed, selection would favor assortative mating, because crosses between the two morphs would not be well camouflaged to either habitat. The divergence in breeding times could have been the mechanism by which assortative mating was brought about. The result would be the two species we now see.

Henry subsequently showed that reproductive isolation between the species is influenced by their courtship songs, as well as the separation of breeding seasons and habitats that the Taubers identified. If the Taubers' theory is correct, the differences in courtship song would have evolved after the establishment of the habitat polymorphism: courtship song, along with breeding season, would have caused assortative mating. In theory, the divergence of song could have occurred before, after, or at the same time as the divergence in breeding season.

The Taubers' interpretation is theoretically plausible and fits in with all the evidence. However, it is impossible to confirm it with certainty. It cannot be proved that the two species never had an allopatric phase in the past and an advocate of the allopatric theory of speciation need not feel compelled to accept the sympatric interpretation. However, the question of what the evidence suggests is at least as interesting as that of what it would force a fanatic to accept. There is no evidence for an allopatric phase, whereas the species' sympatry is now observable. All the conditions for sympatric speciation seem to be met. The Taubers' study gives about as complete a set of evidence for sympatric speciation as is possible.

16.6.3 Phytophagous insects may split sympatrically by host shifts

Rhagoletis pomonella is a tephritid fly and a pest of apples. It lays its eggs in apples and the maggot then ruins the fruit. This was not always so. In North America, *R. pomonella*'s native larval resource is the hawthorn. Only in 1864 were these species first found on apples. Since then it has expanded through the orchards of North America, and has also started to exploit cherries, pears, and roses. These moves to new food plants are called *host shifts*. In the host shift of *R. pomonella*, speciation may be happening before our eyes.

The *R. pomonella* on the different hosts are currently different genetic forms. Females prefer to lay their eggs in the kind of fruit they grew up in: females isolated as they emerge from apples will later choose to lay eggs in apples, given a choice in the laboratory. Likewise, adult males tend to wait on the host species that they grew up in; mating takes place on the fruit before the females oviposit. Thus there is assortative mating: males from apples mate with females from apples, males from hawthorn with females from hawthorn.

The genetic forms are presumably about 100 generations old (given that they first moved on to apples a century ago). Is this long enough for genetic differences between the forms to have built up? Gel electrophoresis shows that the two forms have evolved extensive differences in their enzymes. They also differ genetically in their development time: maggots in apples develop in about 40 days, whereas hawthorn maggots develop in 55–60 days. This difference also acts to increase the reproductive isolation between the forms, because the adults of each are not active at the same time.

Apples and hawthorns differ and selection will therefore probably favor different characters in each form; this may be the reason for their divergence. If it is, selection may also favor prezygotic isolation, and speciation. If flies from the different forms are put together in the laboratory, however, they mate together indiscriminately. Either reinforcement has not operated when it might have been expected, or, alternatively, the differences in behavior and development time in the field may be enough to reduce interbreeding to the level natural selection favors. Selection would then not be acting to reinforce the degree of prezygotic isolation. It is not known which interpretation is correct; we need to know more about the forces maintaining the genetic differences between the forms.

In the case of host shifts, unlike the Taubers' lacewings, we can be practically certain that the initial host shift, and development of a new form, has happened in sympatry. The shift took place in historical time. However, it is not a full example of sympatric speciation because the forms have not fully speciated. Indeed, we do not know whether they will, or whether the current situation, with incomplete speciation, is stable.

How general a process is sympatric speciation by host shifts? A definite answer cannot be given; it has not even been confirmed that sympatric speciation ever does take place by host shifts. But there are interesting hints that the process might be important (section 21.3.2, p. 566). Several phytophagous insect taxa have undergone extensive phylogenetic radiations on plant host taxa. There are, for example, about 750 species of fig wasp, and each breeds on its own species of fig; in Britain alone there are 300 species of leaf miners in the dipteran family Agromyzidae, and 70% of them each feed on only one plant species. It is easy to imagine how these groups could have radiated from a single common ancestor, as successive new species arose by host shifts like the one taking place in the apple maggot fly in the USA. If phytophagous insect species consisted of odd species scattered through the phylogeny of the insects, and feeding on unrelated kinds of food plants, the process would probably have not been operating; but the existence of whole large taxa of host plant specific phytophages does suggest that speciation by host shifts could have contributed to their diversification.

16.7 Reinforcement is suggested by greater prezygotic isolation between a pair of species than in allopatry

Drosophila mojavensis and *D. arizonensis* are two closely related species of fruitfly that coexist in Sonora, Arizona, while each species lives on its own elsewhere. When a male of one species is put with a female of the other, they are less likely to mate than are a pair from the same species: the two species show a degree of prezygotic reproductive isolation. Wasserman and Koepfer measured the degree of mating discrimination in populations taken both from where the two species co-occur and from where only one of the species lives. They found that discrimination against potential mates from the other species was stronger in the flies from regions where both species are found (Table 16.2). The technical term for the observation is *character displacement*. Character displacement means that the two species differ more in sympatry than in allopatry. The term can refer to any character; *D. mojavensis* and *D. arizonensis* show character displacement for prezygotic reproductive isolation. These two species are one example among several in which this result has been found.

Observations like Wasserman and Koepfer's have often been explained by reinforcement. When the two species do not encounter each other (i.e. are allopatric), there will have been no selection on them to discriminate against mates from the other species; but in sympatry, where interbreeding may produce hybrids of reduced fitness, selection will have favored mechanisms to prevent cross-breeding. Prezygotic isolation has been reinforced in sympatry.

Reinforcement may well be the correct explanation; but, as Templeton pointed out, it is not certain. The same result could occur without reinforcement. Fusion is one alternative (and not the only one). Suppose that some pairs of species, while the two are living allopatrically, evolve higher degrees of prezygotic isolation than do other pairs. The geographic ranges of the various pairs of species then change, and they come into contact. In those cases in which reproductive isolation is low, the pair would be more likely to merge back into a single species than would more isolated pairs. The same process will not operate in allopatry, where fusion is impossible. The result,

Table 16.2 Isolation between *Drosophila mojavensis* and *D. arizonensis* taken from areas where both species live (sympatry) and from places where only one of them lives (allopatry). The experiments give either a female of one species a choice of mating with males of the other two species, or vice versa. In the experiment, the number of matings with members of the same (H_s) and with the other species (H_o), and total number of matings (*N*) was measured. An isolation index was calculated, where $I = (H_s - H_o)/N$. If $I = 1$ there was complete isolation (no cross-mating between species); if $I = 0$, there was random breeding between and within species. The numbers should be understood as follows: the top left number, for example, means that *D. mojavensis* females, taken from a place where there are no *arizonensis* (i.e. allopatry), when put with males of both species, show an isolation index of only 0.3. From Wasserman and Koepfer (1977)

D. mojavensis female with *arizonenis* male		*D. arizonensis* female with *mojavensis* male	
D. mojavensis female		*D. arizonensis* female	
Allopatry	$I = 0.3$	Allopatry	$I = 0.9$
Sympatry	$I = 0.94$	Sympatry	$I = 0.8$
D. arizonensis male		*D. mojavensis* male	
Allopatry	$I = 0.54$	Allopatry	$I = 0.78$
Sympatry	$I = 0.6$	Sympatry	$I = 0.92$

as seen by an external observer, would be that sympatric species pairs would show higher reproductive isolation than allopatric pairs: the allopatric pairs include all kinds of species, whereas the pairs of sympatric species have higher isolation—those with lower isolation have all fused. We should then observe character displacement, even though there has been no sympatric reinforcement.

16.8 A study of speciation in *Drosophila*, provides evidence about reinforcement, and other interesting results

Templeton's criticism is to a large extent answered by a study of speciation in *Drosophila*, by Coyne and Orr. Their result emerges from a comparison of prezygotic and postzygotic isolation between many pairs of species; the species pairs differ in how closely related they are. Box 16.3 explains how the key variables were quantified. Some of the pairs of species lived sympatrically, in at least part of their geographic range, others were completely allopatric. Coyne and Orr were able to find 42 independent pairs of species of *Drosophila* for which all four variables (pre- and postzygotic isolation, genetic distance, geographic distribution) had been measured.

Figure 16.12 shows the relationship between genetic distance and reproductive isolation. The first point to make is that reproductive isolation increases with time (that is, with genetic distance). The result makes good evolutionary sense. As time passes from the original split between the two species, genes are substituted in each lineage, genetic distance increases, and as the two species diverge they will come not to recognize, or meet, each other (prezygotic isolation); and if they do mate, the genetic differences between them will tend to result in inviable or sterile hybrids (postzygotic isolation).

A second interesting result is less immediately visible in Figure 16.12. If we

Box 16.3 Quantitative measures of genetic distance, postzygotic isolation, and prezygotic isolation.

Genetic distance is a standard concept, of which the most common measure is Nei's index D. It is calculated like this. We take two populations and measure the frequency of both of the alleles at several loci. Consider an allele A_i at a locus in population 1 and 2, and define its frequency as p_i in population 1 and q_i in population 2. (By the usual convention, i has as many values as there are alleles at a locus: if there are two alleles, $i = 1$ for one of them and $i = 2$ for the other.) First calculate $j_1 = \Sigma p_i^2$, $j_2 = \Sigma q_i^2$, and $j_{12} = \Sigma p_i q_i$,

where j_1 is the probability that two randomly chosen genes in population 1 are identical, j_2 is the same probability for population 2, and j_{12} is the probability that two genes, one drawn randomly from population 1 and the other from population 2, are identical. (j_1 is more easily recognized as the frequency of all homozygotes in population 1, j_2 for population 2.) The same set of three numbers can be calculated for every locus; the next step is to calculate the average for all loci in each of the three cases. Define the averages as follows: $J_1 = j_1$, $J_2 = j_2$, $J_{12} = j_{12}$. Finally, Nei's formula for genetic distance is:

$$D = -\log_e \frac{J_{12}}{\sqrt{J_1}\sqrt{J_2}}$$

D decreases as more alleles are shared between the two populations, making the product J_{12} higher. In the extreme case of two samples from one population, $J_1 = J_2 = J_{12}$, and $D = -\log_e 1 = 0$. As shared alleles are lost, D increases. D has no mathematic upper limit (in the extreme case of two samples that share no alleles, $J_{12} = 0$, $D = -\log_e 0 = \infty$); but in practice, it is found to have values of about 0.1–2 for different species, and of about 1 for different genera.

The formula can also be expressed $D = -\log_e I$, where $I = J_{12}/\sqrt{(J_1 J_2)}$. I is called the genetic identity between the two populations (see Table 16.1).

Because many proteins evolve at least approximately in the manner of a molecular clock (see chapter 7), genetic distance increases approximately in proportion to the time since the two species diverged.

Postzygotic isolation. If two species can be crossed in the laboratory, some or all of their offspring may be inviable or sterile. We saw in chapter 14 that it is often the heterogametic sex that is inviable or sterile, a generalization known as Haldane's rule. There are two sorts of cross between two species: either a male of species 1 and a female of species 2, or a male of species 2 and a female of species 1. According to its sex and the type of cross, an offspring can be one of four types: male and from male species 1 × female species 2; female and from male species 1 × female species 2, etc. For a species pair we count the number of the four types that are sterile or inviable and divide the result by four. The index can therefore have values 0, 0.25, 0.5, 0.75, 1 in order of increasing isolation. For a pair of species that fit Haldane's rule, the index is 0.5.

Prezygotic isolation. Prezygotic isolation is generally measured in a mate choice experiment in which an individual of one species has a choice of mates from its own, and another, species. If it mates with its own species, the mating is called homospecific; if with the other, heterospecific. As an index of prezygotic isolation, Coyne and Orr used

Prezygotic isolation = 1 −

$$\frac{\text{Frequency of heterospecific matings}}{\text{Frequency of homospecific matings}}$$

The index can vary from 1 (complete isolation) to 0 (no isolation) through, in theory, to $-\infty$ for complete disassortative mating. In practice the index varies between 0 and 1 for most pairs of species. (In Table 16.3 a slightly different isolation index was used.)

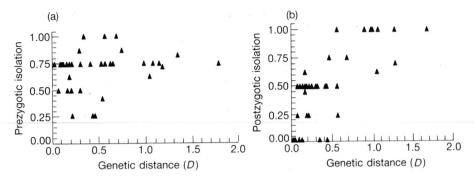

Figure 16.12 Strength of isolation in relation to genetic distance for pairs of species of *Drosophila*. (a) Prezygotic isolation. (b) Postzygotic isolation. Note prezygotic isolation is higher for low distances than is postzygotic isolation. From Coyne and Orr (1989a).

concentrate on closely related pairs of species, prezygotic isolation is higher than postzygotic isolation in those with genetic distance ($D \leq 0.5$). Both prezygotic and postzygotic isolation eventually reach one for high enough genetic distances, but prezygotic isolation does so at a lower value of D. Once the difference has been pointed out, it can be seen in Figure 16.12 (compare the large number of points at low values of D with isolation about 0.75 in the prezygotic isolation figure with the large number at more like 0.5 in the postzygotic isolation figure); and it can be confirmed statistically. The result suggests that prezygotic isolation tends to evolve during speciation before postzygotic isolation. We shall come to a possible explanation later; here we need only notice the result.

Finally, we can compare the isolation of allopatric and sympatric pairs of species. Figure 16.13 shows the degree of prezygotic isolation between pairs of allopatric species and between pairs of sympatric species. The result to notice is that the sympatric pairs are much more isolated from each other. The result is an interspecific analog of the character displacement of *Drosophila mojavensis* and *D. arizonensis*. There, different populations of the

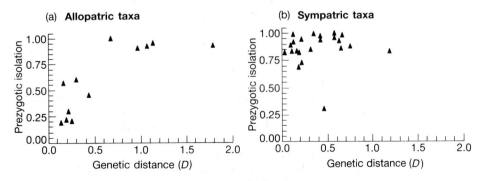

Figure 16.13 Strength of prezygotic isolation in relation to genetic distance for pairs of species of *Drosophila*. (a) For allopatric pairs of species; (b) for sympatric pairs of species. Note prezygotic isolation is higher for sympatric pairs of species. From Coyne and Orr (1989a).

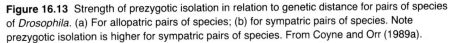

same species were compared; here we are comparing different species. The result again suggests that sympatric species have evolved mechanisms to prevent interbreeding, but Templeton's objection could again apply. The result suggesting that reinforcement has taken place during evolution is Figure 16.14. Prezygotic isolation is higher in sympatric than in allopatric species; but for postzygotic isolation there is no such difference. Templeton's argument that species with low isolation will tend to fuse in sympatry applies equally to postzygotic and prezygotic isolation; it does not predict that prezygotic isolation will be increased in sympatry while postzygotic will not.

By contrast, the theory of reinforcement can explain the result. Suppose that hybrids are initially selected against, though by some small amount such that their disadvantage is not detected in Coyne and Orr's index of postzygotic isolation. Selection would then favor preferential mating with conspecifics, and in sympatry prezygotic isolation will evolve. Postzygotic isolation, however, evolves only as an incidental consequence of the divergence between the two species; natural selection will not act to reinforce postzygotic isolation in fruitflies. As the species diverge, the genes within each will become less well coadapted to each other, and the hybrids become decreasingly viable. After prezygotic isolation has been reinforced, the two species' gene pools will diverge more rapidly, and the evolution of postzygotic isolation therefore follows that of prezygotic isolation.

Such is the explanation of Coyne and Orr's result by reinforcement. Its crucial prediction is that hybrids are disadvantageous even between closely related (low D) pairs of species: it is this disadvantage that drives the whole system. If closely related species in which prezygotic isolation has evolved actually have hybrids as fit as crosses within a species, the argument fails. The prediction is testable, but has not yet been tested; meanwhile, the argument is plausible, but unproven.

A related conclusion from Coyne and Orr's study should also be noted. The x-axis of the graphs have all been drawn as genetic distances. Genetic distance is approximately proportional to time for low values of D (Box 16.3), and the one can be calibrated against the other. Nei has estimated, from a number of studies, that it takes about 5000000 years for D to increase from zero to one. We are now in a position to calculate how long it takes for reproductive isolation to evolve between a pair of *Drosophila* species. The

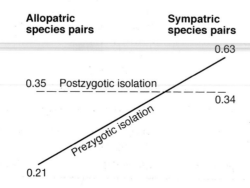

Figure 16.14 Average strength of pre- and postzygotic isolation between pairs of species of *Drosophila*, for allopatric and sympatric pairs. Note that prezygotic isolation is higher among sympatric than among allopatric pairs of species, but postzygotic isolation is the same. From Coyne and Orr (1989a).

average figure for isolation between sympatric pairs of *Drosophila* is about 0.9 (Figure 16.13b); the lower bound, from 95% confidence intervals, is 0.85. Isolation of 0.85 is therefore probably sufficient to keep two sympatric species apart. The relationship between genetic distance and reproductive isolation suggests that isolation of 0.85 evolves at a genetic distance of $D = 0.53$ (for all the data combined), or 0.66 (for allopatric species), or 0.31 (for the sympatric species). Reading these numbers back through Nei's estimate implies that it takes between 1 500 000 and 3 500 000 years for a new *Drosophila* species to evolve.

In a separate study, Coyne and Orr collected genetic results of crosses between closely related species of *Drosophila*. In these crosses, it is possible to make *Drosophila* with various chromosomal combinations to find out on which chromosomes the genes responsible for hybrid sterility are located. Their striking result is that the genes are *always* concentrated on the X chromosomes (Table 16.3, which includes results from four other taxa too). The result is similar to Haldane's rule, that in interspecific crosses, if only one sex of hybrids is sterile, it is usually the heterogametic sex (section 15.2.3, p. 390).

Coyne and Orr suggest these phenomena are related. Experiments concerned with sterile interspecific hybrids are automatically about the early stages of speciation, because before speciation, and long after it has happened, hybrids do not exist—offspring are either normal offspring or no hybrids are produced at all. If the genetic changes during the early stages of

Table 16.3 Experimental genetic interspecific crosses in which the sex chromosomes have a large effect on hybrid sterility. From Coyne and Orr (1989b), who should be referred to for the original references

Cross	Trait affected	Reference
Drosophila species		
pseudoobscura/persimilis	Male and female fertility	Dobzhansky (1936); Orr (1987)
pseudoobscura USA/Bogota	Male fertility	Dobzhansky (1974); Orr (1989)
simulans/mauritiana	Male fertility	Coyne and Kreitman (1984)
simulans/sechellia	Male fertility	Coyne (1986)
mohavensis/arizonensis	Male fertility	Zouros *et al.* (1988)
micromelanica A/B	Male fertility	Sturtevant and Novitski (1941)
littoralis/virilis	Male fertility	Orr and Coyne (1989)
novamexicana/virilis	Male and female fertility	Orr and Coyne (1989)
texana/virilis	Male and female fertility	Orr and Coyne (1989)
lummei/virilis	Female sterility	Orr and Coyne (1989)
buzzattii/serido	Female sterility	Naviera and Fontdevila (1986)
mulleri/aldrichi-2	Female viability	Crow (1942)
texana/montana	Female viability	Patterson and Griffen (1944)
Glossina morsitans subspp. 1/2	Fertility or mating ability	Curtis (1982)
Anopheles arabiensis/gambiae	Fertility or mating ability	Curtis (1982)
Colias eurytrheme/C. philodice	Female viability, fertility, mating vigor	Grula and Taylor (1980)
Cavia rufescens/C. porcellus	Male sterility	Detlefsen (1914)

speciation occur mainly on the X chromosomes, we should expect difficulties to arise first in the heterogametic hybrid offspring (the males in *Drosophila*). They will have a mismatch between the X chromosome and one set of autosomes, and the interactions between these might cause defects.

But why should genetic changes happen first in the X chromosome? Coyne and Orr's preferred explanation is that, for recessive genes, evolution on the X chromosome is faster. Recessive genes are rarely expressed on the autosomes, but they only need to be in one dose to be expressed (and selected for, if advantageous) on the X chromosome. An advantageous recessive gene will therefore be fixed faster on an X chromosome than an autosome. If Coyne and Orr are right, there are two sorts of genetic change that influence reproductive isolation, and they happen at different times in speciation: changes in recessive genes, on the X chromosomes, come first, and other sorts of change (dominant, overdominant, etc.), on the autosomes, follow later, after the first detectable reproductive isolation has evolved.

Coyne and Orr's work, in summary, suggests that selection acts in sympatry to reinforce the prezygotic isolation of closely related species. They provide evidence of the rates (and relative rates) at which prezygotic and postzygotic isolation evolve, and demonstrate that isolation between species increases through time. They have shown that the earliest genetic changes in speciation are on the X chromosome, and suggest this is because natural selection works more effectively on recessive genes on the sex chromosomes.

16.9 Chromosomal changes could potentially lead to speciation

Species differ in their chromosomes as well as in their morphology, genetics, and behavior. The chromosomes of many taxa have been described and species are known to differ in their chromosomal numbers, lengths, and structure. Humans, for example, have 46 chromosomes, whereas chimpanzees, gorillas, and the orangutan have 48, suggesting that a pair of chromosomes have fused in human ancestry. These sorts of observations would not be of peculiar interest—given that species differ in all other respects too—were it not for a theory suggesting that chromosomal changes may be exceptionally important in speciation. The theory runs like this.

Consider two forms of a chromosome, differing (for example) in an inversion. We can call the standard and inverted forms A and A'. We suppose, for simplicity, that A and A' have identical sets of genes and produce the same phenotype (Figure 16.15a). The two kinds of homozygote (AA and A'A') will have identical fitness. The heterozygote, however, is likely to be selected against, because recombinants between the two forms may have double sets of some genes, and lack others (Figure 16.15b), and organisms that lack genes will usually be selected against. We have here a precondition for the evolution of assortative mating. AA types should preferentially mate with other AA types and avoid crossing with A'A'; and vice versa. The usual difficulties with reinforcement will apply, but a gene causing its bearer to mate assortatively with respect to chromosomal genotype will be favored. If assortative mating becomes established, speciation will result.

The argument works for a population that already possesses both chromosomes in reasonable frequencies. In evolution, one of the chromosomes would have existed first, and the second would have originated as a rare

(a) Chromosomal inversion

Chromosome form A

Inverted chromosome form A′

(b) Recombination in inversion heterozygote

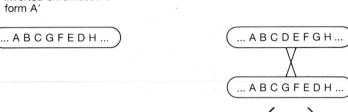

Figure 16.15 (a) A chromosomal inversion has a set of genes inverted. The letters represent genes along the chromosomes. (b) Recombination in a heterozygote for a chromosomal inversion can produce chromosomes that lack some genes and have others in double dose. These forms are probably selected against.

mutation. Suppose the population was initially fixed for A and A′ arose as a mutation. A′ would initially exist only in AA′ heterozygotes. These heterozygotes are selected against, and the most likely evolutionary result would be for A′ to be lost. However, there is a condition under which A′ will not be selected against. If A′ can somehow become established in the form of A′A′ homozygotes then it will have some chance of being maintained (the exact chance depends on the relative frequency of the two chromosomal forms, because selection favors the more common form).

How could A′A′ homozygotes be generated in fairly high frequency? They can be generated by matings between AA′ heterozygotes, but these matings will normally be rare if A′ is a new mutation. The condition needed to establish A′ is inbreeding. Inbreeding enhances speciation both by producing more homozygotes and by the effects of random genetic drift, which is more likely, in an inbred population, to produce one subpopulation with a new chromosomal form in high frequency. If an individual is an AA′ heterozygote, then its close relatives are likely to be too, and were such an individual to mate incestuously the chance of generating A′A′ individuals would be greatly increased. In an isolated subpopulation with a high degree of inbreeding, the A′A′ homozygote has a higher chance of reaching a non-trivial frequency. Once the population possesses both homozygotes, two things can happen. Selection favors either the loss of one of the chromosomal forms, or reproductive isolation between them. If one chromosomal form is lost, it could be either the new form (in which case nothing would have changed) or the old (in which case a chromosomal evolutionary event would have occurred). If speciation proceeds, we should have two new species, one with the old chromosomal form and one with the new.

If the theory is correct, we can predict an association between whether or not the members of a taxonomic group have an inbred population structure, their rate of speciation, and of chromosomal change. Some species live in small social groups, or otherwise have subdivided populations, and in these the degree of inbreeding will be higher than in a species in which individuals breed in a large population. Bush *et al.* tested whether genera whose members live in small family groups have a higher rate of speciation and chromosomal evolution than panmictic genera. They collected evidence for 225 vertebrate genera. For each, they counted the number of species in it and its chromosomal diversity. The two are correlated, as would be predicted by almost any theory. They then argued that the taxa with higher speciation rates tended to have subdivided population structures; mammals, especially primates and horses (though not whales), have high rates of speciation and chromosomal evolution; fish, amphibians, and reptiles have lower rates. Primates and horses are two taxa that usually live in social groups. Bush *et al.* did not systematically compare the variables in all taxa; but the result is still suggestive. Maybe taxa with subdivided population structures do have higher speciation rates, and maybe this is due to the way chromosomal evolution can proceed more rapidly in these kinds of species.

16.10 Asexual species may split when they are subdivided

Two forms of herpes simplex virus (HSV-1 and HSV-2) can infect humans. HSV-1 is transmitted orally and HSV-2 genitally. Gentry *et al.* compared the amino acid sequences of one of the viral proteins (deoxythimidine kinase) in HSV-1 and HSV-2, and in some related viruses, including marmoset herpes virus. Figure 16.16 shows the phylogenetic tree of the viruses; Gentry *et al.* estimated that the two human herpes viruses have diverged relatively recently, at about the same time as humans were diverging from the great apes. This coincidence suggests the following model of speciation.

Humans are peculiar among mammals in their face-to-face sexual behavior; most other mammals typically copulate back-to-front. The relative frequencies of oral–oral, oral–genital, and genital–genital contact may consequently differ between humans and other mammals. In humans, the frequencies of oral–oral and genital–genital contact may be relatively high compared with oral–genital contact. Before the evolution of humans, oral, genital, or oral–genital transmission of the herpes virus might have been equally probable; one population of the virus would then have been exposed to the whole oral–genital environment. But with the evolution of human sexual behavior, the oral and genital transmission routes might have become differentiated. The herpes viruses could then have split into two, one specialized for each route of transmission (Figure 16.16).

The frequencies of oral, genital, and oral–genital contact have not been measured quantitatively in any natural primate; but the assumptions of Gentry *et al.*'s model are plausible and fit in with such anecdotal observations as exist. If they are right, the amount of divergence between the two herpes simplex viral forms implies (by the molecular clock) that the missionary sexual position in the human evolutionary lineage is about 8 million years old.

Gentry *et al.*'s model is, in a way, typical of much work on speciation. It starts with observations on two species, suggests a mechanism by which they might have split, and fits the available facts to the model. Unfortunately, theories for particular species pairs are difficult to test. They refer to events—

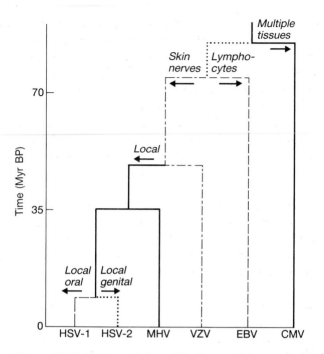

Figure 16.16 Phylogenetic tree of the herpes viruses and related viruses. Time, on the *y*-axis, was inferred from the molecular clock, calibrated by an estimated time of divergence between the New World and Old World monkeys of 35 million years: the times are therefore approximate. HSV-1 and HSV-2, human herpes viruses 1 and 2. MHV, marmoset herpes virus. VZV, varicella roster virus. EBV, Epstein–Barr virus. CMV, cytomegalovirus. Changes in the part of the host that the viruses attack are written on the branch points where they may have evolved. From Gentry *et al.* (1988).

in this case, the sexual behavior of human ancestors 8 million years ago—that we cannot observe. Was the virus speciation allopatric or sympatric? In a sense, it was allopatric because the ancestral viral population would have been split into two separate populations that rarely encountered each other; but it is in a strict sense sympatric because the split took place within the geographic range of the ancestral species. In other words, the distinction is not all that interesting in this case: what matters is that the population was split into two separate populations, and divergence followed. In this case, there is the added complication that the two species mainly reproduce asexually; there is no gene flow to be broken down.

The question of the relative frequencies of different modes of speciation probably will not be settled by counts of individual cases. It is too difficult to establish conclusively what the mode of speciation was in any one case; there are too many unobservable, historical, but crucial elements in the models. Some progress can be made in reconstructing individual cases; but not to the point of certainty. The question also has to be studied theoretically, to establish how plausible speciation is under different conditions. We have a good idea of the main ways in which speciation could take place; but there remains plenty of room for more conclusive work on the question of how a new species can arise on earth—the question that John Herschel once described, in a phrase that inspired Darwin, as "the mystery of mysteries."

16.11 Summary

1 The evolution of a new species happens when one population of interbreeding organisms splits into two separately breeding populations.
2 It has been a matter of controversy whether new species evolve in subpopulations geographically isolated from (allopatric), contiguous with (parapatric), or overlapping (sympatric), the ancestral population.
3 At least in vertebrates, the majority of new species probably evolve allopatrically. All species show extensive geographic variation that could be converted into speciation; and there are no theoretical difficulties for allopatric speciation (whereas there are for parapatric and sympatric speciation).
4 Allopatric speciation may be by subdivision of the species range or by a peripheral isolate—a small population which becomes cut off at the edge of the species range.
5 Parapatric speciation could happen if a steep cline evolved into a hybrid zone and barriers to interbreeding then evolved.
6 Sympatric speciation is most likely if selection first establishes a stable polymorphism and then favors assortative mating within each polymorphic type.
7 Reinforcement is the enhancement of reproductive isolation by natural selection: forms are selected to mate with their own, and not with the other, type. Sympatric speciation requires reinforcement to happen; parapatric speciation usually requires it; allopatric speciation can take place with or without it. Reinforcement is theoretically difficult, which is one reason to suspect that allopatric speciation may be more common than sympatric and parapatric speciation.
8 Reinforcement may explain why pairs of closely related species show greater reproductive isolation where they co-occur sympatrically, than where they are allopatric populations; but the same result could arise without reinforcement.
9 The amount of prezygotic isolation is greater between recently evolved sympatric pairs of species of fruitflies than it is between equivalent pairs of allopatric species; but the amount of postzygotic isolation is similar in sympatric and allopatric pairs of species. Reinforcement is probably the explanation.
10 New species of fruitfly take about 1500 000–3500 000 years to evolve.
11 Chromosomal evolution might produce faster speciation in species with a subdivided population structure, in which inbreeding is more frequent; there is some evidence that chromosomal rates of evolution are correlated with social structure in vertebrate species.

16.12 Further reading

The classic treatises on speciation by Mayr (1942, 1954, 1963) and Dobzhansky (1970) remain good, if dated, introductions. See Mayr (1982a) for his more recent ideas. There have been a series of multi-author books on the topic which are also good, if erratic, introductions: Atchley and Woodruff (1981), Barigozzi (1982), and Otte and Endler (1989). Review papers include: Bush (1975), Templeton (1981), Barton (1988), Harrison (1991), and Coyne (1992).

Mayr's works are the references for allopatric speciation. Murray and Clarke (1980) describe two other examples of ring species, in snails. On the peripheral isolate and founder effect theories, see the general reviews by Barton and Charlesworth (1984), Barton (1989), and the book edited by Giddings *et al.* (1989). Lande's papers (1980, 1986) are important but advanced. On sexual selection, see Dominey (1984) and Endler (1989).

There are several excellent reviews of hybrid zones. See Hewitt (1988, 1989), Barton and Hewitt (1985, 1989), and Harrison (1990). Most other work can be traced through their bibliographies. On mosaic hybrid zones see Harrison and Rand (1989). On crows, see Cook (1975). The general works listed above also discuss the topic. On parapatric speciation see Endler (1977), which includes an important discussion of the biogeographic evidence.

On sympatric speciation: Mayr (1942, 1963, 1982a) is the classic critic. Seger (1985) contains the model discussed in the text. Tauber and Tauber (1989) review the work on *Chrysopa*, the evidence for other insects, and give references to other theoretical models. For host shifts see Bush (1975) and Butlin (1987b) for the question in general; see Feder *et al.* (1988), McPheron *et al.* (1988), and Smith (1988) for further details on *Rhagoletis*.

Lynch (1989) uses biogeographic evidence to assess the relative frequencies of the different modes of speciation.

For Coyne and Orr's study, see Coyne and Orr (1989a,b) and Endler (1989). On reinforcement generally, Butlin (1987a, 1989) provides critical reviews and the key references (see also the general references on speciation particularly on hybrid zones). The comparisons between sympatric and allopatric populations are an example of the general phenomenon of character displacement, on which see Brown and Wilson (1956) and Grant (1972, 1975).

On chromosomal mechanisms of speciation see: Bush *et al.* (1977), White (1978), and Lande (1979, 1985).

The reconstruction of phylogeny

How do we know that humans and chimpanzees have a more recent common ancestor with each other than either has with the amoeba? The quick answer is that humans and chimpanzees look more similar. For phylogenetic inference, the vague expression "look more similar" has to be made more precise. We can divide up the phenotypic appearance of an organism into a series of characters and character states, and humans and chimpanzees look similar because they share many more characters in common than does either with the amoeba. Humans and chimpanzees share such vertebrate characters as brains and backbones, mammalian characters like lactation, and great ape characters such as their distinctive molar teeth and absence of a tail. Humans and amoebas do have features in common—many of the characters of cellular metabolism, for example—but these are equally shared with chimpanzees and do not suggest a separate grouping of human and amoeba.

There is a more fundamental principle in phylogenetic reconstruction than that of shared characters. A phylogeny is more plausible if it requires less, rather than more, evolutionary changes in character states. This is the principle of *parsimony*. To illustrate the principle of parsimony in the case of the human, the chimpanzee, and the amoeba, we need one more species, such as the magnolia. The procedure is to write out all the possible phylogenies for the set of species and then, given a knowledge of the character states in each species, to count the minimum number of events implied by each phylogeny. The best estimate of the true phylogeny is the one requiring the least evolutionary change. Now look at Figure 17.1. The phylogeny in (a) suggests that humans share their most recent common ancestor with magnolias, in (b) it is with amoebas, and in (c) with chimpanzees. The three phylogenies are *unrooted trees*. An unrooted tree specifies the branching relationships between species, but not where the ancestor of all the species lies. The ancestor could have been at any point in the unrooted tree. In phylogeny (c), for example, it could be that the common ancestor of the four species was the human, and from the human ancestor a lineage first split off leading to chimpanzees and then a second split led to the most recently evolved forms, amoebas and magnolias; alternatively, the amoeba might be the common ancestor, and the first split led to magnolias, and the second to humans and chimpanzees; or the common ancestor could be at any point in one of the branches, such as in the central branch, in which case two lineages would have evolved from it, one leading to amoebas and magnolias, the other to humans and chimpanzees. Many full phylogenies, in which the

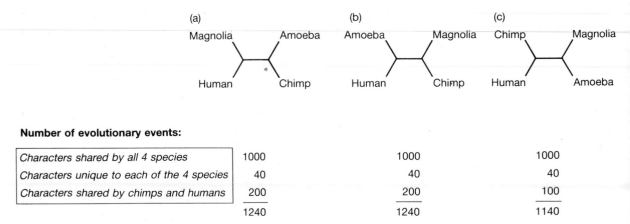

	(a)	(b)	(c)
Characters shared by all 4 species	1000	1000	1000
Characters unique to each of the 4 species	40	40	40
Characters shared by chimps and humans	200	200	100
	1240	1240	1140

Number of evolutionary events:

Figure 17.1 Possible phylogenetic relationships of amoeba, chimpanzee, human, and magnolia. Diagrams (a), (b), and (c) are three unrooted trees; they specify branching, but not ancestral, relationships. The sums for the number of evolutionary events implied by each are made for 1000 characters shared by all four species, 10 characters unique to each species, and 100 characters shared by humans and chimpanzees.

ancestral relationships are specified, are compatible with any one unrooted tree.

How many evolutionary events do each of the unrooted trees require? Suppose we know that 1000 characters are shared in all four species (these will be all the characters, such as the use of DNA, proteins, and many cellular structures, that are common to all eukaryotic life); and that each species has 10 characters unique to itself (leaves, wonderfully perfumed flowers, etc. in the magnolia; an adult brain of about 1500 cm^3 in humans, etc.); and that 100 characters are shared between humans and chimpanzees, but are absent in amoebas and magnolias (backbone, vocal communication, etc.). No other kinds of character are known. We can now count the number of evolutionary events in each phylogeny. In (a) the 1000 common characters could have evolved once each in the common ancestor of all four species and been retained throughout the phylogeny; that would require a total of 1000 evolutionary events (one evolutionary origination for each character). Each species' 10 unique characters can have evolved in the lineage leading to that species, making 40 evolutionary events in all. The 100 characters shared by humans and chimpanzees are more puzzling. The smallest number of events producing the observed distribution of characters is 200. They could have evolved separately in the human lineage and in the chimpanzee lineage (100 events in each, making 200). Or, supposing that the magnolia was the common ancestor of the group of four species, they could have evolved once before the common ancestor of the amoeba, human, and chimpanzee and then been lost in the lineage leading to the amoeba; there are then 100 gains and 100 losses, or 200 events in all. (The same argument can be made if the amoeba is the common ancestor; the characters then have to be lost in the lineage leading to the magnolia.) The total number of events in (a) is therefore 1240. This is the minimum. The tree is logically compatible with any number of events between 1240 and infinity. The minimum number was obtained by supposing the 10 characters unique to the magnolia only evolved in the lineage leading to the magnolia. However, logically they could have evolved and then been lost any even number of times in any other lineage. Perhaps, in the human lineage, leaves were evolved and then lost, scented flowers were evolved and then lost, then leaves were evolved and lost again . . . and so on to infinity. The interesting number is the minimum number of events implied by the tree, not the number of events that logically could have occurred.

The same calculation in (b) also implies 1240 events. In (c), however, the 100 characters shared by humans and chimpanzees only need to have evolved once, in the common ancestor of humans and chimpanzees, and would not have to be lost again; (c) therefore only requires 1140 events. It is the most parsimonious tree. Such, in outline, is the reason for thinking that humans and chimpanzees share a more recent common ancestor with each other than either has with an amoeba or a magnolia.

17.2 The parsimony principle works if evolutionary change is improbable

How can the parsimony principle be justified? Why is a phylogeny requiring fewer evolutionary events a more plausible inference than one requiring more? The parsimony principle is reasonable because evolutionary change is improbable. Suppose we know that a modern species and one of its ancestors both have the same character state (Figure 17.2). Parsimony suggests that all the intermediate stages in the continuous lineage between ancestor and modern species possessed that same character state. As we have seen, an indefinitely large number of changes could logically have occurred between ancestor and descendant. However, a change followed by a reversal of that change is unlikely. Each change requires a gene (or set of genes) to arise by mutation and then to be substituted, either by drift, if the change is neutral,

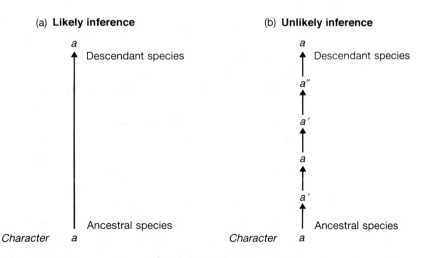

Figure 17.2 The same character is found in both a descendant species and one of its ancestors. It is more likely (a) that the character has remained constant and been passed on by inheritance than (b) that it has changed and reverted to its original state a number of times between ancestor and descendant.

or by selection; both these processes are improbable. It is much more likely that the same character would have been continuously passed on, in much the same form, from ancestor to descendant by simple inheritance. We know this is plausible because it happens every time a parent produces an offspring: the parental characters are passed on. For the characters shared between humans and chimpanzees, the argument is particularly powerful. Chimpanzees and humans share whole complex organ systems like hearts and lungs, eyes, brains, and spinal cords. The initial evolution of each of these characters required improbable mutations, and natural selection operating over millions of generations. It is evolutionarily improbable to the point of near impossibility that the same changes would have evolved independently in the two lineages after their common ancestor. By contrast, there is nothing improbable about postulating that the characters could have been passed on in passive inheritance from the common ancestor of chimpanzees and humans to the modern descendants. In conclusion, it is more likely that a character will be shared by common descent than by independent, convergent evolution. For any set of species, a phylogeny requiring less evolutionary change is more plausible than one requiring more.

17.3 Phylogenetic inference uses two principles: parsimony and distance statistics

So far we have met two principles of phylogenetic inference, and we shall meet them recurrently in this chapter. They are:

1 *Parsimony*. Species are arranged in a phylogeny such that the smallest number of evolutionary changes is required.

2 *Distance*. Species are arranged in a phylogeny such that each species is grouped with the other species that it shares the most characters with.

In the case of the chimpanzee, amoeba, and human, the argument that humans and chimpanzees share a more recent common ancestor because they look more similar to each other, implicitly used measurements of distance (section 14.5, p. 360; Box 16.3, p. 436). The counts of evolutionary events in the three unrooted trees for the human, chimpanzee, amoeba, and magnolia used the principle of parsimony.

In easy cases, like those given, the distance and parsimony principles give the same result, and it might seem that it does not matter which is used. But in other cases they differ. The parsimony principle (as we shall see) is then more reliable, because it has a better theoretical justification. However, distance statistics are an important method in phylogenetic inference, because there are circumstances in which they are almost as reliable as parsimony— and they can often be collected more rapidly. In order to know when any given method can be relied on, it is necessary to understand the underlying principle of phylogenetic inference. Such is the topic of this chapter.

17.4 In most real cases, not all the characters suggest the same phylogeny

In the imaginary example of Figure 17.1, all the characters implied the same phylogeny. There were characters unique to each species, characters shared by all, and characters shared by just two species; the 1140 characters "agreed" in the sense that we could randomly divide them into two sets and still obtain the same answer. Thus if we divided the 1000 + 40 + 100 characters into two sets of 500 + 20 + 50, the phylogeny inferred from each of those two sets would be the same. Now imagine instead that as well as the 100 characters shared between chimpanzees and humans (and absent from magnolias and amoebas) there are 50 characters shared by magnolias and humans (and absent from chimpanzees and amoebas). Now the characters do not agree: if we concentrate on the characters shared between humans and magnolias, the phylogeny (a) in Figure 17.1 is correct; but the characters shared between humans and chimpanzees favor phylogeny (c).

Conflict between characters is usual in real phylogenetic problems. (In cases like those given conflict might not be a problem, but the answer is obvious. It would be more accurate to say that the cases in which there is no character conflict have all been worked out, and the real problems left for phylogenetic research are the cases in which the evidence is conflicting.) The reptiles and birds are an example (Figure 14.4, p. 361): some characters, such as gait, suggest one phylogenetic grouping; other characters, such as skull anatomy, suggest another.

When characters conflict, how should we proceed? One possibility is simply to stay with the principle of parsimony. If there are 100 characters shared between humans and chimpanzees, and 50 shared between humans and magnolias, the phylogeny putting the humans and chimpanzees together will be more parsimonious; it will require less evolutionary events. But there is also a second method. The characters can be analyzed further to see whether some are more reliable indicators of phylogenetic relationships than others. The 150 characters shared by humans with magnolias and chimpanzees could be studied to see whether some could be ruled out; in the best case, it might be possible to rule out all 50 of one class (or all 100 of the other) to leave an unambiguous set of evidence. A full method therefore can have two stages: (a) analyze the characters to identify the most reliable ones for purposes of phylogenetic inference, and then (b) apply statistical techniques, such as parsimony or a distance measurement, to that sifted residue of reliable characters. In practice, character analysis is used more with morphologic evidence and the classical morphologic study of phylogeny uses the full two-stage method; character analysis is less applicable to molecular evidence and the inference of phylogeny using molecular evidence goes straight to the statistical stage. In this chapter, we shall first look at how morphologists analyze characters and then move on to look in more detail at the inference of phylogeny from molecular evidence, using parsimony and distance statistics.

17.5 The characters that are most reliable for phylogenetic inference can be theoretically specified

We met the necessary distinctions when discussing classification in chapter 14 (see Figure 14.7 and Table 14.2, pp. 366–367), but there we assumed that the character analysis had been performed, and were concerned with how to use the different characters in classification. Here we are going to consider how to analyze the characters; but first we should quickly recapitulate the distinctions and their rationale. Any character shared between two species can belong to one of three categories. If it was present in the common ancestor of the two species, it is a homology; if it was not (and therefore evolved convergently) it is an analogy. Homologies, in turn, divide into ancestral and derived homologies. An ancestral homology was present in the common ancestor of the two species, but evolved earlier and is also shared with other, more distantly related species; a derived homology first evolved in the common ancestor of the two species and is not shared with other more distantly related species. If the two species were a human and a salmon, for example, the backbone is a derived homology (it first evolved in the common ancestor of the vertebrates—which is also the common ancestor of these two species), but the presence of a distinct nucleus within all their cells is an ancestral homology (it was present in the common ancestor of the vertebrates, but evolved earlier, in the common ancestor of all eukaryotes). Of the three kinds of character, only shared derived homologies are evidence that the two species share a more recent common ancestor with each other than with any other species (Figure 17.3). Analogies or shared ancestral homologies can be shared by species that do not form phylogenetic groups.

The fact that analogies and ancestral homologies within a group of species can be shared between species that do not share a recent common ancestor leads to an important conclusion. Phenetic similarity does not reliably indicate phylogenetic relationship. Consider a classic example of convergence from the two major groups of mammals, the marsupials and placentals (Figure 17.4). The marsupial and placental saber-toothed carnivores both evolved long, gashing canine teeth and there are striking similarities in skull shape and body form in the marsupial and placental wolves. If we consider the phylogeny of the marsupial wolf, the placental wolf, and the kangaroo, the two wolves are phenetically more similar but the marsupial wolf is phylogenetically closer to the kangaroo than to the placental wolf. That phenetic similarity is misleading in the case of convergence is widely

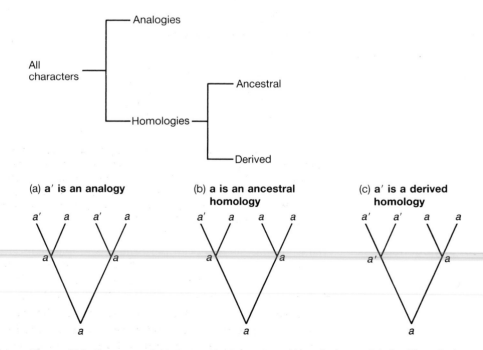

Figure 17.3 Characters divide into analogies, ancestral homologies, and derived homologies. Analogies (a), and ancestral homologies shared between species (b) do not indicate they share a recent common ancestor. Shared derived homologies do (c).

(a) **Saber-toothed carnivores**

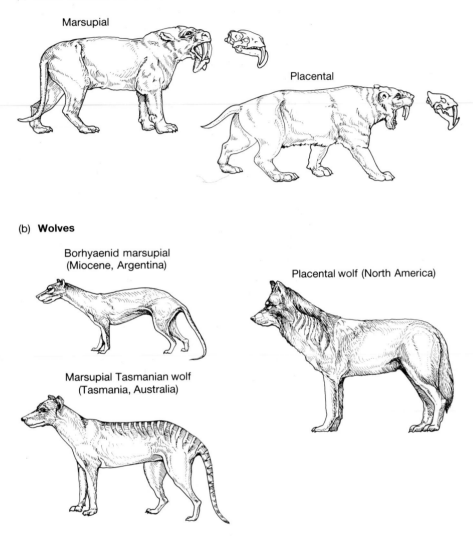

(b) **Wolves**

Figure 17.4 Convergence in marsupial and placental carnivores. (a) The reconstructed bodies, and skulls, of *Thylacosmilus*, a saber-toothed marsupial carnivore that lived in South America in the Pliocene and of *Smilodon*, a saber-toothed placental carnivore from the Pleistocene in North America. (b) *Prothylacynus patagonicus*, a borhyaenid marsupial from the early Miocene in Argentina; *Thylacinus cynocephalus*, the extinct marsupial Tasmanian wolf; and *Canis lupus*, the modern placental wolf.

appreciated; but ancestral homologies cause the same problem, and more insidiously. We have seen it in the reptiles (though they are not the only example): a crocodile looks more like a lizard than a bird, but is phylogenetically closer to a bird than to a lizard. The crocodile and lizard share ancestral homologies—characters that were present in the common ancestor of the bird, crocodile, and lizard—that have been lost in the rapidly evolving bird lineage. The point of these two examples is not that phenetic similarity never indicates phylogenetic relationship, but that it is unreliable; there are cases in which it will undoubtedly give the wrong result. The unreliability of phenetic similarity is the reason to distrust distance statistics in phylogenetic inference, because distance statistics group species according to their phenetic similarity (exactly the same argument applies for genotypic similarity). If we had applied the informal reasoning used at the opening of the chapter for the case of chimpanzee, human, and amoeba to the case of the lizard, crocodile, and bird, we should have blundered.

A final point about derived homologies is that they must all agree on the same phylogeny. Because all characters have evolved in the same phylogenetic tree, all the shared homologies must fall into the same pattern of groups (horizontal transfer of characters between lineages is an exception). If, in a number of characters, the homologies and character polarities have been correctly identified, it is *impossible* for different shared derived characters to suggest incompatible phylogenies. A set of species can no more have multiple phylogenetic relationships than a human family can have more than one family tree. If a human family has two family trees in its possession, at least one of them must be wrong; likewise, if two characters suggest incompatible phylogenies, the analysis for at least one of them must be wrong. The same is not true for analogies and ancestral homologies. Ten different analogies, or 10 different ancestral homologies, can fall into up to 10 different, and conflicting, groupings of species. In summary, the revealed secret of phylogenetic inference is to distinguish the derived homologies from the ancestral homologies and analogies, and use only the derived homologies to reconstruct the phylogeny.

17.6 Homologies can be distinguished from analogies by several criteria

Suppose we have a list of 100 characters that show similarities within a group of species. The characters give conflicting indications about the true phylogeny, with maybe 30 characters implying one phylogeny, 30 another, 20 still another, and the final 20 pointing to various idiosyncratic possibilities. The task is to shorten the list and reduce its ambiguity by eliminating the analogies. How can we detect them? If a shared character is a real homology (i.e. inherited from a common ancestor), it must be the same character in the two species, and we can test whether it is by examining the character in detail to find out whether it is the same in the different groups. First, if a character is homologous it must have the same fundamental structure. The wings of birds and bats, for example, are superficially similar; but they are constructed from different materials and supported by different limb digits (Figure 17.5); they probably evolved independently, and not from a winged common ancestor. Second, homologies must have the same relationships to surrounding characters; homologous bones, for example, should be connected in a similar way with their surrounding bones. Third, they must have the same embryonic development in different groups. A character that looks similar in the adult forms, but develops by a different series of stages, is unlikely to be homologous. One example, which we have met before, is the relationship between a barnacle, a mollusc, such as a limpet, and a crab (Figure 14.2, p. 358). At least superficially, the adult form of a barnacle is more like a limpet than a crab. The relations of barnacles had been uncertain for centuries until John Vaughan Thompson discovered their larvae in 1830; it is very like the larva of several groups of Crustacea, and unlike those of molluscs.

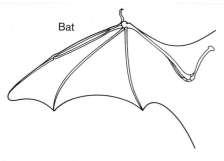

Figure 17.5 The wings of birds and bats are analogies. They are structurally different: the bird wing is supported by digit number 2, the bat wing by digits 2–5. The bird wing is also covered with feathers, the bat's with skin.

Barnacles therefore share a more recent common ancestor with crabs than with limpets. The similarities between the adult barnacle and limpet, such as their hard external armor, attachment to rocks, and feeding through a hole in the shell, are all convergent.

The classic exposés of convergence have mainly used these criteria. There are some further criteria that do not conclusively determine whether a character is a homology or analogy, but can be grounds for suspicion. For example, because convergence is caused by natural selection, neutral characters, which are not acted on by natural selection, are unlikely to form analogies. Many molecular changes may be neutral (see chapter 7); and it was once believed that many morphologic changes were too, but this is no longer widely accepted. The criterion is rarely used in modern morphology. A more convincing argument can be found in the possibility that some characters are adapted to a broader range of environments than others; the more broadly adapted characters will change less often and are more likely to be homologies. An adaptation constituting a deep part of an organism's structure may be more evolutionarily conservative than a more superficial adaptation. A backbone is a very broad adaptation, which is functional in almost all vertebrate environments; at the other extreme, color patterns change rapidly in evolution according to the needs of camouflage and sexual selection. A shared backbone is arguably more likely to be homologous than a shared color pattern (though the breadth of adaptation may not be the only reason for this inference). It has proved difficult to make this sort of argument rigorous and general, but morphologists aim to use the deeper, broader adaptations to reconstruct phylogeny. Finally, evidence may be used from other related groups of species. If some character is known to be reliable in a well-studied group of species, someone setting out to study a new but related group may begin with that character. So there are considerable possibilities for the subtle application of evolutionary knowledge in assessing the value of morphologic evidence.

The possibility remains, however, that no matter how much character analysis is performed, we may not be able to remove all the conflict from the list of characters. Maybe, from that initial list of 100 characters, morphologic research will cut it down to a list of 30 reliable characters, and of them perhaps 20 point to one phylogeny, eight to another, and two to a third. Then we have to fall back on the principle of parsimony. A phylogeny treating the 20 characters as homologies and the other 10 as analogies will require fewer evolutionary events than one treating the eight as homologies and the other 22 as analogies. The analysis by parsimony itself is a criterion of homology, for it tells us which of the 30 characters were most likely to be inherited from common ancestors and which to be convergent. But that does not mean that the final phylogeny is being settled only by the criterion of parsimony. It could be that the largest set of characters in the original list of 100 characters pointed to one phylogeny, and the largest set in the final list of reliable characters pointed to another. The study of the characters, to determine which are the most reliable homologies, is then crucial to the final result.

In summary, a homology can be recognized as a character that has fundamentally the same structure, relationship with surrounding parts, and development, in two species. There are other criteria that may suggest that a character is a homology, such as that it is a deep, or broad adaptation in the design of a body. Finally, the homologies can be recognized in a set of conflicting characters by parsimony.

17.7 Groups of species can be arranged in an unrooted tree

A knowledge of morphologic homologies can proceed a stage further: it can produce an unrooted tree for a series of forms. For example, the mammals are known from the fossil record to have evolved from the reptiles through a long series of intermediates known as the mammal-like reptiles. The transition took place in a series of gradual stages (section 20.1, p. 533), and the fossil series can be arranged in a row, leading from those that are more like reptiles to those that are more like mammals (Figure 17.6).

The position of most of the species in Figure 17.6 has been inferred from their morphologic similarity with the other species. *Diademodon* is placed between *Probelesodon* and *Cynognathus* rather than between (say) *Probainognathus* and *Oligokyphus* because it is more similar to the former two species than to the latter pair. The arrangement of species by homologous similarity tells us which species is most closely related to which, but it does not tell us where the ancestor was: it could be anywhere in the tree. Given an unrooted tree, the next question is where the ancestor (or the root of the tree) was. In this case, the answer is easy. The root is where the oldest fossils are; it is at the bottom of Figure 17.6.

The fruitflies of Hawaii provide a second, and larger example of an unrooted tree. For some reason, there is an extraordinarily large number of species of *Drosophila* in the Hawaiian archipelago. There are probably about 3000 drosophilid species in the world, and about 800 of them appear to be endemic to this archipelago. The phylogeny of one subgroup of the Hawaiian fruitflies is better known than that of any other equivalently large group of living things. It was worked out, by Carson and his colleagues, from

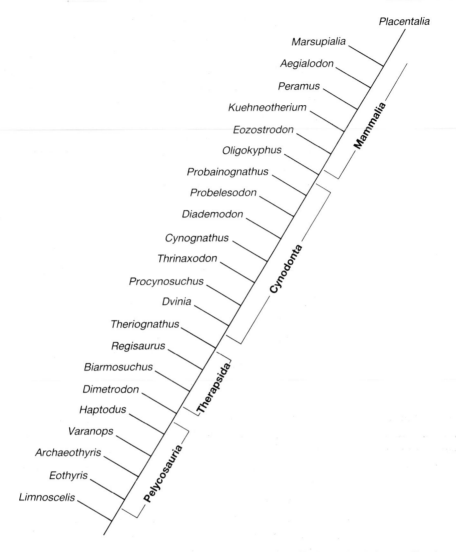

Figure 17.6 A morphoseries of fossil mammal-like reptiles. Each species is intermediate in form between the one below it and the one above it in the series. (Figure 20.2 illustrates some of the forms). From Kemp (1982a).

chromosomal banding patterns. Chromosome bands are clearly visible in fruitflies (section 4.5, p. 69).

The banding patterns differ between species, and it soon becomes obvious that regions of the chromosomes have been inverted during evolution: a segment of genes within a chromosome has been inverted as a whole. The important event, for phylogenetic inference, is for a second inversion to happen across the end of an earlier inversion (Figure 17.7). When this happens, we can infer with near certainty that the unrooted tree is $1 \leftrightarrow 2 \leftrightarrow 3$, not $1 \leftrightarrow 3 \leftrightarrow 2$. If species 1 had evolved directly into species 3, and then 3 into 2, the two inversions in Figure 17.7 would be needed for the evolution of species 3; then, to go to species 2, the exact same two breaks (one at each

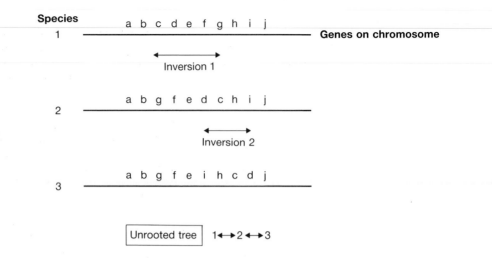

Figure 17.7 Overlapping inversions, in different species, can be used to infer their phylogenetic relations, in the form of an unrooted tree. With this pattern of inversions, the tree must be $1 \leftrightarrow 2 \leftrightarrow 3$, and not $1 \leftrightarrow 3 \leftrightarrow 2$ or $3 \leftrightarrow 1 \leftrightarrow 2$.

end) of the second inversion would have to happen again in reverse—which is much less probable than evolution in the order $1 \leftrightarrow 2 \leftrightarrow 3$. As more species are added, with more overlapping inversions, the improbability of most alternative trees multiplies to the point of practical impossibility.

To apply the technique in practice, we first have to work out the chromosomal banding patterns of the group of species. Then one species is picked, more or less arbitrarily, as the standard against which the other species are compared. Starting with the species that have a chromosomal banding pattern most like the standard, we gradually work outwards through the tree until all have been included. Carson has concentrated on the picture-wing group of Hawaiian drosophilids; there are about 110 species in this group and a recent phylogeny included 103 of these, based on 214 inversions (Figure 17.8). It is a marvellous piece of work. We can get some idea of how certain the inference is from the fact that *none* of the 214 inversions contradict the phylogeny; the characters all agree.

The tree of inversion patterns is unrooted. It is traced outwards from an arbitrary starting point; the ancestor could be anywhere within it. For the Hawaiian fruitflies, two independent lines of evidence have been used to locate the root. One is to look outside the archipelago for the likely ancestor of the group and see what banding pattern it has. The favored ancestors live in South America and are most similar to *Drosophila primaeva* and *D. attigua* (species 1 and 2 on Figure 17.8). These species are therefore probably closest to the root of the tree. The inference is supported by the archipelago's geologic history. Kauai is the oldest island and Hawaii the youngest. The ancestor of the group would probably have colonized Kauai and, if it is still alive, it should still be there because almost every Hawaiian drosophilid is confined to a single island. *Drosophila primaeva* and *D. attigua* live on Kauai.

Unrooted trees are inferred from the homologous similarities of species. An

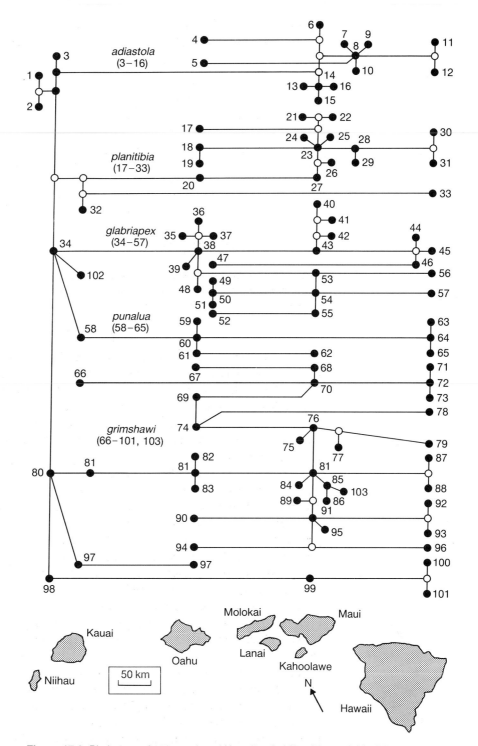

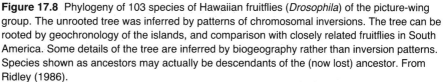

Figure 17.8 Phylogeny of 103 species of Hawaiian fruitflies (*Drosophila*) of the picture-wing group. The unrooted tree was inferred by patterns of chromosomal inversions. The tree can be rooted by geochronology of the islands, and comparison with closely related fruitflies in South America. Some details of the tree are inferred by biogeography rather than inversion patterns. Species shown as ancestors may actually be descendants of the (now lost) ancestor. From Ridley (1986).

unrooted tree is phylogenetically informative in itself, but it becomes more so if we can identify the ancestral species.

17.8 The polarity of character states can be inferred by three main techniques

The second stage in phylogenetic inference, after the distinction between homologies and analogies, is to work out which homologies are ancestral and which derived. This is the problem of *character polarity*. If we have one *Drosophila* with chromosomal state a and another with a', we need to know whether a evolved from a', or the other way round. Then we can root the tree. In this section, we shall discuss the three most important methods: outgroup comparison, the embryological criterion, and the fossil record.

17.8.1 Outgroup comparison

Suppose we know that the amniotes (reptiles, birds, and mammals) share a unique common ancestor; but we do not know the relationships between the different amniotes. For instance, in a set of six amniotic species, such as a mouse, a kangaroo, a bird of paradise, a robin, a crocodile, and a tortoise, does the kangaroo share a more recent common ancestor with a mouse, a bird of paradise, or what? Suppose we have established homologies in various characters, including reproductive physiology. The kangaroo and mouse are viviparous, and the other four species are oviparous. Did the ancestor of the group of six species breed viviparously, in which case viviparity was ancestral and oviparity derived, or did it breed oviparously, in which case evolution went the other way round? By the method of *outgroup comparison*, the answer is found by looking at a closely related species which is known to be phylogenetically outside the group of species we are studying. The character state in that outgroup is likely to have been ancestral in the group under consideration.

In this case, we might look at a salamander, or a frog, or even a fish. They are all near relatives of the amniotes, but not amniotes themselves. These outgroup species breed oviparously. The inference by outgroup comparison, therefore, is that oviparity is ancestral in the amniotes. Viviparity, in the kangaroo and mouse, would then be a shared derived character and oviparity in the other four a shared ancestral character.

In the abstract, there could be two species, species 1 and 3, sharing homology a and two others, species 2 and 4, with homology a' (Figure 17.9). We wish to know whether character a evolved into a', or a' into a. We look at a closely related species and infer that the state there is ancestral in the group of four. If the outgroup had a we should infer that species 2 and 4 share a more recent common ancestor with each other than with any of the other species. The relationships of 1 and 3 remain uncertain, pending further knowledge.

Outgroup comparison works on the assumption that evolution is parsimonious. In Figure 17.9, if the character in the outgroup (a) is ancestral in the group of species 1–4, there must have been at least one evolutionary event in the phylogeny: a transition from a to a' before the ancestor of species 2 and 4. If, having observed a in the outgroup, we reasoned that a' was the ancestral state of species 1–4, we should need at least two events: a change from a' to a somewhere between the outgroup and species 1–4, and then a

(a) **Observations**

Species	1	2	3	4	Outgroup
Character state	a	a'	a	a'	a

(b) **Phylogenetic inference**

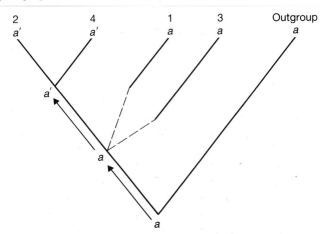

Figure 17.9 (a) Species 1–4 have the character states as given. We wish to know whether *a* or *a'* was the state in their common ancestor. (b) We look at a closely related species, the outgroup. It has state *a*, and we infer that was the state in the ancestor of species 1–4. The dotted lines for species 1 and 3 indicate their branching relationships remain uncertain.

change from *a'* back to *a* in species 1 and 3. If the character state in the outgroup is ancestral, the fewest evolutionary events are required.

Outgroup comparison, like all techniques of phylogenetic inference, is fallible. Often, one possible outgroup will suggest that one character state is ancestral, but another possible outgroup will suggest that a different character state is ancestral. The result will then depend on which outgroup we rely on. The method is most reliable when the closely related species that could be used as outgroups all suggest the same inference; but it is perfectly possible to be led astray by the method in particular cases. The inference should be treated with caution, and if possible tested against other evidence.

Returning to the case of the six amniotes, the inference about viviparity provides evidence that the mouse and kangaroo share a more recent common ancestor with each other than with any of the other four species. It provides no evidence about the relationships between the other four species: they share only an ancestral character. The evidence does not distinguish between the two types of phylogeny in Figure 17.10: the mammals could be a separate branch, or they could have branched off from any one of the other four. The shared ancestral homology of oviparity is not evidence that the four species share a more recent common ancestor with each other than with the mammals. We need further characters to sort out these four species, with the ancestral character state.

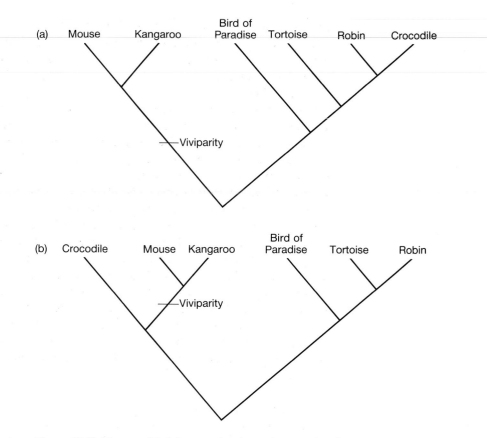

Figure 17.10 Two possible inferences for six amniote species. Once we know that mouse and kangaroo are more closely related, we still know nothing about the relationships of the other four species. Neither do we know whether the mouse and kangaroo (a) share a separate branch, or (b) branched off from any one of the other taxa. Many more phylogenies would be compatible with the evidence.

Before we can use outgroup comparison, we need to know something about the phylogeny. We needed to know that fish and amphibians were outside the Amniota in order to use them as outgroups. In practice this is not a major problem. Outgroup comparison cannot be used when we are absolutely ignorant, but if we know something about the phylogeny of a group (for example that amphibians are not amniotes, but are closely related to them) we can use that knowledge to find out more (in this case, more about the phylogeny within the amniotes).

17.8.2 The embryological criterion

The pre-Darwinian embryologist Karl Ernst von Baer (1792–1876) described, from microscopic observations, the course of development of several animal groups. He summarized his observations in his "embryological laws." For phylogenetic inference, the first law is most important. It states: "the general features of a large group of animals appear earlier in the embryo than the special features." Cartilage, for example, is found in all fish—in cartilaginous fish, such as sharks, rays, and dogfish, as well as in bony fish. Cartilage is a general character; bone is a special character, being found only in bony fish. Von Baer's law predicts that, in bony fish, cartilage will appear earlier in individual development, and will transform into bone. That is what happens (Figure 17.11).

Von Baer's law can be given an evolutionary interpretation and then put to phylogenetic use. The characters von Baer called general are in evolutionary terms ancestral; and his special characters are evolutionarily derived: the successive transformations from general to special forms of a character are evolutionary changes between ancestral and derived character states. By the embryological criterion, cartilage is inferred to be an ancestral state and bone derived: the bone in bony fish evolved from a cartilaginous ancestry. In general, if we have a group of species and a list of homologous characters, then (if they are the kind of characters that undergo development—i.e. not things like chromosomal bands) the relative ancestral and derived character states can be inferred from their order in development.

The embryological criterion only works when von Baer's law is correct (see section 20.5, p. 543). The law's scope has never been systematically studied. It is widely accepted to have some truth, and for this reason it can be used in phylogenetic inference. But it is also known to have exceptions. Clearly, von Baer's law should only be applied where we can be reasonably confident that it is valid.

17.8.3 The fossil record

In the evolution of the mammal-like reptiles into mammals (Figure 17.6), many characters changed (section 20.1, p. 533). Posture evolved from a sprawling to an upright gait; jaw articulation and circulatory physiology also changed. Some, though not all, of these characters leave a fossil record, and we can infer which character states were ancestral and which derived by seeing which is found in the earlier fossils. This is the *paleontological criterion* of character polarity.

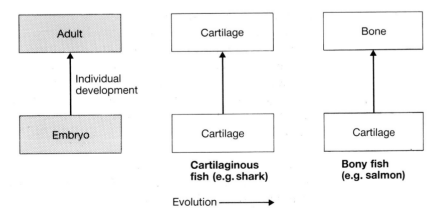

Figure 17.11 The character (cartilage) that appears earlier in the development of bony fish is a general character, found in the wider phylogenetic group of bony and cartilaginous fish. Bone is a more special character, being found only in bony fish. The character state of cartilage is inferred to be ancestral, and the state bone derived.

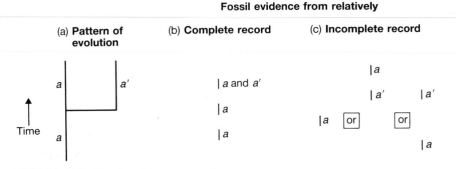

Figure 17.12 (a) The ancestral state of a character must have evolved before its derived state. If the fossil record is relatively complete (b), the ancestral state will be preserved in earlier fossils; but if it is incomplete (c), the derived state may or may not be preserved earlier than the ancestral.

The reasoning could hardly be easier. The ancestral state of a character must have preceded the derived states in fact, and therefore the earlier state in the fossil record is likely to be ancestral. In the case of the mammal-like reptiles, the criterion is reliable, because the fossil record is relatively complete. If the record is less complete, a derived character could be preserved earlier than its ancestral state (Figure 17.12), and the paleontologic inference will be the opposite of the truth.

For a whole fossil series like the mammal-like reptiles, we can be reasonably sure which states are ancestral. At the other extreme, of few fossils and a highly imperfect record, the evidence may be practically worthless. Most real cases lie between, and an intermediate level of confidence is appropriate.

17.8.4 The criteria may conflict

In some cases, the same group of species can be studied by more than one method. When they agree, the inference is proportionally stronger; but they can also disagree. When methods disagree, the procedure to follow is the same as for conflicting characters: we can either resolve the conflicts by the parsimony principle, or suspend judgment.

A third procedure has sometimes been advocated. If it could be shown that some (or one) of the criteria are more reliable than others, then when they conflict we could reject the least reliable class of evidence. Let us consider the tetrapods again as an example. If we ignore the fossil record, and consider the phylogeny of only the modern tetrapods, there is considerable evidence that the mammals share a more recent common ancestor with birds than with any group of reptiles (Figure 17.13a). Birds and mammals are both warm-blooded and there are many changes associated with homeothermy. The exact result depends on how the characters are divided up; but Gardiner, the strongest advocate of a bird–mammal relation, considered 47 characters in modern forms, and argued that most favored his controversial scheme. A critic would reply that the characters shared by birds and mammals are nearly

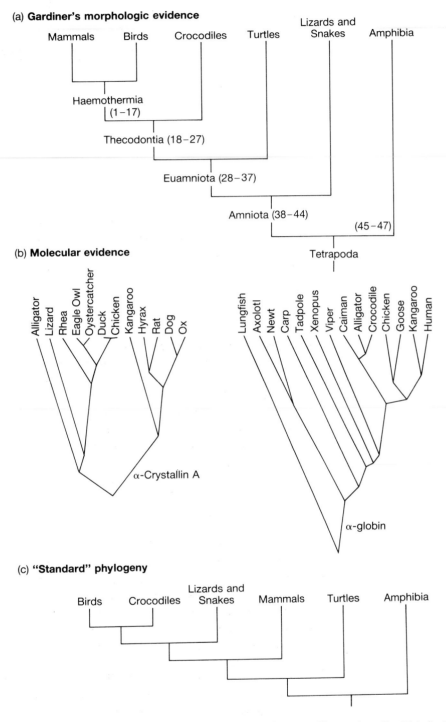

Figure 17.13 Tetrapod phylogenies. (a) According to Gardiner. The numbers (1–47) indicate numbers of inferred shared derived characters in each group; the characters are morphologic and modern. Note the unorthodox grouping together of birds and mammals. (b) Inferred from two molecules: α-globin supports, though not strongly, Gardiner's grouping of mammals with birds; α-crystallin A does not. (However, α-crystallin A itself gives an unorthodox grouping of birds with lizards rather than alligators, and look at the newts! The evidence is not definitive.) (c) Standard relationships, mainly based on skull anatomy and fossil evidence of mammal-like reptiles. From Gardiner (1982) and Bishop and Friday (1988).

all connected with the evolution of homeothermy and are not really in-dependent; perhaps only one convergence is required for all of them. However, the comparative anatomy of modern forms certainly does not strongly suggest that mammals and birds are related to different reptilian groups.

The strong evidence against the bird–mammal group comes from the fossil record. There is a good sequence of fossil intermediates—the mammal-like reptiles—between reptiles and mammals (Figure 17.6). Birds are nowhere in this sequence; indeed, the well known avian fossil *Archaeopteryx* is closely related to the dinosaurs, a quite separate reptilian group. If we include all the fossil evidence then parsimony supports the standard tetrapod groupings (Figure 17.13c). (The matter is not settled, however, as there is also molecular evidence for (and against) a mammal–bird grouping (Figure 17.13b).)

So there is conflict between modern comparative anatomy and the fossil record. If one class of evidence (and such claims have most often been made against fossil evidence) were inherently unreliable, we could resolve the conflict by ignoring it. However, there is no scientific case for excluding whole classes of evidence. Phylogenetic inference is difficult enough as it is, and we handicap ourselves unnecessarily if we exclude any potential line of evidence. When the evidence conflicts, it all has to be weighed together, by the principle of parsimony, to reach a conclusion; and the evidence from fossils, in this case, is extensive enough for most zoologists to over-rule the evidence from molecules and modern morphology.

17.9 Phylogenetic relationships can be inferred by molecular evidence and statistical techniques

The first protein to have its amino acid sequence worked out was insulin, which was sequenced by Sanger in 1954. Protein sequencing is now an automated process, and large amounts of data are available. The sequences of some proteins have been worked out in many species, which makes it possible to use the sequences to reconstruct phylogenies. The molecular evidence is in principle just like morphologic evidence, but the general problems we have met with morphology look rather different in molecules. When confronted by apparently conflicting homologies for morphologic characters (like the wings of birds and bats), the first thing to do is to re-examine the organs, and their embryology, in detail to see whether their similarity really is fundamental, or superficial and analogous. But it would be absurd to do the same for an amino acid in a protein. For example, the amino acid at site 12 of cytochrome *c* is methionine in humans, chimpanzees, and rattlesnakes, but glutamine in all the other species—including many mammals and birds—that have been studied. We cannot dissect the rattlesnake's methionine, or trace its embryonic development, to see whether it is only "superficially" methionine and "more fundamentally" glutamine. It is a methionine molecule, and that is that.

Moreover, we cannot usually assess the reliability of different pieces of molecular evidence by thinking about how natural selection could have acted on them. When morphologists examine a similarity between the organs of two species, they keep a look out for functional convergences—such as the evolution of wings in species that fly. This kind of analysis is possible if we understand the relationship between the structure (the wing) and its function (flight). However, for molecules we frequently lack this understanding. If we knew, for example, that a change from glutamine to methionine at site 12 of cytochrome *c* made functional sense in certain kinds of animal, then the same kind of arguments as appear in morphology could be used for the protein analysis. Otherwise, we have to treat molecules in the way a morphologist would treat an organ of unknown function.

Protein sequences have other distinctive properties. The amount of evidence they provide is large; cytochrome *c* alone, for example, has 104 amino acids, providing 104 pieces of phylogenetic evidence. A typical morphologic study might be based on perhaps 20 or so characters, and it is exceptional for many more than about 50 characters to be used. A second point is that the recognition of independent units of evidence is straightforward. With morphologic evidence, two apparently separate organs may really be a single evolutionary unit. At one extreme, non-independence is obvious; no one would think of treating the right leg and the left leg as two pieces of evidence. But less obvious correlations can also arise as a consequence of developmental processes, which makes it tricky to recognize independent characters. For amino acids, the mutations down the DNA molecule are (with certain exceptions) effectively independent; each site is free to evolve independently of each other site. Thirdly, evolution at different amino acid sites is easily comparable: one change at one site is equivalent to one change at another. But how can we say what amount of evolution in a knee bone is equivalent to any given change in a skull bone?

These four properties of protein sequence data—the impossibility of any deeper analysis of the character, the large amounts of evidence, the recognizability of independent units, and the comparability of evidence—have encouraged the development of statistical techniques to infer phylogenies. The same techniques are in principle just as applicable to morphologic evidence, though here it is always tempting to try to pre-empt statistical analysis and resolve the apparent conflicts by ever-deepening character analysis; morphologic data is also less readily divisible into neat character states for statistical analysis. The fundamental principle of these statistics is parsimony. (The term parsimony is sometimes applied to particular techniques, but here refers generally to all techniques based on the assumption that evolutionary change is improbable.)

The art of using molecules to infer phylogenetic relationships is to pick a molecule that evolves at a rate appropriate to the group of species in question. Different proteins, and stretches of DNA, evolve at different rates (Table 7.1, p. 142), and they can be used like clocks with hands that revolve at different rates. If you use a rapidly evolving molecule for an ancient group, the molecule will have "turned over" many times during the phylogeny and once multiple changes at the same site become common the phylogenetic information in the sequence similarity is lost: a stopwatch with only a second hand would be no use in comparing professors' lecture times. Likewise, slowly evolving molecules are useless for fine phylogenetic resolution because they will not have changed enough.

Mitochondrial DNA and ribosomal RNA are two extremes (Figure 17.14). The DNA that is present in mitochondria evolves rapidly; it has a high rate of mutation, perhaps because mitochondria lack the DNA repair enzymes that are present in the cell nucleus. Mitochondrial DNA is only useful for working out phylogenies of species that have common ancestors less than about 15 million years ago; for older relationships the mitochondrial DNA is effectively randomized. It is therefore used for recently diverged species, such as humans and great apes; for incompletely separate species (such as in hybrid zones); and for the relationships of populations within a species. Ribosomal RNA is at the other extreme. It evolves so slowly it would be useless in a study of the great apes; but it provides powerful evidence for the relationships of bacteria, and of the grand groups in the animal kingdom. Figure 17.14b illustrates the relationships of several large metazoan groups, from the work of Lake. Careful inspection of the phylogeny reveals a zoologic surprise—the arthropods may be paraphyletic. But the main point here is that a slowly evolving molecule is needed to infer phylogenetic relations of this degree of antiquity.

17.10 The parsimony principle with molecular data implicitly makes the same distinctions as are used in morphology

The principle of parsimony, applied to molecular sequences, follows the same procedure as the simple example of the human, chimpanzee, magnolia, and amoeba with which the chapter began. Figure 17.15 illustrates the mechanical procedures for more realistic data. Let us use the inference in the figure to make some points about molecular phylogenies. The first point to notice is that, although Figure 17.15 does not explicitly mention the methods of character analysis described earlier, the same fundamental principles are at work. When a pair of closely related species in a phylogeny share the same nucleotide (or amino acid), it is reasoned to have been present in their common ancestor. In the most parsimonious phylogeny, the characters that are shared by all members of a group and are used to recognize it are characters that were present in the common ancestor of the group: that is, homologies. The homologies and analogies are correctly identified in the tree requiring the least change and are incorrectly identified in all the alternative trees—if, as we are assuming, the principle of parsimony is correct.

Moreover, in the full phylogeny, groups of species are defined by derived, not ancestral, homologies. Look at species 3, 4, and 5 at the top of Figure 17.15. Species 3 and 4 share G at site 4; and species 4 and 5 share T at site 3. The phylogeny at the bottom of the figure groups together species 3 and 4 because the G at site 4 is a derived homology. The T at site 3 is shared with species other than 3, 4, and 5; it is an ancestral homology and therefore does not indicate that species 4 and 5 are closely related. The grouping of species by shared derived homologies follows from the use of parsimony. The alternative to grouping species by minimizing the number of changes would be to group them by maximizing the similarity between species; that is, to use a distance statistic. As it happens, in Figure 17.15 the species are grouped according to similarity, but this does not have to be so. Imagine, for instance, that another five bases are sequenced in all the species, and in species 1, 2, 4, and 5 they are GGGGG whereas in species 3 they are TATAT. By parsimony, in the lineage from node 1 to species 3 there would have been five more

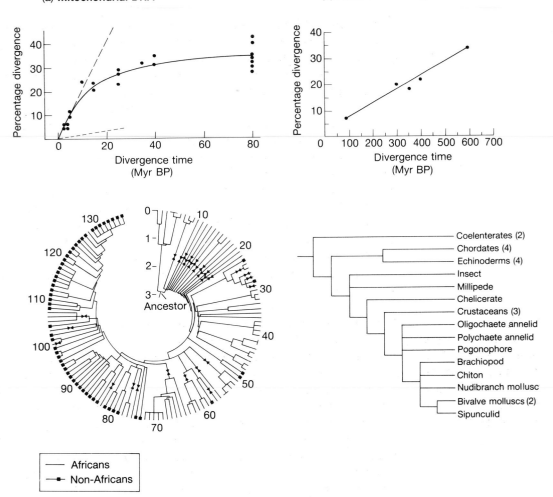

(a) **Mitochondrial DNA**

(b) **Ribosomal DNA gene**

— Africans
—■— Non-Africans

Figure 17.14 (a) (above) Mitochondrial DNA evolves rapidly. No further divergence is
detectable after 15–20 million years of separation. The different points are for mammals,
mainly primates. It tails off (at about 33% divergence) because of multiple changes at the
same site. The dotted line at the bottom shows the rate of evolution in nuclear DNA; it evolves
much more slowly. This graph is for restriction enzyme studies of the whole of the
mitochondrion. (a) (below) Relationships within *Homo* sapiens, as revealed by mitochondrial
DNA. Each of the 135 tips is a mitochondrial DNA type; the 135 types came from 189
individual human beings. The phylogeny suggests that humans originated in Africa and there
have been successive colonizations from that source. The phylogeny is based on sequences
of the control region within the mitochondrion, which evolves four to five times as fast as the
average for the whole mitochondrion depicted above. The 135 types have the following ethnic
sources: western Pygmies (1, 2, 37–48), eastern Pygmies (4–6, 30–32, 65–73), !Kung
(7–22), African Americans (3, 27, 33, 35, 36, 59, 63, 100), Yorubans (24–26, 29, 51, 57, 60,
63, 77, 78, 103, 106, 107), Australian (49), Herero (34, 52–56, 105, 127), Asians (23, 28, 58,
74, 75, 84–88, 90–93, 95, 98, 112, 113, 121–124, 126, 128), Papua New Guineans (50,
79–82, 97, 108–110, 125, 129–135), Hadza (61, 62, 64, 83), Naron (76), Europeans (89, 94,
96, 99, 101, 102, 104, 111, 114–120). The computational procedures for calculating the most
parsimonious tree for 135 units are imperfect and the tree shown is only one possibility among
many. (b) (above) Ribosomal RNA evolves slowly; 33% divergence is just being reached
between taxa with a common ancestor 600 million years ago. (b) (below) Relationships of
major animal groups, as revealed by ribosomal RNA. From Vigilant *et al.* (1991), Mindell and
Honeycutt (1990), and Patterson (1990; data from Lake 1990).

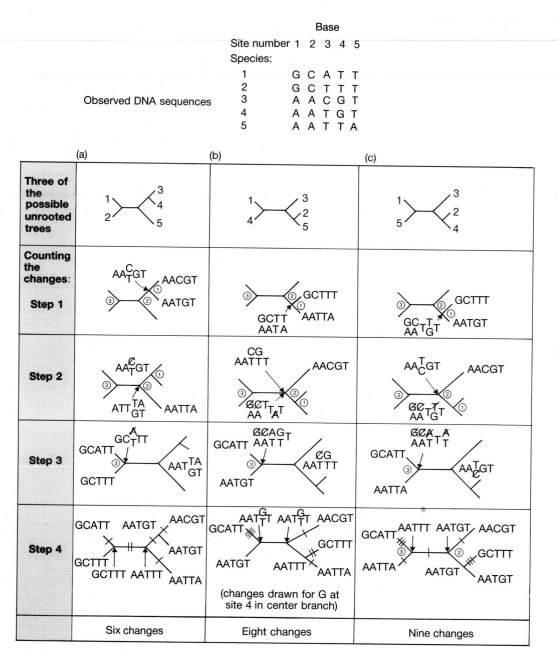

If outgroup has AATTT the tree is

changes; but the phylogeny would not be altered. Now, however, though species 3 is still the nearest relative of species 4 it is no longer the most similar species to it. Species 4 still only differs from species 5 by two evolutionary changes; but it differs from species 3 by six evolutionary changes (the one in Figure 17.15 plus the five from the five further bases). There can be any amount of ancestral similarity between species in a phylogeny, but it is all ignored when the tree is inferred by minimizing the number of changes. Using parsimony is therefore conceptually identical to using shared derived homologies in the method with explicit character analysis.

The parsimony principle is controversial. Its justification was that it infers phylogeny reliably when evolutionary change is improbable (section 17.2). However, when evolutionary change is not improbable, and in particular when evolutionary change is much more probable in some parts of a tree than in other parts, parsimony can give the wrong answer. Many variants of the parsimony principle have been devised to get round the difficulties, though all of the reliable methods assume that a tree requiring less change is more likely than one requiring more.

It has sometimes been claimed that the existence of convergence disproves the parsimony principle. Convergence in some characters and some taxa is common; and estimates of the frequency of convergence are probably underestimates because they are themselves made on the assumption that evolution is relatively parsimonious. However, the parsimony principle does not claim that convergence never happens. In fact, parsimony is a method of inferring phylogeny given that convergence *does* happen: it was because of

Figure 17.15 How to count the minimum number of changes among a set of species. Here there are five species and five variable bases have been sequenced in the DNA of each. The first step is to write out all the possible unrooted trees; only three are given here. Take the tree shown in (a) as an example. The procedure is as follows: (1) pick any species as a starting point: e.g. species 3. Trace back down its lineage to the nearest node. It is node ①. Deduce the possible sequence at the node that would minimize the number of changes below the node. Thus if a base is the same in species 3 and 4, write that base at the node: e.g. site 1 has A in both species and we write A at site 1 at the node. If a site varies, write both alternatives at the node: e.g. site 3 has C and T and we write C or T at the node because the node could have had either and one change would be required either way. (2) Trace back to the next node. It is node ②. Use the sequence in species 5 and at node ① to deduce the sequence at node ② that would require the fewest changes. Again, if a site does not vary, write the base at the node; if it varies, write both alternatives. Also, if a site is variable at node ① and one of the alternatives is in species 5, strike out the other alternative at node ①: e.g. the T at site 3 in species 5 means that node ① must have had T not C at that site. (3) Deduce node ③ from the sequences in species 1 and 2 and (4) use the common bases between the possible sequences at nodes ② and ③ to strike out alternatives when there is one base common to them: e.g. site 3 of node ③ could be A or T from looking at species 1 and 2, but it is T at node ②, so strike out the A at node ③. Now we have the most parsimonious sequences throughout the tree and can count the number of changes required. They are written as lines struck through the branches, and there are six. The same calculations for (b) and (c) give eight and nine changes; in fact (a) requires the least changes for all possible trees connecting the five species; (a) is the correct unrooted tree. The root can be found by looking at the sequence in an outgroup. Here the outgroup has the sequence AATTT, implying the phylogeny at the bottom. In principle, the outgroup could have located the root at any point in the unrooted tree.

character conflicts caused by convergence that the method of parsimony was introduced into phylogenetic inference.

17.11 Some molecular evidence can only be used to infer phylogenetic relationships with distance statistics

The rate at which molecules—both proteins and DNA—can be sequenced is speeding up all the time, but we still have sequences for only small parts of the genome even in well-studied species. A phylogeny based on one or two genes may be biased, and methods that do not require exact sequences and use samples from a larger part of the genome can be useful. There are several versions of these methods. One of the first compared proteins by immunologic techniques, and it led to an important redating of the ancestry of humans—from about 15 million years to about 5 million years ago (Box 17.1). A more recent method uses rates of DNA hybridization. The feature all the methods have in common is that they infer phylogenies by distance statistics.

If the DNA of two individuals is put together in solution, the molecules hybridize—i.e. join together. The strength of the bond is proportional to the similarity of their sequences. Double stranded DNA can be "melted" (i.e. caused to separate) by heating it, and a quick method of assessing the similarity of the whole DNA of two species is to hybridize their DNA and then measure how much lower the melting temperature is of hybrid DNA than of the DNA of a single species. The temperature reduction is proportional to the percentage similarity of the DNA; the percentage similarity of the DNA can then be used as a distance statistic. Sibley and Ahlquist have applied the method extensively to birds, and also to the hominoids (Figure 17.16).

The problem with distance statistics is that they do not distinguish the three different kinds of shared character. If shared ancestral similarities are more common between two species than are shared derived similarities, distance statistics confuse paraphyletic and monophyletic groups. If some members of a group of species (like the reptiles) evolve slowly while other members of the group evolve rapidly, the slowly evolving species are left looking similar even though they do not share a recent common ancestor; distance statistics group species by similarity and will recognize the disparate set of slowly evolving species as a phylogenetic group. However, it can be argued that this general problem does not apply to DNA hybridization. DNA hybridization measures the percentage similarity of the whole (or a large part of) the DNA of two species. The DNA sequences of species may well evolve at a relatively constant rate. Genes and proteins provide reasonably good molecular clocks (see chapter 7), and the same is likely to be true for the DNA as a whole. Then the DNA hybridization method is justified. The problematic cases like reptiles arise when evolution proceeds much more rapidly in one lineage than another; but if evolution has a constant rate for a certain class of characters, distance statistics are reliable.

In summary, there are classes of molecular evidence, of which DNA–DNA hybridization is currently the most important, that can be used to infer phylogenies by distance statistics. Distance statistics suffer the general disadvantage that they do not concentrate on shared derived homologies. They

Box 17.1 When did the hominids first evolve?

Ramapithecus (which is now usually classified in the genus *Sivapithecus*) is a group of fossil apes that lived about 9–12 million years ago. Until the late 1960s, almost all paleoanthropologists thought that *Ramapithecus* was a hominid: that is, it was more closely related to *Homo* than to chimpanzees and gorillas (Figure B17.1a). (Hominoids—formally superfamily Hominoidea—are the group of all great apes, including humans; hominids— formally family Hominidae or subfamily Homininae—are the narrower group of *Homo* and the australopithecines.) *Ramapithecus* and *Homo* apparently shared a number of derived characters: for example, (a) *Homo* has a rounded, parabolic dental arcade, whereas chimpanzees have a more pointed dental arcade; that of *Ramapithecus* was initially thought to be shaped more like *Homo*; (b) *Ramapithecus*' canine teeth were thought to

be relatively diminished compared with its other teeth, as in *Homo* but unlike chimpanzees (in which the canines, especially in males, are large); and (c) most importantly, *Homo* and *Ramapithecus* were thought to share, as a derived condition, a thickened layer of tooth enamel, unlike the thinner layer in other apes (and which was thought to be the condition in the ancestors of the Hominidae). This morphologic and paleontological argument for a relationship between *Homo* and *Ramapithecus* has a classical form: a set of character states are shown to be shared uniquely by these two species, and the characters are derived within the larger group of Hominoidea. The corollary was that the human lineage must have split from the great apes at least 12 million years ago, because *Ramapithecus* is nearer to us than to the great apes.

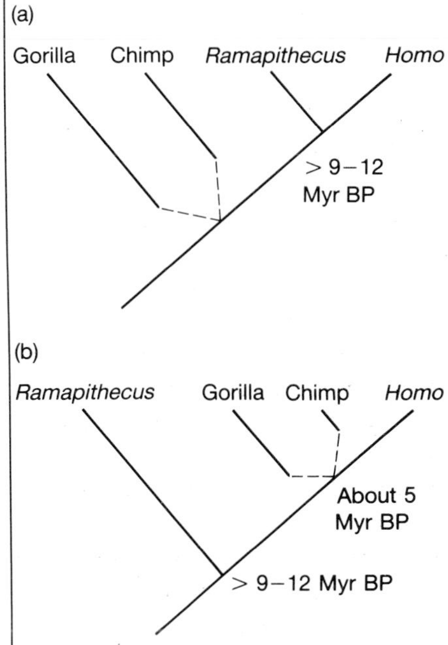

In the early 1960s, Goodman first demonstrated the molecular similarity of humans and other great apes; but the molecular argument for a recent human–ape split was most influentially made in a paper by Sarich and Wilson in 1967. Sarich and Wilson used an immunologic distance measure. They made an antiserum against human albumin by injecting it into rabbits. They then measured how much that antiserum cross-reacted with the albumin of other species, such as chimpanzees, gorillas, and gibbons. The antiserum recognizes the albumins of closely related species, because they are similar to human albumin; but it does not recognize them quite as efficiently as it does human albumin. The degree of cross-reactivity gives a measure of the immunologic distance (ID) between a pair of species. Immunologic distance increases among phylogenetically more distant relatives, and the relative rate test (Box 7.1, p. 144) suggests that it increases at a constant rate through time; immunologic distance is a sort of "molecular clock." The clock can be calibrated using the fossil record for some of the studied species, and the ID can then be used to estimate the divergence time for other pairs of species. The method

(a)

(b)

Fig B17.1 Relationships of *Homo*, other great apes, and *Ramapithecus*, according to (a) original paleontological and morphologic; and (b) molecular (and revised paleontological and morphologic) evidence. (Dotted lines imply uncertainty in the order of the human/chimpanzee/gorilla split.)

continued on page 474

suggests that *Homo* and the other great apes have too short an ID to fit with a pre-*Ramapithecus* divergence: Sarich and Wilson suggested humans and chimpanzees diverged only about 5 million years ago. Subsequent molecular work has supported this. The DNA hybridization results suggest a similar, if perhaps slightly older, figure (see Figure 17.16), and other molecules suggest a figure of 3.75–4 million years; hence 5 million years (or somewhere in the range 4–8 million years) is now a widely accepted figure for the time of origin of the hominid lineage. The corollary is that if *Homo* diverged from the chimpanzee and gorilla 5 million years ago, it cannot be more closely related to *Ramapithecus* than to the living great apes. The phylogeny must be more like Figure B17.1b.

It is always interesting when two independent lines of evidence, from very different fields, are applied to the same question. In this case, the evidence was found to conflict. A controversy began, in which both the molecular and morphologic evidence was challenged (often by experts in the other field). The controversy has now been settled (with a few dissenters) in favor of the original molecular evidence. The morphologic characters previously believed to show a relationship between *Homo* and *Ramapithecus* have all succumbed to reanalysis. The dental arcade of *Ramapithecus* had been wrongly reconstructed (originally by combining parts from two very different specimens), the reduced canine teeth may be because the fossil *Ramapithecus* specimens were female, and Martin in 1985 finally removed the last important character—thickened enamel—by reinterpreting it as an ancestral character. It is now widely thought that *Ramapithecus* is not a close relative of *Homo*; it is probably closely related to orangutans. The controversy does not show that molecular evidence is inherently superior to paleontological evidence. Maybe, on another occasion, paleontologists will successfully challenge a molecular orthodoxy; but the controversy does show how a new kind of evidence can cause an accepted conclusion, based on other kinds of evidence, to be creatively reanalyzed.

Further reading. Andrews (1986), Gribbin and Cherfas (1982), Kay and Simons (1983), Martin (1985), and Sarich and Wilson (1967).

Box 17.1 (*continued*) When did the hominids first evolve?

are most reliable if molecular evolution is clock-like, which in turn is most likely if it is driven by neutral drift.

17.12 Conclusion

Phylogenetic inference uses similarities between species, whether similarities in morphologic characters or in molecular sequences. The simplest form of inference is to reason that, for a group of three species, the two that look most like each other have the most recent common ancestor. This principle can sometimes confuse a group of distantly related, but slowly evolving, species with a group of species sharing a unique common ancestor. A more generally reliable principle is to infer the phylogeny as the tree requiring the smallest number of evolutionary changes. Then a phylogenetic relationship requires not just similarity, but similarity in derived homologies.

In practice, both the less and the more reliable principles are used. They lead, respectively, to distance and to parsimony methods. With morphologic evidence, a distance statistic measures the average phenetic similarity of the species. The parsimony principle divides the total phenetic similarity into similarity due to analogies and to ancestral and derived homologies. Phylogenies inferred using only derived homologies are the most reliable. For

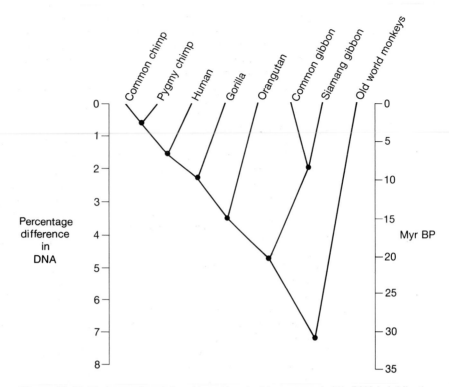

Figure 17.16 Phylogenetic relationships of hominoids, as revealed by DNA hybridization.

molecular evidence, sequence data are usually needed to infer phylogenetic relationships with a parsimony statistic. Then the individual amino acids (or nucleotides) can be treated as characters, and we can reconstruct the phylogeny requiring the minimum number of evolutionary changes. Other kinds of molecular evidence provide instead an average distance measure for each pair of species. Molecular distances are phylogenetically informative if there is a molecular evolutionary clock. There is good enough evidence for such a clock in molecular evolution for it to be used to infer phylogenies.

The reliability of our phylogenetic knowledge is highly variable. For some large taxa, we have little idea what the phylogeny is; but other cases can be more encouraging. Although it is possible that the phylogeny of the Hawaiian fruitflies has been wrongly inferred, at present it looks like one of the more certain facts in evolutionary biology.

17.13 Summary

1 Phylogenetic relationships are inferred using the shared characters of species. The best estimate of the phylogeny of a set of species is the one requiring the smallest number of evolutionary character changes. This is called the principle of parsimony.

2 When different characters imply the same phylogeny, phylogenetic inference is easy; when they do not, methods are needed to unravel the disagreement.

3 Analogies (convergent characters) and ancestral homologies do not reliably indicate phylogenetic groups. Derived homologies do. Phylogenetic inference should rely on shared derived homologies.

4 Analogies can be recognized by close analysis of a character, to reveal fundamental differences between the character in different species. They are also suggested if the character's taxonomic distribution differs from most other characters.

5 An unrooted tree specifies the branching relationships among species, but not the direction of evolution. The root can be identified if character polarities can be inferred.

6 The unrooted tree of Hawaiian fruitflies is the most firmly established phylogeny of any large group of species. It has been reconstructed using chromosomal inversions.

7 Character polarities can be inferred, with varying degrees of certainty, by outgroup comparison, the embryologic criterion, and the fossil record.

8 For molecular characters, convergence cannot be exposed by analysis of the molecule. Phylogenies are inferred using statistical techniques that weigh all characters equally.

9 If sequences are known for the molecules, the best methods make use of the principle of parsimony.

10 Molecular phylogenies can also be inferred by distance statistics; DNA–DNA hybridization is an example.

11 The statistical methods used with molecular evidence, and the character analysis used in morphology, both reconstruct phylogenies from those characters that, according to the principle of parsimony, can be expected to reveal phylogenetic relationships most reliably.

17.14 Further reading

Phylogenetic inference is introduced by, for example, Wiley (1981) and Ridley (1986), who both give references to earlier authorities. The statistical techniques used with molecular evidence are introduced by Li and Grauer (1991), and reviewed by Felsenstein (1982, 1988a), and Swofford and Olsen (1990); see also Miyamoto and Cracraft (1991). Patterson (1987) contains several papers comparing molecular and morphologic evidence. A topic not covered in the text is the question of whether phylogenetic reconstruction revolves a circular argument: see Hull (1967). For the fruitflies of Hawaii, see Carson and Kaneshiro (1976), Carson (1983, 1990), and M. Williamson (1981). For the use of fossils, see Donoghue *et al.* (1989), Norell and Novacek (1992), and Novacek (1992).

On parsimony and likelihood, see Felsenstein's (1982) paper, Friday (1987) and Sober (1984b, 1989). Wilson *et al.* (1985) introduce mitochondrial DNA; for its phylogenetic application see Harrison (1989) and (for humans) Vigilant *et al.* (1991). On Lake's work, see Lake (1987, 1990, 1991), Felsenstein (1988b), Patterson (1990); Mindell and Honeycutt (1990) and Woese (1991) review the uses of ribosomal RNA; Springer and Krakewski (1989) review the use of DNA hybridization and Sibley and Ahlquist (1990) apply it to birds. The journal *Molecular Biology and Evolution* often publishes important papers on phylogenetic inference; Bulmer (1991) is a recent example.

On the phylogeny of tetrapods, see Gardiner (1982), Kemp (1988), Benton

(1988, 1990), and Carroll (1988). For the phylogeny of other taxa not discussed in the chapter, see Fernholm *et al.* (1989) on everything, Martin (1990) and Miyamoto and Goodman (1990) on primates, Willmer (1990) on invertebrates, Hennig (1981) on insects, and Sytsma (1990) on plants.

18 Evolutionary biogeography

18.1 Species have defined geographic distributions

Biogeography is the science that seeks to explain the distribution of species on the surface of the Earth. The geographic distributions of species can be of a number of types. Consider Figure 18.1, which shows the distribution of three species of toucan in the genus *Ramphastos*, in South America. Two of the species, *R. vitellinus* and *R. culminatus*, have *endemic* distributions: they are limited to a particular area. Endemic distributions can be more-or-less widespread, and the extreme case of species that are found on all continents of the globe are called *cosmopolitan*. The pigeon, for example, is found on all continents except Antarctica; on a strict definition, the pigeon might not be allowed to be cosmopolitan, but the term is usually intended less strictly— and the pigeon is called a cosmopolitan species. Other species, like *R. ariel* in Figure 18.1, are not confined to a single area, but are distributed in more than one region with a gap between them: these are called *disjunct* distributions.

Maps like the ones for species in Figure 18.1 can be drawn for a taxonomic group at any Linnaean level: just as species have geographic distributions, so too do genera, families, and orders. The distributions of the higher taxonomic levels are, obviously, more widespread than those of species, but some taxonomically isolated higher groups, with small numbers of species (they are usually examples of living fossils, see also section 18.5), have localized distributions. For example, the tuatara *Sphenodon punctatus* is the only surviving species of a whole order of reptiles (or almost the only survivor—there may be more than one surviving species of *Sphenodon*). Of about 20 orders of reptiles, 16 are completely extinct and four have living survivors. Of those four, three contain the turtles and tortoises, lizards and snakes, and crocodiles respectively. The fourth has only *Sphenodon*, which is now confined to some rocky islands off New Zealand.

When the biogeographers of the nineteenth century looked at the distributions of large numbers of species on the globe, they saw that many species often lived in the same broad areas. They suggested that there are large-scale faunal regions on the Earth. The first map of these faunal regions was drawn for birds by the British ornithologist Philip Lutley Sclater (1829–1913), and Alfred Russel Wallace soon generalized Sclater's regions to other groups of animals. The Earth was thus divided up into six main biogeographic regions (Figure 18.2). Strictly speaking, these are zoogeographic regions, because botanists have tended to draw rather different lines on the map; but different zoologists have also disagreed about the correct borders of the regions. Figure 18.2, therefore, does not illustrate a set of hard-and-fast facts; the regions are

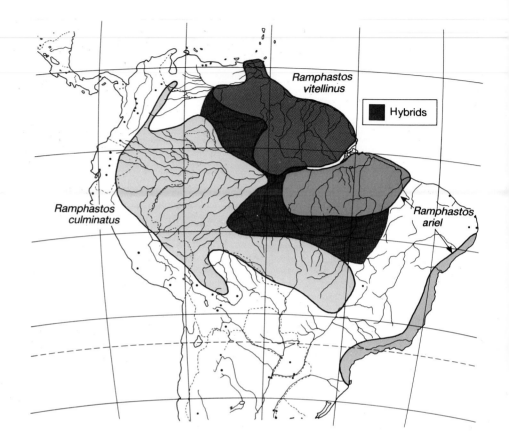

Figure 18.1 The natural distribution of three species of toucan in the genus *Ramphastos* in South America. *Ramphastos vitellinus* and *R. culminatus* have endemic distributions; *R. ariel*'s is disjunct. Adapted from Haffer (1974).

approximate. The regional terms—like Nearctic and Neotropical—are often used in biogeographic discussion.

The degree of similarity between the lists of the species living in two places can be quantified by various *indexes of similarity*. One of the simplest indexes is Simpson's index. If N_1 is the number of taxa in the area with the smaller number of taxa, and N_2 is the number of taxa in the other area, and C is the number of taxa in common between the two regions, then Simpson's index of similarity between the two areas is:

$$\text{Index of similarity} = \frac{C}{N_1}$$

Table 18.1 gives the faunal similarities, for mammalian species, between several regions (they are expressed as percentages: that is, $(C/N_1) \times 100$). The indexes illustrate the division of the Earth into faunal regions, as in Figure 18.2. For example, the faunas of Australia and New Guinea are 93% similar, whereas those of New Guinea and the Philippines are only 64% similar: the Philippines have as high a similarity with Africa as they do with New Guinea.

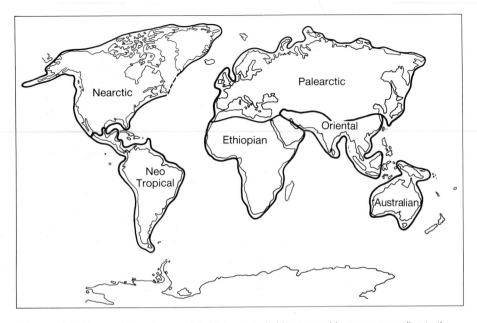

Figure 18.2 The world has been divided into six main biogeographic areas, according to the similarity of their animals. Here is the version of Simpson (1983). The discontinuity between the Australian and Oriental regions is called Wallace's Line.

Table 18.1 Indexes of similarity (%) for the mammalian species of various regions. Data from Flessa *et al.* (1979)

	North America	West Indies	South America	Africa	Madagascar	Eurasia	South East Asian islands	Philippines	New Guinea	Australia
North America										
West Indies	67									
South America	81	73								
Africa	31	27	25							
Madagascar	38	27	35	65						
Eurasia	48	27	36	80	69					
South East Asian islands	37	20	32	82	63	92				
Philippines	40	20	32	88	50	96	100			
New Guinea	36	21	36	64	50	64	79	64		
Australia	22	20	22	67	38	50	61	50	93	

This Indonesian discontinuity, which can be seen in Figure 18.2, is known as *Wallace's line*. It was not properly understood until the discovery of continental drift.

18.2 The ecological characteristics of a species limit its geographic distribution

The distributional limits of a species are set by its ecological attributes. One way of understanding how ecological factors limit a species' distribution is in terms of the species *fundamental niche* and *realized niche*. A species will be able to tolerate a certain range of physical factors—temperature, humidity, and so on—and could in theory live anywhere these tolerance limits were satisfied.

This is its fundamental niche. However, competing species will often occupy part of this range and the competition may be too strong to permit both species to exist. Each species' realized niche will then be smaller than its physiology would make possible: each will occupy a smaller range than it could in the absence of competition. Much ecological research has been carried out to discover the factors—whether physical or biological—that act to limit particular species' distributions in any particular place.

However, there are patterns in the distribution of species that probably cannot be explained by ecological factors alone. It becomes difficult, after looking at the map of one species' distribution after another, to accept that species are always excluded from all the places where they are absent purely by ecological factors that have operated persistently in the past and present; or that the places where a species exists always differ in some crucial ecological respect from the places where the species is absent. Another possibility is that historical factors have been at work. There may be places where a species ecologically could be present, but it is absent because it has never arrived—that is, never migrated to and established itself at. The same factor could work via an ecological competitor. Suppose that a species is equally well adapted to the physical conditions of two places, but is present in one but absent from the other. We then find that it is excluded, by an ecological competitor, from one place but not the other. The distribution could then be determined by the historical accident of which of the two competitors established itself in a place first. The distribution of the species will again then be determined historically—by where the species was at certain times in the past.

In what sense are ecological and historical factors alternatives? If we consider a particular distributional limit of a species, we can ask whether it lies at the limit of the species' ecological tolerance, or whether the species could ecologically survive on the other side of the border but for some historical reason has not come to be living there. It can therefore be meaningful to test between ecological and historical explanations. In most real cases, however, a complete account of a species distribution needs both ecological and historical knowledge. A species cannot live outside its ecological tolerance range: its biogeography therefore cannot contradict its ecology. However, within its ecological tolerances, historical factors may have determined where it is living and where it is not. The two factors will then not be opposed, and the sensible method of analysis is to work out how ecology and history have combined to produce the species distribution.

18.3 Geographic distributions are influenced by dispersal

A number of historical factors have been identified that influence the geographic distributions of species. A species' range will be changed if either the members of the species move in space, or space in some sense moves beneath the species, or the species splits into two species with different geographic ranges. *Dispersal* is the term for the first of these factors. Individual animals and plants move, actively and passively, through space both in order to seek out unoccupied areas and in response to environmental change. When the climate cools, the ranges of species in the northern hemisphere move southwards, and tropical forests fragment into smaller forest patches. It would also be possible for the range of a species to change, when the climate changed, without the movement of individuals: those in the colder regions (for example) might die off, and the range would shrink and move on average to the south. In practice, though, individuals would move southwards as well and extend the species range as they did so. If a species originated in one area and subsequently dispersed to fill out its existing distribution, the place where it originated is called its center of origin.

Various dispersal routes might have been followed in the biogeographic history of a species. Simpson distinguished dispersal by means of *corridors*, *filter bridges*, and *sweepstakes*. Two places are joined by a corridor if they are part of the same land mass: Georgia and Texas, for example. Animals can move easily along a corridor and any two places joined by a corridor will have a high degree of faunal similarity. A filter bridge is a more selective connection between two places, and only some kinds of animal will manage to pass over it. For instance, when the Bering Strait was above water, mammals moved from North America to Asia and vice versa, but no South American mammals moved to Asia and no Asian species moved to South America. The reason is presumably that the land bridges at Alaska and Panama were so far apart, so narrow, and so different in ecology that no species managed to disperse across them. Finally, sweepstakes routes are hazardous or accidental dispersal mechanisms by which animals move from place to place. The standard examples are island hopping and natural rafts. Many land vertebrates live in the Caribbean Islands, and (if their biogeography is correctly explained by dispersal) they might have moved from one island to another, perhaps being carried on some sort of raft.

There is good evidence for the power of dispersal. In 1883, for example, a volcanic eruption covered the Indonesian island of Krakatoa with ash and killed all the plants and animals. Biologists then recorded the recolonization of the island, particularly for birds and plants. The recolonization was astonishingly rapid. Fifty years later, the island was already recovered with tropical forest, which supported 271 plant species and 31 bird species. Invertebrate animals, such as insects, had come too, though their numbers were less closely monitored. The immigrants mainly came from the neighboring islands of Java (40 km away) and Sumatra (80 km away); the birds would have dispersed by active flight and the plants would have been carried as seeds. Dispersal, therefore, in the right circumstances can have a clear effect on the ranges of species.

18.4 Geographic distributions are influenced by climate, such as in the Ice Age

The current geologic age is called the Quaternary; it began 2.5 million years ago (see the Appendix for the ages of geology). During the Quaternary, the climate has mainly been cooler than in the preceding Tertiary; and there have been continuous cycles of increasing and decreasing temperature. Many of the cooler times were glacial periods, and the warmer times interglacials. These climatic changes have happened recently enough for the fossil record in some cases to be revealingly complete. When the weather turns cool, animals tend to migrate southwards and plant ranges contract. At any one site, the local ecology changes to one characteristic of the cooler climate; a change from a temperate to a tundra type of ecosystem has been well

(a)

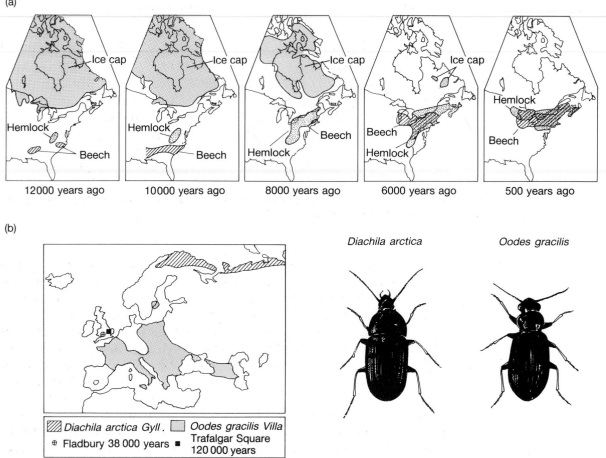

(b)

Figure 18.3 (a) Changing American geographic distribution of beech (*Fagus*) and hemlock (*Tsuga*) as the polar ice cap retreated after the most recent Ice Age. (b) Modern distributions of two beetle species: *Diachila arctica*, which lives well north of Britain; and *Oodes gracilis*, which lives further south. Fossil *D. arctica* have been found from 38 000 years ago in Britain and *O. gracilis* from 120 000 years ago.

documented, from pollen records, in the northern temperate zone through recent Ice Ages.

The change can also be seen in the distribution of single species (Figure 18.3). The most recent Ice Age ended about 10 000 years ago and Figure 18.3 shows how the geographic distributions of hemlock and beech trees moved north through the USA as the temperature warmed up and the ice cap retreated. The same factor has controlled the distribution of certain beetle species in Europe. Some 38 000 years ago the climate was cooler than it is today and Coope has found, in deposits from that time in Britain, fossil beetles belonging to species that are now not found south of northern Scandinavia. Likewise, he has found beetles belonging to species that are now not found north of France in sediments beneath Trafalgar Square in London, from 120 000 years ago when the weather was warmer (Figure

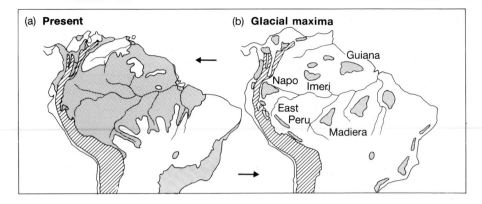

Figure 18.4 Areas of high plant species diversity in modern tropical forest. These are inferred localities of glacial forest refuges.

18.3b). The beetles thus act as indicators of past climates. Another point to note is that the beetles do not show any evolutionary changes in their morphology when the climate changes. As the interglacials and glacials come and go, the beetles do not evolve: they just move north and south.

Events in the Ice Age are also crucial in the hypothesis of *glacial forest refuges*. Tropical forests, like those of Latin America, are now continuous over large (though decreasing) areas. They are also highly diverse in species of plants and animals. There are many hypotheses to explain this diversity, one of which, suggested by Haffer in 1969, is that the forests shrunk into smaller localities during the Ice Ages. Each local area could have evolved different species from the other localities, and thus the number of species would have increased. The main evidence that led Haffer to put forward this idea is that there are small areas within the modern forest in which diversity is higher than elsewhere. These areas of high diversity, on Haffer's interpretation, are the modern vestiges of Ice Age refuges (Figure 18.4). Haffer's hypothesis is controversial. It predicts that speciation should have had a higher than average frequency in the Pleistocene, during the Ice Ages, and speciation in different groups should have been almost simultaneous. Critics argue that the speciation events that generated modern species took place long before the last Ice Age. The detailed work on the phylogenetic and biogeographic history of Amazonian species needed to test the theory is being carried out, though it is not yet conclusive.

18.5 Geographic distributions are influenced by vicariance events, such as continental drift and speciation

A second factor influencing geographic distributions is *continental drift*. The continents have moved over the surface of the globe through geologic time. The positions of the main continents since the Permian have been reconstructed in some detail (Figure 18.5), and these maps immediately suggest the reason for many biogeographic observations. For example, when we looked at the faunal regions of the world (Figure 18.2), we saw the difference between the faunas of the northern and southern Indonesian islands known as Wallace's line. It turns out that the two regions have separate tectonic histories and have only recently come into close geographic

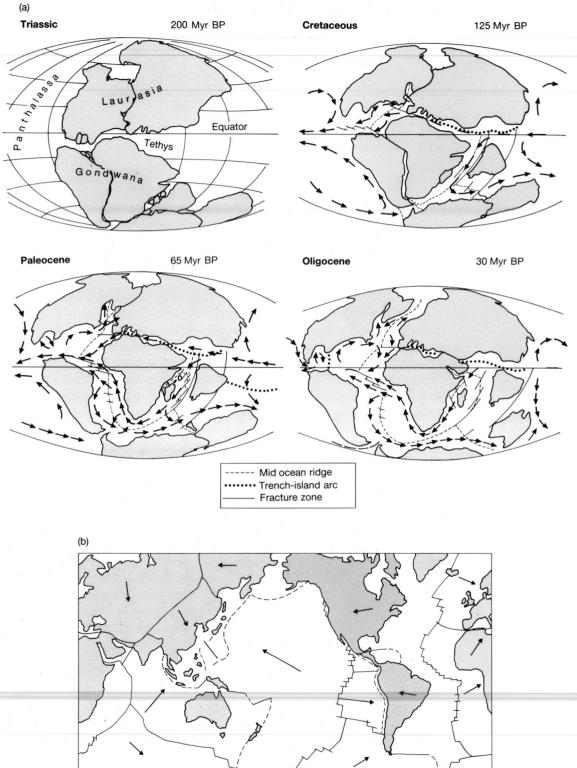

(a)

Triassic   200 Myr BP

Panthalassa

Laurasia

Equator

Tethys

Gondwana

Cretaceous   125 Myr BP

Paleocene   65 Myr BP

Oligocene   30 Myr BP

----- Mid ocean ridge
•••••• Trench-island arc
——— Fracture zone

(b)

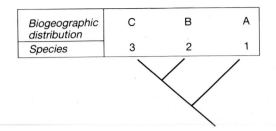

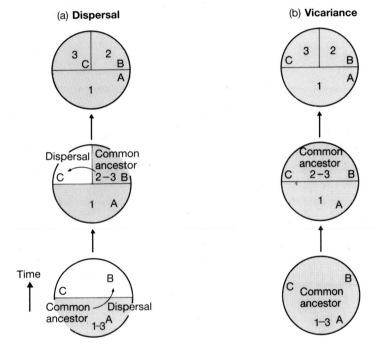

Figure 18.6 Dispersal and range splitting can be alternative hypotheses to explain the biogeography of a group. (a) An ancestral species with center of origin in area A dispersed first to area B and a descendant from there then dispersed to area C. (b) An ancestor occupying A + B + C had its range split first into A and B + C and then the descendant in B + C had its range split.

contact. The patterns of faunal similarity are therefore exactly what we should expect given the fact of continental drift. This kind of argument has been developed in detail in what is called *vicariance biogeography*.

The drifting apart of tectonic plates is the sort of event that could cause speciation (section 16.4.1, p. 412). If the splitting of the land and of the species on it do coincide, the result is two species occupying complementary parts of a formerly continuous area that was occupied by their common ancestor. This is an example of a *vicariance event*. (Vicariance means a splitting in the range

Figure 18.5 (*Opposite*) Continental drift. (a) The movements of the continents during the past 200 million years. (b) The positions of the main tectonic plates today.

of a taxon.) In theory, continental drift is just one process that could split a species' range. It could also be that a species was split by parapatric speciation, or some other process (like mountain building, or the formation of a river) that subdivided an area. In any such case, the result would be a pair of species each occupying a subsection of the ancestral species range.

According to the theory of vicariance biogeography, the distributions of taxonomic groups are determined by vicariance events rather than dispersal (Figure 18.6). The idea can be tested in two ways. The first is to see whether the pattern of splitting in one group matches the geologic history of the region where it lives. The first major piece of vicariance biogeographic research, by Brundin in 1966, was of this kind. He studied the Antarctic chironomid midges. These midges are distributed around the southern hemisphere (Figure 18.7). Brundin reconstructed their phylogeny by standard morphologic techniques (see chapter 17) and then used the species' modern biogeographic distributions to draw a combined picture of their phylogeny

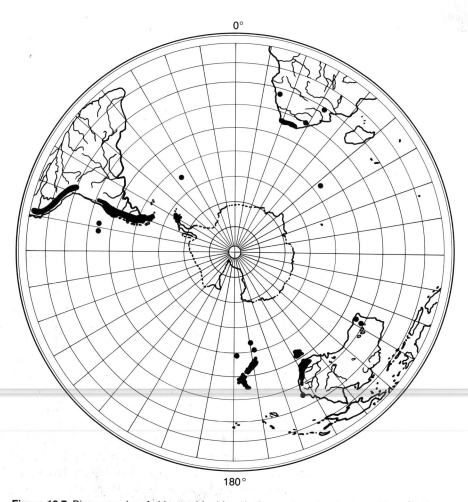

Figure 18.7 Biogeography of chironomid midges in the southern hemisphere. The midges live in the regions indicated in black. From Brundin (1989).

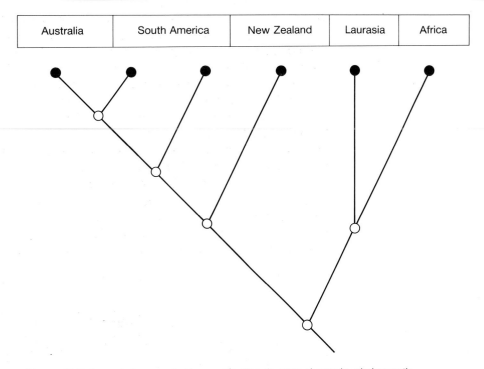

Figure 18.8 Area cladogram of chironomids. The diagram shows the phylogenetic relationships between the midges from different areas: the midges of Australia, for example, are phylogenetically more closely related to those of South America, than those of South Africa. For "Laurasia" see Figure 18.5. From Brundin (1989).

and biogeography, called an *area cladogram* (Figure 18.8). If the successive splits in the phylogeny were driven by successive break-ups of the land, the phylogeny implies a definable sequence of tectonic events. To begin with, the common ancestor of the modern forms would have occupied a large area made up of all their modern distributional zones—which implies the existence, sometime in the past, of a southern supercontinent, Gondwanaland. Gondwanaland would then, Brundin's analysis predicts, have split in the following order. First, South Africa split from east Antartica; then New Zealand split from west Antartica; later on, Australia and east Antartica split from South America and west Antarctica. This prediction can be tested against the geologic evidence, which was only accumulating during and after Brundin's work: the geology turned out to fit Brundin's prediction.

Brundin's test concerns a single taxon. A second test is to compare the relationships between phylogeny and biogeography in many taxa. As continents—or, in general, the habitats occupied by species—move in a particular pattern through time, all the groups of living things inhabiting an area will be affected in a similar manner. If members of each group tend to speciate when their ranges are fragmented, they should all show similar relationships between phylogeny and biogeography: their area cladograms should match. This prediction can be tested (Figure 18.9). In Figure 18.9, the three species of taxon 1 have a particular area cladogram, implying a parti-

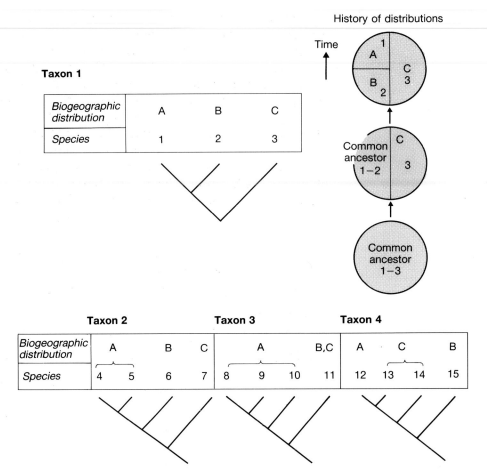

Figure 18.9 Testing vicariance biogeography by comparing area cladograms of four taxa. Phylogenies of four taxa; their species are symbolized by numbers (1, 2, 3) and the places where the species live by letters (A, B, C). Taxa 1–3 have congruent distributions, but taxon 4 is incongruent.

cular biogeographic history for the taxon. Given the area cladogram of taxon 1, we can compare it with the area cladogram of another taxon inhabiting the same region and see whether they are the same. In technical language, vicariance biogeography predicts their area cladograms will be *congruent*. Congruence is a term that can be applied to any sort of branching diagram (in phylogeny or biogeography). If two branching diagrams are congruent, the order of branching in the two does not contradict one another (they do not have to be identical, because one place, or taxon, might be missing from one of the branching diagrams; but the information that is present has to be compatible). In Figure 18.9, taxa 2 and 3 are congruent, and taxon 4 is incongruent, with taxon 1. If the land area A + B + C had first split into A + B and C and then into A and B, the phylogeny of taxa 1–3 all would fit with it: the phylogeny can be understood as a series of vicariant events. Taxon 4 does not fit. If its common ancestor occupied the whole area, its first split suggests that the land first divided not into A + B and C, but into A + C and

B. The congruence of taxa 1–3 conform with the ideas of vicariance biogeography, but taxon 4 does not.

Before these methods were developed—indeed before continental drift was widely accepted—the Venezuelan biogeographer Léon Croizat had established that different taxonomic groups often show correlated distributions. Croizat called them "generalized tracks," the distribution of any one species being its "track." He argued that if different species independently dispersed from centers of origin, they would not end up with correlated distributions. Correlated distributions are more likely to result from common vicariance events, such as continental drift, that split the ranges of several taxa in the same way. Modern vicariance biogeography adds to Croizat's ideas in two ways. One is that we now know more about the details of

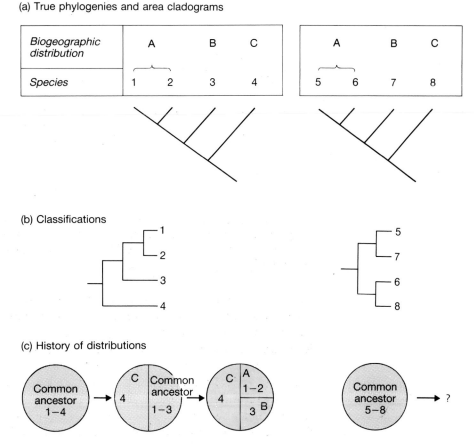

Figure 18.10 (a) True phylogenies and biogeographies of two hypothetical taxa. Species are symbolized by numbers, places by letters. Their actual biogeographies are identical. (b) Taxon 1 has been classified cladistically into monophyletic groups, but taxon 2 has been classified into two polyphyletic groups. (c) Their implied biogeographic histories are now incongruent. Taxon 1 implies a split into first C on the one side and A + B on the other. If taxon 2 were mistaken for a tree of monophyletic groups, it would need migration to explain the biogeographic distributions.

continental drift. The other is the importance of using cladistic groups when testing whether different taxa have congruent distributions.

The analysis in Figure 18.9 is only possible for taxa that are monophyletic in the cladistic sense (Figure 14.7, p. 366). If a set of phylogenetic groups have been classified into a mixture of mono-, para-, and polyphyletic groups, then even if they have experienced the same sequence of range subdivisions, their area cladograms need not be congruent. Look at Figure 18.10. In it are two taxa with congruent phylogenetic and biogeographic relationships; but one of the taxa has been classified into a polyphyletic group, and the biogeographic ranges of the two classificatory taxa are not congruent. There is no reason to expect different polyphyletic groups to have congruent biogeographic patterns, or for polyphyletic groups to be congruent with monophyletic ones. A similar argument could be made for paraphyletic groups. It is therefore essential for vicariance biogeography that taxa are cladistically classified. If the classifications contain a mixture of phenetic and cladistic taxa, any general biogeographic study is liable to become meaningless.

Let us now turn to an example. It is part of a larger study by Patterson. His starting point was a probable area cladogram for the marsupials (Figure 18.11a). Recent marsupials live in Australia and New Guinea, and South and North America (where they are represented by the opossum *Didelphis*). Fossil marsupials can also be found in Europe, making five areas in the complete area cladogram for marsupials.

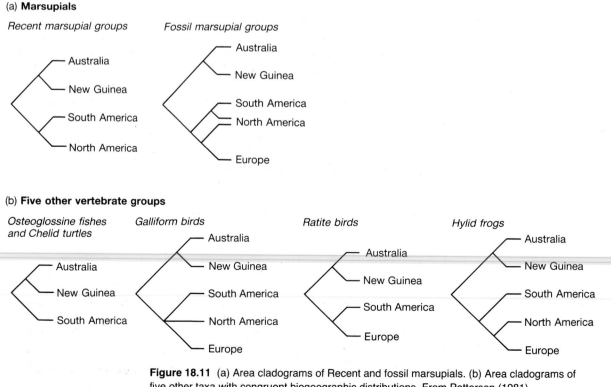

Figure 18.11 (a) Area cladograms of Recent and fossil marsupials. (b) Area cladograms of five other taxa with congruent biogeographic distributions. From Patterson (1981).

Now, marsupials have evolved on the same globe as all other species. If the modern distributions of vertebrates result from a history of range splitting, they should all have shared much the same geologic history and their area cladograms should therefore all be more-or-less congruent. What do the area cladograms for the other vertebrates look like? Figure 18.11b reveals, for five other vertebrate groups, that the vicariance prediction is upheld: all their area cladograms are congruent. The result could in theory be because all the taxa have dispersed in the same order and the same direction, but that arguably requires an implausible series of coincidences. It is more likely that the common pattern is simply due to a shared history of range splits by tectonic events.

Dispersal has probably had some influence on the history of the taxa in Patterson's study. The osteoglossid fish are found in South East Asia as well as Australia, New Guinea, and South America (Figure 18.11b). As it happens, none of the other four taxa are represented in South East Asia. There are three possible explanations for the result. One is that all six taxa used to live in South East Asia and that five of them have since become extinct there; a second is that all six were originally absent from Asia and the osteoglossid fish (in the form of *Scleropages*) arrived there by dispersal. The fossil record could in principle be used to show that a taxon once lived in Asia but is now extinct. But in the absence of any such evidence, Patterson reasons that it is more likely that one group (the osteoglossids) dispersed to Asia than that five groups became extinct there. Finally, it could be that osteoglossids originally had a broader distribution than the other five vertebrate groups, and were ancestrally present in South East Asia. The vicariance of the osteoglossids would then have taken place within a larger range.

Vicariance biogeography has been successful in finding a number of area cladograms that are consistent between different taxa and also consistent with tectonic history. But there are also exceptions. Patterson drew attention to the sciadocerid flies, which are incongruent with other dipteran flies (Figure 18.12). In this case, the area cladograms are contradictory. In other exceptional cases, some taxa are missing from areas where others are present. Incongruency may be caused by dispersal, or by erroneous phylogenies (or biogeographies) for the taxa. Vicariance biogeographers seek to resolve incongruent area cladograms by further study.

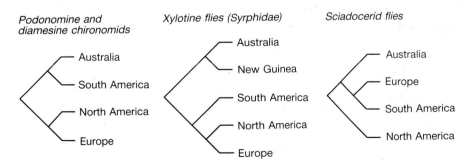

Figure 18.12 Area cladograms of four groups of dipteran flies. The pattern for the sciadocerid flies is incongruent with the rest. From Patterson (1981).

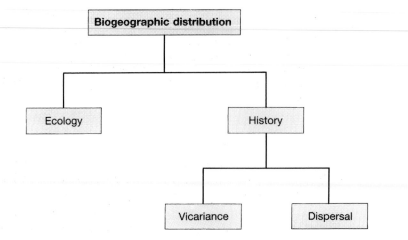

Figure 18.13 Relationship between explanatory dichotomies.

Dispersal and vicariance form a pair of possible explanations one stage below the one between ecological and historical explanations; they form two historical alternatives (Figure 18.13). Much the same point can be made about the distinction between dispersal and vicariance as was made for ecological and historical factors. It is possible that in any particular case, either dispersal or vicariance has been exclusively at work. But it is also possible that both dispersal and range subdivision have operated together, and the challenge is then to work out the relative contributions of the two.

In summary, vicariance biogeography looks for relationships between the phylogeny of a group and the geologic history of the region it inhabits, and it looks for consistent patterns in the biogeography and phylogeny of different taxa. It is a promising area of research, and has the virtues of more rigorous methods and greater generality than some of the earlier dispersalist biogeography. But vicariance biogeography is an empirical claim as well as a set of techniques, and we do not yet know how far it is possible to explain biogeography in terms of the splitting of ranges without any dispersal.

18.6 The Great American Interchange

The processes of continental drift and dispersal have both contributed to the events that take place when two previously separate faunas come into contact. These events are called biotic interchanges, and several are known from the history of life. The most famous is the Great American Interchange. Its deep geologic cause is probably connected with the tectonic processes that have been raising the Andean mountains for the past 15 million years or so. The rate of this mountain building has varied from time to time, but during a period between 4.5 and 2.5 million years ago, it intensified. At the same time—maybe 3 million years ago—the modern Isthmus of Panama rose out of the sea and the South and North American continents reconnected. This connection had dramatic repercussions for the fauna, most noticeably the mammalian fauna, of the southern continent.

North and South America had been connected before, over 50 million years earlier. They may have had similar mammalian inhabitants, but the

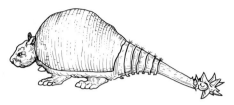

Figure 18.14 A reconstruction of *Doedicurus*, a Pleistocene glyptodont. The glyptodonts were a strange group of armored South American mammals, related to armadillos.

Cretaceous mammals of South America are too poorly known for us to be sure. Then, probably in the late Paleocene, the two halves of the American continent drifted apart. At that time, the modern orders of mammals—the groups such as horses, dogs, cats that are still the dominant land vertebrates—evolved in North America, Africa, and Europe; but South America shows no sign of possessing these forms. It instead evolved its own distinctive mammalian fauna. The South American mammals of the Paleocene and Eocene fall into three groups: marsupials, xenarthrans (armadillos, sloths, anteaters), and ungulates. Armadillos, tree sloths, and opossums still survive in South American forests, but they formerly lived along with many other curious, and now extinct, forms. There were marsupial saber-tooth carnivores (see Figure 17.4, p. 453), ground sloths (the group from which the giant ground sloth *Megatherium* of the Pleistocene evolved), and the most heavily armored mammals that ever lived—the glyptodonts (Figure 18.14), which were first described from Darwin's collections made during the *Beagle* voyage.

New arrivals came in from the outside, on rare occasions, from the early Oligocene on. They probably immigrated by waif dispersal, hopping from island to island before there was a continuous land bridge between the continents. Rodents are a major group which first appeared in the Oligocene. It is so uncertain where they came from that experts still dispute whether the South American rodents are more closely related to African or North American species (though the latter is the more widely favored source). The South American rodents, like the other mammalian groups in that land, also in turn evolved peculiar South American forms, including one called *Telicomys gigantissimus* (in the Pleistocene), which is the biggest rodent ever to have lived and was almost as large as a rhinoceros.

In the late Miocene, about 8–9 million years ago, further small additions to the fauna arrived. These were the procyonids (racoons and allies) who came from North America, and the cricetid rodents. These too almost certainly entered by waif dispersal; it is possible that North and South America had tectonically wandered closer together at that time; but the connection can have been neither close nor lasting, because it was another 6 million years before the South American mammal fauna encountered the full range of the outside world's mammalian types.

Then, about 3 million years ago, the Bolivar Trough finally disappeared and the modern Panamanian land bridge formed. The vegetation on both sides of the bridge was probably savannah, not the tropical rainforest of modern times. Mammals adapted to the similar vegetation of the two sides could move freely both ways, and it was now that the mustelids (skunks), canids

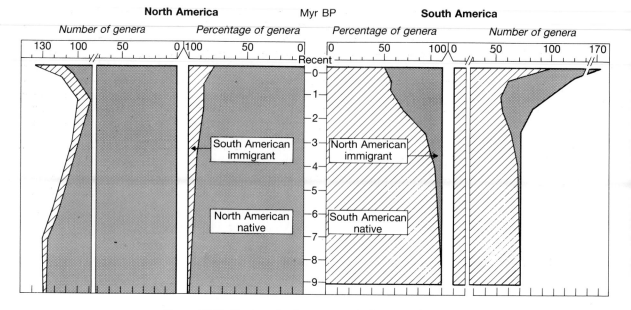

Figure 18.15 Numbers (and percentages) of genera of land mammals in the last 9 million years in North and South America. Immigrant and native genera are distinguished in both places. Note the wave of immigration after about 3 million years ago. From Marshall *et al.* (1982).

(dogs), felids (cats), equids (horses), ursids (bears), and camels invaded South America from the north, while the dasypodids (armadillos), didelphids (opossums), callithricids (marmosets), and edentate anteaters moved rather less dramatically in the opposite direction—in both cases, accompanied by many other less-well-known forms. This extraordinary clash and exchange of faunas is known as the Great American Interchange, and popular biology still portrays it as a competitive rout of the South American mammals by the superior northern forms. There is some truth in that idea; but the increasing quantity of fossil evidence is allowing a more detailed reconstruction of the events.

A study by Marshall *et al.* has examined the time course of the Interchange in detail. They counted the number of mammalian genera in South and in North America at successive times and divided the genera according to where they originally evolved. They then divided the immigrant genera into primary immigrants (genera that evolved in the south and migrated to the north or vice versa) and secondary immigrants (genera descended from primary immigrants). They argued that the primary invasions were roughly equal in both directions, and that the takeover of South America by northern mammals had two other sources: weight of numbers and different rates of speciation after arrival.

Figure 18.15 shows the numbers of genera, expressed both as absolute numbers and as proportions, of mammals in North and South America. On both sides, after 2.5–3 million years ago, an increasing proportion of the mammal genera were immigrants (or descendants of immigrants) from outside. At present, about 50% of South American genera are descended

Table 18.2 Pattern of faunal exchange between North and South America. The table gives the total numbers of mammalian genera of South or North American origin in each region (these are the numbers plotted in Figure 18.15), and breaks down the immigrant genera according to whether they were "primary" (genus itself immigrated) or "secondary" (genus descended from a primary immigrant genus—e.g. a secondary immigrant in North America is a genus that evolved in North America but whose ancestor evolved in South America). The total of the immigrant genera in the bottom rows equals the number of alien genera in the "number of genera" row. Note: (1) the similar proportions of primary immigrant genera moving in each direction; and (2) the much greater number of secondary immigrants in South than in North America. From Marshall *et al.* (1982)

	South American mammals						North American mammals			
Time period (Myr ago)	9–5	5–3	3–2	2–1	1–0.3	0.3–Recent	9.5–4.5	4.5–2	2–0.7	0.7–Recent
Duration (Myr)	4	2	1	1	0.7	0.3	5	2.5	1.3	0.7
Number of genera										
North American	1	4	10	29	49	61	128	99	90	102
South American	72	68	62	55	58	59	3	8	11	12
Total	73	72	72	84	107	120	131	107	101	114
Numbers of immigrant genera										
Primary	1	1	2	10	18	20	2	6	8	9
Secondary	0	3	8	19	31	41	1	2	3	3

from originally North American mammals; the proportion of southern mammals in the North is much lower, being less than 20%. The numbers become more revealing when we break them down further (Table 18.2). We can begin by counting the total number of genera before the Interchange; the total number is higher in the North, perhaps because of the continent's greater area. (It is an important principle of island biogeography that a larger area supports a larger number of species.) Marshall *et al.* then counted the numbers of North American mammals moving south, and South American mammals moving north, and expressed them both as proportions of the total pool. They found that the proportions are about the same: about 10% of available North American mammals invaded South America (for example, as Table 18.2 shows, there were about 100 endemic mammalian genera in North America and about 10 of them migrated south) and about 10% of South American mammals moved north (for example, there were about 60 South American mammals 3 million years ago and the number of South American genera in the north increased by about six between 4.5 and 1 million years ago). The greater absolute numbers moving south is mainly due to the larger number of mammals in the north to begin with.

The pattern of primary immigration is thus similar in both directions. Something like 10% of the genera from each side successfully invaded the other. But when we look at the subsequent proliferation of the immigrants, the pattern diverged markedly (Table 18.2). By the Recent period, a total of 12 (the nine in Table 18.2 is the number alive—three others had arrived and then became extinct) immigrant southern mammal genera had produced only three new genera, while the 21 immigrant northern mammalian genera

Table 18.3 Relative brain sizes (expressed as encephalization quotient (EQ), see section 21.3.5, p. 570, which increases with increasing brain size) of North and of South American ungulates in the Cainozoic. From Jerison (1973)

Time (Myr ago)	Ungulate brain size (EQ)	
	South America	North America
65–22	0.44	0.38
	($n = 9$)	($n = 22$)
22–2	0.47	0.63
	($n = 17$)	($n = 13$)

in the south produced 49. In the Recent period the trend has continued. The North American mammals showed their superiority, therefore, not in the original invasion, but in their relative success afterwards.

Why did the North American mammals prove superior? The increase in number of originally North American genera can be seen across a wide range of mammal types, which suggests they had some general advantage. There are several ideas why. One is that the North American mammals had lived a more competitive life, in a larger continent with more species, than the isolated southern mammals. The "arms race" of competition had progressed further in the north. The idea can be illustrated by Jerison's study of brain size (section 21.3.5, p. 570). In North American mammals, brain sizes, relative to body size, increased with time in both predators and prey in the past 65 million years. Jerison's interpretation is that brain size increased as predators and prey grew increasingly intelligent, in an escalating improvement of offensive and defensive behavior; the pattern of brain evolution fits his interpretation (see Figure 21.11, p. 571). However, in South American mammals, no such increase seems to have happened (Table 18.3). Arguably, then, when the North American mammals invaded the south they had been prepared by 50 million years or so of more demanding competition. They possessed advanced armaments, probably not only in intelligence, that enabled them to over-run the southern mammals.

Alternatively, as Marshall and his co-authors suggest, the North American mammals may have enjoyed some advantage in the environmental change of the past 3 million years. The Andean upthrust sheltered the Americas from the Pacific, creating a rain shadow east of the mountains. In South America, dryer pampas or even semi-desert replaced the moist savannah and forest. Quite why such a change should benefit the North American mammals at the expense of the South American forms is unclear; but such a change would be likely to benefit one of the two groups more than the other. It was a large change, and was therefore probably influential in the faunal replacements of the time.

The Great American Interchange is one of the most dramatic case studies in historical biogeography. The mammalian faunas of North and South America have only been connected, by a narrow isthmus, for less than 3 million years. Yet 50% of the mammalian genera in the south are now of northern origin, and such wonderful animals as that rhinoceros-sized rodent, the giant ground sloth, and the saber-toothed borhyaenid were somehow involved in the general destruction of species during the Interchange. That the events of the Interchange were at least partly due to competition is very plausible: but to demonstrate it is a taller task.

18.7 Conclusion

Evolutionary biogeographers have particularly been interested in the historical processes that have shaped the geographic distributions of species—though they by no means rule out the well-documented influence of modern ecology. They have mainly studied two sorts of historical process: movement and range-splitting. Species undoubtedly do move, by dispersal, and when a new corridor appears on the Earth allowing a new encounter of faunas, it can precipitate dramatic evolutionary events. The Great American Interchange is a famous example. It is not easy to disentangle the exact causes that were at work, but the data allow a plausible inference that the faunal changes were substantially influenced by both weight of numbers and competition.

The recent biogeographic work on vicariance events illustrates well how a discipline can be invigorated by new, formal techniques. It is possible to write the modern distributions of large taxa on a phylogeny, and use the information to answer two questions: (1) whether the order of phylogenetic splitting is consistent with the tectonic history of the areas occupied by the taxa; and (2) whether other large taxa show the same congruent biogeographic patterns. It is unlikely that all biogeographic distributions can be accounted for by vicariance events, for it is known that dispersal can be influential in cases like the recolonization of Krakatoa and the movement of species during the Ice Age. However, the relative importance of vicariance and disperal, and of ecology and history, are both topics that can in principle be settled by empirical research.

18.8 Summary

1 Species, and higher taxa, have geographic distributions, which biogeographers aim to explain.

2 The similarity of the flora or fauna of two regions can be measured by indexes of similarity. The world can be divided up into six main faunal regions.

3 The ecological properties of a species set limits on where it can live.

4 The distributions of species are influenced by historical accidents, where species happened to be at certain times, as well as their ecologic tolerances.

5 The range of species may be altered by dispersal (when a species moves in space) and by continental drift (when movement of the land subdivides the ranges of species). The splitting of a species range is called vicariance.

6 When climates cooled in the Ice Age, the range of species in the northern hemisphere moved to the south. The range of tropical forests may have been fragmented into localized refuges.

7 An area cladogram shows the phylogenetic relationships between the members of a taxon in different areas.

8 Vicariance biogeography suggests that geographic distributions are determined mainly by vicariance events, not dispersal. It predicts that the area

cladogram of a taxon should match the geologic history of the area, and the area cladograms of different taxa in an area should be similar. Both predictions are supported by evidence.

9 In the encounter between the North and South American faunas when the Isthmus of Panama formed 3 million years ago, similar proportions of mammals initially moved in both directions; but the immigrant North American mammals in the south proliferated at a greater rate.

18.9 Further reading   Cox and Moore (1985) is an introduction, Brown and Gibson (1983) is a comprehensive textbook, and Myers and Giller (1988) is a symposium on modern research topics. The *American Zoologist*, vol. 22, pp. 347–471 (1982) contains another such symposium, including a general paper about ecologic and historical factors by Endler (1982a).

Simpson (1980a, 1983) explains how the great faunal regions of the Earth were discovered, as well as the importance of movement, and how to measure faunal similarity. The ecologic influences on distribution can be read about in any ecology text, such as Begon *et al.* (1990). Baker (1978) is a thorough account of animal migration.

On Ice Age biogeography, see Pielou (1991), and the general references, particularly the chapter in Cox and Moore (1985). See Graham and Grimm (1990) for the influence of climate; on the beetles, see Coope (1978, 1979). On forest refuges, see Haffer (1969), Lynch's chapter in Myers and Giller (1988), Endler (1982b) and other chapters in the same volume, and Mayr and O'Hara's (1986) reply.

On vicariance biogeography: the chapter by Brundin in Myers and Giller (1988); Wiley (1981, 1989); Nelson and Rosen (1981), which contains the chapter by Patterson; *Systematic Zoology*, vol. 37, pp. 219–419 (1988). See also Humphries and Parenti (1986). For Croizat's biogeography, see Croizat *et al.* (1974), or the larger work of Croizat (1962).

Vermeij (1991) is a general study of biotic interchanges; as is an issue of *Paleobiology*, vol. 17, pp. 201–324 (1991), which contains a paper by Webb on the Great American Interchange. Stehli and Webb (1985) is a book about the Great American Interchange. Simpson (1980b) describes the South American mammals. On the brain size difference, see Jerison (1973) and a popular essay by Gould (1977b, chapter 23). Part of another of Gould's popular essays (1983a, chapter 27) is about the Interchange.

Part 5
Paleobiology and Macroevolution

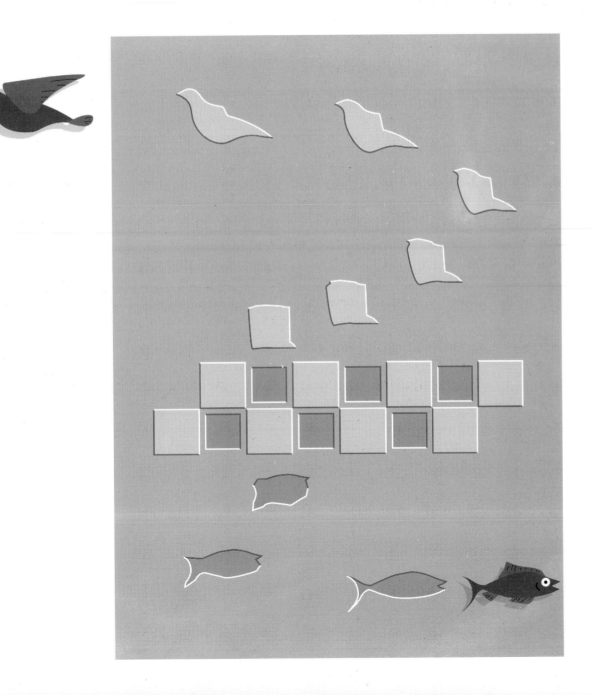

Paleobiology and Macroevolution

Part 5 is about *macroevolution*. Macroevolutionary changes are large: the kinds of event that can be studied in the fossil record, such as the origin of new organs, or body plans, or of new higher taxa (that is, taxa above the species level). These large-scale changes can be distinguished from *microevolution*, which refers to changes in gene frequencies within a population. The conventional dividing line between macro- and microevolution is at speciation: events below that level are microevolution and above it are macroevolution. The distinction is as much between the subject matter of evolutionary genetics and adaptation (parts 2 and 3) and paleobiology (this part) as between different magnitudes of change. We begin with evolutionary rates in chapter 19; we see how they are measured and consider one controversy—about the relative rates of evolution during, and between, speciation events—in detail. That controversy introduces us to a much larger question: of whether macroevolution is microevolution extrapolated over a long time scale or whether it takes place by different, though not incompatible, mechanisms from microevolution. We stay with that issue in chapter 20, which is about macroevolutionary change. The chapter presents evidence that higher groups, such as the mammals, evolved by natural selection. We then consider the different ways that development may change in evolution, because changes in developmental mechanisms are probably the typical means by which large changes in morphology evolve. Chapter 21 considers two ideas. One, coevolution, is well supported and the other, species selection, is more conjectural; but they are both possible explanations for evolutionary trends—that is, long-term net evolutionary changes in a character. The theory of species selection introduces us to the importance of extinction in determining patterns in the fossil record and chapter 22 considers why extinctions happen. We look at one theory—the Red Queen—which applies the idea of coevolution to explain extinctions, and at mass extinctions, which are more probably caused by physical catastrophes. Mass extinctions, by clearing the ground, are thought to have been important influences in the replacements of one higher taxon by another in the history of life. This part also contains an Appendix which explains the elementary facts (including the terminology of geologic time) about the fossil record that the four main chapters assume.

Rates of evolution

19.1 How to measure the rate of evolution of horse teeth

The six to eight modern species of the horse family (Equidae) are the modern descendants of a well-known evolutionary lineage in the fossil record. The record extends back through forms such as *Dinohippus* and *Mesohippus* to *Hyracotherium*, which lived 55 million years ago and was once called *Eohippus*. Horses have characteristic teeth, adapted to grind up plant material, and fossilized teeth provide the main evidence that has been used to trace the history of horses. Early members of the lineage were smaller on average than later forms (section 21.1, p. 559), and their teeth also tended to be smaller and had different shapes from the teeth of their descendants. The ancestor–descendant relationships of the several equid species are known with reasonable, if imperfect, confidence and it is therefore possible to estimate the rate of evolutionary change in their teeth by direct measurement.

Horse teeth are classic subject matter in the study of evolutionary rates, ever since the early work of Stirton (1947), Haldane (1949b), Romer (1949), and Simpson (1949a). The most comprehensive modern work is by MacFadden. He measured four properties of 408 tooth specimens, from 26 inferred ancestor–descendant pairs of species (Figure 19.1). The measure of rate used by MacFadden, and many other paleontologists, was first suggested by Haldane. For a character that has been measured and has an average value of x_1 in earlier specimens and x_2 in later specimens, where the two sets of specimens come from times Δt years apart, the evolutionary rate (r) is

$$r = \frac{\ln x_2 - \ln x_1}{\Delta t}$$

A formal unit for the rate of evolution, the "darwin" is often used, 1 darwin is defined as a change in the character by a factor of e in 1 million years ($e \simeq 2.718$, the factor for natural logarithms). The reason for transforming the measurements logarithmically is to remove spurious scaling effects. If logarithms were not taken, the rate of evolution of a character would appear to speed up when it became larger even if its proportional rate of change remained constant. With logarithmically transformed measurements, it is also possible to compare rates of change in species of very different size, such as mice and elephants.

The 26 ancestor–descendant species pairs and four dental characters measured by MacFadden produced $4 \times 26 = 104$ estimates of evolutionary rates (Figure 19.1). The different tooth characters show different patterns, with height (M1MSTHT in Figure 19.1), for instance, evolving rapidly between *Parahippus* and *Merychippus*, while the other

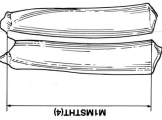

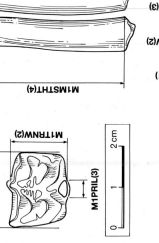

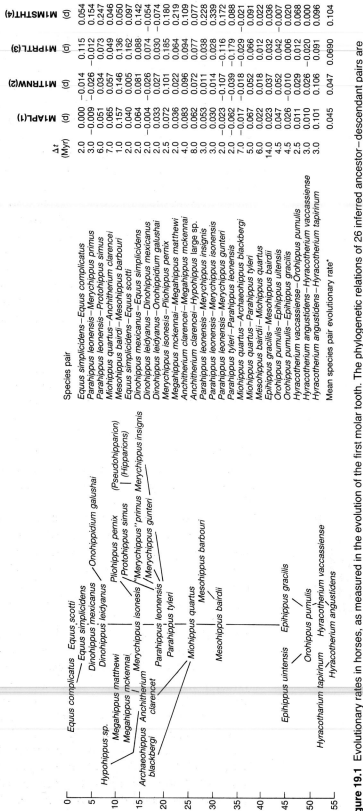

Species pair	Δt (Myr)	M1APL(1) (d)	M1TRNW(2) (d)	M1PRTL(3) (d)	M1MSTHT(4) (d)
Equus simplicidens–Equus complicatus	2.0	0.000	−0.014	0.115	0.054
Parahippus leonensis–Merychippus primus	3.0	−0.009	−0.026	−0.012	0.154
Parahippus leonensis–Protohippus simus	6.0	0.051	0.034	0.073	0.247
Miohippus quartus–Anchitherium clarencei	7.0	0.065	0.057	0.049	0.046
Mesohippus bairdi–Mesohippus barbouri	1.0	0.157	0.146	0.136	0.050
Equus simplicidens–Equus scotti	2.0	0.040	0.005	0.162	0.097
Dinohippus mexicanus–Equus simplicidens	2.0	0.064	0.081	0.088	0.142
Dinohippus leidyanus–Dinohippus mexicanus	2.0	−0.004	−0.026	0.074	−0.054
Dinohippus leidyanus–Onohippidium galushai	2.0	−0.033	0.027	0.030	−0.074
Merychippus isonesis–Pliohippus pernix	2.5	0.072	0.101	0.185	0.180
Megahippus mckennai–Megahippus matthewi	2.0	0.036	0.022	0.064	0.219
Anchitherium clarencei–Megahippus mckennai	4.0	0.083	0.096	0.094	0.109
Parahippus leonensis–Merychippus insignis	8.0	0.062	0.072	0.077	0.077
Parahippus leonensis–Merychippus isonensis	3.0	0.053	0.011	0.038	0.228
Parahippus leonensis–Merychippus gunteri	3.0	0.030	0.014	0.028	0.339
Parahippus tyleri–Parahippus leonensis	2.0	−0.023	−0.107	−0.116	0.172
Miohippus quartus–Archaeohippus blackbergi	2.0	−0.062	−0.039	−0.179	0.088
Miohippus quartus–Parahippus tyleri	7.0	−0.017	−0.018	−0.029	−0.021
Mesohippus bairdii–Miohippus quartus	5.0	0.022	0.052	0.066	0.091
Epihippus gracilis–Mesohippus bairdii	6.0	0.022	0.018	0.012	0.022
Orohippus pumulis–Epihippus uitensis	14.0	0.023	0.037	0.032	0.036
Orohippus pumulis–Epihippus gracilis	4.5	0.047	0.052	0.042	−0.007
Hyracotherium vaccassiense–Orohippus pumulis	4.5	0.026	−0.010	0.006	0.020
Hyracotherium angustidens–Hyracotherium vaccassiense	2.5	0.011	0.029	0.012	0.068
Hyracotherium angustidens–Hyracotherium tapirinum	3.0	0.010	0.026	−0.020	0.000
	3.0	0.101	0.106	0.091	0.096
Mean species pair evolutionary rate*		0.045	0.047	0.0690	0.104

Figure 19.1 Evolutionary rates in horses, as measured in the evolution of the first molar tooth. The phylogenetic relations of 26 inferred ancestor–descendant pairs are shown, and for each lineage the rate of evolution in darwins (d) is given in the table, for four properties of the first molar. It is not important to study the numbers in detail! They are meant only to illustrate the results that come out of a study of evolutionary rates. *Absolute values used to calculate. From MacFadden (1988a).

characters were evolving at normal rates. But the detailed pattern of the numbers is not important here; they could perhaps be interpreted in terms of the grinding functions of the teeth and the diets eaten by individual horse species. The approximate absolute values of the rates are worth bearing in mind. The values in Figure 19.1 are mainly about 0.05–0.1 darwins, or about a 15–30% change per million years. Each value in Figure 19.1 is an average value for a lineage connecting an ancestral–descendant species pair, and it does not imply that evolution had a steady rate throughout the lineage, of 0.05 darwins every million years. An average is not a constant, and the rates for short periods may have been very different from the long-term average. However, as average figures, the values in Figure 19.1 are fairly typical for the fossil record, being neither exceptionally fast or slow. Mammals evolving rapidly show rates of more like 1–10 darwins over short periods. Simpson, who did more than anyone to stimulate the study of fossil evolutionary rates, noticed that rates vary between taxa, characters, and times, and he invented the terms bradytelic, horotelic, and tachytelic to describe slow, typical, and rapid evolution, respectively; horse evolution as such is horotelic.

19.2 How do population genetic, and fossil, rates of evolution compare?

Rates of evolution in the fossil record have been measured for many characters, in many species, at many different geologic times. A compilation by Gingerich in 1983 included 521 different estimates, of which 409 were for the fossil record. The estimates vary, between 0 and 39 darwins in fossil lineages. The main problem of evolutionary rates is to understand why they differ between times and taxa in the patterns observed.

Before we come to that problem, we can ask a more general question. Are the rates of change seen in the fossil record consistent with the mechanisms of evolutionary change studied by population geneticists? Population genetics identifies two main mechanisms of evolution, natural selection and random drift, though it is questionable whether drift is important in morphologic evolution (section 7.11, p. 164). For changes like those of tooth size in the history of horses, we cannot confirm directly that selection was the cause; we should have to show that larger toothed horses gave rise to larger toothed offspring than average (i.e. the character was inherited) and produced more offspring than average (i.e. selection favored it). That kind of study is impossible with fossils. However, we can at least find out whether the results of research in the two areas are consistent. We can first ask whether there is any contradiction between the rates of evolution observed in population genetic work, such as artificial selection experiments, and those observed in fossils. If, for example, the fossil rates are significantly higher, it would suggest that selection alone cannot be the only cause of evolution; some other more rapid factor would be needed. In fact, it turns out, the rates of evolution in artificial selection experiments are far higher than those measured in fossils. Evolution under artificial selection has proceeded about five orders of magnitude faster than in the fossil record (Table 19.1). We can conclude that the known mechanisms of population genetics can comfortably accommodate the fossil observations.

Table 19.1 Gingerich's summary of evolutionary rates. The summary is large but not complete; it is based on 521 different measurements. Gingerich divided the measurements into four classes. The importance of the column for time intervals will become apparent in section 19.3. From Gingerich (1983)

Domain	Sample size	Evolutionary rate (d) Range	Geometric mean	Time interval Range	Geometric mean
I Selection experiments	8	12 000–200 000	58 700	1.5–10 yr	3.7 yr
II Colonization	104	0–79 700	370	70–300 yr	170 yr
III Post-Pleistocene Mammalia	46	0.11–32.0	3.7	1000–10 000 yr	8200 yr
IV Fossil Invertebrata and Vertebrata	363	0–26.2	0.08	8000 yr–350 Myr	2.8 Myr
Fossil Invertebrata alone	135	0–3.7	0.07	0.3–350 Myr	7.9 Myr
Fossil Vertebrata alone	228	0–26.2	0.08	8000 yr–98 Myr	1.6 Myr
I to IV combined	521	0–200 000	0.73	1.5 yr–350 Myr	0.2 Myr

d, darwins; yr, years; Myr, million years ago.

Strictly speaking, this does not confirm that the fossil changes *were* driven by selection and (perhaps) drift. However, it does show that the observations are consistent. For this reason, and because no other mechanisms of evolution are known, no one seriously doubts that the microevolutionary processes of chapters 4–10—even if operating indirectly (tooth sizes might increase because of selection for larger body size, for instance)—are fundamentally responsible for all evolution in the history of life. There are, as we shall see (section 19.8), some senses in which it is debatable whether macroevolution can be reduced to microevolution; but those debates are not concerned with the most fundamental question of whether selection, mutation, and drift are the causes of evolution.

19.3 Why do evolutionary rates vary?

Paleobiologists have studied a number of generalizations about evolutionary rates. For example, it has been suggested that mammals evolve faster than molluscs; that structurally more complex forms evolve more rapidly than simpler forms; and that species usually change more rapidly during, rather than between, speciation events. We shall examine the latter of these issues in more detail later. But before doing so, a number of general points should be noticed about the study of evolutionary rates.

Gingerich observed, in his compilation of evolutionary rates, an inverse relationship between the rate and the time interval over which it was measured: the observed cases of rapid evolution have tended to be for shorter intervals than the cases of slower evolution (Figure 19.2). The relationship is unlikely to be real or natural. Nothing in evolutionary theory constrains rapid

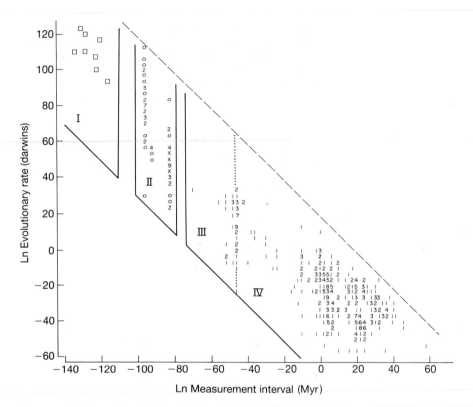

Figure 19.2 The relationship between estimated logarithmically transformed evolutionary rate and the time interval used, for the 521 studies summarized in Table 19.1. The relationship is negative. For the meaning of samples I, II, III, and IV see Table 19.1. The digits higher than 1 on the graph mean that number of cases fell on that spot (x for numbers higher than 9). From Gingerich (1983).

evolution, at these speeds, to take place in short intervals and slower evolution in longer intervals. At the molecular level, by way of comparison, the rates of evolution seem to be fairly constant over all time periods (e.g. Figure 7.2, p. 143). The relationship found by Gingerich is therefore probably an artifact, reflecting some kind of preselection of the fossil series used to measure evolutionary rates. The inverse relationship between rate and interval in Gingerich's data sets I (artificial selection—see Figure 19.2) and II (postcolonization) is probably due to higher selection coefficients in artificial selection; but that is unlikely to be the explanation within the fossil data sets (III and IV). There, the most likely reason is that the evolutionary rate of a character can only be measured if the character is recognizably the same character throughout the measuring period. Once a character changes more than a certain amount, it will start to look like a different character; it may then not be possible to measure its evolutionary rate in the same way. Eventually a leg starts to look like a wing, or a flipper. If the rates of change produced in artificial selection experiments persisted for millions of years, the character would probably be changed beyond recognition. More generally, there could be a range of changes that we are predisposed to measure; if the change is greater, the character will not appear to be the same character throughout and if it is smaller, the change may appear too trivial to notice. The remnant of changes that we measure will then show the negative relationship between interval and rate found by Gingerich: cases of rapid change over long periods and of slow change over short periods will be excluded.

This interpretation, if it is correct, matters for some kinds of generalizations about evolutionary rates, but not for others. It does not invalidate the measurements themselves. In the 14 million years between *Epihippus gracilis* and *Mesohippus bairdii*, horse teeth evolved at a rate of 0.023–0.036 darwins (Figure 19.1), and that is that. All questions about individual measurements, and comparisons between them, remain valid. It is for the more general patterns that Gingerich's result should make us suspicious. The samples may be biased. The generalization that mammals evolve faster than molluscs, for example, is reflected in Gingerich's data: he found that vertebrates as a whole tended to evolve faster than invertebrates (compare the mean rates of evolution for the two groups in Table 19.1). Now, it remains true that *in the samples measured* vertebrates did evolve 1.14 times as fast as invertebrates; but there is a danger that the rapidly evolving invertebrates and slowly evolving vertebrates have been excluded. The time intervals in fact suggest that the samples are biased (compare the mean time intervals for vertebrates and invertebrates in Table 19.1). The difference in rates can be more than accounted for by the shorter time intervals used to estimate the vertebrate evolutionary rates. When Gingerich corrected for the difference in intervals (by extrapolation from Figure 19.2) he deduced that invertebrates actually evolve faster than vertebrates. That particular correction may or may not be appropriate, but anyhow it is advisable to look at the time intervals in comparisons of evolutionary rates.

These problems do not mean it is impossible to compare rates of evolution between different taxa, or between different kinds of taxa. However, they do need to be taken into account. We shall concentrate on a question that should not be influenced by the perceptual difficulties implied by Gingerich's result;

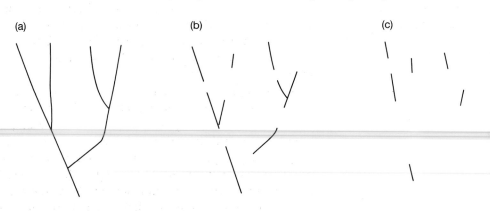

Figure 19.3 The record of gradual evolution is eroded if the fossil record is incomplete. (a) The real pattern of evolution. (b) In a fairly complete record, some of the gradual change can still be seen, but (c) as the record becomes more incomplete, the evidence of gradual change is lost.

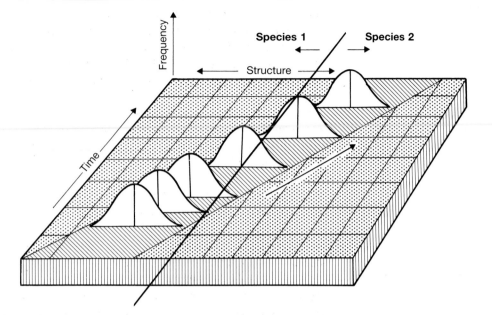

Figure 19.4 Evolution by phyletic gradualism. The figure is slightly modified from Eldredge and Gould (1972), who had in turn reproduced it from a paleontologic textbook, to illustrate how paleontologists thought in term of phyletic gradualism.

it is also outstandingly the most lively modern controversy about evolutionary rates: the theory of punctuated equilibrium.

19.4 The theory of punctuated equilibrium originally applied the idea of allopatric speciation to predict the pattern of change in the fossil record

In 1972, Eldredge and Gould published a famous essay. It argued that paleontologists had misinterpreted neo-Darwinism. The fossil record had posed an apparent problem for Darwin because it does not show smooth evolutionary transitions. A common pattern is for a species to appear suddenly, to persist for a period, and then to become extinct. A related species may then arise, but with little sign of any transitional forms between the putative ancestor and descendant. Darwin explained this pattern by the incompleteness of the fossil record; if evolution really was gradual, but most of the record has been lost, the result would be the pattern we now see (Figure 19.3). (The Appendix discusses the incompleteness of the fossil record.)

Paleontologists duly studied the record in more detail to see whether they could find examples of gradual change. According to Eldredge and Gould, paleontologists came single-mindedly to pursue examples of *phyletic speciation*, in which new species arise by transformation, rather than splitting, of a lineage, and interpreted all observations in terms of this model (Figure 19.4); deviations from it were attributed to gaps in the fossil record. Eldredge and Gould summarized this evolutionary model as follows:

> We refer to it here as "phyletic gradualism" and identify the following as its tenets:
>
> (1) New species arise by the transformation of an ancestral population into its modified descendants.

(2) The transformation is even and slow.

(3) The transformation involves large numbers, usually the entire ancestral population.

(4) The transformation occurs over all or a large part of the ancestral species' geographic range.

In chapter 16, we discussed a number of theories of speciation. Eldredge and Gould drew on one of them in particular—the peripheral isolate model—and used it to suggest that the fossil record should show a pattern different from phyletic gradualism. If new species arise allopatrically and in small isolated populations, then the fossil record at any one site may not reveal the speciation event. If the site preserves the record of the ancestral species, the descendant species will be evolving elsewhere. The newly evolving species will not be preserved at the same site as its ancestor. It may even not be preserved at all: fossils may not form in the peripheral region and the small size of the intermediate population makes it less likely to leave fossils. The new species will only leave fossils at the same site as its ancestor if it reinvades the same area. Reinvasion could happen if the descendant either was outcompeting its ancestor or was sufficiently different to coexist ecologically. Either way, the new species would be fully formed by the time it turned up as fossils in the same place as its ancestor. The transitional forms would be unrecorded not because of the incompleteness of the fossil record at that site but because the interesting evolution took place elsewhere. The reason why the transitional forms are absent is again the incompleteness of the fossil record—but not in the same way as the theory of phyletic gradualism suggested.

Eldredge and Gould summarized their view of the pattern of evolution, which they called *punctuated equilibrium*, as follows:

(1) New species arise by the splitting of lineages.

(2) New species develop rapidly.

(3) A small sub-population of the ancestral form gives rise to the new species.

(4) The new species originates in a very small part of the ancestral species' geographic range—in an isolated area at the periphery of the range.

Eldredge and Gould believe that phyletic speciation (Figure 19.4) does not happen. (In chapter 15 we discussed the cladistic species concept, according to which phyletic speciation is impossible by definition. That is not the issue here. Here we are concerned with whether change happens within a lineage as a matter of fact: whether, if it does, the lineage should be taxonomically subdivided is another matter.)

The key difference between the theories of phyletic gradualism and punctuated equilibrium concerns the relative rates of evolution during, and between, speciation events. Under phyletic gradualism, evolution has a similar range of rates during the two times; under punctuated equilibrium, it is much faster at the time of splits (Figure 19.5). Between the punctuations are long periods with little or no evolutionary change, a condition Eldredge and Gould call *stasis*. Phyletic gradualism, as understood here, does not deny the existence of splitting: it would be too absurd to do so, because there are over

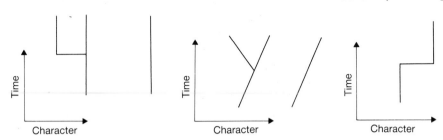

(a) **Punctuated equilibrium** (b) **Phyletic gradualism** (c) **Punctuated change without split in lineage**

Figure 19.5 The crucial difference between punctuated equilibrium and phyletic gradualism concerns the rate of evolution at, and between, splitting events. (a) Punctuated equilibrium; (b) phyletic gradualism. (c) On a strict interpretation of punctuated equilibrium, sudden change without splitting contradicts the theory (section 19.6.2).

a million species now, descended from a unique origin of life. However, it differs from punctuated equilibrium in that it predicts evolutionary change between, as well as during, speciation events.

19.5 What is the evidence for punctuated equilibrium and for phyletic gradualism?

19.5.1 A satisfactory test requires a complete stratigraphic record and biometric evidence

It is indeed common to see an apparently punctuated pattern of evolution in the fossil record. However, this fact does not strongly count in favor of the theory of punctuated equilibrium. The fossil record is incomplete and the punctuated pattern will therefore appear in most fossil samples whether the underlying evolutionary pattern is gradual or punctuated. Two crucial conditions must be met by any test of the ideas. One is that the stratigraphic sequence should be relatively complete (i.e. sediments should have been laid down fairly continuously). The other is that the fossils should be measured biometrically. It is futile to reason with taxonomic evidence: if it is observed that species a persists for a certain amount of time and then a related species b suddenly appears and in turn persists for some time, the underlying pattern of evolution could be either gradual or punctuated. Taxonomists, quite rightly, include a range of forms within a single species, and the observation that a single species persists for a certain amount of time tells us nothing about whether its morphology is changing gradually, or is constant in form.

The question is not simply whether "either" punctuated equilibrium "or" phyletic gradualism is right. The two theories represent extreme points in two continuous dimensions: the pattern of evolution itself in any one lineage, and the relative frequency of the patterns in different lineages, can occupy any point between these extremes. Thus, according to the theory of punctuated equilibrium, the majority (perhaps more than 90%) of evolutionary lineages should show a punctuated pattern whereas a phyletic gradualist might claim the opposite. Nature itself could lie anywhere between the two. Likewise there are plenty of patterns between the punctuated and gradual types of change (Figure 19.6) and nature could show any of them. Punctuated equilibrium and phyletic gradualism are not the only alternatives to be tested

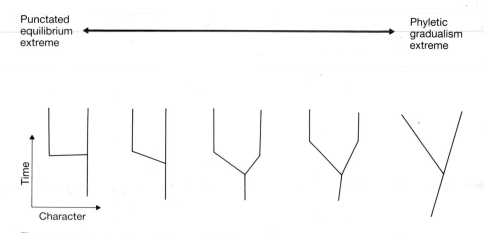

Figure 19.6 Punctuated equilibrium and phyletic gradualism are extremes of a continuum.

between. It is therefore not revealing to try and test "between" gradual and punctuated evolution; it is more appropriate to find out what the frequency of the different patterns is.

Moreover, the two extreme positions are not schools of thought that are being advocated by two opposed camps of evolutionary biologists. They are disembodied theories, not positions that large numbers of people are committed to. Some individual paleobiologists do think that the majority of cases fit the punctuational pattern; but the same cannot be said for phyletic gradualism. It is even possible that no phyletic gradualists exist. We shall return to the question of what neo-Darwinism predicts about the pattern of evolutionary rates. The question of what the frequency distribution of evolutionary rates is merits an answer—independently of what patterns people believe in.

It is too early to draw a general conclusion. The number of biometric studies, using relatively complete stratigraphic sequences, remains small. Some appear to show gradual evolution, others punctuated equilibrium; but they are all controversial. We shall confine ourselves to only two examples, to illustrate the two patterns and the kinds of evidence that are available.

19.5.2 The Plio-Pleistocene snails of Lake Turkana show a punctuated equilibrial pattern of evolution

Williamson has studied in detail the evolution of the aquatic snails fossilized in the sediments of Lake Turkana, in Kenya. He concentrated on the fossils deposited 1–5 million years ago, during the Pliocene and Pleistocene (these strata are also of great interest in human evolution). The sedimentary record is exceptionally complete—the "completenesses" of Williamson's two main samples has been estimated as 45 and 73% (Appendix, Table A2). Williamson's was a thorough study. He measured between five and 24 (19 on average) characters in 3300 specimens from 13 species lineages. His result was that the snails, in all 13 lineages, showed no change for prolonged periods, with occasional bursts of punctuational change. The abrupt changes were concentrated in intervals of about 5000 to 50 000 years, much shorter than the

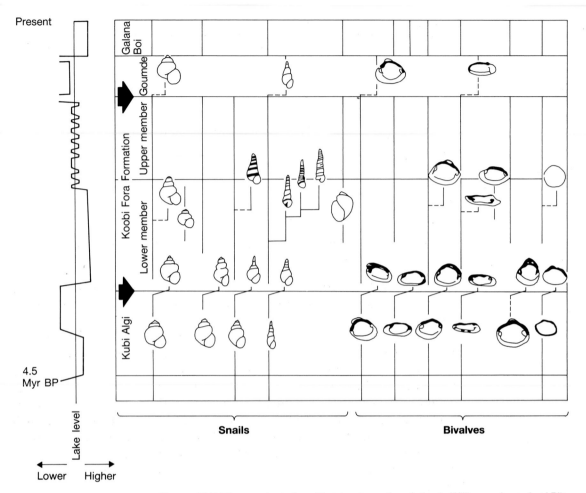

Figure 19.7 The punctuated equilibrial pattern of evolution in Williamson's study of Plio-Pleistocene snails at Lake Turkana. Some 13 separate lineages were studied; time goes up the page. At the left, the sea level is shown, and the main stratigraphic divisions. Note the punctuational changes are often simultaneous, and may coincide with changes in sea level. The x-axis is notional, not biometric. From Williamson (1981a).

long periods of constancy. The biometric data are unpublished, but Figure 19.7 is a diagram of the pattern of change.

The species all tended to change simultaneously, at times when the sea level changed: the main simultaneous punctuational events occurred at the times of the rise and fall of the water level at the beginning and end of the Koobi Fora Formation. As the water level lowered, larger lakes would have fragmented into archipelagos of smaller lakes, and the populations of snails would have had their geographic ranges fragmented into many smaller, isolated populations at the same time. The combination of environmental change and isolated populations are just the conditions for allopatric speciation, and taxonomists do indeed count the forms after the regressions as new species. The snails of Lake Turkana, therefore, illustrate the essential properties of the punctuated equilibrium model.

Williamson's study has, however, been criticized. In particular, the changes in the snails may not have been evolutionary changes, but "ecophenotypic switches". A snail's adult phenotype depends on the environment it grows up in, as can be shown by experimentally rearing snails under different environmental treatments. Some treatments can switch the snail phenotype between strikingly different forms. The effect is important because the forms can be different enough to look like different species; there are even examples of living snails, complete with formal names and descriptions, that are now known to be ecophenotypic variants of a single species. To change one "species" into the other is just a matter of rearing the eggs under appropriate conditions.

It is not known whether the separate "species" in Williamson's study are real species or ecophenotypic variants. Some experts on molluscs have remarked that the species do show the kinds of morphologic differences known from living ecophenotypic variants; there is some reason for caution at least. If Williamson's punctuations are ecophenotypic switches it would hardly be surprising that they took place rapidly, and they would not illustrate Eldredge and Gould's theory. Williamson and his followers, however, believe that the forms are good species. Apart from this controversial point, his study provides a clear example of evolution by punctuated equilibrium.

19.5.3 Ordovician trilobites show gradual evolutionary change

The extinct arthropod group of trilobites is classified by external morphologic features like the number of pygidial ribs (the pygidium is the tail region of a trilobite's body). Sheldon has made a rigorous biometric study of their evolution at a site in Wales. He measured the number of pygidial ribs in 3458 specimens from eight generic lineages, taken from seven stratigraphic sections. The total time period spanned by the sections is about 3 million years.

In all eight genera, the average number of pygidial ribs increased through time, and in all eight the evolution was gradual (Figure 19.8); a population at any one time was usually intermediate between the samples before and after it. There is one possible artifactual explanation for the result, and Sheldon was able to rule it out. A gradual increase in the number of pygidial ribs would result if two populations, one with a higher number of ribs than the other, were mixed together, with successively later samples having increasing proportions of the high rib number population. Sheldon argued this was not the case because, with rare exceptions, his samples did not show bimodal frequency distributions, as they should if they contained a mixture of two distinct populations. The study looks like a good illustration of gradual evolution.

Eldredge and Gould, however, disagree. They claim that the amount of change in these trilobites is too small to be significant; it could almost count as an example of stasis. Maynard Smith then pointed out that a trilobite example (from Eldredge's own work) originally used by Eldredge and Gould to support their ideas showed a comparable amount of evolutionary change. What is sauce for the gradual goose is sauce for the punctuated gander.

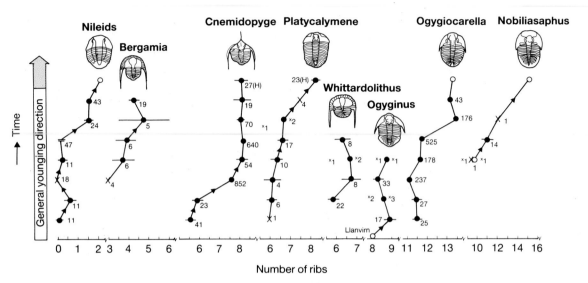

Figure 19.8 Gradual evolution in Sheldon's study of Ordovician Welsh trilobites. In eight lineages, the pattern of change is gradual rather than punctuated. Time goes up the page (total time span 3 million years) and the biometric variable (number of ribs) is on the bottom. From Sheldon (1987).

Either Eldredge's trilobites do not show punctuations, and Sheldon's do not show gradualism; or Eldredge's is a case of punctuation, and Sheldon's trilobites do show gradual evolution.

19.5.4 Conclusion

The empirical question, of what the frequencies are of different patterns of evolutionary rates, remains open for the future. Eldredge and Gould have inspired paleobiologists to collect data to new standards, and the first studies with fairly complete stratigraphy and biometric measurements are now starting to appear. Some, like Williamson's snails, illustrate a punctuated equilibrial pattern of evolution; others, like Sheldon's trilobites, show a pattern of phyletic gradualism. We shall need at least several, and probably many more, such studies before a rational conclusion can be drawn.

19.6 The theory of punctuated equilibrium has developed since 1972

There have been a number of developments since Eldredge and Gould first stated their idea. In some modern discussions, the original theory has been confused, in one great melting pot of controversy, with various later claims which have, and have not, been made by Eldredge, Gould, and their followers. Here we shall discuss the independent ideas separately. We shall concentrate on four questions.

19.6.1 Punctuated evolution may be due to allopatric speciation, or to other speciation mechanisms, such as the breakdown of homeostatic constraints, or macromutation

The original model of punctuations attributed the rapid change at speciation to allopatric events in peripheral isolates. A number of more-or-less well

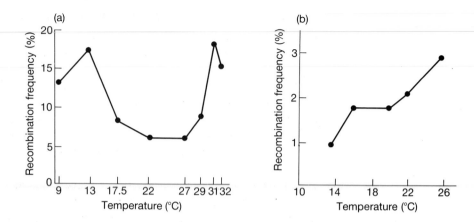

Figure 19.9 Organisms can be subjected to stress by keeping them at abnormal temperatures. Here are two examples of how recombination rate increases at stressful, unnatural temperatures: (a) recombination rate in the purple–black region of chromosome 2 in the fruitfly *Drosophila melanogaster*; (b) recombination rate between *dpy-5* and *unc-15* in the nematode worm *Caenorhabditis elegans*.

worked out suggestions have been made about how evolution might happen in a small peripherally isolated population. One possibility is that evolution might proceed as normal, under selection and drift; that would be enough to generate the punctuated equilibrial pattern. Alternatively, the population might evolve relatively rapidly, it has been suggested, simply because it is small; we discussed the idea in chapter 16 (section 16.4.4, p. 417) and saw that there are no good reasons to expect any general relationship between population size and evolutionary rate.

Another possibility is that the peripheral population may be living under abnormal conditions. In itself, this need make no difference. However, stabilizing selection—in the form of canalizing selection in this case—favors the evolution of homeostatic mechanisms (section 13.9, p. 336); under extreme conditions, outside the normal experience of the species, these mechanisms may break down and an unusual array of phenotypes may then be expressed. The generation of new phenotypes alone has no straightforward, permanent evolutionary effect: new genotypes would be needed for that. Maybe genetic variation that was previously disguised by homeostatic mechanisms could be expressed in the new conditions. The course of selection could then be altered. Moreover, strange genotypic effects can arise under abnormal conditions; recombination rates, for instance, can increase under stress (Figure 19.9).

The breakdown of homeostatic constraints is not the only unorthodox hypothesis about punctuated equilibrium. Gould revived the idea, particularly associated with Goldschmidt, that major evolutionary breakthroughs could be by macromutation. Macromutations may perhaps tend to arise in genes controlling regulatory, or developmental, processes rather than in structural genes. Gould duly suggested that new species and major groups might arise by regulatory macromutation, which is a different type of evolution from the natural selection of small mutations. Macroevolution would then proceed by non-adaptive mechanisms, and adaptive evolution would be confined to minor adjustments within populations. (We return to this topic in chapter 20.)

These are no more than hypotheses; but we should be aware of them. The main idea behind them is that natural selection and ordinary variation produce only small-scale, microevolutionary change; major changes are achieved in the revolutionary crucible of small isolated populations: only there can the new genotypes of the future be forged.

19.6.2 Does rapid change always take place at speciation?

Eldredge and Gould not only claimed that evolution proceeds in alternating periods of rapid change and stasis: they also suggested that the rapid evolution would be accompanied by speciation; when evolution is rapid, the lineage splits. For much, though not all, real fossil evidence, this is virtually a taxonomic tautology. Few lineages will be preserved so completely that all the intermediates survive, and whenever there is rapid change a new species will usually be named whether or not the lineage has split. Even if the record is complete, the possibility that a sudden change is due to immigration has to be ruled out. (Sheldon's trilobites are a possible exception: if fossils only from the beginning and end of one of the lineages had been found, they would probably have been put in different species.) Lineage splits are the crucial issue here. What reason do we have to think that rapid evolution will always cause, or only be caused by, a split in the lineage?

In logic, evolutionary changes do not have to be concentrated in the times when a lineage splits; but Eldredge and Gould gave a more biological argument to suggest that it often will be. Evolutionary change, they suggested, does not happen in whole populations: it happens in geographically isolated subpopulations. Then, as the isolated subpopulation changes, it will split from the unchanged ancestral rump population because they will now have different phenotypes.

But is it true that large evolutionary changes cannot happen in whole, rather than isolated, populations? Examples of evolutionary transitions in large, central populations are known (see chapter 5), and Turner has pointed to cases of rapid change, without splitting, in butterfly species with mimetic polymorphism. Maybe it could be argued that these are exceptional. It is a safe general statement that we do not know the relative frequency of rapid evolution in small isolated subpopulations and in large, complete populations. This being so, it would be premature to draw any conclusion ruling out one or the other. Evolution in large populations is reasonable in theory and has been observed in fact.

Real speciation (in which the ancestral species splits) and phyletic speciation (in which a lineage changes enough to be given a new name but has not split) are usually indistinguishable in the fossil record. But Eldredge and Gould's idea that rapid evolution is confined to the times of lineage splits is an interesting hypothesis. It has some theoretical appeal, though as a limited rather than absolute generalization (because exceptions are known); decisive evidence for or against it is lacking.

19.6.3 Is stasis due to stabilizing selection or constraints?

In the fossil record, stasis—evolutionary lineages that persist for long periods without change—is common. Stasis is not a new discovery, but it has taken on a new importance since Eldredge and Gould's paper. Two main ideas have been put forward to explain it. They are *stabilizing selection* and *constraint*; we have met effectively the same pair before (section 13.9, p. 337). Constraint means that species do not change because they lack the necessary genetic variation. It has been suggested that in the sorts of peripherally isolated populations where speciation may take place, latent or new genetic variation is expressed; genetic constraint is lifted and evolutionary change becomes possible. We can thus distinguish two extreme theories to explain stasis and rapid speciation: species may remain constant because they are under stabilizing selection and then change rapidly when there are new selective conditions; or they may stay constant while they are constrained and change rapidly when those constraints break down under exceptional conditions. In the real world, any mixture of these processes could operate.

Stabilizing selection does not have to mean that the environment is constant. In chapter 18 (section 18.4, p. 484), we saw how beetle populations migrate south and north as Ice Ages and interglacials come and go. The beetles remain constant in form, but move about as the conditions change: the beetle population as a geographic whole could then be experiencing stabilizing selection even though the environment at any one place was not constant.

The question of whether evolutionary constancy is caused by stabilizing selection or constraint is a particular form of the general question of what causes phenotypic patterns in nature, and the same general points can be made. The hypothesis that a character is not changing because it has no genetic variation can be tested. It predicts that the character's heritability is zero. The heritability of fossil characters cannot be measured, but that of similar characters in modern species can be. For example, Williamson suggested that the Lake Turkana snails might not have changed because of developmental constraints. Various population geneticists replied that heritabilities have been measured in snail characters like those studied by Williamson, and significantly positive values found. (The counter-argument, that Pleistocene snails had different genetics from their modern descendants, is desperate.) Stabilizing selection is well documented, whereas a lack of genetic variation has never been shown to be the reason why any character remains unchanged. Most characters show significant heritability; those that do not are puzzling, because the nature of the constraint is not understood.

The evidence of geographic variation is another reason to doubt the importance of constraint. At any one locality, species show a limited range of variability; but they vary markedly from place to place and the existence of ring species shows that geographic variation within a species can be large enough to produce speciation (section 16.4, p. 412). Williamson, however, denied this argument. He suggested that geographic variation is limited, and inadequate to account for speciation; and he attributed the limitation to developmental constraint. But he did not support his case with appropriate evidence. Variation within a species must, as a necessary consequence of taxonomy, be narrower than the variation of a group of species. The case where populations of a species are arranged geographically in a ring provides the only clear test case for his claim. The fact that these rings lead to speciation in all known cases strongly suggests that ordinary geographic variation alone can explain speciation.

In conclusion, there is a debate over why species can persist in the fossil record for long periods of time without changing. Stabilizing selection is a well-documented and plausible explanation. Constraint is an undocumented but still possible explanation that should not be ruled out before investigation begins.

19.7 Darwin was not a phyletic gradualist

We have now dealt with the main scientific questions in the punctuated equilibrium controversy. However, the full controversy also includes two historical issues. Gould has claimed that the theory of punctuated equilibrium contradicts both Darwin's own ideas and also those of neo-Darwinism. Darwin certainly did emphasize repeatedly that evolution is slow and gradual; and Gould thought that Darwin was advocating (of Eldredge and Gould's two opposed theories) the theory of phyletic gradualism. It is more likely, however, as Dawkins has argued, that Darwin meant something crucially different by gradual evolution. Darwin did not make his remarks about gradualism particularly in the context of evolutionary rates at and between speciation events. When he did discuss that subject, he said things that sound quite like punctuated equilibrium, such as (from *On The Origin of Species*):

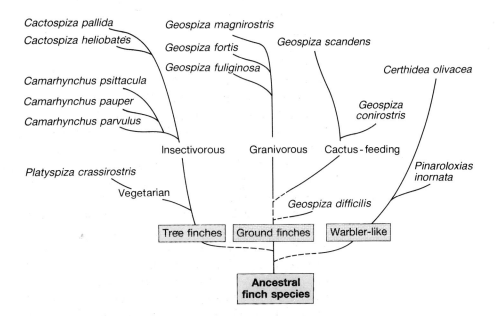

Figure 19.10 Possible phylogeny of Darwin's finches, according to Lack. The dotted lines indicate uncertainty. Other phylogenies have been suggested too. Was there time enough for the evolution of 14 species, by selection within a population, since the Galápagos were colonized by the ancestral finch maybe 570 000 years ago? From Lack (1947).

Many species once formed never undergo any further change...; and the periods during which species have undergone modification, though long as measured by years, have probably been short in comparison with the periods during which they retained the same form.

Darwin's theory was not gradualist in the sense of demanding similar evolutionary rates at, and between, speciation events. The sense in which his theory *is* strongly gradualist concerns the evolution of adaptations, particularly of complex adaptations. Adaptations cannot evolve in sudden jumps (section 13.3, p. 326). If they are to evolve at all, it must be in a large number of small stages; and this process will be slow and gradual. Darwin's theory is gradualist about the evolution of complex adaptations, not about the pattern of evolutionary rates. Darwin's own statements consistently suggest that he was talking about adaptation, rather than speciation, when he described evolution as slow and gradual. Therefore, both the structure of Darwin's theory and the literary evidence from Darwin's writings count against Gould's claim. Darwin was not a phyletic gradualist: but he did believe that adaptations are built up in many small evolutionary stages.

19.8 The theory of punctuated equilibrium does not render the modern synthesis "effectively dead"; the relationship between microevolution and macroevolution is an open question

Eldredge and Gould in 1972 originally accepted a neo-Darwinian theory of speciation and used it to update the paleobiological study of evolutionary rates. Around 1980, though, Gould moved on to a more radical position, which has been actively discussed ever since. He turned the theory of punctuated equilibrium back on the modern synthesis and (he claimed) toppled that synthesis off its throne. In 1980, he described the synthetic theory of evolution as "effectively dead."

According to Gould, the modern synthesis asserted that evolution at all levels takes place by the same process: natural selection among alleles of different fitnesses. Microevolution within populations, speciation, and the origin of higher groups, are all thought to be driven by natural selection among alleles within populations. In this sense the modern synthesis is narrow and *extrapolationist* (or *reductionist*): narrow because it recognizes only one cause of evolution; extrapolationist because it explains all evolution by extrapolating from microevolution; reductionist, in a philosophic sense, because it "reduces" macroevolution to microevolution.

Let us consider an example. We have seen (section 9.1, p. 211) how natural selection operates on the beaks of Darwin's finches. The evidence there was for natural selection within a species demonstrated that individuals with larger beaks are favored when the seeds and fruit that they eat are large, whereas smaller beaks are favored when the food size is smaller. In 1976–1977 (and two subsequent periods), Grant measured the strength of selection on the finches' beaks, and its evolutionary results (Figure 9.10, p. 229). Now, there are 14 Galápagos finch species and they mainly differ in their beak and body proportions (Figure 19.10). We can therefore calculate whether the kind of selection he observed would be enough to account for the origin of all the finches in the Galápagos in the time available.

How long would it take for the process studied by Grant in 1976–1977 to convert one species of finch into another? During the 1977 drought, the beak size of *Geospiza fortis* on the island of Daphne Major increased by 4%. *Geospiza magnirostris* is a close relative of *G. fortis*; the two species differ mainly in body and beak proportions and they coexist on many islands of the Galápagos. From the average difference in beak size between *G. fortis* and *G. magnirostris* on Daphne Major, Grant estimated that 23 bouts of evolution of the 1977 type would be enough to turn *G. fortis* into *G. magnirostris*. On other islands *G. fortis* is larger than on Daphne Major and, using one of the larger *G. fortis* populations as a starting point, only 12–15 such events would be needed.

And how much time is available? The Galápagos are volcanic islands. The oldest rocks are 4.2 million years old, and all the islands had appeared by 1–0.5 million years ago. It has been estimated that the common ancestor of Darwin's finches arrived from South America about 570 000 years ago: the radiation of the 14 finch species has occurred in about half a million years. Using either the low estimate of 12–15 events (where an event is a bout of evolution such as occurred in 1977 on Daphne Major) or the higher estimate of 25 events, we can see that even if only one such event took place per century, the evolutionary divergence between the two species could still have been accomplished well within the time available. In fact, the real evolutionary transition probably did not happen that way. The 1982–1983 El Niño event reversed the evolution of 1977 and there was probably not a steady transition from one population into another, but frequent reversals of accumulated small changes. In any case, it is unlikely that *G. fortis* simply changed into *G. magnirostris* (or vice versa); they probably both diverged from another common ancestor.

The rough calculation is not intended to represent the exact history of the birds. It illustrates instead how we can extrapolate from natural selection operating within a species to explain the diversification of the finches from a single common ancestor about 570 000 years ago to the present 14 species. If the extrapolation is correct, the reason for the speciation in the finches was the same process as has been observed in the present: natural selection for changes in beak shape, which were probably in turn due to changes in food types through time and between islands. Arguments of this kind are common in the theory of evolution: we shall meet another in the first section of the next chapter, where natural selection over long periods will be used to explain the major evolutionary transition from the mammals to the reptiles. All these arguments are extrapolationist.

The theory of punctuated equilibrium, in its more radical form, suggests that large-scale evolution is not simply microevolution extended over long periods. If speciation, for example, happens only under extraordinary, revolutionary circumstances, it would not be a simple extrapolation of normal evolution within a population: speciation would be *uncoupled* from microevolution. The theory of punctuated equilibrium would then contradict the modern synthesis's extrapolationism. Gould suggested that several evolutionary processes might operate, each with its own taxonomic level for its operations. He calls his idea *pluralist* (because it employs more than one causal process), and suggests that, once pluralism is accepted, the narrow modern synthesis would benefit from *hierarchic expansion* (as different processes would be studied at the micro- and macroevolutionary levels).

We shall see in the next three chapters how paleobiologists have also challenged extrapolationism at higher levels than speciation. It has been suggested that the origin of major groups, and macroevolutionary trends, may be driven by non-adaptive processes, or distinctive developmental reorganization, or by selection at higher levels than within a population. This is an influential and substantial line of research, but how much of a challenge is it to the modern synthesis?

Let us consider another interpretation of the modern synthesis, rather different from Gould's. The modern synthesis, it can be argued, does not predict that evolution has only one, consistent rate: the classic treatment of evolutionary rates was by Simpson and he reasoned that rates may be slow, average, or fast (section 19.1). Simpson was a pluralist about evolutionary rates. Seen against this background, the theory of punctuated equilibrium is novel mainly for being narrow. According to the modern synthesis, evolutionary rates may vary from slow to fast, and no consistent pattern had been identified. According to Eldredge and Gould, evolution is either fast (punctuation) or slow (equilibrium), and the rapid evolution is concentrated in the times of branching, the slow evolution between the branches. If punctuated equilibrium turns out to be empirically true, it will therefore refute the pluralist as well as the gradualist view of rates; but the way in which it differs from the modern synthesis is arguably different from what Gould claims.

And what about the extrapolationism of the modern synthesis? With the alternative interpretation, the modern synthesis is extrapolationist only in explaining adaptations. For other kinds of (non-adaptive) evolution, there is no reason why several processes should not operate, and at various taxonomic levels. Gould's interpretation of the modern synthesis would then imply that it claimed all evolution to be adaptive; but the modern synthesis never ruled out non-adaptive evolution. Individuals such as Cain did from time to time question its existence, but that is different from ruling it out in theory, and in any case Cain's was not a mainstream position. About non-adaptive evolution, the modern synthesis was again pluralist: nothing in the modern synthesis rules it out and there are whole areas of research (like molecular evolution) that are concerned with it.

In summary, the two interpretations of the modern synthesis differ on the

Table 19.2 Two interpretations of the modern synthesis

	Gould's 1980 development of the theory of punctuated equilibrium	Modern synthesis	
		Gould's interpretation	Alternative interpretation
Macroevolutionary changes in adaptations	Pluralist, with uncoupling of micro- and macroevolutionary levels	Extrapolationist	Extrapolationist
Macroevolutionary changes in all characters		Extrapolationist	Pluralist
Pattern of evolutionary rates	Punctuated equilibrium	Phyletic gradualism	Pluralist

evolution of adaptive and non-adaptive characters and on the pattern of evolutionary rates (Table 19.2). For the practicing evolutionary biologist, it is most important to be aware of the two interpretations: according to one (Gould's), the modern synthesis takes a narrow and reductionist view of all evolution and favors the phyletic gradualist theory of rates; according to the alternative, it is reductionist only about adaptive characters and is pluralist about both non-adaptive evolution and the pattern of rates.

19.9 Evolutionary rates can be measured for non-continuous character changes and the evolution of "living fossil" lungfish illustrates how

The measurement of evolutionary rates in darwins is appropriate for metric changes, such as a character evolving to be longer or shorter; but for larger changes, such as from a leg to a wing, this method ceases to be useful (section 19.2). However, it is still possible to measure rates of evolution for larger changes. Here we describe two methods. The first is a famous early study of evolutionary rates: Westoll's work on lungfish.

Lungfish (Dipnoi) form one of the four main divisions of fishes (see Figure 13.3, p. 330) and are an ancient group dating back over 300 million years; there are only six modern species. The modern forms are examples of *living fossils*, species that have changed little from their fossil ancestors in the distant past. They should therefore show, at least recently, slow rates of evolution; Westoll investigated this question quantitatively. He distinguished 21 different characters of fossil Dipnoi; they are all skeletal characters. For each of the 21, he distinguished a number of character states (like the character states discussed earlier for classification and phylogenetic inference). The 21 characters showed between three and eight different states. Character number 11, for example, was "degree of fusion of bones along supraorbital canal"; Westoll distinguished the five different states listed below.

4 Irregular, more or less random fusions.

3 Tendency for fusions to be in twos, especially in some parts of canal.

2 Still stronger tendency to fusion, rarely in threes or fours in specific sections.

1 Three or four elements (K to M) generally fuse, but numerous irregularities.

0 Three or four elements ($K-L_2$ or $K-M$) always fuse.

(Letters like K and L_2 refer to particular, identifiable bones.) The highest state (4) is the most primitive condition of the character, and 3, 2, 1, and 0 are successively later, more derived states. Westoll made an analogous list of states for all 21 characters. These character states are not the sort of metric changes for which evolutionary rates can be measured in darwins (section 19.1). The fusion of two bones into one is a discrete, not a continuous, evolutionary change.

For each fossil, Westoll calculated a total score, made up of its total for all 21 characters. The most advanced possible lungfish, with 21 characters in the most advanced state, would therefore have a total score of zero; the score for the most primitive possible lungfish, which had the highest scores for all 21, would have been 100. The rate of change in the score measures the rate of evolution of the group. The numbers assigned to the characters states are arbitrary, but they can still be used to portray evolutionary rates. (By the way,

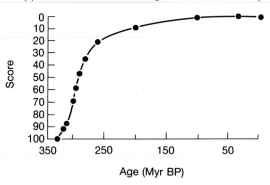

(a) **Modernization of a lungfish character complex**

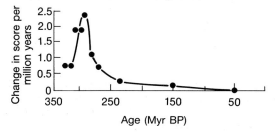

(b) **Rate of evolution in lungfishes**

Figure 19.11 (a) Evolution in lungfish, shown as the total score for each fossil; the text explains the scoring. Score = 100 for the most primitive form, 0 for the most advanced. The rate of evolution is the slope of the graph: when the graph is flat, evolution is not happening. (b) Rate of evolution. The graph is derived from (a) and shows the rate of change of the score through time. Lungfish have been living fossils since about 200 million years ago. Modified from Westoll (1949).

the lungfish in Westoll's study are not a simple sequence of ancestors and descendants, in the way that the horses in MacFadden's work probably were. Westoll's rates are for evolution within the Dipnoi as a whole and are not rates of change down a single evolutionary lineage; there would have been many lineages within the Dipnoi as a whole.)

Figure 19.11 is Westoll's result. Dipnoi, it reveals, have not always been living fossils. Around 300 million years ago they were evolving rapidly, but since about 250–200 million years ago, their evolution has slowed right down, and their description as living fossils is accurate for the modern forms.

The obvious biological question is why evolutionary change came almost to a stop in lungfish 200 million years ago. Lungfish are not the only examples of living fossils; other examples include the brachiopod *Lingula* and the horseshoe crab *Limulus* (Figure 19.12). There are many particular conjectures about why these groups have changed so little, but no general theory. The question is an instance of the general question of why there should be evolutionary stasis. Their stability may be due to stabilizing selection or absence of genetic variation. Some living fossil species live in relatively isolated habitats, with no apparent competitors; and if their habitats have been stable there will have been no pressure for them to change. There is no

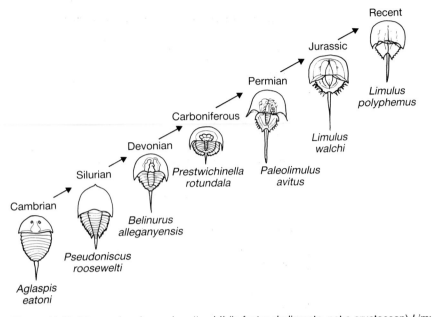

Figure 19.12 The modern horseshoe "crab" (in fact a chelicerate, not a crustacean) *Limulus polyphemus*, which lives along the east coast of the USA, is a living fossil: it is morphologically very similar to forms that lived about 200 million years ago, and not all that different from Cambrian species. From Newell (1959).

evidence that living fossils have peculiar genetic systems that might prevent evolutionary change, for instance by constraining the degree of new genetic variation. We can measure the amount of protein polymorphism in modern living fossil species by gel electrophoresis, and in *Limulus polyphemus* it is not noticeably low.

But the point of the example here is methodologic, not biological. It is to show how evolutionary rates can be studied quantitatively in characters whose evolutionary changes are not simply metric. The characters can be divided into discrete states; the states assigned arbitrary scores; and the changes in those scores measured through time.

The quantification is mainly useful for purposes of illustration. Figure 19.11 neatly shows how rates of evolution have varied through time in a way that a table of raw character data could not. But the division of characters into states, and the assignment of scores to states, is arbitrary. The five states for supraorbital canal bone fusion could just as well have been scored 40, 16, 15, 14, 0 or 2, 8, 17, 39, 40 as it was 4, 3, 2, 1, 0. It would therefore be meaningless to compare exact numerical rates of change between characters, or between taxa. The scores are incommensurable. The approximate shape of a graph like Figure 19.11 could be compared with another such graph for another group; but there would be no point in asking why one group changed at, say, 2.1 units per million years and another at 1.3 units per million years. The scores are not intended for that kind of analysis. But as an illustration of how rates of change in lungfish have risen and fallen and declined to a virtual standstill, Westoll's analysis is a classic.

19.10 Taxonomic data can be used to describe the rate of evolution of higher taxonomic groups

For a question such as "Do mammals evolve faster then bivalve molluscs?", a further kind of measurement of evolutionary rate can be used. The question could be tackled by either of the methods we have discussed so far. We could calculate the evolutionary rates for individual characters, measured either metrically or in discrete states, in mammals and bivalves, and then compare them; but there is a danger the comparisons might be meaningless. Any differences in evolutionary rate may reflect only the way we measure (for example) teeth in mammals and (for example) shell shape in bivalves, and nothing real. However, there may be some purposes—comparisons with rates of nucleotide change, perhaps—for which the measurements might be useful.

The third method uses taxonomic evidence. When taxonomists divide a set of organisms into a number of species, they make their judgment according to the degree of phenetic differences among the forms. Their judgment will not usually be based on a single character, but on several characters, integrated in the taxonomist's mind into a single dimension of taxonomic similarity. Thus if there were two separate but comparable evolutionary lineages, and in the same time interval a taxonomist divided one lineage into two species and the other into three, it would suggest that the latter lineage had evolved at a higher rate (Figure 19.13). This comparison uses a *taxonomic rate* of evolution.

A taxonomic rate of evolution offers an abstract measure of how rapidly change is taking place in a group of species. The exact meaning of a taxonomic rate is less easy to specify than for an evolutionary rate of a single character; the taxonomic rate has a relatively imprecise meaning. It can be said in their defence that they summarize evidence from more than one character, and have greater generality. How reliable a taxonomic rate is depends on how reliable is the judgment of the taxonomist who had divided the lineage up into species and genera.

Taxonomic rates of evolution are expressed in two main ways. One is the number of species, or genera (or whatever taxon) per million years. Table 19.3 gives some examples from the two taxa that we have discussed in this

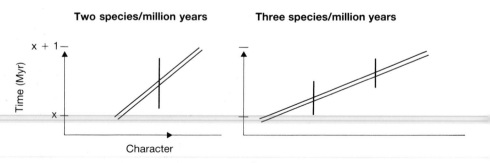

Figure 19.13 Taxonomic measurement of evolutionary rate. If a taxonomist has divided one group into two species, and another group into three species, in the same time interval, the latter group shows a 50% higher taxonomic rate of evolutionary change. The diagram illustrates only the logic of the argument. In real data, there would be gaps in the lineages; and the real pattern of evolution could have been smooth or jerky, with any number of branches in addition to the lineages shown here.

Table 19.3 Taxonomic rates of evolution in mammals and lungfish. Early in their evolution, lungfish evolved about as rapidly as mammals; but they have subsequently slowed down. *Hyracotherium–Equus* is the horse lineage discussed in section 19.1. From Simpson (1953)

Group or line	Average duration of genus (Myr)
Hyracotherium-Equus	7.7
Lungfish	
Devonian	7
Permo-Carboniferous	34
Mesozoic	115

chapter: horses and lungfish. The rates for lungfish reillustrate how that group initially had a high rate of evolution, which then slowed down so much that they became living fossils.

The same data, but for a group made up of a larger number of lineages, can also be expressed as a *survivorship curve* (Figure 19.14). Survivorship curves are constructed by taking a sample of a number (such as 100) of mammalian genera and measuring how long each one lasts in the fossil record. (Any taxonomic level within the mammals can be used; genera are an example.) The survivorship curve plots the number of genera surviving for different times: most of the genera are still surviving after a short time, but as time passes the members of the original sample drop out one by one. The slope of a survivorship curve measures the rate of evolution of the group. If the group is evolving fast, the survivorship curve falls rapidly as a taxonomist does not recognize a species as surviving for long; the curve is more drawn out for a slowly changing group. (Survivorship curves are more familiar for populations of individuals. See section 4.1, p. 61. An actuarial survivorship curve plots the survival of a sample of individuals through time, but the same type of graph can be plotted for species and other taxonomic groups.)

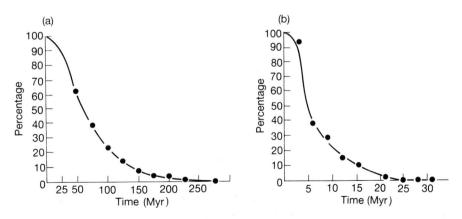

Figure 19.14 Survivorship curves for (a) bivalves and (b) carnivores (Mammalia). The curves express the numbers of genera surviving for different amounts of time. Note bivalve genera tend to last longer than carnivore genera; the average duration of a bivalve genus is 78 million years, of a carnivore, 8.1 million years. From Simpson (1953).

There is a similar problem in comparing taxonomic evolutionary rates between groups as there is for single characters. A bivalve taxonomist may be a good judge of a bivalve genus, and a mammal taxonomist of a mammal genus, but it is still difficult to know how to interpret any differences in the rates of turnover of the two sorts of genera: it could just reflect some difference in the way the two taxonomists work. Within a group, survivorship curves can have revealing features, as we shall see in chapter 22. For now, however, we only need to know that taxonomic evidence can provide another sort of measurement of evolutionary rates, along with measurements of single characters.

19.11 Conclusion

This chapter has introduced three methods of measuring rates of evolution. For single characters that show continuous variation, we can measure the metric rate of change and express it in darwins. For characters with discrete states, we can give each state an arbitrary score and measure the rate of change of the score. A cruder measure is provided by the duration of species in the fossil record, because taxonomists will recognize a higher turnover of species in groups that evolve rapidly. All three methods have their particular uses and applications.

Paleobiologists have used measures of evolutionary rate to study a number of general questions. This chapter has concentrated on the question of what the relative rates of evolution are during, and between, speciation events. The controversy between punctuated equilibrium and phyletic gradualism as a whole has involved several large issues in evolutionary biology; but the empirical controversy, concerning the characteristic pattern of evolution in fossils, remains open for further work.

19.12 Summary

1 Evolutionary rates of fossil characters can be measured as simple rates of change through times; logarithms are often taken of the character measurements and the rate expressed in darwins. Evolution in horse teeth is a classic example.

2 Rates of evolution measured in the fossil record are slower than those produced by artificial selection in the laboratory.

3 Evolutionary rates vary between different geologic times, taxa, and types of taxa. The science of evolutionary rates is mainly concerned with explanation of the pattern of evolutionary rates.

4 Among published measurements of evolutionary rates, the rate and the time interval over which it was measured are inversely related: faster evolution is seen in shorter intervals. The reason is probably that the sample of rate measurements is biased.

5 Eldredge and Gould stimulated a controversy about evolutionary rates by their suggestion that rates have a strict pattern (called punctuated equilibrium): evolution is fast at times of splitting (speciation) and comes to a halt between splits. The opposite pattern, in which evolution has a constant tempo, they called phyletic gradualism.

6 It is difficult to discover the pattern of evolutionary rates at, and between, speciation events because the fossil record is incomplete.

7 There is some controversial evidence for punctuated equilibrium, such as

Williamson's study of Lake Turkana snails, and some for phyletic gradualism, such as Sheldon's study of Ordovician trilobites. No general empirical conclusion is yet possible.

8 The theory of punctuated equilibrium has been developed in radical, if speculative, directions: it has been suggested that stasis is due to constraints on genetic variability and speciation to the breakdown of these constraints, or even macromutation. There is evidence for stabilizing and directional selection, but little or none for constraint and revolutionary speciation.

9 Neither Darwin's theory nor neo-Darwinism necessarily predict pure phyletic gradualism. They were gradualist about the evolution of adaptation. It is historically controversial whether Darwin and the neo-Darwinians mainly held gradualist, or pluralist, views about the rates of evolution during, and between, speciation events.

10 Evolutionary processes and rates can be examined at all taxonomic levels from evolution within populations, through speciation, to the origin of the higher groups. Evolution may have characteristic mechanisms and rates at different levels, or the same set of rates and processes may operate equally at all levels.

11 For large changes, like from a limb into a wing, evolutionary rates cannot be measured as a continuous variable. The character can instead be divided into states. The evolutionary rate can then be studied in the rate of change between states. Westoll studied the evolution of lungfish by this method.

12 The number of species in a lineage per million years is a complex measure of evolutionary rate, called a taxonomic rate of evolution. Taxonomic rates can be expressed as survivorship curves.

19.13 Further reading

Simpson (1953) remains a good introduction to the study of evolutionary rates. More recent research is discussed in a special issue of *Paleobiology*, vol. 9, pp. 326–428 (1983), and of *Special Papers in Paleontology*, vol. 33, pp. 1–203 (1985), in Raup and Jablonski (1986), Campbell and Day (1987), and single-author works such as Stanley (1979) and Levinton (1988). Fenster and Sorhannus (1991) is a review, and Eldredge (1985b) a more popular introduction.

MacFadden (1987, 1988a,b) discusses his work on horse evolution; Gingerich (1983) is the survey of evolutionary rates discussed in sections 19.2 and 19.3; see also Gould (1984) and Gingerich's (1984) reply. For generalizations about evolutionary rates, see the general references above; and Kauffman (1978) on rates of evolution in molluscs and mammals, and Schopf *et al.* (1975) on the relationship with structural complexity.

The literature on punctuated equilibrium is now vast. Eldredge and Gould (1972) and Gould and Eldredge (1977) are the key original works; see Brown (1987) on their evidence. On snails, and developmental constraints, see Williamson (1981a,b) and the comment in *Nature*, vol. 296, pp. 608–612 (1982) and in Fryer *et al.* (1985). Sheldon (1987) is the reference for trilobites; for the controversy, see Maynard Smith (1987c), Eldredge and Gould (1988), and Maynard Smith (1988). Many more case studies are discussed in the multi-author works cited. For later developments of the theory: Gould (1980b, 1982a,b) and Williamson (1981b). Eldredge (1989) steered a different course.

For rapid change without speciation, see MacLeod (1991). For what the population geneticists think, see Charlesworth *et al.* (1982), Turner (1986), and Maynard Smith (1983), and Hoffmann and Parsons (1991) for the effect of environmental stress. See Grant (1986, 1991) on the finches, and Carson (1992) and Christie *et al.* (1992) for evidence of earlier islands in the Galápagos. On history, see (in addition to Gould) Rhodes (1983) and Dawkins (1986).

Eldredge and Stanley (1984) is about living fossils. Westoll (1949) and Simpson (1953) are the references for lungfish. On taxonomic rates, see Simpson (1953), Stanley (1979), and Gilinsky (1988).

Macroevolutionary change

20.1 The mammals evolved from the reptiles in a long series of small changes

The mammals are a distinct group of vertebrates in many respects. They have warm blood and a constant body temperature, and the high metabolic rate and homeostatic mechanisms that go with it; they have a characteristic mode of locomotion, or gait, in which the body is held upright with the legs underneath (in contrast to the sprawling gait of reptiles, such as lizards, in which the legs stick out sideways); mammalian brains are large; the mammalian method of reproduction, including lactation, is also distinctive; the active metabolism of mammals demands efficient feeding, and mammals have powerful jaws and a set of relatively durable teeth, differentiated into a number of tooth types. Therefore, when the mammals evolved from the reptiles, there had to be changes on a large scale in many characters. It was a *macroevolutionary* change. How did this transition take place?

Not all of the distinctively mammalian characters are preserved in the fossil record. The earliest mammalian fossils are a group called the morganucodontids, such as *Kuehneotherium*, from the Triassic–Jurassic border almost 200 million years ago. Whether *Kuehneotherium* was viviparous and lactated is not known directly. But it is possible to see that it had a mammalian jaw, gait, and tooth structure, and therefore probably also had warm blooded physiology. The origin of the mammals can be traced back before the morganucodontids, through a series of reptilian groups called the *mammal-like reptiles*. Mammal-like reptiles is an informal name for three main groups. They extend in a reasonably complete record of fossils from the mid Carboniferous to the end of the Triassic (c. 232–195 million years ago), when the earliest mammals appear. The most interesting feature of their evolutionary history is the pattern they suggest for the origin of a major new group.

The characters that can most clearly be reconstructed in fossils are those concerned with locomotion and feeding, because these are simply related to the form of preserved bones and teeth. The reptilian jaw contrasts in many respects with the mammalian (Figure 20.1): mammalian teeth have a complex, multicusped structure and they are differentiated down the jaw into canines, molars, and so on, whereas reptilian teeth form a relatively undifferentiated row and have a simpler structure; the mammalian jaw has cheek muscles which enable it to close more powerfully and accurately than the reptilian jaw, which has its muscles at the back of the jaw, where it also articulates. Let us see how the jaw, and gait, changed in the evolution of the mammal-like reptiles.

The evolution of the mammal-like reptiles proceeded in three main phases,

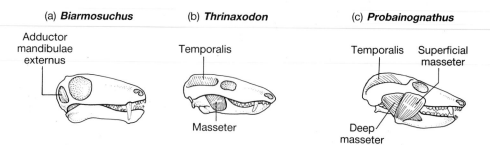

(a) *Biarmosuchus*   (b) *Thrinaxodon*   (c) *Probainognathus*

Adductor mandibulae externus

Temporalis

Masseter

Temporalis   Superficial masseter

Deep masseter

Figure 20.1 Articulation of jaw in mammal-like reptiles. In the early form *Biarmosuchus* (a), the jaw muscle is at the basal articulation. In the evolution of mammals, the muscles move forward, and by *Thrinaxodon* (b), and advanced mammal-like reptile, the masseter has split into two. The superficial masseter joins a characteristic region of the upper jaw, and the presence of the advanced jaw condition can be recognized from the jaw alone (see Figure 20.2, in which these three forms are again illustrated, but without the muscles drawn in). Note also the increasing complexity and serial differentiation of the teeth. For the different gaits of reptiles and mammals, see Figure 14.4, p. 361. From Carroll (1988).

corresponding to the three main taxonomic divisions. The first phase sees the evolution of the *pelycosaurs*. *Archaeothyris*, which lived about 300 million years ago, is an exemplary pelycosaur (Figure 20.2). It was a lizard-like animal, about 50 cm long and differing little from its reptilian contemporaries. One important difference was an opening in the cheek region, on both sides of the face. Each opening is called a temporal fenestra, and in the living animal a muscle passed through it. The muscle acted to close the jaw, and the opening up of the temporal fenestrae is the first sign of the more powerful jaw mechanism of the mammals. (The temporal fenestra, by the way, is the defining character of the mammal-like reptiles.) A better known pelycosaur was *Dimetrodon*, with its extraordinary sails on its back; its gait, like those of all pelycosaurs, was a reptilian sprawling gait (Figure 20.2). The pelycosaurs, through their 50 million year history, evolved into three main groups, and most of them went extinct quite suddenly about 260 million years ago. A few of the sphenacodontids survived and it was from an unknown line within the sphenacodontids that the second main group of mammal-like reptiles evolved.

The second phase of mammal-like reptilian evolution belonged to the *therapsids*. They underwent a remarkably similar pattern of evolution to the pelycosaurs. Their temporal fenestrae were larger than those of the pelycosaurs, and they show the first sign of holding their legs in an upright position. In the late Permian a group within the therapsids in turn gave rise to the third, final phase: the *cynodonts*. Cynodont jaws resemble modern

Figure 20.2 (*Opposite*) The evolutionary radiation of the mammal-like reptiles. There were three main phases: pelycosaurs (sphenacodontids and ophiacodontids in this picture); therapsids; and cynodonts. Within each phase, there were many smaller evolutionary lineages. Some fossil forms are illustrated. Note again the evolution of the more powerful and precision action mammalian jaw, and the change from a sprawling gait in *Dimetrodon* to an upright gain in *Probelesodon*. From Kemp (1982b).

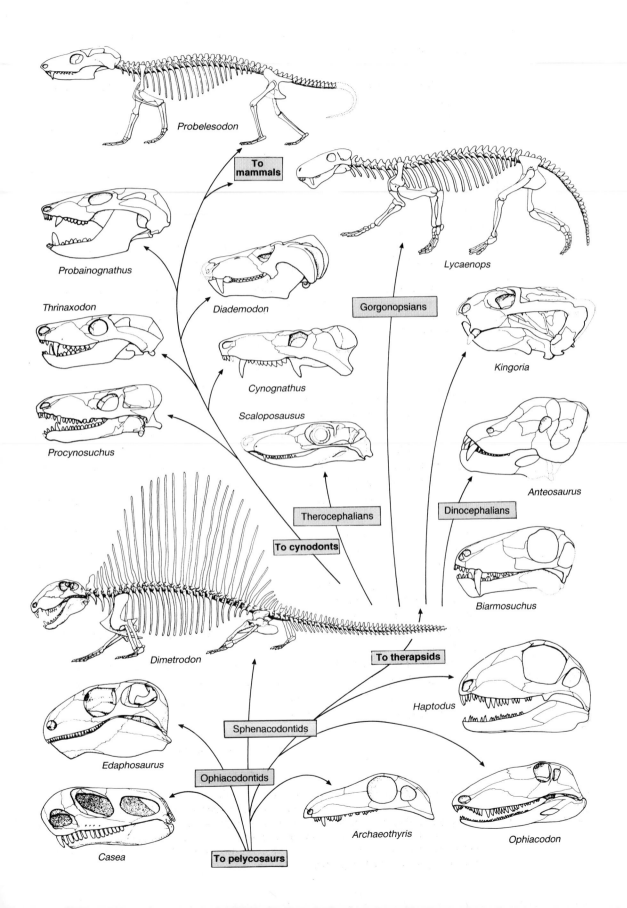

Probelesodon

To mammals

Lycaenops

Probainognathus

Thrinaxodon

Diademodon

Gorgonopsians

Kingoria

Procynosuchus

Cynognathus

Scaloposausus

Anteosaurus

Therocephalians

Dinocephalians

To cynodonts

Biarmosuchus

Dimetrodon

To therapsids

Haptodus

Edaphosaurus

Sphenacodontids

Ophiacodontids

Casea

Archaeothyris

Ophiacodon

To pelycosaurs

mammal jaws more closely and their teeth are multicusped and differentiated down the jaw. It was from a line of cynodonts, the exact identity of which is uncertain, that the ancestors of the modern mammals such as *Kuehneotherium* evolved.

20.2 The mammal-like reptiles illustrate the neo-Darwinian theory of macroevolution

The fossil record for the origin of mammals is superior to that for the origin of any other major group. It is therefore an important test case for general theories about how major evolutionary transitions take place. We should draw two important conclusions from the historical narrative. One is that the changes from reptilian to mammalian characters evolved in gradual stages. The other is that the large-scale differences between mammals and reptiles concern adaptations. The mammals have a high energy, high metabolic rate kind of physiology, with locomotory adaptations for rapid movements (upright rather than sprawling gait) and adaptations for powerful and efficient feeding (the mammalian teeth and jaw articulation). These are surely adaptive changes, which would have been brought about by natural selection. The general evolutionary model suggested by the mammal-like reptiles, therefore, is one of cumulative action of natural selection over a long (40 million year) period; the accumulation of many small-scale changes resulted in the large-scale change from reptile to mammal.

From the previous chapter, this can be recognized as an extrapolative theory. Macroevolution is taking place by the same process—natural selection and adaptive improvement—as has been observed within species and at speciation; but the process is operating over a much longer period. The extrapolative model is not the only model for the evolution of major groups, but it is the most important and the only one that can be illustrated with detailed fossil evidence. It can also, in a sense, be thought of as the "neo-Darwinian" theory of macroevolution. In population genetics, neo-Darwinism explains microevolution by changes in the frequencies of pre-existing variants: most characters show variation and the character evolves as its frequency distribution is altered by selection. No extraordinary kinds of variation are needed. Likewise, in a neo-Darwinian account, macroevolution can be explained without recourse to extraordinary kinds of variation: natural selection, on ordinary variation and mutation, is enough. That is how the mammals evolved from the reptiles.

There is a potential verbal confusion in the meaning of "neo-Darwinian". The pattern for the mammal-like reptiles may not be the universal mode of evolution. Other large groups may have evolved by non-adaptive, or sudden, processes. Their evolution would then in a sense not have been neo-Darwinian. It is important to understand what this means. The neo-Darwinian extrapolative theory of macroevolution is not the same as neo-Darwinism as a whole. Neo-Darwinism is used to refer to various sets of ideas, some of which are more fundamental than others. As a theory of adaptation, neo-Darwinism asserts that adaptations only evolve by natural selection. This is a most fundamental claim, because we have no theory other than natural selection to explain adaptation; without it, we have only miracles to fall back on. The extrapolative theory of macroevolution is a weaker neo-Darwinian claim: if it turns out to be wrong, plenty of scientific alternatives—all compatible with the neo-Darwinian theory of natural selection—still remain; the extrapolative theory of macroevolution can therefore easily be jettisoned if it ever turns out to be wrong in fact. If it is shown that a major evolutionary transition is sudden, or non-adaptive, neo-Darwinism in the fundamental sense certainly has not been refuted. Provided that we do not confuse the neo-Darwinian theory of adaptation with the neo-Darwinian theory of macroevolution, no problem arises.

Notice that the extrapolative model does not imply that there is a single line of evolution from reptile to mammal, with each new step being added on to the one before. Within each phase of mammal-like reptilian evolution there was a radiation into many evolutionary lines (Figure 20.2), and most of them became extinct without issue. But if we retrospectively trace back the line from the mammals to the reptiles of 300 million years ago, we pass through a series of stages in which the mammalian adaptations emerge. Within that line, there would have been steps both toward and away from the mammalian condition. All that matters for the extrapolative model is that, whatever else was going on, the many mammalian adaptations evolved in stages.

20.3 Morphologic transformations are generally accomplished by developmental changes

Macroevolution has mainly been studied morphologically, because we have more taxonomic and fossil evidence for morphology than for other kinds of characters, such as physiology or chromosomes. Morphologic structures are produced by growth, and their form emerges from the process of development. Thus evolutionary changes in the form of an organ are frequently developmental: they are produced by changes in the rate or timing of developmental events. An organ may evolve to be larger if its growth speeds up, and it may change shape if the growth rate of one of its parts speeds up relative to other parts. We shall see in this section that there are several types of evolutionary change in development; but we can begin by looking at one particular type: *recapitulation*. Recapitulation, or the "Biogenetic law," is a bold and influential idea; it is often associated with Haeckel, but had many other adherents in the nineteenth century. Many ideas about the relationship of development and evolution have grown up in opposition to the theory of recapitulation. Let us see what it means.

According to the theory of recapitulation, the stages of an organism's development correspond to the species' phylogenetic history: in a phrase, "ontogeny recapitulates phylogeny." Each stage in development corresponds to (i.e. recapitulates) an ancestral stage in the evolutionary history of the species. The transitory appearance of structures resembling gill slits in the development of humans, and other mammals, is a striking example. Mammals evolved from an ancestral fish stage and their embryonic gill slits recapitulate the piscine ancestry. Another example, often quoted in the nineteenth century, is seen in the tail shapes of fish (Figure 20.3). During the development of an evolutionarily advanced individual fish species, such as the flatfish *Pleuronectes*, the tail has a diphycercal stage in the larva, and develops through a heterocercal stage, to the homocercal form of the adult. Now, it is also possible to find different species of fish possessing these three tail types in the adult: lungfish, sturgeon, and salmon, on the right of Figure 20.3 are examples. The lungfish is thought to most resemble an early

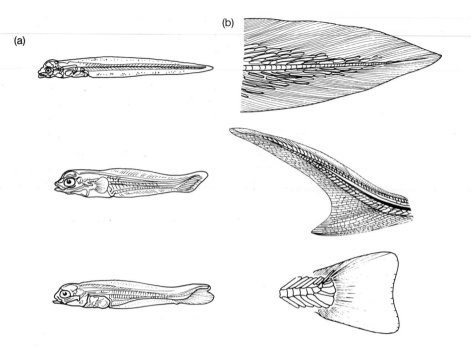

Figure 20.3 Recapitulation, illustrated by fish tails. (a) The development of a modern teleost, the flatfish *Pleuronectes*, passes through (starting at the top) a diphycercal stage, to a stage in which the upper lobe of the tail is larger (heterocercal), and the adult has a tail with equal sized lobes (homocercal). (b) Adult forms in order of evolution of tail form, from top to bottom: lungfish (diphycercal), sturgeon (heterocercal), salmon (homocercal).

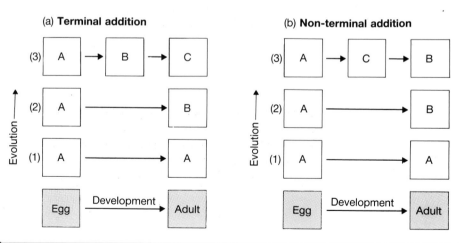

Figure 20.4 (a) Evolution by terminal addition. The stages of an individual's development are symbolized by alphabetic letters: under terminal addition, new stages are added only to the end of the life cycle. (b) Evolution by non-terminal addition. A new evolutionary stage has been added in early development, not on to the end of the life cycle in the adult.

fish, the sturgeon to be a later stage, and the salmon to be the most recently evolved form. Thus evolution has proceeded by adding on successive new stages to the end of development. Let us symbolize the diphycercal, heterocercal, and homocercal tails by A, B, and C, respectively. The development of the early fish advanced to stage A and then stopped. Then, in evolution, a new stage was added on to the end: the development of the fish at the second stage was AB; the final type of development was ABC. Gould has called this mode of evolution *terminal addition* (Figure 20.4a).

In logic, evolution does not inevitably or automatically have to proceed by terminal addition. New, or modified, characters logically could be intruded at earlier developmental stages (Figure 20.4b). There are indeed many examples of specialized larval forms that are not recapitulated ancestral stages (the zoea of crabs, the Müller's larva of echinoderms, and the caterpillar of Lepidoptera, almost certainly, are examples); they probably evolved by modification of the larva. These exceptions notwithstanding, recapitulation is noticeably common. Evolution has often proceeded by terminal addition. The process is common enough to demand an explanation, and we shall look at

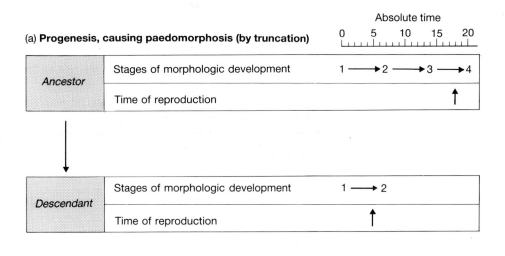

Figure 20.5 Paedomorphosis, in which a descendant species reproduces at a morphologic stage that was juvenile in its ancestors, can be caused by (a) progenesis, in which reproduction is earlier in absolute time, or (b) neoteny, in which reproduction is at the same age but somatic development has slowed down.

some below. But first we shall look at the other kinds of developmental change.

There are two types of reason why recapitulation breaks down. One, as was just mentioned, is the evolution of new early developmental stages. The other is a kind of *heterochrony*. During development, the different parts of the body appear at their own particular times. The relative time of maturation of the reproductive (germ) cells and the rest of the body (soma) is an important case in point: the body becomes reproductively mature at a certain stage of somatic development. Heterochrony means a change in the timing or rate at which the different body parts develop relative to each other; in the case of germ and somatic cell lines, it means the individual reproduces at an earlier (or later) stage of somatic development. A mutation that alters the rate at which a cell line develops relative to other cell lines is a heterochronic mutation. When reproduction occurs earlier, the organism will not recapitulate its ancestry, because the new adult form is not added on after the ancestral adult form.

Two processes can lead to this result. If reproduction occurs at an earlier stage of somatic development, then either somatic development is proceeding at the same rate in absolute time as always, and germ line development has speeded up; or reproduction happens at the same absolute age, and somatic development has slowed down (Figure 20.5). The morphologic result is the same either way: reproduction is seen in what was ancestrally a juvenile morphologic stage. We thus need terms both for the morphologic result and the processes producing them. The morphologic result (reproduction in ancestral juvenile form) is called *paedomorphosis*; and the two ways of generating it are called *progenesis* (speeding up the germ line) and *neoteny* (slowing down the soma). Likewise, there are two possible causes for recapitulation—for a shift in the time of reproduction to a somatic stage corresponding to the post-adult of the ancestor. In all, we distinguish two morphologic results, with two possible causes each, for any pair of cell lines (Table 20.1). (Some modern analyses use more than four terms; but four are enough here.)

Table 20.1 Categories of heterochrony. In modern work, the term "peramorphosis" is sometimes substituted for "recapitulation." From Gould (1977a)

Developmental timing		Name of evolutionary process	Morphologic result
Somatic features	Reproductive organs		
Accelerated	Unchanged	Acceleration	Recapitulation (by acceleration)
Unchanged	Accelerated	Progenesis	Paedomorphosis (by truncation)
Retarded	Unchanged	Neoteny	Paedomorphosis (by retardation)
Unchanged	Retarded	Hypermorphosis	Recapitulation (by prolongation)

20.4 Examples exist of the different types of heterochronic evolution

Examples exist of all the types of developmental change in Table 20.1. Let us look at two cases of neoteny, one in modern species and the other in fossils, to see how the inference can be made. Among modern species, the classic example of neoteny is the Mexican axolotl, *Ambystoma mexicanum*. The axolotl is an aquatic salamander. Most salamanders have an aquatic larval stage, which breathes through gills; the larva later emerges from the water as a metamorphosed terrestrial adult form, with lungs instead of gills. The Mexican axolotl, however, remains in the water all its life and retains its external gills for respiration. It reproduces while it has this juvenile morphology. It takes only a quite straightforward experiment to make a Mexican axolotl grow up into a conventional adult salamander (it can be done, among several ways, by injection of thyroid extract). These experiments practically confirm that the timing of reproduction has moved earlier in development during the axolotl's evolution. Otherwise there would be no reason for it to possess all the unexpressed adaptive information of a terrestrial form.

So the Mexican axolotl is paedomorphic: but is it neotenous or progenetic? Its age of breeding (and the body size at which it breeds) is not abnormally early (or small) for a salamander. It is therefore a reasonable inference that the time of reproduction has stayed roughly constant, while somatic development has slowed down: the axolotl is neotenous. Humans are probably also neotenous. As adults, we are morphologically similar to the juvenile forms of great apes. Our age of breeding, however, has not shifted earlier—indeed, it has shifted later, which suggests that our somatic development has slowed down even more than our reproductive development.

In fossils, we do not know the time of breeding; but an inference can be made from the size of the forms. Consider three fossil cockles (Figure 20.6). The juvenile *Cardium plicatum* (a) has fewer ribs than the adult (b). A later descendant, *C. fittoni* (c), has a rib pattern like the juvenile of *C. plicatum*. *Cardium fittoni* is therefore paedomorphic: but is it neotenous or progenetic? An answer is suggested by the sizes. The juvenile *C. plicatum* is about 5 mm, the adult about 17 mm. *Cardium fittoni* is about 35 mm long. If both species grow at about the same rate in absolute time, then the juvenile morphology of *C. fittoni* must be neotenous: the development of the ribbing pattern has slowed down in absolute time.

If the adult *C. fittoni* had been about 5 mm long, the stronger inference

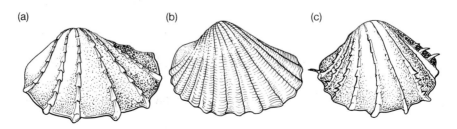

Figure 20.6 Neoteny in fossil cockles. (a) *Cardium plicatum* juvenile, 5 mm in length; (b) same species, adult 17 mm long; (c) *C. fittoni*, 35 mm long, and descended from *C. plicatum*.

would be that is was progenetic. It would then have the form of the ancestral juvenile and its size too—suggesting it bred at an earlier age than its ancestor. The rule of inference with fossils, therefore, is that if the adult descendant looks like the juvenile ancestor (i.e. it is paedomorphic), then if the descendant is as large as (or larger than) the ancestor it is probably neotenous, whereas if it is the size of a juvenile ancestor it is probably progenetic. Clearly, the inference could sometimes be wrong. It makes two main assumptions: (1) that size is proportional to age of breeding; and (2) that we can tell that the juvenile-formed descendant species actually is an adult (our evidence is that no larger specimens have been found—which will be stronger for a richer fossil record). Either assumption could be wrong; but that means only that the inference is uncertain, not that it is unreasonable.

20.5 The question of the relative frequencies of the different types of developmental change is interesting but not yet answered

We have considered a number of different types of developmental change in evolution. What are the relative frequencies of these processes? Are the four kinds of heterochrony equally frequent? Or is one more common than the others? Maybe the different types of evolution are correlated with different ecological conditions, with different types of development, or with different kinds of evolutionary change.

One study by McNamara illustrates the kind of comparison that can be made. For 44 trilobites, he divided the species into those that were inferred to show progenesis, neoteny, hypermorphosis, and acceleration, and displayed the frequency of the different types of change. His main result was a difference in the frequency distribution between the Cambrian and Postcambrian: progenesis seems to have been common in the Cambrian and neoteny afterwards (Figure 20.7). No explanation has been suggested. Work like McNamara's will become more common, but at present we have only limited factual knowledge. However, evolutionary biologists have from time to time discussed the possibility that one or other mode of evolution is particularly important. Let us look at some of these ideas.

We can take recapitulation first. Many explanations for it were offered in the nineteenth century. Only a few of them remain interesting today, and they apply equally well to a related but crucially different phenomenon called *von Baer's law* (section 17.8.2, p. 462), which states that "the general features of a large group of animals appear earlier in the embryo than the special features." In other words, the early developmental stages of a group of animals are more similar than are the later stages. In different mammalian species, for example, the early embryonic stages are more similar than the adult forms.

What is the relationship between recapitulation and von Baer's law? Consider first a case in which there is recapitulation: related evolutionary lineages must then have accumulated their new characters in the adult, and von Baer's law will also be satisfied (Figure 20.8a): if recapitulation is true, von Baer's law will be too. But what about if recapitulation does not happen? Is von Baer's law then violated also? We have seen two reasons why recapitulation may not happen: one was the modification of early stages (Figure 20.8b) and the other was paedomorphic heterochrony (Figure 20.8c). In either case, the ontogenetic stages are not in phylogenetic order and the

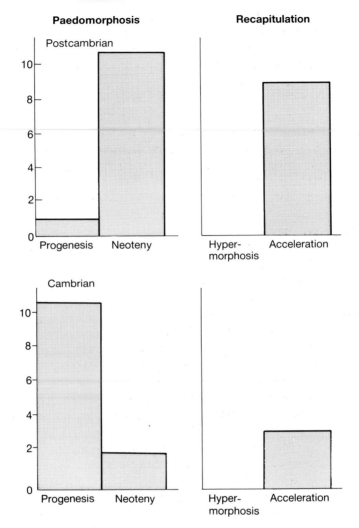

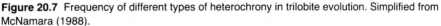

Figure 20.7 Frequency of different types of heterochrony in trilobite evolution. Simplified from McNamara (1988).

principle of recapitulation is wrong; the ancestor of the Mexican axolotl, for example, was a terrestrial form, which the axolotl does not recapitulate. Similarly, a modified early stage is not a recapitulated ancestral form. Von Baer's law, however, is violated by one of these processes but not the other. It requires the relative rarity of modifications to early stages; but it does not rule out heterochrony. It is still true of the axolotl and a salamander, for example, that their earlier developmental stages are more similar: the truncation of adult stages does not alter the similarity of earlier stages. In summary, recapitulatory evolution leads to von Baer's law, but non-recapitulatory evolution may or may not. Paedomorphosis is consistent with the law; but the modification of larval stages violates it.

Von Baer's law is a weaker claim than the theory of recapitulation, and stands a better chance of being correct; indeed, many biologists believe that

(a) Terminal addition   Shows recapitulation: fits von Baer's law

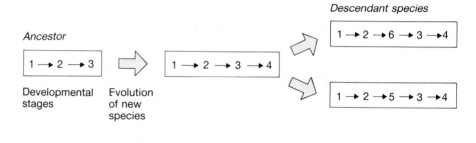

(b) Non-terminal addition of new characters   No recapitulation: violates von Baer's law

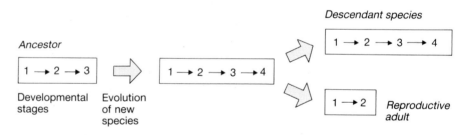

(c) Heterochronic paedomorphosis   No recapitulation: fits von Baer's law

Figure 20.8 (a) Evolution by terminal addition (as in Figure 20.4). Both von Baer's law and the principle of recapitulation are correct: the two species are more similar at earlier than at later developmental stages and the ontogeny of each recapitulates its phylogeny. (b) Non-terminal addition of new characters: both von Baer's law and recapitulation are violated. (c) Heterochronic evolution (paedomorphosis): von Baer's law is correct, but the principle of recapitulation is violated in the lower species in the figure.

the law is a sufficiently good generalization for it to need an explanation. If the law is correct, it must be because evolutionary change happens most often in the later stages of development; early stages must be evolutionarily more conservative. Why should this be? One possibility is a special case of the general argument against the evolutionary importance of macromutations (Box 7.3, p. 170; section 13.3, p. 326). A mutation that influences early development is more likely to have a large phenotypic effect than one that influences a later stage. Development (excepting metamorphoses) is continuous and cumulative, and changes early on will have ever magnifying and ramifying consequences later in the adult. Some of these consequences may well be deleterious. A late-acting mutation has a higher probability of being a fine-tuning change, with small phenotypic effect. Fine-tuning, in turn, has a higher probability of being selectively advantageous; changes late in development should therefore be more common than early changes. Hence the relative frequencies of changes at different stages and their manifestation as von Baer's law. The argument does not require that macromutations are never selectively advantageous, only that small mutations are more likely to be.

A second (and not incompatible) theory reasons from external, ecological selective forces rather than the magnitude of mutations. Early developmental stages are shielded from the environment. In many species, the mother broods her offspring, but even in species in which the eggs are abandoned the embryo is provided with yolk to support its early growth. As the embryo grows up, it increasingly has to fend for itself. Selection is relatively unlikely to demand changes in an embryo being brooded inside its mother; and the embryo will tend to retain its ancestral form. Once it is exposed to the external environment, it will be selected to adapt to any change in conditions. Early stages will therefore (as von Baer's law describes) tend to be more similar, and later stages will accumulate their own adaptations in different species. The same reason may explain why some groups fit von Baer's law better than do others. In the nineteenth century, Balfour argued that species with brooding will tend to retain their ancestral forms in the embryo; but species with less yolk and independent development will tend to evolve individual embryonic adaptations. Species that brood their young should then fit the law better than species that discharge eggs. Many examples of specialized larval forms are from groups in which the larva has to fend for itself; but Balfour's idea still needs to be systematically tested. In summary, there are two theories to explain why evolutionary changes should be more common in the adult than earlier in development: one from the magnitude of mutations acting at different developmental stages, the other from external selective circumstances.

There is another idea about the evolutionary importance of paedomorphosis. It has been argued that paedomorphosis is specially important in the origin of higher taxa—in the kind of evolutionary breakthroughs that generate new classes or phyla. The argument, if correct, would "uncouple" macro- and microevolution: paedomorphosis might be relatively unimportant in ordinary microevolution, and even in speciation: its importance would be in the origin of groups at higher levels. Then macroevolution would have a different characteristic process from microevolution.

The first component of the argument is empirical. For many large groups of animals, the adults appear to resemble an early developmental stage of a possible ancestor. The chordates provide the classic example. They must have evolved from one of the invertebrate groups, but the distinctive chordate characters—a backbone or notochord, a dorsal nerve chord, and segmented muscles—are not found in any of the invertebrates. They are found in the larvae of a group called the tunicates, or sea-squirts. Tunicates are included in the chordates because of their chordate larva, but adult tunicates lack any chordate characters. Adult tunicates look like perfectly good invertebrates. It has therefore been argued that the ancestral chordate resembled a larval tunicate, and the chordates originated by paedomorphosis. Paedomorphosis has subsequently been suggested in the origin of many animal groups. De Beer, for example, in his important work on the subject, *Embryology and evolution* (1931, renamed *Embryos and ancestors*, 1940), suggested that the following groups had originated from paedomorphic ancestors: ctenophores, siphonophores, hexacorals, proparian trilobites, cladocerans, copepods, insects, graptolites, pteropods, appendicularians, chordates, ratite birds, hominids.

This is a long list. If each case was correct, paedomorphosis would definitely have been a major, perhaps *the* major, process in the origin of higher taxa. We shall consider in a moment how convincing the evidence is; but let us first ask why the trend might exist. The classic explanation was that paedomorphosis enables an "escape from specialization" (in Hardy's phrase). Within a higher taxon, the argument said, most changes are in the adult; and the species become increasingly specialized to narrow niches by these adult adaptations. A return to the juvenile stage means that the slate of evolutionary specialization can be wiped clean, and new large niches may then be invaded. The idea is debatable at best; but we shall not discuss it in detail here because the phenomenon it is intended to explain is too uncertain. None of the cases in de Beer's list, not even the origin of the chordates, is well confirmed. The trouble is that, even if the chordates did evolve from a form like the modern tunicate larva, that does not prove it was a larval form *then*, at the time of the origin. As Gould and others have pointed out, 500 million years ago the tunicate larva could have been an adult form. During the subsequent millions of years that adult could have evolved a new sessile (and invertebrated) adult stage by terminal addition. The origin of the chordates would then have been from an ancestral tunicate, but not by paedomorphosis. With unique events, happening millions of years ago, it is difficult to rule out such a possibility. The importance of paedomorphosis in the origin of higher taxa remains only a conjecture.

In summary, theoretical work has outstripped the empirical in the study of the relative frequency of different types of developmental change. Apart from McNamara's compilation, no quantitative work has been done on the frequency of the different types of heterochrony. Moreover, no systematic study has been done on the relationship between heterochrony and macro-evolution. However, that has not stopped work on provisional theories of why evolutionary change should be concentrated in the adult, and of why paedomorphosis should be associated with macroevolution. How well these theories will be factually supported remains to be determined.

20.6 Heterochronic change can occur between different somatic cell lines

The examples of heterochrony we have discussed so far have concerned the relative timing of reproduction and somatic development as a whole. But within the soma, many different cell lines develop and the form of the adult can be altered by heterochronic changes between them. This kind of heterochrony can be studied by two of the classic methods of studying morphologic evolution. The simpler of the two was largely invented by Huxley and is called *allometry*. A typical allometric graph plots a character such as testis size on the *y*-axis and body size on the *x*-axis (see Figure 13.7,

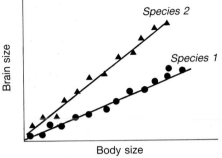

Figure 20.9 Imaginary graph of allometric relationships between brain and body size in two species. The points on the graph for each species could either be for many different individuals, or for one individual as it grows up. Species 2 has a larger brain at each body size, and this could be because the growth of the brain cell lines have been heterochronically speeded up relative to the growth of other cell lines.

p. 341). The points on these graphs can be for the same individual measured at different ages, for different individuals of a species (in which case the scatter will mainly be due to variation in age), or for different species in a higher taxon.

For the study of heterochrony, the most interesting allometric graphs show the intraspecific variation (or individual growth) for two or more species. If brain size increases more rapidly relative to body size in one species than in another, that is probably because the growth of the brain cell lines has been speeded up relative to the other cell lines in the body (Figure 20.9). In other words, there has been a heterochronic change between the species. Heterochrony was introduced above for the somatic and germ lines of the body; but it can operate between any two cell lines, or sets of cell lines. The term allometry, on the other hand, is customarily applied to the relationships between the sizes of two organs; but no point of principle depends on the distinction between the size of an organ, such as the brain, and a reproductive character such as numbers of offspring. We can see that allometry and heterochrony are, in an abstract sense, two ways of talking about the same thing if we look at how a case of heterochrony between germ and somatic lines can be expressed on an allometric graph. The allometric graph now has body size (or age) on the x-axis and cumulative number of offspring produced on the y-axis (Figure 20.10). If reproduction were speeded

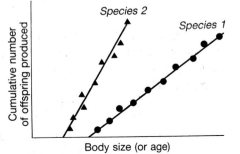

Figure 20.10 Allometric relationship between the size of an organism and its cumulative number of offspring produced. If reproduction is speeded up (progenesis), the allometric graph will shift to the left. Heterochrony often alters the shape of allometric graphs.

up (progenesis) the reproduction curve would be shifted to the left, as shown in Figure 20.10. In general, we can think of the two axes on the graph as representing two cell lines, or sets of cell lines; when there is a heterochronic change in the rate at which the two cell lines develop, the allometric graph will change shape.

Allometry is an important method for describing morphologic evolution. The body shape of a species can usually be partly represented in an allometric graph, and allometry provides a method of representing formally the shape differences between species. The axis of an allometric graph may or may not represent a true growth gradient in individual development. If it does, then the shifts in the graph slope between species will portray the changes in the development regulatory system that produced the evolution. Even if the variable on an axis does not correspond to an actual developmental mechanism, it will to some extent be correlated with one, and the graph for two species gives some idea of the underlying developmental changes.

Huxley's method works easily for two characters; and its many dimensioned extension is obvious (just add a third axis for the third character, etc.). Multidimensional measurement, however, is clumsy for

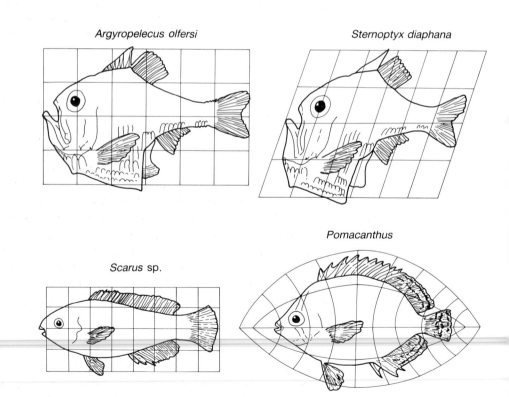

Figure 20.11 A D'Arcy Thompson transformational diagram. The shapes of two pairs of species of fish have been plotted on Cartesian grids. *Argyropelecus olfersi* could have evolved from *Sternoptyx diaphana* by changes in growth patterns corresponding to the distortions of the axes, or the direction of evolution could have been in the other direction, or they evolved from common ancestral species. From D'Arcy Thompson (1942).

complex shapes. For complex shapes, *D'Arcy Thompson's transformations* can be more illuminating (Figure 20.11). D'Arcy Thompson found that related species superficially looking very different could in some cases be represented as simple Cartesian transformations of one another. We met the most thoroughly worked out modern example in an earlier chapter (Raup's analysis of snail shell shapes: Figure 13.6, p. 339). With some simplification, the axes on the fish grids in Figure 20.11 or the snails of Figure 13.6 can be thought of as growth gradients. The evolutionary change between the species would then have been produced by a genetic change in the regulatory mechanisms controlling those gradients.

The work of D'Arcy Thompson and his followers shows that apparently complex changes may turn out, if looked at in the right way, to be fairly simple; regulatory changes in growth gradients are probably the means of many evolutionary changes in shape.

20.7 Changes in regulatory genes can produce macromutational "monsters"

If a developmental change is to have any evolutionary importance, it must have a genetic basis. Genes that influence development are often referred to as regulatory genes. Little is known about them genetically, but we can deduce something about their evolution. The main, if obvious, deduction is that the general theory of evolution should apply to evolutionary changes in development in much the same way as it applies to any other character, such as structural molecules, enzymes, or behavior. Therefore, evolutionary changes in development will probably proceed by small genetic changes. Growth systems have proved amenable to quantitative genetic study (chapter 9) just like any other character. Small genetic changes in growth patterns can be produced, and these are presumably the source of most developmental change in evolution.

However, as we have seen, mutations with effects in early stages of development can have large phenotypic consequences. The *homeotic mutations* of fruitflies (*Drosophila*) are clear examples. Homeotic mutations typically transpose part of the body from one region to another. The mutant *bithorax* is a well-known case. It has two rear thoraxes instead of one front and one rear, giving it two pairs of wings instead of a pair of wings and a pair of halteres. *Antennapedia* is another example, in which legs grow out of the antennal sockets instead of antennae. At an abstract level, it is easy to imagine how homeotic mutations work. There is presumably a set of genes encoding the growth of a leg and another set specifying where these leg genes are switched on. Mutations in the position specifying genes could result in the genes encoding for leg growth being switched on in the wrong place.

Homeotic mutations are examples of macromutations, or "monsters." They are mutations of large phenotypic effect. Their mere existence demonstrates that macromutations can arise. We can imagine that even more monstrous phenotypes might be generated by mutations in genes controlling early development, though mutations with too radical effects would be lethal and never emerge from the egg stage. But how important are these developmental macromutations in evolution?

There are two main viewpoints. The standard neo-Darwinian argument, due to Fisher, is that developmental macromutations are evolutionarily inconsequential. They may arise from time to time, but will always be selectively disadvantageous because they are introducing a gross change into a fairly well-adjusted machine: they will therefore soon disappear from the population. A further point is worth making about whether macromutations can happen. It is necessary to make a distinction. The homeotic mutations of fruitflies illustrate that macromutations can happen; but they produce their large effect by reshuffling existing information. They produce new forms showing large phenotypic changes, but they do not produce large adaptive innovations. No one can deny that macromutations happen, but there is good reason to deny that macromutations generating novel adaptations ever arise. The many adaptations by which a mammal differs from a reptile, such as in tooth shape, jaw, locomotion, and circulatory physiology, could have evolved in small stages, but it is improbable to the point of impossibility that they all appeared in one big mutation in an ordinary reptile. A macromutation producing a new complex set of adaptations would require a miracle. The argument only applies against *adaptive* macromutations, not phenotypic macromutations in general. A phenotypic macromutation that shuffles the parts of a reptile to produce any old monster might happen; an adaptive macromutation that shuffled the parts of a reptile to produce a mammal all at once would not.

On the other hand, Goldschmidt and his followers have suggested that regulatory gene macromutations are the basis of many evolutionary innovations. The mutations are often known as *hopeful monsters*. The idea is that radically new phenotypes are produced by mutations in regulatory genes, and these mutations are the mechanism by which new major groups evolve. The origin of major groups, according to Goldschmidt, takes place by a different process from microevolution: here is another example of the (hypothetical) uncoupling of micro- and macroevolution. Microevolution may be by natural selection of small genetic variants; but major groups (Goldschmidt said) evolve by a different process—via hopeful monsters. The idea has some modern supporters and, as a rare theoretical possibility, it cannot be ruled out. Maybe selectively advantageous macromutations do sometimes arise. However, there is no evidence that they do, and theory suggests they would not.

At the beginning of the chapter, we saw how the mammals evolved by gradual and adaptive evolution. They provide a counter-example to Goldschmidt's theory—and to any other theory suggesting major groups evolve by a sudden saltation from their ancestors. The origin of no other major group is understood so well, but for birds we also have evidence of fossil intermediate stages and gradual evolution. The transition from reptiles to birds was adaptive, because most of the characters of birds are adaptations for flying; it must therefore have proceeded in many stages. It is more difficult to rule out macromutations in the origin of groups, such as many invertebrates, for which there is no appropriate fossil evidence. But to say that a theory cannot be ruled out by negative evidence is very different from saying that it is correct. The theory of evolution by hopeful monsters should be no more than borne in mind as a hypothesis; it is implausible in theory and unsupported in fact.

20.8 Conclusions: developmental change and evolution

The relationship between development and evolution is one of the classic topics in evolutionary biology. The nineteenth century generalization, that new evolutionary stages are added on to the end of the life cycle, was no longer believed when the modern synthesis was being put together in the 1920s. Then for about half a century, many evolutionary biologists showed less interest in the subject; but now it is again an active field of research. Work is being done on the frequency, and circumstances, of the different types of developmental change; on many individual case studies of heterochronic evolution in modern and fossil species; and on the genetics of developmental change. The best understood macroevolutionary transitions suggest that macroevolution occurs by extrapolated microevolution; the evidence does not suggest that any kinds of developmental change characteristically happen any more frequently in the origin of higher groups than in smaller-scale evolution.

20.9 Higher taxa rise, fall, and replace one another

After the origin of a higher taxon, the number of species in it usually increases; so too does the number of genera, families, and other taxonomic levels. The pattern of increase is called *radiation* and its exact form can differ between groups. In the mammals, for example, the major subgroups—dogs, cats, primates, ungulates—all originated relatively suddenly in a burst of evolution during the early Paleocene; once established, the number of groups remained constant to the present. In bivalves, by contrast, there has been a steady increase in the number of families (Figure 20.12). Eventually, after a phase of expansion, a higher taxon goes into decline and ends with extinction.

What factors explain the timing and pattern of a group's rise and fall? Why,

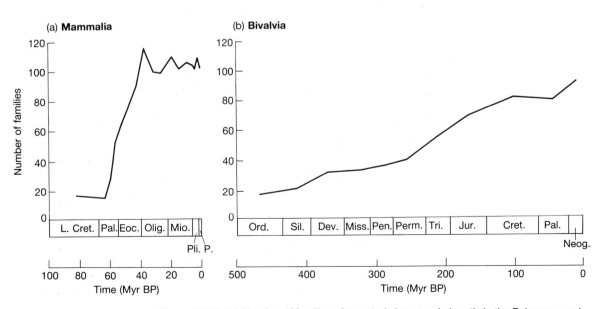

Figure 20.12 (a) Number of families of mammals increased abruptly in the Paleocene and Eocene, after which it has remained constant. (b) Number of families of bivalves has increased steadily through time. From Stanley (1979).

for example, did the mammals remain a minor group for about 150 million years after their origin and then radiate rapidly 50–60 million years ago? The obvious answer is that until the Paleocene the ecological niches later to be occupied by mammals were already filled by dinosaurs: the mammals radiated as the ecological replacements for the dinosaurs. The rise and fall of a higher taxon has often been connected with the fall of one set of ecologically equivalent species and the later rise of another set.

Why should one higher taxon replace another higher taxon of ecologically similar species? Two theories can be tested. One (competitive displacement) says that the second group outcompeted the first, and drove it extinct. The other (independent replacement) says that the first group declined and became extinct for some reason—environmental change, perhaps—unrelated to the presence of the second group, and the second group only radiated after the first had been cleared away. The pattern of change in diversity of the two groups provides the best evidence to test between the two theories (Figure 20.13). If the first group declines before the second expands, it suggests competition was not influential; whereas if the first group declines in proportion to the increase in the second, it suggests competition.

The test is not foolproof. In practice, environmental change could produce the pattern of competitive replacement (Figure 20.13b) if the conditions favoring the second group gradually appear while those favoring the first group disappear. Or the environmental conditions could themselves influence the relative competitive power of the two groups. The environment might change from an earlier condition under which the first group outcompeted the second to a later condition under which the second outcompeted the first. A pattern of gradual replacement therefore does not rule out factors other than ecological competition. But the test (Figure 20.13) is still of some interest. If the first group does become extinct clearly before

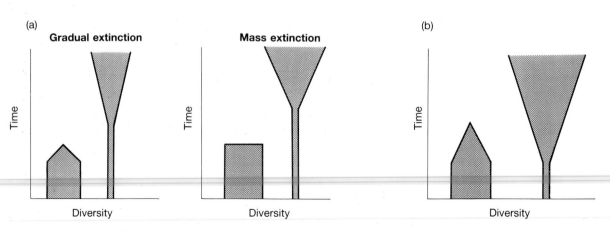

Figure 20.13 The exact pattern of replacement of one group by another suggests whether or not competition was at work. (a) If the initially dominant group declines before the second group expands, it suggests that the replacement was not caused by competitive displacement. The dominant group may decline either gradually or catastrophically. Whereas (b) if the dominant group declines as the other group gains at its expense, competition and relative adaptation are more likely to have influenced the replacement.

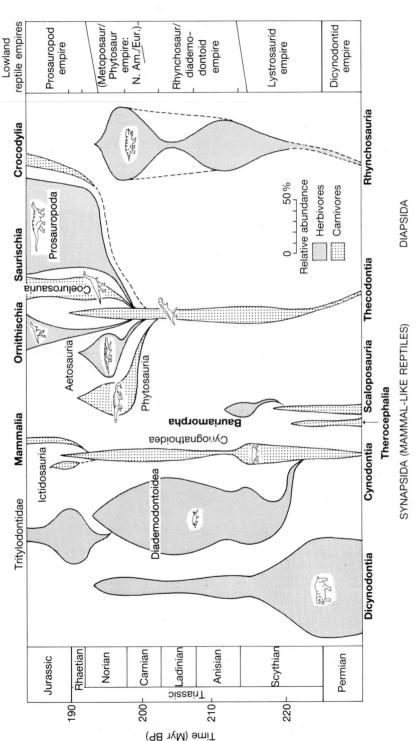

Figure 20.14 Patterns of diversity among the main terrestrial reptile groups of the Triassic. The width of the stippled area for each group corresponds to the abundance of the group. Dinosaurs are the saurischians and ornithischians. Note the replacements of the diademodontoids and rhynchosaurs by the dinosaurs (Ornithischia and Saurischia) in the Norian, and of dicynodonts by diademodontoids in the Scythian. See Figure 20.2 for some of the mammal-like reptile forms. From Benton (1983a).

the rise of the second, it is difficult to explain the replacement by competition. And if the takeover is correlated in time and the two groups are ecologically analogous, competition is at least suggested.

Let us look at some examples. The mammal-like reptiles are one. In the Permo-Triassic period, the mammal-like reptiles were the dominant group of large terrestrial vertebrates; there were herbivorous, carnivorous, and smaller insectivorous taxa. By the end of the Triassic, they had been replaced in most of these niches by the dinosaurs. It had been argued that the dinosaurs were competitively superior (in Mesozoic conditions) to the mammalian lineage and that was the reason for the replacement. But the most detailed study, by Benton, of the diversity changes of the two groups does not closely support that idea; the evidence is ambiguous (Figure 20.14). The principal Triassic herbivores were a group of mammal-like reptiles (Diademodontoidea, a group of cynodonts) and another unrelated group of reptiles called rhynchosaurs. The herbivorous prosauropod dinosaurs radiated at about the time the two groups became extinct, but there is little evidence of a parallel rise and fall of the groups. The decline of the Diademodontoidea was almost complete by the time the dinosaurs expanded significantly. A more suggestive example of competitive displacement can be found lower down the figure. The dicynodonts did decline more or less as the diademodontids expanded: maybe this was due to competition. But the main conclusion from this first example is that the evidence for the competitive displacement of the mammal-like reptiles by the dinosaurs is unconvincing.

A second example is the replacement through the Cainozoic of the perissodactyl mammals (horses and their relatives) by the artiodactyls (cattle, deer, and their relatives). Perissodactyls such as horses and zebra do still exist, but the group was relatively much more abundant in the Eocene. The artiodactyls and perissodactyls are ecologically similar groups of large-bodied herbivorous grazers, and the rise of the artiodactyls has been explained by

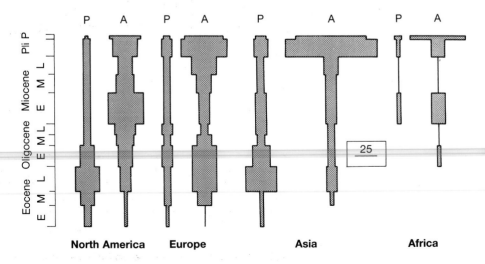

Figure 20.15 The diversities of perissodactyl (P) and artiodactyl (A) mammals in four continents. Diversities are numbers of genera. From Cifelli (1981).

their competitive superiority, due to the evolution of such factors as their special kind of teeth (called selenodont molars) and ruminant digestion. However, again the facts are not clear-cut. Cifelli broke down the data for the diversity of the two groups for four separate continents. He found that the classic pattern of a correlated rise and fall existed only in North America: in Europe, Africa, and Asia the artiodactyls do not appear to have proliferated at the expense of the perissodactyls (Figure 20.15). It is possible, therefore, that there was a competitive displacement in North America; but the evidence for a consistent competitive superiority of artiodactyls to perissodactyls is poor.

The angiosperms (flowering plants) have replaced the conifers and ferns as the dominant terrestrial plants (Figure 20.16). The rise of the angiosperms and their replacement of the gymnosperms has been attributed to a number of factors. Angiosperms can grow more rapidly than gymnosperms under many conditions, which gives them a competitive advantage. Pollination and seed dispersal by insects is mainly confined to flowering plants, and it may have enabled them to colonize new habitats more rapidly. Specialized pollinator relationships in angiosperms may also have increased their

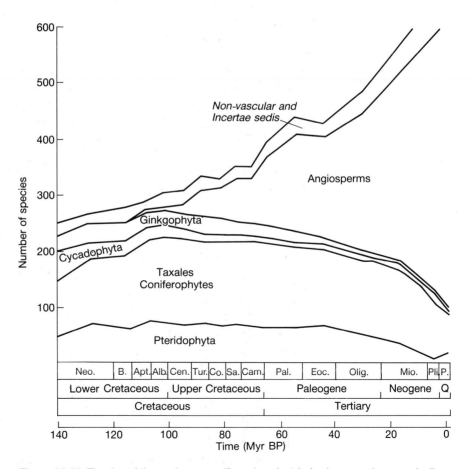

Figure 20.16 The rise of the angiosperms (flowering plants). Angiosperms have gradually expanded in diversity since the Cretaceous. Gymnosperms (Coniferophytes, Cycadophyta, Ginkgophyta) have declined, as have pteridophytes. From Niklas (1986).

speciation rate (see section 21.3.2, p. 566). Figure 20.16 suggests that the increase of the flowering plants may well have been at the expense of the gymnosperms: they do show a parallel increase and decline, rather than a decline of the one preceding the increase of the other.

A final example of a faunal replacement shows how both factors may operate in the same case. In the Paleozoic, the dominant sessile filter-feeding invertebrate group was the brachiopods; they still survive today, but have largely been replaced by another taxonomic group, the bivalve molluscs. Gould and Calloway plotted the diversity changes of the two groups during the crucial period around the Permo-Triassic (Figure 20.17). The bivalves, we can see, were increasing relative to the brachiopods during the Permian; they then suffered less than the brachiopods in the mass extinction (see chapter 22) at the end of the Permian, and since then the bivalves have expanded at a higher rate. It is possible, therefore, that the replacement was due both to a competitive superiority of bivalves and to the better fortunes of the bivalves in the extraordinary conditions of the Permian mass extinction. (It is also possible that the bivalves' better survival in the mass extinction was also due to their competitive superiority. That, however, would depend on what events took place at that time, and we have no evidence either way.) Mass extinctions, by causing a temporary reduction in the level of competition, may lead to phases in which ecological replacements are particularly likely to occur.

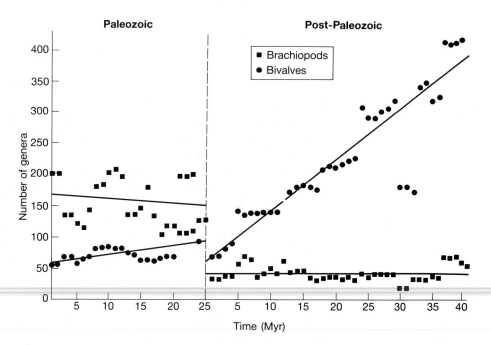

Figure 20.17 Replacement of brachiopods by bivalves through the Permo-Triassic mass extinction. The bivalves were gaining before the mass extinction, and survived through it relatively better; which suggests (see Figure 20.13) that relative adaptation and the fortunes of survival in the mass extinction may both have influenced the replacement. Time is for the 25 million years preceding, and the 40 million years following, the Permo-Triassic boundary. From Gould and Calloway (1980).

In summary, we have considered two reasons why one higher taxon may replace another, and how to test which has operated. One is competitive displacement; it has often been put forward as a hypothesis, but is difficult to support with evidence. The rises and falls of higher taxa suggests that replacements often happen after the earlier group has gone extinct: the second group then radiates into the empty ecologic space.

20.10 Summary

1 The evolution of mammals from reptiles is an example of adaptive evolution and the fossil record reveals that it proceeded in a series of stages, through various groups of mammal-like reptiles.

2 The neo-Darwinian theory of the origin of higher taxa suggests that they evolve in many small adaptive stages, by natural selection. Macroevolution is microevolution extrapolated over long periods.

3 Morphologic transitions often take place by changes in the rate, and timing, of developmental processes.

4 A formerly popular idea—the principle of recapitulation—held that successive new evolutionary stages can only be added on to the adult stage; an individual then develops by "climbing up its family tree."

5 In fact, new evolutionary stages may be added at any stage of an individual's development. The relative timing and rate of different development processes can also shift, in a process called heterochrony.

6 The time of reproduction may be shifted earlier relative to somatic development (paedomorphosis) either because somatic development is slowed down (neoteny, e.g. the axolotl) or reproduction is accelerated (progenesis). Heterochrony can also shift reproduction to a relatively later stage of somatic development (by "acceleration" or by "hypermorphosis").

7 Evolutionary biologists are interested in the relative frequencies of evolutionary changes early and late in development, and the frequencies are of different types of heterochrony.

8 Two theories have been put forward to explain a possible tendency for new evolutionary stages to be added more often in the adult than earlier in development: (a) early-acting mutations will be more likely to be macromutations; and (b) the forces of natural selection from the external environment are more intense on the independent adult than the protected early embryo.

9 It has been argued that new higher taxa often evolve by paedomorphosis from the larval stage of ancestors. The possible origin of the chordates from an animal like the tunicate larva is an example. The reason could be that larval stages are evolutionarily more flexible. The argument as a whole is questionable and its factual support uncertain.

10 Changes in the relative sizes of two organs can be studied by allometric graphs.

11 Changes in complex shapes can be studied in D'Arcy Thompson's transformational diagrams.

12 Mutations influencing developmental processes can produce macromutational phenotypic effects. Homeotic mutations in fruitflies, such as *antennapedia* and *bithorax*, are examples. Developmental macromutations are usually (or even always) deleterious and evolutionarily inconsequential.

13 Goldschmidt, and his followers, have argued that developmental

macromutations are the source of new large evolutionary changes: there is no evidence for this idea, and it is theoretically implausible, if not impossible.

15 Large-scale evolutionary replacements of one higher taxon by another may be due to competitive ecological displacement, or to the extinction of the earlier group for an unrelated reason followed by the radiation of the second group. The two ideas can be tested by the exact timing of the rise and fall of the two groups.

20.10 Further reading

Simpson (1953) and Rensch (1959), for example, give classic neo-Darwinian accounts of macroevolutionary morphologic changes. On the mammal-like reptiles see Kemp (1982a,b, 1985), Hotton *et al.* (1986), Bramble and Jenkins (1989), Benton (1990), and Carroll (1988). See Wellnhofer (1990) on *Archaeopteryx*.

There is a large modern literature on the relation between developmental change and macroevolution. Gould's (1977a) book discusses the history of recapitulatory ideas and modern work on heterochrony: much of the more recent work has been inspired by Gould. Raff and Kaufman (1983) is a useful text. On heterochrony, there is a recent book by McKinney and McNamara (1991) and a conference edited by McKinney (1988) which contains general chapters by Gould (on its significance), McKinney (on analysis), and McNamara (on frequency); McNamara (1988) discusses further fossil examples of heterochrony. There is a popular article by McNamara (1989), several chapters in the volume edited by Bonner (1982), and Slatkin (1987) discusses its genetics. See also the special issue of the journal *Bioessays* 14, 209–290; 1992. For the classic writers on recapitulation, von Baer's law, and paedomorphosis, see Gould (1977a), and Ridley (1985) for Balfour's argument.

On allometry see chapter 13 of this volume. D'Arcy Thompson's book (1942) is one of the great books of biology; see Medawar and le Gros Clark (1945), Medawar (1958), Gould (1971), Bookstein (1977). Development is not the only idea in macroevolutionary morphology: for others, see Nitecki (1990) in general and Margulis and Fester (1991), on symbiosis.

On replacements, see the general articles by Benton (1983b, 1987), as well as the particular studies discussed in the text (Benton, 1983a; Cifelli, 1981; Knoll, 1986; Bond, 1989; Gould and Calloway, 1980). Radiation itself is a further topic; see the authors in Ross and Allman (1990), and see Levinton (1988) and Gould (1989) on the number of higher groups through geologic time.

21 Macroevolutionary trends, coevolution, species selection

21.1 Cope's law states that animals tend to evolve larger size

The fossil record of horses, which we used in chapter 19 to illustrate how evolutionary rates can be measured, also shows another important paleo-biological phenomenon. If the body size of horses is plotted through time, we see that more recent horses are bigger (Figure 21.1). The Eocene ancestors of modern horses were about the size of a dog; since then, in the lineage showing the largest increase, horses have evolved to become as much as 10 times heavier. The *evolutionary trend* toward size increase can be seen in two ways. If we look at the average size of all horse species at any one time, the average has increased. It looks as if the trend has arisen mainly because the lower size limit has been constant while the range of sizes has increased through time. There are small horse species, with body mass of about 50 kg, at all times, and a steadily increasing number of larger species; the result is an increase in the average for the whole family. The trend can also be seen within some ancestor–descendant lineages. Four or five of the later lineages in the figure show increases in size; but the trend clearly is not universal: the lineage to *Nannipus* shows a decrease in size and for the first 30 million years or so there was no net change at all. Evolutionary trends toward an increase in size are common in the fossil record, and the tendency is often called Cope's law or Cope's rule, after the paleontologist Edward Drinker Cope.

Cope's law is not a "law" in the sense that it has no exceptions. Some horse lineages showed a size increase, some did not, others showed a size decrease; the lineages showing size increases are only a part of all the evidence. No one has ever done a systematic study of size increases in all (or even a sample of) taxonomic groups at many geologic times; and it is not possible to make any exact general statement about its frequency. One example, however, can give an idea of the sort of results that are found. Hallam measured shell size in 41 Jurassic lineages of bivalves and 19 of ammonites; he reported that "a majority, almost the same in both groups (69% of bivalves, 68% of ammonites) have increased in size by a factor of approximately two or more, and a significant minority have trebled, quadrupled or even quintupled in size." Figure 21.2 illustrates 10 of the lineages.

An evolutionary trend is a directional evolutionary change that persisted for a long enough period to be detectable in the fossil record. (In Gould's words, a trend is a "sustained biostratigraphic character gradient.") This chapter is about why evolutionary trends happen. As we discuss different explanations for trends, it is important to keep in mind the distinction between the two kinds of trend (Figure 21.3). The simplest kind of trend is a

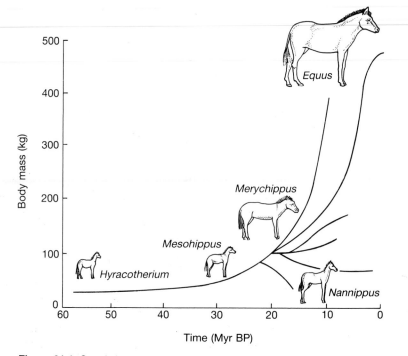

Figure 21.1 Cope's law in horses. The individual lineages are inferred ancestor–descendant lineages, though the real pattern would have had many more. The graph is based on 40 species of North American horse, and includes representatives of most of the main evolutionary groups. From MacFadden (1988b).

change in a single lineage; i.e. a sequence of ancestor–descendant populations. The second kind is a trend in the average state of a character in a group of related lineages. This chapter will mainly be about two explanations for evolutionary trends: natural selection and a process called species selection. Natural selection can explain trends within a lineage, and also in the average for a group of lineages if they are produced by a trend in the majority of their constituent lineages. Species selection does not explain trends within a lineage, but it can produce a trend in the average for a group of lineages even when the majority of its constituent lineages are not showing the trend.

21.2 Several general theories can explain evolutionary trends

In addition to natural selection and species selection, a third possible explanation has been put forward to explain evolutionary trends like Cope's law: namely, chance. Just as random drift of gene frequencies has some probability of causing one gene to be fixed (see chapter 6), so there is a theoretical probability that random changes between morphologic states would produce a trend. Suppose, for example, that an individual's body size makes no difference to the number of offspring it leaves: small, medium-sized, and large individuals all have the same reproductive success. The mean body size of the population in the next generation would be equally likely to increase or decrease; on average it will stay the same. If, by chance, it increased, then in

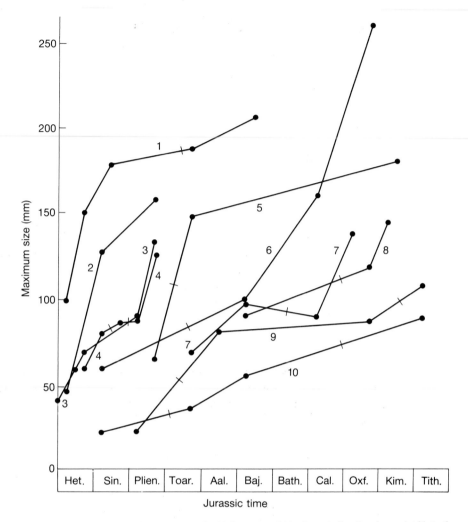

Figure 21.2 Patterns of size increase in 10 lineages of bivalves during the Jurassic. Note the *y*-axis is maximum size of an individual in the lineage. The names on the *x*-axis are the successive divisions of Jurassic time. Cross bars in a lineage indicate a change in species or subspecies. From Hallam (1975).

the second generation it might again decrease, stay the same, or increase. There is a chance that body size will increase through enough generations for a trend to result. The process can be simulated: Figure 21.4 illustrates some randomly evolving lineages, and at least one of them shows an apparent evolutionary trend.

Chance is a possibility to be kept in mind when interpreting evolutionary trends. It is more plausible for short than prolonged trends, and for unique trends than for trends that take place independently in many lineages. Chance is also a relatively unlikely explanation for morphologic evolution (section 7.11, p. 164). In the case of an evolutionary size increase within a lineage, for example, body size could influence fitness by its relation with competitive power in both sexes, with fecundity in females, with

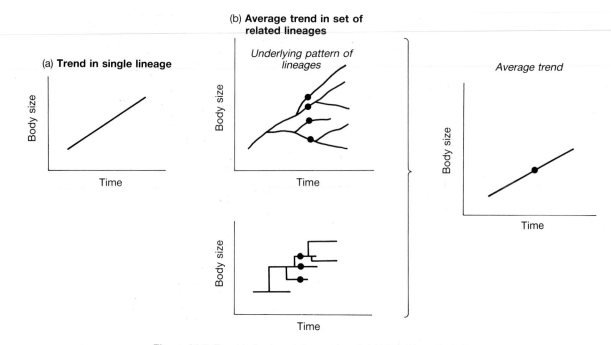

Figure 21.3 Two kinds of evolutionary trend. (a) Trend in a single lineage—a single series of ancestor–descendant populations. (b) Trend in average state of a character (such as body size) in a group of lineages through time. In the graph on the right, each point for body size is the average for several lineages.

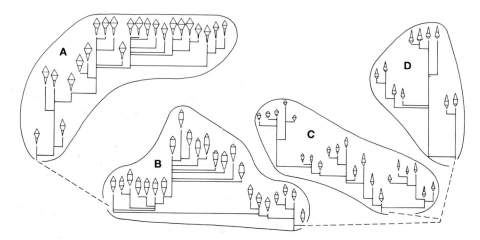

Figure 21.4 Part of a simulated random pattern of evolution. Time goes up the page. In the simulation, change in form occurs only at branch points, and is equally likely to increase or decrease the size of a character. Form is indicated in the diagram at the top of each lineage. There are three main characters, a "head," a "thorax," and a "tail" (actually, five variables in the simulation control the three observed characters). Four "taxonomic" groups are recognizable (labeled A–D). Taxon C shows a trend to decreasing size through time. From Raup (1977).

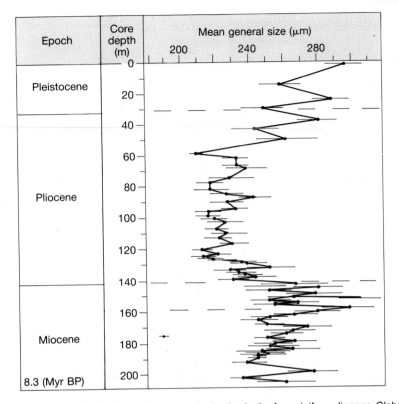

Figure 21.5 Evolutionary changes in body size in the foraminiferan lineage *Globorotalia conoidea–G. inflata*. (The same lineage is classified as *G. conoidea* in the Miocene, but *G. inflata* from the Pleistocene.) Note the two rounds of increased size, with a decrease between in the Pliocene. From Malmgren and Kennett (1981).

thermoregulation, and other factors: it is implausible that changes in body size do not affect fitness. Although some trends may be due to chance, we shall concentrate on other explanations.

Evolutionary trends could be driven by directional change in the *physical environment*. However, the process is difficult to demonstrate convincingly with fossil evidence. Figure 21.5 shows the change in size of a foraminiferan lineage from the Miocene to the recent past. Size increases in the late Miocene, then decreases in the early Pliocene, and increases again in the late Pliocene and Pleistocene. Malmgren and Kennett suggested the trends might have been driven by changes in sea temperature: there were periods of cooling in the late Miocene and late Plio–Pleistocene, and a period of warming at the beginning of the Pliocene. However, they stress that the interpretation is uncertain, because "very little is known about environmental–morphological relations in foraminifera": and even if there is a reason why larger foraminiferans are favored when the sea cools down, it would still be difficult to rule out the possibility that the trend was really driven by changes in some other factor, such as the foraminifer's biological competitors.

If the physical environment is constant, natural selection can still produce long-term evolutionary change, by either of two mechanisms. First, evolutionary change may be limited by the availability of beneficial mutations. As each beneficial mutation arises, it will be fixed; but if they arise quite rarely, evolutionary change will be spread out over a long time. We should then see the change through time as an evolutionary trend. There would be evolutionary *progress* over time. In practice, this theory is as difficult—or more difficult—to test with fossils as the theory of continued directional change in the physical environment. Moreover, it is unlikely to be a general explanation; the rate of evolution is probably not usually limited by the amount of genetic variation, because most characters are genetically variable in natural populations. But it is a theoretical possibility.

The final possibility, and theoretically the most important one, is that a long-term trend results when each successive stage in evolution sets up selection for the evolution of a new stage. Direct competition between individuals within a species is a simple example, such as in sexual selection (section 11.4, p. 283). If the largest individual has an advantage in reproductive competition, body size should theoretically increase without limit. In a real case, there will be limits due to other factors and a long-term trend would be most likely to result if there were successive rounds of selection for increase in size, interrupted by times when other factors put a brake on the process. But whatever the details, an open-ended process in which evolution occurs in response to other members of the species has the potential to produce a trend. Interspecific relationships can set up the same kind of selection pressure, in cases in which evolution in one species happens in response to evolutionary changes in another. This is called coevolution. Coevolution is a large subject in its own right, and we shall be looking at it as a whole in addition to seeing how it might explain evolutionary trends.

21.3 Coevolution happens when two or more species influence each other's evolution

21.3.1 Coadaptation alone suggests, but is not conclusive evidence for, coevolution

Figure 21.6a shows an ant (*Formica fusca*) feeding on the caterpillar of the lycaenid butterfly *Glaucopsyche lygdamus*. The ant is not eating the caterpillar; it is drinking "honeydew" from a special organ (Newcomer's organ), the sole purpose of which seems to be to provide food for ants. The reason why the caterpillars feed the ants has been the subject of several hypotheses, and in 1981 Pierce and Mead carried out an experiment which suggests that the caterpillars, at least in *G. lygdamus*, gain by being protected from parasites.

The caterpillars are parasitized by braconid wasps and tachinid flies. Alone, they are almost defenceless against these lethal parasites; but the tending ants will fight off parasites from their caterpillars (Figure 21.6b). Pierce and Mead experimentally prevented ants from tending caterpillars, and measured the rates of parasitism in the experimentally unprotected and in normally protected (control) caterpillars. Their results show that ants reduce the rate of parasitism in *G. lygdamus* (Table 21.1). The ants and caterpillars are therefore closely adapted to each other; the ants gain food, and the caterpillars gain protection. They form a kind of interspecific *coadaptation*. (Here the term refers to the mutual adaptation of two species; it has also been used to

Figure 21.6 (a) The ant (*Formica fusca*) is tending a caterpillar of the lycaenid butterfly species *Glaucopsyche lygdamus*. The ant is drinking honeydew, secreted from a special organ. (b) *Formica fusca* defending a caterpillar of *G. lygdamus* against a parasitic braconid wasp: the ant has seized the wasp in its mandibles. Bars indicate 1 mm. (Photographs by Naomi Pierce.)

Table 21.1 Caterpillars of the lycaenid butterfly *Glaucopsyche lygdamus* are more likely to be parasitized if they are not tended by ants. Ants were experimentally excluded from some caterpillars, and the rate of parasitism on these and untreated control caterpillars was measured. The two sites are in Gunnis County, Colorado. Parasites were wasps and flies. *N* is the sample size. From Pierce and Mead (1981)

Site	Caterpillars without ants		Caterpillars with ants (control)	
	Percentage parasitized	*N*	Percentage parasitized	*N*
Gold Basin	42	38	18	57
Naked Hills	48	27	23	39

describe the mutual adaptation of genotypes (section 8.2, p. 186) and of parts (section 13.3, p. 327) within an organism.) Relationships like that between ants and lycaenids are called *mutualism*; many examples of mutualism are known, and they provide some of the most charming details in natural history.

How could the coadaptation between ant and lycaenid have evolved? The morphologic structure and the behavior patterns of both ant and caterpillar appear to have evolved in relation to each other. It seems likely that, after the ancestors of the two species had become associated, natural selection would have favored mutually adapted changes in each species. Changes in one species, such as to increase honeydew production, would favor changes in the other (to increase protection) as the caterpillars became more beneficial to the ants. This kind of reciprocal influence is what is meant by *coevolution*: each species exerts selection pressures on, and evolves in response to, the other species. The two lineages evolve together (Figure 21.7).

Logically, however, the observation of coadaptation between two species,

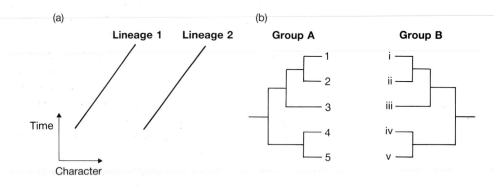

Figure 21.7 Coevolution means that two separate lineages mutually influence each other's evolution. The two lineages tend to (a) change together, and (b) speciate together. Lineages 1 and 2 could be, for example, an ant lineage and a lycaenid butterfly lineage.

such an ant and a caterpillar, is not enough to confirm that the two have coevolved. As Janzen pointed out, it could be that the two lineages have been evolving independently, and it just turned out at some stage that the two forms were mutually adapted to each other. The ancestors of *G. lygdamus* might have evolved their Newcomer's organs for some other reason than feeding *Formica* and the ants might have evolved anti-wasp behavior patterns for some other reason than defending caterpillars; when the two came together they were coadapted. To demonstrate coevolution requires showing not only two forms at one time are coadapted but also that their ancestors evolved together, exerting selective forces on each other.

This is a tall order. In practice, biologists tend to assume that interspecific coadaptations are due to a long history of coevolution unless a convincing alternative hypothesis can be put forward. Janzen's stricture is logically correct, but difficult to live up to in practical biology. Further evidence that a coadapted system arose by coevolution can come from comparison with related species. The relationship between *G. lygdamus* and *Formica* is not unique. Lycaenids and ants have evolved a large number of relationships in different species and this suggests the two groups have been evolving together for some time.

21.3.2 Coevolution should be distinguished from sequential evolution

In a paper that is perhaps the most influential modern discussion of coevolution, Ehrlich and Raven (1964) listed the food plants of the main butterfly taxa. Each family of butterflies feeds on a restricted range of plants, but these plants are in many cases not phylogenetically closely related. Ehrlich and Raven explained the diet patterns mainly in terms of plant biochemistry. Plants produce natural insecticides—chemicals like alkaloids that can poison herbivorous (phytophagous) insects. Insects, in the manner of pest species evolving resistance to artificial pesticides (section 5.7, p. 102), may evolve resistance to these chemicals, perhaps by means of detoxifying mechanisms. When the detoxifying mechanism arises, it will open up a new array of food supplies, consisting of all those plants that produce the now harmless chemical to protect themselves. The insects can feed on them, and will diversify to exploit the resource. The result will be that each insect group can feed on a range of food plants, the range being set by the capabilities of the insect's detoxifying mechanisms. The range of food plants will form a biochemical group, but need not form a phylogenetic group because unrelated plants could use the same defensive chemicals. Ehrlich and Raven's pattern of butterfly–plant relationships could arise as a result.

In turn, there will be selection on the plants to evolve improved insecticides. Plant–insect coevolution should therefore consist of cycles, as plant groups are drawn into, and removed from, the diets of insect groups, and the insects "move" between plant types according to their biochemical abilities. The biochemical arms race between plants and insects should persistently favor new mechanisms on both sides, and might therefore have promoted the diversification of insects and of angiosperms (section 16.6.3, p. 433; section 20.9, p. 555). In Ehrlich and Raven's words "the fantastic diversification of modern insects has developed in large measure as the result of a stepwise pattern of coevolutionary stages superimposed on the changing pattern of angiosperm variation."

The defining property of coevolution is the reciprocal selective influence between the two parties, over a prolonged period. For true coevolution between insects and plants, insects must influence plant evolution, and plants influence insect evolution. Another possibility, would be for one party to influence the evolution of the other, but not vice versa. Plants and insects appear to influence each other's fitnesses strongly, and it is therefore widely believed that they have coevolved. Some biologists, however, have doubted how strong the reciprocal influences are. Jermy, for instance, suggests that plants and insects show *sequential evolution* rather than coevolution: plants

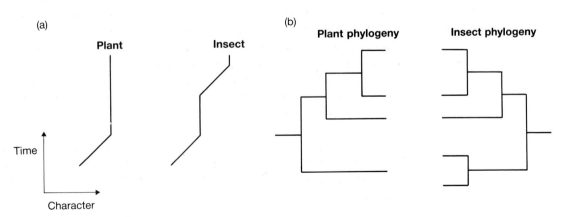

Figure 21.8 Sequential evolution means that change in one lineage selects for change in the other lineage, but not vice versa. Jermy hypothesizes this is true for insects and plants. Plant evolution then influences insects, but a change in an insect lineage does not select for a change in its food plants. The pattern of change in (a) lineages, and (b) phylogenies differs from strict coevolution (compare with Figure 21.7). (a) Change in plants coevolves with changes in insects, but changes (for some reason other than changes in plants) in insects do not cause changes in plants. (b) When plants speciate, so do insects, but when insects speciate, it has no effect on plants.

influence the evolution of insects, but insects have less effect on evolution in plants (Figure 21.8). He offers a number of reasons. One is that many insects eat only one type of plant, whereas plants are eaten by many insects. When a plant changes, its insects will all have to change to keep up; but when one insect species changes, it alone will exert only a small selective pressure on its food plant. It is not yet known whether coevolution, or sequential, or even independent evolution is the better description of insect–plant relationships. The challenge for research is to test between these possibilities by combining ecological studies of the reciprocal influences of plants and insects with phylogenetic evidence for the species involved.

21.3.3 Coevolutionary relationships will often be diffuse

The clearest examples of coevolution come from ecologically coupled pairs of species. In practice each species will experience selective pressures and exert them on many other species. The evolution of a species will be an aggregate response to all its mutualists and competitors, and any evolutionary change may not be easy to explain in terms of any one competitor. The process is called *diffuse coevolution*. It undoubtedly operates in nature; indeed, it may be the main force shaping the evolution of communities of species. But it is difficult to study, and its importance is consequently controversial.

21.3.4 The phylogenetic branching of parasites and hosts may be simultaneous

Step-by-step coevolution is particularly likely to take place between parasites and their hosts. They can have specific and close relationships, and it is easy to imagine how a change in a parasite, which improves its ability to penetrate its hosts, will reciprocally set up selection for a change in the host. If the range of genetic variants in parasite and host is limited, coevolution can be cyclic (section 11.3.5, p. 275); but if new mutants continually arise, the parasite and host may undergo unending coupled changes that may or may not be directional according to the type of mutations that arise.

A further coevolutionary result can be seen in the classifications of parasites and their hosts. They often form the kind of mirror images we expect for coevolving taxa (Figure 21.7). The correspondence suggests that speciation in parasites and their hosts is approximately simultaneous, a relationship called Fahrenholz's rule. Recent work by Timm, and by Hafner and Nadler, on the rodent family Geomyidae (the pocket gophers) and their ectoparasitic lice (Mallophaga) makes a good example. Hafner and Nadler took gel electrophoretic samples from 31 gene loci in eight species of pocket gopher and from 11 loci in 10 parasitic lice, and used the data to reconstruct their phylogenies (Figure 21.9). The phylogenies are nearly mirror images; but there are some deviations. The mirror pattern is attributed to *cospeciation*, and the deviations to *host switching*. In Figure 21.9, for example, if we concentrate on the split labeled D in the phylogeny of the lice, one branch after the split must originally have been a parasite of the gopher *Geomys bursarius*. That ancestral louse then split twice: two descendants "switched" to the hosts *Thomomys talpoides* and *T. bottae*, while the louse *Geomydoecus ewingi* retained the ancestral host *Geomys bursarius*. It is therefore possible to count the

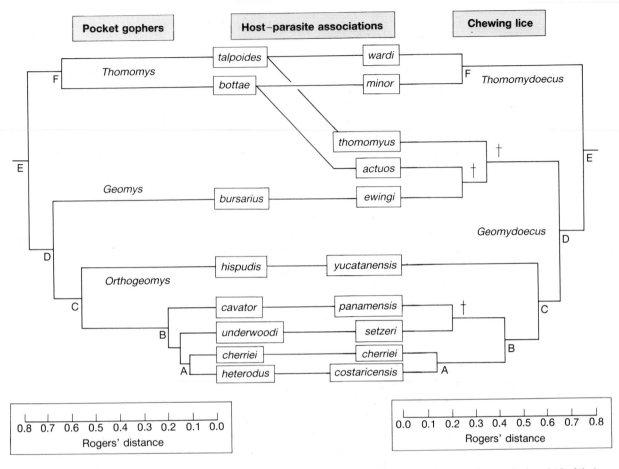

Figure 21.9 The phylogenies of eight species of pocket gophers (Geomyidae) and 10 of their mallophagan parasites. The phylogenies were reconstructed by gel electrophoresis, which gives a measure of distance between species that is probably proportional to time since common ancestry. The phylogenies mainly form mirror images, both in the branch pattern and the time of splitting. Some cases of probable host switching can also be identified (parasite splits marked by daggers (†) are not reflected in the hosts). The pattern illustrates Fahrenholz's rule. From Hafner and Nadler (1988).

frequencies of cospeciation and host switching by inspecting phylogenies like those in Figure 21.9. The figure has another interesting feature: the branch lengths are correlated between parasite and host. This can be seen by comparing the length of a branch on one side with the length of its mirror image on the other side. The branch lengths are expressed as distances (Box 16.3, p. 436), and measure the molecular difference between the two groups connected by the branches. If we assume that the molecules evolve according to the molecular clock, distance will be proportional to time since divergence. The correlated branch lengths between pocket gophers and lice therefore suggests that the host and parasite species tended to speciate at the same time.

Why should host and parasite speciate synchronously? Probably because the same circumstances favor speciation in both groups. Suppose, for example, that pocket gophers speciate geographically. When the host population becomes geographically subdivided, their parasites' populations will automatically be subdivided too. (Mallophaga have limited independent powers of dispersal.) The conditions for speciation in the parasite are then set up whenever the host is speciating. The process could be reinforced if each parasite needs a distinct set of adaptations to penetrate an individual host species; parasites will then be selected to breed only with other parasites of the same host species, in order not to mix the parasitic adaptations to different host species.

The process requires that the parasite cannot move independently of its host. Timm has found evidence that a comparable group of lice that can move independently do not show mirror image phylogenies. The dispersal abilities of parasites should generally influence whether they speciate simultaneously with their hosts. But the parasitic lice in Figure 21.9 move with their hosts; for them, therefore, the observed mirror image phylogenies of host and parasite make sense.

21.3.5 Coevolution can proceed in an "arms race"

A graph of brain size against body size for many vertebrate species reveals that larger vertebrates have larger brains (Figure 21.10). Brain size is usually measured as the relative *encephalization quotient*; that is, the deviation from the line for all species in Figure 21.10. A large brain is one above the line: it is larger than would be expected for an animal of that body size. The idea is that brain size is determined by two factors: the main one is body size; and there is a secondary factor of intelligence. The more intelligent animals, in a loose sense, are those that deviate further above the line; they have greater relative encephalization. Let us provisionally accept here that the encephalization quotient is an index of intelligence: then we can consider, from the work of

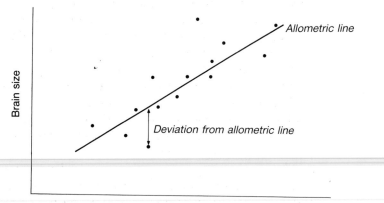

Figure 21.10 Relative brain size may be measured as relative encephalization by the deviation of a species' brain size from the allometric line for many species. Relative encephalization measures whether a species has a brain larger or smaller than would be expected for an animal with its body size. The species illustrated here has a relatively small brain.

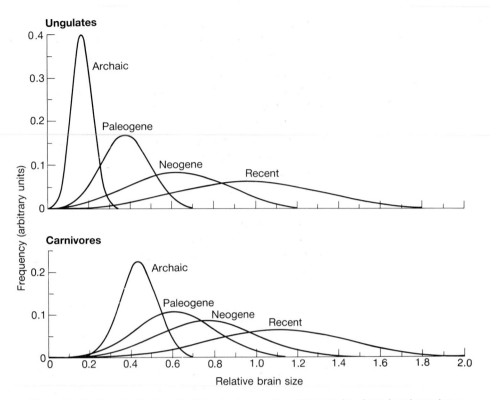

Figure 21.11 The distribution of relative brain sizes for carnivores (predators) and ungulates (prey) through the Cainozoic. There was an increase in brain size through time, and at any one time carnivores had bigger brains than ungulates. From Jerison (1973).

Jerison, a possible example of coevolution between prey and predator. (We met a related part of Jerison's general study when we discussed the Great American Interchange in section 18.6, p. 498.)

In Cainozoic mammals, predators typically have relatively larger brains than their prey. This relationship can be seen in Figure 21.11; but the same figure also shows another, more interesting fact. The relative brain sizes of both predators and prey have increased through time. In order to estimate encephalization in a fossil, it is necessary to estimate body size, and the method used by Jerison has been criticized. The result is therefore uncertain, but Jerison's explanation is nevertheless interesting. He suggests that natural selection has favored higher intelligence both in the prey, to escape predators, and in predators, to catch prey. There has been a coevolutionary *arms race* between predator and prey, leading to ever larger brain sizes in both. The selective forces might well be truly coevolutionary, with each party exerting a reciprocal selective pressure on the other: as predators become cleverer at catching prey, the prey are selected to avoid them more intelligently, and vice versa. In this case, the evidence remains inconclusive; but it is difficult to demonstrate coevolutionary relationships in fossils and Jerison's work is well worth noticing as a rare example in which coevolution is a plausible explanation.

21.3.6 Coevolutionary arms races can result in evolutionary escalation

Vermeij has applied the same argument as was used by Jerison, but much more generally. Vermeij suggests that predators and prey typically show an evolutionary pattern that he calls *escalation*. By escalation, he means that life has become more dangerous over evolutionary time: predators have evolved more powerful weapons and prey have evolved more powerful defences against them. Vermeij distinguishes escalation from evolutionary progress. If evolution is progressive, organisms will become better adapted to their surroundings through evolutionary time; if it is escalatory, the improvement in predatory adaptations may be matched by improvements in prey defences, and neither ends up any better off. The two concepts are easy to distinguish by a thought experiment. If evolution is progressive in predators (for example), then later predators would be better at catching their prey than were earlier predators. If, however, evolution is escalatory, later predators will be no better than their ancestors at catching their contemporary prey types; but if transported in a time machine and set loose on the prey hunted by ancestral predators, they should cut through them like a modern jet fighter in a dog-fight with a biplane from the First World War.

In his book *Evolution and Escalation*, Vermeij discusses both biogeographic and paleontological evidence for escalation. His evidence comes mainly from molluscs in shallow water marine environments: the molluscs are abundant as fossils; and the nature of a shell itself can reveal how strongly adapted a species was, as a predator or a prey. More strongly defended shells have properties such as general thickening, or thickening concentrated around their apertures; burrowing species, or those that cement themselves down, are better defended than those that lie loose on the bottom surface. The degree of escalation in a habitat can be estimated by the proportions of more and less powerfully defended prey, and equivalent measures of the power of predators. Vermeij's biogeographic evidence suggests that at present some places are more dangerous than others for molluscs: the faunas of freshwater habitats, for instance, show fewer signs of escalation that shallow marine ones, and within shallow marine environments, escalation is highest in the western Pacific and Indian Oceans, intermediate in the east Pacific, and lowest in the Atlantic.

We can look at Vermeij's fossil evidence for escalatory coevolution in both predatory and prey species. The evidence for predators is less convincing than for prey, but it is worth looking at both classes of evidence. They reveal what has to be shown to demonstrate escalatory evolution. Let us take the predators first. The earliest predatory molluscivores in the fossil record were whole body ingestors who simply swallowed molluscan prey whole; this is an unspecialized method of eating a mollusc and for that reason perhaps not a particularly menacing one. By the Devonian, specialized shell breakers have appeared—in the case of fish this can be inferred from their teeth—and by the late Devonian there were already 10 fish families of specialized shell breakers. Decapods, which crush shells in their claws, proliferated in the Triassic; gastropods that drill holes in other shelled species appear in the late Triassic; and starfishes, with their specialized suckers and pullers that enable them to prize bivalved shells apart, first appear in the Jurassic. By the late Mesozoic, many more shell breaking fish forms had appeared. This narrative is consistent with an increase in the number, and specialization, of molluscivores through time; however, it is not persuasive evidence for escalation. The total number of forms recorded as fossils has increased through time, and there are inevitably trends through time toward increased numbers of many things. It is not possible to show that life had become more dangerous for molluscan prey by showing that there are more predators later on in the fossil record. There may also be more non-molluscivorous predators and weakly-molluscivorous predators as well as more specialized molluscivores; to show escalation, we have to show the numbers of escalated predatory forms has *proportionally* increased. For the prey species, Vermeij has found some evidence of that kind.

The most obvious measure for escalation in prey defences would perhaps be shell thickness, but Vermeij's best evidence is for other characters. Consider first some evidence on shell repair. When a mollusc is non-lethally attacked, it repairs the damage to its shell and the repair pattern can be observed in the shell. Proportions of shells showing signs of repair have been measured in several fossil faunas, and the trend appears to be toward increasing amounts of repair over time (Figure 21.12). This Vermeij interprets as meaning that the prey have been suffering higher frequencies of predatory attacks over evolutionary time. (Logically it could also mean—though this is

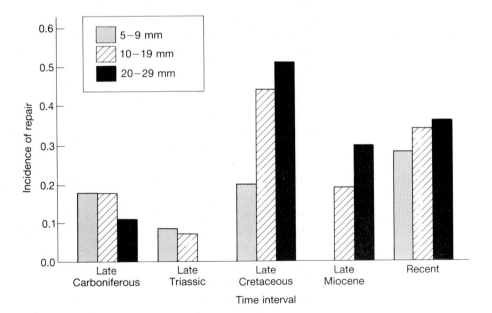

Figure 21.12 Incidence of repair at five successive time periods, for shells divided into three size classes. Note that the incidence of repair is higher in more recent times, and that larger shells show a relatively high incidence of repair compared with small shells in more recent times, as compared with earlier times. Large shells tend to be more resistant to breakage than small ones; one interpretation of this second trend is that more recent predators have become stronger, and therefore able to injure large-shelled animals. From Vermeij (1987).

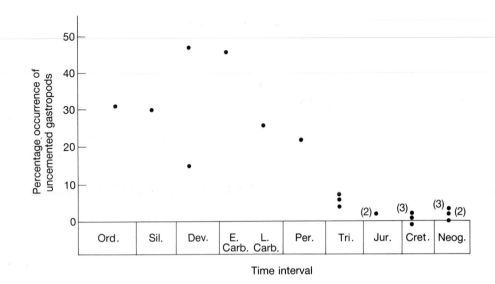

Figure 21.13 The incidence of sessile or sedentary uncemented gastropods through time. Note that the proportion decreases. Each point is for one fossil assemblage. For each assemblage, Vermeij divided the gastropods into different types (burrowers, sessile, attached forms, etc.), adding up to 100%. This graph gives the proportion of gastropods that lie unattached on the bottom surface. Neogene is the Miocene and the Pliocene. From Vermeij (1987).

perhaps unlikely—that the predators have de-escalated from forms that destroyed their prey to forms that sometimes merely injured them!) Vermeij's best evidence that molluscan prey defences have escalated comes from the proportion of different types of gastropod shells. Shells that lie loose, we have seen, are less well defended than burrowers and attached shells, and if we look at the proportions of the different types through time, we see that the loosely attached forms have decreased (Figure 21.13). Vermeij also found limited evidence that the better defended burrowers increased from comprising about 5–10% of genera in the Late Carboniferous–Late Triassic to about 37% in the Late Cretaceous and 62–75% in modern formations. Internal thickening or narrowing of the aperture is another form of escalated defence and these types, too, have proportionally increased through time (Figure 21.14). We can conclude that more recent molluscs are more strongly defended than were their earlier ancestors.

Vermeij's evidence would not be persuasive enough to convince a skeptic. The evidence is not abundant; it is noisy; the patterns are not all consistent; and it is difficult to prove that the fossil record is unbiased. For example, could there be some unidentified reason why shell forms with narrower apertures became relatively more likely to be preserved as time passed? Biases of this general sort (in which some forms are more likely to be preserved at some times than others) do exist in the fossil record. However, in the absence of any particular argument that the trends are due to sampling biases, we can give the evidence the benefit of the doubt. It does then suggest that some escalation of prey defences and predatory weapons has taken place in evolution.

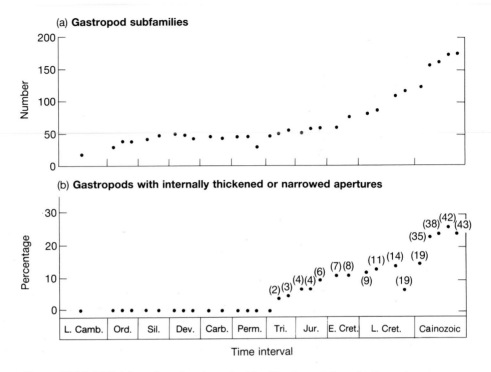

Figure 21.14 (a) Total number of gastropod subfamilies through time. (b) Proportion of subfamilies with members that have evolved internally thickened or narrowed apertures, a character that probably evolved as a defence against predators. Because the total number of subfamilies has increased, it is necessary (as here) to plot the *proportion*, not the total *number* of subfamilies, in order to show a trend. From Vermeij (1987).

21.4 Natural selection can explain trends like Cope's law

The ideas in sections 21.2 and 21.3 all explain evolutionary trends by natural selection within a population. In the case of a trend to larger size, larger individuals must be leaving more offspring at any one time in an evolving lineage (Figure 21.15). Either physical or biological factors could be at work. Animals tend to be larger nearer the poles than the Equator (see, for example, Figure 16.2, p. 411), for elementary reasons of thermoregulation: larger animals suffer less heat loss because there surface area is lower per unit of volume. Likewise, climatic cooling could select for larger body size. The biological factor most likely to select for increased body size is competition, particularly direct physical competition. Large size can also increase efficiency in predators, and in seed eating finches (section 9.1, p. 211), or enhance the defences of prey. These kinds of competition can operate within a species, or between a number of different species. Depending on the conditions, it can lead to a kind of escalatory arms race in which ever larger size is favored—up to a limit. These are no more than hypotheses; but they do illustrate how net directional selection within a population (perhaps interrupted by periods of relaxed, or reversed, selection pressure) could result in a long-term evolutionary trend. Generally, when faced with a long-term evolutionary trend, natural selection is a hypothesis to consider. Natural selection in the

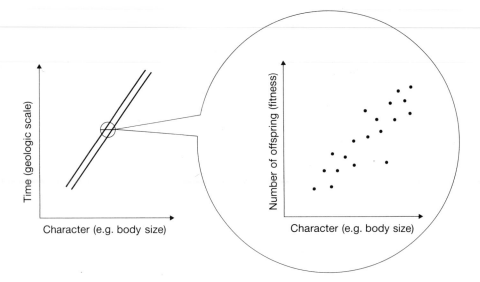

Figure 21.15 The general condition for a macroevolutionary trend to be driven by natural selection is that at all times during the trend, individuals that deviate from their population average in the direction of the trend should leave more offspring. (Whether the pattern of evolution is gradual or punctuated is irrelevant here.)

form of coevolution is particularly likely because it is open ended, which enables it to favor a consistent trend over a long time.

21.5 Species selection

21.5.1 Species selection takes place by differences in the extinction and survival rates of whole species

We shall now explore the theoretical possibility that evolutionary trends, instead of being caused by natural selection, are driven by the differential survival and proliferation (speciation) of whole species. We shall discuss the process in detail, but here is an abstract outline to begin with. In terms of Figure 21.3, species selection can only produce trends in the average state for a group of lineages: it cannot work in single lineages. Imagine, therefore, that there are a set of different species lineages, and that natural selection has established (and continues to favor) some adaptations in some lineages, and other adaptations in others. The key point is that the various adaptations may influence the chance of extinction, and of speciation, of the lineages they exist in. Some kinds of species will then have higher changes of survival, or of splitting, than others. The kinds of species with higher survival and speciation rates will become more numerous. The group of lineages will show a trend, toward the kind of character states in the proliferating species. The trend is driven by *species selection*. Modern thought on the subject has mainly been inspired by a paper by Eldredge and Gould in 1972. Stanley introduced the term "species selection."

Let us consider an evolutionary trend toward increasing body size as a hypothetical example. Species selection can generate the trend in two ways. One, discussed by Eldredge and Gould, works by differential extinction rates. If species whose members have larger body size become extinct at a lower rate than species with smaller-sized members, there will be a trend to an

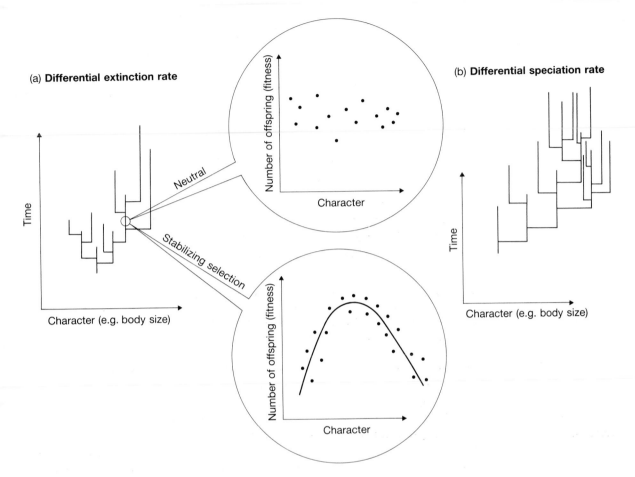

Figure 21.16 Species selection by differences among lineages in (a) extinction rate, and (b) speciation rate. There is a trend through time toward more species with large body size. In (a) species with larger body sizes have lower extinction rates (last longer) than species with smaller body size. Speciation is equally likely to produce a new species with a smaller or a larger body size than its ancestor; it also has a constant rate through time. In (b) species with larger body sizes have a higher speciation rate than species with smaller body size. Speciation is equally likely to produce a descendant species with smaller or larger body size than its ancestor. Each species has the same longevity (extinction rate). In both cases, natural selection within the lineage is not favoring individuals with larger body sizes. Compare with Figure 21.15. (Whether the pattern of evolution is gradual or punctuated is irrelevant here too.)

increased average body size for the set of lineages (Figure 21.16a). The other, emphasized by Vrba, works by the speciation rate. If species with smaller-sized members for some reason split at a lower rate, a trend towards increasing numbers of species with larger members will result and the average size of the set of species will increase (Figure 21.16b). The two kinds of species selection are analogous to natural selection working by survival and fecundity respectively.

The trends in Figure 21.16 are not produced by natural selection. Larger individuals in a population at any time do not have higher fitness; and the average size of each species is constant between successive generations. The role of natural selection is to produce a larger average body size in some species than in others. Body size is presumably an adaptation in each species and different average body sizes are being favored in each species' particular ecological niche. However, once the average body size of a species has been set by natural selection, that body size can have consequences for the rates of speciation and extinction. The species with larger body sized members might have higher speciation rates because the populations are more subdivided, or have lower extinction rates because of a greater durability of larger organisms in the face of environmental fluctuations. These are mere hypotheses and they may be right or wrong in particular cases: it could alternatively be, for example, that larger body sized species have higher extinction rates, not lower, perhaps because their population sizes are smaller. Species selection would then produce a trend to smaller size. We shall meet some more plausible hypotheses when we come to real examples, but the important abstract point is that the different ecological adaptations of different species may have different effects for speciation and extinction rates; when they do, species selection will tend to produce an evolutionary trend.

The conditions necessary for species selection to operate are as follows:

1 The character showing the trend, such as body size, is associated with the extinction rate, or speciation rate, or both. The association also has to be in the right direction: for example, if the trend is to increased size then species with larger-sized individuals must have lower, not higher, extinction rates.

2 The rates of speciation and extinction are uncoupled from what natural selection favors within a population.

3 The character shows heritability through speciation events. Species with larger-sized members must tend to produce new species with larger average

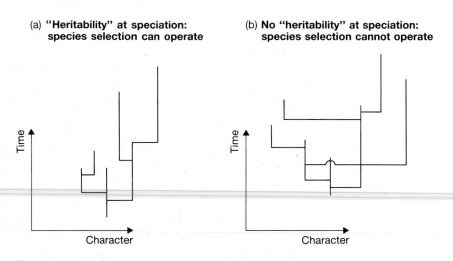

(a) **"Heritability" at speciation: species selection can operate**

(b) **No "heritability" at speciation: species selection cannot operate**

Time

Character

Time

Character

Figure 21.17 (a) Species selection requires that the attributes of ancestral and descendant species are similar: there has to be a species level analog of heritability in ordinary natural selection. (b) If descendants were uncorrelated with their ancestors, lower extinction rates or higher speciation rates of certain kinds of species would not result in a trend toward more of them.

body sizes than do species with smaller-sized members (Figure 21.17). This condition is surely valid, but is worth stating to complete the formal analogy with natural selection.

Natural selection requires there to be variation in fitness associated with a character, inheritance, and reproduction (section 4.2, p. 63): species selection requires the same properties at the species level. Condition 1 is analogous to variation in fitness (and condition 2 says it is due to properties of species, not individuals), condition 3 is analogous to inheritance, and the reproduction of species happens when its constituents reproduce themselves every generation. In so far as these conditions are satisfied, species selection will generate long-term evolutionary trends.

These conditions provide the means to test whether species selection is operating. We can assume that condition 3 is met. The practical research on species selection has mainly consisted of attempts to demonstrate condition 1. A proper demonstration of species selection should also satisfy condition 2; it should show that any association of a character with the speciation or extinction rate is not simply due to natural selection. This is an important point. We can see why it is important by analogy with a microevolutionary study of natural selection. Evolution within a population is produced by selection and mutation. Mutation is uncoupled from natural selection; it produces the variants among which selection chooses. If (in theory) mutation is directed, it can add to the short-term trends produced by natural selection, or even produce trends of its own; however, in practice this possibility can be ignored because we have good reason to think that mutation is undirected (section 4.6.2, p. 78). Short-term trends can reliably be attributed to selection.

With species selection, natural selection within each lineage is analogous to mutation; it produces the variants that species selection works upon. But in this case, we cannot ignore the possibility that natural selection produces short-term trends—because it does. Indeed, for a real trend, such as an increase in average body size, natural selection is a very plausible hypothesis. Similarly, in a major evolutionary transition such as that from reptiles to mammals (section 20.1, p. 533) it is unlikely that new species with less powerful jaws, or with less efficient homeothermy, were arising at the same rate as species with more powerful jaws and improved warm bloodedness; natural selection would have been working on these characters. Species selection only produces non-adaptive trends.

The strongest evidence that species selection produced a trend, in the average state of a character in a set of lineages, would show three things. It would show that speciation rates were higher, or extinction rates lower, in the lineages that deviated from the average in the direction of the trend; it would show that the trend, and pattern of apparent extinction and speciation rates, was not caused by a bias in the fossil record; and it would show that the trend, and pattern of extinction and speciation rates, was not being controlled by natural selection within a species. These are demanding conditions, and have yet to be met in any factual study of selection. However, the theory is interesting, and even imperfect evidence can be suggestive. Let us now turn to the evidence. (Box 21.1 describes some further conceptual properties of species selection.)

Box 21.1 Four conceptual points about species selection.

1 *Species selection is not only a theory of trends.* Just as natural selection can be stabilizing or disruptive, or change direction from generation to generation, so species selection does not always have to operate in a consistent direction. Species selection operates whenever there is a relation between the characters of species and their rates of proliferation and extinction. Relations of this kind may sometimes produce trends, but the process has broader application.

2 *Species selection does not depend on the theory of speciation by punctuated equilibrium* (see chapter 19). Eldredge and Gould originally discussed species selection as an extension of the theory of punctuated equilibrium, but all the assumptions of species selection can be met equally well, whether speciation is sudden or gradual. To see the point, simply imagine all the evolutionary lineages in a diagram of species selection (Figure 21.16) redrawn as smooth slopes rather than as right angles.

3 *Species selection is not an explanation for adaptation.* It is manifestly impotent as an adaptive force. Building up adaptations requires the cumulative action of selection over a long period: probably millions of gene substitutions were needed in the evolution of the eye, and each substitution requires selection between alternative genes over many generations. Natural selection adjusts gene frequencies over the time scale of individual reproduction and death, species selection adjusts species frequencies on the time scale of speciation and extinction events. The rate of evolution by species selection will tend to be slower than by natural selection by a factor of a million or so. If genes could only incrementally change in frequency when speciation happened, the eye might not yet have evolved. It would take too long. Natural selection, however, is capable of building up complex adaptations.

4 *Species selection should not be confused with the theory of group selection* (section 12.2.5, p. 310). Group selection was put forward as an explanation of adaptation; it proposed that adaptations exist for the good of groups, rather than individuals. It is generally rejected because group selection is so much less powerful than individual selection. Likewise, because species selection is millions of times slower than natural selection, natural selection would dominate evolution whenever the two

came into conflict. If the theory of species selection were making a claim like that of the group selectionists—i.e. that adaptations exist for the good of the species—it would encounter the same problem and be rejected for the same reason. But it makes no such claim. It makes only the more modest claim that if different adaptations, each fixed by natural selection for individual advantage, have different consequences for speciation and extinction rates, then that can affect the species composition of lineages through time. Species selection will not produce individuals that act against their individual interest.

Suppose, for example, that species selection were favoring species with larger body size, and the number of species with larger body-sized members were increasing through time. This would not mean that within a species small size was best for the individual, nor that the increase in numbers of large-sized species was resulting as individuals sacrificed themselves by growing an individually disadvantageous large size for the good of their species. It simply means that natural selection favors small size in some species, and large size in others, and for some reason the species with larger body-sized members speciate at a higher rate or have a lower extinction rate.

In chapter 12, gene, cell, organism, kin, and group selection were competing theories of adaptation, and it was argued that most adaptations benefit organisms. The theory of species selection accepts that natural selection generally produces adaptations that benefit individual organisms, and is only concerned with the consequences of different individual level adaptations for the rate of proliferation of lineages. Species selection, therefore, does not challenge natural selection as the explanation of adaptation. If the different attributes of different species have consequences for extinction and speciation rates, species selection will operate: if they do not, there will be no variation in fitness at the species level and species selection will be powerless. In fact many individual adaptations do affect things like breeding structure, mate choice, and population size, which probably influence speciation and extinction rates—species selection therefore probably does operate in the history of life.

21.5.2 Extinction and speciation rates differ between gastropod molluscs with planktotrophic and direct development

Different molluscs grow up in different ways. In gastropod snails, planktotrophic and direct development are two of the main types of development. With planktotrophic development, the egg is released into the surface waters of the ocean and develops into a planktotrophic larval form which feeds and disperses for a time before it settles and metamorphoses into the relatively sessile adult form. With direct development, the eggs and young grow up near, or inside, the parental snail. Various ecological trends are known among modern forms, such as that planktotrophic development is more common among shallow than deep water species, and among tropical than polar species.

The relationship between larval type and speciation and extinction rates

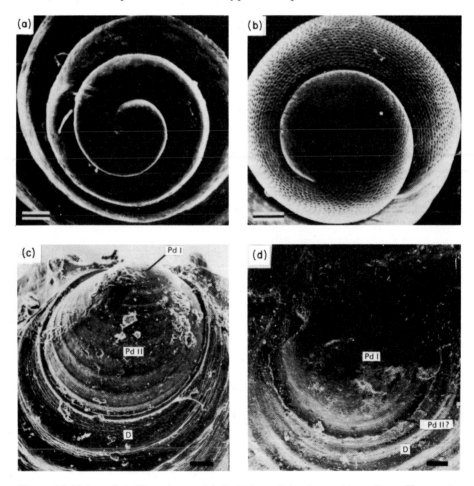

Figure 21.18 Larval shell form is correlated with type of development in molluscs. The species in (a) and (b) are modern gastropods, (c) and (d) are Late Cretaceous fossil bivalves. Note the relative sizes of the regions labeled Pd I and Pd II. (a) is *Rissoa guerina*, and known to be planktotrophic; (b) is *Barleeia rubra*, and is known not to be planktotrophic; (c) is *Uddenia texana* and has small Pd I and large Pd II like (a) and is inferred to have been planktotrophic; (d) is *Vetericardiella crenalirata*; it has large Pd I and small Pd II like (b) and is inferred to have been non-planktotrophic. Pd, prodissoconch. From Jablonski and Lutz (1983).

can be studied in fossil gastropods. Larval types in fossils are inferred by analogy with modern species. These kinds of inference were pioneered in the work of Thorson, and several criteria have now been used. Figure 21.18 shows one, which uses the size of regions in the larval shell. Modern species with planktotrophic development typically have small, yolk-poor eggs; in the larval shell, a region called prodissoconch I tends to be small and another region called prodissoconch II is larger: species with direct development have the reversed condition (Figure 21.18). These morphologic regions can be distinguished in fossil larval shells, by the scanning electron microscope, and it is a reasonable inference that shell form is correlated with development type in the same way as modern forms.

What is the relation between larval type and extinction rate? Several studies have found that species with planktotrophic larva have lower extinction rates (Figure 21.19). As the figure shows, they also have wider geographic ranges. This may be the reason for their lower extinction rate, because a species with a wider range is less vulnerable to local circumstances. Or it may mean merely that planktotrophic forms have a higher chance of being preserved as fossils than have directly developing forms, because their wider distribution increases the chance that in one site the conditions will be right for fossilization. The difference could then be just a bias in the fossil record.

The relationship with speciation rate is more interesting. Hansen predicted

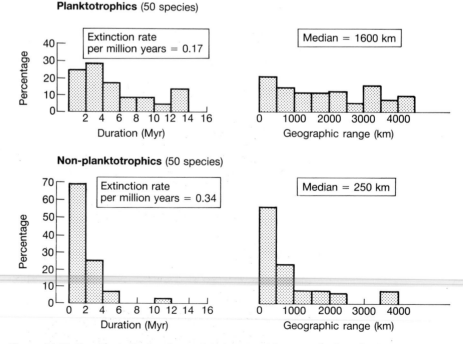

Figure 21.19 Duration in the fossil record and geographic ranges for Late Cretaceous gastropods from North America. Species with planktotrophic larvae last longer in the fossil record (i.e. they have lower extinction rates) and also have wider geographic ranges. Extinction rate is given as the chance that a species lineage will go extinct per million years. From Jablonski and Lutz (1983).

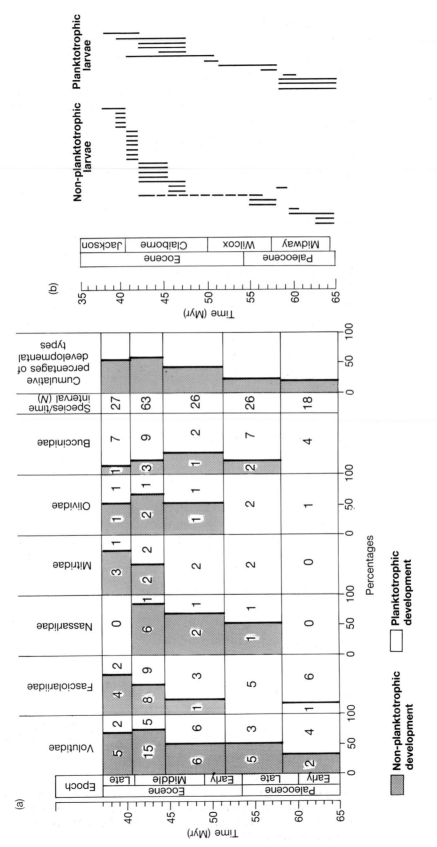

Figure 21.20 (a) The proportion of gastropod species with planktotrophic, as opposed to non-planktotrophic, development increases through the Paleocene and Eocene. The effect exists within several of the six families, and as an average for all six families. (b) Detailed observations for Volutidae, from the Gulf Coast of the USA (Volutidae are furthest left of the six families in (a)). Note the proliferation of non-planktotrophic species. The extinction rate of planktotrophic species appears to be lower—which counters the trend toward more non-planktotrophically developing species. Small discrepancies between the numbers in (a) and the number of lineages in (b) may be due to extra data in the later paper (a). From Hansen (1978, 1983).

that snails with direct development will speciate more rapidly than plankto-trophic forms; non-planktotrophic snails will be more likely to be geographi-cally localized and isolated, which makes allopatric speciation easier. He used this idea to explain an observed trend in snails of the early Tertiary (Figure 21.20). The proportion of planktotrophically developing species declined through the Paleocene and Eocene. Now, if pure species selection by extinction rates was in control, the trend ought to be the opposite (plankto-trophically developing species have lower extinction rates): that process, therefore, was not producing the trend.

Two alternatives are left. Natural selection could have been favoring direct development within the majority of lineages. Hansen "suggested" this was not true (though he gave no evidence). The period was a time of global cooling (see also section 22.6.3, p. 612), which might favor direct develop-ment, given the latitudinal trend mentioned earlier. Hansen says the decline in planktotrophically developing forms preceded the global cooling; con-crete evidence, however, rather than a vague statement would be needed to persuade a skeptic. The second alternative is species selection, working on the speciation rate. According to Hansen, the speciation rates in directly developing forms were higher. So the trend to an increase in directly developing species could be due to species selection.

The results published so far are not conclusive. However, Hansen's theory is consistent with what evidence there is. Species selection, by the speciation rate, may have caused the decline in planktotrophically developing forms in the early Tertiary. His work is particularly interesting because the relationship of developmental type with speciation rate is not open to any obvious alternative interpretation in terms of bias in the fossil record. It could still be due to some non-obvious bias, but the obvious bias would have produced the opposite trend from the one observed: planktotrophic forms are more likely to leave fossils (and so should show a higher speciation rate) than directly developing forms. Hansen's evidence is currently the best candidate for an evolutionary trend produced by species selection.

21.5.3 The taxonomic pattern of asexual reproduction suggests that it increases extinction rates

Species selection can also be tested with taxonomic evidence in modern species: the clearest example comes from sexual and asexual reproduction. It has been known since a paper by Weismann in 1886 that asexual reproduction has a peculiar taxonomic distribution. It is mainly confined to a number of scattered and small taxa: it is typically found in an odd species, or perhaps a whole genus, within a larger taxonomic group that mainly reproduces sexually. There are only two large taxa in all which members reproduce asexually: the bdelloid rotifers and an obscure group of gastrotrich invertebrates called the Chaetonoidea.

So the taxonomic pattern of asexuality appears "spindly" in the phylogenetic tree of life; only small twigs, not whole branches, are asexual. This fact is usually interpreted to mean that asexual reproduction increases the extinction rate. There are good reasons to think that asexual reproduction does increase the chance of extinction (sections 11.3.3 and 11.3.4, pp. 271–

275) and, if it does, a group that had lost sex would be likely to become extinct before it had lasted long enough to evolve into a higher taxonomic group. This would be an instance of species selection, because a character (asexual reproduction) is influencing the extinction rate; other things being equal, we should expect evolutionary trends, within lineages, toward increasing numbers of sexually reproducing species.

Williams noticed a problem in the argument. Suppose that taxa with asexual reproduction actually have the same probability of extinction as sexual taxa. The phylogenetic patterns of asexual and sexual reproduction should then be similar. However, that expectation is only the mean of a random process; there is some chance that one of the characters will have a more spindly distribution than the other. There is a lower, but still finite, chance that one will have an extremely spindly distribution. We cannot, therefore, prove that asexual reproduction increases the extinction rate simply by inspection of its taxonomic distribution; almost any taxonomic distribution could arise by chance. If we look at a large number of characters, all of which have the same chance of driving their bearers extinct, some will by chance have spindly distributions, others will by chance occupy a few phylogenetic main branches. Williams' question was, how do we know that asexual reproduction's spindly taxonomic distribution is not a preselected extreme from a random set of patterns?

Maynard Smith partly answered this question by comparing the taxonomic pattern of asexual reproduction with that of haplodiploidy. In species with haplodiploidy, females are produced by normal sex, but males are haploid and produced asexually; males contain half their mother's genome. Moreover, haplodiploidy is, like full asexual reproduction, relatively rare in nature. Haplodiploidy is therefore a comparable kind of reproduction to asexuality, and Maynard Smith used it as a point of comparison, to see how extreme the taxonomic pattern of asexual reproduction is. Haplodiploidy, it turns out, has a quite different pattern from asexuality. Haplodiploidy is found in whole higher groups, such as Hymenoptera and Thysanoptera, as well as most of the order of rotifers called the Monogonta. It is also found in a whole tribe of homopteran bugs, in several families of mites (Acari), and in smaller taxa, such as the beetle *Micromalthus debilis*. Maynard Smith argues that haplodiploidy is a normal character, which has no particular effect on the extinction rate. It is found in large groups, which of course incorporate many smaller groups. The comparison between the spindly pattern of asexuality and the normal pattern illustrated by haplodiploidy suggests that asexual reproduction really does increase the extinction rate.

From taxonomic evidence alone, we cannot generally distinguish whether a spindly distribution is caused by a high extinction rate or a low speciation rate or any mixture of the two. In the case of sex, we have a strong theoretical prediction that asexuality will increase the extinction rate, and this prediction can be tested taxonomically. We have no such theory for speciation rate. Indeed, the need for reproductive isolation places some limit on the speciation rate in sexual forms, and asexual groups might "speciate" (in a phenetic, or perhaps ecological, sense—see chapter 15) at a higher rate than equivalent sexual groups.

In summary, taxonomic evidence can be used to test the relation between a character and the extinction or the speciation rates of the species that possess it. In the case of reproductive mode, it can be predicted that sexual species should have lower extinction rates than asexual species. The evidence supports the prediction.

21.5.4 Species selection is an interesting idea but so far there is little evidence for it

Some, probably most, characters may influence the extinction and speciation rates of the species that possess them. Other things being equal, species with characters that lead to high speciation rates and low extinction rates should become more abundant over geologic time. In reality, however, other things will not be equal, and the influence of species selection may be reduced by natural selection acting within lineages in the short term. The importance of species selection is therefore an empirical question. So far, little work has been done and any conclusion would be premature. The work does show, however, what conditions a convincing study must meet. It is not enough to show that a character is associated with the speciation or the extinction rate: if it is, it could just be due to natural selection within each species. To demonstrate species selection, we must also show that the trend is not due to natural selection. In fossil evidence, it must also be shown that any trend is not simply due to the bias of the fossil record. In taxonomic evidence, it must be shown that any trend is not merely random. These conditions will have to be satisfied if we are to gain a rigorous understanding of how important species selection has been in evolution.

21.6 The whole history of life may show evolutionary trends on the grand scale

This chapter has been mainly concerned with two kinds of evolutionary trends, namely trends in the average for a character within a single lineage, and trends in the average for a group of related lineages; the initial examples of Cope's law illustrated both types of trend (Figures 21.1 and 21.2). Hansen's work on the proportion of gastropod species with planktotrophic and direct development is also concerned with a trend in a group of species. We can finish the chapter by looking at some trends in even larger groups of lineages: indeed, in the group made up of all the lineages in the whole history of life.

The history of life, from its origin to the present, probably shows a grand-scale analog of Cope's law. Life probably originated as a replicating molecular system. Since then the range of the sizes of living things has increased while the smallest size has remained about constant: there are still replicating molecules. As the range has expanded with a constant lower limit, the average size of a living thing must have increased through time. Even if we take as our starting point a single-celled prokaryote, there will still have been a trend to increased average size in the history of life. The trend in average size cannot be plotted exactly, because we do not know the average size of a living organism in the world today; one estimate, by Vogel, puts the average size somewhere between 1 mm and 1 cm. The estimate depends crucially on the abundance of very small bacteria, and in any case we cannot plot a graph of average size against time because no equivalent estimates have been made for other stages in the history of life. Bonner has instead plotted the size of

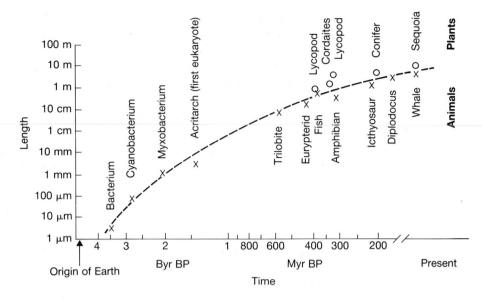

Figure 21.21 A graph of a rough estimate of the maximum size of living thing at different times in the history of life. Both axes are logarithmic. From Bonner (1988).

the largest known species against time, and the average size has probably mirrored the same trend, at a much lower scale (Figure 21.21).

Figure 21.21 is an example of an evolutionary trend on the largest possible scale. It is a trend in the state of all life through time. Bonner has also tried to quantify a number of similar macroevolutionary trends. Figure 21.22 suggests a trend towards increased complexity in living things. Organismic complexity is notoriously difficult to measure, and Bonner used the number of cell types in an organism as an approximate criterion. Rough estimates of the number of cell types in different sized modern species suggests that larger species have more cell types and it is a reasonable inference that complexity (by this measure) will have shown a comparable increase in the history of life to that shown by size. But whatever the quantitative pattern, it is almost necessarily true that complexity must have on average increased, because the first living things would have been at the simple extreme of the range of complexity seen in modern species.

What is the explanation for these very large scale trends? Maynard Smith has pointed out that the increase in size and complexity is almost inevitable, because life started small and simple and there was nowhere to go except up. Bonner favors a similar theory; he calls it "pioneering" (it is essentially the same as Darwin's principle of divergence (section 14.8, p. 377)). Because earlier living things were small and simple, species could always escape from competition by evolving to be larger and more complex. The ecological niche that could be occupied by organisms that were larger and more complex than any other organisms existing at the time was always vacant, and selection could have favored pioneers that evolved to occupy it. The net effect of the process will be for ever larger and more complex organisms to arise. The principle of divergence forces species to explore extreme niches. If Bonner is correct, biological competition, and the divergence it produces, has been the

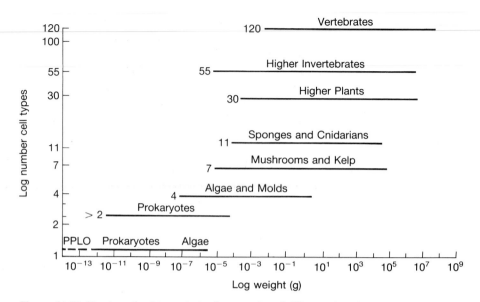

Figure 21.22 Number of cell types in modern species of different sizes. Larger individuals tend to be more complex, as measured by number of cell types. PPLO is a mycoplasma. From Bonner (1988).

main motor of the pattern of increasing size and complexity—of progress, some would say—in the history of life.

21.7 Summary

1 The fossil record reveals many examples of evolutionary trends, in which a lineage undergoes directional evolution over a long period.

2 Trends can be seen within a single lineage, or as the average of several lineages within a taxon, or as the average for the whole of life.

3 Trends could in theory arise if advantageous mutations appeared only rarely, but this is unlikely and untestable; alternatively trends could be driven by competition within a species, if selection constantly favors forms that deviate from the average; or they could be driven by changes in the physical environment.

4 Coevolution is a plausible general theory of evolutionary trends. Coevolution occurs when two or more lineages reciprocally influence each other's evolution. It may or may not generate trends according to the nature of the reciprocal influences.

5 Coadaptation between species, such as in any example of mutualism, is probably but not necessarily the result of coevolution.

6 Insects and plants influence each other's evolution, by the evolution of insecticides in plants and of detoxification and avoidance mechanisms in insects.

7 Speciation in parasites and hosts is likely to be simultaneous, if the parasites have limited independent powers of dispersal. The phylogeny of the two form mirror images.

8 Coevolutionary arms races between predators and prey produce escalatory long-term evolutionary trends; they can be seen in the evolution of brain size in mammals and of armor and weapons in molluscs and their predators.

9 If natural selection favors one form of a character in one species and another form in another, and if the different forms of the character cause species to have differing speciation or extinction rates, then there may be a trend toward an increase in the kind of species with higher speciation, or lower extinction rate. The process is called species selection.

10 Associations between characters and extinction or speciation rates have been studied in fossils, and in taxonomic evidence for modern species. In Tertiary snails, species with planktonic development have lower speciation rates than do species with direct development; this may have caused a trend to an increasing proportion of directly developing forms.

11 Trends in the whole history of life can be measured by working out the average state for a character such as body size for all the species alive at a series of times. Body size, and complexity, have probably increased from the origin of life to the present.

21.8 Further reading

The volume edited by McNamara (1990) is all about evolutionary trends in fossils. Simpson (1953) is a classic treatment of the subject; see also Simpson (1949) and Stanley (1979). For trends in horse evolution, see MacFadden (1987, 1988b), and Hallam (1975) for the Jurassic molluscs. The volume edited by Hallam (1977) discusses many examples of macroevolutionary trends and transitions, and includes a chapter by Raup (1977) on random models. See Cronin and Schneider (1990) for the effect of climate.

On coevolution in general, see Futuyma and Slatkin (1983), Nitecki (1983), Kim (1986), Howe and Westley (1988), and Thompson (1989). Ehrlich and Raven (1964) is the classic paper on plant–insect coevolution; Mitter *et al.* (1991) and Pellmyr (1992) are recent modern accounts. The lycaenid butterfly example is the work of Pierce and Mead (1981). See Pierce (1987) on this symbiosis generally. See Janzen (1980) on the inference of coevolution from coadaptation. For parallel phylogenies, see Brooks and McLennan (1991). Dawkins and Krebs (1979) discuss arms race coevolution. Vermeij (1987) is the important work on evolutionary escalation. See Jerison (1973) on brain evolution; Gould (1977b, chapter 23) popularizes Jerison's work. See McMahon and Bonner (1983) for the way selection acts on body size.

On species selection, the original exposition was by Eldredge and Gould (1972). Stanley (1975) coined the term. For the paleontological example, see Hansen (1983), and Vrba (1987, 1989) for further possibilities: for the taxonomic example, see Williams (1975) and Maynard Smith (1978); Reaka and Manning (1987) give another example, from modern stomatopods, which supports Hansen's fossil result. Weismann's (1886) paper was translated in volume 1 of Weismann (1891–1892). Lauder (1981, 1990) describes another method for studying species selection. For further general discussion, see (in addition): Williams (1966, 1992), Gould and Eldredge (1977), Dawkins (1982), Damuth (1985), Eldredge (1985a), Gilinsky (1986; and the reply following it), the exchange between Maynard Smith (1987c, 1988) and Eldredge and Gould (1988), and Hoffman (1989).

On grand-scale trends, see Bonner (1988), Pringle (1951), and Maynard Smith (1972, pp. 92–99). Gould *et al.* (1987) and Kitchell & MacLeod (1988) discuss directionality in the history of life.

Extinction and mass extinction

22.1 Modern extinctions can be studied by observation, but the fossil record can be used to study general ideas about the causes of extinction

Extinction is a relatively recent human discovery. It dates from the late eighteenth and early nineteenth centuries. Fossils had been known about long before then; but when a fossil was found that differed from any known species, it could still have been alive in some unexplored region of the globe. As the global flora and fauna became better and better known through the eighteenth century, it became increasingly likely that some fossil forms were no longer alive; by the end of the century, several naturalists accepted that some marine invertebrate groups, such as the ammonites, were extinct. However, the best known taxa—the vertebrates, and mammals in particular—posed a special problem. Fossil bones are preserved as isolated, disarticulated fragments, and it is even more difficult to show that a single bone does not belong to any modern species, than it is for a complete specimen (like a shell). The decisive work is usually credited to Cuvier. Cuvier reconstructed, with new standards of rigor, whole skeletons from bone fragments; and it is easier to see whether the whole skeleton, rather than just the disarticulated bones, of a vertebrate belongs to any living species. The most convincing cases of extinction were for gigantic forms like mastodons: it was hardly plausible that the explorers would have overlooked them.

Why do species become extinct? Some modern extinctions have been witnessed closely enough for the cause to be known with certainty. The enormous Steller's sea cow was discovered by a shipwrecked German naturalist called Georg Steller in 1742, but he was the only naturalist ever to see it alive. The animals were completely tame—Steller records how he could stroke them—and by 1769 they had been hunted to extinction. The extinction of modern species by analogous human means is all too easy to observe now. Local extinctions can be studied ecologically, in the wild and by experiment. For example, in a famous set of experiments, Huffaker cultured together the predatory mite *Typhlodromus occidentalis* and (as prey) the mite *Eotetranychus sexmaculatus*, in the laboratory. Under simple conditions, the predatory mite predictably ate its prey to extinction, and then became extinct itself. But in a more varied environment, including refuges in which the predator could not easily catch its prey, prey and predator could persist together for longer (Figure 22.1). The experiment thus reveals the influence of environmental complexity on the chance of extinction in a predator–prey system. Many possible causes of extinction, biological or physical, can be investigated in the laboratory in this manner.

For fossils, we do not have any evidence as direct as sailors shooting sea cows, though for recent extinctions the inference can be quite strong. The

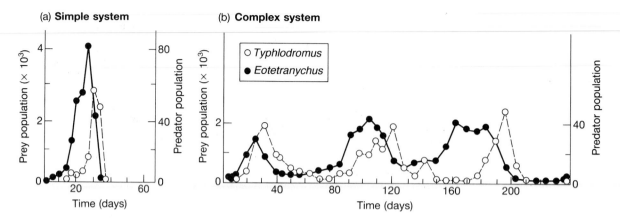

Figure 22.1 Population cycles of the predatory mite *Typhlodromus occidentalis* and its prey *Eotetranychus sexmaculatus*. (a) Under simple conditions, the predator eats its prey to extinction, and then crashes to extinction itself. (b) Under more complex conditions, with environmental refuges, the predator and prey populations show continual cycles. From Huffaker (1958).

most recent Ice Age, for example, which was at its peak about 18 000 years ago, almost certainly caused many local extinctions. If a species disappears before the advancing ice cap and does not return, there is little doubt what the cause of the extinction was. The tulip tree and hemlock are only two of the species lost from the European flora at that time, though both survived in North America. However, as we move further back in time, the uncertainty mounts rapidly. We saw, in the well-studied case of the Great American Interchange, how uncertain the evidence is about the cause of the many extinctions; and the Interchange took place only 2 million years ago and has left a good fossil record (section 18.6, p. 494).

The difficulty of studying the cause of individual extinctions in the fossil record does not mean that fossils cannot be used to study extinction. Modern species can be used to test particular ecological theories about individual extinctions, and fossils for investigating general theories about the pattern of extinctions. We shall look at two kinds of general theory: biological, by competitive coevolution, and physical. Biological extinctions can happen, for example, when predators, or parasites, drive their prey (or hosts) extinct, when superior competitors eliminate inferior competitors, or when a species' food supply disappears. We shall concentrate on these factors in the first half of the chapter, and move on in the second half to mass extinctions, which are more likely to have had physical causes. We shall consider there such factors as astronomical and geologic catastrophes that wipe out most of the species in a whole region, and changes in sea level and climate. The division of the chapter into two halves is not meant to suggest that evolution alternates between mass extinctions, dominated by physical factors, and periods of biological coevolution in between. Indeed, we shall see at the end that there is no clear distinction between mass extinctions and in-between phases: extinction rates vary continuously through time. All the extinction factors,

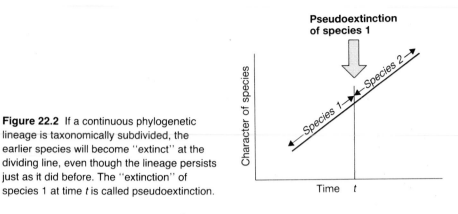

Figure 22.2 If a continuous phylogenetic lineage is taxonomically subdivided, the earlier species will become "extinct" at the dividing line, even though the lineage persists just as it did before. The "extinction" of species 1 at time *t* is called pseudoextinction.

therefore, have probably operated side by side, though with varying frequencies, through most of the history of life.

The extinction of a species means that its members have ceased to reproduce. They have died out and left no descendants. For modern species, the meaning is unambiguous, but for fossils true extinction has to be distinguished from a kind of taxonomic artifact called *pseudoextinction* (Figure 22.2). As a lineage evolves, later forms may look sufficiently different from earlier ones that a taxonomist who only has a few specimens will classify them as different species, even though there is a continuous breeding lineage. We have discussed before (section 15.3, p. 399) the purely taxonomic question of whether lineages that undergo evolutionary change ought to be split into more than one species. Both viewpoints have their supporters, but either way there can be no doubt that this kind of taxonomic extinction is conceptually different from the literal death of a reproducing lineage. For the study of extinction, pseudoextinction is a nuisance. There are no interesting theoretical questions about it, and students of extinction have to keep an eye open for it, in order to be able to exclude it. All the theories of extinction that we shall discuss in this chapter are concerned with true extinction: the extinction of a lineage.

22.2 The probability that a species will become extinct is approximately independent of its length of existence

We met taxonomic survivorship curves in chapter 19 as a means of studying evolutionary rates (section 19.10, p. 528). Here we shall use them to demonstrate a property of extinction rates. To plot a taxonomic survivorship curve, take a higher taxonomic group, such as a family or order, measure the duration in the fossil record of its member species, and plot the number (or per cent) of species that survive for each duration (Figure 19.14, p. 529). In 1973, Van Valen published a large study, based on measurements of the durations of 24 000 taxa; he plotted the graphs after taking logarithms of the numbers surviving and found that survivorship is generally linear on the log scale, or log linear (Figure 22.3). It has been questioned just how linear his results really are, and many of them are not for species, but for genera or even families within a higher taxon; however, there is a strong enough suggestion of log linearity in taxonomic survivorship curves for us to ask what it means.

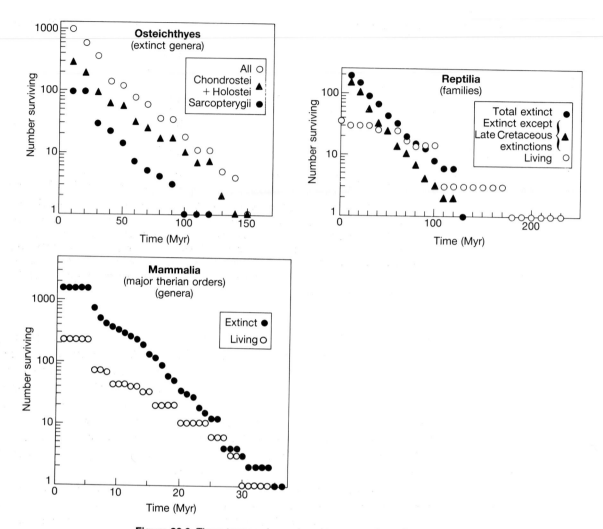

Figure 22.3 Three taxonomic survivorship curves: they plot the number of genera (for Mammalia and Osteichthyes) or families (for Reptilia) surviving for various durations in the fossil record. Note that the lines, with this logarithmic *y*-axis, are approximately straight. Osteichthyes are bony fish and the two groups drawn on that graph are subgroups of bony fish. From Van Valen (1973).

The answer is that it means species do not evolve to become any better (or worse) at avoiding extinction. The straight line on the logarithmic survivorship curve means that the chance a species will become extinct is independent of its "age." Of the species that survived to a time *t* million years after their origin, a certain proportion become extinct by time *t* + 1 million years; then, of the survivors to time *t* + 1, the same proportion become extinct by time *t* + 2, and so on. There is an exponential decay of species, with a constant proportion of the survivors becoming extinct in the next age unit. It could have been otherwise. If, for example, evolution is progressive, the probability of extinction might decrease with time, as the level of adaptation improves;

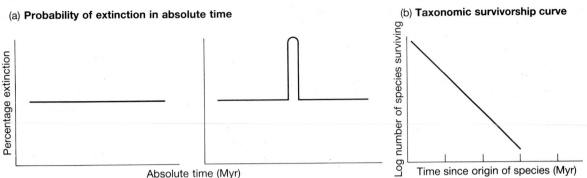

Figure 22.4 A linear survivorship curve need not imply that the probability of species extinction is constant through the lifetime of the higher taxon. The graphs in (a) and (b) have crucially different *x*-axes. (a) Extinction probabilities in absolute time: constant and non-constant. (These correspond to interpretations (1) and (2) respectively in the text.) Both will produce (b) a log linear taxonomic survivorship curve, provided the chance of extinction at any time is independent of the age of the taxon. The *x*-axis in (b) has the origin of every species at time 0 and plots and numbers surviving for various durations.

later species would then last longer. Alternatively, if evolution is escalatory (section 21.3.6, p. 572), the level of individual adaptation under normal conditions relative to competitors may not change in any particular direction, but as the species come to invest in heavier armaments it might become more vulnerable to environmental stresses and extinction rate might increase with age. Or, as a purely theoretical possibility, every species could have a fixed lifespan and go extinct a certain time *t* after its origin. The importance of Van Valen's result is that it suggests that no such process leading to an increase or decrease in the chance of extinction generally operates.

We can distinguish two extreme interpretations of the log linear survivorship curves, though in any real taxonomic group both could be operating in any mixture (Figure 22.4). Interpretation (1) is that each species has a constant chance of becoming extinct each year. We can call this constant extinction. Alternatively, (2) the chance a species becomes extinct could vary between years, but be the same for all species, independently of the length of their existence. The two interpretations can be illustrated by analogy with a human actuarial survivorship curve. A log linear survivorship curve means that the chance of dying at any age is a constant: the chance of dying at age 20 years equals the chance of dying at age 21 years, 22 years, and so on. The two interpretations are then as follows: (1) each individual has a constant chance of dying each year, a chance *d* in 1993, the same chance *d* in 1994, in 1995, and so on (we also assume there is no "baby boom" effect—people are born at random times); or (2) the chance of dying varies between years. It might be very low, perhaps 0.001, every year through the 1990s, but then shoot up to maybe 0.8 in the year 2000, if there is a plague or some other disaster. If the disaster hits young and old alike, the survivorship curve will again be log linear. To get a non-linear curve, the chance of death must vary with age. For example, if the plague kills more old people than young, the survivorship would slope downwards. A log linear survivorship in a human population would mean that people do not become better (or worse) at avoiding death as they grow older: a log linear taxonomic survivorship curve means that species do not become better or worse at avoiding extinction as they persist through time. Extinction factors—whatever they are—hit species independently of the length of their existence.

22.3 Competitive coevolution can have various forms

What can we deduce from the taxonomic survivorship curves about the factors that cause extinction? A first division of extinction factors is into physical and biological. Physical factors could be catastrophic or more gradual in action. A geologic catastrophe might, for instance, hit an area once every million years and wipe out a certain proportion of the species. Or it could be that there is a steadier elimination of species by a more constant physical, perhaps climatic, change. There is no puzzle about what pattern of extinction rates results in either case: in the first case they are low with sudden increases at the times of the catastrophes; in the second they are intermediate in level and less variable. Either could lead to log linear survivorship curves on interpretation (2) above.

The biological factor most likely to be causing extinction is competitive coevolution (section 21.3, p. 564). As ecological competitors, or parasites and hosts, evolve against each other, if one competitor fails to evolve an adaptive improvement to keep up with its antagonists, it may become extinct. If a host species evolves a new kind of immunity, then if the parasite does not soon evolve a way of penetrating the defence, it will become extinct. What is the pattern of extinction rates likely to be for this process?

Stenseth and Maynard Smith have discussed the question theoretically in terms of what they called *lag load*. The lag load has the standard form for loads (section 7.3, p. 148). We imagine that at any time there is an optimum state that the members of a species could have; a genotype with that optimum has fitness w_{opt}; it is defined such that $w_{opt} = 1$. But the genotype that is needed to have that optimum fitness keeps changing, because of the changes in competing species. Each species will evolve towards its optimum (i.e. to have all its members with the optimum genotype); but will lag a certain distance behind it. The actual mean fitness of the population is $\bar{w}$, a number usually less than one; $\bar{w} = 1$ if all members of the population should have the optimum genotype. The lag load L is then defined as:

$$L = \frac{w_{opt} - \bar{w}}{w_{opt}}$$

As the lag load L increases, the rate of evolution of the species will increase, because it is subject to stronger selection pressure. The lag load also controls the chance that the species will become extinct. As L increases, the species lags further and further behind its competitors and its chance of extinction goes up. As L decreases toward zero, its chance of extinction also decreases.

Competitive coevolution, in this model, can have four forms. The first two are unstable. If a species lags behind its competitors and does not evolve fast enough to keep up with them, it will fall further behind until it becomes extinct; this is called the contractionary mode of evolution. Secondly, a species could be ahead of its competitors and out-evolve them; it could theoretically expand until it had an infinite number of descendants; this is called the expansionary mode. Clearly, the expansionary mode could not continue for long. In real systems, the main possibility is that a species could alternate between periods of expansionary and contractionary evolution as the fortunes of natural selection favored one species and then another among a group of competitors. The third evolutionary mode is called stationary: in this the competing species evolve to a set of optimal states and then stay there; the lag load reduces to zero, no species changes, and no species becomes extinct. In practice, it would mean that biological coevolution would always take the species to a stable equilibrium, from which it could be perturbed by physical events; after a perturbation, it would evolve back to a stationary equilibrium again. At the stationary equilibrium, coevolution is not driving species extinct; any extinctions would be due to physical processes. This may be a common kind of coevolution. If there is one optimum form for an organism to have in order to compete with members of other species, its species will evolve to it and stay there.

The fourth result is also an equilibrium, but it is dynamic. It is the *Red Queen* equilibrium. Instead of evolving to an optimal state and then staying there, this result arises when there are always new adaptive possibilities opening up, and the species continually evolves toward them (host–parasite coevolution, section 11.3.5, p. 275, could be an example). The species all have constant positive values for their lag loads, and the system therefore neither expands nor contracts. Van Valen originally suggested the idea to explain the log linear survivorship curves he had documented. His name for it (the Red Queen hypothesis) alludes to the Red Queen's remark in Lewis Carroll's *Alice Through the Looking Glass*: "here, you see, it takes all the running you can do, to keep in the same place." The analogy for running is coevolutionary change. In the Red Queen mode of coevolution, natural selection continually operates on each species to keep up with improvements made by competing species; each species' environment deteriorates as its competitors evolve new adaptations. This deterioration is the cause of extinctions in the model. On average, a group of competing species have balanced levels of adaptation, and they all lag behind their best possible states. At any one time, one species may experience some random run of bad reproductive luck, and become extinct. Competitive coevolution will result in a log linear survivorship curve if the rate of environmental deterioration is roughly constant through time: if the species' competitive environments deteriorate in fits and starts, the survivorship curve will be non-linear.

Why should the rate of environmental deterioration be approximately constant? Van Valen reasoned that it would follow from the zero-sum nature of competitive ecological interactions. The total resources available, he thought, will stay approximately constant. If one species adaptively improves, it will temporarily at least be able to take more of the resources; its population will expand. This increase will be experienced by its competitors as an equivalent decrease in the resources available to them. The selection pressure on them to improve will increase, by an amount proportional to the loss in resources caused by the competitor's improvement. They will then tend to improve their competitive abilities, and make up the ground lost to the competitor. Then, on average, the lag load of each species will be constant. The justification may be correct, but it is debatable. Resource levels, for example, have probably changed through evolutionary time and competitors may not always compete for a constant sized pie. If resource levels increase, Van Valen's argument might predict that extinction rates would decrease, and the change in the resource level would cause a change in the extinction rate. The point here is only that Van Valen's argument is not theoretically watertight. However, it is plausible and may even identify the general mode of coevolution.

The value of Stenseth and Maynard Smith's analysis is mainly theoretical (though it should ultimately help in testing too). By considering all the possible coevolutionary results we have seen that the Red Queen equilibrium is by no means inevitable when species are coevolving. Indeed, a constant lag load is a "knife edge" result, and it could easily be tripped into an expansionary or contractionary mode by the fortunes of natural selection; and those three are not the only possibilities, for coevolution may simply proceed to stationary standstill. The Red Queen equilibrium requires an exact balance between species such that the improvements of one are exactly balanced by the aggregate improvements of its competitors. If the balance is inexact, the equilibrium breaks down.

In summary, a set of competing species that coevolve with one another can generate four kinds of pattern of extinction: expansionary, contractionary, stationary, and the Red Queen equilibrium. Van Valen suggested the Red Queen result would follow from the zero-sum nature of a competition for constant resources. He invoked it to explain the log linearity of taxonomic survivorship curves.

22.4 The Red Queen hypothesis can be tested by plotting extinction rates against absolute time

So much for the Red Queen in theory: how can we test it? Although Van Valen put forward the hypothesis to explain the log linearity of survivorship curves, it is not the only explanation for the result. The Red·Queen is a type (1) interpretation: it explains the log linear survivorship curve by a constant chance of extinction through time. At the Red Queen coevolutionary equilibrium, each species has about the same chance of becoming extinct every year, and the ones that do are replaced by the same number of new species; the system remains in a dynamic balance. However, a linear survivorship curve can also arise if extinction rates are not constant (interpretation (2)). Therefore, the shape of survivorship curves is at best a poor test of the Red Queen hypothesis.

The possibility of testing the Red Queen hypothesis depends on whether we adopt a "strong" or a "weak" version of the hypothesis. According to the strong version, most extinctions are accounted for by the actions of the Red Queen; the Red Queen then not only contributes to the shape of the survivorship curves but is the full reason why they are log linear. This strong version is testable, as we shall see. The weak version states only that the Red Queen is the normal mode of biological coevolution and accounts for most extinctions caused by biological factors, but that in reality many other extinctions are caused by physical factors. Then the log linear survivorship curves could be due to some mixture of the Red Queen theory, producing constant extinction probabilities, and physical factors, producing fluctuating extinction probabilities. The weak version of the Red Queen hypothesis is a reasonable claim about the real world, but is practically untestable. (Moreover, it dethrones the Red Queen from its original explanatory position. It is no longer much of an explanation for the log linear survivorship curves, because if physical factors are allowed in they can explain the curves.)

The strong version can be tested by plotting the extinction probability for members of a taxon against absolute time (Figure 22.4). Absolute time means that the x-axis has dates on it, not time since the origin of the species. Then we can see whether the extinction probability is constant, as the Red Queen theory implies, or wanders up and down in the manner of a type (2) interpretation of the log linear survivorship curves. If the graph is not flat, the Red Queen is wrong. If it is flat, then logically we have not proved the Red Queen to be right, but we have good evidence for it. The Red Queen is much the most plausible known process that produces constant extinction probabilities within a taxon. If extinctions are mainly caused by some non-Red Queen mechanism, such as physical factors, it is implausible that each species' environment would deteriorate at a constant rate relative to the species adaptive improvements. Changes in the physical environment are more likely to generate a wobbly extinction probability through time. Likewise, coevolution in any of three modes other than the Red Queen in Stenseth and Maynard Smith's model would not give a constant extinction probability through time: it is therefore a strong prediction of the Red Queen.

If, however, we adopt the weak interpretation and say that the Red Queen process interacts with physical factors to produce the observed pattern of extinctions, then we cannot make a testable prediction about extinction probabilities in absolute time. The weak theory would fit any pattern of extinction probabilities. The only hope for a test would be if we could recognize which extinctions were caused by physical factors and then exclude them to test whether the residue of biological extinctions had a flat probability through time. But at present we cannot do that.

Let us now turn to the evidence. What does it suggest about the strong version of the Red Queen? Van Valen plotted some graphs of extinction probability against absolute time in his 1973 work, and they were quite variable; but he followed it up with a larger study in 1985, and concluded that extinction probabilities fitted the Red Queen's prediction of constancy surprisingly well. (Constant here does not mean fixed; it means the average chance of extinction for members of a taxon is probabilistically constant—it may wander up or down at random in the short term but has the same average over the long term.) We can look at the results of two even larger studies, though neither was carried out for the purpose to which it is being put here (Figure 22.5). Gilinsky and Bambach examined the extinction probabilities through time for 99 taxa; 14% deviated from a flat line; that is, there was no net slope up or down. They did not test for scatter in the short term (reasonably enough, because it is statistically far from clear how to do so and they were mainly interested in whether the lines were flat or showed a net slope up or down). They illustrated (Figure 22.5a) the result for one of the taxa; it shows a flat line with one outlier, which is probably a mass extinction,

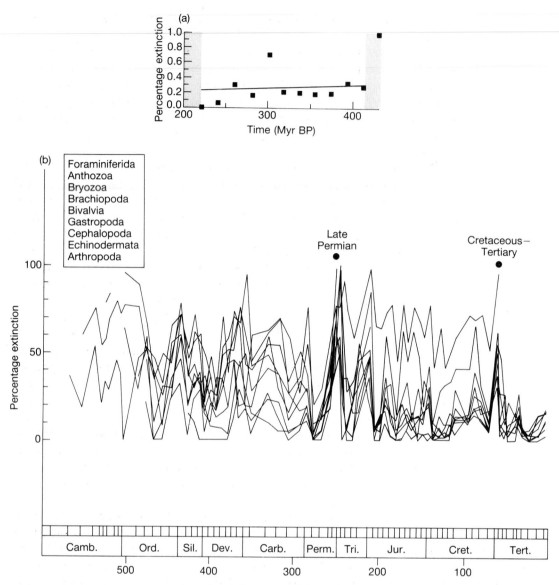

Figure 22.5 (a) Proportions of Rugosa (a group of corals), becoming extinct through geologic time. The first and last time interval (the stippled area of the figure) are excluded from the statistical analysis of the slope because the proportion of species becoming extinct in the final geologic interval of any taxon has to be one, and is likely to be low, even zero, in the first interval. The in-between points can test whether the probability of extinction is constant. (b) Nine invertebrate taxa show cyclic changes in probability of extinction through time. From Gilinsky and Bambach (1987) and Raup and Boyajian (1988).

but does not reveal the extent of scatter around the line. The other study, based on 20 000 genera, by Raup and Boyajian, shows how variable the chances of extinction can be (Figure 22.5b). This result suggests that the strong version of the Red Queen is not well supported. If Raup and Boyajian's data had been aggregated into 20 million year data points, and the scatter left off, they would look as flat as Figure 22.5a. The Red Queen, in its strong version, does not merely predict that data averaged into 20 million year points will show no net trend up or down through 400 million years; it predicts near constancy in the evolutionary history of small groups of competing species—over a time period of a few million years. Raup and Boyajian's study is not an ideal test either, because the data are aggregated into large taxonomic groups, whereas the Red Queen applies to groups of coevolving species; but it is the best evidence available and is not supportive. The fluctuations in extinction probability may have physical or biological causes, but they appear to be there. However, these results will not be the final word. Every year, new evidence from the fossil record comes in, and there are many statistical methods of analyzing it (we shall see below how the cyclic pattern seen in Figure 22.5b has been criticized in other, analogous studies). The properly cautious conclusion, therefore, is that the evidence does not support the hypothesis that extinction probabilities are constant in the history of a taxon: it would be premature to say the evidence rules it out.

In summary, the log linearity of survivorship curves is a well-documented result. It implies that species do not grow better, or worse, at avoiding extinction as they persist in time; old species have the same chance of extinction as young ones. Van Valen suggested the log linearity of survivorship curves was caused by competitive coevolution between species, in the Red Queen mode: species have a probabilistically constant chance of becoming extinct each year because of ecological competition. The Red Queen may be the normal mode of competitive coevolution, though this is uncertain because evolution could also be contractionary, expansionary, or stationary. To test whether the chance of extinction is constant, they have to be plotted for a taxon in absolute time. Much less work has been done than on survivorship curves; but the initial results, taken as a whole, do not suggest that the chance of extinction is constant through the history of a taxon.

22.5 At certain moments in the history of life, there have been mass extinctions of many species

The geologists of the nineteenth century who worked out the main eras of the Earth's history did so by looking for characteristic fossil faunas that lasted for a noticeable time (or rather, depth) in the sediments. Different characteristic faunas were recognized as different time periods. They recognized three large scale faunal types and named them Paleozoic, Mesozoic, and Cainozoic, and there were shorter-term characteristic faunas within each of the three. Two major faunal transitions divide the three eras: the Permo-Triassic boundary, between the Paleozoic and Mesozoic, and Cretaceous–Tertiary boundary, between the Mesozoic and Cainozoic. In the nineteenth century, it was not known whether these two transitions corresponded to long gaps in the sedimentary record, or to the massive and rapid extinction of one fauna and its replacement by another. We now know, from radioisotope dating, that the second interpretation is correct. At the end of the Permian, about 225 million years ago, and the end of the Cretaceous, 65 million years ago, there were *mass extinctions*.

And they were massive. Raup has estimated that at the end of the Permian, as many as 96% of marine species became extinct. The Cretaceous extinction was less severe, but still perhaps 60–75% of marine species became extinct.

These mass extinctions pose immediate questions of description and cause. How long did they last? Were they global, or local? Did they affect all groups equally? We shall begin with the Cretaceous–Tertiary mass extinction, which has been most thoroughly studied; but it will soon become apparent that the Cretaceous extinction may be but one instance of a more general phenomenon, and its explanation should therefore be considered in a broader context.

22.5.1 The best-studied mass extinction occurred at the Cretaceous–Tertiary boundary

The mass extinction at the end of the Cretaceous has been found in all regions of the globe, and affected more-or-less every group of plants and animals (Figure 22.6). The fossil record of small, abundant, microfossil groups such as Foraminifera provides the best evidence for the fine-scale pattern of the extinction, but the demise of larger groups provides its drama. Some groups, such as the dinosaurs and ammonites, were finally driven extinct; most large groups were drastically reduced in diversity, though some odd groups, such as crocodiles, were not noticeably affected at all. The obvious question is: why did it happen?

In 1980, Alvarez *et al.* published an influential observation. They sampled the rocks of the Cretaceous–Tertiary border from Gubbio in Italy, and found exceptionally large concentrations of rare earth elements, particularly iridium (Figure 22.7). These elements also have high concentrations in extraterrestrial objects; and Alvarez and his colleagues explained the biological mass extinction, and the geochemical *iridium anomaly*, by the collision of a large asteroid with the Earth. Since then, similar iridium anomalies have been found in Cretaceous–Tertiary boundary rocks at several other sites. Some geologists have argued that the iridium anomaly could have had a terrestrial cause, by volcanic eruptions; but asteroids are the most widely accepted explanation.

The exact means by which such an impact could have precipitated the mass extinction has been considered in detail by Alvarez, and other authors. Alvarez *et al.* originally suggested that the impact would have thrown up a global dust cloud, which would have blocked out sunlight for several years until it settled again. When Krakatoa erupted in 1883, it ejected an estimated $18 \, km^3$ of matter into the atmosphere; and this took 2.5 years to fall down again. From the iridium concentrations at Gubbio, and in known asteroids, Alvarez *et al.* estimated the size of their hypothesized asteroid at a diameter of about 10 km. Such an asteroid, whose kinetic energy they described as "approximately equivalent to that of 10^8 megatons of TNT," would have produced an explosion about 1000 times as large as the eruption of Krakatoa.

The loss of sunlight alone would have been enough to cause the extinctions, but the impact could have had other destructive side-effects too. Global warming, acid rain, extreme volcanism, and perhaps an associated global fire, are some of the possibilities. We need not enter into the geologic and climatic ideas here; it is enough to note that an impact on the scale suggested by Alvarez *et al.* would surely have been capable of causing the mass extinction at the end of the Cretaceous.

The main evidence for the impact remains the iridium anomaly. More recently, other kinds of geologic evidence have been added. Zhao and Bada found certain amino acids (racemic isovaline and α-amino-isobutyric acid) that are said to be characteristic of asteroids rather than the Earth, in anomalously high concentrations in Cretaceous–Tertiary rocks from Stevens Klint (Denmark). There might also be direct physical evidence of the impact. Rocks, such as shocked tektites and quartzes, which are suggestive of a high velocity collision, have been found from several Cretaceous–Tertiary sites; but the most convincing evidence would be the impact crater itself. It has not been found. A 10 km wide asteroid should have produced a crater of perhaps 100–200 km in diameter, and something that big would be difficult to miss. Maybe it is at the bottom of an ocean; maybe it has been tectonically subducted; maybe the extinctions were caused by two medium-sized asteroids or a shower of smaller asteroids rather than one large one. About 120 impact craters are known from around the globe today; most of them are too small to have been produced by one asteroid of the size calculated by Alvarez *et al*. None has been convincingly dated to the Cretaceous–Tertiary boundary, though further chronologic work may conform that one 35 km impact structure in Iowa, called the Manson structure, does date from this time.

The fossil record can also test Alvarez *et al*.'s theory. If the theory is correct, the extinctions at the Cretaceous–Tertiary border should have been sudden, concentrated in a short interval of time, and not preceded by any decline through the Cretaceous; they should be synchronous in different taxa and geographic localities; and they should coincide with the iridium anomaly. This is a highly testable and stimulating set of predictions.

There are some problems in the evidence. You might think you could simply peer into the fossil record and observe whether extinctions were sudden or gradual, synchronous or spread out in time. In reality it is not so easy. How, for instance, do we observe the exact time of an extinction? The last appearance of a species in the fossil record will usually precede its final, true extinction (and a species certainly cannot appear *after* its true extinction). The species' population may decline before it finally disappears, which would reduce the chance of leaving fossils; and even if the population is constant, its chance of fossilization will still be much less than one. Species will therefore appear to go extinct in the fossil record before they actually did, and this "push backwards" is greater for forms that are less likely to leave fossils. The tendency has been called the Signor–Lipps effect.

The Signor–Lipps effect is only one difficulty. It can also be difficult to correlate events at different geographic localities, because absolute dates are often unavailable. The incompleteness of the fossil record also introduces uncertainty: a species may appear to become extinct suddenly at what is really a gap in the sedimentary record (see Figure 22.6a for a controversial example). Or, on the other hand, a sudden extinction can be smoothed out by the secondary reworking of the fossils—fossils may be moved up and down after their original deposition. For all these reasons, evidence from the fossil record is controversial when used to show either sudden or gradual, or synchronous or asynchronous, extinction patterns.

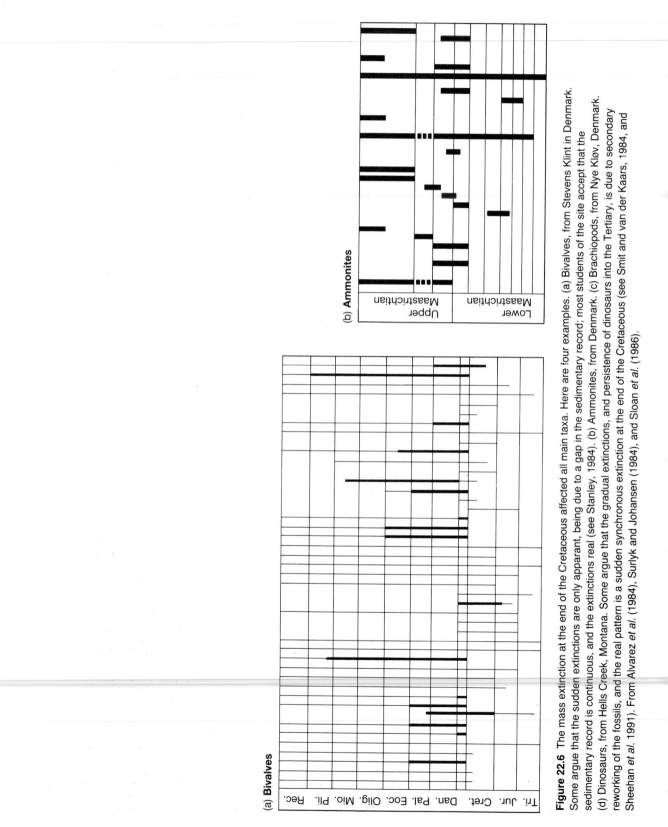

Figure 22.6 The mass extinction at the end of the Cretaceous affected all main taxa. Here are four examples. (a) Bivalves, from Stevens Klint in Denmark. Some argue that the sudden extinctions are only apparant, being due to a gap in the sedimentary record; most students of the site accept that the sedimentary record is continuous, and the extinctions real (see Stanley, 1984). (b) Ammonites, from Nye Kløv, Denmark. (c) Brachiopods, from Denmark. (d) Dinosaurs, from Hells Creek, Montana. Some argue that the gradual extinctions, and persistence of dinosaurs into the Tertiary, is due to secondary reworking of the fossils, and the real pattern is a sudden synchronous extinction at the end of the Cretaceous (see Smit and van der Kaars, 1984, and Sheehan *et al.* 1991). From Alvarez *et al.* (1984), Surlyk and Johansen (1984), and Sloan *et al.* (1986).

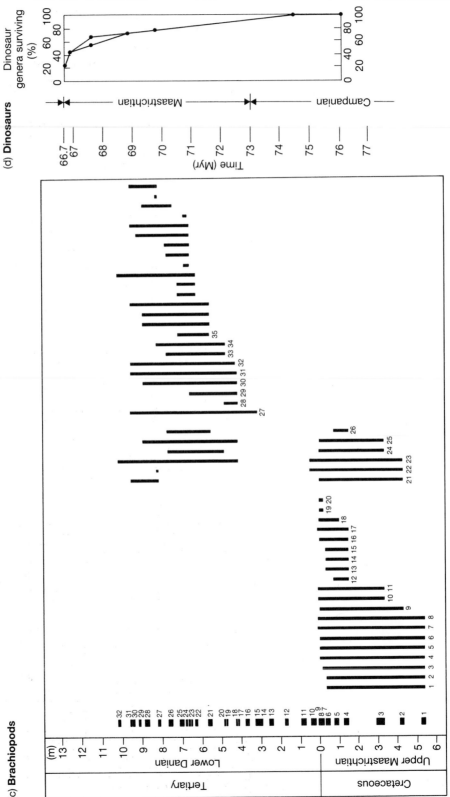

Figure 22.6 *(continued)*

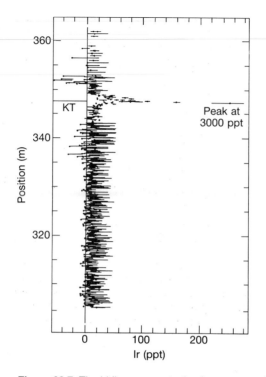

Figure 22.7 The iridium concentration increases suddenly by two to three orders of magnitude at the Cretaceous–Tertiary (marked KT) boundary rocks at Gubbio, Italy. From Alvarez *et al.* (1990).

So there are problems; but that does not mean the evidence cannot be used (Figure 22.6). There is an increasing amount of evidence that is consistent with a sudden and synchronous mass extinction. Supporters of Alvarez *et al.*'s theory accept the evidence; critics distrust it for the reasons we have just considered. The evidence is not all consistent. Figure 22.8 shows the results of a detailed test for synchroneity. The Cretaceous–Tertiary mass extinction can be observed in small planktotrophic fossils from all over the globe, and we can ask whether they all happened at exactly the same time. Radioisotope dating is too uncertain to test whether events 65 million years ago happened at the same time or as much as a few million years apart. Officer and Drake, however, made use of magnetic time zones (see Appendix) and found that the Cretaceous–Tertiary transition take place in different time zones in different places (Figure 22.8): the events may not have been as synchronous as the catastrophic Alvarez *et al.* hypothesis suggests. Officer and Drake estimate, for example, that the dinosaur extinction at the San Juan Basin in New Mexico dates about 400 000 years after the iridium spike at Gubbio. Even this evidence is not definitive, however, because it can be argued that the asynchrony is due to secondary reworking of the fossil material, such that it has been shifted from its original rocks to later strata.

Alvarez *et al.*'s asteroid theory also predicts a different duration of the Cretaceous–Tertiary extinction from a less catastrophic theory. If caused by an asteroidal impact, the mass extinction should have been almost

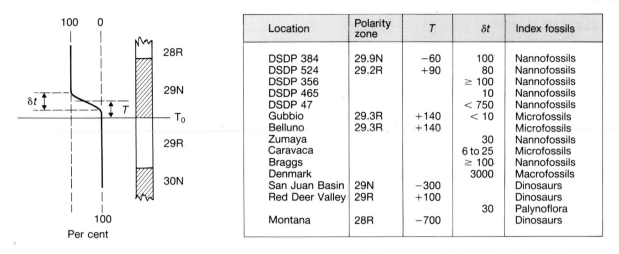

Location	Polarity zone	T	δt	Index fossils
DSDP 384	29.9N	−60	100	Nannofossils
DSDP 524	29.2R	+90	80	Nannofossils
DSDP 356			≥ 100	Nannofossils
DSDP 465			10	Nannofossils
DSDP 47			< 750	Nannofossils
Gubbio	29.3R	+140	< 10	Microfossils
Belluno	29.3R	+140		Microfossils
Zumaya			30	Nannofossils
Caravaca			6 to 25	Microfossils
Braggs			≥ 100	Nannofossils
Denmark			3000	Macrofossils
San Juan Basin	29N	−300		Dinosaurs
Red Deer Valley	29R	+100		Dinosaurs
			30	Palynoflora
Montana	28R	−700		Dinosaurs

Figure 22.8 Officer and Drake's summary of Cretaceous–Tertiary transition times (T, in thousands of years) and transition time intervals (δt, in thousands of years) for various sites with Cretaceous–Tertiary boundary rocks. T is expressed relative to the time of the switchover from magnetic time interval 29N to 29R, which is variously dated at 64.03, 64.09, or 66.56 million years before the present. Microfossils and nannofossils are fossils of very small planktotrophic organisms. From Officer and Drake (1983).

instantaneous, and the effect would have worked through to all ecosystems within a few years at most. All the extinctions should have been completed in an interval below the minimum resolution of the fossil record, which would be much more than 100 years for such ancient events. Almost any biological or non-catastrophic physical theory will predict the extinctions should be more spread out over time. The temporal spread of the extinctions, and indeed the iridium spikes, therefore provides another test. Alvarez and his supporters duly argue that the fossil record suggests such an instantaneous extinction pattern (see Figure 22.6). On the other hand, Officer and Drake (1985) estimated that the iridium spikes themselves were spread out over a time interval of at least 10 000–100 000 years; but their estimates crucially depend on the degree of reworking by bioturbation (animals, like worms, burrowing up and down in the sediments and moving sediments with them) and are uncertain.

Let us summarize the argument so far. The evidence for an iridium anomaly at the Cretaceous–Tertiary boundary is generally accepted. Alvarez *et al.* attributed it to a global collision with an extraterrestrial meteor, and they have many supporters. However, the iridium anomaly has also been explained by terrestrial processes such as volcanic eruptions. Either way, the iridium anomaly was probably caused by the same process as caused the mass extinction. This section has also been concerned with the way fossil evidence can be used to test between theoretical causes of the mass extinction. Agreement can hardly be expected in so rapidly moving and controversial a subject, but the different authors do agree about what should be looked for in the evidence. The crucial matters are the time courses of extinctions, and physical geologic changes, within each site and between different sites. There is good evidence for both suddenness and synchroneity of extinctions; but it is not complete enough to persuade everyone. We shall reconsider the cause of the Cretaceous mass extinction again below; but we have now gone far enough with this event in isolation. The Cretaceous mass extinction is not the only mass extinction in the fossil record, and the other ones may throw some light on possible general causes.

22.5.2 Mass extinctions may occur at regular 26 million year intervals in the history of life

Sepkoski has compiled, from the published literature, the distributions in the fossil record of all the families of marine organisms. Sepkoski's compilation is not the first of its kind, but it is the most up to date and is now the most widely used. The data set is expanding all the time, and in 1984, he had evidence for 3500 families. He, with Raup, used it to examine the grand pattern of extinctions through the past 250 million years. The data set has more and less inclusive versions according to which relatively poor parts are excluded; for the initial study the 3500 families were reduced to a sample of 567 families—still a large number. The striking result was that the probability of extinction appeared to vary cyclically, with a wavelength of about 26 million years (Figure 22.9a). The initial study suggested as many as 12 peaks of extinction; as Figure 22.9 shows, some peaks are clearer than others. Raup and Sepkoski have repeated their analysis from time to time, as new data have come in. An analysis in 1986 with 2160 families also showed a cyclical extinction pattern (Figure 22.9b) as did a sample of 20 000 genera, broken down into taxonomic groups in 1988 (Figure 22.5b, discussed above).

Raup and Sepkoski's factual claim has proved highly controversial. The time scale they used, the taxonomic data, and the statistical methods, are all open to criticism. The absolute time of the geologic periods is not known with certainty at this fine level, and more than one chronology is widely used. Raup and Sepkoski demonstrated periodicity using the "Harland" and "Odin" time scales and found more strongly significant results with the former. Hoffman re-examined their evidence using several more time scales and showed that the periodicity depends on the time scale, which is worrying—we should have more confidence in the result if it did not depend on any particular time scale. A statistical analysis by Stigler and Wagner also suggested the Harland time scale is particularly likely to generate cycles with a wavelength in the 25–30 million year range.

The fossil evidence has been criticized too. Boucot, an expert on brachiopods, estimated that as many as 75% of the "extinctions" recorded by Sepkoski were pseudoextinctions; they happened merely because taxonomists had subdivided continuously changing lineages. A related criticism has been made by Patterson and Smith, who are experts on fossil fish and echinoderms, respectively. They looked at the families of these two groups in Sepkoski's data and estimated that only 25% of them were monophyletic, the remainder being paraphyletic (see Figure 14.7, p. 366). Moreover, the periodicity seemed to be concentrated in the paraphyletic families; the subsample of monophyletic families did not show periodicity. Neither Boucot's nor Patterson and Smith's

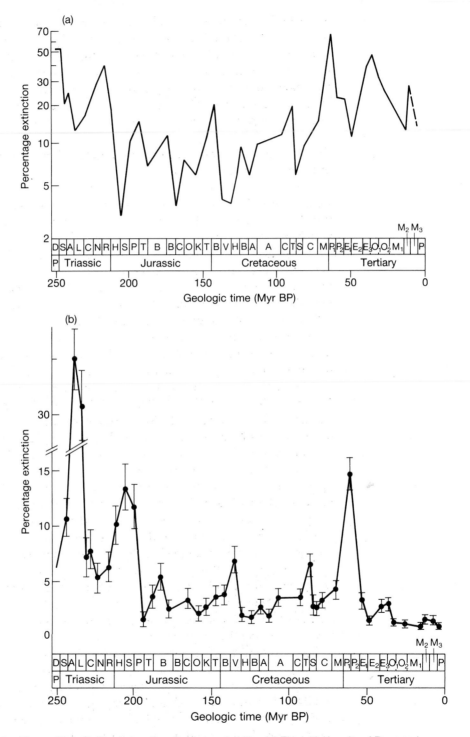

Figure 22.9 Cyclic change in extinction probability. (a) The initial results of Raup and Sepkoski in 1984 for 567 marine families. (b) 1986 result for 2160 families. From Raup and Sepkoski (1984, 1986).

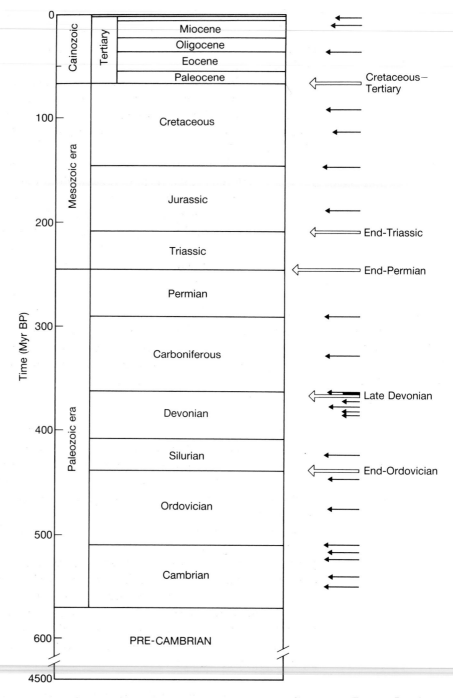

Figure 22.10 Five mass extinctions are particularly clear: Cretaceous, Triassic, Permian, Devonian, and Ordovician. Arrows represent the intensity of extinctions. From Raup (1991).

result invalidates Raup and Sepkoski's claim, but they do raise the suspicion that the periodicity is due to some artifact of phenetic taxonomy rather than a real extinction pattern.

We can draw two conclusions. One is that there may be a strongly cyclic pattern in the probability of extinction. If there is, something peculiar, and previously unrecognized, must have been influencing the history of life on the grand scale. However, the evidence for cycles is plagued by geologic, statistical, and taxonomic problems and is not an established result. Therefore, when we consider below the causes of mass extinctions, we should have cyclicity in mind as a possibility. If a theory of mass extinctions can account for a cyclic pattern, it may turn out to be a virtue; but the evidence for cycles is not strong enough presently to rule out a theory that lacks the property.

We can pause here to notice another pattern in Figure 22.9b. The average probability of extinction has gradually decreased from the past to the present. There are two possible explanations. First (and this is the more likely), that the fossil record becomes more complete nearer the present; in recent time a family is therefore less likely to become extinct merely because it is not preserved in the fossil record. Alternatively, however, the decline could be a sign of evolutionary progress; the more recent taxa could have lower chances of extinction.

22.6 Several theories have been suggested to explain why mass extinctions happen

22.6.1 It is difficult to test the theories because the facts are not settled

We can proceed in either (or both) of two ways. One is to take Raup and Sepkoski's result at face value—we then provisionally accept that there is a 26 million year cycle of mass extinctions, and look for an explanation. The other is to be more skeptical and doubt whether the cyclic pattern is well-enough documented to merit thinking about. However, even those who prefer the skeptical position cannot doubt that there are *some* mass extinctions. About five are widely accepted (Figure 22.10): three of them are the tallest peaks in Raup and Sepkoski's time series (Cretaceous, Triassic, and Permian) and two others precede the Permian (Devonian and Ordovician). Thus even if cyclicity is not accepted, we need an explanation for a recurrent phenomenon. Mass extinctions, we have seen, may either be regular and cyclic, or rare and erratic features in the history of life.

Which position we take on the time distribution of mass extinctions will influence the kind of explanation we look for. Physical processes, in particular, are more likely to have 26 million year periodicities than are biological processes. Some large-scale biological convulsion could perhaps have caused one, or a few scattered, mass extinctions; but it is difficult to imagine biological cycles on the time scales of 26 million years that could generate mass extinctions. Here, we shall consider several of the more important theories; some are more plausible for cyclic, others for erratic, patterns of mass extinction.

22.6.2 The evidence does not suggest that mass extinctions other than the Cretaceous–Tertiary event are driven by periodic asteroid impacts

Three kinds of evidence for an asteroidal collision with the Earth have been offered: the geochemical signature (iridium anomaly), the characteristic signs of an impact (shocked quartz), and impact craters. Only the Cretaceous–Tertiary boundary has been extensively investigated in the first two ways. Some iridium measurements have been made for other mass extinctions

Table 22.1 Iridium anomalies at times other than the Cretaceous–Tertiary boundary. Most times were for mass extinctions, though one may not be. Magnitudes for iridium spikes are expressed relative to the average for the type of rock before and afterward. Compare Figure 22.7 for the iridium anomaly at the end of the Cretaceous. It is doubted whether any of the iridium anomalies in the table have extraterrestrial causes. From Orth (1989) who gives references for all the sources, except Holser *et al.* (1989)

Period	Extinction	Iridium
Eocene–Oligocene	Mass	Small spike
Eocene–Oligocene	Mass	Small spike
Eocene–Oligocene	Mass	Small spike
Permo-Triassic	Mass	Spike
Permo-Triassic	Mass	Normal
Permo-Triassic	Mass	10 times spike
Late Devonian	Mass	3–7 times spike
Late Devonian	Mass	Small spike
Ordovician–Silurian	Mass	Normal
Ordovician–Silurian	Mass	Normal
Precambrian–Cambrian	Mass?	Small spike
Precambrian–Cambrian	Normal?	Small spike

(Table 22.1): some have revealed small increases; but most do not. Small increases, of about one order of magnitude, may be better explained by terrestrial processes that concentrate iridium, rather than by an asteroid collision. The iridium spike at Gubbio is much larger, by three to four orders of magnitude (see Figure 22.7). There is no good evidence for an extraterrestrial collision for any mass extinction except the one at the end of the Cretaceous. The largest mass extinction, at the end of the Permian, has no striking iridium anomaly and almost certainly had a different cause from the Cretaceous–Tertiary event.

The fossil record provides further reason to doubt that the 26 million year pattern (if it is real) is driven by periodic collisions with large asteroids. The record for the Cretaceous–Tertiary event, we saw (see Figure 21.6), is consistent with a global and synchronous extinction. Other mass extinctions have been less thoroughly studied, but there are strong hints that some do not show the same pattern. Hallam, for example, analyzed two of Raup and Sepkoski's mass extinctions from the Jurassic, and found that they were regional, rather than global, in extent. The Plienisbachian extinction took place in North, but not in South, American bivalves; and the extinction at the Jurassic–Cretaceous boundary can be seen clearly in samples from all over Europe, but not from Argentina and Chile. In both cases, small and local causes were probably operating.

22.6.3 Mass extinctions may have been caused by changes of climate and sea level

Some mass extinctions have coincided with times of climatic change, especially cooling; and climatic change is the sort of physical factor that might be expected to cause extinctions. However, the evidence for the association is imperfect. The five principal mass extinctions illustrate the point. The mass extinction in the late Ordovician coincides with a relatively sudden cooling. The one in the late Devonian coincides with a sudden cooling according to some authors, but not according to others. For the Permo-Triassic extinction, some authors have found an association with a bout of cooling, or with salinity changes in the ocean that would probably indicate a change in climate. No evidence of cooling has been found associated with the late Triassic extinction; and the Cretaceous–Tertiary extinction has been argued both ways. We should consider not only whether extinctions are associated with cooling, but also whether cooling is associated with extinction. There are periods of cooling, such as a bout of glaciation at the end of the Carboniferous, without an exceptional number of extinctions. The theory therefore is not supported by the evidence; it convincingly fits only one of five mass extinctions and makes erroneous predictions elsewhere.

Even if climatic change is not the general cause of mass extinctions, it may operate during smaller extinction events. This has been suggested for some more recent extinctions. A round of extinctions in the late Eocene 36 million years ago coincides with a time of climatic cooling, when average

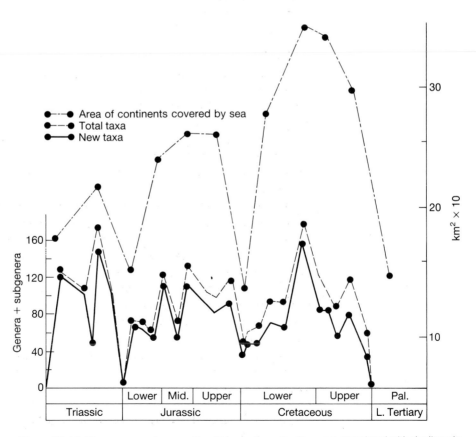

Figure 22.11 Three successive rounds of ammonite extinction were correlated with declines in the sea level (expressed here as area of continents covered by sea: when the sea is higher, it covers more of continental land). From Hallam (1983).

temperatures may have decreased by about 4°C. And Stanley, who has particularly argued that climatic cooling has caused extinctions, has documented an extinction event 2.4–3 million years ago in molluscs that took place when the climate cooled down. The hypothesis, however, still needs to be tested systematically with all the evidence; most of the evidence consists of pointing to cases in which the climatic change and extinction happen to be associated. Further evidence for the influence of climate on extinctions comes from the selective pattern of extinctions. In some extinction events, species that are more tolerant of cool climates survived better than did warm-adapted species (see section 22.8.1 below).

Asteroidal collisions would cause large short-term climatic changes, as would intense global volcanic eruptions. The climatic theory is not necessarily an alternative to these two ideas. Climatic change may also be connected with another theory of mass extinction: changes in sea level. The times of lower sea level (called marine regressions) in the Mesozoic appear to have coincided with times of mass extinction (Figure 22.11). The correlation is better than for the climatic theory, but again there are regressions that do not coincide with mass extinctions. In recent times, particularly in the Pleistocene, fluctuations of sea level have also not coincided with extinctions, though it is possible to argue round the exceptions. Changes in sea level should be taken into account when studying extinctions.

In summary, there is evidence that climatic and sea level changes are associated with extinctions. Both factors could in principle cause extinctions. They are likely to be connected because of the way the Earth's oceans and climate interact with each other. However, the evidence for an association with mass extinctions is unpersuasive. Proper statistical testing is needed, to see whether the majority of the evidence supports the idea, or whether a few confirmatory instances have been selected from a wide range of data.

22.6.4 Extinctions may be connected with changes in the shape of continents

Valentine and Moores have suggested that changing land and shore areas, due to plate tectonic wandering, should be associated with global species diversity and extinction rates. The number of species in a locality is usually proportional to its geographic area, for good ecological reasons, and a reduction in area should therefore produce extinctions. The best fit with the pattern of mass extinctions is at the end of the Permian. If, as has been suggested, the continents coalesced at the end of the Permian to produce the supercontinent Pangaea (Figure 18.5, p. 486), the length of continental perimeter would have been reduced for simple geometric reasons. Many fossils are marine invertebrates that inhabited the shoreline habitats. As the area of their habitat decreased, the diversity of these species would have decreased too: hence, in Valentine and Moores' theory, the Permian mass extinction. The idea is difficult to test because of uncertainty about the exact continental positions at times so distant in the past; the theory therefore remains poorly supported by evidence. Changes in habitat area are less clear for other mass extinctions; but the theory should operate generally. As continents wander, habitat areas will change, and species diversity can be expected to alter accordingly. Thus, Valentine and Moores have identified a potentially general influence on extinction rates, and its most plausible application so far is to the extinction at the end of the Permian.

22.7 Summary for mass extinctions

Mass extinctions are real events in the history of life. The exact number of mass extinctions that have occurred since the Cambrian is probably somewhere between the five clear events and the 23 or so implied by a 26 million year cycle and a 600 million year span; but whatever the number is, we are dealing with a recurrent phenomenon. Mass extinctions are more likely to have a physical than a biological cause. For none of the mass extinctions is there universal agreement as to what the cause was. The best studied mass extinction happened at the Cretaceous–Tertiary boundary, and for this there is a growing body of evidence that it was caused by a collision of the Earth with an asteroid. This idea has clear implications for the pattern of fossil extinctions, and those implications are the subject of active research.

The causes of other mass extinctions are under research too, but they are even less certain. If the 26 million year cycle of mass extinctions is not an artifact, then a good theory of mass extinctions should be able to account for a regular, and not just a recurrent and erratic, event; but that is not a pressing demand with the present quality of the evidence for cycles. Several geologic processes, including volcanism, changes in sea level, climate, and continental geometry, may contribute to mass extinctions. Table 22.2 gives a recent summary of the explanations for various mass extinctions that were favored at an expert symposium in 1989. The summary is not definitive, however! This subject moves fast, and it is not appropriate to hold firm opinions.

Other theories have been put forward from time to time; and the large number of hypotheses for the extinction of the dinosaurs is notorious. Many are frivolous, and the points to be made about them are too obvious to need to be printed. A good hypothesis for the extinction of the dinosaurs must be general enough to explain the whole mass extinction at the end of the Cretaceous, because it is implausible that one process was driving the

Table 22.2 Mass extinctions and their possible causes, suggested by various authorities. Simplified from Donovan (1989)

Eocene to Oligocene	Stepwise extinction associated with severe cooling, glaciation, and changes of oceanographic circulation, driven by the development of the circum-Antarctic current
End Cretaceous	Bolide impact producing catastrophic environmental disturbance
Late Triassic	Possibly related to increased rainfall with implied regression
End Permian	Gradual reduction in diversity produced by sustained period of refrigeration, associated with widespread regression and reduction in area of warm, shallow seas
Late Devonian	Global cooling associated with (causing?) widespread anoxia of epeiric seas
Late Ordovician	Controlled by the growth and decay of the Gondwanan ice sheet following a sustained period of environmental stability associated with high sea level

dinosaurs extinct while other processes were independently disposing of the many other groups that suffered extinction at exactly the same time. Any causal hypothesis needs to be tuned to the generality of the facts it is meant to explain, and thus further work on the fossil record for mass extinctions should continue to influence thinking about the cause, or causes, or these extraordinary catastrophes.

22.8 Some kinds of species may survive mass extinctions better than others

The traumatic environmental circumstances—whatever they were—of mass extinctions must have imposed exceptional selective forces, in kind as well as degree. Living things cannot become adapted, by natural selection, to conditions that recur on the time scale of at least 26 million years, and we should not expect organisms to possess adaptations for survival in mass extinctions. However, it is likely that some taxa will have attributes enabling them to survive better than others. Species will probably differ in their chances of surviving mass extinctions, just as they do for extinction generally. In chapter 21, we met the idea under the title of species selection. The work we shall discuss here is for the special case of species selection in the exceptional conditions of mass extinctions.

22.8.1 Species with different temperature adaptations may have survived mass extinction differentially

Stanley, as we have seen, has argued that mass extinctions are caused by climatic cooling. A major part of his evidence is the selective pattern of extinctions at those times. In the late Eocene extinction, for instance, warm water molluscan species suffered more extinctions than polar species. Likewise, many foraminiferid groups became extinct, but the globigerines (a cool water group) survived and became the dominant Oligocene foraminifers. There is other anecdotal evidence that tropical species suffer more in mass extinctions; but it is always important to check that a few confirmatory instances have not been selected from evidence that points in all directions. The evidence also needs to be corrected for the greater species diversity and the narrower geographic range of individual species at the tropics. Tropical species will have a higher total extinction rate than polar species even if extinction hits species at random, because there are more tropical species. The narrower geographic range of tropical species increases their apparent extinction rate, because they are less likely to be preserved in the fossil record. Thus the evidence that temperature adaptations influence survival through mass extinctions is suggestive, but not yet rigorous enough to be persuasive.

22.8.2 The selectivity of extinction at the Cretaceous–Tertiary boundary differed from the pattern in the late Cretaceous

Jablonski has carried out a more systematic study, to find out whether extinctions were selective at the end of the Cretaceous. He looked at three properties of various gastropods and bivalves: the species richness of the taxon, the geographic range of the constituent species, and whether larval development was direct or planktotrophic (i.e. whether the young develop on the sea bottom near the adult or there is a planktotrophic dispersal and feeding stage).

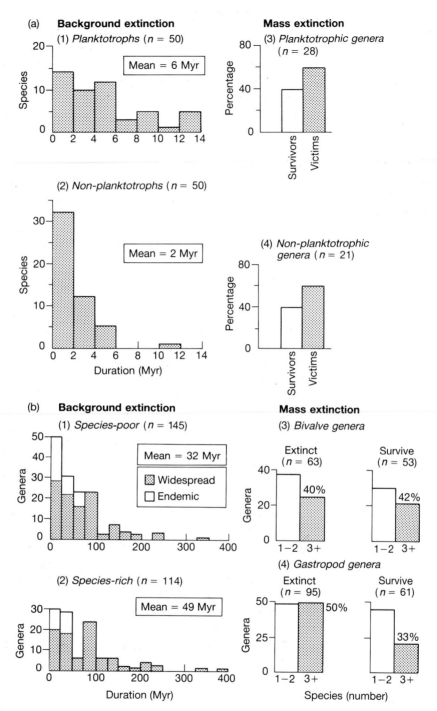

Figure 22.12 (a) Relation between extinction rate and developmental mode in gastropods at two times: ("background extinction") and during mass extinctions. Normal times are times between mass extinctions. Species with planktotrophic development have lower extinction rates in normal times, but were no more likely to survive the Late Cretaceous mass extinction than species with direct development. (b) Relation between extinction rate and species richness of a taxon. In normal times, species-poor taxa have higher extinction rates, i.e. shorter durations (compare the mean duration in the two cases). But in the mass extinction, there was no tendency for species-rich genera (containing three or more species) to be more likely to survive than species-poor genera (one or two species) in bivalves; and in gastropods, the species-rich genera actually made up fewer of the survivors than the species-poor ones. From Jablonski (1986).

His study is interesting for the contrast between two times: the Cretaceous–Tertiary mass extinction and the more normal times of the later Cretaceous, before the mass extinction began. In the normal period, species richness, geographic range, and planktotrophic (as opposed to direct) development were all correlated with lower extinction rate. We met the part of this result for normal periods in chapter 21 (section 21.5.2, p. 581) and noticed that the results may all be due to biases in the fossil record. Groups with more species, wider ranges, and planktotrophic development (which leads to a wider range) are all more likely to leave fossils and therefore will show lower "extinction" probabilities. Alternatively, however—as Jablonski suggests—the trends may be real. A taxon with a narrower geographic range will be more vulnerable to local disturbances, for example, which should increase its extinction rate.

Jablonski's interesting result is that during the mass extinction two of the three correlations disappeared (Figure 22.12). In the Cretaceous mass extinction, species-rich taxa had the same chance of extinction as species-poor ones, and planktotrophic species had the same chance of extinction as directly developing ones. Only broad geographic range continued to be associated with lower extinction rate. The extinction seems to have been so massive as to have taken out groups almost at random. At any rate, the relationships between the characters of a taxon and its extinction probability were significantly altered: in normal times, planktotrophically developing and species-rich taxa have lower probabilities of extinction than directly developing and species-poor taxa; in the Cretaceous–Tertiary mass extinction the difference disappeared.

22.9 A two-geared engine of macroevolution?

The different relationships between a taxon's extinction rate and its characters led Jablonski to suggest that there are two *macroevolutionary regimes*. He argued that evolution alternates between "normal" periods and mass extinctions, and the kinds of taxa favored in the two periods differ. Mass extinctions, he suggests, are extraordinary and intense and remove whole taxa in a predominantly random way. In the normal periods between, planktonically developing forms (for example) will proliferate relative to directly developing forms by species selection, as planktotrophic development reduces extinction rate in those quiet times.

Jablonski's idea is attractive. It seems likely that the relation between the characters possessed by taxa and their extinction rates will differ between mass and "normal" extinctions, because the circumstances are probably so different that the same character would not assist in avoiding extinction in both. The idea has little factual support, however; even the evidence he has collected is open to the criticism that it reflects mainly the biases of the fossil record. However, we can pose one more question. Are there, as Jablonski suggests, two radically different macroevolutionary regimes? Or is there a continuum, with Jablonski's two regimes as its two extremes? The question has been tackled by looking at the frequency distribution of extinction rates for the whole history of life (Figure 22.13a,b). If there are two regimes, then the frequency distribution should be bimodal, with one peak for the "normal" low extinction rate and another higher peak for the mass extinctions. If extinction rates have one long-term average for the history of

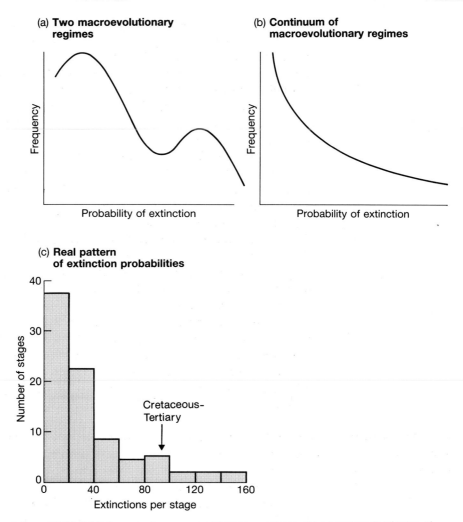

Figure 22.13 (a) If there are two macroevolutionary regimes, the frequency distribution of extinction probabilities in the fossil record should be bimodal. Whereas (b) if there is a continuum of macroevolutionary regimes, it will be a continuous Poisson distribution. (c) The actual distribution for 2316 marine animal families in the 79 generally recognized divisions of geologic time since the Cambrian is continuous. The extinction intensity at the Cretaceous–Tertiary boundary is indicated for comparison. From Raup (1986).

life, and the actual rate changes continuously between low and high values, the frequency distribution would probably be a Poisson distribution (the Poisson distribution is the frequency distribution for the number of event per time unit for a random process). The real pattern turns out to show little deviation from the random Poisson pattern (Figure 22.13c): there is a small secondary peak, but it is entirely due to the Cretaceous–Tertiary mass extinction and is statistically insignificant. The evidence therefore counts against the strong prediction of biomodality.

As ecological and geologic conditions change through time, they place varying demands on living species. Sometimes the environment will be harsh and extinction rates will rise; sometimes there will be a period of calm and extinction rates will decline. The species that survive best in the different environments will vary, both because of their adaptations to the environment and because of chance relationships between adaptations and the probability of extinction (by the mechanism of species selection). Mass extinctions, therefore, may have been important influences in the replacements of one taxon by another in the history of life (section 20.9, p. 551). After the late Triassic extinction, the dinosaurs rose to become the dominant large terrestrial vertebrates, replacing the formerly dominant mammal-like reptiles (Figure 20.14, p. 553). Then, between the Cretaceous and the Tertiary, the tables were turned: the dinosaurs were extinguished and the mammals took over.

The replacement may have been competitive, and have taken place over a long period: the evidence cannot rule out the possibility. However, it is possible that it was determined by the relative survival of the two groups in the Cretaceous–Tertiary mass extinction. There could have been two causes. If the mass extinction was indiscriminate, and survival a matter of luck, then the mammals survived to radiate into the empty ecological space after the mass extinction simply because they were luckier than the dinosaurs. Alternatively, the characters of the two groups might have influenced their chance of survival. Species selection in a mass extinction might have favored a group of small-bodied animals, which were adapted to survive in marginal habitats. While the dinosaurs were dominant on land, mammals persisted as small, mainly nocturnal, species; maybe mammals were better equipped than dinosaurs to survive the extraordinary selective conditions of the Cretaceous catastrophe.

Through the history of life on Earth, the changing geologic conditions have probably acted to sift among the groups of plants and animals and to generate the grand patterns of taxonomic rises, falls, and replacements—the patterns Medawar has called "evolutionary dynastics." But the continuous variation of extinction rates (Figure 22.13) suggests that the conditions change continuously, among an infinite variety of states, rather than swinging back and forth between only two kinds of macroevolutionary regime.

22.10 Summary

1 The ecological causes of particular species' extinctions are best studied in modern, not fossil, forms.

2 The real extinction of lineages should be distinguished from pseudo-extinction, which is due to the taxonomic subdivision of a continuous lineage.

3 The extinction rates of species are independent of the length of their existence: a species does not become more likely to reach extinction as time passes. Taxonomic survivorship curves are logarithmically linear.

4 Van Valen explained the log linearity of survivorship curves by his Red Queen hypothesis. It suggests that (1) each species' environment deteriorates as competing species evolve new, superior adaptations; (2) the competing species improve at a constant rate relative to each other; and (3) the constant deterioration in the environment causes the chance of extinction of any one of them to be probabilistically constant.

5 Coevolution within groups of competing species need not have the Red

Queen mode. Rates of change, and relative competitive abilities, need not be constant.

6 The Red Queen hypothesis could be tested by seeing whether extinction rates within taxa are constant in absolute time. Little conclusive work of this kind has been done.

7 At various times in the history of life, an exceptionally large number of species have become extinct together; these are called mass extinctions. The best studied mass extinction occurred at the boundary of the Cretaceous and Tertiary, 65 million years ago.

8 Alvarez *et al.*'s discovery of anomalously large concentrations of the rare earth element iridium at the Cretaceous–Tertiary boundary rocks at Gubbio, in Italy, suggested that the mass extinction may have been caused by the collision of an asteroid, about 10 km in diameter, with the Earth.

9 The impact theory of mass extinctions predicts that extinctions in different taxa should be sudden, synchronous, and global; the evidence for the Cretaceous–Tertiary extinction mainly fits the prediction, though other interpretations of the pattern are possible.

10 A survey, by Raup and Sepkoski, of extinction rates in the history of life, suggests extinction rates vary cyclically; mass extinctions recur about every 26 million years.

11 The evidence used by Raup and Sepkoski has been challenged taxonomically, statistically, and geochronologically. The result is controversial and uncertain.

12 Asteroidal collisions, volcanic eruption, climatic cooling, sea level changes, and changes in habitat area caused by continental drift are the five potentially general theories of mass extinction. The evidence does not suggest mass extinctions are generally caused by asteroidal collisions; the effects of climatic sea level changes need to be tested systematically; and the effect of continental drift is difficult to test at present.

13 Different kinds of species may suffer differentially in mass extinctions. The relationship between the characters of taxa and their extinction rates changed between the late Cretaceous and the Cretaceous–Tertiary mass extinction.

14 Mass extinctions may represent an endpoint of a continuum of macro-evolutionary extinction regimes: the kind of taxa favored in macroevolution will depend on the prevailing conditions.

22.11 Further reading

The popular books by Eldredge (1991), Raup (1991), and Stanley (1987) cover much of the material in this chapter. Gould's popular monthly column in *Natural History* (collected in Gould 1977b, 1980a, 1983a, 1985, 1991) contains a number of excellent essays on extinction. At a more scholarly level, there are several symposia: Elliot (1986), Raup and Jablonski (1986), Larwood (1988), Donovan (1989), and *Philosophical Transactions of the Royal Society of London, B*, vol. 325, pp. 239–488 (1989). The modern work can be followed, mainly in the journals *Science*, *Nature*, and *Paleobiology*. Raup (1986) and Marshall (1988) are general review papers; Hoffman (1989) discusses the subject too.

A textbook such as Begon *et al.* (1990) discusses the ecology of modern extinctions. References for Van Valen's ideas are: Van Valen (1973, 1985),

Gilinsky and Bambach (1987), Hoffman (1991), Hoffman and Kitchell (1984), McCune (1982), Raup (1978), Stenseth and Maynard Smith (1984).

On the Cretaceous–Tertiary mass extinction, see the pair of papers in *Scientific American* by Alvarez and Asaro (1990) and Courtillot (1990). The key original Alvarez papers are Alvarez *et al.* (1980, 1984); Grieve (1990) is the source for impact craters. Officer and Drake (1983, 1985) and Officer *et al.* (1987) give a different perspective. The end-Permian extinction is discussed by Erwin (1989, 1990). On cycles, the key papers by Raup and Sepkoski are (1984, 1986, 1988). Raup and Boyajian (1988) is a more recent update. Geochronologic and statistical critics include Stigler and Wagner (1987, 1988) and Hoffman (1985), taxonomic critics include Patterson and Smith (1987) and Boucot (1988).

On the effect of climate, in addition to Eldredge (1991) and Stanley (1987) cited above, see Crowley and North (1988), Stanley (1984), and Shackleton (1986) for late Eocene climatic change. Hallam (1983) discusses the effect of sea level; also Officer *et al.* (1987).

Jablonski (1986) is the main reference for the section on selectivity. For the frequency distribution of extinction rates, see Raup (1986) and McKinney (1987). On replacement, see the general references in chapter 20, and for dinosaurs and mammals see Gould (1983a, chapter 30) and Van Valen and Sloan (1977).

Appendix: the fossil record

The study of fossils—paleontology—is a large science, and its scope extends far beyond the theory of evolution. This appendix provides background material only for those parts of paleontology that are most important in evolutionary biology.

Fossilization

A fossil is any trace of past life. The most obvious fossils are body parts, such as shells, bones, and teeth; but fossils also include remains of the activity of living things, such as burrows or footprints (called trace fossils), and of the organic chemicals they form (chemical fossils). For any organism to leave a fossil requires a series of events, each of which is unlikely. We can consider these events for the case of hard parts; analogous points apply for trace and chemical fossils. When the organism dies, its soft parts will either be eaten by scavengers or be decayed by microbial action. For this reason, organisms that consist mainly of soft parts (such as worms and plants) are much less likely to leave fossils than are organisms that have hard parts. The fossil record is biased in favor of species that possess skeletons. Only an organism's hard parts stand much chance of fossilization; but in most cases these too will be destroyed. Hard parts may be crushed by rocks, stones, or wave action, or broken up by scavengers. If the hard parts survive, the next stage in fossilization is for them to be buried in sediment at the bottom of a water column: for only sedimentary rocks contain fossils. (Geologists distinguish three main rock types: igneous rocks, formed from volcanic action; sedimentary rocks, formed from sediments; and metamorphic rocks, formed deep in the Earth's crust by the metamorphosis of other rock types—when sedimentary rocks undergo metamorphosis, any fossils are lost.)

Animals that normally live within sediments are more likely to be buried in sediment before being destroyed, and these animals are therefore more likely to leave fossils than are species that live elsewhere. Likewise, species that live on the surface of the sediment (i.e. on the sea bottom) are more likely to be fossilized than are species that swim in the water column. Terrestrial species are least likely of all to be fossilized. The further a species lives from sediments, the less likely it is to be fossilized. For most of the delicate kinds of animal that live on the sea bottom, such as feather stars and worms, practically the only way they may come to leave fossils is by catastrophic burial: for example, in a slide of sediment from shallower water into the depths, that carries with it and buries some soft skeletoned animals. Feather stars, for instance, are known to decay into nothing within 48 hours of death on the sea bottom; they therefore have to be buried rapidly to have any chance of fossilization.

Once an organism's remains have been buried in the sediment, they can remain there, potentially, for an indefinitely long period of time. As new sediment piles on top of older sediment, the lower sediments are compacted: water is squeezed out and the sedimentary particles forced closer together. The organism's hard parts may be destroyed or deformed in the process. As the sediments compact, they are gradually turned into sedimentary rock. They may subsequently be moved up, down, or around the globe by tectonic movement, and can be re-exposed in a terrestrial area. Any fossils they contain can then be picked up, or dug up, on land (a fossil is, etymologically, anything that is dug up). Sediments may also be lost by tectonic subduction and geologic metamorphosis.

Over geologic time, the original hard parts of an organism will be transformed while lying in the sedimentary rock. Minerals from the surrounding rock slowly impregnate the bones, or shell, of the fossil, changing its chemical composition. Calcareous skeletons also change chemically. There are two forms of calcium carbonate: aragonite and calcite; aragonite is less stable and becomes rare in older fossils. Calcite may be replaced in some fossils by silica or pyrite. In extreme cases, the calcite may be dissolved away completely, and the space filled in by other material: the fossil then acts as a mold, or cast, for the new material. The remains still reveal the shape of the organism's hard parts.

Fossilization is an improbable eventuality. It is more probable for some kinds of species than others, and for some parts of an organism than for other parts. After burial in sediment, the fossils slowly transform through time; but the transformed remains, if they are preserved, can still tell us (after expert interpretation) much about the original living form.

Geologic time

Geologic ages

The geologic history of the Earth is conventionally divided into a series of eras, periods, and epochs, as in Figure A1. The main divisions were recognized by nineteenth century geologists on the basis of characteristic fossil faunas. The times of transition between two eras are times of transition between different characteristic fossil faunas: the fossils of the Permian, for instance, characteristically differ from those of the Triassic and at the Permo-Triassic boundary there is a relatively sudden transition between the fossil faunas. In the nineteenth century, it was not known whether the transition times corresponded to mass extinctions and sudden replacements, or to long gaps in the fossil record while a slower replacement was proceeding. It is now known they were mass extinctions in short periods. To demonstrate this, techniques to establish absolute times were necessary. For the smaller divisions of geologic time, such as epochs, there is no exact agreement on the absolute dates and more than one geologic time scale exists.

Absolute time

Geologists date events in the past both by relative and absolute techniques. An absolute time is a date expressed in years (or millions of years); a relative

Era	Period	Epoch	Myr BP (approx.)
Cainozoic	Quaternary	Recent	0.01
		Pleistocene	1.8 —
	Neogene (Tertiary)	Pliocene	5 —
		Miocene	24 —
	Paleogene (Tertiary)	Oligocene	37 —
		Eocene	54 —
		Paleocene	65 —
Mesozoic	Cretaceous		144 —
	Jurassic		213 —
	Triassic		248 —
Paleozoic	Permian		286 —
	Carboniferous	Pennsylvanian	320 —
		Mississippian	360 —
	Devonian		408 —
	Silurian		438 —
	Ordovician		505 —
	Cambrian		590 =

Figure A1 The geologic time scale. From Harland *et al.* (1982).

time is a time relative to some other known event: the times on Figure A1 are absolute. They were established from the radioactive decay of elements. For example, the isotope of rubidium ^{87}Rb decays into an isotope of strontium, ^{87}Sr. The decay is very slow: it has a half life of about 48.6 billion years: that is, half of an initial sample of ^{87}Rb will have decayed into ^{87}Sr in 48.6 billion years (about 10 times the age of the Earth).

Radioactive decay proceeds at an exponentially constant rate. Exponential decay means that a constant proportion of the initial material decays in each time unit. For example, suppose we start with 10 units and one-tenth of them decay per time interval; in the first time interval one unit will decay, and we shall have nine units left. In the second time interval, a proportion equal to one-tenth of the remaining nine units (i.e. 0.9 units) will decay; and we shall be left with 8.1 units. In the third time interval, a further one-tenth of the 8.1 units will decay, leaving 7.29 units (8.1 − 0.81) at the beginning of the fourth time interval and so on. In radioactive decay, the proportion of the isotope that decays each year is called the decay constant (λ), and for $^{87}Rb/^{87}Sr$ $\lambda = 1.42 \times 10^{-11}$ per year. Therefore, whatever the amount of ^{87}Rb that is present at any time, a proportion equal to 1.42×10^{-11} of it will decay into ^{87}Sr in the next year.

To estimate the age of a rock by the radioisotope technique, we need to be able to make two measurements and validate one assumption. The two measurements are the isotope composition of the rock now and when it was formed. The proportions of ^{87}Rb and ^{87}Sr are obviously measureable now. The composition of the rock was originally fixed when it crystallized as an igneous rock from liquid magma, and the ratios of ^{87}Rb and ^{87}Sr in modern magma can be measured: the ratio is a good estimate of the isotope ratio when the rock first formed. The isotope ratio will slowly change from the original ratio as ^{87}Rb radioactively decays into ^{87}Sr. In order to estimate the age of the rock from the change in the isotope ratio, we must assume that all the change in the ratio is due to radioactive decay. For the case of $^{87}Rb/^{87}Sr$, the assumption is probably valid. Neither isotope seeps into, or leaks out of, the rock, and the ratios are therefore solely determined by time and radioactive decay. For some other radioisotopes, the assumption is less well met. Uranium, for example, can be oxidized into a mobile form and move among rocks (though the problem in this case can be dealt with by combining two uranium decay schemes, such that the time is inferred from the ratio of two lead isotopes and the concentration of uranium does not matter). Table A1 lists the main radioisotopes used in geochronology. The decay of ^{40}K, for example, is a geochronologically useful decay scheme. When a volcano erupts the heat volatalizes all the ^{40}Ar out of the volcanic lava and ash, but not the ^{40}K. When the volcanic dust cools, therefore, it contains (of ^{40}K and ^{40}Ar) only ^{40}K. The ^{40}K then decays into both ^{40}Ca and ^{40}Ar. In practice, there is so

Table A1 Radioactive decay systems used in geochronology

Radioactive isotope	Decay constant ($\times 10^{-11}$ per year)	Half life (years)	Radiogenic isotope
^{14}C	1.2×10^{7}	5.73×10^{3}	^{14}N
^{40}K	$5.81 + 47.2$	1.3×10^{9}	$^{40}Ar + ^{40}Ca^{*}$
^{87}Rb	1.42	4.86×10^{10}	^{88}Sr
^{147}Sm	0.654	1.06×10^{11}	^{143}Nd
^{232}Th	4.95	1.39×10^{10}	^{208}Pb
^{235}U	98.485	7×10^{8}	^{207}Pb
^{238}U	15.5125	4.4×10^{9}	^{206}Pb

* ^{40}K decays into both ^{40}Ar and ^{40}Ca, with the two decay constants given; the half life is for the sum of the two.

much ^{40}Ca in the rock from other sources that it is not convenient to use it for dating purposes, but all the ^{40}Ar in the rock will have been produced by the decay of ^{40}K. The decay is so slow that it is not practical to use it for rocks less than about 100000 years old; but there are other radioisotopes for shorter times. The decay of ^{14}C into nitrogen, for example, has a half life of only 5730 years.

The exact age of a rock is calculated as follows. Take ^{87}Rb/^{87}Sr as an example. Let N_0 be the number of ^{87}Rb atoms in the sample of rock when it was formed, and N be the number today. Then $N = N_0 e^{\lambda t}$, where λ is the decay constant and t is the age of the rock. Take logs and $t = (1/\lambda)\ln N_0/N$. It is practically easier to measure the quantity of the isotope generated by decay. Therefore, let N_R be the number of ^{87}Sr atoms generated by radiocative decay up to any time. Because each ^{87}Sr atom has been generated by the decay of one ^{87}Rb atom, $N_R = N_0 - N$. This can be substituted into the formula for time:

$$t = \frac{1}{\lambda} \ln \frac{N + N_R}{N}$$

For example, if 3% of the original ^{87}Rb in a rock has decayed into ^{87}Sr, then the age of the rock is calculated as

$$t = \frac{1}{1.42 \times 10^{-11}} \ln(1.03) = 2.08 \times 10^9 \text{ years}$$

(or about 2 billion years).

(The decay constant and the half life of a radioisotope are simply related. When half the original ^{87}Rb has decayed into ^{87}Sr, the number of ^{87}Sr atoms formed must equal the number of ^{87}Rb atoms that have decayed. $N = N_R$. Substitute that into the formula for time, and $t_{1/2} = \ln 2/\lambda = 0.693/\lambda$.)

In summary, if we know the isotope ratio in the rock when it was formed and in a modern sample, if we can reasonably assume that the change in the ratio between then and now was only caused by radioactive decay, we can estimate the absolute age of the rock.

The radioisotope method can only be used for rocks that contain the isotopes. For ^{14}C, fossils can be dated directly, because carbon is found in living material, and the ratio of ^{14}C to ^{14}N in the fossil can be used to infer its age. Some other radioisotopes are found in corals or shells, but many (such as ^{87}Rb or ^{40}K) are only found in igneous rocks. In order to date a deposit of fossils, it is then necessary to infer that the fossils were laid down at about the same time as an associated igneous rock that can be absolutely dated. If the fossiliferous sediments are laid down on top of an igneous rock, it can be inferred that the fossils are no older than the date of the igneous rock (on the principle that younger rocks lie on top of older ones). If an igneous rock has been intruded into a sedimentary rock, it can be inferred that the sediments are older than the igneous rock (because igneous rocks only intrude into existing sedimentary rocks). In the best case, a fossiliferous sediment will lie on top of one older set of igneous rocks, and have another younger igneous rock intruded in it. Then the fossils can be dated to a time between the age of the two igneous rocks.

Relative time

The inference of the age of fossils from that of surrounding igneous rocks is an example of relative time measurement. If we know the relative date of a rock or fossil, it means we know its date relative to that of another rock or fossil: we have a statement of the form "rock A was laid down before/at the same time as/after rock B." Some of the procedures for finding relative times are as follows. At any one site, more recent sediments are deposited on top of older sediments. Fossils lower down a sedimentary column are therefore likely to be older (sometimes a large geologic convulsion, such as a volcanic explosion, may turn a sedimentary column upside down, but it can be detected when it has happened). The date of any one fossil deposit relative to those at different sites can also usually be estimated. It is done by comparing the fossil composition of the site, for some common fossils such as ammonites or foraminifers, with a standard reference collection. For these reference fossils, the fossils deposited at one place and time will be much the same as those being deposited at another place. They therefore show that two sites

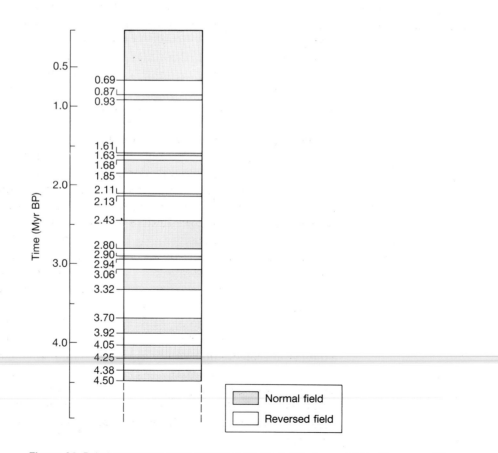

Figure A2 Polarity reversals of the Earth's magnetic field in the past 4.5 million years. The picture may not be complete: it is possible that further, shorter events have still to be discovered.

had the same relative date. This kind of study is called correlation; the paleontologist is said to be correlating the two sites.

Magnetic time zones supply a similar principle. The magnetic field of the Earth has reversed its polarity at intervals through geologic history. When the magnetic field is as it is now (compasses point North) it is called normal; when it has the opposite polarity it is called reversed. The history of the Earth is an alternating sequence of normal and reversed time zones. (The reason for the reversals is not known for sure—though there are plenty of hypotheses.) Polarity switches have been more common at some times than others; Figure A2 gives an idea of their frequency in recent times. All the rocks in the globe at any one time have the same polarity, and the polarity at the time rocks were formed can be detected. The polarities can then be used in fine scale time resolution. If two rocks are known to have been formed at similar times, but we are not sure whether they are exact or only near contemporaries, magnetic polarities can provide the answer. If the rocks have different polarities they cannot have been exact contemporaries.

The completeness and resolving power of the fossil record

It is important for a number of questions in evolutionary biology to know how complete the fossil record is. When estimating rates of evolution (see chapter 19), we need to know whether apparently sudden evolutionary changes really are bouts of rapid evolution or whether they just mean that a long series of intermediate stages left no record as fossils. In mass extinctions (see chapter 22), we are interested in whether extinctions in the fossil record at different sites were exactly synchronous (i.e. they happened within a few decades of each other) or were more crudely synchronous (within one or two million years of each other). The problem of how complete the record is can be tackled in two stages. In a series of fossils from successive deposits at a site, the first thing a geologist can look for is any evidence of a break, or non-uniformity, in the sedimentary deposition: for instance, a change from one form of rock to another, suggesting that the earlier rocks were laid down under different conditions from the later; there could also have been a hiatus between them. It would be a reasonable hypothesis that changes in the form, or composition, of fossils between the different kinds of rock reflect changes in fossilizing conditions, and a possible gap in the record between the two rocks, rather than continuous evolutionary change. Within an apparently uniform sedimentary rock, the apparent events in the record are more likely to be real.

But even a fairly uniform rock will not contain deposits from every year when it was laid down. There will be gaps, of shorter or longer duration. It is possible to estimate how complete the record is if we have radioisotope dates for the top and bottom of the sequence. The rate of deposition of sediments in various environments, such as the ocean bottom or river estuaries, can be measured over a human time scale. We can then extrapolate the short-term rate over the time of the fossiliferous sedimentary sequence to estimate how complete it is:

$$\text{Completeness} = \frac{\text{Observed thickness}}{\text{Short-term rate of deposition} \times \text{timespan of rock}}$$

Table A2 Completeness of some important fossil sequences. Simplified from Schindel (1982a,b)

Subject	Authority	Sediment thickness (m)	Temporal scope (Myr)	Percentage stratigraphic completeness
Eocene mammals	Gingerich	520	3.5	28
Permian foraminiferans	Ozawa	200	12	4
Neogene radiolarians	Kellogg	5	1.9	2
Neogene foraminiferans	Malmgren & Kennett	200	8.3	23
Jurassic ammonites	Raup & Crick	14	1+	3
Pennsylvanian snails	Schindel	1300	27	34
Plio–Pleistocene molluscs (a)*	Williamson	265	0.4	45
(b)*		110	0.1	73

* (a) Entire Koobi Fora Formation; (b) Lower member, Koobi Fora Formation.

Completeness is then a number between zero and one. If sediments were laid down continuously at the short-term rate, the rock sequence would have a completeness of one; in so far as there were periods without deposition, completeness will be reduced below one. The estimates also have to be corrected for compaction.

For various studies that have been used to test between punctuated equilibrium and phyletic gradualism (see chapter 19), Schindel has estimated the completeness of the record. His results are in Table A2. He used a compaction factor of two in his estimates (i.e. he assumed that compaction had reduced the thickness of the rock to one-half of what it would be if sediments had just piled up at the short-term rate). The main interest of the table is to show how completeness can be quantitatively estimated. However, some other points are worth noticing. For example, even though these are some of the most complete records available, their completeness can be low: several are less than 10%. Also, it can be seen that the completenesses of older fossil sequences tend to be lower. This is useful background information when we use the actual fossil records to test theories about the pattern of evolutionary rates.

Further reading

There are many books about the fossil record. For example: Black (1989), Cvancara (1990), Donovan (1991), Fortey (1982), Paul (1980), Shipman (1981), Simpson (1983), and Smith (1981); also the relevant chapter in Hoffman (1989). On potassium-argon dating, see Dalrymple and Lanphere (1969). On completeness, see Sadler (1981) and Schindel (1982a,b). See also Raup and Jablonski (1986).

Glossary

adaptation   Feature of an organism enabling it to survive and reproduce in its natural environment better than if it lacked the feature.

adaptive landscape   A graph of the average *fitness* of a population in relation to the frequencies of *genotypes* in it: peaks on the landscape correspond to genotypic frequencies at which the average fitness is high, valleys to genotypic frequencies at which the average fitness is low. Also called a fitness surface.

allele   A variant of a single *gene*, inherited at a particular genetic *locus*; it is a particular sequence of *nucleotides*, coding for *messenger RNA*.

allometry   The relationship between the size of an organism and the size of any of its parts: for example, there is an allometric relationship between brain size and body size, such that (in this case) animals with bigger bodies have bigger brains. Allometric relationships can be studied during the growth of a single organism, between different organisms within a *species*, or between organisms in different species.

allopatric speciation   Speciation via geographically separated *populations*.

amino acid   Unit molecular building block of *proteins*: a protein is a chain of amino acids in a certain sequence. There are 20 main amino acids in the proteins of living things, and the properties of a protein are determined by its particular amino acid sequence.

analogy   Character shared by a set of species but not present in their common ancestor. A convergently evolved *character*. (Compare with *homology*.)

anatomy   (1) The structure itself of an organism, or one of its parts. (2) The science that studies those structures.

ancestral homology   *Homology* that evolved before the common ancestor of a set of *species*, and is present in other species outside that set of species. (Compare with *derived homology*.)

artificial selection   Selective breeding, carried out by humans, to alter a *population*. The forms of most domesticated and agricultural species have been produced by artificial selection; it is also an important experimental technique for studying evolution.

asexual reproduction   Production of offspring by virgin birth, i.e. without sexual fertilization of the eggs.

assortative mating   Tendency of like to mate with like. It can be for a certain *genotype* (e.g. individuals with genotype *AA* tend to mate with other individuals of genotype *AA*) or *phenotype* (e.g. tall individuals mate with other tall individuals).

atomistic (as applied to a theory of inheritance) Inheritance in which the entities controlling heredity are relatively distinct, permanent, and capable of independent action; *Mendelian inheritance* is an atomistic theory because, in it, inheritance is controlled by distinct *genes*.

autosome Any *chromosome* other than a sex chromosome.

base The *DNA* is a chain of *nucleotide* units; each unit consists of a backbone made of a sugar and a phosphate group, with a nitrogenous base attached. The base in a unit is one of adenine (A), guanine (G), cytosine (C), or thymine (T). In *RNA*, uracil (U) is used instead of thymine. A and G belong to the chemical class called *purines*; C, T, and U are *pyrimidines*.

Batesian mimicry A kind of *mimicry* in which one non-poisonous species (the Batesian mimic) mimics another poisonous species.

Biogenetic law Name given by Haeckel to *recapitulation*.

biological species concept Concept of *species*, according to which a species is a set of organisms that can interbreed among each other. (Compare with *cladistic species concept, ecological species concept, phenetic species concept, recognition species concept*.)

biometrics Quantitative study of *characters* of organisms.

blending inheritance The historically influential but factually erroneous theory that organisms contain a blend of their parents' hereditary factors and pass that blend on to their offspring. (Compare with *Mendelian inheritance*. See section 2.5, p. 32.)

character Any recognizable trait, feature, or property of an organism.

chloroplast Structure (or *organelle*) found in some cells of plants; their function is photosynthesis.

chromosomal inversion *See inversion*.

chromosome Structure in the cell nucleus that carries the DNA. At certain times in the cell cycle they are visible as string-like entities. Chromosomes consist of the DNA with various proteins, particularly histones, bound to it.

clade Set of species descended from a common ancestral species. Synonym of *monophyletic group*.

cladism Phylogenetic classification. The members of a group in a cladistic classification share a more recent common ancestor with each other than with the members of any other group. A group at any level in the classificatory hierarchy, such as "family," is formed by combining a subgroup (at the next lowest level, perhaps the genus in this case) with that other subgroup it shares its most recent common ancestor with. (Compare with *evolutionary classification* and *phenetic classification*.)

cladistic species concept Concept of species, according to which a species is a *lineage* of populations between two phylogenetic branch points (or speciation events). (Compare with *biological species concept, ecological species concept, phenetic species concept, recognition species concept*.)

classification Arrangement of organisms into hierarchic groups. Modern biological classifications are *Linnaean* and classify organisms into species, genus, family, order, class, phylum, kingdom, and certain intermediate categoric levels. *Cladism, evolutionary classification*, and *phenetic classification* are three methods of classification.

cline A geographic gradient in the frequency of a gene, or in the average value of a character.

clock   *See molecular clock.*

clone   A set of genetically identical organisms asexually reproduced from one ancestral organism.

coadaptation   Beneficial interaction between a number of: (1) genes at different loci within an organism; (2) different parts of an organism; or (3) organisms belonging to different species.

codon   Triplet of *bases* (or nucleotides) in the *DNA* coding for one *amino acid*. The relationship between codons and amino acids is given by the *genetic code*. The triplet of bases that is complementary to a codon is called an anti-codon; conventionally, the triplet in the *messenger RNA* is called the codon and the triplet in the *transfer RNA* is called the anti-codon.

coevolution   Evolution in two or more species in which the evolutionary changes of each species influence the evolution of the other species.

comparative biology   Study of patterns among more than one species.

comparative method   Study of *adaptation* by comparing many species.

concerted evolution   Tendency of the different genes in a *gene family* to evolve in concert; i.e. each gene *locus* in the family comes to have the same genetic variant.

convergence   Process by which a similar character evolves independently in two species. Also, a synonym for *analogy*, i.e. an instance of a convergently evolved character: a similar character in two species that was not present in their common ancestor.

Cope's rule (law)   Evolutionary increase in body size over geologic time in a lineage of populations.

creationism   The theory that species have separate origins and never change after their origin. Most versions for creationism are religiously inspired and suggest that the origin of species is by supernatural action.

crossing over   The process during *meiosis* in which the chromosomes of a *diploid* pair exchange genetic material. It is visible under the light microscope. At a genetic level, it produces *recombination*.

cytoplasm   Region of *eukaryotic* cell outside the nucleus.

Darwinism   Darwin's theory, that species originated by evolution from other species and that evolution is mainly driven by natural selection. Differs from *neo-Darwinism* mainly in that Darwin did not know about *Mendelian inheritance*.

derived homology   *Homology* that first evolved in the common ancestor of a set of species and is unique to them. (Compare with *ancestral homology*.)

diploid   Having two sets of *genes* and two sets of *chromosomes* (one from the mother, one from the father). Many common species, including humans, are diploid. (Compare with *haploid* and *polyploid*.)

directional selection   Selection causing a consistent directional change in the form of a population through time, for example selection for larger body size.

disruptive selection   Selection favoring forms that deviate in either direction from the population average. Selection favors forms that are larger or smaller than average, but works against the average forms between.

distance   In *taxonomy*, refers to the quantitatively measured difference between the phenetic appearance of two groups of individuals, such as populations or species (phenetic distance), or the difference in their gene frequencies (genetic distance).

DNA   Deoxyribonucleic acid: the molecule that controls inheritance.

dominance (genetic)   An allele (*A*) is dominant if the phenotype of the heterozygotes *Aa* is the same as the homozygote *AA*. The allele *a* does not influence the heterozygote's phenotype and is called *recessive*. An allele may be partly, rather than fully, dominant: then the heterozygous phenotype is nearer to, rather than identical with, the homozygote of the dominant allele.

drift   Synonym of *genetic drift*.

duplication   The occurrence of a second copy of a particular sequence of DNA: the duplicate sequence may appear next to the original, or be copied elsewhere into the *genome*. When the duplicated sequence is a *gene*, the event is called gene duplication.

ecological genetics   Study of evolution in action in nature, by a combination of fieldwork and laboratory genetics.

ecological species concept   Concept of species, according to which a species is a set of organisms adapted to a particular, discrete set of resources (or niche) in the environment. (Compare with *biological species concept, cladistic species concept, phenetic species concept, recognition species concept*.)

electrophoresis   Method of distinguishing entities according to their motility in an electric field. In evolutionary biology, it has been mainly used to distinguish different forms of proteins. The electrophoretic motility of a molecule is influenced by its size and electric charge.

epistasis   An interaction between the *genes* at two or more loci, such that the phenotype differs from what would be expected if the loci were expressed independently.

eukaryote   Comprised of cells with distinct nuclei. Almost all multicellular organisms are eukaryotic. (Compare with *prokaryote*.)

evolution   Darwin defined it as "descent with modification." It is the change in a lineage of populations between generations.

evolutionary classification   Method of classification using both *cladistic* and *phenetic* classificatory principles. To be exact, it permits *paraphyletic* groups (which are allowed in phenetic but not in cladistic classification) and *monophyletic* groups (which are allowed in both cladistic and phenetic classification), but excludes *polyphyletic* groups (which are banned from cladistic classification but permitted in phenetic classification).

exon   The nucleotide sequences of some genes consist of parts that code for amino acids, with other parts that do not code for amino acids interspersed among them. The coding parts, which are translated, are called exons; the interspersed non-coding parts are called introns.

fitness   The average number of offspring produced by individuals with a certain genotype, relative to the number produced by individuals with other genotypes. When genotypes differ in fitness because of their effects on survival, fitness can be measured as the ratio of a genotype's frequency among the adults divided by its frequency among individuals at birth.

fixation   A gene has achieved fixation when its frequency has reached 100%

in the population.

fixed (1) In *population genetics*, a gene is fixed when it has a frequency of 100%. (2) In *creationism*, species are described as fixed in the sense that they are believed not to change their form, or appearance, through time.

founder effect The loss of genetic variation when a new colony is formed by a very small number of individuals from a larger population.

frequency-dependent selection Selection in which the *fitness* of a genotype (or phenotype) depends on its frequency in the population.

gamete Reproductive cell: sperm and eggs in animals, pollen and ovules in plants.

gene Sequence of *nucleotides* coding for a *protein* (or, in some case, part of a protein).

gene duplication *See duplication.*

gene family Set of related *genes* occupying various *loci* in the *DNA*, almost certainly formed by *duplication* of an ancestral gene, and having a recognizably similar sequence. The globin gene family is an example.

gene flow The movement of *genes* into, or through, a population by interbreeding, or by migration and interbreeding.

gene frequency The frequency in the population of a particular gene relative to other genes at its *locus*. Expressed as a proportion (between 0 and 1) or percentage (between 0 and 100%).

gene pool All the *genes* in a *population* at a particular time.

genetic code The code relating *nucleotide* triplets in the *messenger RNA* (or *DNA*) to amino acids in the proteins. It has been decoded (see Table 2.1, p. 26).

genetic distance *See distance.*

genetic drift Random changes in *gene frequencies* in a *population*.

genetic load A reduction in the average *fitness* of the members of a *population* because of the deleterious genes, or gene combinations, in the population. It has many particular forms, such as mutational load, segregational load, and recombinational load.

genetic locus *See locus.*

genome The full set of *DNA* in a cell or organism.

genotype The set of two *genes* at a *locus* possessed by an individual.

geographic isolation *See reproductive isolation.*

geographic speciation *See allopatric speciation.*

germ plasm The reproductive cells in an organism: the cells that produce the *gametes*. All the cells in an organism can be divided into the *soma* (the cells that ultimately die) and the germ cells (that are perpetuated by reproduction).

group selection Selection operating between groups of individuals rather than between individuals. It would produce attributes beneficial to a group in competition with other groups, rather than attributes beneficial to individuals.

haploid Condition of having only one set of *genes* or *chromosomes*. In normally *diploid* organisms such as humans, only the *gametes* are haploid.

haplotype Set of *genes* at more than one *locus* inherited by an individual from one of its parents. It is the multilocus analog of an *allele*.

Hardy–Weinberg ratio Ratio of genotype frequencies that evolve when mating is random and neither *selection* nor *drift* are operating. For two alleles (*A* and *a*) with frequencies *p* and *q*, there are three genotypes *AA*, *Aa*, and *aa*; and the Hardy–Weinberg ratio for the three is p^2 *AA* : $2pq$ *Aa* : q^2 *aa*. It is the starting point for much of the theory of population genetics.

heritability Broadly, the proportion of variation (more strictly *variance*) in a phenotypic character in a *population* that is due to individual differences in *genotype*. Narrowly, the proportion of variation (more strictly *variance*) in a phenotypic character in a population that is due to individual genetic differences that will be inherited by the offspring.

heterogametic Sex with two different *sex chromosomes* (males in mammals, because they are XY). (Compare with *homogametic*.)

heterozygosity Proportion of individuals in a *population* that are *heterozygotes* (for most purposes).

heterozygote Individual having two different *alleles* at a genetic *locus*.

heterozygote advantage Condition in which the *fitness* of a *heterozygote* is higher than the fitness of either *homozygote*.

homeostasis (developmental) Self-regulating process in development, such that the organism grows up to have much the same form independently of the external influences it experiences while growing up.

homeotic mutation *Mutation* causing one structure of an organism to grow in the place appropriate to another. For example, in the mutation called *antennapedia* in the fruitfly, a foot grows in the antennal socket.

homogametic Sex with two of the same kind of *sex chromosomes* (females in mammals, because they are XX). (Compare with *heterogametic*.)

homology Character shared by a set of *species* and present in their common ancestor. (Compare with *analogy*.)

homozygote Individual having two copies of the same *allele* at a genetic *locus*.

hybrid Offspring of a cross between two *species*.

idealism Philosophic theory that there are fundamental non-material ideas, plans, or forms underlying the phenomena we observe in nature. It has been historically influential in classification.

inheritance of acquired characters Historically influential but factually erroneous theory that an individual inherits characters that its parents acquired during their lifetimes.

intron The *nucleotide* sequences of some *genes* consist of parts that code for *amino acids*, with other parts that do not code for amino acids interspersed among them. The interspersed non-coding parts, which are not translated, are called introns; the coding parts are called *exons*.

inversion An event (or the product of the event) in which a sequence of *nucleotides* in the *DNA* are reversed, or inverted. Sometimes inversions are visible in the structure of the *chromosomes*.

isolating mechanism Any mechanism, such as a difference between *species* in courtship behavior or breeding season, that results in *reproductive isolation* between the species.

isolation Synonym for *reproductive isolation*.

Lamarckian inheritance Historically misleading synonym for *inheritance of acquired characters*.

larva (and **larval stage**) Prereproductive stage of many animals; the term is used particularly when the immature stage has a different form from the adult.

lineage An ancestor–descendant sequence of (1) *populations*; (2) cells; or (3) *genes*.

linkage disequilibrium Condition in which the *haplotype* frequencies in a *population* deviate from the values they would have if the *genes* at each *locus* were combined at random. (When there is no deviation, the population is said to be in linkage equilibrium.)

linked Of *genes*, present on the same *chromosome*.

Linnaean classification Hierarchic method of naming classificatory groups, invented by the eighteenth century Swedish naturalist Carl von Linné, or Linnaeus. Each individual is assigned to a species, genus, family, order, class, phylum, and kingdom, and some intermediate classificatory levels. Species are referred to by a Linnaean binomial of its genus and species, such as *Magnolia grandiflora*. Universally used by educated persons.

locus The location in the *DNA* occupied by a particular *gene*.

macroevolution *Evolution* on the grand scale: the term refers to events above the *species* level; the origin of a new higher group, such as the vertebrates, would be an example of a macroevolutionary event.

macromutation *Mutation* of large phenotypic effect; one that produces a *phenotype* well outside the range of variation previously existing in the *population*.

mean The average of a set of numbers. For example, the mean of 6, 4, and 8 is $(6 + 4 + 8)/3 = 6$.

meiosis Special kind of cell division that occurs during the reproduction of *diploid* organisms to produce the *gametes*. The double set of *genes* and *chromosomes* of the normal diploid cells is reduced during meiosis to a single *haploid* set. *Crossing over* and therefore *recombination* occur during a phase of meiosis.

Mendelian inheritance The mode of inheritance of all *diploid* species, and therefore of nearly all multicellular organisms. Inheritance is controlled by *genes*, which are passed on to the offspring in the same form as they were inherited from the previous generation. At each *locus*, an individual has two genes, one inherited from its father and the other from its mother. The two genes are represented in equal proportions in its *gametes*.

messenger RNA (mRNA) Kind of *RNA* produced by *transcription* from the *DNA* and which acts as the message that is decoded to form *proteins*.

microevolution Evolutionary changes on the small scale, such as changes in *gene frequencies* within a *population*.

mimicry A case in which one species looks more-or-less similar to another species. See *Batesian mimicry*, *Müllerian mimicry*.

mitochondrion A kind of organelle in *eukaryotic* cells; mitochondria burn the digested products of food to produce energy. They contain *DNA* coding for some mitochondrial *proteins*.

mitosis Cell division. All cell division in multicellular organisms is by mitosis except for the special division called *meiosis* that generates the gametes.

modern synthesis Synthesis of *natural selection* and *Mendelian inheritance*. Also called *neo-Darwinism*.

molecular clock Theory that molecules evolve at an approximately constant rate. The difference between the form of a molecule in two species is then proportional to the time since the species diverged from a common ancestor, and molecules become of great value in the inference of *phylogeny*.

monomorphic Population in which all individuals have the same genotype. Opposite of *polymorphism*.

monophyletic group Set of *species* containing a common ancestor and all its descendants.

morphology The study of the form, shape, and structure of organisms.

Müllerian mimicry A kind of *mimicry* in which two poisonous species evolve to look like each other.

natural selection Process by which the forms of organisms in a population that are best adapted to the environment increase in frequency relative to less well-adapted forms over a number of generations.

neo-Darwinism (1) Darwin's theory of *natural selection* plus *Mendelian inheritance*. (2) The larger body of evolutionary thought that was inspired by the unification of natural selection and Mendelism. A synonym of the *modern synthesis*.

neutral drift Synonym of *genetic drift*.

neutral mutation *Mutation* with the same *fitness* as the other *allele*, or alleles, at its *locus*.

neutral theory (and neutralism) Theory that most *evolution* at the molecular level occurs by *genetic drift*.

nucleotide Unit building block of *DNA* and *RNA*; a nucleotide consists of a sugar and phosphate backbone with a *base* attached.

nucleus Region of *eukaryotic* cells containing the *DNA*.

numerical taxonomy In general, any method of *taxonomy* using numerical measurements; in particular, it often refers to *phenetic classification* using large numbers of quantitatively measured *characters*.

organelle Any of a number of distinct small structures found in the cytoplasm (and so outside the nucleus) of *eukaryotic* cells, e.g. *mitochondrion*, *chloroplast*.

orthogenesis The erroneous idea that species tend to evolve in a fixed direction because of some inherent force driving them to do so.

ovule Female *gamete* of plants.

paleobiology Biological study of fossils.

paleontology Scientific study of fossils.

panmixis Random mating throughout a *population*.

parapatric speciation Speciation in which the new species forms from a population contiguous with the ancestral species' geographic range.

paraphyletic group Set of species containing an ancestral species together with some, but not all, of its descendants. The species included in the group are those that have continued to resemble the ancestor; the excluded species have evolved rapidly and no longer resemble their ancestor.

parsimony Principle of phylogenetic reconstruction in which the *phylogeny* of a group of *species* is inferred to be the branching pattern requiring the smallest number of evolutionary changes.

parthenogenesis Reproduction by virgin birth: a term for *asexual reproduction*.

particulate (as property of theory of inheritance) Synonym of *atomistic*.

peripatric speciation Synonym of *peripheral isolate speciation*.

peripheral isolate speciation A form of *allopatric speciation* in which the new *species* is formed from a small *population* isolated at the edge of the ancestral population's geographic range. Also called *peripatric speciation*.

phenetic classification Method of classification in which *species* are grouped together with other species that they most closely resemble phenotypically.

phenetic species concept Concept of *species*, according to which a species is a set of organisms that are phenetically similar to one another. (Compare with *biological species concept, cladistic species concept, ecological species concept,* and *recognition species concept.*)

phenotype The *characters* of an organism, whether due to the *genotype* or environment.

phylogeny "Tree of life": branching relationships among *species*, showing which species shares its most recent common ancestor with which other species.

plan of nature Philosophic theory that nature is organized according to a plan. It has been influential in *classification*, and is a kind of *idealism*.

plasmid A genetic element that exists (or can exist) independently of the main *DNA* in the cell. In bacteria, plasmids can exist as small loops of DNA and be passed between cells independently.

Poisson distribution Frequency distribution for number of events per unit time, when the number of events is determined randomly.

polymorphism Condition in which a *population* possesses more than one *allele* at a *locus*. Sometimes it is defined as the condition of having more than one allele with a frequency of over 5% in the population.

polyphyletic group Set of *species* descended from more than one common ancestor. The ultimate common ancestor of all the species in the group is not a member of the polyphyletic group.

polyploid An individual containing more than two sets of *genes* and *chromosomes*.

population A group of organisms, usually a group of sexual organisms that interbreed and share a *gene pool*.

population genetics Study of processes influencing *gene frequencies*.

postzygotic isolation *Reproductive isolation* in which a *zygote* is successfully formed but then either fails to develop or develops into a sterile adult. Donkeys and horses are postzygotically isolated from each other: a male donkey and a female horse can mate to produce a mule, but the mule is sterile.

prezygotic isolation *Reproductive isolation* in which the two *species* never reach the stage of successful mating, and thus no *zygote* is formed. Examples would be species with different breeding seasons or courtship displays, and which therefore never recognize each other as potential mates.

prokaryote Cell without a distinct nucleus. Bacteria and some other simple organisms are prokaryotic. (Compare with *eukaryote*.)

protein A molecule made up of a sequence of *amino acids*. Many of the important molecules in a living thing are proteins: all enzymes, for example, are proteins.

pseudogene A sequence of *nucleotides* in the *DNA* that resembles a *gene* but is non-functional for some reason.

purine A kind of *base*; in the *DNA*, adenine (A) and guanine (G) are purines.

pyrimidine A kind of *base*; in *DNA*, cytosine (C) and thymine (T), and in RNA, cytosine (C) and uracil (U) are pyrimidines.

quantitative character A *character* showing continuous variation in a *population*.

random drift Synonym of *genetic drift*.

random mating Mating pattern in which the probability of mating with another individual of a particular genotype (or phenotype) equals the frequency of that genotype (or phenotype) in the population.

recapitulation Partly or wholly erroneous theory that an individual, during its development, passes through a series of stages corresponding to its successive evolutionary ancestors. An individual thus develops by "climbing up its family tree."

recessive An *allele* (*a*) is recessive if the phenotype of the heterozygote *Aa* is the same as the homozygote (*AA*) for the alternative allele *A* and different from the homozygote (*aa*) for the recessive allele. The allele *A* controls the heterozygote's phenotype and is called *dominant*. An allele may be partly, rather than fully, recessive: then the heterozygous phenotype is nearer to, rather than identical with, the homozygote for the dominant allele.

recognition species concept Concept of *species*, according to which a species is a set of organisms that recognize one another as potential mates; they have a shared mate recognition system. (Compare with *biological species concept*, *cladistic species concept*, *ecological species concept*, and *phenetic species concept*.)

recombination Event, occurring by the *crossing over* of *chromosomes* during *meiosis*, in which *DNA* is exchanged between a pair of chromosomes. Thus two *genes* that were previously unlinked, being on separate chromosomes, can become *linked* because of recombination; and vice versa: linked genes may become unlinked.

reinforcement Increase in *reproductive isolation* between incipient *species* by *natural selection*. Natural selection in most species can only directly favor an increase in *prezygotic isolation*; reinforcement therefore amounts to selection for *assortative mating* between the incipiently speciating forms.

reproductive isolation Two populations, or individuals of opposite sex, are reproductively isolated from each other if they cannot together produce fertile offspring.

ribosomal RNA (rRNA) Kind of *RNA* that constitutes the ribosomes and provides the site for *translation*.

ribosome Site of *protein* synthesis (or *translation*) in the cell, and mainly consisting of *ribosomal RNA*.

ring species Situation in which two reproductively isolated populations (*see reproductive isolation*), living in the same region, are connected by a geographic ring of populations that can interbreed.

RNA Ribonucleic acid. *Messenger RNA, ribosomal RNA,* and *transfer RNA* are its three main forms. They act as the intermediaries by which the hereditary code of *DNA* is converted into proteins. In some viruses, RNA is itself the hereditary molecule.

selection Synonym of *natural selection*.

selectionism Theory that some classes of evolutionary events, such as molecular or phenotypic changes, have mainly been caused by *natural selection*.

separate creation *See creationism.*

sex chromosome *Chromosome* that influences sex determination. In mammals, including humans, the X and Y chromosomes are the sex chromosomes (females are XX, males XY). (Compare with *autosome*.)

sexual selection *Selection* on mating behavior, either through competition among members of one sex (usually males) for access to members of the other sex or through choice by members of one sex (usually females) of certain members of the other sex. Sexual selection works on the *fitness* of a *genotype* relative to other genotypes in individuals of the same sex, whereas natural selection works on the fitness of a *genotype* relative to the whole *population*.

soma The mortal cell lines in a body. (Compare *germ plasm*.)

spacer region Sequence of *nucleotides* in the *DNA* between coding *genes*.

species An important classificatory category, which can be variously defined by the *biological species concept, cladistic species concept, ecological species concept, phenetic species concept,* and *recognition species concept*. The biological species concept, according to which a species is a set of interbreeding organisms, is the most widely used definition, at least by biologists who study vertebrates. A particular species is referred to by a *Linnaean* binomial, such as *Homo sapiens* for human beings.

stabilizing selection *Selection* tending to keep the form of a *population* constant: individuals with the mean value for a *character* have high *fitness*, those with extreme values have low fitness.

stepped cline *Cline* with a sudden change in *gene* (or character) *frequency*.

substitution The evolutionary replacement of one allele by another in a population.

sympatric speciation Speciation via populations with overlapping geographic ranges.

systematics A near synonym of *taxonomy*.

taxon (taxa) Any named taxonomic group, such as the family Felidae, or the genus *Homo*, or species *Homo sapiens*.

taxonomy Theory and practice of biological *classification*.

tetrapod A member of the group made up of amphibians, reptiles, birds, and mammals.

transcription Process by which *messenger RNA* is read off the *DNA* forming a *gene*.

transfer RNA (tRNA) Kind of *RNA* that brings the *amino acids* to the *ribosomes* to make *proteins*. There are 20 kinds of transfer RNA molecules, one for each of the 20 main amino acids. A transfer RNA molecule has an amino acid attached to it, and has the anti-*codon* corresponding to that amino acid in another part of its structure. In protein synthesis, each codon in the *messenger RNA* combines with the appropriate tRNA's anti-*codon*, and the amino acids are thus arranged in order to make the protein.

transformism Evolutionary theory of Lamarck, in which changes occur within a *lineage* of *populations*, but in which lineages to do not split (i.e. there is no *speciation*, at least not in the sense of the *cladistic species concept*) and do not become extinct.

transition *Mutation* changing a *purine* into the other purine, or a *pyrimidine* into the other pyrimidine (i.e. changes from adenosine to guanine or vice versa and changes from cytosine to thymine or vice versa).

translation The process by which a *protein* is manufactured at a *ribosome*, using *messenger RNA* code and *transfer RNA* to supply the *amino acids*.

transversion Mutation changing a *purine* into a *pyrimidine* or vice versa (i.e. changes from adenine or guanine to cytosine or thymine and changes from cytosine or thymine to adenosine or guanine).

typology (1) Definition of classificatory groups by *phenetic* similarity to a "type" specimen. A *species*, for example, might be defined as all individuals less than x phenetic units from the species' type. (2) Theory that distinct "types" exist in nature, perhaps because they are part of some *plan of nature*. (See also *idealism*.) The type of the species is then the most important form of it, and variants around that type are noise or mistakes. *Neo-Darwinism* opposes typology because in a *gene pool* no one variant is any more important than any others.

unequal crossing over *Crossing over* in which the two *chromosomes* do not exchange equal lengths of *DNA*: one receives more than the other.

variance A measure of how variable a set of numbers are. Technically, it is the sum of squared deviations from the *mean* divided by n (the number of numbers in the sample). Thus to find the variance of the set of numbers, 4, 6, and 8, we first calculate the mean, which is 6; then sum the squared deviations from the mean $(4 - 6)^2 + (6 - 6)^2 + (8 - 6)^2$, which comes to 8; and divide by n (which is 3 here). The variance of the three numbers is $\frac{8}{3} = 2.67$. The more variable a set of numbers are, the higher the variance. The variance of a set of identical numbers (such as 6, 6, and 6) is zero.

virus A kind of intracellular parasite, consisting of nucleic acid surrounded by *protein*, that can only replicate inside a living cell. (Less formally, on Medawar's definition, "a piece of bad news wrapped in a protein.")

vitamin A member of a chemically heterogeneous class of organic compounds that is essential, in small quantities, for life.

wild type The normal form of members of a species, as distinct from a rare mutant form.

zygote The cell formed by the fertilization of male and female gametes.

References

Alberts, B., Bray, D., Lewis, J., Raff, M., Roberts, K. & Watson, J.D. (1989) *Molecular Biology of the Cell*, 2nd edn. Garland, New York.

Alexander, R.D. & Borgia, G. (1978) Group selection, altruism, and the levels of organization of life. *Ann. Rev. Ecol. System.* **9**: 449–474.

Alexander, R.McN. (1985) The ideal and the feasible: physical constraints on evolution. *Biol J. Linn. Soc.* **26**: 345–358.

Alvarez, L.W., Alvarez, W., Asaro, F. & Michel, H.V. (1980) Extraterrestrial cause for the Cretaceous–Tertiary extinction. *Science* 208: 1095–1108.

Alvarez, W. & Asaro, F. (1990) An extraterrestrial impact. *Sci. Am.* **263** (October): 78–84.

Alvarez, W., Asaro, F. & Montanari, A. (1990) Iridium profile for 10 million years across the Cretaceous–Tertiary boundary at Gubbio (Italy). *Science* **250**: 1700–1702.

Alvarez, W., Kauffman, E.G., Surlyk, F., Alvarez, L.W., Asaro, F. & Michel, H.V. (1984) Impact theory of mass extinctions and the invertebrate fossil record. *Science* **233**: 1135–1141.

Andrews, P. (1986) Molecular evidence for catarrhine evolution. In: Wood, B., Martin, L. & Andrews, P. (eds) *Major Topics in Primate and Human Evolution*, pp. 107–129. Cambridge University Press, Cambridge, UK.

Antonovics, J., Bradshaw, A.D. & Turner, J.R.G. (1971) Heavy metal tolerance in plants. *Adv. Ecol. Res.* **7**: 1–85.

Antonovics, J., Ellstrand, N.C. & Brandon, R.N. (1988) Genetic variation and environmental variation: expectations and experiments. In: Gottlieb, L.D. & Jain, S.K. (eds) *Plant Evolutionary Biology*, pp. 275–303. Chapman & Hall, London.

Arnheim, N. (1983) Concerted evolution of multigene families. In: Nei, M. & Koehn, R.K. (eds) *Evolution of Genes and Proteins*, pp. 38–61. Sinauer Associates, Sunderland, Massachusetts.

Atchley, W.R. & Woodruff, D.S. (eds) (1981) *Evolution and Speciation.* Cambridge University Press, New York.

Baker, R.R. (1978) *The Evolutionary Ecology of Animal Migration.* Hodder & Stoughton, London.

Baker, R.R. & Parker, G.A. (1979) The evolution of bird coloration. *Phil. Trans. R. Soc. Lond.* B **287**: 63–130.

Barigozzi, C. (ed.) (1982) *Mechanisms of Speciation.* A.R. Liss, New York.

Barthélemy-Madaule, M. (1982) *Lamarck the Mythical Precursor.* MIT Press, Cambridge, Massachusetts.

Barton, N.H. (1988) Speciation. In: Myers, A.A. & Giller, P.S. (eds) *Analytical Biogeography*, pp. 185–218. Chapman & Hall, London.

Barton, N.H. (1989) Founder effect speciation. In: Otte, D. & Endler, J.A. (eds) *Speciation and its Consequences*, pp. 229–256. Sinauer Associates, Sunderland, Massachusetts.

Barton, N.H. & Charlesworth, B. (1984) Genetic revolutions, founder effects, and speciation. *Ann. Rev. Ecol. System.* **15**: 133–164.

Barton, N.H. & Hewitt, G.M. (1985) Analysis of hybrid zones. *Ann. Rev. Ecol. System.* **16**: 113–148.

Barton, N.H. & Hewitt, G.M. (1989) Adaptation, speciation, and hybrid zones. *Nature* **341**: 497–503.

Barton, N.H. & Turelli, M. (1989) Evolutionary quantitative genetics: how much do we know? *Ann. Rev. Genet.* **23**: 337–370.

Beer, G.R. de (1931) *Embryology and Evolution.* Oxford University Press, Oxford, UK.

Beer, G.R. de (1940) *Embryos and Ancestors.* (2nd edition, 1958.) Oxford University Press, Oxford, UK.

Begon, M., Harper, J.L. & Townsend, C.R. (1990) *Ecology: Individuals, Populations, and Communities*, 2nd edn. Blackwell Scientific Publications, Cambridge, Massachusetts.

Bell, G. (1982) *The Masterpiece of Nature.* University of California Press, Berkeley.

Bendall, D.S. (ed.) (1983) *Evolution from Molecules to Men.* Cambridge University Press, Cambridge, UK.

Benson, S.A., Partridge, L. & Morgan, M.J. (1988) Is bacterial evolution random or selective? *Nature* **336**: 21–22.

Bentley, D. & Hoy, R.R. (1974) The neurobiology of cricket song. *Sci. Am.* **231** (August): 34–44.

Benton, M.J. (1983a) Dinosaur success in the Triassic: a noncompetitive ecological model. *Quart. Rev. Biol.* **58**: 29–55.

Benton, M.J. (1983b) Large-scale replacements in the history of life. *Nature* **302**: 16–17.

Benton, M.J. (1987) Progress and competition in macroevolution. *Biol. Rev.* **62**: 305–338.

Benton, M.J. (ed.) (1988) *The Phylogeny and Classification of the Tetrapods.* Oxford University Press, Oxford, UK.

Benton, M.J. (1990) *Vertebrate Paleontology: Biology and*

Evolution. Unwin Hyman, Boston.

Berlin, B. (1973) Folk systematics in relation to biological classification and nomenclature. *Ann. Rev. Ecol. System.* **4**: 259–271.

Bernstein, H., Hopf, F.A. & Michod, R.E. (1988) Is meiotic recombination an adaptation for repairing DNA, producing genetic variation, or both? In: Michod, R.E. & Levin, B.R. (eds) *The Evolution of Sex*, pp. 139–160. Sinauer Associates, Sunderland, Massachusetts.

Bertram, B. (1978) *Pride of Lions.* Scribner, New York.

Bishop, M.J. & Friday, A.E. (1988) Estimating the interrelationships of tetrapod groups on the basis of molecular sequence data. In: Benton, M.J. (ed.) *The Phylogeny and Classification of the Tetrapods*, vol. 1, pp. 33–58. Oxford University Press, Oxford, UK.

Black, R.M. (1989) *The Elements of Palaeontology*, 2nd edn. Cambridge University Press, Cambridge, UK.

Blair, W.F. (1964) Isolating mechanisms and interspecies interactions in anuran amphibians. *Quart. Rev. Biol.* **39**: 334–344.

Bodmer, W.F. (1983) Gene clusters and genome evolution. In: Bendall, D.S. (ed.) *Evolution from Molecules to Men*, pp. 197–208. Cambridge University Press, Cambridge, UK.

Bodmer, W.F. & Cavalli-Sforza, L.L. (1976) *Genetics, Evolution, and Man.* W.H. Freeman, San Francisco.

Bond, W.J. (1989) The tortoise and the hare: ecology of angiosperm dominance and gymnosperm persistence. *Biol. J. Linn. Soc.* **36**: 227–249.

Bonner, J.T. (ed.) (1982) *Evolution and Development.* Dahlem Workshop. Springer-Verlag, Berlin.

Bonner, J.T. (1988) *The Evolution of Complexity.* Princeton University Press, Princeton.

Bookstein, F.L. (1977) The study of shape transformation after D'Arcy Thompson. *Mathematical Biosciences* **34**: 177–219.

Boucot, A.J. (1988) Periodic extinctions within the Cainozoic. *Nature* **331**: 395–396.

Bowler, P.J. (1989) *Evolution: the History of an Idea*, revised edn. University of California Press, Berkeley.

Box, J.F. (1978) *R.A. Fisher. The Life of a Scientist.* John Wiley, New York.

Bradbury, J.W. & Andersson, M. (eds) (1987) *Sexual Selection: Testing the Alternatives.* Dahlem Workshop. John Wiley, Chichester, UK.

Bradshaw, A.D. (1971) Plant evolution in extreme environments. In: Creed, E.R. (ed.) *Ecological Genetics and Evolution*, pp. 20–50. Appleton-Century-Croft, New York.

Brakefield, P.M. (1987) Industrial melanism: do we have the answers? *Trends Ecol. Evol.* **2**: 117–122.

Bramble, D.M. & Jenkins, F.A. (1989) Structural and functional integration across the reptile–mammal boundary: the locomotor system. In: Wake, D.B. & Roth, G. (eds) *Complex Organismal Functions: integration and evolution in vertebrates*, pp. 133–146. Dahlem Workshop. John Wiley, Chichester, UK.

Brandon, R.N. (1990) *Adaptation and Environment.* Princeton University Press, Princeton.

Brenner, S. (1992) Dicing with Darwin. *Current Biology* **2**: 167–168.

Brown, D.D., Wensink, P.C. & Jordan, E. (1972) A comparison of the ribosomal DNAs of *Xenopus laevis* and *Xenopus mulleri*: the evolution of tandem genes. *J. Mol. Biol.* **63**: 57–73.

Brooks, D.R. & McLennan, D.A. (1991) *Phylogeny, Ecology, and Behavior.* University of Chicago Press, Chicago.

Brooks, D.R. & Wiley, E.O. (1984) Evolution as an entropic phenomenon. In: Pollard, J.W. (ed.) *Evolutionary Theory: Paths into the future*, pp. 141–171. John Wiley, New York.

Brooks, D.R. & Wiley, E.O. (1988) *Evolution as Entropy: Toward a Unified Theory of Biology*, 2nd edn. University of Chicago Press, Chicago.

Brown, J.H. & Gibson, A.C. (1983) *Biogeography.* Mosby, St. Louis, Missouri.

Brown, W.L. (1987) Punctuated equilibrium excused: the original examples fail to support it. *Biol. J. Linn. Soc.* **31**: 383–404.

Brown, W.L. & Wilson, E.O. (1956) Character displacement. *System. Zool.* **5**: 49–64.

Browne, J. (1983) *The Secular Ark: Studies in the History of Biogeography.* Yale University Press, New Haven.

Brundin, L. (1988) Phylogenetic biogeography. In: Myers, A.A. & Giller, P.S. (eds) *Analytical Biogeography*, pp. 343–369. Chapman & Hall, London.

Bull, J.J. & Charnov, E.L. (1988) How fundamental are Fisherian sex ratios? *Oxford Surv. Evol. Biol.* **5**: 96–135.

Bulmer, M. (1980) *The Mathematical Theory of Quantitative Genetics.* Oxford University Press, Oxford, UK.

Bulmer, M. (1987) Coevolution of codon usage and tRNA abundance. *Nature* **325**: 728–730.

Bulmer, M. (1988) Estimating the variability of substitution rates. *Genetics* **123**: 615–619.

Bulmer, M.G. (1989) Maintenance of genetic variability by mutation–selection balance: a child's guide through the jungle. *Genome* **31**: 761–767.

Bulmer, M. (1991) Use of the method of generalized least squares in reconstructing phylogenies from sequence data. *Mol. Biol. Evol.* **8**: 868–883.

Bulmer, M., Wolfe, K.H. & Sharp, P.M. (1991) Synonymous nucleotide substitution rates in mammalian genes: implications for the molecular clock and the relationships of mammalian orders. *Proc. Nat. Acad. Sci. USA* **88**: 5974–5978.

Burke, T. (1989) DNA fingerprinting and other methods for the study of mating success. *Trends Ecol. Evol.* **4**: 139–144.

Burkhardt, F. & Smith, S. (eds) (1985–) *The Correspondence of Charles Darwin*, vol. 1–. Cambridge University Press, Cambridge, UK.

Burkhardt, R.W. (1977) *The Spirit of System: Lamarck and Evolutionary Biology.* Harvard University Press, Cambridge, Massachusetts.

Bush, G.L. (1975) Modes of animal speciation. *Ann. Rev.*

Ecol. System. **6**: 339–364.

Bush, G.L., Case, S.M., Wilson, A.C. & Patton, J.L. (1977) Rapid speciation and chromosomal evolution in mammals. *Proc. Nat. Acad. Sci. USA* **74**:3942–3946.

Buss, L.W. (1987) *The Evolution of Individuality*. Princeton University Press, Princeton.

Butlin, R.K. (1987a) Speciation by reinforcement. *Trends Ecol. Evol.* **2**: 8–13.

Butlin, R.K. (1987b) A new approach to sympatric speciation. *Trends Ecol. Evol.* **2**: 310–311.

Butlin, R.K. (1989) Reinforcement of premating isolation. In: Otte, D. & Endler, J.A. (eds) *Speciation and its Consequences*, pp. 158–179. Sinauer Associates, Sunderland, Massachusetts.

Cain, A.J. (1954) *Animal Species and their Evolution*. Hutchinson, London.

Cain, A.J. (1964) The perfection of animals. In: Carthy, J.D. & Duddington, C.L. (eds) *Viewpoints in Biology*, vol. 3, pp. 36–63. Butterworths, London. (Reprinted in *Biol. J. Linn. Soc.* (1989) **36**: 3–29.)

Cain, A.J. & Sheppard, P.M. (1954) Natural selection in *Cepaea*. *Genetics* **39**: 89–116.

Cairns, J., Overbaugh, J. & Miller, S. (1988) The origin of mutants. *Nature* **335**: 142–145.

Camin, J.H. & Ehrlich, P.R. (1958) Natural selection in water snakes (*Natrix sipedon* L.) on islands in Lake Erie. *Evolution* **12**: 504–511.

Campbell, K.S.W. & Day, M.F. (eds) (1987) *Rates of Evolution*. Allen & Unwin, London.

Carroll, R.L. (1988) *Vertebrate Paleontology and Evolution*. W.H. Freeman, New York.

Carson, H.L. (1983) Chromosomal sequences and interisland colonizations in Hawaiian *Drosophila*. *Genetics* **103**: 465–482.

Carson, H.L. (1990) Evolutionary process as studied in population genetics: clues from phylogeny. *Oxford Surv. Evol. Biol.* **7**: 129–156.

Carson, H.L. (1992) The Galápagos that were. *Nature* **355**: 202–203.

Carson, H.L. & Kaneshiro, K.Y. (1976) *Drosophila* of Hawaii: systematics and ecological genetics. *Ann. Rev. Ecol. System.* **7**: 311–345.

Cavalier-Smith, T. (ed.) (1985) *The Evolution of Genome Size*. John Wiley, New York.

Cavalli-Sforza, L.L. & Bodmer, W.F. (1971) *Genetics of Human Populations*. W.H. Freeman, San Francisco.

Charlesworth, B. (1980) *Evolution in Age-structured Populations*. Cambridge University Press, Cambridge, UK.

Charlesworth, B. & Langley, C.H. (1991) Population genetics of transposable elements in *Drosophila*. In Selander, R.K., Clark, A.G. & Whittam, T.S. (eds) *Evolution at the Molecular Level*, pp. 150–176. Sinauer Associates, Sunderland, Massachusetts.

Charlesworth, B., Lande, R. & Slatkin, M. (1982) A neo-Darwinian commentary on macroevolution. *Evolution* **36**: 474–498.

Charlesworth, D., Charlesworth, B., Bull, J.J. *et al.* (1988)

The origin of mutants disputed. *Nature* **336**: 525–528.

Charnov, E.L. (1982) *The Theory of Sex Allocation*. Princeton University Press, Princeton.

Christie, D.M., Duncan, R.A., McBirney, A.R. *et al.* (1992) Drowned islands downstream from the Galápagos hotspot imply extended speciation times. *Nature* **355**: 246–248.

Cifelli, R.L. (1981) Patterns of evolution among the Artiodactyla and Perissodactyla (Mammalia). *Evolution* **35**: 433–440.

Clark, A.G. & Kao, T-H (1991) Excess synonymous substitutions at shared polymorphic sites among self-incompatibility alleles of Solanaceae. *Proc. Natl. Acad. Sci. USA* **88**: 9823–9827.

Clark, R.W. (1969) *JBS: The Life and Work of J.B.S. Haldane*. Coward-McCann, New York.

Clarke, B.C. (1979) The evolution of genetic diversity. *Proc. R. Soc. Lond. B* **205**: 453–474.

Clarke, C.A. & Sheppard, P.M. (1969) Further studies on the genetics of the mimetic butterfly *Papilio memnon*. *Phil. Trans. R. Soc. Lond. B* **263**: 35–70.

Clarke, C.A., Sheppard, P.M. & Thornton, I.W.B. (1968) The genetics of the mimetic butterfly *Papilio memnon*. *Phil. Trans. R. Soc. Lond. B* **254**: 37–89.

Clarke, G.M. & McKenzie, J.A. (1987) Developmental stability of insecticide resistant phenotypes in blowfly: a result of canalizing selection. *Nature* **325**: 345–346.

Clayton, G.A. & Robertson, A. (1955) Mutation and quantitative variation. *Am. Nat.* **89**: 151–158.

Clegg, M.T. & O'Brien, S.J. (eds) (1990) *Molecular Evolution*. Wiley–Liss, New York.

Clutton-Brock, T.H. & Harvey, P.H. (1984) Comparative approaches to investigating adaptation. In: Krebs, J.R. & Davies, N.B. (eds) *Behavioural Ecology*, pp. 7–29. Blackwell Scientific Publications, Oxford, UK.

Clutton-Brock, T.H., Albon, S.D. & Guinness, F.E. (1984) Maternal dominance, breeding success, and birth sex ratios in red deer. *Nature* **308**: 358–360.

Cook, A. (1975) Changes in the carrion/hooded crow hybrid zone and the possible importance of climate. *Bird Study* **22**: 165–168.

Coope, G.R. (1978) Constancy of insect species versus inconstancy of Quaternary environments. In: Mound, L.A. & Waloff, N. (eds) *Diversity of Insect Faunas*, pp. 176–187. Blackwell Scientific Publications, Oxford, UK.

Coope, G.R. (1979) Late Cenozoic fossil Coleoptera: evolution, biogeography, and ecology. *Ann. Rev. Ecol. System.* **10**: 247–267.

Corsi, P. (1988) *The Age of Lamarck*. University of California Press, Berkeley, California.

Cosmides, L.M. & Tooby, J. (1981) Cytoplasmic inheritance and intragenomic conflict. *J. Theoret. Biol.* **89**: 83–129.

Courtillot, V.E. (1990) A volcanic eruption. *Sci. Am.* **263** (October): 85–92.

Cox, C.B. & Moore, P.D. (1985) *Biogeography*, 4th edn. Blackwell Scientific Publications, Cambridge, Massachusetts.

Coyne, J.A. (1992) Genetics and speciation. *Nature*, **355**: 511–515.

Coyne, J.A. & Orr, H.A. (1989a) Patterns of speciation in Drosophila. *Evolution* **43**: 362–381.

Coyne, J.A. & Orr, H.A. (1989b) Two rules of speciation. In: Otte, D. & Endler, J.A. (eds) *Speciation and its Consequences*, pp. 180–207. Sinauer Associates, Sunderland, Massachusetts.

Coyne, J.A., Orr, H.A. & Futuyma, D.J. (1989) Do we need a new species concept? *System. Zool.* **37**: 190–200.

Creed, E.R., Lees, D.R. & Bulmer, M.G. (1980) Pre-adult viability differences of melanic *Biston betularia* (L.) (Lepidoptera). *Biol. J. Linn. Soc.* **13**: 25–62.

Croizat, L. (1962) *Space, Time, Form: the Biological Synthesis*. Published by the author, Caracas. (Available from Wheldon & Wesley, Lytton Lodge, Codicote, Herts, UK.)

Croizat, L., Nelson, G. & Rosen, D.E. (1974) Centers of origin and related concepts. *System. Zool.* **23**: 265–287.

Cronin, T.M. & Schneider, C.E. (1990) Climatic influences on species: evidence from the fossil record. *Trends Ecol. Evol.* **5**: 275–279.

Crow, J.F. (1979) Genes that violate Mendel's rules. *Sci. Am.* **240** (February): 104–113.

Crow, J.F. (1986) *Basic Concepts in Population, Quantitative, and Evolutionary Genetics*. W.H. Freeman, New York.

Crow, J.F. & Kimura, M. (1970) *An Introduction to Population Genetics Theory*. Harper & Row, New York.

Crowley, T.J. & North, G.R. (1988) Abrupt climate change and extinction events in Earth history. *Science* **240**: 996–1002.

Crowson, R.A. (1970) *Classification and Biology*. Atherton Press, New York.

Curio, E. (1973) Towards a methodology for teleonomy. *Experientia* **29**: 1045–1058.

Curtis, C.F., Cook, L.M. & Wood, R.J. (1978) Selection for and against insecticide resistance and possible methods of inhibiting the evolution of resistance in mosquitoes. *Ecol. Entomol.* **3**: 273–287.

Cushing, D.H. (1975) *Marine Ecology and Fisheries*. Cambridge University Press, Cambridge, UK.

Cvancara, A.M. (1990) *Sleuthing Fossils*. John Wiley, New York.

Czelusniak, J., Goodman, M., Hewett-Emmett, D., Weiss, M.L., Venta, P.J. & Tashian, R.E. (1982) Phylogenetic origins and adaptive evolution of avian and mammalian haemoglobin genes. *Nature* **298**: 297–300.

Dalrymple, G.B. & Lanphere, M.A. (1969) *Potassium–Argon Dating: Principles, Techniques, and Applications to Geochronology*. W.H. Freeman, San Francisco.

Damuth, J. (1985) Selection among "species": a formulation in terms of natural functional units. *Evolution* **39**: 1132–1146.

Danciger, E., Dafni, N., Bernstein, Y., *et al.* (1986) Human Cu/Zn superoxide dismutase gene family. *Proc. Natl. Acad. Sci. USA* **83**: 3619–3623.

Darwin, C.R. (1859) *On the Origin of Species*. John Murray, London.

Darwin, C.R. (1871) *The Descent of Man, and Selection in Relation to Sex*. John Murray, London. (Paperback edition: Princeton University Press, Princeton.)

Darwin, C.R. (1872) *The Expression of the Emotions in Man and Animals*. John Murray, London.

Darwin, F. (ed.) (1887) *The Life and Letters of Charles Darwin*, vols 1–3. John Murray, London.

Darwin, F. & Seward, A.C. (eds) (1903) *More letters of Charles Darwin*. 2 vols. John Murray, London.

Dawkins, R. (1976) *The Selfish Gene*. Oxford University Press, Oxford, UK. (Second edition, 1989.)

Dawkins, R. (1982) *The Extended Phenotype*. W.H. Freeman, Oxford, UK. (Paperback edition: Oxford University Press, Oxford, UK.)

Dawkins, R. (1986) *The Blind Watchmaker*. W.W. Norton, New York and Longman, London.

Dawkins, R. & Krebs, J.R. (1979) Arms races between and within species. *Proc. R. Soc. Lond. B* **205**: 489–511.

Dean, G. (1972) *The Porphyrias*. Pitman, London.

Desmond, A. & Moore, J. (1992) *Darwin*. Warner Books, New York.

Devonshire, A.L. & Field, L.M. (1991) Gene amplification and insecticide resistance. *Ann. Rev. Entomol.* **36**: 1–23.

Diamond, J. (1990a) Alone in a crowded universe. *Natural History* June 1990. 30–34.

Diamond, J. (1990b) Biological effects of ghosts. *Nature* **345**: 769–770.

Dobzhansky, T. (1937) *Genetics and the Origin of Species*. Columbia University Press, New York.

Dobzhansky, T. (1951) *Genetics and the origin of species*. 3rd edn. Columbia University Press, New York.

Dobzhansky, T. (1970) *Genetics of the Evolutionary Process*. Columbia University Press, New York.

Dobzhansky, T. & Pavlovsky, O. (1957) An experimental study of interaction between genetic drift and natural selection. *Evolution* **11**: 311–319.

Dominey, W.J. (1984) Effects of sexual selection and life history on speciation: species flocks in African cichlids, and Hawaiian *Drosophila*. In: Echelle, A.A. & Kornfield, I. (eds) *Evolution of Fish Species Flocks*. University of Maine at Orono Press, Orono, Maine.

Donoghue, M.J., Doyle, J.A., Gauthier, J., Kluge, A.G. & Rowe, T. (1989) The importance of fossils in phylogeny reconstruction. *Ann. Rev. Ecol. System.* **20**: 431–460.

Donovan, S.K. (ed.) (1989) *Mass Extinctions: Processes and Evidence*. Columbia University Press, New York.

Donovan, S.K. (ed.) (1991) *The Process of Fossilization*. Columbia University Press, New York.

Doolittle, W.F. & Sapienza, C. (1980) Selfish genes, the phenotype paradigm, and genome evolution. *Nature* **284**: 601–603.

Dover, G. A. (1986) Molecular drive in multigene families: how biological novelties arise, spread, and are assimilated. *Trends Genet.* **2**: 159–165.

Dudley, J.W. (1977) 76 generations of selection for oil and protein percentage in maize. In: Pollack, E., Kempthorne,

O. & Bailey, T.B. (eds) *Proceedings of the International Conference on Quantitative Genetics*, pp. 459–473. Iowa State University Press, Ames, Iowa.

Dybas, H.S. & Lloyd, M. (1962) Isolation by habitat in two synchronized species of periodical cicadas (Homoptera: Cicadidae: *Magicicada*). *Ecology* 43: 444–459.

Eberhard, W.G. (1980) Evolutionary consequences of intracellular competition. *Quart. Rev. Biol.* 55: 231–249.

Eberhard, W.G. (1985) *Sexual Selection and Animal Genitalia.* Harvard University Press, Cambridge, Massachusetts.

Eberhard, W.G. (1990) Evolution in bacterial plasmids and levels of selection. *Quart. Rev. Biol.* 65: 3–22.

Edwards, A.W.F. (1977) *Foundations of Mathematical Genetics.* Cambridge University Press, Cambridge, UK.

Ehrlich, P.R. & Raven, P.H. (1964) Butterflies and plants: a study in coevolution. *Evolution* 18: 586–608.

Ehrlich, P.R. & Raven, P.H. (1969) Differentiation of populations. *Science* 165: 1228–1232.

Eldredge, N. (1982) *The Monkey Business: a Scientist Looks at Creationism.* Washington Square Press, New York.

Eldredge, N. (1985a) *Unfinished Synthesis: Biological Hierarchies and Modern Evolutionary Thought.* Oxford University Press, New York.

Eldredge, N. (1985b) *Timeframes.* Simon & Schuster, New York.

Eldredge, N. (1989) *Macroevolutionary Dynamics, Species, Niches, and Adaptive Peaks.* McGraw-Hill, New York.

Eldredge, N. (1991) *The Miner's Canary: Unraveling the Mysteries of Extinction.* Simon & Schuster, New York.

Eldredge, N. & Gould, S.J. (1972) Punctuated equilibria: an alternative to phyletic gradualism. In: Schopf, T.J.M. (ed.) *Models in Paleobiology*, pp. 82–115. Freeman, Cooper, & Co, San Francisco.

Eldredge, N. & Gould, S.J. (1988) Punctuated equilibrium prevails. *Nature* 332: 211–212.

Eldredge, N. & Stanley, S.M. (eds) (1984) *Living Fossils.* Springer-Verlag, New York.

Elliott, D.K. (ed.) (1986) *Dynamics of Extinction.* John Wiley, New York.

Endler, J.A. (1977) *Geographic Variation. Speciation and Clines* Princeton University Press, Princeton.

Endler, J.A. (1982a) Problems in distinguishing historical from ecological factors in biogeography. *Am. Zool.* 22: 441–452.

Endler, J.A. (1982b) Pleistocene forest refuges: fact or fancy? In: Prance, G.T. (ed.) *Biological Diversification in the Tropics*, pp. 641–657. Columbia University Press, New York.

Endler, J.A. (1986) *Natural Selection in the Wild.* Princeton University Press, Princeton.

Endler, J.A. (1989) Conceptual and other problems in speciation. In: Otte, D. & Endler, J.A. (eds) *Speciation and its Consequences*, pp. 625–648. Sinauer Associates, Sunderland, Massachusetts.

Erwin, D.H. (1989) The end-Permian mass extinction: what really happened and did it matter? *Trends Ecol. Evol.* 4: 225–229.

Erwin, D.H. (1990) The End-Permian mass extinction. *Ann. Rev. Ecol. System.* 21: 69–92.

Ewens, W.J. (1979) *Mathematical Population Genetics.* Springer-Verlag, Berlin.

Falconer, D.S. (1981) *Introduction to Quantitative Genetics*, 2nd edn. Longman, London.

Feder, J.L., Chilcote, C.A. & Bush, G.L. (1988) Genetic differentiation between sympatric host races of the apple maggot fly *Rhagoletis pomonella*. *Nature* 336: 61–64.

Felsenstein, J. (1982) Numerical methods for inferring phylogenetic trees. *Quart. Rev. Biol.* 57: 379–404.

Felsenstein, J. (1985) Recombination and sex: is Maynard Smith necessary? In: Greenwood, P.J., Harvey, P.H. & Slatkin, M. (eds) *Evolution: Essays in Honour of John Maynard Smith*, pp. 209–220. Cambridge University Press, Cambridge, UK.

Felsenstein, J. (1988a) Phylogenies from molecular sequences: inference and reliability. *Ann. Rev. Genet.* 22: 521–565.

Felsenstein, J. (1988b) Perils of molecular introspection. *Nature* 335: 118.

Fenster, E.J. & Sorhannus, U. (1991) On the measurement of morphological rates of evolution: a review. *Evol. Biol.* 25: 375–410.

Fernholm, B., Bremer, K. & Jörnvall, H. (eds) (1989) *The Hierarchy of Life.* Excerpta Medica, Amsterdam.

Fisher, R.A. (1918) The correlation between relatives under the supposition of Mendelian inheritance. *Trans. R. Soc. Edinburgh* 52: 399–433.

Fisher, R.A. (1930) *The Genetical Theory of Natural Selection.* Oxford University Press, Oxford, UK.

Fisher, R.A. & Ford, E.B. (1947) The spread of a gene in natural conditions in a colony of the moth *Panaxia dominula* L. *Heredity* 1: 143–174.

Fitch, W.M. (1976) Molecular evolutionary clocks. In: Ayala, F.J. (ed.) *Molecular Evolution*, pp. 160–178. Sinauer Associates, Sunderland, Massachusetts.

Flessa, K.W., Barnett, S.G., Cornue, D.B., *et al.* (1979) Geologic implications of the relationship between mammalian faunal similarity and geographic distance. *Geology* 7: 15–18.

Ford, E.B. (1975) *Ecological Genetics*, 4th edn. Chapman & Hall, London.

Fortey, R. (1982) *Fossils: the Key to the Past.* Van Nostrand Reinhold, New York.

Frazzetta, T.H. (1975) *Complex Adaptations in Evolving Populations.* Sinauer Associates, Sunderland, Massachusetts.

Friday, A.E. (1987) Models of evolutionary change and the estimation of evolutionary trees. *Oxford Surv. Evol. Biol.* 4: 61–88.

Fryer, G., Greenwood, P.H. & Peake, J.F. (1985) The demonstration of speciation in fossil molluscs and living fishes. *Biol. J. Linn. Soc.* 26: 325–336.

Futuyma, D.J. (1983) *Science on Trial: the Case for Evolution.* Pantheon Books, New York.

Futuyma, D.J. & Slatkin, M. (eds) (1983) *Coevolution.* Sinauer Associates, Sunderland, Massachusetts.

Gale, J.S. (1990) *Theoretical Population Genetics*. Unwin Hyman, Boston.

Gardiner, B.G. (1982) Tetrapod classification. *Zool. J. Linn. Soc.* **74**: 207–232.

Gentry, G.A., Lowe, M., Alford, G. & Nevins, R. (1988) Sequence analyses of herpesviral enzymes suggest an ancient origin for human sexual behavior. *Proc. Natl. Acad. Sci. USA* **85**: 2658–2661.

Ghiselin, M.T. (1966) On psychologism in the logic of taxonomic controversies. *System. Zool.* **15**: 207–215.

Ghiselin, M.T. (1969) *The Triumph of the Darwinian Method*. University of California Press, Berkeley.

Ghiselin, M.T. (1974) *The Economy of Nature and the Evolution of Sex*. University of California Press, Berkeley.

Ghiselin, M.T. (1984) 'Definition', 'character', and other equivocal terms. *System. Zool.* **33**: 104–110.

Ghiselin, M.T. (1987) Species concepts, individuality, and objectivity. *Biol. Phil.* **2**: 127–143.

Gibbs, H.L & Grant, P.R. (1987) Oscillating selection in Darwin's finches. *Nature* **327**: 511–513.

Giddings, L.V., Kaneshiro, K.Y. & Anderson, W.W. (eds) (1989) *Genetics, Speciation, and the Founder Principle*. Oxford University Press, New York.

Giddings, L.V. & Templeton, A.R. (1983) Behavioral phylogenies and the direction of evolution. *Science* **220**: 372–378.

Gilinsky, N.L. (1986) Species selection as a causal process. *Evol. Biol.* **20**: 249–273.

Gilinsky, N.L. (1988) Survivorship in the Bivalvia: comparing living and extinct genera and families. *Paleobiology* **14**: 370–386.

Gilinsky, N.L. & Bambach, R.K. (1987) Asymmetrical patterns of origination and extinction in higher taxa. *Paleobiology* **13**: 427–445.

Gillespie, J.H. (1992) *The Causes of Molecular Evolution*. Oxford University Press, New York.

Gingerich, P.D. (1983) Rates of evolution: effects of time and temporal scaling. *Science* **222**: 159–161.

Gingerich, P.D. (1984) [reply to Gould 1984, q.v.] *Science* **226**: 995–996.

Gingerich, P.D., Smith, B.H. & Simons, E.L. (1990) Hind limbs of Eocene *Basilosaurus*: evidence of feet in whales. *Science* **249**: 154–157.

Glass, B. (ed.) (1980) *The Roving Naturalist: Travel Letters of Theodosius Dobzhansky*. American Philosophical Society, Philadephia.

Golding, G.B. (1987) Nonrandom patterns of mutation are reflected in evolutionary divergence and may cause some of the unusual patterns observed in sequences. In: Loeschcke, V. (ed.) *Genetic Constraints on Adaptive Evolution*, pp. 151–172. Springer-Verlag, Berlin.

Goldschmidt, R.B. (1940) *The Material Basis of Evolution*. Yale University Press, New Haven, Connecticut.

Gomendio, M., Clutton-Brock, T.H., Albon, S.D., Guinness, F.E. & Simpson, M.J. (1990) Mammalian sex ratios and variation in costs of rearing sons and daughters. *Nature* **343**: 261–263.

Goodman, M. (1963) Man's place in the phylogeny of the primates as reflected in serum proteins. In: Washburn, S.L. (ed.) *Classification and Human Evolution*, pp. 204–234. Aldine, Chicago.

Goto, H.E. (1982) *Animal Taxonomy*. Edward Arnold, London.

Gould, S.J. (1966) Allometry and size in ontogeny and phylogeny. *Biol. Rev.* **41**: 587–640.

Gould, S.J. (1971) D'Arcy Thompson and the science of form. *New Lit. Hist.* **2**: 229–258.

Gould, S.J. (1977a) *Ontogeny and Phylogeny*. Harvard University Press, Cambridge, Massachusetts.

Gould, S.J. (1977b) *Ever Since Darwin*. W. W. Norton, New York.

Gould, S.J. (1980a) *The Panda's Thumb*. W. W. Norton, New York.

Gould, S.J. (1980b) Is a new and general theory of evolution emerging? *Paleobiology* **6**: 119–130.

Gould, S.J. (1982a) Darwinism and the expansion of evolutionary theory. *Science* **216**: 380–387.

Gould, S.J. (1982b) The meaning of punctuated equilibrium and its role in validating a hierarchical approach to macroevolution. In: Milkman, R. (ed.) *Perspectives on Evolution*, pp. 83–104. Sinauer Associates, Sunderland, Massachusetts.

Gould, S.J. (1983a) *Hen's Teeth and Horse's Toes*. W. W. Norton, New York.

Gould, S.J. (1983b) Irrelevance, submission, and partnership: the changing role of palaeontology in Darwin's three centennials, and a modest proposal for macroevolution. In: Bendall, D.S. (ed.) *Evolution from Molecules to Men*, pp. 347–366. Cambridge University Press, Cambridge, UK.

Gould, S.J. (1984) Gingerich's smooth curve of evolutionary rate: a psychological and mathematical artifact. *Science* **226**: 994–995.

Gould, S.J. (1985) *The Flamingo's Smile*. W. W. Norton, New York.

Gould, S.J. (1989) *Wonderful Life*. W. W. Norton, New York.

Gould, S.J. (1991) *Bully for Brontosaurus*. W. W. Norton, New York.

Gould, S.J. & Calloway, C.B. (1980) Clams and brachiopods—ships that pass in the night. *Paleobiology* **6**: 383–396.

Gould, S.J. & Eldredge, N. (1977) Punctuated equilibria: the tempo and mode of evolution reconsidered. *Paleobiology* **3**: 115–151.

Gould, S.J., Gilinsky, N.L. & German, R.Z. (1987) Asymmetry of lineages and the direction of evolutionary time. *Science* **236**: 1437–1441.

Gould, S.J. & Johnston, R.F. (1972) Geographic variation. *Ann. Rev. Ecol. System.* **3**: 457–498.

Gould, S.J. & Lewontin, R.C. (1979) The spandrels of San Marco and the panglossian paradigm: a critique of the adaptationist program. *Proc. R. Soc. Lond. B* **205**: 581–598.

Gould, S.J. & Vrba, E.S. (1982) Exaptation—a missing term in the science of form. *Paleobiology* **8**: 4–15.

Grafen, A. (1984) Natural selection, kin selection, and group selection. In: Krebs, J.R. & Davies, N.B. (eds) *Behavioural Ecology: an Evolutionary Approach*, pp. 62–84. Blackwell Scientific Pultications, Cambridge, Massachusetts.

Grafen, A. (1985) A geometric view of relatedness. *Oxford Surv. Evol. Biol.* **2**: 28–89.

Grafen, A. (1988) A centrosomal theory of the short term evolutionary maintenance of sexual reproduction. *J. Theoret. Biol.* **131**: 163–173.

Grafen, A. (1989) The phylogenetic regression. *Phil. Trans. R. Soc. Lond. B* **326**: 119–156.

Grafen, A. (1990a) Sexual selection unhandicapped by the Fisher process. *J. Theoret. Biol.* **144**: 473–516.

Grafen, A. (1990b) Biological signals as handicaps. *J. Theoret. Biol.* **144**: 517–546.

Graham, R.W. & Grimm, E.C. (1990) Effects of global climatic change on the patterns of terrestrial biological communities. *Trends Ecol. Evol.* **5**: 289–292.

Grant, P.R. (1972) Convergent and divergent character displacement. *Biol. J. Linn. Soc.* **4**: 39–68.

Grant, P.R. (1975) The classic case of character displacement. *Evol. Biol.* **8**: 237–237.

Grant, P.R. (1986) *Ecology and Evolution of Darwin's Finches*. Princeton University Press, Princeton.

Grant, P.R. (1991) Natural selection and Darwin's finches. *Sci. Am.* **265** (October): 82–87.

Grantham, R., Perrin, P. & Mouchiroud, D. (1986) Patterns in codon usage of different species. *Oxford Surv. Evol. Biol.* **3**: 48–81.

Gribbin, J. & Cherfas, J. (1982) *The Monkey Puzzle*. Bodley Head, London.

Grieve, R.A.F. (1990) Impact cratering on the Earth. *Sci. Am.* **262** (April): 66–73.

Haffer, J. (1969) Speciation in Amazonian forest birds. *Science* **165**: 131–137.

Haffer, J. (1974) Avian speciation in tropical South America. *Publ. Nuttall Ornithol. Club* **14**: 1–390.

Hafner, M.S. & Nadler, S.A. (1988) Phylogenetic trees support the coevolution of parasites and hosts. *Nature* **332**: 258–259.

Haig, D. & Grafen, A. (1991) Genetic scrambling as a defence against meiotic drive. *J. Theoret. Biol.* **153**: 531–558.

Haldane, J.B.S. (1924) A mathematical theory of natural and artificial selection. Part I. *Trans. Cambridge Phil. Soc.* **23**: 19–41.

Haldane, J.B.S. (1932) *The Causes of Evolution*. Longman, London. (Reprints by Cornell University Press, New York (1966) and by Princeton University Press, Princeton (1990).)

Haldane, J.B.S. (1949a) Disease and evolution. *La Ricercha Scientifica*, **19**: (suppl.) 1–11.

Haldane, J.B.S. (1949b) Suggestions as to quantitative measurement of rates of evolution. *Evolution* **3**: 51–56.

Haldane, J.B.S. (1957) The cost of natural selection. *J. Genet.* **55**: 511–524.

Haldane, J.B.S. (1958) The theory of selection for melanism in Lepidoptera. *Proc. R. Soc. Lond.* **145**: 303–306.

Haldane, J.B.S. (1963) A defense of beanbag genetics. *Persp. Biol. Med.* **7**: 343–359.

Hallam, A. (1975) Evolutionary size increases and longevity in Jurassic bivalves and ammonites. *Nature* **258**: 493–496.

Hallam, A. (ed.) (1977) *Patterns of Evolution*. Elsevier, Amsterdam.

Hallam, A. (1983) Plate tectonics and evolution. In: Bendall, D.S. (ed.) *Evolution From Molecules to Men*, pp. 367–386. Cambridge University Press, Cambridge, UK.

Hallam, A. (1986) The Pliensbachian and Tithonian extinction events. *Nature* **319**: 765–768.

Hamilton, W.D. (1964) The genetical theory of social evolution, I and II. *J. Theoret. Biol.* **6**: 1–52.

Hamilton, W.D. (1967) Extraordinary sex ratios. *Science* **156**: 477–488.

Hamilton, W.D. (1972) Altruism and related phenomena, mainly in the social insects. *Ann. Rev. Ecol. System.* **3**: 193–232.

Hamilton, W.D. (1990) Mate choice near or·far. *Am. Zool.* **30**: 341–352.

Hamilton, W.D. & Zuk, M. (1982) Heritable true fitness and bright birds: a role for parasites. *Science* **218**: 384–387.

Hamilton, W.D., Axelrod, R. & Tanese, R. (1990) Sexual selection as an adaptation to resist parasites (a review). *Proc. Natl. Acad. Sci. USA* **87**: 3566–3573.

Hansen, T.A. (1978) Larval dispersal and species longevity in Lower Tertiary neogastropods. *Science* **199**: 885–887.

Hansen, T.A. (1983) Modes of larval development and rates of speciation in early Tertiary neogastropods. *Science* **220**: 501–502.

Harcourt, A.H., Harvey, P.H., Larson, S.G. & Short, R.V. (1981) Testis weight, body weight, and breeding system in primates. *Nature* **293**: 55–57.

Hardison, R.C. (1991) Evolution of globin gene families. In: Selander, R.K., Clark, A.G. & Whittam, T.S. (eds) *Evolution at the Molecular Level*, pp. 272–289. Sinauer Associates, Sunderland, Massachusetts.

Harland, W.B., Cox, A.V., Llewellyn, P.G., Pickton, C.A.G., Smith, A.G. & Walters, R. (1982) *A Geologic Time Scale*. Cambridge University Press, Cambridge, UK.

Harris, H. (1966) Enzyme polymorphisms in man. *Proc. R. Soc. Lond. B* **164**: 298–310.

Harrison, R.G. (1989) Animal mitochondrial DNA as a genetic marker in population and evolutionary biology. *Trends Ecol. Evol.* **4**: 6–11.

Harrison, R.G. (1990) Hybrid zones: windows on evolutionary processes. *Oxford Surv. Evol. Biol.* **7**: 129–156.

Harrison, R.G. (1991) Molecular changes at speciation. *Ann. Rev. Ecol. System.* **22**: 281–308.

Harrison, R.G. & Rand, D.M. (1989) Mosaic hybrid zones and the nature of species boundaries. In: Otte, D. & Endler, J. (eds) *Speciation and its Consequences*, pp. 111–

133. Sinauer Associates, Sunderland, Massachusetts.

Hartl, D.L. (1988) *A Primer of Population Genetics*, 2nd edn. Sinauer Associates, Sunderland, Massachusetts.

Hartl, D.L. & Clark, A.G. (1989) *Principles of Population Genetics*, 2nd edn. Sinauer Associates, Sunderland, Massachusetts.

Harvey, P.H., Kavanagh, M. & Clutton-Brock, T.H. (1978) Sexual dimorphism in primate teeth. *J. Zool.* **186**: 475–486.

Harvey, P.H. & Krebs, J.R. (1990) Comparing brains. *Science* **249**: 140–146.

Harvey, P.H. & Pagel, M.D. (1991) *The Comparative Method in Evolutionary Biology*. Oxford University Press, Oxford, UK.

Hayden, M. (1981) *Huntington's chorea*. Springer-Verlag, Berlin.

Hedrick, P.W. (1983) *Genetics of Populations*. Science Books International, Boston.

Hedrick, P.W. (1986) Genetic polymorphism in heterogeneous environments: a decade later. *Ann. Rev. Ecol. System.* **17**: 535–566.

Hedrick, P.W., Jain, S. & Holden, L. (1978) Multilocus systems in evolution. *Evol. Biol.* **11**: 101–184.

Hedrick, P.W., Klitz, W., Robinson, W.P., Kuhner, M.K. & Thomson, G. (1991a). Evolutionary genetics of HLA. In: Selander, R.K., Clark, A.G. & Whittam, T.S. (eds) *Evolution at the Molecular Level*, pp. 248–271. Sinauer Associates, Sunderland, Massachusetts.

Hedrick, P.W., Whittam, T.S. & Parham, P. (1991b) Heterozygosity at individual amino acid sites: extremely high levels for the *HLA-A* and *-B* genes. *Proc. Natl. Acad. Sci. USA* **88**: 5897–5901.

Hennig, W. (1966) *Phylogenetic Systematics*. University of Illinois Press, Urbana.

Hennig, W. (1981) *Insect Phylogeny*. John Wiley, Chichester, UK.

Henry, C.S. (1985) Sibling species, call differences, and speciation in green lacewings (Neuroptera: Chrysopidae: *Chrysoperla*). *Evolution* **39**: 965–984.

Herre, E.A. (1985) Sex ratio adjustment in fig wasps. *Science* **228**: 896–898.

Hewitt, G.M. (1988) Hybrid zones—natural laboratories for evolutionary studies. *Trends Ecol. Evol.* **3**: 158–167.

Hewitt, G.M. (1989) The subdivision of species by hybrid zones. In: Otte, D. & Endler, J.A. (eds) *Speciation and its Consequences*, pp. 85–110. Sinauer Associates, Sunderland, Massachusetts.

Hill, W.G. (ed.) (1984) *Quantitative Genetics*, vols 1 and 2 (Benchmark Papers in Genetics series). Van Nostrand Reinhold, New York.

Hoffman, A. (1985) Patterns of family extinction depend on definition and geological timescale. *Nature* **315**: 359–362.

Hoffman, A. (1989) *Arguments on Evolution: a Palaeontologist's Perspective*. Oxford University Press, New York.

Hoffman, A. (1991) Testing the Red Queen hypothesis. *J. Evol. Biol.* **4**: 1–7.

Hoffman, A. & Kitchell, J.A. (1984) Evolution in a pelagic planktic system: a paleobiologic test of models of multispecies evolution. *Paleobiology* **10**: 9–33.

Hoffmann, A.A. & Parsons, P.A. (1991) *Evolutionary Genetics and Environmental Stress*. Oxford University Press, New York.

Holman, E.W. (1987) Recognizability of sexual and asexual species of rotifers. *System. Zool.* **36**: 381–386.

Holser, W.T., *et al.* (1989) A unique geochemical record at the Permo-Triassic boundary. *Nature* **337**: 39–44.

Horner, H.A. & Macgregor, H.C. (1985) Normal development in newts (*Triturus*) and its arrest as a consequence of an unusual chromosomal situation. *J. Herpetol.* **19**: 261–270.

Hotton, N.H., MacLean, P.D., Roth, J.J. & Roth, E.C. (eds) (1986) *Ecology and Biology of Mammal-like Reptiles*. Smithsonian Institution, Washington, DC.

Howe, H.F. & Westley, L.C. (1988) *Ecological Relationships of Plants and Animals*. Oxford University Press, New York.

Hudson, R.R., Kreitman, M. & Aguadè, M. (1987) A test of neutral molecular evolution based on nucleotide data. *Genetics* **116**: 153–159.

Huffaker, C.B. (1958) Experimental studies on predation: dispersion factors and predator–prey oscillations. *Hilgardia* **27**: 343–383.

Hughes, A.H. & Nei, M. (1988) Pattern of nucleotide substitution at major histocompatibility complex class I loci reveals overdominant selection. *Nature* **335**: 167–170.

Hughes, A.H. & Nei, M. (1989) Nucleotide substitution at major histocompatibility complex class II loci: evidence for overdominant selection. *Proc. Natl. Acad. Sci. USA* **86**: 958–962.

Hull, D.L. (1965) The effect of essentialism on taxonomy. *Brit. J. Phil. Sci.* **15**: 314–326; **16**: 1–18.

Hull, D.L. (1967) Certainty and circularity in evolutionary taxonomy. *Evolution* **21**: 174–189.

Hull, D.L. (1978) A matter of individuality. *Phil. Sci.* **45**: 335–360.

Hull, D.L. (1988) *Science as a Process*. University of Chicago Press, Chicago.

Humphries, C.J. & Parenti, L.R. (1986) *Cladistic Biogeography*. Oxford University Press, Oxford, UK.

Hunt, H.R., Hoppert, C.A. & Rosen, S. (1955) Genetic factors in experimental rat caries. In: Sognnaes, R.F. (ed.) *Advances in Experimental Caries Research*, pp. 66–81. American Association for the Advancement of Science, Washington, DC.

Huxley, J.S. (1932) *Problems of Relative Growth*. Methuen, London.

Huxley, J.S. (ed.) (1940) *The New Systematics*. Oxford University Press, Oxford, UK.

Huxley, J.S. (1942) *Evolution: the Modern Synthesis*. Allen & Unwin, London.

Huxley, J.S. (1970) *Memories*. Allen & Unwin, London.

Huxley, J.S. (1973) *Memories II*. Allen & Unwin, London.

Jablonski, D. (1986) Background and mass extinctions: the alternation of macroevolutionary regimes. *Science* **231**: 129–133.

Jablonski, D. & Lutz, R.A. (1983) Larval ecology of marine

benthic invertebrates: paleobiological implications. *Biol. Rev.* **58**: 21–89.

Jackson, J.F. & Pounds, J.A. (1979) Comments on assessing the dedifferentiating effects of gene flow. *System. Zool.* **28**: 78–85.

Janzen, D.H. (1980) When is it coevolution? *Evolution* **34**: 611–612.

Janzen, D.H. (1983) Dispersal of seeds by vertebrate guts. In: Futuyma, D.J. & Slatkin, M. (eds) *Coevolution*, pp. 232–262. Sinauer Associates, Sunderland, Massachusetts.

Janzen, D.H. & Martin, P.S. (1982) Neotropical anachronisms: the fruits the gomphotheres ate. *Science* **215**: 19–27.

Jeffreys, A.J., Harris, S., Barrie, P.A., Wood, D., Blanchetot, A. & Adams, S.M. (1983) Evolution of gene families: the globin genes. In: Bendall, D.S. (ed.) *Evolution From Molecules to Men*, pp. 175–195. Cambridge University Press, Cambridge, UK.

Jeffreys, A.J., MacLoed, A., Tamaki, K., Neil, D.L. & Monkton, D.G. (1991) Minisatellite repeat coding as a digital approach to DNA typing. *Nature* **354**: 204–209.

Jeffreys, A.J., Royle, N.J., Wilson, V. & Wong, Z. (1988) Spontaneous mutation rate to new length alleles at tandem-repetitive hypervariable loci in human DNA. *Nature* **332**: 278–281.

Jeffreys, A.J., Wilson, V. & Thein, S.L. (1985) Hypervariable minisatellite regions in human DNA. *Nature* **314**: 67–73.

Jepsen, G.L., Mayr, E. & Simpson, G.G. (eds) (1949) *Genetics Paleontology, and Evolution*. Princeton University Press, Princeton.

Jerison, H.J. (1973) *Evolution of the Brain and Intelligence*. Academic Press, New York.

Jermy, T. (1984) Evolution of insect/host relationships. *Am. Nat.* **124**: 609–630.

John, B. & Miklos, G.L.G. (1988) *The Eukaryote Genome in Development and Evolution*. Allen & Unwin, London.

Johnson, C. (1976) *Introduction to Natural Selection*. University Park Press, Baltimore.

Johnson, C. (1979) An overview of selection theory and analysis. In: Thompson, J.N. & Thoday, J.M. (eds) *Quantitative Genetic Variation*, pp. 111–119. Academic Press, New York.

Johnson, L.A.S. (1970) Rainbow's end. *System. Zool.* **19**: 203–239.

Johnston, R.F. & Selander, R.K. (1971) Evolution in the house sparrow. II. Adaptive differentiation in North American populations. *Evolution* **25**: 1–28.

Jones, D.A. (1989) 50 years of studying the scarlet tiger moth. *Trends Ecol. Evol.* **4**: 298–301.

Jones, J.S. (1982) More to melanism than meets the eye. *Nature* **300**: 109–110.

Jones, J.S. (1987) An asymmetrical view of fitness. *Nature* **325**: 298–299.

Jones, J.S., Leith, B. & Rawlings, P. (1977) Polymorphism in *Cepaea*: a problem with too many solutions? *Ann. Rev. Ecol. System.* **8**: 109–143.

Jones, J.S. & Probert, R.F. (1980) Habitat selection main-

tains a deleterious allele in a heterogeneous enviroment. *Nature* **287**: 632–633.

Kaneshiro, K.Y. (1989) The dynamics of sexual selection and founder effects in species formation. In: Giddings, L.V., Kaneshiro, K.Y. & Anderson, W.W. (eds) *Genetics, Speciation, and the Founder Principle*, pp. 279–296. Oxford University Press, New York.

Kaplan, N.L., Hudson, R.R. & Langley, C.H. (1989) The "hitchhiking effect" revisited. *Genetics* **123**: 887–899.

Karn, M.N. & Penrose, L.S. (1951) Birth weight and gestation time in relation to maternal age, parity, and infant survival. *Ann. Eugen.* **16**: 147–164.

Kauffman, E.G. (1978) Evolutionary rates and patterns among Cretaceous bivalves. *Phil. Trans. R. Soc. Lond. B* **284**: 277–304.

Kay, R.F. & Simons, E.L. (1983) A reassessment of the relationship between later Miocene and subsequent Hominoidea. In: Ciochin, R.L. & Corruccini, R.S. (eds) *New Interpretations of Ape and Human Ancestry*, pp. 577–624. Plenum Press, New York.

Keeler, E.F. (1983) *A feeling for the Organism: the Life and Work of Barbara McClintock*. W.H. Freeman, San Francisco.

Kemp, T.S. (1982a) *Mammal-like Reptiles and the Origin of Mammals*. Academic Press, London.

Kemp, T.S. (1982b) The reptiles that became mammals. *New Sci.* **93**: 581–584. (Reprinted in Cherfas, J.J. (ed.) (1982) *Darwin Up to Date*, pp. 31–34. New Scientist, London.

Kemp, T.S. (1985) Synapsid reptiles and the origin of higher taxa. *Special Paper. Paleontol.* **33**: 175–184.

Kemp, T.S. (1988) Haemothermia or Archosauria? The interrelationships of mammals, birds, and crocodiles. *Zool. J. Linn. Soc.* **92**: 67–104.

Kessler, S. (1966) Selection for and against ethological isolation between *Drosophila pseudoobscura* and *Drosophila persimilis*. *Evolution* **20**: 634–645.

Kettlewell, H.B.D. (1973) *The Evolution of Melanism*. Oxford University Press, Oxford, UK.

Kidwell, M.G. (1972) Genetic change of recombination value in *Drosophila melanogaster*. I. Artificial selection for high and low recombination and some properties of recombination-modifying genes. *Genetics* **70**: 419–432.

Kim, K.C. (ed.) (1986) *Coevolution of Parasitic Arthropods and Mammals*. Academic Press, New York.

Kimura, M. (1965) A stochastic model concerning the maintenance of genetic variability in quantitative characters. *Proc. Natl. Acad. Sci. USA* **68**: 984–986.

Kimura, M. (1968) Evolutionary rate at the molecular level. *Nature* **217**: 624–626

Kimura, M. (1979) The neutral theory of molecular evolution. *Sci. Am.* **241** (November): 98–126.

Kimura, M. (1983) *The Neutral Theory of Molecular Evolution*. Cambridge University Press, Cambridge, UK.

Kimura, M. (1991) Recent developments of the neutral theory viewed from the Wrightian tradition of theoretical population genetics. *Proc. Natl. Acad. Sci.* **88**: 5969–5973.

Kimura, M. & Crow, J.F. (1964) The number of alleles that can be maintained in a finite population. *Genetics* **49**:

725–738.

King, J.L. (1967) Continuously distributed factors affecting fitness. *Genetics* **55**: 483–492.

King, L. & Jukes, T. (1969) Non-Darwinian evolution. *Science* **164**: 788–789.

Kirkpatrick, M. & Jenkins, C.D. (1989) Genetic segregation and the maintenance of sexual reproduction. *Nature* **339**: 300–301.

Kirkpatrick, M. & Ryan, M.J. (1991) The evolution of mating preferences and the paradox of the lek. *Nature* **350**: 33–38.

Kitchell, J.A. & MacLeod, N. (1988) Macroevolutionary interpretations of symmetry and synchroneity in the fossil record. *Science* **240**: 1190–1193.

Klekowski, E.J. & Godfrey, P.J. (1989) Ageing and mutation in plants. *Nature* **340**: 389–391.

Knoll, A.H. (1986) Patterns of change in plant communities through geological time. In: Diamond, J. & Case, T. (eds) *Community Ecology*, pp. 126–141. Harper & Row, New York.

Kondrashov, A.S. (1988) Deleterious mutations and the evolution of sexual reproduction. *Nature* **336**: 435–440.

Kondrashov, A.S. & Turelli, M. (1992) Deleterious mutations, apparent stabilizing selection, and the maintenance of quantitative variation. *Genetics* **132**: 603–618.

Krebs, J.R. & Davies, N.B. (1993) *An Introduction to Behavioural Ecology*, 3rd edn. Blackwell Scientific Publications, Cambridge, Massachusetts.

Kreitman, M. (1983) Nucleotide polymorphism at the alcohol dehydrogenase locus of *Drosophila melanogaster*. *Nature* **304**: 412–417.

Kreitman, M. (1987) Molecular population genetics. *Oxford Surv. Evol. Biol.* **4**: 38–60.

Lack, D. (1947) *Darwin's Finches*. Cambridge University Press, Cambridge, UK.

Lake, J.A. (1987) A rate-dependent technique for analysis of nucleic acid sequences: evolutionary parsimony. *Mol. Biol. Evol.* **4**: 167–191.

Lake, J.A. (1990) Origin of the Metazoa. *Proc. Natl. Acad. Sci. USA* **87**: 763–766.

Lake, J.A. (1991) Tracing origins with molecular sequences: metazoans and eukaryotic beginnings. *Trends Biochem. Sci.* **16**: 46–50.

Lamarck, J-B. (1809) *Philosophie Zoologique*. Paris.

Lande, R. (1976) The maintenance of genetic variability by mutation in a polygenic character with linked loci. *Genet. Res.* **26**: 221–235.

Lande, R. (1979) Effective deme size during long-term evolution estimated from rates of chromosomal rearrangement. *Evolution* **33**: 234–251.

Lande, R. (1980) Genetic variation and phenotypic evolution during allopatric speciation. *Am. Nat.* **116**: 463–479.

Lande, R. (1985) The fixation of chromosomal rearrangements in a subdivided population with local extinction and recolonization. *Heredity* **54**: 323–332.

Lande, R. (1986) The dynamics of peak shifts and the pattern of morphological evolution. *Paleobiology* **12**:

343–354.

Langley, C.H. (1977) Nonrandom associations between allozymes in natural populations of *Drosophila melanogaster*. In: Christiansen, F.B. & Fenchel, T.M. (eds) *Measuring Selection in Natural Populations*, pp. 265–273. Springer-Verlag, Berlin.

Larson, E.J. (1989) *Trial and Error. The American Controversy over Creation and Evolution*. Updated edn. Oxford University Press, New York.

Larwood, G.P. (1988) *Extinction and Survival in the Fossil Record*. Systematics Association Special Volume, no. 34. Oxford University Press, Oxford, UK.

Latter, B.D.H. (1970) Selection in finite populations with multiple alleles. II. Centrepetal selection, mutation, and isoallelic variation. *Genetics* **66**: 165–186.

Lauder, G.V. (1981) Form and function: Structural analysis in evolutionary morphology. *Paleobiology* **7**: 430–442.

Lauder, G.V. (1991) Functional morphology and systematics: studying functional patterns in an historical context. *Ann. Rev. Ecol. System.* **21**: 317–340.

Law, R. (1991) Fishing in evolutionary waters. *New Sci.* 2 March: 35–37.

Leary, R.F. & Allendorf, F.W. (1989) Fluctuating asymmetry as an indicator of stress: implications for conservation biology. *Trends Ecol. Evol.* **4**: 214–217.

Lee, B.T.O. & Parsons, P.A. (1968) Selection, prediction, and response. *Biol. Rev.* **43**: 139–174.

Lees, D.R. (1971) Industrial melanism: genetic adaptation of animals to air pollution. In: Bishop, J.A. & Cook, L.M. (eds) *Genetic Consequences of Man Made Change*, pp. 129–176. Academic Press, London.

Levene, H. (1953) Genetic equilibrium when more than one niche is available. *Am. Nat.* **87**: 331–333.

Leverich, W.J. & Levin, D.A. (1979) Age-specific survivorship and reproduction in *Phlox drummondii*. *Am. Nat.* **113**: 881–903.

Levin, D.A. (1978) The origin of isolating mechanisms in flowering plants. *Evol. Biol.* **11**: 185–317.

Levin, D.A. (1979) The nature of plant species. *Science* **204**: 381–384.

Levinton, J. (1988) *Genetics, Paleontology, and Macroevolution*. Cambridge University Press, Cambridge, UK.

Lewin, B. (1990) *Genes IV*. Oxford University Press, New York.

Lewontin, R.C. (1970) The units of selection. *Ann. Rev. Ecol. System.* **1**: 1–18.

Lewontin, R.C. (1974) *The Genetic Basis of Evolutionary Change*. Columbia University Press, New York.

Lewontin, R.C. (1980) Adaptation. *Encyclopaedia Einaudi*. Milano, Italy. (Reprinted in Sober, E. (ed.) (1984) *Conceptual Issues in Evolutionary Biology*, pp. 235–251. MIT Press, Cambridge, Massachusetts.)

Lewontin, R.C. (1982) *Human Diversity*. Scientific American Library, New York.

Lewontin, R.C. (1985a) Population genetics. In: Greenwood, P.J., Harvey, P.H. & Slatkin, M. (eds) *Evolution: Essays in Honour of John Maynard Smith*, pp.

3–18. Cambridge University Press, Cambridge, UK.

Lewontin, R.C. (1985b) Population genetics. *Ann. Rev. Genet.* **19**: 81–102.

Lewontin, R.C. (1986) How important is population genetics for an understanding of evolution? *Am. Zool.* **26**: 811–820.

Lewontin, R.C. & Hubby, J.L. (1966) A molecular approach to the study of genic heterozygosity in natural populations. II. Amount of variation and degree of heterozygosity in natural populations of *Drosophila pseudoobscura*. *Genetics* **54**: 595–605.

Lewontin, R.C., Moore, J.A., Provine, W.B. & Wallace, B. (eds) (1981) *Dobzhansky's Genetics of Natural Populations I-XLIII*. Columbia University Press, New York.

Li, C.C. (1976) *First Course in Population Genetics*. Boxwood Press, Pacific Grove, California.

Li, W-H. (1977) *Stochastic Models in Population Genetics*. (Benchmark Papers in Genetics, vol. 7.) Dowden, Hutchinson & Ross, Stroudsburg, Pennsylvania.

Li, W-H. & Grauer, D. (1991) *Fundamentals of Molecular Evolution*. Sinauer Associates, Sunderland, Massachusetts.

Li, W-H., Tanimura, M. & Sharp, P.M. (1987) An evaluation of the molecular clock hypothesis using mammalian DNA sequences. *J. Mol. Evol.* **25**: 330–342.

Li, W-H., Wu, C-I. & Luo, C-C. (1985) A new method for estimating synonymous and nonsynonymous rates of nucleotide substitution considering the relative likelihood of nucleotide and codon changes. *Mol. Biol. Evol.* **2**: 150–174.

Liebert, A.G. & Brakefield, P.M. (1987) Behavioural studies on the peppered moth *Biston betularia* and a discussion of the role of pollution and lichens in industrial melanism. *Biol. J. Linn. Soc.* **31**: 129–150.

Loeschcke, V. (ed.) (1987) *Genetic Constraints on Adaptive Evolution*. Springer-Verlag, Berlin.

Luria, S.E. (1973) *Life—the Unfinished Experiment*. Scribner, New York.

Luria, S.E. & Delbrück, M. (1943) Mutations of bacteria from virus sensitivity to virus resistance. *Genetic* **28**: 491–511.

Lynch, J.D. (1989) The gauge of speciation: on the frequencies of modes of speciation. In: Otte, D. & Endler, J.A. (eds) *Speciation and its Consequences*, pp. 527–553. Sinauer Associates, Sunderland, Massachusetts.

MacFadden, B.J. (1987) Fossil horses from "Eohippus" (*Hyracotherium*) to *Equus*: scaling, Cope's law, and the evolution of body size. *Paleobiology* **12**: 355–369.

MacFadden, B.J. (1988a) Fossil horses from "Eohippus" (*Hyracotherium*) to *Equus*, 2: rates of dental evolution revisited. *Biol. J. Linn. Soc.* **35**: 37–48.

MacFadden, B.J. (1988b) Horses, the fossil record, and evolution: a current perspective. *Evol. Biol.* **22**: 131–158.

Macgregor, H.C. & Horner, H.A. (1980) Heteromorphism for chromosome 1, a requirement for normal development in crested newts. *Chromosoma* **76**: 111–122.

Macleod, N. (1991) Punctuated anagenesis and the importance of stratigraphy to paleobiology. *Paleobiology* **17**: 167–188.

Mallett, J. (1989) The evolution of insecticide resistance: have the insects won? *Trends Ecol. Evol.* **4**: 336–340.

Malmgren, B.A. & Kennett, J.P. (1981) Phyletic gradualism in a Late Cenozoic planktonic foraminiferal lineage; DSDP site 284, southwest Pacific. *Paleobiology* 7: 230–240.

Mani, G.S. (1982) A theoretical analysis of the morph frequency variation in the peppered moth over England and Wales. *Biol. J. Linn. Soc.* **17**: 259–267.

Margulis, L. (1981) *Symbiosis in Cell Evolution*. W.H. Freeman, San Francisco.

Margulis, L. & Fester, R. (eds) (1991) *Symbiosis as a Source of Evolutionary Innovation*. MIT Press, Cambridge, Massachusetts.

Marshall, L.G. (1988) Extinction. In: Myers, A.A. & Giller, P.S. (eds) *Analytic Biogeography*, pp. 219–254. Chapman & Hall, London.

Marshall, L.G., Webb, S.D., Sepkoski, J.J. & Raup, D.M. (1982) Mammalian evolution and the Great American Interchange. *Science* **215**: 1351–1357.

Martin, L. (1985) Significance of enamel thickness in hominoid evolution. *Nature* **314**: 260–263.

Martin, R.D. (1990) *Primate Origins and Evolution*. Chapman & Hall, London.

Mather, K. (1943) Polygenic inheritance and natural selection. *Biol. Rev.* **18**: 32–64.

May, A.W. (1967) Fecundity of Atlantic cod. *J. Fish. Res. Board Can.* **24**: 1531–1551.

Maynard Smith, J. (1968) 'Haldane's dilemma' and the rate of evolution. *Nature* **219**: 1114–1116.

Maynard Smith, J. (1968) *Mathematical Ideas in Biology*. Cambridge University Press, Cambridge, UK.

Maynard Smith, J. (1972) *On Evolution*. Aldine Atherton, Chicago.

Maynard Smith, J. (1976) Group selection. *Quart. Rev. Biol.* **51**: 277–283.

Maynard Smith, J. (1978a) *The Evolution of Sex*. Cambridge University Press, Cambridge, UK.

Maynard Smith, J. (1978b) Optimization theory in evolution. *Ann. Rev. Ecol. System.* **9**: 31–56.

Maynard Smith, J. (1983) The genetics of stasis and punctuation. *Ann. Rev. Genet.* **7**: 11–25.

Maynard Smith, J. (1986) *The Problems of Biology*. Oxford University Press, Oxford, UK.

Maynard Smith, J. (1987a) J.B.S. Haldane *Oxford Surv. Evol. Biol.* **4**: 1–9.

Maynard Smith, J. (1987b) How to model evolution. In: Dupré, J. (ed.) *The Latest on the Best*, pp. 119–131. MIT Press, Cambridge, Massachusetts.

Maynard Smith, J. (1987c) Darwin stays unpunctured. *Nature* **330**: 516.

Maynard Smith, J. (1988) Punctuation in perspective. *Nature* **332**: 311–312.

Maynard Smith, J. (1989) *Evolutionary Genetics*. Oxford University Press, Oxford, UK.

Maynard Smith, J. (1991) Theories of sexual selection. *Trends Ecol. Evol.* **6**: 146–151.

Maynard Smith, J., Burian, R., Kauffman, S., *et al.* (1985) Developmental constraints and evolution. *Quart. Rev. Biol.* **60**: 265–287.

Maynard Smith, J. & Haigh, J. (1974) The hitch-hiking effect of a favourable gene. *Genet. Res.* **23**: 23–35.

Maynard Smith, J. & Vida, G. (eds) (1990) *Organizational Constraints on the Dynamics of Evolution*. Manchester University Press, Manchester, UK.

Mayr, E. (1942) *Systematics and the Origin of Species*. Columbia University Press, New York.

Mayr, E. (1954) Change of genetic environment and evolution. In: Huxley, J.S., Hardy, A.C. & Ford, E.B. (eds) *Evolution as a Process*, pp. 157–180. Allen & Unwin, London.

Mayr, E. (1963) *Animal Species and Evolution*. Harvard University Press, Cambridge, Massachusetts.

Mayr, E. (1969) *Principles of Systematic Zoology*. McGraw-Hill, New York.

Mayr, E. (1976) *Evolution and the Diversity of Life*. Harvard University Press, Cambridge, Massachusetts.

Mayr, E. (1981) Biological classification: toward a synthesis of opposing methodologies. *Science* **214**: 510–516.

Mayr, E. (1982a) Processes of speciation in animals. In: Barigozzi, C. (ed.) *Mechanisms of Speciation*, pp. 1–19. Alan Liss, New York.

Mayr, E. (1982b) *The Growth of Biological Thought: Diversity, Evolution, and Inheritance*. Harvard University Press, Cambridge, Massachusetts.

Mayr, E. (1988) The why and how of species. *Biol. Phil.* **3**: 431–441.

Mayr, E. (1991) *One Long Argument*. Harvard University Press, Cambridge, Massachusetts.

Mayr, E. & O'Hara, R.J. (1986) The biogeographic evidence supporting the Pleistocene refuge hypothesis. *Evolution* **40**: 55–67.

Mayr, E. & Provine, W.B. (eds) (1980) *The Evolutionary Synthesis*. Harvard University Press, Cambridge, Massachusetts.

McCune, A.R. (1982) On the fallacy of constant extinction rates. *Evolution* **36**: 610–614.

McDonald, J.H. & Kreitman, M. (1991) Adaptive evolution at the *Adh* locus in *Drosophila*. *Nature* **351**: 652–654.

McKinney, M.L. (1987) Taxonomic selectivity and continuous variation in mass and background extinctions of marine taxa. *Nature* **325**: 143–145.

McKinney, M.L. (ed.) (1988) *Heterochrony in Evolution: an Interdisciplinary Approach*. Plenum Press, New York.

McKinney, M.L. & McNamara, K. (1991) *Heterochrony: the Evolution of Ontogeny*. Plenum Press, New York.

McMahon, T.A. & Bonner, J.T. (1983) *On Size and Life*. Scientific American Library, New York.

McNamara, K.J. (1988) Heterochrony and the evolution of echinoids. In: Paul, C.R.C. & Smith, A.B. (eds) *Echinoderm Phylogeny and Evolutionary Biology*, pp. 149–163. Oxford University Press, Oxford, UK.

McNamara, K.J. (1989) The great evolutionary handicap. *New Sci.* **123**: 10 September: 47–51.

McNamara, K.J. (ed.) (1990) *Evolutionary Trends*. University of Arizona Press, Tucson, Arizona.

McPheron, B.A., Smith, D.C. & Berlocher, S.H. (1988) Genetic differences between host races of *Rhagoletis pomonella*. *Nature* **336**: 64–66.

Medawar, P.B. (1958) Postscript: D'Arcy Thompson and *Growth and Form*. In: D'Arcy Thompson, R. *D'Arcy Thompson: the Scholar Naturalist*. Oxford University Press, London, UK (Reprinted in Medawar, P.B. (1982) *Pluto's Republic*. Oxford University Press, Oxford, UK.)

Medawar, P.B. & le Gros Clark, W.E. (eds) (1945) *Essays on Growth and Form*. Oxford University Press, Oxford, UK.

Michod, R.E. & Levin, B.R. (eds) (1988) *The Evolution of Sex*. Sinauer Associates, Sunderland, Massachusetts.

Milkman, R.D. (1967) Heterosis as a major cause of heterozygosity in nature. *Genetics* **55**: 493–495.

Mindell, D.P. & Honeycutt, R.L. (1990) Ribosomal RNA in vertebrates and phylogenetic applications. *Ann. Rev. Ecol. System.* **21**: 541–566.

Mishler, B.D. & Brandon, R.N. (1987) Individuality, pluralism, and the phylogenetic species concept. *Biol. Phil.* **2**: 397–414.

Mishler, B.D. & Donoghue, M.J. (1982) Species concepts: a case for pluralism. *System. Zool.* **31**: 491–503.

Mitchell-Olds, T. & Rutledge, J.J. (1986) Quantitative genetics in natural plant populations: a review of the theory. *Am. Nat.* **127**: 379–402.

Mitter, C., Farrell, B. & Futuyma, D.J. (1991) Phylogenetic studies of insect–plant interactions: insights into the generation of diversity. *Trends Ecol. Evol.* **6**: 290–293.

Mivart, St G.J. (1871) *The Genesis of Species*. John Murray, London.

Miyamoto, M.M. & Cracraft, J. (eds) (1991) *Phylogenetic Analysis of DNA Sequences*. Oxford University Press, New York.

Miyamoto, M.M. & Goodman, M. (1990) DNA systematics and evolution of primates. *Ann. Rev. Ecol. System.* **21**: 197–220.

Miyata, T., Hayashida, H., Kikuno, R., *et al.* (1982) Molecular clock of silent substitution: at least six-fold preponderance of silent changes in mitochondrial genes over those in nuclear genes. *J. Mol. Evol.* **19**: 28–35.

Møller, A.P. (1988) Female choice selects for male sexual tail ornaments in the monogamous swallow. *Nature* **332**: 640–642.

Møller, A.P. (1989) Viability costs of male tail ornaments in a swallow. *Nature* **339**: 132–134.

Møller, A.P. (1990) Effects of a haematophagous mite on the barn swallow (*Hirundo rustica*): a test of the Hamilton and Zuk hypothesis. *Evolution* **44**: 771–784.

Moore, J.A. (1949) Patterns of evolution in the genus *Rana*. In: Jepsen, G.L., Mayr, E. & Simpson, G.G. (eds), *Genetics, Paleontology, and Evolution*, pp. 315–338. Princeton University Press, Princeton.

Moore, W.S. (1977) An evaluation of narrow hybrid zones in vertebrates. *Quart. Rev. Biol.* **53**: 263–277.

Muller, H.J. (1959) One hundred years without Darwinism

are enough. *School Sci. Math.* **49**: 314–318.

Murray, J. & Clarke, B. (1980) The genus *Partula* on Moorea: speciation in progress. *Proc. R. Soc. Lond. B* **211**: 83–117.

Myers, A.A. & Giller, P.S. (eds) (1988) *Analytical Biogeography.* Chapman & Hall, London.

Nei, M. (1987) *Molecular Evolutionary Genetics.* Columbia University Press, New York.

Nei, M. & Koehn, R.K. (eds) (1983) *Evolution of Genes and Proteins.* Sinauer Associates, Sunderland, Massachusetts.

Nelkin, D. (1977) *Science Textbook Controversies and the Politics of Equal Time.* MIT Press, Cambridge, Massachusetts.

Nelkin, D. (1982) *The Creation Controversy: Science or Scripture in the Schools.* W.W. Norton, New York.

Nelson, G. & Rosen, D.E. (eds) (1981) *Vicariance Biogeography: a Critique.* Columbia University Press, New York.

Neufeld, P.J. & Colman, N. (1990) Genetic fingerprinting. *Sci. Am.* **262** (May): 46–53.

Nevo, E. (1988) Genetic diversity in nature. *Evol. Biol.* **23**: 217–246.

Newell, N.D. (1959) The nature of the fossil record. *Proc. Amer. Philosoph. Soc.* **103**: 264–285.

Niklas, K.J. (1986) Large-scale changes in animal and plant terrestrial communities. In: Raup, D.M. & Jablonski, D. (eds) *Patterns and Processes in the History of Life.* Dahlem Workshop. John Wiley, Chichester, UK.

Nitecki, M.H. (ed.) (1983) *Coevolution.* University of Chicago Press, Chicago.

Nitecki, M.H. (ed.) (1990) *Evolutionary Innovations.* University of Chicago Press, Chicago.

Nordenskiöld, E. (1929) *The History of Biology.* Knopf, New York.

Nordström, K. & Austin, S.J. (1989) Mechanisms that contribute to the stable segregation of plasmids. *Ann. Rev. Genet.* **23**: 37–69.

Norell, M.A. & Novacek, M.J. (1992) The fossil record and evolution: comparing cladistic and paleontological evidence for vertebrate history. *Science* **255**: 1690–1693.

Novacek, M.J. (1992) Fossils, typologies, missing data, and the higher level phylogeny of Eutherian mammals. *Systematic Biology* **41**: 58–73.

Novick, A. (1955) Mutagens and antimutagens. *Brookhaven Symp. Biol.* **8**: 201–215.

Ochman, H., Jones, J.S. & Selander, R.K. (1983) Molecular area effects in *Cepaea. Proc. Natl. Acad. Sci. USA* **80**: 4189–4193.

Officer, C.B. & Drake, C.L. (1983) The Cretaceous–Tertiary transition. *Science* **219**: 1383–1390.

Officer, C.B. & Drake, C.L. (1985) Terminal Cretaceous environmental events. *Science* **227**: 1161–1167.

Officer, C.B., Hallam, A., Drake, C.L. & Devine, J.D. (1987) Late Cretaceous and paroxysmal Cretaceous/Tertiary extinctions. *Nature* **326**: 143–149.

Ohta, T. (1974) Mutation pressure as the main cause of molecular evolution and polymorphism. *Nature* **252**: 351–354.

Ohta, T. (1977) Extension of the neutral mutation drift hypothesis. In: Kimura, M. (ed.) *Molecular Evolution and Polymorphism*, pp. 148–167. National Institute of Genetics, Mishima, Japan.

Ohta, T. (1988) Multigene and supergene families. *Oxford Surv. Evol. Biol.* **5**: 41–65.

Orgel, L.E. & Crick, F.H.C. (1980) Selfish DNA: the ultimate parasite. *Nature* **284**: 604–607.

Orth, C.J. (1989) Geochemistry of the bio-event horizons. In: Donovan, S.K. (ed.) *Mass Extinctions: Processes and Evidence*, pp. 37–72. Columbia University Press, New York.

Ospovat, D. (1981) *The Development of Darwin's Theory.* Cambridge University Press, Cambridge, UK.

Oster, G.F. & Wilson, E.O. (1978) *Caste and Ecology in the Social Insects.* Princeton University Press, Princeton.

Otte, D. & Endler, J.A. (eds) (1989) *Speciation and its Consequences.* Sinauer Associates, Sunderland, Massachusetts.

Parker, G.A. & Maynard Smith, J. (1990) Optimality theory in evolutionary biology. *Nature* **348**: 27–33.

Paterson, H.E.H. (1978) More evidence against speciation by reinforcement. *S. Afr. J. Sci.* **74**: 369–371.

Paterson, H.E.H. (1985) The recognition concept of species. In: Vrba, E.S. (ed.) *Species and Speciation*, pp. 21–29. (Transvaal Museum Monograph, no. 4.) Transvaal Museum, Pretoria, South Africa.

Patterson, C. (1981) Methods of paleobiogeography. In: Nelson, G. & Rosen, D.E. (eds) *Vicariance Biogeography: a Critique*, pp. 446–489. Columbia University Press, New York.

Patterson, C. (ed.) (1987) *Molecules and Morphology in Evolution: Conflict or Compromise.* Cambridge University Press, Cambridge, UK.

Patterson, C. (1990) Reassessing relationships. *Nature* **344**: 199–200.

Patterson, C. & Smith, A.B. (1987) Is the periodicity of extinctions a taxonomic artefact? *Nature* **330**: 248–251.

Paul, C.R.C. (1980) *The Natural History of Fossils.* Weidenfeld & Nicolson, London.

Pellmyr, O. (1992) Evolution of insect pollination and angiosperm diversification. *Trends Ecol. Evol.* **7**: 46–49.

Penny, D., Foulds, L.R. & Hendy, M.D. (1982) Testing the theory of evolution by comparing the phylogenetic trees constructed from five different protein sequences. *Nature* **297**: 197–200.

Pielou, E.C. (1991) *After the Ice Age: the Return of Life to Glaciated North America.* University of Chicago Press, Chicago.

Pierce, N.E. (1987) The evolution and biogeography of associations between lycaenid butterflies and ants. *Oxford Surv. Evol. Biol.* **4**: 89–116.

Pierce, N.E. & Mead, P.S. (1981) Parasitoids as selective agents in the symbiosis between lycaenid butterfly larvae and ants. *Science* **211**: 1185–1187.

Pinker, S. & Bloom, P. (1990) Natural language and natural selection. *Behav. Brain Sci.* **13**: 707–784.

Policansky, D. (1982) Sex change in plants and animals.

Ann. Rev. Ecol. System **13**: 471–495.

Pollack, E., Kempthorne, O. & Bailey, T.B. (eds) (1977) *Proceedings of the International Conference on Quanititative Genetics*. Iowa State University Press, Ames, Iowa.

Potts, W.K. & Wakeland, E.K. (1990) Evolution of diversity at the major histocompatibility complex. *Trends Ecol. Evol.* **5**: 181–187.

Primack, R.B. & Kang, H. (1989) Measuring fitness and natural selection in wild plant populations. *Ann. Rev. Ecol. System.* **10**: 367–396.

Pringle, J.W.S. (1951) On the parallel between learning and evolution. *Behaviour* **3**: 174–215.

Provine, W.B. (1971) *The Origins of Theoretical Population Genetics*. University of Chicago Press, Chicago.

Provine, W.B. (1985) The R.A. Fisher–Sewall Wright controversy and its influence upon modern evolutionary biology. *Oxford Surv. Evol Biol.* **2**: 197–219.

Provine, W.B. (1986a) *Sewall Wright and Evolutionary Biology*. University of Chicago Press, Chicago.

Provine, W.B. (1986b) *Evolution: Selected Papers*. University of Chicago Press, Chicago.

Raff, R.A. & Kaufman, T.C. (1983) *Embryos, Genes, and Evolution*. Macmillan, New York.

Raup, D.M. (1966) Geometric analysis of shell coiling. *J. Paleontol.* **40**: 1178–1190.

Raup, D.M. (1977) Stochastic models in evolutionary palaeontology. In: Hallam, A. (ed.) *Patterns of Evolution*, pp. 59–78. Elsevier, Amsterdam.

Raup, D.M. (1978) Cohort analysis of generic survivorship. *Paleobiology* **4**: 1–15.

Raup, D.M. (1979) Size of the Permo-Triassic bottleneck and its evolutionary implications. *Science* **206**: 217–218.

Raup, D.M. (1986) Biological extinction in earth history. *Science* **231**: 1528–1533.

Raup, D.M. (1991) *Extinction: Bad Genes or Bad Luck?* W.W. Norton, New York.

Raup, D.M. & Boyajian, G.E. (1988) Patterns of generic extinction in the fossil record. *Paleobiology* **14**: 109–125.

Raup, D.M. & Jablonski, D. (eds) (1986) *Patterns and Processes in the History of Life*. Dahlem Workshop. John Wiley, Chichester, UK.

Raup, D.M. & Sepkoski, J.J. (1984) Periodicity of extinctions in the geologic past. *Proc. Natl. Acad. Sci. USA* **81**: 801–805.

Raup, D.M. & Sepkoski, J.J. (1986) Periodic extinction of families and genera. *Science* **231**: 833–836.

Raup, D.M. & Sepkoski, J.J. (1988) Testing for periodicity of extinction. *Science* **241**: 94–96.

Raven, P.H. (1976) Systematics and plant population biology. *System. Bot.* **1**: 284–316.

Raymond, M., Callaghan, A., Fort, P. & Pasteur, N. (1991) Worldwide migration of amplified resistance genes in mosquitoes. *Nature* **350**: 151–153.

Read, A.F. (1988) Sexual selection and the role of parasites. *Trends Ecol. Evol.* **3**: 97–102.

Read, A.F. & Harvey, P.H. (1989) Reassessment of comparative evidence for Hamilton & Zuk theory on the evolution of secondary sexual characters. *Nature* **339**: 618–620.

Reaka, M.L. & Manning, R.B. (1987) The significance of body size, dispersal potential, and habitat for rates of morphological evolution in stomatopod Crustacea. *Smithsonian Contributions to Zoology*, no. 448, 46 pp.

Rendel, J.M. (1967) *Canalisation and Gene Control*. Logos, London.

Rensch, B. (1959) *Evolution Above the Species Level*. Methuen, London.

Rhodes, F.H.T. (1983) Gradualism, punctuated equilibrium, and the *On the Origin of Species*. *Nature* **305**: 269–272.

Ricker, W.E. (1981) Changes in the average size and average age of Pacific salmon. *Can. J. Fish. Aquat. Sci.* **38**: 1636–1656.

Ridley, M. (1982) Coadaptation and the inadequacy of natural selection. *Brit. J. His. Sci.* **15**: 45–68.

Ridley, M. (1985) Embryology and classical zoology in Great Britain. In: Horder, T.J. Witkowski, J.A. & Wylie, C.C. (eds) *A History of Embryology* (Symposium 8, British Society for Developmental Biology), pp. 35–67. Cambridge University Press, Cambridge, UK.

Ridley, M. (1986) *Evolution and Classification: the Reformation of Cladism*. Longman, London.

Ridley, M. (1989) The cladistic solution to the species problem. *Biol. Phil.* **4**: 1–16.

Robson, G.C. & Richards, O.W. (1936) *The Variations of Animals in Nature*. Longman, London.

Roff, D.A. & Mousseau, T.A. (1987) Quantitative genetics and fitness: lessons from *Drosophila*. *Heredity* **58**: 103–118.

Romer, A.S. (1949) Time series and trends in animal evolution. In: Jepsen, G.L., Mayr, E. & Simpson, G.G. (eds) *Genetics, Paleontology, and Evolution*, pp. 103–120. Princeton University Press, Princeton.

Ross, R.M. & Allman, W.D. (eds) (1990) *Causes of Evolution*. University of Chicago Press, Chicago.

Rudwick, M.J.S. (1964) The inference of function from structure in fossils. *Brit. J. Phil. Sci.* **15**: 27–40.

Ruse, M. (1979) *The Darwinian Revolution: Science Red in Tooth and Claw*. University of Chicago Press, Chicago.

Ruse, M. (1982) *Darwinism Defended: a Guide to the Evolution Controversies*. Addison-Wesley, Reading, Massachusetts.

Sadler, P.M. (1981) Sediment accumulation rates and the completeness of the fossil record. *J. Geol.* **89**: 569–584.

Salthe, S.N. (1985) *Evolving Hierarchical Systems: their Structure and Representation*. Columbia University Press, New York.

Sarich, V. & Wilson, A.C. (1967) Immunological time scale for human evolution. *Science* **158**: 1200–1203.

Schaller, G.B. (1972) *The Serengeti Lion: a Study of Predator–Prey Relations*. University of Chicago Press, Chicago.

Scharloo, W. (1987) Constraints in selective response. In: Loeschcke, V. (ed.) *Genetic Constraints on Adaptive Evolution*, pp. 125–149. Springer-Verlag, Berlin.

Scharloo, W. (1991) Canalization: genetic and developmental aspects. *Ann. Rev. Ecol. System* **22**: 65–94.

Scherer, S. (1990) The protein molecular clock. *Evol. Biol.*

24: 83–106.

Schimenti, J.C. & Duncan, C.H. (1984) Ruminant globin gene structures suggest an evolutionary role for Alu-type repeats. *Nucleic Acid. Res.* **12**: 1641–1655.

Schimenti, J.C. & Duncan, C.H. (1985) Concerted evolution of the cow e² and e⁴ β-globin genes. *Mol. Biol. Evol.* **2**: 505–513.

Schindel, D.E. (1982a) The gaps in the fossil record. *Nature* **297**: 282–284.

Schindel, D.E. (1982b) Resolution analysis: a new approach to the gaps in the fossil record. *Paleobiology* **8**: 340–353.

Schmitt, J. & Antonovics, J. (1986) Experimental studies on the evolutionary significance of sexual reproduction. IV. Effect of neighbor relatedness and aphid infestation on seedling performance. *Evolution* **40**: 830–836.

Schopf, T.J.M., Raup, D.M., Gould, S.J. & Simberloff, D.S. (1975) Genomic versus morphologic rates of evolution: influence of morphologic complexity. *Paleobiology* **1**: 63–70.

Schram, F.R. (1986) *Crustacea.* Oxford University Press, New York.

Scott, J.A. (1986) *The Butterflies of North America.* Stanford University Press, Stanford.

Seger, J. (1985) Intraspecific resource competition as a cause of sympatric speciation. In: Greenwood, P.J., Harvey, P.H. & Slatkin, M. (eds) *Evolution*, pp. 43–53. Cambridge University Press, Cambridge, UK.

Selander, R.K., Clark, A.G., Whittam, T.S. (eds) (1991) *Evolution at the Molecular Level.* Sinauer Associates, Sunderland, Massachusetts.

Shackleton, N.J. (1986) Paleogene stable isotope events. *Palaeogeog. Palaeoclimatol. Palaeoecol.* **57**: 91–102.

Sheehan, P.M., Fastovsky, D.E., Hoffmann, R.G., Berghaus, C.B. & Gabriel, D.L. (1991) Sudden extinction of the dinosaurs: latest Cretaceous, Upper Great Plains, U.S.A. *Science* **254**: 835–839.

Sheldon, P.R. (1987) Parallel gradualistic evolution of Ordovician trilobites. *Nature* **330**: 561–563.

Sheppard, P.M. (1975) *Natural Selection and Heredity*, 5th edn. Hutchinson, London.

Shipman, P. (1981) *The Life History of a Fossil: an Introduction to Taphonomy and Paleoecology.* Harvard University Press, Cambridge, Massachusetts.

Sibley, C.G. & Alquist, J.E. (1987) DNA hybridization evidence of hominoid phylogeny: results from an expanded data set. *J. Mol. Evol.* **26**: 99–121.

Sibley, C.G. & Alquist, J.E. (1990) *Phylogeny and Classification of Birds: a Study in Molecular Evolution.* Yale University Press, New Haven.

Signor, P.W. & Lipps, J.H. (1982) Sampling bias, gradual extinction patterns and catastrophes in the fossil record. In: Silver, L.T. & Schultz, P.H. (eds) *Geological Implications of Impacts of Large Asteroids and Comets on the Earth*, pp. 291–296. Geological Society of America, Boulder, Colorado.

Silberglied, R.E., Ainello, A. & Windsor, D.M. (1980) Disruptive coloration in butterflies: lack of support in *Anartia fatima. Science* **209**: 617–619.

Simpson, G.G. (1944) *Tempo and Mode in Evolution.* Columbia University Press, New York.

Simpson, G.G. (1949a) Rates of evolution in animals. In: Jepsen, G.L., Simpson, G.G. & Mayr, E. (eds) *Genetics, Paleontology, and Evolution*, pp. 205–228. Princeton University Press, Princeton.

Simpson, G.G. (1949b) *The Meaning of Evolution.* Yale University Press, New Haven, Connecticut.

Simpson, G.G. (1953) *The Major Features of Evolution.* Columbia University Press, New York.

Simpson, G.G. (1961a) One hundred years without Darwin are enough. *Teach. Coll. Record* **60**: 617–626. (Reprinted in Simpson, G.G. (1964) *This View of Life.* Harcourt, Brace & World, New York.)

Simpson, G.G. (1961b) *Principles of Animal Taxonomy.* Columbia University Press, New York.

Simpson, G.G. (1978) *Concession to the Improbable.* Yale University Press, New Haven, Connecticut.

Simpson, G.G. (1980a) *Why and How: Some Problems and Methods in Historical Biology.* Pergamon Press, Oxford, UK.

Simpson, G.G. (1980b) *Splendid Isolation.* Yale University Press, New Haven, Connecticut.

Simpson, G.G. (1983) *Fossils and the History of Life.* Scientific American Library, New York.

Sims, S.H., Macgregor, H.C., Pellat, P.S. & Horner, H.A. (1984) Chromosome 1 in crested and marbled newts (*Triturus*). *Chromosoma* **89**: 169–185.

Slatkin, M. (1985) Gene flow in natural populations. *Ann. Rev. Ecol. System.* **16**: 393–430.

Slatkin, M. (1987) Quantitative genetics of heterochrony. *Evolution* **41**: 799–811.

Sloan, R.E., Rigby, J.K., Van Valen, L.M. & Gabriel, D. (1986) Gradual dinosaur extinction and simultaneous ungulate radiation in the Hell Creek Formation. *Science* **232**: 629–633.

Smit, J. & van der Kaars, S. (1984) Terminal Cretaceous extinctions in the Hell Creek area, Montana: compatible with catastrophic extinction. *Science* **223**: 1177–1179.

Smith, D.C. (1988) Heritable divergence of *Rhagoletis pomonella* host races by seasonal asynchrony. *Nature* **336**: 66–67.

Smith, D.G. (ed.) (1981) *The Cambridge Encyclopedia of the Earth Sciences.* Cambridge University Press, Cambridge, UK.

Smith, H.G. & Montgomerie, R. (1991) Sexual selection and the tail ornaments of North American barn swallows. *Behav. Ecol. Sociobiol.* **28**: 195–201.

Sneath, P.H.A. & Sokal, R.R. (1973) *Numerical taxonomy*, 2nd edn. W.H. Freeman, San Francisco.

Sober, E. (1984a) *The Nature of Selection.* MIT Press, Cambridge, Massachusetts.

Sober, E. (ed.) (1984b) *Conceptual Issues in Evolutionary Biology.* MIT Press, Cambridge, Massachusetts.

Sober, E. (1989) *Reconstructing the Past.* MIT Press, Cambridge, Massachusetts.

Sokal, R.R. (1966) Numerical taxonomy. *Sci. Am.* **215** (December): 106–116.

Spiess, E.B. (1989) *Genes in Populations*, 2nd edn. John Wiley, New York.

Springer, M. & Krakewski, C. (1989) DNA hybridization in animal taxonomy: a critique from first principles. *Quart. Rev. Biol.* **64**: 291–318.

Stanley, S.M. (1975) A theory of evolution above the species level. *Proc. Natl. Acad. Sci. USA* **72**: 646–650.

Stanley, S.M. (1979) *Macroevolution*. W.H. Freeman, San Francisco.

Stanley, S.M. (1984) Mass extinctions in the ocean. *Sci. Am.* **250** (June): 46–54.

Stanley, S.M. (1987) *Extinction*. Scientific American Library, New York.

Stehli, F.G. & Webb, S.D. (eds) (1985) *The Great American Biological Interchange*. Plenum Press, New York.

Stenseth, N.C. & Maynard Smith, J. (1984) Coevolution in ecosystems: Red Queen evolution or stasis? *Evolution* **38**: 870–880.

Stigler, S.M. & Wagner, M.J. (1987) A substantial bias in nonparametric tests for periodicity in geophysical data. *Science* **238**: 940–944.

Stigler, S.M. & Wagner, M.J. (1988) [reponse to Raup & Sepkoski (1988), q.v.] *Science* **241**: 96–99.

Stirton, R.A. (1947) Observations on evolutionary rates in hypsodonty. *Evolution* **1**: 32–41.

Strobeck, C., Maynard Smith, J. & Charlesworth, B. (1976) The effects of hitch-hiking on a gene for recombination. *Genetics* **82**: 547–558.

Surlyk, F. & Johansen, M.B. (1984) End-Cretaceous brachiopod extinctions in the chalk of Denmark. *Science* **223**: 1174–1177.

Suzuki, D.T., Griffiths, A.J.F., Miller, J.H. & Lewontin, R.C. (1989) *An Introduction to Genetic Analysis*, 4th edn. W.H. Freeman, New York.

Sved, J.A., Reed, T.E. & Bodmer, W.F. (1967) The number of balanced polymorphisms that can be maintained in a natural population. *Genetics* **55**: 469–481.

Swofford, D.L. & Olsen, G.J. (1990) Phylogeny reconstruction. In: Hillis, D.M. & Moritz, C. (eds) *Molecular Systematics*, pp. 411–501. Sinauer Associates, Sunderland, Massachusetts.

Sytsma, K. (1990) DNA and morphology: inference of plant phylogeny. *Trends Ecol. System.* **5**: 104–110.

Tauber, C.A. & Tauber, M.J. (1989) Sympatric speciation in insects: perception and perspective. In: Otte, D. & Endler, J.A. (eds) *Speciation and its Consequences*, pp. 307–344. Sinauer Associates, Sunderland, Massachusetts.

Taylor, C.E. (1986) Genetics and evolution of resistance to insecticides. *Biol. J. Linn. Soc.* **27**: 103–112.

Templeton, A.R. (1981) Mechanisms of speciation—a population genetic approach. *Ann. Rev. Ecol. System.* **12**: 23–48.

Thoday, J.M. & Gibson, J.B. (1962) Isolation by disruptive selection. *Nature* **193**: 1164–1166.

Thompson, D'Arcy W. (1942) *On Growth and Form*. Cambridge University Press. Cambridge, UK.

Thompson, J.N. (1989) Concepts of coevolution. *Trends Ecol. Evol.* **4**: 179–183.

Thompson, J.N. & Thoday, J.M. (1979) *Quantitative Genetic Variation*. Academic Press, New York.

Thomson, G. (1977) The effect of a selected locus on linked neutral loci. *Genetics* **85**: 753–788.

Timm, R.M. (1983) Fahrenholz's rule and resource tracking: a study of host–parasite coevolution. In: Nitecki, M.H. (ed.) *Coevolution*, pp. 225–265. University of Chicago Press, Chicago.

Travis, J. (1989) The role of optimizing selection in natural populations. *Ann. Rev. Ecol. System.* **20**: 279–296.

Trivers, R.L. (1985) *Social Evolution*. Benjamin/Cummings, Palo Alto, California.

Trivers, R.L. & Hare, H. (1976) Haplodiploidy and the evolution of social insects. *Science* **191**: 249–263.

Trivers, R.L. & Willard, D.E. (1973). Natural selection of parental ability to vary the sex ratio of offspring. *Science* **179**: 90–92.

Turelli, M. (1984) Heritable genetic variation via mutation–selection balance: Lerch's zeta meets the abdominal bristle. *Theoret. Popn. Biol.* **25**: 138–193.

Turelli, M. (1985) Effects of pleiotropy on predictions concerning mutation selection balance for polygenic traits. *Genetics* **111**: 165–195.

Turelli, M. (1986) Gaussian versus non-Gaussian genetic analyses of polygenic mutation–selection balance. In: Karlin, S & Nevo, E. (eds) *Evolutionary Processes and Theory*, pp. 607–628. Academic Press, New York.

Turelli, M. (1988) Population genetic models for polygenic variation and evolution. In: Weir, B.S., Eisen, E.J., Goodman, M.M. & Namkoong, R.J. (eds) *Proceeding of the Second International Conference on Quantitative Genetics*, pp. 601–618. Sinauer Associates, Sunderland, Massachusetts.

Turelli, M. & Hoffmann, A.A. (1991) Rapid spread of an inherited incompatibility factor in California *Drosophila*. *Nature* **353**: 440–442.

Turner, J.R.G. (1967) Why does the genotype not congeal? *Evolution* **21**: 645–656.

Turner, J.R.G. (1977) Butterfly mimcry: the genetical evolution of an adaptation. *Evol. Biol.* **11**: 163–206.

Turner, J.R.G. (1984) Mimicry: the palatability spectrum and its consequences. In: Vane-Wright, R.I. & Ackery, P.R. (eds) *The Biology of Butterflies*, pp. 141–161. Academic Press, London.

Turner, J.R.G. (1985) Fisher's evolutionary faith and the challenge of mimicry. *Oxford Surv. Evol. Biol.* **2**: 159–196.

Turner, J.R.G. (1986) The evolution of mimicry: a solution to the problem of punctuated equilibrium. In: Raup, D.M. & Jablonski, D. (eds) *Patterns and Processes in the History of Life*, pp. 42–66. Dahlem Workshop. John Wiley, Chichester, UK.

Turner, J.R.G. (1987) Random genetic drift, R.A. Fisher, and the Oxford School of Ecological Genetics. In: Gigerenzer, G., Krüger, L. & Morgan, M. (eds) *The*

Probabilistic Revolution. Vol.2. Ideas in the Sciences, pp. 313–354. MIT Press, Cambridge, Massachusetts.

Valentine, J.W. & Moores, E.M. (1972) Global tectonics and the fossil record. *J. Geol.* **80**: 167–184.

Van Valen, L.M. (1971) Adaptive zones and the orders of mammals. *Evolution* **25**: 420–428.

Van Valen, L.M. (1973) A new evolutionary law. *Evol. Theory*, **1**: 1–30.

Van Valen, L.M. (1976) Ecological species, multispecies, and oaks. *Taxon*, **25**: 233–239.

Van Valen, L.M. (1985) How constant is extinction? *Evol. Theory* **7**: 93–106.

Van Valen, L.M. & Sloan, R.E. (1977) Ecology and the extinction of the dinosaurs. *Evol. Theory* **2**: 37–64.

Vermeij, G.J. (1987) *Evolution and Escalation*. Princeton University Press, Princeton.

Vermeij, G.J. (1991) When biotas meet: understanding biotic interchange. *Science* **253**: 1099–1104.

Vigilant, L., Stoneking, M., Harpending, H., Hawkes, K. & Wilson, A.C. (1991) African populations and the evolution of human mitochondrial DNA. *Science* **253**: 1503–1507.

Vogel, S. (1988) *Life's Devices*. Princeton University Press, Princeton.

Vrba, E.S. (1987) Ecology in relation to speciation rates: some case histories of Miocene–Recent mammal clades. *Evol. Ecol.* **1**: 283–300.

Vrba, E.S. (1989) Levels of selection and sorting with special reference to the species level. *Oxford Surv. Evol. Biol.* **6**: 111–168.

Waage, J.K. (1979) Dual function of the damselfly penis: sperm removal and tranfer. *Science* **203**: 916–918.

Waddington, C.H. (1957) *The Strategy of the Genes*. Allen & Unwin, London.

Wade, M.J. (1976) Group selection among laboratory populations of *Tribolium*. *Proc. Natl. Acad. Sci. USA* **73**: 4604–4607.

Wade, M.J. (1978) A critical review of the models of group selection. *Quart. Rev. Biol.* **53**: 101–114.

Wahls, W.P., Wallace, L.J. & Moore, P.D. (1990) Hypervariable minisatellite DNA is a hotspot for homologous recombination in human cells. *Cell* **60**: 95–103.

Wake, D.B. & Roth, G. (eds) (1989) *Complex Organismal Functions: Integration and Evolution in Vertebrates*. Dahlem Workshop. John Wiley, Chichester, UK.

Wallace, B. (1968) *Topics in Population Genetics*. W. W. Norton, New York.

Wallace, B. (1981) *Basic Population Genetics*. Columbia University Press, New York.

Walsh, J.B. (1987) Sequence-dependent gene conversion: can duplicated genes diverge fast enough to escape conversion? *Genetics* **117**: 543–557.

Ward, R.D. & Skibinski, D.O.F. (1985) Observed relationships between protein heterozygosity and protein genetic distance and comparisons with neutral expectations. *Genet. Res.* **45**: 315–340.

Wasserman, M. & Koepfer, H.R. (1977) Character displace-ment for sexual isolation between *Drosophila mojavensis* and *Drosophila arizonensis*. *Evolution* **31**: 812–823.

Watson, J.D., Hopkins, N.H., Roberts, S.W., Steitz, J.A. & Weiner, A.M. (1988) *Molecular Biology of the Gene*, 4th edn. Benjamin/Cummings, Menlo Park, California.

Weaver, R.F. & Hedrick, P.W. (1991) *Genetics*. W. C. Brown Publishers, Dubuque, Iowa.

Weir, B.S., Eisen, E.J., Goodman, M.M. & Namkoong, G. (eds) (1988) *Proceedings of the Second International Conference on Quantitative Genetics*. Sinauer Associates, Sunderland, Massachusetts.

Weismann, A. (1891–2) *Essays upon Heredity and Kindred Biological Problems*, 2 vols. Oxford University Press, London.

Wellnhofer, P. (1990) Archaeopteryx. *Sci. Am.* **262** (May): 70–77.

Werren, J.H. (1980) Sex ratio adaptations to local mate competition in a parasitic wasp. *Science* **208**: 1157–1159.

White, M.J.D. (1973) *Animal Cytology and Evolution*, 3rd edn. Cambridge University Press, Cambridge, UK.

White, M.J.D. (1978) *Modes of Speciation*. W.H. Freeman, San Francisco.

Whitham, T.G. & Slobodchikoff, C.N. (1981) Evolution by individuals, plant–herbivore interactions, and mosaics of genetic variability: the adaptive significance of somatic mutations in plants. *Oecologia* **49**: 287–292.

Wiley, E.O. (1981) *Phylogenetics*. John Wiley, New York.

Wiley, E.O. (1988) Vicariance biogeography. *Ann. Rev. Ecol. System* **19**: 513–542.

Williams, G.C. (1966) *Adaptation and Natural Selection*. Princeton University Press, Princeton.

Williams, G.C. (1975) *Sex and Evolution*. Princeton University Press, Princeton.

Williams, G.C. (1985) A defense of reductionism in evolutionary biology. *Oxford Surv. Evol. Biol.* **2**: 1–27.

Williams, G.C. (1992) *Natural Selection: Domains, Levels, and Challenges*. Oxford University Press, New York.

Williamson, M. (1981) *Island Populations*. Oxford University Press, Oxford, UK.

Williamson, P.G. (1981a) Paleontological documentation of speciation in Cainozoic molluscs from Turkana Basin. *Nature* **293**: 437–443.

Williamson, P.G. (1981b) Morphological stasis and developmental constraint: real problems for neo-Darwinism. *Nature* **294**: 214–215.

Willmer, P.G. (1990) *Invertebrate Relationships*. Cambridge University Press, Cambridge, UK.

Wills, C. (1981) *Genetic Variability*. Oxford University Press, New York.

Wilson, A.C., Carlson, S.S. & White, T.J. (1977) Biochemical evolution. *Ann. Rev. Biochem.* **46**: 573–639.

Wilson, A.C., *et al.* (1985) Mitochondrial DNA and two perspectives on population genetics. *Biol. J. Linn. Soc.* **26**: 375–400.

Wilson, D.S. (1980) *The Natural Selection of Populations and Communities*. Benjamin/Cummings, Menlo Park, California.

Wilson, D.S. (1983) The group selection controversy:

history and current status. *Ann. Rev. Ecol. System.* **14**: 159–187.

Wolpert, L. (1990) The evolution of development. *Biol. J. Linn. Soc.* **39**: 109–124.

Woese, C.R. (1991) The use of ribosomal RNA in reconstructing evolutionary relationships among bacteria. In: Selander, R.K., Clark, A.G. & Whittam, T.S. (eds) *Evolution at the Molecular Level*, pp. 1–24. Sinauer Associates, Sunderland, Massachusetts.

Woolfenden, G.E. & Fitzpatrick, J.W. (1984) *The Florida Scrub Jay*. Princeton University Press, Princeton.

Wright, S. (1931) Evolution in Mendelian populations. *Genetics* 97–159.

Wright, S. (1932) The roles of mutation, inbreeding, cross-breeding, and selection in evolution. *Proc. VI Intern. Cong. Genetics* **1**: 356–366.

Wright, S. (1968) *Evolution and Genetics of Populations, vol. 1. Genetic and Biometric Foundations*. University of Chicago Press, Chicago.

Wright, S. (1969) *Evolution and Genetics of Populations, vol. 2. The Theory of Gene Frequencies*. University of Chicago Press, Chicago.

Wright, S. (1970) Random drift and the shifting balance theory of evolution. In: Kojima, K. (ed.) *Mathematical Topics in Population Genetics*, pp. 1–31. Springer-Verlag, Berlin.

Wright, S. (1977) *Evolution and Genetics of Populations, vol. 3. Experimental Results and Evolutionary Deductions*. University of Chicago Press, Chicago.

Wright, S. (1978) *Evolution and Genetics of Populations, vol. 4. Variability Within and Among Natural Populations*. University of Chicago Press, Chicago.

Wright, S. (1980) Genic and organismic selection. *Evolution* **34**: 825–843.

Wynne-Edwards, V.C. (1962) *Animal Dispersion in Relation to Social Behaviour*. Oliver & Boyd, Edinburgh, UK.

Wynne-Edwards, V.C. (1986) *Evolution through Group Selection*. Blackwell Scientific Publications, Oxford, UK.

Zahavi, A. (1975) Mate selection—a selection for a handicap. *J. Theoret. Biol.* **53**: 205–214.

Zetterberg, J.P. (ed.) (1983) *Evolution Versus Creationism: the Public Education Controversy*. Oryx Press, Phoenix.

Zhao, M. & Bada, J.L. (1989) Extraterrestrial amino acids in Cretaceous/Tertiary boundary sediments at Stevens Klint, Denmark. *Nature* **339**: 463–465.

Index